Volume II: Cultures

African Americans[1]

Eric J. Bailey

ALTERNATIVE NAMES

"Black" Americans, "Blacks", Afro-Americans and Afro-Caribbeans.

LOCATION AND LINGUISTIC AFFILIATION

In 1996, 53% of African Americans lived in the South, making up 19% of that region's population. Nationwide, 55% resided in the central cities of metropolitan areas. The 10 counties with the most African American residents on July 1, 1996, were Cook, Illinois (1.4 million); Los Angeles (1.0 million); Kings, New York (900,000); Wayne, Michigan (900,000); Philadelphia (600,000); Harris, Texas (600,000); Bronx, New York (500,000); Queens, New York (500,000); Dade, Florida (400,000); and Baltimore City, Maryland (400,000) (U.S. Census Bureau, 1998).

With regard to linguistic affiliation, African Americans' primary language pattern is English. English is in the Germanic branch of the Indo-European language family. Yet similar to all American populations, African Americans adapted their language patterns and dialects according to their sociocultural environment. African Americans' language patterns reflect influences from Africa, Britain, Creole populations, the Caribbean islands, rural, urban, southern regions of the United States, and Canada.

Often identified with such terms as African American English, "Black" English, "Black" English Vernacular, Ebonics, and, most recently, "Spoken Soul," African American language patterns are a reflection of their dynamic culture. The languages and dialects regularly spoken in the African American community are: Spoken Soul (considered by some a dialect of American English, and by others a language distinct and separate from American English); the U.S. Language of Wider Communication (LWC), a.k.a. "Standard American English"; Non-standard American English; and Arabic, Spanish, Swahili, Creole (and other foreign languages, but these are the main ones) (Rickford & Rickford, 2000; Smitherman, 2000, p. 20).

Depending on time, place, setting, and audience, all or some of the various languages and dialects may be used in the African American community. Most middle-class and professional African Americans speak the LWC, as well as some aspect of Spoken Soul at least some of the time. Most working-class African Americans speak varying degrees of the LWC, Non-standard English, and Spoken Soul (Smitherman, 2000, p. 20).

OVERVIEW OF THE CULTURE

The African American family as a unit has a historical continuity that began not with the American experience but in Africa, long before the intrusion of Europe into the continent. In the process of adapting to their new environments, West Africans merged their cultural traditions with European and Native American traditions. Although some of the cultural traditions have changed or been Americanized, the family unit remains constant.

The characteristics of the African American family today include a bilateral orientation—an equal recognition of the male and female lines of descent but favoring the mother's kin; extended kin groups; respect for elders; and a high value placed on children and motherhood (Aschenbrenner, 1973; Stack, 1974). Other cultural characteristics include: individual moral "strength" as a human quality; an emphasis on family occasions and rituals; strong belief in spiritualism (Aschenbrenner, 1973; Stack, 1974); reliance upon extended familial network for social, economic, and health care issues; strong orientation toward religious beliefs, activities, and organizations; outwardly expressed emotions; emphasis on nurturing children and participating in many rites of passage; preference for group activities as opposed to individual activities; preference for oral communication and oral history to share news and information; admiration of art, dance, and music; and preference for women and men sharing roles and responsibilities.

As early as the 1500s and 1600s, the ancestors of African Americans were forcibly transported to

South America, the Caribbean, and North America. African Americans are primarily descendants of West African people who mostly shared a common history, place of origin, language, food preferences, values, and health beliefs. With regard to health beliefs and health care practices, African Americans are believed to have retained many of the preventive and treatment practices associated with indigenous West African cultures, primarily because these methods were perceived to be most useful. Jacques (1976), Jordan (1975), Jackson (1985), Baer (1985), Spector (1985), Goodson (1987), Mbiti (1975), Harvey (1988), Tinling (1967), and Tallant (1990) all contend that African Americans continued to utilize folk and herbal medical practices as a result of communication difficulties with Europeans and the fears of European physicians.

There is documented evidence that traditional West African health beliefs and herbal remedies were handed down and maintained by various African American populations throughout North America (Bailey, 1991a; Spector, 1985). For example, oral histories of former slave women from all over the South contain frequent and sometimes elaborate descriptions of the wide variety of plants that constituted the material base from which the slave medical practice operated (Goodson, 1987, p. 200).

Specifically, female African American slave doctors used drugs derived from plants to prevent and cure worms, malaria, croup, pneumonia, colds, teething, and measles. Sometimes, they used the root, and other times they selected the leaf, bark, fruit, or gum resin to boil into a tea or make into a poultice or wear in a bag around the neck (Goodson, 1987, p. 200). Not only did African American slave doctors have this medical knowledge of plants, but also many other slaves knew how to diagnose and treat illnesses. In fact, a number of medicine chests have been discovered filled with the popular preparations of the day: calomel, blue mass pills, castor oil, ipecac, tarter emetic, and various tinctures (Ewell, 1813; Postell, 1951).

During the 1700s and 1800s in the United States, African American health beliefs and practices continued to show similarities with the West Africans' health beliefs and practices. As more African Americans migrated to northern cities during the mid-1800s and early 1900s, they carried a repertoire of health care beliefs and practices. African American health beliefs and practices became a composite, containing elements from a variety of sources: European folklore, Greek classical medicine, modern scientific medicine, vodoun religion, Christianity, and particularly African folklore (Bailey, 1991a; Hill, 1976).

To better understand the relationship between culture and health care issues associated with the African American population, we must examine the current social, economic, and demographic data from the U.S. Census. Although recently there has been much discussion on the lack of reporting and inaccuracy of the U.S. Census data and the fact that many African Americans are very skeptical of any governmental data agency, the sociodemographic data show some very intriguing and positive trends associated with the African American population.

According to the U.S. Census Bureau (1998) on October 1, 1996, there were an estimated 33.7 million African Americans in the United States, constituting 12.7% of the total population. The African American population is young, with an estimated median age on November 1, 1997, of 29.8 years, nearly 8 years younger than the median for the non-Hispanic "White" population. It is projected that the African American population will grow more than twice as fast as the European American population between 1995 and 2050. By the middle of the next century, the African American population should nearly double its present size to 61 million (U.S. Census Bureau, 1998).

To critically assess the health status and health issues associated with African Americans, one must use a culturally relativistic approach. This entry defines African American culture, highlights the cultural historical origins of their health beliefs and values, examines their cultural health-seeking patterns, and discusses the impact of alternative and complementary medicine on African Americans' ethnomedical beliefs.

The persistence of traditional African American health beliefs and practices provides a meaningful alternative to mainstream formal medicine for many African Americans because of its role in maintaining a sense of ethnic identity. It also indicates a pattern of adaptation to social and economic conditions both within the African American community and in the larger society.

THE CONTEXT OF HEALTH: ENVIRONMENTAL, ECONOMIC, SOCIAL, AND POLITICAL FACTORS

Medical anthropologists are trained to view the health status of each individual and population from a

The Context of Health

biopsychosociocultural perspective. Our approach is holistic; that is, medical anthropologists believe that all factors contribute to the well-being or sickness of a person/population. In order to understand how this holistic approach applies to African Americans, I will: highlight the culture of African Americans and then discuss how culture relates to African American health; examine the major cultural, historical, and political issues that may have had an impact on the health-seeking process of the African American population; and examine psychosocial environmental issues such as religion and caregiving and how these factors relate to African American health.

African Americans learn certain health beliefs and practices from their extended familial networks. When a health care crisis occurs, African Americans tend to seek health care information from their extended familial networks or a lay health professional first, and then opt for professional care.

African Americans are acutely aware of nonverbal body language that indicates that a health care provider is comfortable or not comfortable in treating an African American person. If the African American patient perceives the nonverbal language as positive, he or she will most likely return for care. However, if the African American patient perceives the nonverbal body language as negative, he or she will most likely not return for care and not adhere to the prescribed regimen.

Like all Americans, African Americans are connected to the social, economic, political, and health care fabric of U.S. society. Yet once an economic or health care crisis occurs in U.S. society, African Americans often are affected first, primarily because a third of the population is considered the working poor.

Culture adds meaning to reality. African Americans respect and honor those who train, graduate, and serve in the health professional fields. Often, African American health care professionals feel a commitment to serve their community. Because of the low numbers of African American physicians and health care specialists, African Americans frequently choose to work in underserved minority communities as opposed to communities with relatively low numbers of minorities. African Americans' commitment to work and to serve those who are underserved adds meaning to their profession in the health care field.

Culture is differently shared. African Americans perceive health and the health care system differently. Although African Americans have many common traits and patterns, there remains a high degree of diversity within the African American population. For instance, intragenerational health care issues differ among African Americans as they relate to perceptions of the health care system and health belief system. Additional sociodemographic factors—such as gender, region of the country, income level, and educational level—may also cause differing perceptions of health care issues in the African American population.

Cultural History

Another component of the comprehensive, holistic approach to health is cultural history. The cultural historical approach helps us to better understand issues such as African Americans' past health-seeking process; African Americans' current perceptions of the national and local health care systems; African Americans' lack of participation in clinical trials; and African Americans' alternative medical practices.

The significance of the cultural historical approach can best be illustrated in an article entitled, "African American Suspicion of the Healthcare System is Justified: What Do We Do about It?" Dula (1994) uses historical and contemporary examples of health care issues that contributed to the mistrust and suspicion that many African Americans have for the U.S. health care system. According to Dula (1994), real abuses in experimentation have contributed greatly to the high level of distrust within the African American population for decades.

Suspicion of the medical system in the United States has its origin in slavery medical practices. Dula (1994) states that slaves were sources for medical experimentation and research by doctors, instructional material for medical students, and that enslaved albinos and Siamese twins were displayed as freaks at medical society meetings. African Americans contributed to scientific progress, usually without consent or benefit to themselves, and sometimes without benefit even to science (Dula, 1994, p. 348).

Another cultural historical case is the Tuskegee experiment. Dula (1994) states that the Tuskegee experiment (1932)—in which 400 African American men participated in a government-sponsored study to find the effects of untreated syphilis—was the ultimate proof that European American health policy-makers, indeed, deserved mistrust. Jones (1981) and Dalton (1989) support Dula's claim. Thomas and Quinn (1991), who stated the following, also support Dula: "The history of the Tuskegee Syphilis Study, with its failure to educate the participants and treat them adequately, helped to lay the foundation for Blacks' pervasive sense of distrust of public health authorities today" (p. 1499).

Religion and Spirituality

Another important factor in the holistic approach to health is the psychosocial component. For the purpose of our discussion, this section refers to psychosocial factors in a broad-based sense, encompassing such issues as mental health, religiosity, and spirituality. In particular, religiosity and spirituality play a significant role in the well-being of the African American population.

The church, one of the symbols of religious involvement, has long been one of the major social and cultural institutions that define how people should see themselves and direct their behavior (Berry & Blassingame, 1982). Taylor and Chatters (1991) emphasize that the teachings of the church regarding human nature and human relationships may foster either mentally healthy attitudes or destructive, neurotic attitudes in its members. Inevitably, a church teaches its members, either directly or indirectly, how to deal with aggression, anger, pride, sexuality, competition, social relations, child-raising, and marital relations (Taylor & Chatters, 1991).

Within the African American community, the significance of the church to community life may be attributed, in part, to the church's position as one of the few indigenous institutions in African American communities—that is, built, financed, and controlled by African Americans. Moreover, African American churches serve as an outlet for social expression, a forum for the discussion of political and social issues, and a training ground for potential community leaders (Chatters, Levin, & Ellison, 1998; McRae et al., 1998; Neighbors, Musick, & Williams, 1998).

In 1991, Taylor and Chatters reported the results from a National Survey of Black Americans ($N = 2,094$) with regard to their religious affiliations. Taylor and Chatters (1991) found results such as: 40 different religious affiliations were reported; one half of the respondents indicated that they were Baptist (52.1%); 11% were Methodist; 6% were Roman Catholic; 3% were Holiness; 2% were Jehovah's Witness; 15% identified one or more of the 35 remaining denominations; one out of ten respondents indicated no religious preference; and eight respondents indicated that they were either atheist or agnostic.

In a study to examine the significance of religiosity and health outcomes among African Americans, Picot, Debanne, Namazi, and Wykle (1997) conducted face-to-face interviews with 136 African American and 255 European American caregivers of community-dwelling elders. Results revealed that African American caregivers perceived higher levels of rewards from participating in religious services than did European American caregivers. Additionally, the relationship between race and perceived rewards was mediated by comfort from religion and prayer. African Americans are taught to depend on a supreme being, the Lord, and to take their burdens to the Lord and leave them there (Picot et al., 1997). Picot et al. (1997) emphasized the importance of religious beliefs and health outcomes among all groups, particularly the effects within the African American population.

MEDICAL PRACTITIONERS

The types and sources of treatment actions among African Americans are likely to vary according to gender, class, region of the United States, and degree of assimilation into mainstream society. In the United States, most ethnic groups have the option of selecting from a variety of sources: (1) self; (2) alternative or native health practitioners; and (3) formal health professionals.

If the illness is perceived as naturalistic (i.e., due to inadequate rest, poor nutrition, or germs), African Americans tend to use initially one and/or a combination of the following treatment actions: self-care, alternative (indigenous) health practitioner, and formal health professionals (Jackson, 1981; Jacques, 1976; Leininger, 1985; Spector, 1985). A discussion of self-care will presented be later in this entry.

In addition to the variety of home remedies or patent medicines practiced among African Americans, there are primarily two types of alternative indigenous health practitioners serving the African American community: (1) independent practitioners and (2) cultic practitioners (Baer, 1985, p. 327). These two types of alternative indigenous health practitioners operate as individuals or are affiliated with some sort of occult-supply store, either as the owner, an employee, or someone who rents office space. The cultic practitioner is affiliated with a religious group and practices in both public and private settings. The multiplicity of African American alternative indigenous health practitioners today stems from the role adaptability of traditional African American healers of the past (Bailey, 2000a,b).

For instance, one type of African American alternative indigenous health practitioner, the neighborhood prophet/Old Lady, does not dispense medicine but merely advises clients about concocting herbal medicines. Rather than selling or giving a herbal remedy, the Old Lady tells the client to use it in varying proportions to treat the perceived illness. In addition, she gives advice on various emotional, personal, and domestic problems. She does not receive monetary gifts for her service, only gifts of food or expressions of gratitude (Jordan, 1979, p. 38). The neighborhood prophet/Old Lady treats the individual's mind, body, and spirit in an attempt to return the individual to harmony with nature.

This indigenous alternative health practitioner, as well as many others, have successfully matched their holistic approach to treating illness/disease with the model or perception of treating illness/disease among many African Americans. Although there are no specific utilization data to document the total number of African Americans seeking care from these alternative indigenous health practitioners, their presence and growth within the alternative health delivery network can no longer be denied and overlooked.

Formal health practitioners are referred to as licensed medical doctors (MDs) and doctors of osteopathic medicine (DOs) in licensed allopathic (MD-staffed) and osteopathic hospitals. The U.S. Department of Health and Human Services (1986) regularly documents the number of physician contacts by the general public. Interestingly, the national health care utilization data have always showed a distinctive difference between African Americans' and European Americans' use of formal health practitioners. For example, African Americans sought care from a doctor's surgery less (47%) than did European Americans (57.7%). African Americans were also less likely to contact a doctor by phone (9.2%) than were European Americans (13.4%). Yet, African Americans were reported to have experienced more physician contact in hospital outpatient departments (21.4%) than European Americans (13.7%) and to have more physician contact at home (5.8%) than European Americans (2.5%) (U.S. Department of Health and Human Services, 1986).

Most health care researchers would contend that an obvious explanation for the difference in physician utilization between African Americans and European Americans is economics. There is no doubt that the cost of seeking care and a lack of health insurance cause many African Americans to delay or not seek care from formal health practitioners. However, some researchers contend that there are other social and cultural factors that have truly caused the disparity in seeking care from formal health practitioners.

In particular, Blendon, Aiken, Freeman, and Correy's (1989) study of 10,130 persons living in the continental United States found that even African Americans above the poverty line have less access to medical care than their European American counterparts. The researchers contend that ethnic-related differences in health care arrangements and lifestyle were the most significant factors in the disparity between African American and European American health care utilization.

For example, African Americans are more likely than European Americans to report that during their last visit, their physician did not inquire sufficiently about pain, did not tell them how long it would take for a prescribed medicine to work, did not explain the seriousness of the illness or injury, and did not discuss tests or examination findings. In addition, fewer than three-fifths of African Americans were completely satisfied with the care provided during their last hospitalization, compared with over three-fourths of European Americans (Blendon et al., 1989). It is apparent that there are differences not only in access but also in the perception of the care provided for African Americans and European Americans. African Americans seem to adhere to ethnomedical beliefs and practices, and they use alternative health care therapies more extensively than do European Americans (Blendon et al., 1989).

Unnatural illnesses are perceived to be caused by a supernatural spirit, magic, sorcery, voodoo or some other personalistic agent (Bailey, 1991a; Hill, 1976; Snow, 1974). The literature on African American illness causation suggests that the African American healing tradition considers the universe a place where the forces of good and evil, God and Satan, struggle for control (Gregg & Curry, 1994, p. 522). Therefore, unnatural illnesses are perceived to be a struggle between the forces of good and evil.

Within this belief system, religion or spirituality is of utmost importance to illness perceived to be caused by unnatural means. Even though religion and spirituality are most commonly reported as important sources of social support for African Americans, they can also be used for dealing with unnatural illnesses in a more positive way (Ford, Tilley, & McDonald, 1998; Gregg & Curry, 1994; Jackson, Jackson, & Nixon, 1970; Stolley & Koenig, 1997; Taylor & Chatters, 1986). Taylor and Chatters (1986), Stolley and Koenig (1997), and many other researchers have shown that African Americans' use of religion and spirituality for natural as well as unnatural illnesses is their particular method of treating an illness/disease within their health belief system.

Another means of treating an unnatural illness and sometimes natural illness is with voodoo "rooting." As a form of religion, voodoo is a complex of African belief and ritual governing, in large measure, the religious life of African natives (Jordan, 1975). Harvey (1988) states that "as a belief system which combined historical conceptions with practices that were acceptable in a hostile social environment, voodoo is a striking example of a cultural adaptive mechanism used by members of an oppressed group as a survival technique."

The first is the mystic component, which deals with the supernatural such as spells and spirits. The second component is that part of voodoo that deals with psychological support of the individual, and the third part is herbal and folk medicine. African American voodoo prospers, particularly in the South, primarily because it fills a void left by inaccess and denial of medical care by formal American health practitioners (Jordan, 1979, p. 38).

In conclusion, African Americans tend to consult alternative indigenous health practitioners for unnatural and natural causation of illnesses primarily because of their attempt to cope with health problems within the context of one's resources and social and cultural environment; their belief that alternative indigenous health practitioners have some control over the forces that cause illness/diseases in a person's life, whereas Westernized medical physicians cannot heal certain cases of illness and misfortune; and lower monetary expense associated with such treatments (Cockerham, 1986, p. 88; Hill, 1976, p. 14).

CLASSIFICATION OF ILLNESS, THEORIES OF ILLNESS, AND TREATMENT OF ILLNESS

Health

African Americans' perception of what constitutes "healthy" encompasses a relatively high tolerance of discomforts from symptoms (Jackson, 1981). Studies have shown that African Americans often ignore minor discomforts as backache, upset stomach, or headache until they reach such proportions that they interfere with the business of living (Jackson, 1981). In fact, African Americans, particularly the elderly, often perceive their health status as "poorer" than do European Americans (Manuel, 1985). Furthermore, since one's self-esteem, sanity, and survival are at risk, in a state of health (good or bad), one becomes adaptive, flexible, alert, and able to use a wide range of strategies effectively to endure and reach his/her optimal potential. In other words, until the degree of discomfort impairs daily activities or is acknowledged by the individual's sociocultural network, seeking health care, particularly from mainstream health care facilities, is not warranted.

Illness

Illness is defined as sickness of body or mind; disease is defined as a harmful departure from the normal state of a person or other organism (Green & Ottoson, 1994, p. 697). Illnesses are classified as either naturalistic or personalistic (Snow, 1974). Naturalistic agents identify what caused an illness, whereas personalistic agents recognize who caused the illness (Chino & Vollweiler, 1986). Such impersonal agents as inadequate rest, poor nutrition, and germs cause illnesses. The etiology of illnesses falls into three general categories: environmental hazards, divine punishment, and impaired sociocultural relationships.

In some cases, serious and life-threatening illnesses are perceived to be sent by God (personalistic agent) or some other personalistic agent as punishment for sin (Hill, 1976; Snow, 1974). For example, many African Americans who suspect they have a terminal illness, may delay medical diagnosis. During this delay and/or denial period, many African Americans turn to powers considered greater than themselves to fathom the reason for the disease, thereby accepting terminal illness as "God's will" and believing that nothing more can be done.

Examples of naturalistic and personalistic causative agents are described in the following studies. First, a study investigating the treatment patterns of hypertension among 285 African Americans in the Detroit metropolitan area contended that African Americans' beliefs about the etiology of hypertension were based on naturalistic agents (Bailey, 1991b). Bailey reported that informants considered inadequate rest, poor nutrition, weather disturbances, and imbalances in hot and cold properties as naturalistic agents affecting their blood pressure. Informants described "richy" foods such as heavily salted greens, pork, and sweets as naturalistic agents. Moreover, cold weather was described as a naturalistic agent that can affect the viscosity of one's blood pressure (Bailey, 1991b, p. 294). These findings were directly comparable with those of other studies on folk symptomatology (Blumhagen, 1982; Dressler, 1982; Garro, 1986; Snow, 1974).

Second, a study investigating the narratives of 26 African American women with advanced breast cancer found that these women attempted to relate the meaning of their cancer to an indigenous model of health and disease (Mathews, Lannin, & Mitchell, 1994; Lannin et al., 1998). The women interviewed ranged in age from 39 to 83 years, with the majority being over the age of 50.

Cancer was seen by patients to be the worst of all diseases because they believe it is always fatal and is essentially incurable. Consequently, for many of the women interviewed, cancer resembled illnesses that did not respond to conventional categories or cures in the indigenous medical system (Mathews et al., 1994, p. 795). Thus, the only likelihood of finding a cure for cancer was "to turn it over to God." One of the informants said the following: "Cancer is a horrible disease. It just eats you up. The only one powerful enough to overcome it is the Lord. You just have to trust in Jesus to do battle for you and save you from this horrible affliction" (Mathews et al., 1994, p. 795). Mathews et al. contend that here the battle metaphor is used to portray a struggle between God, as the all-powerful force for good, and cancer, as consummate evil.

Interestingly, Mathews et al. (1994) suggest that too often in the past health care professionals have assumed that patients who delay seeking treatment for cancer or who fail to utilize the screening services available either lack knowledge, are too poor to access services, deny reality, or are excessively fatalistic. In actuality, these patients have well-worked-out ideas about their own health and about their disease (p. 799). Patients, in this case African Americans, also have well-worked-out ideas about treatment strategy depending on whether the illness is perceived as naturalistic or personalistic.

Self-Care

Self-care includes positive steps taken by individuals to either prevent disease or promote general health status through health promotion or lifestyle modification; medical self-care for the identification or treatment of minor symptoms of ill health or self-management of chronic health conditions; and steps taken by laypersons to compensate or adjust for functional limitations affecting routine activities of daily living.

The use of home remedies or herbs is an example of a self-care strategy. The basic assumption behind the use of herbs/home remedies links the natural organic properties of herbs with the natural healing capabilities of human beings. Herbalists use these organic substances in an effort to neutralize or eliminate one's body of harmful substances that impair its power to heal itself (Lust, 1974, p. 8). According to herbalists, any herb, if mixed and used properly, can effectively treat any natural illness.

Herbs from the woods are used in many ways. Herb teas are prepared to treat pain and reduce fevers. Sassafras tea frequently is used to treat colds, and placing raw onions on the feet and wrapping them in warm blankets is used to break a fever (Lust, 1974, pp. 196–197).

Spector (1985) has identified some African American home remedies as successful in the treatment of disease. A few selected examples are: a method for treating colds is hot lemon water and honey; when congestion is present in the chest and the person is coughing, the person can be wrapped with warm flannel after his/her chest is rubbed with hot camphorated oil; hot toddies are used to treat colds and congestion (these drinks consist of hot tea with honey, lemon, peppermint, and a dash of brandy or whatever alcoholic beverage the person likes and has available); and Vick's vapor rub is swallowed.

In general, self-care strategies such as home remedies and herbs are convenient and effective sources of therapy (Davis, 1977). The three major reasons for their continual use in the African American community are that folk remedies may be the only alternatives to costly treatment of acute illness by the health care system; folk remedies have been given the stamp of approval by generations of African American caregivers; and the loving care, attention, and overall nurturance that accompanies the use of remedies cannot be overlooked (Davis, 1997, p. 433).

Furthermore, Davis (1997) states that employing folk remedies in treatment follows closely the culture of a particular group of people and is replete with memories, comfort, and familiarity. In fact, folk remedies may be so enmeshed in tradition that not to perform them is tantamount to sacrilege.

HEALTH THROUGH THE LIFE CYCLE

Pregnancy and Birth

According to the Committee on the Status of "Black" Americans of the National Research Council (1989), birthrates among teenage African American women have been dropping since 1960s. However, because the total number of African American adolescent women has increased by 20%, there were substantial increases in the total births to African American teenagers, despite the declining birthrates. In addition, because birthrates to African American teenagers remain two to three times higher than those for European Americans, a higher proportion of all African American births occur among teenage mothers; in 1984, 20% of all African American births were to teenagers, compared with 11.1% among European Americans (National Research Council, 1989, p. 412). Interestingly, African American teenage women represent only 14% of the U.S. adolescent female population whereas European American adolescent girls have the overwhelming majority of all teenage births.

The higher birthrate for African American teenagers can be accounted for by earlier initiation of sexual intercourse (on average two years earlier than European Americans); less use of contraception; less likelihood of abortion; and the almost universal decision to keep and rear children who are born, rather than offer them for adoption (National Research Council, 1989, p. 412). This pattern of keeping and rearing children is reflected in U.S. Census data.

For example, in 1993, about four in ten African American preschoolers were cared for by grandparents or other relatives besides their fathers while their mothers worked, compared with only about two in ten European American children. Care by grandparents was especially important to African American families, accounting for one-fifth of all arrangements used for preschoolers (U.S. Census Bureau, 1998).

Infancy

In 1991, the mortality rate for African American infants (17.6 per 1,000 live births) was 2.4 times that for European American infants (7.3 per 1,000 live births). Between 1970 and 1991, the mortality rate decreased more for European American infants (59%) than for African American infants (46%), widening the gap in infant mortality between the two populations.

The national average figures for infant mortality do not show all the disparities between African Americans and European Ameircans in infant mortality rates across the United States. African American infant mortality rates show considerable variation by region. For example, during 1982–1984, the African American infant mortality rate was lowest in the Mountain and Pacific states (15.4 and 16.2 deaths per 1,000 live births, respectively), and highest in the East North Central states (21.7), particularly Illinois (23.3) and Michigan (23.7) (National Research Council, 1989, p. 399).

The marked gap in infant mortality rates between European Americans and African Americans mirrors the more than twofold difference in the rates of low birthweight and very low birthweight between the two groups. African Americans are twice as likely as European Americans to have low-birthweight infants: the African American rate is 12.4 per 1,000 live births, and the European American rate is 5.6 (National Research Council, 1989, p. 400).

Childhood

In 1996, there were 8.1 million African American families, 46% of them married-couple families. The majority of African American families (57%) had children. Families with children averaged two children apiece (U.S. Census Bureau, 1998).

Overall, African American children have benefited from the impressive health gains for all American children since 1950 (National Research Council, 1989, p. 405). The rate of death from all causes for children aged 1–4 was 139.4 per 1,000 live births in 1950 and 51.4 per 1,000 live births in 1985. For children aged 5–14, the comparable rates were 60.1 per 1,000 live births in 1950 and 26.3 per 1,000 live births in 1985 (National Center for Health Statistics, 1997, p. 80). However, African American children have not shared equally in the overall health gains, and their death rates are much higher than those for European American children.

Interestingly, the leading cause of death among children is injury. Accidents cause three times as many deaths than do either of the next two leading causes of childhood death (cancer and congenital anomalies). For African American children, the highest rates of injuries occur in or near the home (National Research Council, 1989, p. 405). The National Research Council states that injuries are related to socioeconomic status: poor children are very likely to live in areas in which heavy traffic patterns lead to pedestrian injury. Within the house, unrepaired stairwells and inadequate or absent screens or window guards expose children to the risk of falls. Missing smoke detectors along with defective heaters and other household appliances pose fire hazards. Poor homes are also more likely to contain toxic substances, such as chemicals for pest control, or peeling lead paint (National Research Council, 1989, p. 406).

Adolescence

Females. Adolescence is a period of significant life transition, during which children cross the bridge into adulthood (Adams, Scholenborn, Moss, Warren, & Kann, 1995). Behaviors established during this period are often carried into adult life. Adams et al. (1995) state that health behaviors established prior to adulthood can significantly influence health and longevity, both in the short term and later in life.

In 1992, the Centers for Disease Control and Prevention conducted the National Health Interview Survey—Youth Risk Behavior Survey (NHIS—YRBS). The NHIS—YRBS was a collaborative effort of the Division of Health Interview Statistics, the National Center for Health Statistics, and the Division of Adolescent and School Health of the National Center for Chronic Disease Prevention and Health Promotion. Of the 13,789 youth identified as eligible for the NHIS—YRBS, 10,645 completed questionnaires, representing an estimated 77.2% of eligible respondents (Adams et al., 1995).

With regard to weight control, Adams et al. (1995) found that African American adolescent females (36%) were less likely than European American adolescent females (46.1%) to perceive themselves as being overweight. In addition, African American female teens (41.5%) were less likely than European American females (59.5%) and Hispanic females (55.1%) to report any attempt to lose or keep from gaining weight in the past week. Overall, the data show that proportionately, fewer African American female and male youths considered themselves overweight compared with European American female and male youths.

The importance of these national health data on the perceptions of weight control among African American adolescent females relates to the issue of obesity and its health consequences. Obesity rates among African American women are significantly higher than for European American women (Kumanyika, 1994).

Moreover, chronic disease profiles of minority populations indicate that African Americans have higher than average cardiovascular disease-related mortality (Kumanyika, 1991; Otten et al., 1990). These health trends and mortality suggest an urgent need for obesity prevention programs for African American adolescent females.

Males. Much of what has been written and researched about African American adolescent males' health and social behavioral outcomes has been negative and downright gloomy. The health and social data that are often cited in the literature and reported in the media are the rates of homicide and sexually transmitted diseases.

According to the National Center for Health Statistics (1997), the homicide rate for African American males (72.5 per 100,000) were eight times higher than for European American males (9.4 per 100,000) and nearly five times higher than for African American females (13.9 per 100,000) as for European females (3.0 per 100,000). Among African American youth, the homicide rate (77.9 per 100,000) is eight to nine times that for European American (9.6) and Asian American youths (8.8) (National Center for Health Statistics, 1994, p. 29). The cause of this disparity in homicide rates is multifaceted—in other words, factors such as personal situational issues, societal discrimination, cultural historical racism, persistent unemployment, peer pressure, and lack of formal education all contribute to the high rate of homicide among African American adolescent males.

With regard to African American male sexual behavior and the risk for HIV infection, the cause for the high prevalence rate of HIV infection is also multifaceted. Whitehead (1997), for example, found that broader historical and sociocultural issues associated with African American adolescent males and middle-aged men contributed to the disproportionately high rate of HIV infection.

According to Whitehead, adolescent African American males need a masculinity transformation primarily because the existing constructs of ideal masculinity transformation fragment masculinity. In other words, the process of becoming a male in our gender-based society is incorrect. Our society teaches young males inappropriate ways of achieving adulthood.

The goal of masculinity transformation is to achieve a sense of masculine gender identity as a whole (Whitehead, 1977, p. 436). Masculine transformation is a strategy of empowerment that moves away from notions of masculinity that focus on gaining economic capacity to achieve social (with male peers) or economic status. Furthermore, Whitehead states that masculine transformation emphasizes community service, goal-setting, and discipline in achieving goals, and integrates body, mind, and spirit.

The cultural relativistic strategy resulting from Whitehead's masculinity transformation involves peer training in which older, more mature males, including low-income males, work with preadolescent and adolescent boys and young adult men. It helps men on their road to masculine wholeness and works to help younger men overcome their fragmented masculinity (p. 437). Thus, Whitehead recommends masculine transformation, along with the usual HIV/AIDS materials, to effectively address HIV/AIDS among African American adolescent males in low-income American communities.

Adulthood

According to the National Center for Health Statistics (1997), the 10 leading causes of death (from highest incidence to lowest) for African Americans in 1995 were diseases of the heart; malignant neoplasms (cancers); cerebrovascular diseases (stroke); human immunodeficiency virus infection (HIV); unintentional injuries; homicide and legal intervention; diabetes mellitus; pneumonia and influenza; chronic obstructive pulmonary diseases; and certain conditions originating in the perinatal period (National Center for Health Statistics, 1997, p. 117). African American men and women, however, differ in the order and rank of the leading causes of death.

For example, HIV infection ranked third among leading causes of death for African American men. With regard to African American women, cerebrovascular disease and diabetes mellitus ranked third and fourth, respectively, among leading causes of death. Moreover, ranking ninth for leading causes of death among African American women were nephritis, nephritic syndrome, and nephrosis (lupus).

The Aged

African American elders are a diverse group, and it is important to recognize this group's heterogeneity (Brangman, 1995). No typical African American elder exists. They can vary from an elder living in the rural South to an elder in an urban area in the Northeast. They may have been born in the northern or southern parts of the United States or be members of a subgroup, as are immigrants from various parts of the Caribbean, such as Jamaica or Haiti. Their history, religious, educational, socioeconomic, and marital statuses, and cultural backgrounds must be taken as a starting point for understanding the individual while avoiding overgeneralizations and stereotypes (Brangman, 1995, p. 16; Mouton, Johnson, & Cole, 1995).

Alzheimer's disease (AD), also called primary degenerative dementia, is a deterioration of mental capacity. AD accounts for over half of all dementias. Death rates from AD increase with age. For Americans aged 65–74 years, the death rate is nearly 10 deaths per 100,000 people (National Center for Health Statistics, 1996). Age-adjusted rates are nearly two times higher for the European American population than for the African American population.

To determine the medical, social, and cultural pattern of AD among African Americans, Gorelick et al. (1994) studied 113 African American AD patients and 79 African American vascular dementia (VaD) patients in Chicago. They found that the typical demographic and background profile of their African American AD and VaD patients was that of a woman born in the southern portion of the United States who had lived on a farm until her mid-teenage years before moving to Chicago. These women currently lived in a house or apartment with other family members, were retired, widowed, and Protestant, and had some form of medical insurance. AD patients were generally older than VaD patients (76.4 vs. 71.8 years) and remained in their state of birth for a greater length of time before moving to Chicago than VaD patients (26.3 vs. 18.8 years). This study represents one of the few research initiatives that exclusively targeted African American patients with AD and VaD. Moreover, Martin and Panicucci's (1996) study of 40 elderly African American women's health behaviors and beliefs highlighted the difference in this study's results versus stereotypical beliefs associated with elderly African American women. Findings revealed that Southern, community-living African American older women generally have a high level of adherence to commonly recommended health promotion/disease prevention habits.

Martin and Panicucci stipulated that the most likely explanation for the high level of adherence may stem from cultural and religious doctrines that discourage certain unhealthy practices such as excessive alcohol consumption, cigarette smoking, and ineffective coping outlets. Because study findings indicate that African American older women want to maintain their health, increased attention must be directed to the importance of primary prevention behaviors as an assertion of control over one's future health, well-being, and quality of life (Martin & Panicucci, 1996, p. 47).

Martin and Panicucci (1996) suggest that nurses must set into place interventions that enhance perceptions of relevance regarding the lifestyle practice of regular exercise. Regular community-based exercise classes in churches, residential settings, senior centers, or other frequently attended community sites would increase the likelihood of regular participation in exercise.

Finally, it is clear that more and more African Americans are living longer and are in better health. It is imperative that health care administrators develop and plan culturally sensitive health care services for older African Americans from various social and cultural backgrounds.

Dying and Death

Through the legacy of slavery, segregation, and discrimination, African Americans have developed a perspective on death that is unique to this ethnic group (Mouton, 2000). End-of-life decision-making practices draw on religious characterizations of death. The majority of African Americans adhere to a Christian belief system, with 83% claiming a Protestant affiliation (the majority being Baptist) and 14% claiming Catholic affiliation (Ellison & Sherkat, 1990). These religious beliefs recognize a transcendent soul that rises to heaven upon death. African Americans also have a strong belief in a heaven not of this earth. Furthermore, most African Americans subscribe to an African American theology, a religious doctrine that views God as the fighting God of the Old Testament. It is important to realize that not all African Americans view death from a religious construct. However, religion does seem to play a significant role in African Americans view of death and dying (Mouton, 2000, pp. 74–75).

As a whole, African Americans are less likely than European Americans to approve of euthanasia. For example, in a study comparing attitudes toward life-prolonging treatment among 139 patients from a general medicine clinic, only 63% of African Americans approved of stopping life-prolonging treatments compared with 89% of European Americans (Caralis, Davis, Wright, & Marcial, 1993).

African American attitudes to postmortem issues also are drawn from their shared historical, religious, and social experiences (Mouton, 2000). Overall, African Americans are less likely than European Americans to agree to organ donation. Most have a strong belief in the resurrection of the body, and this belief inhibits their willingness to donate organs because they will need their bodies to be intact on Judgment Day (Mouton, 2000, p. 80).

The effects of history and religious views, combined with individual life experiences and beliefs, influence African Americans' attitudes about the end of life. Clinicians need to bring to the table a degree of cultural sensitivity to eliminate some of the mistrust and fear that some African Americans might have when making end-of-life decisions. Addressing these issues might alleviate some of the apprehension that African Americans feel concerning end-of-life care (Mouton, 2000, p. 80).

NOTE

1. Excerpts from this article are reprinted from Eric Bailey's book, *Medical Anthropology and African American Health*, with permission from Greenwood Publishing.

REFERENCES

Adams, P., Scholenborn, C., Moss, A., Warren, C., & Kann, L. (1995). *Health-risk behaviors among our nation's youth: United States, 1992*. Washington, DC: U.S. Department of Health and Human Services.

Aschenbrenner, J. (1973). Extended families among Black Americans. *Journal of Comparative Family Studies, 4*, 257–268.

Baer, H. (1985). Toward a systematic typology of black folk healers. *Phylon, 43*, 327–343.

Bailey, E. (1991a). *Urban African American health care*. Lanham, MD: University Press of America.

Bailey, E. (1991b). Hypertension: An analysis of Detroit African American health care treatment patterns. *Human Organization, 50*, 287–296.

Bailey, E. (2002a). *Medical anthropology and African American health*. Westport, CT: Bergin & Garvey.

Bailey, E. (2002b). *African American alternative medicine: Using alternative medicine to prevent and control chronic diseases*. Westport, CT: Praeger.

Berry, M., & Blassingame, J. (1982). *Long memory: The black experience in America*. New York: Oxford University Press.

Blendon, R., Aiken, L., Freeman, H., & Correy, C. (1989). Access to medical care for black and white Americans. *Journal of the American Medical Association, 261*, 278–281.

Blumhagen, D. (1982). The meaning of hypertension. In N. Chrisman & T. Maretzki (Eds.), *Clinically applied anthropology* (pp. 297–324). Boston: D. Reidel.

Braithwaite, R. & Taylor, S. (Eds.). (1992). *Health issues in the black community*. San Francisco: Jossey-Bass.

Brangman, S. (1995). African American elders: Implications for health care providers. *Clinics in Geriatric Medicine, 11*, 15–23.

Caralis, P. V., Davis, B., Wright, K., & Marcial, E. (1993). The influence of ethnicity and race on attitudes toward advance directives, life-prolonging treatments, and euthanasia. *Journal of Clinical Ethics, 4*(2), 155–165.

Chino, H., & Vollweiller, L. (1986). Etiological beliefs of middle-income anglo Americans seeking clinical help. *Human Organization, 45*, 245–254.

Chatters, L., Levin, J., & Ellison, C. (1998). Public health and health education in Faith communities. *Health Education and Behavior, 25*, 689–699.

Cockerham, W. (1986). *Medical sociology*. Englewood Cliffs, NJ: Prentice Hall.

Dalton, H. (1989). AIDS in blackface. *Daedalus: Journal of American Academy Arts and Science*, Summer 205–228.

Davis, R. (1977). Understanding ethnic women's experiences with pharmacopeia. *Health Care for Women International, 18*, 425–437.

Dressler, W. (1982). *Hypertension and culture change: Acculturation and disease in the West Indies*. New York: Redgrave.

Dula, A. (1994). African American suspicion of the healthcare system is justified: What do we do about it? *Cambridge Quarterly of Health Care Ethics, 3*, 347–357.

Ellison, C. G., & Sherkat, S. E. (1990). Patterns of religious mobility among African Americans. *Sociological Quarterly, 4*, 551–566.

Ewell, J. (1813). *Planters' and mariners' medical companion*. Philadelphia, PA: Scholarly Press.

Ford, M., Tilley, B., & McDonald, P. (1988). Social support among African American adults with diabetes, Part 2: A review. *Journal of the National Medical Association, 90*, 425–432.

Garro, L. (1986). *Cultural models of high blood pressure*. Paper presented at the 85th American Anthropological Association, Philadelphia, PA.

Goodson, M. (1987). Medical–botanical contributions of African slave women to American medicine. *Western Journal of Black Studies, 2*, 198–203.

Gorelick, P. B., Freels, S., Harris, B. A., Dollear, T., Billingsley, M., & Brown, N. (1994). Epidemiology of vascular and Alzheimer's dementia among African Americans in Chicago, IL. *Neurology, 44*, 1391–1396.

Green, L., & Ottoson, J. (1994). *Community health* (7th ed.). St. Louis, MO: Mosby.

Gregg, J., & Curry, R. (1994). Explanatory models for cancer among African American women at two Atlanta neighborhood health centers: The implications for a cancer screening program. *Social Science & Medicine, 39*, 519–526.

Harvey, W. (1988). Voodoo and Santeria: Traditional healing techniques in Haiti and Cuba. In C. Zeichner (Ed.), *Modern and traditional health care in developing societies* (pp. 101–114). New York: University Press of America.

Hill, C. (1976). Folk medical belief system in the American south: Some practical considerations. *Southern Medicine*, 11–17.

Jackson, J. (1981). Urban black Americans. In A. Harwood (Ed.), *Ethnicity and medical care* (pp. 37–129). Cambridge, MA: Harvard University Press.

Jackson, J. (1985). Race, national origin, ethnicity, and aging. In R. Binstock & E. Shanas (Eds.), *Handbook of aging and the social sciences* (pp. 264–303). New York: Van Nostrand.

Jackson, J., Jackson, O., & Nixon, W. (1970). Medicine in the black community. *California Medicine, 113*, 57–61.

Jacques, G. (1976). Cultural health traditions: A black perspective. In M. Branch & P. Paxton (Eds.), *Providing safe nursing care for ethnic people of color* (pp. 115–123). New York: Appleton-Century Crofts.

Jones, J. H. (1981). *Bad blood: The Tuskegee syphilis experiment: A tragedy of race and medicine*. New York: Free Press

Jordan, W. (1975). Voodoo medicine. In R. Williams (Ed.), *Textbook of black-related diseases* (pp. 716–738). New York: McGraw-Hill.

Jordan, W. (1979). The roots and practice of voodoo medicine in America. *Urban Health, 8*, 38–41.

Kumanyika, S. (1994). Racial and ethnic issues in diet and cancer epidemiology. *Advanced Experimental Medical Biology, 354*, 59–70.

Kumanyika, S., & Adams-Campbell, L. (1991). Obesity, diet, and psychosocial factors contributing to cardiovascular disease in blacks. *Cardiovascular Clinics, 21*, 47–73.

Lannin, D., Mathews, H., Mitchell, J., Swanson, M., Swanson, F., & Edwards, M. (1988). Influence of socioeconomic and cultural factors on racial differences in late-stage presentation of breast cancer. *Journal of the American Medical Association, 279*, 1081–1807.

Leininger, M. (1985). Southern rural black and white American lifeways with focus on care and health phenomena. In M. Leininger (Ed.), *Qualitative research methods in nursing* (pp. 195–216). New York: Grune & Stratton.

Lust, J. (1974). *The herb book*. New York: Bantam Books.

Manuel, R. (1985). *Demographics and the black elderly*. Paper presented at the Gerontological Meetings, Chicago, IL.

Martin, J., & Panicucci, C. (1996). Health-related practices and priorities: The health behaviors and beliefs of community-living black older women. *Journal of Gerontological Nursing, 22*, 41–48.

Mathews, H., Lannin, D., & Mitchell, J. (1994). Coming to terms with advanced breast cancer: Black women's narratives from eastern North Carolina. *Social Science & Medicine, 38*, 789–800.

Mbiti, J. (1975). *Introduction to African religion*. Ibadan, Nigeria: Institute for Urban Studies.

McRae, M. Carey, P., & Anderson-Scott, R. (1998). Black churches as therapeutic systems: A group process perspective. *Health Education and Behavior, 25*, 778–789.

Mouton, C. (2000). Cultural and religious issues for African Americans. In K. Braun, J. Pietsch, & P. Blanchette (Eds.), *Cultural issues in end-of-life decision making* (pp. 71–82). Thousand Oaks, CA: Sage.

Mouton, C., Johnson, M., & Cole, D. (1995). Ethical considerations with African American elders. *Ethnogeriatrics, 11*, 113–129.

National Center for Health Statistics. (1996). Mortality trends for Alzheimer's disease: Fact sheets. In *Vital and Health Statistics* (Series 20, No. 28). Washington, DC: Centers for Disease Control and Prevention.

National Center for Health Statistics. (1997). *Health, United States 1996–97 and injury chartbook* (DHHS publication No. PHS 97-1232). Hyattsville, MD: US Government Printing Office.

National Research Council, Committee on the Status of Black Americans. (1989). Black Americans' health. In G. D. Jaynes & R. M. Williams, Jr. (Eds.), *A common destiny: Blacks and American Society* (pp. 391–450). Washington, DC: National Academy Press.

Neighbors, H., Musick, M., & Williams, D. (1998). The African American minister as a source of help for serious personal crises: Bridge or barrier to mental health care? *Health Education and Behavior, 25*, 759–777.

Otten, M. Jr., Teutsch, S., Williamson, D., & Marks, J. (1990). The effect of known risk factors on the excess mortality of black adults in United States. *Journal of the American Medical Association, 264*, 571–573.

Picot, S., Debanne, S., Namazi, K., & Wykle, M. (1997). Religiosity and perceived rewards of black and white caregivers. *The Gerontologist, 37*, 89–101.

Postell, W. (1951). *The health of slaves on southern plantations.* Baton Rouge: Louisiana State University Press.

Rickford, J., & Rickford, R. (2000). *Spoken soul: The story of black English.* New York: Wiley.

Smitherman, G. (2000). *Talkin that talk: Language, culture, and education in African America.* New York: Routledge.

Snow, L. (1974). Folk medical beliefs and their implications for care of patients. *Annals of Internal Medicine, 81*, 82–96.

Spector, R. (1985). *Cultural diversity in health and illness.* New York: Appleton-Century Crofts.

Stack, C. (1974). *All our kin: Strategies for survival in a black community.* New York: Harper & Row.

Stolley, J., & Koenig, H. (1997). Religion/spirituality and health among elderly African Americans and Hispanics. *Journal of Psychosocial Nursing, 35*, 32–38.

Tallant, R. (1990). *Voodoo in New Orleans.* Gretna, LA: Pelican.

Thomas, S., & Quinn, S. (1991). The Tuskegee syphilis study, 1932 to 1972: Implications for health education and AIDS risk education programs in the black community. *American Journal of Public Health, 81*, 1498–1503.

Tinling, D. (1967). Voodoo, root work, and medicine. *Psychosomatic Medicine, 29*, 483–490.

Taylor, R., & Chatters, L. (1986). Church-based informal support among elderly blacks. *The Gerontologist, 26*, 637–643.

Taylor, R. J., & Chatters, L. M. (1991). Religious life. In J. Jackson (Ed.), *Life in black America* (pp. 105–123). Newbury Park, CA: Sage.

U.S. Census Bureau. (1998). *Census bureau facts for features.* Washington, DC: U.S. Census Bureau's Public Information Office.

U.S. Department of Health & Human Services. (1986). *Black and minority health: Report of the Secretary's task force: IV.* Washington, DC: U.S. Government Printing Office.

Whitehead, T. (1997). Urban low-income African American men, HIV/AIDS, and gender identity. *Medical Anthropological Quarterly, 11*, 411–447.

Amish

Lawrence P. Greksa and Jill E. Korbin

ALTERNATIVE NAMES

Old Order Amish, Plain People.

LOCATION AND LINGUISTIC AFFILIATION

The Old Order Amish, the focus of this entry, live in more than 200 settlements in over 20 states of the United States and one Canadian province (Ontario). Approximately three-fourths of all Amish live in Ohio, Pennsylvania, and Indiana. The Old Order Amish speak Pennsylvania Dutch (a German dialect) within the group, use High German in their church services, and, with the exception of preschool children, are generally fluent in English. All of these languages are in the Indo-European language family.

OVERVIEW OF THE CULTURE

The Old Order Amish are an Anabaptist religious isolate. The Anabaptist (or Swiss Brethren) movement arose in Europe in the early 16th century. The Anabaptists believed in adult rather than infant baptism and refused to bear arms, both of which resulted in them being severely persecuted. *Martyr's Mirror*, a book found in most Amish homes, documents how hundreds of Amish were brutally executed for their religious beliefs.

In 1693 the group now known as the Amish separated from the Mennonites, one of the early Anabaptist groups, because they believed in a stricter adherence to the doctrine of *meidung*, or a total shunning of excommunicated church members. Following this separation, the Amish migrated throughout the German-speaking parts of Europe. They were highly regarded as farmers, but were severely persecuted for their Anabaptist beliefs.

To escape religious persecution, the Amish migrated to North America between about 1727 and 1860. There are no longer any Amish in Europe. In 1865, approximately one third of the Amish population split off from the more liberal Amish majority. This offshoot minority was given the name Old Order Amish in recognition of the fact that they wished to retain the old *Ordnung* (order of behavior), or set of orally transmitted rules that govern the behavior of the Amish.

There are currently over 150,000 Old Order Amish residing in the United States and Canada. The primary unit of organization for the Old Order Amish is the congregation, or church district, each composed of an average of 30 households and approximately 150 people. Each congregation is led by a bishop, with the assistance of two to three ministers and one deacon. Religious leaders are chosen by lot. Bi-weekly worship services are held in homes, and there is no separate church building. Each church district has its own *Ordnung*, which it reaffirms twice a year during communion. The *Ordnung* consists of both rules that are common to all Old Order Amish and rules that are specific to each congregation. If a member consistently violates the rules of the *Ordnung*, a hierarchy of responses is initiated, with the highest level of response being excommunication in association with *meidung*. At the most extreme, *meidung* requires all members of the congregation (and by extension all Amish), to have absolutely no contact with the shunned individual. However, any shunned person who repents is reincorporated into the community. The severity of the *meidung* has been decreasing in recent years.

Because each church district has its own *Ordnung*, there is variability from church to church. There is no higher level of religious organization above the church district, although church districts may be affiliated with one another based on similarity of their *Ordnungs*. The term settlement is used to describe a group of congregations located within the same geographic region.

Religion is the core organizing principle for the Amish and is embedded in every aspect of Amish life. A distinction between religious and non-religious affairs is meaningless for the Amish. Amish life is guided by several key principles, including adherence to adult baptism; *Gelassenheit* (acting with humility and simplicity at all times); conviction that true grace can only be achieved if one lives in isolation from the non-Amish world; an ethic of absolute non-violence; a belief that mutual aid is a key ingredient in maintaining the integrity of the church; and a stance that states have no authority in religious matters. Separation from the world is fostered by the utilization of distinctive symbols, such as 18th century European peasant clothing, horse and buggy travel, and rejection of electricity from power lines. The Amish recognize that separation from the world requires the existence of strong community ties and, in particular, providing each other with assistance when needed. One of the better known examples of mutual aid is a communal barn-raising, but in fact, mutual aid is involved in virtually all aspects of daily life. For example, while families are expected to be self-sufficient in paying for usual health care costs, the church will assist in paying large hospital bills.

The Old Order Amish are often thought of as a static society living the lifestyle of 17th or 18th century farmers. They are, in fact, a dynamic society, with a history of carefully incorporating new technology that they decide is essential for economic competitiveness. However, the Amish are selective, refusing to accept anything that they feel might threaten their core beliefs. It must be remembered that this selection process occurs separately within each church district, resulting in variability among Amish communities in the degree of change that has been accepted.

The Old Order Amish have been undergoing a transition over the past 40–50 years from an economic system based primarily on small family-owned farms to one based on wage labor. This transition appears to be primarily due to the joint effects of a rapid rate of population increase in conjunction with an increase in the cost of farm land in the vicinity of the major settlements. The magnitude of this transition varies substantially between settlements. Some Amish wage laborers work primarily with other Amish men, either in Amish-owned shops or on Amish construction crews, but an increasing number of men now work in factories where they have intensive contact with the non-Amish (variously referred to by the Amish as "Yankees" or "English").

THE CONTEXT OF HEALTH: ENVIRONMENTAL, ECONOMIC, SOCIAL, AND POLITICAL FACTORS

The Old Order Amish originated from a relatively small founding population and each major settlement has remained largely genetically isolated from both other Old Order Amish settlements and the surrounding U.S. and Canadian populations for a little over 200 years. As a result, a number of distinctive recessive disorders have developed among the Old Order Amish, with some of them being unique to particular settlements. Other than these genetic disorders, the general pattern of illness and causes of mortality among the Old Order Amish are similar to those for the United States as a whole, except that accident rates related to agriculture and other manual labor occupations are probably somewhat higher.

Cost is one of the primary factors considered by the Old Order Amish when making decisions about health care utilization. Two aspects of cost are particularly important in the Amish context. First, because the Amish do not own or drive cars and rely on horse-and-buggy transportation, estimates of costs for doctor visits and hospitalization must include the need to hire a driver. This can be very expensive, sometimes equaling the cost of the physician visit. Second, due to Amish beliefs in self-sufficiency, separation from the world, and mutual aid, the Amish have generally rejected any kind of formal assistance that comes from outside the Amish community, including commercial health insurance. Obtaining commercial health insurance goes against the principle of separation from the world, implies an unwillingness to accept God's will, and operates against the principle of providing mutual aid in times of crisis. However, since most Old Order Amish have no compunction about using modern biomedicine, but at the same time reject commercial health insurance, this means that illness can potentially result in very high medical bills. This is particularly true for chronic physical conditions and for mental illnesses, which often require long (sometimes lifetime) periods of treatment, often in association with expensive medications.

The Old Order Amish traditionally have relied on personal savings and various mechanisms of mutual assistance within the immediate and larger Old Order Amish community to meet their medical expenses. This includes the use of Amish Aid health care plans that were established in the 1960s in response to the rising cost of health care and with the explicit goal of providing an alternative to commercial health insurance. Although the Amish Hospital Aid plans are essentially health insurance plans, they are not viewed in the same negative way as commercial health insurance plans because they involve Amish mutual aid rather than assistance purchased from non-Amish. Amish Hospital Aid plans are thus acceptable to many, although not all, Old Order Amish. In recent years there has been a trend in some settlements of at least some of the men working in non-Amish commercial establishments to utilize the commercial health insurance provided by employers as part of their benefits package.

At least as important to the Amish in their selection of health care services as economic factors is the ability to obtain services from an agency that they feel can be trusted to respect Amish culture. If a provider is well thought of or makes a positive impression, news of this will spread in the community, and referrals will occur through personal networks.

The role of the Amish bishop in all aspects of Amish life, including the utilization of health care, cannot be overstated. He is the spiritual leader of the church district as well as the mediator between church members and the outside world. The well-being of the community is an enormous responsibility that bishops take quite seriously. Bishops either can be a barrier to or can facilitate access to health care. If a bishop promotes the view that preventive health care, such as childhood immunization, indicates a lack of faith in God and God's will, members of his church will be less likely to seek that care. If, on the other hand, the bishop is not opposed to such care, then individual members of the church are free to follow their own beliefs and preferences. Bishops can be a barrier to mental health services if they believe that mental illness is the fault of the individual, particularly if they believe it is due to shortcomings of the individual in his or her relationship to God, church, and community. Bishops can also be a barrier to specific agencies if they have heard that an Amish person has had unsatisfactory experiences with that specific agency. On the other hand, bishops can be an enormous asset to service delivery if they feel that a particular mental health care center is effective and can be trusted to respect Amish culture.

The most important point for health providers to understand about Old Order Amish society is that it is dynamic and that there is greater heterogeneity between and within congregations than might be expected by outward manifestations of conformity in dress and transportation.

The Old Order Amish diet reflects its Germanic ancestry and is thus characterized by large servings and an emphasis on meats, starches, and a wide array of pastries. Almost all Old Order Amish families, as recently as 50–60 years ago, produced almost all their food themselves, with the exception of basic staples such as flour, sugar, and salt. All the food produced on the family farm was (and still is) grown with the minimum use of artificial fertilizers, if any at all. The transition from small-scale farming to wage labor appears to be associated with increased dietary diversity (due to the increasing purchase of foods, both raw and processed), but without any substantial change in basic dietary patterns. Most Amish families still have gardens for producing vegetables that will be canned and used throughout the year, but these foods now provide a relatively small proportion of food intake in most families. Instead, most families now purchase a large majority of their foods in the same grocery stores as their non-Amish neighbors. Similarly, whereas eating at restaurants of any kind was rare in the past, generally only occurring on long trips where no other option was available, many Amish families now routinely purchase foods at a variety of fast food and family restaurants.

Medical Practitioners

Amish individuals rely on a variety of health care practitioners. The Amish obtain health care from biomedical practitioners, from a variety of complementary and alternative medicine providers, and through the use of home remedies. As long as core religious beliefs are not violated, individual preferences for health care providers are allowed and respected. Biomedical care is sought for both mental and physical illness, though the level of acceptance of biomedical care, particularly for mental health, varies across church districts. Many individuals simultaneously consider and utilize both biomedical and one or more different complementary and alternative (CAM) treatments for virtually all illnesses. There is also a reported tendency for the Old Order Amish not to utilize preventive health services to any great extent, although there is some suggestion that this is changing with increased access to insurance and recognition of the benefits of some preventive care, for example, childhood immunization.

CAM health care is sought either because it is believed to be efficacious or because of its lower cost. Practitioners ranging from reflexologists to chiropractors to herbal and vitamin practitioners may be consulted for health problems. A unique practitioner found in the past (but apparently no longer used) amongst the Amish was the *braucher*, whose healing was based on repeating secret verses and charms that were passed on orally from braucher to braucher. Brauching was performed by both men and women and the braucher did not need to be in the same location as the ill person. Although many options are available, when an illness persists, recourse is usually toward biomedical care. Individuals may travel substantial distances, across states and to Mexico or Canada, to seek care from practitioners, both biomedical and CAM, whom they believe will provide a cure and/or who are more affordable.

Classification of Illness, Theories of Illness, and Treatment of Illness

Health among the Amish is generally associated with an ability to perform one's work and the ability to eat well. Biomedical paradigms, classifications, and theories about illness and treatment are widely accepted among the Amish, keeping in mind that there is both individual variability and variability across church districts. Biological causes generally predominate in etiological explanations of physical illness. With respect to mental illness, most Old Order Amish distinguish between mental disorders that have a biological basis, and therefore can be treated with medication, and those that are not rooted in biology, and therefore require counseling. Amish families are very likely to accept biomedical treatments, regardless of cost, that restore normal functioning, but are likely to strongly resist treatments primarily designed to extend life when there is no hope of restoring normal functioning. The latter, but not the former, is seen as interfering with God's will by most Amish.

Sexuality and Reproduction

The fertility patterns of most populations, and certainly most modernized populations, are strongly influenced by parity-dependent behaviors that limit family size, particularly the cessation of fertility (stopping behaviors) once a couple has attained its desired family size. Populations whose fertility patterns are not influenced to any great extent by such behaviors, but are instead primarily a function of the biological capacities of individuals to reproduce (fecundity) are referred to as "natural fertility" populations. The Old Order Amish possess very strong religious proscriptions against birth control and thus qualify as a natural fertility population. Unlike most natural fertility populations, however, the Old Order Amish are a healthy and well-nourished population that fully utilizes the available biomedical health care system. As a result, Old Order Amish females tend to have very high fecundities, resulting in very high fertility rates, with an average completed fertility (births to women who have completed the reproductive life span) of 7–8 children. Although there is no evidence that the Amish practice stopping behaviors, there is some evidence that at least some couples practice behaviors that influence the length of birth intervals (spacing behaviors). Such behaviors could eventually have an impact on total fertility rates but are not having a measurable impact at the present time. The transition to wage labor has thus far only resulted in a minor decrease in fertility.

Those few adults who either do not marry or are biologically incapable of having offspring (about 3%, similar to other populations) are considered unfortunate and sometimes have a hard time accepting this reality but are not stigmatized by Amish society. Fertility treatments and medical intervention are sometimes sought by infertile couples.

HEALTH THROUGH THE LIFE CYCLE

Pregnancy and Birth

Children are considered a gift from God. The Amish understand biological procreation, but view God's will as involved in the number, gender, and health (including miscarriages) of their offspring. Abortion is strictly prohibited. Use of prenatal care varies by parity. In general, prenatal care in the biomedical sector is initiated earlier with first pregnancies and later with subsequent pregnancies. If, however, the pregnancy appears to be problematic (for example, if it involves bleeding), biomedical care is sought without regard to parity. Some Amish communities have created free-standing birth centers, generally staffed by a nurse with a physician on call. Birth centers were established to limit the cost of childbirth so that young couples would not limit family size due to the high cost of hospital births. Birth centers also provide an atmosphere congruent with Amish preferences for minimal intervention in birth and cost considerations. Women and families are free to decide whether to use birth centers or hospitals for childbirth. The use of midwives, once common, is now rare.

Amish families are not opposed to the use of biomedical technology in childbirth, as long as it is congruent with Amish values and lifestyles. Fetal monitors are accepted, for example, because they may facilitate a better outcome for infant and mother. Prenatal tests, on the other hand, are considered a wasteful use of funds because all children will be accepted, regardless of any problems identified prenatally.

Infancy

Infants are both breast- and bottle-fed, with both types of feeding usually occuring on demand. Breast-feeding remains the ideal or preferred method of infant feeding, though women who elect to bottle feed are not criticized for this decision. There are no data on the length of breast feeding, but exclusive breast feeding seldom lasts for more than 3 or 4 months, at which time Amish families begin to provide soft supplementary foods, such as mashed potatoes, to infants from the dinner table. There are some reports that Amish babies often sleep in the same bed as their parents for the first few months but the frequency of this behavior is not clear. However, most Amish parents do move their babies' cribs into their bedrooms for the first several months after birth, to facilitate infant care and feeding.

Infants are viewed as not yet having the ability to distinguish between right and wrong, and therefore, they should never be punished in any way for any act. Crying is a sign that an infant needs comfort and is never cause for discipline. Amish infants are primarily cared for by their parents, but older children, particularly older female children, will generally play an important role in child care, even in infancy.

There has been resistance in the past, particularly in the more conservative Amish groups, to child vaccinations, in the belief that vaccinations are attempts to thwart God's will. This has occasionally resulted in outbreaks of disease within Amish communities. There may still be some resistance to vaccinations (as well as biomedical health care in general) in the most conservative groups, but the majority of Amish parents recognize the value of vaccinations. Additionally, while the Amish have their own schools, some parents elect to send their children to non-Amish schools or to kindergarten prior to beginning Amish school in the first grade. These parents comply with immunizations that are mandatory for school attendance.

The attitude of Amish parents toward pediatric care for their children largely reflects their attitudes about preventive medical care for themselves. In other words, most Amish parents would not hesitate to take their child to the family doctor or the local emergency room for any disorder that they or their community recognized from past experience (e.g., respiratory disorder) could most effectively be treated with antibiotics or other medical treatments. On the other hand, few Amish parents would probably feel it necessary to take their child to a pediatrician for an annual health exam.

Childhood

Amish parents recognize that one of their most important functions is to raise their children to accept adult baptism and lead a good Amish life. Parents recognize that they must provide a good model of behavior for their children at all times and believe that they must be constantly vigilant to correct the behavior of their children whenever it deviates from the expected norms of the Amish community. Amish parents see the road to becoming Amish (and thus to salvation) as being straight and narrow. It is their responsibility to keep their children on this track. In particular, children are taught to be obedient, to respect authority, to work hard, and that the well-being of the group always takes precedence over their own well-being. The latter belief is summarized in a saying found in many schools: "JOY: Jesus first, Others second, Yourself last".

Amish parents believe that it is their moral obligation to firmly and consistently correct their children. This sometimes is viewed as requiring physical punishment (spanking). Spanking is seen as a necessary (even if disagreeable to the parent) tool for teaching obedience and raising good Christian children. However, any corrective action, including spanking, must not be performed in anger. Amish parents recognize that any corrective behavior performed in anger will not be an effective learning event, which should be the only goal of such behavior.

Toilet training generally begins at about 2 years of age and is approached like all other stages of life for children. Children are gently and calmly told what should occur. They are not blamed for failing but rather are praised for succeeding. The general attitude is that this is what all children do, so you will too.

Since work is seen as central to both good health and being a good Christian, Amish children are assigned age-appropriate chores from an early age. Children are thus incorporated into the work ethic of the family and community, and are seen by others, as well as by themselves, as contributing to the general family welfare. There was never any difficulty assigning useful and meaningful tasks when most Amish were family farmers. Some Amish are concerned that one negative consequence of the transition to wage labor has been the loss of meaningful chores for the young, leading to potentially dangerous idle hands.

Adolescence

Amish children attend school through the 8th grade. The Amish have their own schools, but families may choose whether to send their children to public or Amish school. Upon completing school, all boys and many girls enter the workforce. Boys with fathers who farm will often assist with farm work, but many boys now work in Amish shops or on construction crews, or for non-Amish businesses. Girls generally work as domestics in both Amish and non-Amish homes. Sometime within in their late teens or early twenties, Amish youth must decide if they will accept adult baptism to join the Amish church.

Adolescence is often a difficult time for parents and youth. This is the stage during which youth must decide if they will make a lifelong commitment to join the Amish church, and during which they locate their spouse for life. During this period, a small proportion of Amish youth develop problems with alcohol, drugs, and tobacco. Although the parents of these youth are often portrayed as passively allowing their youth to misbehave, this is not an accurate description of what generally occurs. In the vast majority of cases, these problems resolve and the youth become baptized in the church. Currently over 80% join the Amish church. For a small, but unknown number of youth, these problems persist into adulthood.

Adulthood

There are no special health or medical issues that arise during adulthood for most Amish. However, due to the unique genetic history of the Amish, there is a high prevalence of a number of genetic disorders whose symptoms often include a reduction in both cognitive and physical functioning. Given the Amish belief in self-sufficiency and in taking care of their own, such adults are very rarely placed into a medical or alternative care setting. They are instead cared for at home. As a result, it is not at all unusual for an Amish family to have an adult child living at home who cannot take care of him or herself. This adult child will be cared for by the parents until they are no longer physically capable of doing so or, frequently, until they die. At that time, the adult child becomes the responsibility of his or her siblings. Such an adult is only placed into a nursing or other medical facility if the family cannot provide adequate health care.

The Aged

Elders are accorded respect by the Amish. Elder individuals try to live as independently as possible, often moving into a small but separate house connected to one of their children's homes (*grossdawdy* house). In the past, this move would occur when the parents turned over their farm to one of their children, often the youngest male. Once the elderly person is no longer capable of living independently, they will generally either move in with one of their children or their children will take turns taking care of them. In very rare cases, the elderly are placed into a health care facility. Such cases generally only occur if the family is convinced that appropriate care can only be obtained in a nursing home.

Dying and Death

The Amish have well-established rituals associated with death, which is seen as an expected life transition, and associated with eternal salvation. As a result, death appears to be associated with less stress than in many societies. Life should not be prolonged by the use of heroic measures, and family members should be allowed, if at all possible, to die at home. If a hospital death is unavoidable, the dying person should be surrounded by family and church members. Once death occurs, family members are supported and relieved of their usual tasks and obligations. For example, neighbors prepare the home by clearing one room for a viewing and the other rooms for benches (which are normally used for church services) for those who come to pay their respects. If death occurs in the hospital, these preparations will generally occur before the family even returns home. Neighbors and church members also perform all necessary household and farm tasks. Viewings and funerals are clearly recognized as symbols of community togetherness and are thus generally well attended, so much so that it is sometimes necessary to hold memorial services simultaneously at more than one location. There is generally an open-casket viewing for 1–2 days after death, after which the individual will be buried in one of the local family cemeteries. Non-Amish morticians prepare the body for burial and deliver it to the home, but the actual burial is performed by the church and community.

BIBLIOGRAPHY

Acheson, L. S. (1994). Perinatal, infant, and child death rates among the Old Order Amish. *American Journal of Epidemiology, 139*, 173–183.

Bryer, K. B. (1979). The Amish way of death: A study of family support systems. *American Psychologist 34*, 255–261.

Campanella, K., Korbin, J. E., & Acheson, L. (1993). Pregnancy and childbirth among the Amish. *Social Science and Medicine, 36*, 333–342.

Fuchs, J. A., Levinson, R. M., Stoddard, R. R., Mullet, M. E., & Jones, D. H. (1990). Health risk factors among the Amish: Results of a survey. *Health Education Quarterly, 17*, 197–211.

Greksa, L. P. (2002). Population growth and fertility patterns in an Old Order Amish settlement. *Annals of Human Biology, 29*, 192–201.

Greksa, L. P., & Korbin, J. E. (1997). Influence of changing occupational patterns on the use of commercial health insurance by the Old Order Amish. *Journal of Multicultural Nursing and Health, 3*, 13–18.

Hamman, R. F., Barancik, J. I., & Lilienfeld, A. M. (1981). Patterns of mortality in the Old Order Amish. I. Background and major causes of death. *American Journal of Epidemiology, 114*, 845–861.

Hewner, S. (1997). Biocultural approaches to health and mortality in an Old Order Amish community. *Collegium Anthropologicum, 2* (1), 67–82.

Hostetler, J. A. (1963). Folk and scientific medicine in Amish society. *Human Organization, 22*, 269–275.

Hostetler, J. A. (1993). *Amish society* (4th ed.). Baltimore: Johns Hopkins University Press.

Hostetler, J. A., & Huntington, G. E. (1971). *Children in Amish Society: Socialization and community education*. New York: Holt, Rinehart and Winston.

Huntington, G. E. (1988). The Amish family. In C. H. Mindel, R. W. Habenstein, & R. Wright, Jr. (Eds.), *Ethnic families in America: Patterns and variations* (3rd ed. pp. 367–399). Englewood Cliffs, NJ: Prentice Hall.

Kraybill, D. B. (1989). *The riddle of Amish culture*. Baltimore: Johns Hopkins University Press.

Kreps, G. M., Donnermeyer, J. F., & Kreps, M. W. (1994). The changing occupational structure of Amish males. *Rural Sociology, 59*(4), 708–719.

McKusick, V. A. (Ed.). (1978). *Medical genetic studies of the Amish.* Baltimore: Johns Hopkins University Press.

Miller, L. (1981). The role of a braucher-chiropractor in an Amish community. *The Mennonite Quarterly Review, 55*(2), 157–171.

Nolt, S. M. (1992). *A history of the Amish.* Intercourse, PA: Good Books.

Olshan, M. A. (1991). The opening of Amish society: Cottage industry as Trojan horse. *Human Organization, 50,* 378–384.

Palmer, C. V. (1992). The health beliefs and practices of an Old Order Amish family. *Journal of the American Academy of Nurse Practitioners, 4,* 117–122.

Tripp-Reimer, T., Sorofman, B., Lauer, G., Martin, M., & Afifi, L. (1988). To be different from the world: Patterns of elder care among Iowa Old Order Amish. *Journal of Cross-Cultural Gerontology, 3,* 185–195.

Waltman, G. H. (1996). Amish health care beliefs and practices. In M. C. Julia (Ed.), *Multicultural awareness in the health care professions* (pp. 23–41). Boston: Allyn & Bacon.

Yoder, K. K. (1997). Nursing intervention considerations among Amish older persons. *Journal of Multicultural Nursing and Health, 3,* 48–52.

Argentine Toba

Claudia R. Valeggia and Florencia Tola

ALTERNATIVE NAMES

There are no current alternative names for the Toba, although Spanish colonizers used the collective term *Guaycurú* to refer to many indigenous communities inhabiting the Gran Chaco. The name *frentones* ("large foreheads" in Spanish) was in widespread use for eastern Toba bands in the early centuries of contact. The Toba call themselves *Qom* or *Qom'pi* (people).

LOCATION AND LINGUISTIC AFFILIATION

The Toba have inhabited mostly the southeastern and central areas of the Argentine Gran Chaco. The Gran Chaco is a vast region spanning 1,000,000 km² through Western Paraguay, Eastern Bolivia, and Northeastern Argentina. It is characterized by a patchwork of savannah grasslands and semi-arid forests, with forests along riverbanks. A marked East–West gradient of rainfall makes the western area considerably drier than the eastern. Seasonal changes in temperature are pronounced. Minimum daily temperatures below 10°C, with occasional frosts, can occur between April and September, whereas maximum daily temperatures above 33°C are frequent and are concentrated between September and March.

At present, Toba communities are found mostly in the provinces of Chaco and Formosa, although peri-urban settlements around major cities in the provinces of Santa Fe and Buenos Aires are increasing in population numbers as well.

The Toba belong to the Guaycurú linguistic family, which also encompasses the Pilagá, the Mocoví, and the Mbayá (Caduveo) (Mason, 1963). However, the number of languages and dialects in what is collectively known as the "Toba" language is still controversial (Braunstein, personal communication). There are at least four mutually unintelligible languages spoken by Toba groups in the Gran Chaco. For example, even when sharing the same Guaycurúan language root, eastern and western Argentine Toba do not understand each other when they meet (Mendoza, 2002).

OVERVIEW OF THE CULTURE

European soldiers wrote about the Gran Chaco Indians as early as the mid-1500s (Schmidel, 1970). However, it was not until the early 1900s that ethnographic work began to be published on the Gran Chaco Indians (Boggiani, 1899; Karsten, 1926; Métraux, 1937; Nordenskiöld, 1912). All Gran Chaco indigenous groups used to share similar subsistence economies despite considerable language variation. The groups were traditionally nomadic or semi-nomadic hunter–gatherers, showing occasional horticulture (Braunstein & Miller, 1999; Mendoza & Wright, 1989). Division of labor was manifest among the Toba (Karsten, 1967). Female gathering played a major role in Chaco economies, complementing the almost exclusively male activities of hunting, fishing, and honey collecting (Braunstein & Miller, 1999; Gordillo, 1995).

The Toba organized themselves in bands composed of groups of extended families. Traditional leadership was limited to extended family heads (Miller, 1980). Shamans stood out as healing specialists and intermediaries between the natural and the supernatural worlds and often also acted as leaders (Métraux, 1946; Miller, 1967, 1980). Monogamy was the main mating pattern. Toba women have had an independent and influential position in their society as a consequence of their central role in the family economy (Braunstein, 1983; Karsten, 1967).

The Gran Chaco Indians successfully resisted Spanish colonization and Argentine expansion policies until the late 1800s. Until the 1930s most communities still relied on foraging for their subsistence. During the last century, disruptions to their traditional lifestyle and ecological deterioration of the habitat forced many communities to migrate to urban centers and become sedentarized. At present, indigenous communities in the Gran Chaco fall along an acculturation continuum ranging from the more traditional, living in rural, isolated areas, to the more Westernized, living on the periphery of most non-indigenous towns in the Gran Chaco and in the cities of Rosario and Buenos Aires (Miller, 1999).

THE CONTEXT OF HEALTH: ENVIRONMENTAL, ECONOMIC, SOCIAL, AND POLITICAL FACTORS

Social, Economic, and Political Context

The Argentine Toba are experiencing a dramatic transition from their original lifestyle to the one offered by non-indigenous communities. The severe degradation of their original environment, together with overpopulation and overwhelming socioeconomic pressures, have considerably diminished the possibility of retaining the traditional subsistence model. The Toba have been described as "an egalitarian society with an immediate-return economy" (Mendoza, 2002). Nowadays, Toba communities present varying degrees of economic dependence on the non-indigenous sectors. Rural communities, located in isolated areas by the Pilcomayo and Bermejo rivers, still rely heavily on the forest and wetlands for their subsistence. Hunting, fishing, and gathering items represent up to 75% of their diet during the wet season (Gordillo, 1995). During the dry season (winter), they rely on temporary jobs in nearby towns and on subsidies to large families from the provincial government. Families settled in peri-urban or urban communities only hunt, fish, and gather opportunistically, when they have access to transportation to the forest or the river. They subsist on the wages of the few men with public employment, on the unstable salaries of temporary jobs, and on governmental subsidies. Older women, usually accompanied by young children, may gather food and other goods from the non-indigenous population by asking door to door or simply sitting on the doorsteps of markets and food stores. Most women do not work for a salary and their activities revolve around childcare and household chores. Some women weave baskets or string bags, which they sell as handicrafts in the non-indigenous towns. The percentage of Toba families, both rural and urban, with unmet basic needs varies between 75% and 100%, depending on the province (Costanzó et al., 2001; Delucchi, Fontan, Grichener, & Wassner, 1996).

Integration of the Toba into Argentine social and political life has been extremely difficult. Education policy, as a basic premise of social equity, has not achieved much success. Schools are not integrated, even in urban settings. Furthermore, in the province of Formosa, some schools follow the "aboriginal modality." In these schools, non-indigenous teachers teach a shorter version of the "regular" curriculum, while bilingual indigenous teachers offer native language writing and reading courses. Although this schooling modality was intended to bridge the language gap in the classroom, it is being seriously criticized by Toba delegates, who argue that their children's opportunity to an equal education is being radically curtailed (Alegre & Francia, 2001). Up to this date, no Toba person in the country has achieved a professional (or equivalent) degree.

The political participation of all Chacoan Indian groups has been reduced to negotiation of their vote. Voting is compulsory in Argentina. Around national and local election times, political parties gain votes by offering food and clothes. However, there is an increasing tendency in urban settlements to form civil associations that can legally request community development funds from national and international organizations. For example, a civil association formed in a Toba village in the province of Formosa is devoting its efforts to enforce the implementation of the *Ley Integral del Aborigen* (Aboriginal Integral Law), sanctioned in 1984 but never properly enforced (Alegre & Francia, 2001).

Health Situation

There are no current demographic statistics on the Toba, but the Institute of Aboriginal Communities of Formosa estimates a total of 70,000 Toba people living in Argentina. As recognized Argentine citizens, the Toba have access to free medical services at public hospitals and health centers. However, it is a common complaint that indigenous people are constantly discriminated against at these places. Complaints from Toba people range from being ignored at hospitals to being abused and mistreated. The provincial governments have acknowledged the lack of communication between indigenous and non-indigenous sectors in the arena of public health. In order to alleviate this situation they implemented a plan that includes training health agents at the local communities. These agents are in charge of monitoring the health of a given number of families in their own villages, reporting illnesses and new pregnancies, administering medicines, and promoting health education. The success of this strategy remains to be evaluated, but preliminary results seem very promising.

Epidemiological statistics for each life cycle stage are given below. Briefly, the current health situation of the Toba is that of poor communities in developing countries. Infant and child mortality are high and mainly a consequence of preventable causes, such as diarrhea, dehydration, and respiratory infections. Tuberculosis is prevalent across all ages and its incidence is more pronounced in rural communities. Cardiovascular problems and obesity-related diseases are becoming increasingly common among sedentarized, urban Toba.

MEDICAL PRACTITIONERS

Among the Toba there are different specialists who are in charge of restoring health (Karsten, 1932; Métraux, 1946, 1967). The main figures associated with medical therapy in the larger part of Toba communities are the shaman (*pi'oGonaq*) and the "curandero" (*ratanataGaq*). Nevertheless, this distinction is somewhat artificial because the functions fulfilled by both often coincide in the same person.

In general terms, the shaman has the capability of curing and killing through the invocation and collaboration of his auxiliary spirits (*nataq*). The shaman's techniques include sucking out the illness that has materialized in an object, blowing on the area of the extraction, and praying to auxiliary spirits. The shaman's calling is restricted to those who have inherited from their father or grandfather the power that allows them to contact non-human entities from different spaces of the cosmos. However, certain people can develop ties with non-human beings who later endow them with the power to cure and kill and, therefore, to become a shaman (Métraux, 1967; Tola, 2001; Wright, 1997).

In contrast, the "curandero" is not characterized by the "cure–kill" ambivalence of his therapeutic powers. His knowledge allows him only to cure with plants, whose medicinal powers are known by the Toba through experimentation. In general, this category of medical practitioners includes persons that participate in the Toba Evangelic cults present in the communities (Miller, 1967; Wright, 1990). Their healings are often done by the extraction of evil spirits out of the sick body through collective prayers.

There are other specialists whose knowledge is more limited and who carry out only the healing of specific illnesses, such as those affecting infants and children. These practitioners acquire their knowledge after a personal experience related to the illness and the proximity to death. The acquisition of curative faculties often relates to the presence in dreams or in moments near death of some non-human being who gives to the ill person the power of curing one specific illness. Unlike shamans, the "curandero" and these specialists had at some time a relation with powerful non-human beings, but they are not in permanent contact with them neither in dreams nor when the healing takes place.

CLASSIFICATION OF ILLNESS, THEORIES OF ILLNESS, AND TREATMENT OF ILLNESS

Sources of illness include spells attributed to the will of shamans, non-human entities, or sorcerers, as well as to the violation of restrictions imposed during specific moments of the life cycle (Métraux, 1937; Nordenskiöld, 1912). Illnesses caused by shamans can only be cured by other shamans. The healer extracts from the body of the victim the object sent by the aggressor shaman. Once the source of evil has been eradicated, the shaman introduces the object into his own body by friction, transforming it to a source of power for him. The non-human entities can generate illnesses in people by materializing in the forest, in dreams, or during the night. The type of illness varies depending on the entity. In general, if the illness is not cured by a shaman whose spirit can talk to the insulting entity, the ill person dies (Métraux, 1967; Wright, 1997). Sometimes, the illness is related to "natural" conditions (e.g., a cold, the flu, or a cough) or to visible causes (e.g., a cut or a wound). These problems do not need the immediate consultation of a shaman. However, if they persist, they are attributed to the actions of shamans, non-human beings, or sorcerers regardless of their initial empirical cause.

When the traditional therapy fails to restore health, the person or his/her relatives may resort to Western medicine physicians or to collective prayer at their local Evangelic church. The general scheme of consultation consists in alternating between the shamanic, evangelic, and biomedical therapies. Physicians often complain about the delay in presenting the sick person to the health center or hospital, and the consequent advancement of the illness. They also consider visits to the "curandero" or to the shaman obstacles to the biomedical treatment. Toba people see illnesses as the result of an intentional damaging action of human and non-human origin. Their view rests on a perception of the person in relation to others (humans and nonhumans) and of the body as an entity that is permeable to symbolic actions. In contrast, the physiological and anatomical knowledge on which occidental medicine is based consecrates the material body and the subject's autonomy. These differences in the representations of illness and the body often cause confrontations between the shamanic and biomedical therapies and produce a constant conflict of interpretations and representations for the present-day Toba.

SEXUALITY AND REPRODUCTION

The concept of sex and sexuality among the Toba cannot be dissociated from more general social rules and beliefs such as pre- and post-partum taboos, the social consequences of violating restrictions, the importance of the gestational process in the infant's health, and ideas concerning family responsibility. After their first menstruation, girls had to submit to a ritual of initiation after which they waited some years before starting families of their own. This waiting period did not mean sexual abstinence, but rather was a time without childbearing responsibilities. Sexual freedom was, and still is, the hallmark of this period (Karsten, 1932; Métraux, 1946). In the past, adolescents who became pregnant without having a stable partner resorted to abortion or infanticide. Since these practices cannot be carried out nowadays, it is common to find a very high number of young unmarried mothers.

Gestation is considered a gradual process in which, through successive semen depositions into the woman's body, the couple begins to engender a new offspring (Karsten, 1932; Métraux, 1946). In order to start a pregnancy, the woman has to receive a *baby spirit* from nonhuman beings (nowadays summarized by the image of God). During the first 4 months of pregnancy, the fetus is formed by the union of sperm and intrauterine menstrual blood. In fact, the same menstrual blood that was not discarded (because menstruation was interrupted by conception) contributes to the fetus's formation in the womb (Idoyaga Molina, 1976/77; Métraux, 1937; Tola, 1999, 2001). This representation of conception and gestation illustrates the idea of reproduction as a process requiring the participation of both parents. This model also emphasizes that conception is only possible through the intervention of a nonhuman element, the presence of the *baby spirit*. The absence of this spirit is one of the causes of infertility, which is considered to be mainly the woman's fault.

HEALTH THROUGH THE LIFE CYCLE

Pregnancy and Birth

Records from prenatal visits to local health centers indicate few pregnancy complications. The most frequently recorded pathologies are pre-eclampsia, eclampsia, and severe anemia, although no quantitative analysis of these problems has been undertaken. An analysis of weight gain during pregnancy in one Toba community indicated an adequate weight gain (mean = 9.9 ± 4.0 kg, range = 5.0–27.2 kg) (Valeggia & Ellison, n.d.).

The majority of births (72–95%, depending on the community) occur in hospitals. Official statistics indicate that in the most remote areas this percentage decreases to about 54% (Programa NacyDef, 2000). However, there are rural communities where all births take place at home with the assistance of experienced midwives. Post-natal records indicate that the incidence of pre-term births among Toba women is as high as among non-indigenous women in the area.

Information about changes in fertility patterns among Toba people is extremely scarce. A survey of reproductive histories of peri-urban Toba women indicated that age at first birth has been declining steadily during the last few decades (Valeggia & Ellison, n.d.). The mean age at first birth declined from 21.5 (±4.5) years for women born in the 1930s and 1940s to 15.5 (±1.0) years for women born in the 1980s.

Abortion and infanticide were common (Karsten, 1967; Vitar, 1999). Abortion was more frequently practiced among unmarried or widowed women and was provoked by mechanical means; that is, by striking the abdomen until the miscarriage occurred (Karsten, 1967). Since abortion is illegal in Argentina, the incidence of this practice at present is difficult to assess. However, hospitalizations due to incomplete abortions are among the most frequent reasons for hospitalization, mainly among women 25–34 years of age (Departamento de Información y Estadísticas, 1999). Women also drink herbal teas that are said to "make a delayed menstruation come right away." The effectiveness of these herbs is unknown. Infanticide was practiced when infants were small, weak, or had an obvious birth defect. The second infant of a twin birth was also put to death, alleging evil intervention. At present, since most Toba follow a Christian religion in which infanticide is interdicted, direct infanticide is virtually unheard of. Yet, as in other societies, some child neglect cases that eventually lead to child death are suspected to be covert forms of infanticide (Gelles & Lancaster, 1987).

Interestingly, interbirth intervals, which now average 28 months (Valeggia & Ellison, 2001a), do not appear to be significantly shorter than those reported in early writings (Karsten, 1967). The Toba respect a postpartum sexual taboo by which couples do not engage in sexual intercourse until the child is able to walk by him/herself, which occurs around the 13–18 month postpartum (Tola, 2000). Given that they have, on average, 10 months of postpartum amenorrhea (Valeggia & Ellison, 2001a), this taboo may represent a social mechanism to space births.

Infancy

Birth weight is within normal ranges in both urban and rural villages. A study of infant growth found that only 3% of infants born in a peri-urban village weighed less than 2,500g, while 85% weighed between 2,500 and 4,000g and 12% weighed more than 4,000g (Valeggia & Ellison, n.d.). The mean birth weight for this population was $3,380 \pm 498$g (range 2,025–4,400g). A summary of vital statistics for the year 2000 showed that in the more rural communities, 8% of infants were born weighing less that 2,500g, 80% were born between 2,500 and 4,000g, and 12% weighed more than 4,000g at birth (Programa NacyDef, 2000).

There are no gender preferences among the Toba and both girls and boys receive much attention during their first year of life. The growth of infants is very good during the first year of life. In fact, the mean weight-for-age falls above two reference curves until 11 months of age, showing a peak around 4 months (Faulkner, Valeggia, & Ellison, 2000; Valeggia, Faulkner, & Ellison, 2001). Growth slows down progressively during the second year.

All infants are breastfed from birth until 2–3 years of age or until the mother becomes pregnant again. Breastfeeding can be defined as "on demand." On average, babies are put to the breast three to four times per hour, even during their second year (Valeggia & Ellison, 2001a). Supplements to breast milk are introduced around the fifth month of life and typically consist of broths, noodles, and rice. In urban settings, formula feeding is starting to replace semi-solid food as the first supplement of choice. The mother is the principal caretaker during infancy (Cohn, Valeggia, & Ellison, 2001). On occasion and for brief periods, the father and older siblings can act as surrogates.

Infant mortality during the first year of life varies from 18.6 per 1,000 in peri-urban villages to 61.8 per 1,000 in rural communities (Programa NacyDef, 2000; Torres, Cabutti, & Palatnik, 1973). The most commonly cited causes of infant death in peri-urban settings are upper-respiratory infections and gastrointestinal infections (diarrhea and dehydration). In more isolated areas, peri-natal deaths appear to be the main cause of infant death (Costanzó et al., 2001).

Childhood

The Toba do not have set ceremonies to mark the beginning and end of childhood. Once children are weaned they are considered to be independent and they are no longer in permanent physical contact with the mother. Girls begin to perform some light household chores and child caretaking when they are 3 years old. However, most helping activities are performed when the girl is between 7 and 15 years old. Girls' helping behavior includes involvement in domestic work (e.g., cooking, cleaning, tending the fire, washing), economic work (e.g., weaving baskets, selling handicrafts), or child caretaking. Young boys are not expected to help, neither in household chores nor in childcare (Bove, Valeggia, & Ellison, 2002).

Toba parents have a very permissive attitude toward children. Children are seldom reprimanded or scolded and are encouraged to learn through experience. As soon as they start walking confidently, they join mixed-age peer groups. Play is unsupervised and children usually play at different locations within the village.

Early childhood (1–3 years old) is the life stage in which Toba children are most vulnerable to problems of malnutrition. A survey carried out by a pediatric hospital in the province of Santa Fe indicated a worsening in the grade of malnutrition in Toba children in successive hospitalizations for other pathologies (Gomez, Morales, Aride, Balonchar, & Jofre, 1998). During the second year of life the mean weight-for-age declines and it falls significantly below Argentine growth reference curves (Faulkner et al., 2000). According to Faulkner et al.'s study, a sharp increase in the percentage of malnourished children occurs around 15 months of age. The authors suggest that this dramatic weight loss may be due to a delay in introducing supplements, coupled with the unavailability of appropriate, nutrient-rich weaning foods.

In 1994, the Department of Epidemiological Surveillance of the Province of Formosa evaluated the nutritional status of indigenous children in the province (Ranaivoarisoa & Ventura, 1998). Compared with non-indigenous children and with national growth curves, Toba children 6–9 years old showed considerable growth faltering. Of the surveyed children, 11% were one standard deviation below the national average and 8.5% were two standard deviations below that average. The authors also pointed out that Toba children in rural communities were in better nutritional status than their urban counterparts, suggesting better diet quality in rural settings.

In the year 2000, mortality rates for children 1–4 years old averaged between 9 per 1,000 and 50 per 1,000 among communities (Programa NacyDef, 2000). The main causes of death for rural communities were tuberculosis, other acute respiratory infections, and malnutrition (Costanzó et al., 2001). In peri-urban settings, early childhood deaths are associated with acute respiratory and gastrointestinal infections. Among the most frequent reasons for medical consultation for young children are acute diarrhea and gastroenteritis, upper respiratory infections, and skin infections, mainly pyoderma resulting from scabies (Stevens, personal communication).

Adolescence

Adolescent girls start their reproductive lives at approximately 15 years of age. Vital statistics reports show high rates of adolescent pregnancy. In certain Formosan villages, as many as 44% of the Toba women giving birth in the year 2000 were 19 years old or younger (Programa NacyDef, 2000), with 3% being younger than 15 years old. These rates were similar regardless of the location and mode of subsistence of the community. Usually, adolescent mothers and their infants remain in the maternal home until the second or third child. The custom of fostering away children born to young girls is fairly common, particularly in urban settings.

Adolescent prostitution is common in peri-urban and urban settings. As many as 20% of the adolescent girls participating in a reproductive history survey conducted in a peri-urban village indicated that they worked as prostitutes in a nearby truck stop (Valeggia, unpublished data). During this survey, the girls indicated that they seldom used condoms during their sexual encounters. Prostitution, although illegal in Argentina, is not stigmatized among the Toba and it is taken as a temporary job.

Health information on adolescent boys is very scarce. With the traditional customs of hunting and fishing severely curtailed, and integration into the non-indigenous society being difficult, many adolescents turn to alcohol. Alcoholism, in turn, leads to violent accidents and death. Violent deaths were, in fact, one of the main causes of adolescent mortality in the last few years in the province of Formosa (Departamento de Información y Estadísticas, 1999). In some villages, adolescents and young adult men find support in the Evangelic church, where they participate actively in ceremonies. School attendance is still very low among adolescent boys, although it is higher than that of girls. While girls are rapidly incorporated into a strong and supportive female network, boys tend to be left by themselves.

Adulthood

There are virtually no written reports on the health of adult Toba before Spanish colonization. In general, they were described as strong, robust people. Early writings by physical anthropologists noted that the Toba, together with the Patagonian Indians, were among the tallest South American Indians, with an average height of around 169 cm for men and 156 cm for women (Lehmann-Nitsche, 1908; Paulotti, 1948). Interestingly, current adult height is not significantly different from these figures.

At present, very few hospitals discriminate health reports based on ethnicity. The data presented here were obtained from districts in which the indigenous population represents the majority of people (Departamento de Información y Estadísticas, 1999). Still, the data should be regarded as tentative.

Tuberculosis is widespread among Toba adults, particularly those living in more isolated communities. The incidence of Chagas' disease increases east to west in the Gran Chaco, favored by the dryer climate and the prevalence of traditional mud and palm huts in rural areas. Among the urban Toba, the main morbidity causes include hypertension, urinary infections (mainly women), gall bladder calculi, and gastrointestinal infections. As many as 80% of adult members of a rural community in the province of Chaco presented with pterygium and hypervascularization of the ocular conjunctiva (Torres et al., 1973).

Although the incidence of gonorrhea is known to be high among Chaco indigenous groups, other sexually transmitted diseases are dramatically underreported. In a survey conducted in 1998, 75% of the adult women in a Toba community of Formosa reported current or past symptoms of vaginitis and other urogenital infections (Valeggia, unpublished data). Records from city hospitals in Formosa and Chaco show a very low incidence of HIV infection among Gran Chaco indigenous people (Cravero, personal communication), but HIV tests are not performed routinely and underreporting might be significant.

The Aged

Aged people had an important role in Toba society in the past. They were the ones teaching traditions to the young and all important community decisions, including marriages, had to be approved by a council of elders. Post-reproductive Toba women also contributed considerably to the community economy, carrying most of the burden of forest-gathering. Today, changes in lifestyle are also accompanied by a change in the role of elders in the community. It is a common complaint of old people that the young no longer respect them and that that is the cause of a gradual loss of their ethnic identity.

It is difficult to accurately assess the age of older Toba people. Even though most have identification documents, these are not reliable sources of dates of birth. In any event, demographic data from some villages show a dramatic drop after age 55 for both men and women. The main cause of hospitalization for older Toba is pneumonia and other respiratory infections. Mortality records from the Formosan Ministry of Human Development indicate that the main causes of mortality among older Toba adults are tuberculosis, pneumonia, and various types of tumors.

Dying and Death

The Toba do not consider illness and death as natural processes of the living organism, but rather think they are related to the intentional action of another human being or of some nonhuman entity. These actions cause a degeneration of the body that leads to death if they are not countered by the therapeutic methods of shamans and healers (see above). At a corporal level, death is produced when the image-soul of the person (*lki'i*) definitively leaves the body. After this moment and during a variable period of time (from 1 month to several years), the *lki'i* remains on earth near the person's family and his/her body. During the first month after the death of a relative some rules and restrictions must be respected to avoid the spirit of the dead person appearing in dreams or during incursions to the forest. This spirit might attempt to take away the spirit of a member of his/her family. For this reason, when a person dies, the closest relatives often appear oblivious to the death of the person. The dead person's belongings are buried or burned after the burial and the house in which he or she died is destroyed. Moreover, relatives avoid even talking about the dead, pronouncing his or her name, or crying for them. These attitudes or actions are observed by the spirit of the dead,

Changing Health Patterns

The Toba are undergoing a rapid nutritional transition from a hyperproteic diet to a hypercaloric one. Entire communities that used to rely on foraging or home-based cultivation are now depending on the processed foods available in city stores. Their diets are based on what is relatively inexpensive, such as processed sugars, starches, and fats. In addition, these changes are occurring together with an increasingly sedentary lifestyle. This pattern has become increasingly common in all of Latin America (Peña & Bacallao, 1997; Uauy, Albala, & Kain, 2001).

A serious consequence of these lifestyle changes is an increase in the rate of obesity in urban and peri-urban communities. Forty percent of the adult women in a peri-urban setting were overweight or obese (Body Mass Index >26 kg/m^2) in 1998–99 (Valeggia & Ellison, 2001b). Although there is no systematic data on changes in incidence of cardiovascular disease or diabetes in this population, hypertension, and gall bladder problems are ranked high within the 10 most-cited reasons for doctor's appointments at local health centers and hospitals (Departamento de Información y Estadísticas, 1999). As is the case in many Native American groups in North America, the Gran Chaco Indians may be more sensitive to the metabolic derangement associated with obesity. The presence of obese adults and undernourished children in the same household represents a serious public health challenge that will require a review of nutrition intervention programs and a culturally sensitive health education plan (Valeggia & Ellison, 2001b).

References

Alegre, I., & Francia, T. (2001). *Historias nunca contadas.* Buenos Aires, Argentina: Ediciones del Tatú.

Boggiani, G. (1899). *Cartografía lingüística del Chaco. Estudio crítico sobre un artículo del Dr. D. G. Brinton.Rev. del Instituto Paraguayo 3* (Compendio de etnografía paraguaya moderna): Asunción Paraguay.

Bove, R. M., Valeggia, C. R., & Ellison, P. T. (2002). *Girl helpers and time allocation of nursing women among the Toba of Argentina. Human Nature,* 13(4), 457–472.

Braunstein, J. (1983). *Algunos rasgos de la organización social de los indígenas del Gran Chaco* (Publication No. 2). Buenos Aires, Argentina: Universidad de Buenos Aires, Instituto de Ciencias Antropológicas, Trabajos de Etnología.

Braunstein, J., & Miller, E. (1999). Ethnohistorical introduction. In E. Miller (Ed.), *Peoples of the Gran Chaco.* Westport, CT: Bergin & Garvey.

Cohn, M., Valeggia, C., & Ellison, P. T. (2001). Child care-taking and maternal activities in a Toba community, Formosa, Argentina. *American Journal of Physical Anthropology* (Suppl. 32), 51.

Costanzó, J., Villaroel, M., Kayser, A., Barrios Centurion, R., Mendoza, L., & Gimenez, M. (2001). *Diagnóstico de salud area programática Pozo de Maza—Año 2000.* Formosa, Residencia de Medicina General, Ministerio de Desarrollo Humano, Provincia de Formosa.

Delucchi, M., Fontan, M., Grichener, S., & Wassner, M. (1996). *Proyecto de saneamiento básico integral: Barrio Namqom, Formosa.* Formosa, Argentina: Convenio SDS-UNICEF.

Departamento de Información y Estadísticas (1999). *Principales causas de morbilidad según grupos etáreos y sexo.* Formosa, Argentina: Ministerio de Desarrollo Humano.

Faulkner, K. M., Valeggia, C., & Ellison, P. T. (2000). Infant growth status in a Toba community of Formosa, Argentina. *Social Biology and Human Affairs,* 65(1), 8–19

Gelles, R. J., & Lancaster, J. (Eds.) (1987). *Child abuse and neglect: Biosocial dimensions.* New York: de Gruyter.

Gomez, T., Morales, L., Aride, I. G., Balonchar, S., & Jofre, C. S. (1998). *Cultura de la alimentación Toba.* Simposio, "Los pueblos indígenas y la salud", Academia Nacional de Medicina Sociedad Argentina de Pediatría.

Gordillo, G. (1995). La subordinación y sus mediaciones: Dinámica cazadora-recolectora, relaciones de producción, capital comercial y estado entre los Tobas del oeste de Formosa. In H. Trinchero (Ed.), *Producción doméstica y capital: Estudios desde la antropología económica.* Buenos Aires, Argentina: Biblos.

Idoyaga Molina, A. (1976/1977). Aproximación hermenéutica a las nociones de concepción, gravidez y alumbramiento entre los pilagá del Chaco Central. *Scripta Ethnológica,* 4(2), 78–98.

Karsten, R. (1926). *The civilization of the South American Indians.* London: Kegan Paul, Trench, Trubner and Co.

Karsten, R. (1932). Indian tribes of the Argentine and Bolivian Chaco: Ethnological Studies. *Societas Scientiarum Fennica,* 4(1), 10–236.

Karsten, R. (1967). *The Toba Indians of the Bolivian Gran Chaco.* Oosterhout, NB: Anthropological Publications.

Lehmann-Nitsche, R. (1908). Estudios antropológicos sobre los Chiriguanos, Chorotes, Matacos y Tobas (Chaco Occidental). *Anales del Museo Nacional de La Plata.*

Mason, J. A. (1963). The languages of South American Indians. In J. Steward (Ed.), *Handbook of South American Indians.* New York: Cooper Square.

Mendoza, M. (2002). Band mobility and leadership among the Western Toba hunter-gatherers of the Gran Chaco in Argentina. New York: Edwin Mellen Press.

Mendoza, M., & Wright, P. G. (1989). Sociocultural and economic elements of the adaptation systems of the Argentine Toba: The Nacilamolek and Taksek cases of Formosa Province. *Archeological Approaches to Cultural Identity,* 243–257.

Métraux, A. (1937). Etudes d'ethnographie Toba-Pilagá (Gran Chaco). *Anthropos,* 32.

Métraux, A. (1946). Ethnography of the Gran Chaco. In J. Steward (Ed.), *Handbook of South American Indians*. Vol. 5, US Government Printing Office, Smithsonian Institution Bureau of American Ethnology, Washington, DC.

Métraux, A. (1967). *Religions et magies indiennes d'Amérique du Sud*. Paris: Gallimard.

Miller, E. (Ed.). (1999). *Peoples of the Gran Chaco. Native peoples of the Americas*. Westport, CT: Bergin & Garvey.

Miller, E. S. (1967). *Pentecostalism among the Argentine Toba*. Doctoral thesis, University of Pittsburgh, Pittsburgh, CA.

Miller, E. S. (1980). *Harmony and dissonance in Argentine Toba society*. New Haven, CT: US Human Relations Area Files.

Nordenskiöld, E. (1912). La vie des indiens dans le Chaco. *Revue de Géographie, 6*(3), 4–130.

Paulotti, O. L. (1948). Los Toba: Contribución a la somatología de los indígenas del Chaco. *RUNA 1*, 9–96.

Peña, M., & Bacallao, J. (1997). *Obesity and poverty: A new public health challenge*. Washington, D.C., Pan American Sanitary Bureau, Regional Office of the World Health Organization, Scientific Publication No. 576.

Programa NacyDef. (2000). *Informe interno sobre estadísticas vitales de la Provincia de Formosa para el año 2000*. Formosa, Argentina: Ministerio de Desarrollo Humano, Departamento de Vigilancia Epidemiológica.

Ranaivoarisoa, M. Y., & Ventura, C. (1998). *Evaluación del estado socio-nutricional de los aborígenes formoseños a partir del censo de talla de escolares de primer grado*. Simposio, "Los pueblos indígenas y la salud", Academia Nacional de Medicina, Sociedad Argentina de Pediatría.

Schmidel, U. (1970). *Viaje al Río de la Plata, 1534–1554*. Buenos Aires: Plus Ultra.

Tola, F. (1999). Fluidos corporales y roles paternos en el proceso de gestación entre los Tobas orientales de Formosa. *Papeles de Trabajo, 8*, 197–221.

Tola, F. (2000). La restricción sexual en la lactancia y la "lucha entre hermanos" en un grupo Toba de Formosa. *Anales de la Sociedad Científica Argentina, 228*(2), 27–38.

Tola, F. (2001). Relaciones de poder y apropiación del 'otro' en relatos sobre iniciaciones shamánicas en el chaco argentino. *Journal de la Société des Américanistes* (in press).

Torres, E. O., Cabutti, N. F. D., & Palatnik, M. (1973). *Aspectos biomédicos*. Genética de la población Toba del Chaco Argentino, Universidad Nacional de La Plata (UNLP).

Uauy, R., Albala, C., & Kain, J. (2001). Obesity trends in Latin America: Transiting from under- to overweight. *Journal of Nutrition, 131*(3), 893S–899S.

Valeggia, C., & Ellison, P. T. (2001a). Lactation, energetics, and postpartum fecundity. In P. T. Ellison (Ed.), *Reproductive ecology and human evolution*. New York: de Gruyter.

Valeggia, C., & Ellison, P. T. (2001b, Spring). Nutrition, breastfeeding, and fertility: Changing lifestyles and policy implications. *DRCLAS News*.

Valeggia, C., & Ellison, P. T. (n.d.). *Lactational amenorrhea in well nourished Toba women of Formosa, Argentina*. Manuscript submitted for publication.

Valeggia, C., Faulkner, K. M., & Ellison, P. T. (2001). Crecimiento en lactantes de una comunidad Toba de Formosa. *Archivos Argentinos de Pediatría* (in press).

Vitar, B. (1999). Prácticas abortivas entre las indígenas chaqueñas en el siglo XVII. In C. M. Cóceres (Ed.), *Ethnohistoria/CD-ROM*. Buenos Aires Argentina: Equipo NAyA (Noticias de Antropología y Arqueología).

Wright, P. G. (1990). Crisis, enfermedad y poder en la Iglesia Cuadrangular Toba. *Cristianismo y Sociedad, 28*(3), 15–37.

Wright, P. G. (1997). *Being-in-the-dream: Postcolonial explorations in Toba ontology*. Doctoral dissertation, Temple University, Philadelphia, PA.

Badaga

Paul Hockings

ALTERNATIVE NAME

Burgher (early 19th century only).

LOCATION AND LINGUISTIC AFFILIATION

The Badagas are peasant farmers only found on the Nilgiri Hills, a small district in the northwest of Tamil Nadu State, India, where they have lived for the past four centuries. They inhabit over 400 villages which are not multi-caste; these are mainly between 5,500 and 6,800 ft in elevation, and lie about 11°N of the Equator. The people speak Badaga, a language of the Dravidian family closely related to Kannada (Kanarese) and Kurumba; all three are in the South Dravidian subfamily.

OVERVIEW OF THE CULTURE

A community of refugees from the plains to the north, the Badagas had to cut fields and village sites out of the forests, mainly during the period 1565–1800. Some swidden cultivation continued until the 1870s. In fields near the villages they grew millets, barley, wheat, and various European vegetables; also cows, buffalo, and poultry were kept. During the 20th century potato, cabbage, and tea became major cash crops, and many educated Badagas moved into urban, professional jobs. Today, except for some 3,000–4,000 Christians, the Badaga community consists of about 150,000 Hindus (2003), of whom a small minority are Lingayats.

Most villages have only several hundred inhabitants, some much less. Each village is surrounded by fields, and usually includes one or two Hindu temples in addition to the several rows of houses and a few cowsheds. These usually lie along the slope of a hill on its leeward side, as protection from the westerly monsoon. Most villages have piped water coming to communal taps, but not long ago the water supply was a nearby stream or at best an open channel running into the village from a stream. Recent water shortages caused by irrigating fields for the first time have led to some villages depending on water trucks. Each village has a green, used for grazing calves and as a danceground or playground, and for certain ceremonies.

The Badaga society was traditionally a chiefdom, and they are still nominally under a paramount chief. This is a hereditary position, always held by the chief of one particular village. Below him are four regional headmen, each traditionally in charge of all the Badaga and Kota villages in one quarter of the Nilgiri Plateau. At the most local level a village has its own headman, and a number of contiguous villages make up a commune with its headman too. At each level—village, commune, region, and Nilgiri Plateau—there is a council for Badaga affairs, its juridical authority now greatly undermined by modern lawcourts and the Indian legal system. Prior to the 20th century disputes would have been settled at one or another level of this council system; major land disputes, ceremonial improprieties, and cases of murder probably reaching the Nilgiri-wide council.

The community is divided into a number of ranked phratries which are mainly endogamous. A conservative Lingayat group forms the top phratry, the Wodeyas, while the headmen's former servants, the Toreyas, are at the bottom. Between these two extremes there are one other phratry of vegetarians and three more of meat-eaters. The Christian Badagas, springing from the first conversion in 1858, now constitute a separate meat-eating phratry. Each phratry is made up of several exogamous clans, which in turn are made up of various lineages.

THE CONTEXT OF HEALTH: ENVIRONMENTAL, ECONOMIC, SOCIAL, AND POLITICAL FACTORS

A modern biomedical listing of the most prevalent diseases among Badagas recently would include pneumonia, typhoid, dysentery and diarrhea, diphtheria, smallpox, rickets, intestinal parasites, tuberculosis, anemia, food deficiency diseases, conjunctivitis, and various skin diseases.

There is now an urbanized middle class consisting of educated Badagas who work in a wide variety of professions. Thousands of Badagas have graduated from South Indian colleges and universities, including medical schools. Modern biomedicine, practiced at several local hospitals and numerous clinics, is only one attractive profession, for doctors and nurses: many other graduates have been drawn toward law, teaching, administration, banking, plantation agriculture, and other specialized professional callings. As well as a hospital in each of the four neighboring towns, large villages have clinics staffed by the Tamil Nadu government medical department. Family planning campaigns have reached into most villages in the past 30 years. The relatively high rate of literacy in the late 20th century meant that thousands could and did read medical advice columns in regional newspapers or magazines, either in Tamil or in English.

MEDICAL PRACTITIONERS

Although today the only practitioners found in the Badaga community are likely to be doctors and nurses trained in biomedicine and working in the state's health-care system, until perhaps 25 years ago there were therapists still functioning in an indigenous, mainly herbal system. While most of these were unschooled general practitioners, there were also some specialists in many of the villages. The commonest was the midwife, always an experienced older woman; a big village might have several. In addition one could encounter exorcists who were necessary in instances of ghost possession; bone-setters; and some specialists who only handled one type of illness, such as fevers. Although by the end of the 20th century the therapists had disappeared, some of the other specialized practitioners remained, for their services might still be needed. These thus included midwives (now with some medical training), exorcists, and perhaps specialists handling a particular kind of illness.

CLASSIFICATION OF ILLNESS, THEORIES OF ILLNESS, AND TREATMENT OF ILLNESS

Since there are no Badaga medical colleges and no textbooks (this having been a non-literate society until the late 19th century), there has been no formalization of medical theory except by the present author.

Indigenous belief was based on a lack of knowledge of the internal organs, especially in the vegetarian phratries where nobody had ever butchered an animal. As a consequence, it was not known that the heart existed, and other anatomical knowledge was similarly inadequate. The following outline of human physiology was given to me by an elderly male therapist who had never been to school.

The chest contains one organ, the *nenju* (which would in biomedical terms encompass both heart and lungs), which is of no great importance in the Badaga understanding. For them the central organ is the belly. There are some 16 organs in the body and nearly all are somehow connected with the belly: this is responsible not only for digestion but for blood circulation, reproduction, breathing, smell, even sights and speech. Food goes through the mouth into a tube, and then passes down to a stomach. As with cattle (!) there is a second stomach adjoining it, where the body stores liquids, which reach it from the mouth by a separate tube. A third tube links the two stomachs. The mixing of food from these two stomachs occurs in the belly, which lies around and below the navel. Here blood is prepared from the mixed food and liquid, and is sent to all parts of the body through arteries. The actual digestion is assisted by worms about the thickness of a pencil, which live in the belly and eat the food there. They are essential to keep a person alive. An important unit within the belly is the colon (*karu*), which must remain erect, otherwise the person will die. When breathing occurs, the air passes through the nostrils, the mouth, the eyes, and the ears, by a single tube to the *nenju*, and then on to the belly, where it is needed by the worms for digestion. The evidence for this state of affairs is in flatulence, and in the occasional discovery of worms in someone's stool. The head is important for sense and understanding, and it is recognized that the brain does the thinking. It is not, however, linked with the eyes, ears, mouth, or any other organ. Speech originates in the belly, and the tongue helps in forming the words. The larynx has nothing to do with speech, being simply the gateway through which the several tubes pass carrying air, food, and liquid. The existence of a skeleton is recognized, and it is seen as God's gift to help people stand upright and work.

A biomedical model would recognize heuristically several rather separate subsystems: the nervous system, the reproductive system, the digestive system, the circulatory system, the skeletal system, the lymphatic system, etc. The Badagas in general only see one system operating, but there is a somewhat separate subsystem for reproduction. God has distinguished between the two sexes by giving facial hair to men, breasts to women, etc., and semen only to men but menses only to women. Inside the belly of women, but not men, there is a bag where the baby grows. The fetus is formed entirely from a man's semen, and so copulation is essential for a woman to conceive. After digestion has occurred in the pregnant woman's belly, blood goes to the fetus and it grows on this. After birth milk flows from the belly to the breasts, and it is sucked out through the nipples, which are thought of as the outer part of the *nenju*.

Badagas do not recognize any particular god as causing all disease, or being able to cure it. Five diseases, however, are brought by five "sister goddesses." These include the dreaded smallpox, as well as measles, chicken-pox, rubella, and acute conjunctivitis. Cholera is also caused by an angry god, but is thought to be contracted in the plains. Mental illness, like a form of hysteria, is caused by an ill-intentioned ghost or bad spirit getting into the patient, usually a woman. Non-Badaga sorcerers or shamans, commonly said to be Kurumbas, can cause an evil spirit to enter a person. Other mental problems may also result from the sins of one's ancestors. People sometimes think of "bad blood" in parents as causing abnormalities in their children. Some illnesses are caused by personal uncleanliness, or bad diet. Eating too quickly, or having insufficient dairy products, are thought to cause the body to "over-heat" and thus bring illness; buttermilk is a common cure. Over-exertion, including sexual excesses, can bring illness, usually because the colon or an artery gets dislodged, it is said.

Among unweaned infants, the odor that accompanies coitus, or the "bad wind" that emanates from a corpse or from childbirth, can bring sickness and even death. Bites, stings, and burns are recognized as being caused by obvious external agents; but depigmentation of the skin is thought to be caused by "the bite of the blindworm" (which is actually harmless). Some more modern theories of causation may have their origin in scientific medicine. For instance, some people recognize that tiny germs cause tooth decay; and in the case of plague it is thought that mosquitoes carry the disease from rats to people. (This may be the effect of a mosquito eradication program on the lower slopes of the hills.)

As with any medical system, the techniques of treatment depend on the diagnosis. As there are many dozens of recognized ailments, there are dozens of distinct treatments, usually involving the swallowing of a herbal medicine made up by the therapist from freshly gathered wild herbs from the neighborhood. The general term in Badaga for "a medicine" is *maddu*, but it should be understood that this word also encompasses "magical potion," "opium," or "poison."

SEXUALITY AND REPRODUCTION

It is by God's grace that only women produce milk. Semen is likewise produced by the belly, whence it passes to the penis during coitus. The function of the testicles is simply to act as a sort of counterweight to aid with the erection. Since the fetus is formed from the father's semen, copulation is essential before a woman can conceive. Her subsequent role is to nurture the developing fetus with blood from her belly.

Adolescents may engage in some premarital sexual behavior, for example in cowsheds or in the forest, but modern requirements of schooling tend to reduce the opportunities for this. In previous times young people might spend much of the day watching the fields or the herds, and many opportunities for casual sexual relations occurred. Today girls expect to retain their virginity until marriage. At the same time, the usual age at marriage has increased from about 12–13 for girls to around 20 or more, and young men now marry in their early 20s too. The latter are now likely to have their first experiences with urban prostitutes, or sometimes with their elder brothers' wives.

HEALTH THROUGH THE LIFE CYCLE

Pregnancy and Birth

If a woman has not menstruated for 40 days she will be taken to a midwife or other therapist, who will tell from the breasts whether the woman is pregnant, as the nipples become darker. Other signs are if she feels lazy about work, even about walking, experiences nausea, does not want to eat, or requires unusual foods. If pregnant, she should have no intercourse during the first 3 months, and thereafter seldom. The above signs tend to disappear after 60 days. Thereafter there are certain restrictions on the couple. The man should grow a beard and moustache, but should not kill a snake lest its spirit enter the fetus. The woman should not eat red onion or garden marrows, or the baby will have itches when born. The woman must not wash her clothes or put on clean ones, must not cross any large river, or the hamlet boundary, must not go to any festival where she might see something interesting. These restrictions only last till the 90th day. After that time the woman can become active again and eat what she likes. By now, she can feel something in her stomach. From the start of the fifth month the fetus moves occasionally.

In the fourth, sixth, and seventh months the woman can be carefree, and in the even numbered months, the fourth, sixth and eighth, she may cross the hamlet boundary and visit her father's house (Badaga villages are exogamous). If her eyes turn red, she gets diarrhea, feels uneasy, and her lips itch, then these are signs that the body is "heating": excess of "heat" may cause a stillbirth in the seventh or eighth month. Therefore the woman is given buttermilk or butter. Buffalo milk freshly milked into a pot in which there is some lime juice is another cure, if drunk immediately. This is given to her every morning for two or three weeks before breakfast.

In an odd-numbered month of the first pregnancy, generally the fifth or seventh, the couple go through a thread-tying ceremony on an auspicious day that confirms their marriage. The relationship has become stable with the pregnancy, although this is not mentioned during the event. Thereafter divorce becomes more difficult. Supposedly the couple sleep side by side that night, with a stick between them to symbolize the coming child, and they have coitus.

In the ninth month the woman is usually at her father's house, and this is where birth should occur. In the ninth or tenth month she will give birth in the presence of a midwife and a few other women, including some young ones interested in learning the craft. (Small girls are usually not invited.) If the pregnancy takes longer there may be twins; one of these is likely to be a weaker child. But both are cared for. In cases of a particularly difficult birth the woman is given a mixture of dried ginger, pepper, cumin, clove, crude sugar, sweet flag, coffee powder, and clarified butter, mixed in particular proportions with water. It is said that long ago a cesarian section might be performed sometimes, using the small knife with which the umbilicus is cut. Some midwives also know how to reach into the womb and remove a dead fetus. Should the woman die in childbirth, the fetus must be cut out before her funeral.

Male relatives, including the husband, must wait outside during the birth. If there are difficulties they will offer advice or go and fetch other women. Because the act of birth was considered impure it would not occur traditionally inside the house but on the veranda. Now it generally does occur in one of the outer rooms, but almost never in a hospital. The door is kept closed: the woman's body is naked during delivery, though a cloth may be draped over her shoulders for warmth. In earlier times the grindstone was in front of the woman, to hold onto. Now it is not unusual for the woman to lie down to deliver, in the modern way. Commonly the woman kneels on the floor, with a girl in front to support her. Her legs are apart, and the midwife is behind her to catch the newborn before it touches the floor. Gingelly oil is sometimes applied to the vagina to ease the delivery, and the midwife will check to make sure the umbilical cord is not around the baby's neck. Some can put a hand inside to help the baby out, but only as a last resort in a difficult case. If the feel of the stomach surface tells the midwife the baby is in the wrong position, then the mother is made to place her head on the pounding stone, which is set into the floor, and five or six women support her as she balances there, upside down with her legs in the air. The knowledgeable midwife then shakes the woman's legs until the baby comes into the right position for birth.

The newborn is laid on sacking on the floor. The umbilicus is tied with a piece of string (formerly a thread lampwick) some 8 cm from the belly. This curtails bleeding, and then the cord is cut with a penknife or razor blade, or formerly with a reaping blade. A couple of handfuls of cold water are poured on the infant's shoulders or dabbed on with a cloth. In particular the left shoulder is wetted if a boy, or the crown of the head if a girl, to make the child strong for carrying loads. It is then bathed in very hot water. Groundnut oil (formerly castor oil) is put all over the body. The mother also takes a bath in very hot water, and then sits to drink some cold milk. Some women now take a sip of brandy. A little opium in a cup of tea or coffee is a cheap alternative. A further dose of opium may be given on the following two days (it was a traditional Badaga product). The new mother is also given a little palm jaggery (crude sugar) and three or four cloves of garlic; one piece is put into each of her ears. The child is then given back to the mother, its face to her right breast (right is the auspicious side). It is given a drop of groundnut oil to drink, and then begins to suck. A lime is then cut in half, and on each cut surface someone puts fresh cowdung, a blade of Bermuda grass, and some clarified butter. An old lady takes the first half and runs it down the right side of the baby, not quite touching it, as she says: "This child belongs to both the father and the mother" (who remain unnamed). Thus the joint parentage is asserted. The half lime is sucked by the mother, touched to her child's mouth, and thrown away. The same thing is done with the other half lime, running it down the left side of the baby. Lime juice is thought to ward off evil spirits. The afterbirth comes half an hour later. In modern practice, some midwives, if properly trained, pull the afterbirth out. It is carried outside and buried somewhere distant from the village. A man may dig the hole but would not touch the polluting afterbirth.

Infancy

After the child has sucked the breast for the first time, he or she is given a small piece of bezoar. After a few days some burnt rhizome of the sweet flag is also fed to the baby. After a year or two the infant might be taken on a pilgrimage to a nearby temple, where during an annual festival mendicants were given consecrated pieces of banana mixed with crude sugar. This they chewed a little, then spat into the hands or mouths of devotees, who either ate it themselves or fed it to their children because it was believed to cure all disease. The main food of the infant is mother's milk, but the child will be weaned at about one year. If the mother eats too many sweet things, she may transmit small worms to her baby through the breast milk. She avoids going near any corpse, or the smell of another woman's birth, or the odor of anyone other than her husband in the early morning, or hearing any frightening story, for fear that any of these "bad winds" may cause her flow of milk to stop. Anyone who comes to look at the baby just after having had coitus, and without bathing first, puts the baby at severe risk; as does someone who has just come from a great distance, if he does not first rest outside the house for awhile.

Three or four days after the birth the mother begins to bathe her baby, perhaps with help from her own mother. The baby is bundled up in old rags, and always kept warm. Care is taken about keeping the face clean, mainly so that a cat will not harm the baby in some way. During the first few weeks the cranium is elongated by manipulation, using oil, though the practice is now rare. Early anthropologists, not knowing of this, commented about the anomalous dolichocephaly of a people whose origin was known to be among the mesaticephalic and brachycephalic peasants of Mysore.

Until the ninth day a baby is thought to have no understanding, though it has vision. After the 40th day mothers try to make the baby laugh, and also take something away to make it cry: this to determine that the child will not be dumb. While a child might be born dumb it cannot be born deaf; so until the fifth or sixth month people are careful not to make a loud noise near the child which could induce deafness. After the third month of its life, the mother will resume her normal work duties.

Childhood

At about the fifth month the child is made to sit up, surrounded by blankets. In the sixth or seventh month he will start to crawl, and will then be allowed to sleep on the floor. People encourage him to crawl, stand, and walk. At around 12–18 months children begin to talk. For common things like sleep, mat, rice gruel, etc. there are baby words that people use. It is primarily the old people in the household who educate their grandchildren and great-grandchildren. The parents may be out working in the fields much of the day. Now that most children are going to school, this interaction with the elderly mainly occurs in the evenings.

Small children are given relative freedom in their actions, and are usually disciplined with threats: the coming of sorcerers, ghosts, mendicants, demons, or vaccinators to take them away are common threats. Otherwise kindness and outright bribery are lavished on small children. Now between the ages of 5 and 7 years they find themselves being sent off to school. Virtually every village larger than a hamlet has at least one; some have several, and recently private English-language elementary schools have sprung up here and there, as Badagas place much stock in good education today. Children become enculturated partly through school activities, partly through doing things in the village with their friends, and partly from tales their elders tell them. As elsewhere in India, small children look to their older siblings for guidance and advice. Outdoor group games are popular with the very young, including hopscotch, string figures, and board games.

The progress of childhood is marked by a series of ceremonies: naming, before the 40th day of life; head shaving in a temple, within a year of the birth; ear boring, often done at the same time; the first tooth ceremony, if a boy gets his first tooth in the upper jaw; nostril piercing, done in a girl's ninth, eleventh, or thirteenth year on an auspicious day; tattooing, formerly done on girls at about the start of puberty; milking initiation for boys, which is no longer done now that most people have no cattle; and initiation for those boys who are of the Lingayat sect. Otherwise there is no observance for a boy's puberty, but complex ones for a girl's.

Adolescence

Menarche usually comes at the age of 14 or 15, sometimes earlier. Girls at about puberty used to be tattooed on the brow, shoulders, and forearms with distinctively Badaga patterns. The tattooing was done, without any ceremony, with a thorn from one of two local plants or, more recently, with pins or needles, using soot scraped from the bottom of a pot for color. Traditionally, a girl's initiation into adulthood was essentially her marriage ceremony, which would occur at about menarche. Nowadays, child marriages do not occur. There are no maladies recognized as specific to the period of adolescence, and no bodily mutilations mark any male initiation ceremony.

Adulthood

Adult women might add further tattoos to the shoulders, forearms, or back of the hands while they were in the menstrual hut. Each village of adequate size would have its own hut, or several small hamlets would share one, until the mid-20th century. There is no theory of why menstruation occurs: it is simply God's will. A woman should be segregated from food preparation activities for six days.

The Aged

Old people are shown respect, but Badaga proverbs suggest that people are well aware of the infirmities that come with advanced age: "A man above sixty years is said to have half-sense"; "If the lord speaks, the whole village quakes; if an old man speaks there is a sound of babbling"; "The activity of a 20-year-old is pleasure-loving; the activity of an 80-year-old is that of old age." Cynical though all of these may sound, respect is enjoined: "If a person cannot see, hear or walk, as he is such an old man, you must first ask his permission to do anything." It is believed that if a man wears gold his lifetime will lengthen; whereas if he sees two crows in sexual congress then he will die within a year. If he feels he is too young to die, then he can countermand the omen by going up a hill near his home and shouting out to the villagers that he—mentioning his name—is dead. (A woman might also persuade a man to do this for her.) Making sure no one sees him, this act will have the effect of persuading some neighbors to go to his house to pay respects to the corpse; and this process of misleading the mourners is supposed to prolong life.

Dying and Death

When death is thought to be approaching, messages are sent out to other villages inviting relatives and friends of the dying person to come and bless him or her. They bring gifts of grain and milk, offer their blessings, and hope to receive some from the dying person too. Only very close relatives will wait there for death to occur.

Once a person has died, the village headman is called, and a senior member of the family tells him, "This corpse is for you." Accordingly, it is the headman who arranges a communal funeral for the deceased and not his own family, though they will meet much of the expense. Another noteworthy feature of the Badaga funeral is that priests are not involved ceremonially. Most Badagas cremate their dead, but the high-status Lingayat groups and the Christians bury their dead in cemeteries.

The funeral is the most complex of all Badaga ceremonies, and sometimes used to take several days to complete. Now it is always done in one day. It has to symbolically handle the public-health problem of a corpse in the house; the emotional trauma felt by still living members of the bereaved family who must all adjust their lives; and the social trauma occasioned by the loss of a usually elderly village resident. A full account of the Badaga mortuary rites and their symbolism is included in Hockings (2001).

Changing Health Patterns

Although several hospitals have been in the district since the mid-19th century, Badagas rarely went near them unless a patient was clearly dying and beyond all hope. Until the discovery of antisepsis in the 1870s, such hospitals were themselves a serious threat to health anyway. By the time this Badaga system was studied by the author during 1963–1972, only a handful of elderly practitioners remained alive in the Badaga villages. Once the prime informant, K. Sithamma, died (July 25, 1976), there were scarcely any other active general therapists, as Badagas were generally availing themselves of biomedical procedures in the local clinics and hospitals, or in some cases using Ayurvedic (q.v.) medications and therapists, and their midwives were going through a short program of training. But for the one book published on Badaga medicine, the system would quickly have become forgotten.

Bibliography

Note: There are only one book and one article on the Badaga medical system:

Blasco, F., & Fauvel, M. T. (1977). Plantes médicinales des Nilgiri. *Journal d'Agriculture tropicale et de botanique appliquée; Travaux d'ethnobotanique et d'ethnozoologie, 24*, 23–39.

Hockings, P. (1980). *Sex and disease in a mountain community*. New Delhi: Vikas; Columbia, MO: South Asia Books.

Related texts:

Hockings, P. (1988). *Counsel from the ancients, a study of Badaga proverbs, prayers, omens and curses*. Berlin and New York: Mouton de Gruyter.

Hockings, P. (1989). The cultural ecology of the Nilgiris District. In P. Hockings (Ed.), *Blue Mountains: The ethnography and biogeography of a South Indian region* (pp. 360–376). New Delhi and New York: Oxford University Press.

Hockings, P. (1999). *Kindreds of the earth: Badaga household structure and demography*. New Delhi and Thousand Oaks, CA: Sage.

Hockings, P. (2001). *Mortuary ritual of the Badagas of Southern India*. (*Fieldiana, Anthropology*, n.s., 32.) Chicago: Field Museum of Natural History.

Bangladeshis

Darryl J. Holman and Kathleen A. O'Connor

ALTERNATIVE NAME

Bangladeshis are often called Bengalis.

LOCATION AND LINGUISTIC AFFILIATION

Bangladeshis make up the largest ethnic group of The People's Republic of Bangladesh, located in Southern Asia. Most Bangladeshis live in the Bengal Basin, an alluvial plain of the Jamuna River, the Padma River, and the Meghna River. Bangladesh shares borders with India to the west, north, and east, and a small southeastern border with Myanmar (Burma); the southern border is the Bay of Bengal.

The term *Bangladeshi* refers to both a national identity (a citizen of Bangladesh), and a member of the Bengali-speaking majority of Bangladesh. The latter use excludes about 250,000 Biharis and about a million tribal people located in hilly regions along the eastern borders of Bangladesh (Bertocci, 1984). A defining characteristic of Bengali culture is the language Bangla (or Bengali), so that a *Greater Bengal culture* (Basu & Amin, 2000) includes West Bengal, India, and parts of Assam, India.

Bangla is the second largest language spoken in South Asia, after Hindi-Urdu. The language falls on the Indo-Aryan branch of the Indo-European family of languages. Bangla is closely related to other South Asian languages such as Assamese, Hindi, and Nepali. Written Bangla is derived from Sanskrit (Maloney, 1974).

OVERVIEW OF THE CULTURE

The Bangla-speaking majority of Bangladesh numbered about 126 million in 1999. About 87% of Bangladeshis are Sunni Muslims, 12% Hindu, and 1% Christian, Buddhist, and animists (Bangladesh Bureau of Statistics, 2001).

Bangladeshis share deep cultural roots with many other people in South Asia. These shared ideas and traditions transcend the major religions that have swept through the Bengal Basin, including Hinduism in the first millennium BC, followed by Buddhism, and finally Islam beginning about the 12th century (Maloney, 1974).

European trade with the region began at the end of the 17th century. By the mid-18th century, the British had taken political control of the region, and a period of colonial rule began that lasted through the mid-20th century. Colonial rule brought about large-scale political and economic reorganization. The region changed from being a producer of export goods to a producer of raw materials used by the industries of Great Britain. The transportation infrastructure underwent substantial development, but political changes, agricultural policies, and land reforms drained the region of much of its prosperity (Bose, 1967).

British political rule ended in 1947, when India and Pakistan attained independence. Pakistan was formed as a single Islamic nation that included present-day Pakistan and present-day Bangladesh (formerly, East Pakistan). The 2,000 km separation between the two regions, along with linguistic and sociocultural differences, led to political tensions, and finally secession of East Pakistan in March 1971. A brutal war between the two regions followed that led to the formation of Bangladesh (Islam, 1978). The new country was under military rule from 1975 to 1990, and has been under civilian rule, with considerable political unrest, since.

A male-headed family is the basic social unit for Bangladeshis. Patrilineally related households are grouped into a common homestead called a *bari*. The extended family that makes up the *bari* functions as a single economic unit, headed by a senior couple, their sons, daughter-in-laws, and unmarried daughters. The senior male has authority over members of the *bari*, and his wife maintains authority over her son's wives. Newly married couples usually establish a new household within the male's father's *bari* (Maloney, Aziz, & Sarker, 1981; Rob & Cernada, 1992).

Muslim Bangladeshis practice *purdah* (literally, veil or curtain), including the seclusion of women from public observation, wearing of concealing clothing, and use of screens or curtains to hide women within the *bari*.

The degree of observance varies widely by region, education, and socioeconomic condition. The practice of *purdah* is believed to protect women from evil and maintains family respectability. Some adherents see *purdah* as a symbol of piety and purity, that provides religious fulfillment. On the other hand, *purdah* can severely restrict women's mobility, and limits access to education, occupational opportunity, and health care (Maloney et al., 1981).

Bangladeshi Muslim society is not highly stratified, in accordance with the egalitarian principles of Islam. The Hindu caste system, while present, does not figure prominently in Hindu communities because most Bangladeshi Hindus belong to lower castes or outcaste groups. Additionally, the generally low socioeconomic conditions in rural areas have led to a reduction in caste consciousness (Maloney et al., 1981).

The country of Bangladesh ranks 73 out of 90 in the United Nations Human Poverty Index (UNDP, 2001). The population density of the region was 900 people per km^2 in 2001, making it the most densely populated non-island area of the world. Economic development over the last 50 years has been hampered by frequent natural disasters and political turmoil (Begum, 2001). Roughly, 7% of the urban population lives in slums; 79% of Bangladeshis live in a rural setting (Bangladesh Bureau of Statistics, 2001), where most households have small plots of land or are landless (Turner & Ali, 1996).

People in rural areas typically are involved in agriculture through sharecropping or wage-labor on larger farms. The primary crop is rice (77% of cultivated land), wheat (6%), and jute (4%). Agriculture makes up about one third of the employment; about 8% are employed in manufacturing, 9% are employed in the business sales sector, and 20% are employed in other service industries. Something over one third of Bangladeshis are unemployed. Children from ages 10–14 make up 12% of the total labor force. Government statistics indicate that almost half of Bangladeshis were below the poverty line in 1996 (Bangladesh Bureau of Statistics, 2001). Independent estimates suggest a figure closer to 87% (Turner & Ali, 1996).

Six years of primary education is compulsory for Bangladeshi children. Even so, economic need results in many children dropping out of school early. Recent government-sponsored Food for Education programs raised primary school retention rates to 70% (Mputu, 2001). Government statistics place literacy at 68% for males and 51% for females. Begum (2001) places literacy at 32%, and suggests the proportion functionally literate is probably lower. A study that tested people for basic skills found 28% could read, 13% could write, and 37% had oral mathematical skills (Greaney, Khandker, & Alam, 1998). Literacy among people living in slums was less than 15% in 1997 (Bangladesh Bureau of Statistics, 2001).

THE CONTEXT OF HEALTH: ENVIRONMENTAL, ECONOMIC, SOCIAL, AND POLITICAL FACTORS

An important consideration in understanding the health literature for Bangladeshis is that a single organization, the International Centre for Diarrhoeal Disease Research, Bangladesh (ICDDR, B), produces most of the health-related research from Bangladesh (Rahman, Laz, & Fukui, 1999). The ICDDR, B has been conducting intensive demographic and medical research since the early 1960s. Most of the research has taken place within Matlab *thana*, a rural administrative unit located about 55 km southeast of Dhaka, with a population of about 210,000 people. A continuous demographic surveillance has recorded all demographic events (births, deaths, marriages, divorces, and migrations) within Matlab since the late 1960s (van Ginneken, Bairagi, de Francisco, Sarder, & Vaughn, 1998). The area is subdivided into two regions for research purposes. The first is a Maternal, Child Health, and Family Planning (MCHFP) area in which numerous health, nutrition, disease, and family planning interventions have been implemented, beginning in 1977. Many of the interventions have had significant effects on health, fertility, and mortality, so that statistics from the MCHFP area cannot be considered representative of rural Bangladeshis. The second region within Matlab is a comparison area where individuals receive limited diarrhea-related health services but otherwise are not part of the health intervention studies. Findings from the comparison region are more representative of rural Bangladeshis, although they may be affected by diffusion of ideas and knowledge from the MCHFP area.

Bangladeshis have been a high-fertility, high mortality population for at least the last 50 years. The annual growth rate was 2.0% from 1978 to 1998 (WHO, 1999); in 1998, the growth rate was 1.5%, reflecting a recent dramatic decline in fertility (Bangladesh Bureau of Statistics, 2001). The total fertility rate in 1999 was 3.0 children compared with 6.7 children in 1978 (WHO, 1999).

The life expectancy at birth was 60 years (rural) and 62 years (urban) in 1998. By contrast, life expectancy was 51 years in 1966 (Strong, 1992). In both urban and rural settings, males had a slightly greater life expectancy than females. The infant mortality rate in 1998 was among the highest in the world at 6.6% (rural) and 4.7% (urban). Twenty one percent of Bangladeshis do not survive to the age of 40 (UNDP, 2001).

The alluvial ecosystem of the Bengal Basin provides both moisture and rich soil necessary for intensive agriculture. The ecosystem may have ample capacity to support the population, and some suggest that the widespread poverty largely results from sociopolitical conditions (Hartmann & Boyce, 1979). The wet, tropical conditions, as well as frequent natural disasters, contribute to high rates of infectious diseases such as cholera and shigella (Siddique et al., 1992). Cholera is endemic and is the leading single cause of mortality in Bangladesh, accounting for about 11% of deaths annually. Untreated surface water has traditionally been used for all household functions. Since the 1970s, major efforts have been undertaken to install tubewells to provide safe drinking water. By 2000, an estimated 80% of people had access to non-surface sources of drinking water. By 1998, 40% of rural households had installed sanitary toilets, compared with only 6% in 1991 (Bangladesh Bureau of Statistics, 2001).

Bangladeshis have endured repeated natural disasters over the last 50 years. Most of the deadliest cyclones in history have affected Bangladesh (Frank & Hussain, 1971), including a single cyclone in 1970 that killed half a million Bangladeshis (Chacko, 1991) and one in 1991 that killed 130,000 Bangladeshis. Flooding occurs annually in Bangladesh during the rainy season, causing property damage from erosion (Haque, 1988), an increase in diarrheal disease, respiratory infections, and deaths (Siddique, Baqui, Eusof, & Zaman, 1991).

The primary health infrastructure in rural areas consists of traditional medical practitioners with little formal training. Access to allopathic physicians is limited. One registered physician for every 8,600 people was reported in 1981 (Bhardwaj & Paul, 1986), and one physician for every 5,000 people in 1996 (Begum, 2001). The entire country had 91 general patient hospitals in 1997 (Bangladesh Bureau of Statistics, 2001). The overall health system ranking for Bangladesh in 1997 was 131 out of 191 (WHO, 2000).

The health-care system has improved since the 1970s, but overall health remains poor (Vaughan, Karim, & Buse, 2000). The present poor health conditions seem to result from a complex of factors including high population density, poor sanitation infrastructure, inadequate health care, inadequate education, multiple natural disasters, the war with Pakistan, and political instability since 1971. Military spending by the government of Bangladesh, until recently, exceeded health sector spending (Begum, 2001; UNDP, 2001). The government has recently reorganized the health care system and prioritized health (Vaughan et al., 2000).

MEDICAL PRACTITIONERS

Medicine is practiced at many levels in Bangladesh, including self-care, care from a non-practitioner who has gained some medical knowledge, care from paid and unpaid practitioners without licenses, and licensed professionals. A number of specialties exist, such as midwives, bone-setters, and dental practitioners (Ashraf, Chowdhury, & Streefland, 1982). Additionally, health advice and remedies are available at local "chemists" (pharmacists).

Practitioners of a number of medical traditions coexist in Bangladeshi society. Traditions are sometimes combined so that a practitioner may draw from elements of indigenous traditions and particular religious practices. Frequently, allopathic treatment is used simultaneously with traditional cures (Stewart, Parker, Chakraborty, & Begum, 1994).

A study of health practitioners in Matlab *thana* in the late 1970s enumerated 1,300 nongovernmental health practitioners, 1,500 traditional midwives, and three physicians in a government hospital (Sarder & Chen, 1981). *Kobiraj* practitioners accounted for 15% of health practitioners. These practitioners follow Āyurvedic medical tradition based on Sanskrit texts belonging to Hindu scriptures dating to before the second century AD (Basham, 1976; Gupta, 1976). *Kobiraj* are unlicensed and trained by apprenticeship. Combinations of herbal medicines, minerals, and dietary restrictions are used to cure disease and for such things as preventing conception (Aziz & Maloney, 1985). Just over half of the practitioners are men; over half have no formal education. Only 10% practice full time (Sarder & Chen, 1981).

Allopathic practitioners made up about 15% of practitioners in the Sarder and Chen (1981) study. Only one in ten allopathic practitioner was licensed after formal training in medical school. Unlicensed practitioners usually acquired their skills by apprenticeship. Most allopathic practitioners were men, and about half practice full time.

Homeopathic practitioners made up 3% of the health practitioners. This system is based on an 18th-century German medical tradition, but is often viewed as an indigenous medical system because of spiritual elements and concepts that are analogous to some indigenous systems (Leslie, 1978). In the Sarder and Chen (1981) study, one in five practitioners was registered, all were male, about half practiced full time, and most had formal education. Practitioners were typically trained by apprenticeship, although some registered practitioners had attended medical school. Homeopathic practitioners have increasingly been prescribing antibiotics and other Western medicines (Maloney et al., 1981).

A small percentage of practitioners follow the Yunani (or Unani) system (Sarder & Chen, 1981). The Yunani system is based on Aristotelian medicine, codified in Arabic and Persian texts, and was brought into the culture by Islam (Leslie, 1978). The practitioners are called *hakim* which is Arabic for "a learned man" (Basham, 1976). *Hakim* usually are men.

The most abundantly practiced system of medicine (61% of all practitioners in Sarder & Chen, 1981) was *totka*. This system is a mixture of Āyurvedic, Yunani, and shamanistic schools with no single uniform concept of illness, although supernatural causes are common. The practitioners (*totkas*) are not licensed and have no registration requirements; they train through an apprenticeship system. Two thirds have no formal education, and most practitioners are women. A practitioner specializes in one or two types of illness, and only a small percentage practice full time (Sarder & Chen, 1981).

There are other more specialized medical practitioners in Bangladeshi society. A tradition called *boneji* is practiced by elderly women as an art of concocting medicines from herbs and other substances such as honey and fruit juice (Mushtaque, Chowdhury, & Kabir, 1991). The *Badhi* are a low-caste Hindu people who sell herbal medicines out of boats. A *hāzām* specializes in performing circumcisions on boys. *Hazurs* are ritual healers associated with Mosques (Zeitlyn & Rowshan, 1997). Hindus may consult an ascetic saint (*saādhu*) for health problems. Muslims and Hindus may make use of mystic healers (*fakīr*) or magic healers (*ōjhā*) (Maloney et al., 1981).

Traditional birth attendants (*dhorunis* or *dais*) make up one of the largest groups of "health specialists," outnumbering all other health practitioners enumerated in the Sarder and Chen (1981) study. The traditional *dai* is an older women in the community, often a widow, with little formal education (Croley, Haider, Begum, & Gustafson, 1966). The term *dai* is sometimes used to denote a midwife that is associated with government family planning efforts (Islam, 1982). Rozario (1995) argues that the role of the traditional birth attendant is not one of a medical practitioner. Rather than managing and facilitating a birth, the birth attendant's role is limited to the removal of pollution associated with the placenta and blood of childbirth.

In addition to assisting with births, *dais* may also provide abortion services, provide contraceptive services, and assist women with fertility problems. Indigenous abortion practices are well known in Bangladesh. Most traditional practitioners are *dais* with no allopathic training (Bhuiya, Aziz, & Chowdhury, 2001; Rozario, 1995).

In the absence of a practitioner, medical services may be sought from any person who has some knowledge of healing skills. Aziz (1977) reports that a medical specialist was not consulted prior to death in Matlab 35% of the time.

CLASSIFICATION OF ILLNESS, THEORIES OF ILLNESS, AND TREATMENT OF ILLNESS

Nearly all Bangladeshis, regardless of religion, believe that disease occurs according to the will of a God or gods (Ashraf et al., 1982). Additional awareness of illness involves elements of pathogenicity, evil supernatural forces, environmental exposures, personal behavior, and, to some extent, heredity.

Some diseases are thought to be caused by exposure to "cold" in the environment. One source of respiratory illnesses, for example, is exposure to cold temperatures, prolonged exposure to cold water, or even cold mud (Stewart et al., 1994). Bangladeshis classify food as being intrinsically hot (e.g., meat, fish, eggs) or intrinsically cold (e.g., rice, many vegetables) (Maloney et al., 1981; Sarkar, 1982). Additionally, foods can turn cold when they become stale or are leftover. Some foods are believed to play a role in disease promotion, particularly cold foods, which are considered to cause respiratory disease (Stewart et al., 1994). Hot foods are often associated with health, vitality, and sexuality (Maloney et al., 1981).

The concept of a disease pathogen is not widely recognized by Bangladeshis without formal education. Some traditional explanations for disease approximate a pathogenic explanation. For example, some forms of diarrhea are believed to be caused by eating inappropriate substances (e.g., mud) or rotten food. Breast milk is recognized as a disease vector. A mother who is exposed to "cold" can pass the disease on to a breastfeeding child (Stewart et al., 1994). "Polluted" breast milk is understood to be the cause of diarrhea in breastfeeding children (Mushtaque et al., 1991; Zeitlyn & Rowshan, 1997).

Bangladeshis recognize heredity as playing a role in some illnesses such as asthma (Stewart et al., 1994). An important category of hereditary effects is the behaviors of a child's parents. Behavior of both parents can influence a child's health prior to conception, during conception, and throughout pregnancy. Parents' behavior can result in illness of the child, overall bad health, or a defective personality. The moral behavior of a mother, for example, is a common explanation for a child's respiratory illness (Stewart et al., 1994).

Evil influences are a common source of illness for rural Bangladeshis. The sources include *bhut* (ghosts), the "evil eye," and "bad spirits (or winds)" (*batash*). These influences are held responsible for many maladies such as tetanus, preeclampsia, and spontaneous abortion (Ashraf et al., 1982). Daily life includes numerous ritual practices designed to prevent attacks of *bhut* (Stewart et al., 1994). Zeitlyn and Rowshan (1997) provide several examples of rituals that protect a mother or her child from evil spirits. Mothers may express a few drops of breast milk before the baby suckles. A box of matches or the bone of an animal sacrificed in a Muslim festival may be placed underneath the bedding of an infant. Another treatment involves a healer cutting a mother's arm in several places and then "sweeping" a mother and her child from head to toe. Amulets (*tabij*) are widely used by both men and women as a means to improve bad health, preserve good health, influence fertility, or ward off evil influences. The practice transcends religions and is probably of ancient origin. An amulet requires a religious functionary to initially empower it; Muslims can recite verses to temporarily bolster the amulet's effectiveness (Maloney et al., 1981).

SEXUALITY AND REPRODUCTION

Bangladeshi beliefs about sexuality and reproduction reflect ancient cultural traditions that precede the major religions of the area. These include proscriptions and prescriptions for pregnant mothers, beliefs about colostrum, and rituals for newborn babies. Additional beliefs and practices shared with groups in South Asia are based on Āyurvedic texts including ideas about sexuality and ritual pollution (Maloney, 1977; Maloney et al., 1981).

The blood of menstruation is considered polluting, and is believed to negatively affect crops, animals, and the health of individuals. Hindu women are barred from the kitchen, their husband's bed, and religious performance while menstruating (Sarkar, 1982). Muslim women will not enter the fields, come near farm animals, visit the sick, serve food, or participate in religious activities while menstruating (Aziz & Maloney, 1985). Following menstruation, a woman must bath before resuming religious activities. We found in 1993 that some women utilized injectable contraceptives at the start of Ramadan as a means to prevent menstruation during the holy month (unpublished data).

Semen is believed to be made of blood (Muslim) or cerebral tissue (Hindu), so that it should not be wasted. Excessive intercourse is discouraged; masturbation is considered unnatural and believed to damage physical and mental health. Excessive depletion of semen (including female "semen") is believed to have negative consequences such as impotence, weakness, loss of sexual drive, and to cause venereal diseases (Maloney et al., 1981; Sarkar, 1982).

Coital frequency shows a strong age-related pattern from about 11 episodes per month for married individuals less than 18 years old to five episodes after age 40 (Ruzicka & Bhatia, 1982). Couples will not engage in sexual intercourse during menstruation. Coitus is also taboo throughout the third trimester of pregnancy. Following a delivery, couples are expected to practice abstinence for 40 days (Muslim) or a number of days based on caste (Hindu) (Hadi, 2000; Sarkar, 1982). Breastfeeding does not seem to limit coital frequency to the extent seen in other cultures (Ruzicka & Bhatia, 1982). Both Hindu and Muslim Bangladeshis are expected to abstain from coitus on certain religious holidays. Bathing is a requirement following coitus for both religions (Sarkar, 1982).

Premarital sexual activity is prohibited by social ideals and religious beliefs. Even so, Aziz and Maloney (1985) estimate that half of all youths have premarital sexual intercourse. These activities must be hidden from others, as Islamic law provides for severe punishments for couples that are caught. Likewise, sexual relations outside marriage are not permissible in Bangladeshi society (Sarkar, 1982). The frequency of the practice is difficult to determine, but males apparently have more opportunity than females for such activities. Punishment for individuals discovered having an illicit relationship is more severe for women (Aziz & Maloney, 1985).

Bangladeshi culture is strongly pronatal. Children are essential for continuing a patrilineal descent group, and required as part of religious duty. For Bangladeshi men and woman, parenthood is a fundamental part of marital life (Sarker, Rahman, Chowdhury, Nasrin, & Tariq, 1996). Traditionally, fate or "dependence on God" was invoked in response to questions about the ideal number of children. The increased use of contraception and the lower total fertility rates seen through the late 1990s suggest that this cultural ideal has been undergoing significant change.

The birth of at least one son is important to Bangladeshis. Son preference arises from the patrilineal social organization and patrilocal household structure. Sons perpetuate the lineage, maintain an economically viable *bari*, provide for their parents in old age, and arrange for funerals and spiritual welfare of the parents after their death (Aziz & Maloney, 1985). Son preference shows in the pattern of child mortality, which deviates from most other societies in that boys have slightly lower mortality rates than girls. Son preference is clearly observed in fertility differences and contraceptive practices that change according to the sex composition of the offspring (Chowdhury & Bairagi, 1990). Infanticide does not appear to be commonly used as a means of ensuring the birth of a son (Maloney et al., 1981).

At least through the last half of the 20th century, Bangladeshis have had high fertility. The total fertility rates (TFRs—estimates of average family size for women who survive to menopause) were 8.6 children for 1960–62, and 6.9 children for 1966–68 (Sirageldin, Norris, & Ahmad, 1975). The trend continued as TFRs dropped from 6.3 children in 1975 to 5.1 children in 1989. Beginning in the 1990s fertility declined sharply, reaching 3.3 children by 1997 (Basu & Amin, 2000). The causes of this decline are manifold, and include health improvements, improvements in education, and the success of family planning programs. Basu and Amin (2000) argue that the sharp fertility declines seen through the 1990s can be understood as a series of historical, cultural, and political circumstances in Greater Bengal that left the region amenable to such change.

Bangladeshis have long known about and used various contraceptive methods, although the extent to which traditional methods have been used in the past is not known. Maloney et al. (1981) provide a list of 23 herbs and substances recommended as contraceptives by indigenous medical practitioners. By 1981, more women were using modern contraceptive methods than traditional methods (including rhythm and withdrawal) (Kabir, Uddin, Chowdhury, & Ahmed, 1986). Bangladeshis use contraception to space births as well as to stop reproduction (Khan, Smith, Akbar, & Koenig, 1989).

Induced abortion is well known in Bangladeshi society. One estimate is that 800,000 induced abortions were performed in 1978, about one in eight pregnancies (Measham et al., 1981); another study found that induced abortion was used to terminate one in 20 pregnancies (Maloney et al., 1981). Some forms of induced abortion are illegal in Bangladesh. Early menstrual regulation, defined as any chemical or mechanical process used to induce menstruation within a few weeks of a missed period, was decriminalized and condoned as an acceptable practice by the Bangladesh government in 1979 (Dixon-Mueller, 1988). Most Bangladeshis consider induced abortion morally and socially wrong; many Muslims believe it to be contrary to Islamic principles (Maloney et al., 1981). Menstrual regulation has wider acceptance, and may not always be considered abortion; rather, it ensures "non-pregnancy" by inducing menses (Dixon-Mueller, 1988). Additionally, menstrual regulation is consistent with some Islamic interpretations that fetal life begins after the fourth month (Aziz & Maloney, 1985).

Traditional methods of induced abortion involve insertion of plant material, usually a root, into the uterus (Islam, 1982). Bhuiya et al. (2001) observed this for 85% of traditional abortions; the remainder were homeopathic remedies. Maloney et al. (1981) list 44 plants, drugs, and other materials used by homeopathic, Āyurvedic, and *kobiraj* practitioners to induce abortions. Allopathic abortion practitioners have become more common in recent years. Bhuiya et al. (2001) found that 44% of all induced abortions were performed by menstrual regulation and 27.5% by other allopathic procedures.

HEALTH THROUGH THE LIFE CYCLE

Stages of the Bangladeshi life cycle can be classified under a number of schema, reflecting different cultural and religious traditions that have been practiced in the region. Aziz and Maloney (1985) documented nine stages in the Bengali life cycle. Stages are largely determined by an individual's age, although behavior and physical characteristics may play a role in defining an individual's stage. Except for a few stages, rituals of passage are not commonly used to denote or celebrate a transition from one stage to another. Alternative systems of life stages come from Bengali religious traditions. Islamic stages of life that pertain to ideals of training, religious duty, and spiritual well-being are recognized by Muslims. Likewise, stages of life based on ancient Sanskrit texts are part of the Hindu tradition.

Pregnancy and Birth

Pregnancy is viewed as a natural state, rather than a medical condition. Fetal life is recognized as a formal stage by Islamic tradition and the Āyurvedic texts. The life of the fetus begins at 4 months, corresponding to the time when the fetus begins movement (Aziz & Maloney, 1985). This stage is understood to be a period of high vulnerability. The nature and future health of a newly conceived individual is closely tied to the behavior of parents prior to, during, and after coitus. Children conceived on certain days, during certain times of the day, or phases of the moon are believed to take on particular (usually negative) personality traits, deformities, or blindness (Maloney et al., 1981).

Bangladeshi women experience food aversions, smell aversions, and cravings during pregnancy. Pregnancy-related nausea and vomiting occur in about half of all pregnancies. Rates are highest (62%) among young women and lowest (36%) among older women. Symptoms occur any time of the day, but are most frequent in the morning and least frequent in the evening. The most common food aversions during pregnancy are fish, rice, and goat (O'Connor & Holman, unpublished data).

A number of foods are prohibited or avoided during pregnancy because of possible health effects on the fetus. Some fish are believed to cause epilepsy or malformation in a child. Pineapples are avoided because they are considered an abortifacient. Excessive salt, ginger, and chilies are believed by some to be unhealthy for the fetus. The most common cravings are sour foods such as unripe mango or lemon (Maloney et al., 1981), which by Āyurvedic tradition also eliminate toxic substances and the unwanted heat of pregnancy (Nichter, 1989).

About 12% of recognized pregnancies terminate spontaneously prior to birth (Ruzicka & Chowdhury, 1987). This rate is higher than those in well-nourished populations, and depends on a woman's age, nutritional status, and the season of conception (Mostafa, Wojtyniak, Fauveau, & Bhuiyan, 1991; Pebley, Huffman, Chowdhury, & Stupp, 1985).

Infancy

Bangladeshis recognize infancy as a life stage from birth to roughly 5 years of age. Infancy is a dangerous phase of life. In 1978, nearly 14% of Bangladeshi infants died in infancy compared with 1.4% in the United States. By 1998, the proportion had dropped to 8% of Bangladeshi infants dying in infancy, compared with 0.7% in the United States (WHO, 1999). The most common causes of death for infants (Matlab comparison area from 1995 to 1999, ICDDR, B 1996a, 1996b, 1998, 1999, 2000, 2001) were neonatal complications (other than tetanus), respiratory infections, and diarrheal disease. Factors associated with elevated postneonatal mortality (deaths from 4 to 54 weeks of age) are households that do not use a latrine and households with more than 10 individuals (Rahman, Rahman, Wojtyniak, & Aziz, 1985). Infants born to teenage mothers, mothers of low socioeconomic status, mothers with low education, and first-born infants experience higher neonatal and postneonatal mortality (Alam, 2000). Mortality by age 5 is 11.2% (rural) and 6.2% (urban) (Bangladesh Bureau of Statistics, 2001).

Breastfeeding is almost universal in rural Bangladesh, but following birth, an infant may not breastfeed for several days. During this spell, the infant is given prelacteal foods such as rice water and honey (Rizvi, 1993). In the past, delayed initiation of breastfeeding came out of the belief, found throughout South Asia, that colostrum is harmful to the infant. Recent public health programs have promoted early breastfeeding out of concern over disease transmission from contaminated prelacteal foods and to ensure infants receive the health benefits of colostrum. These programs may be responsible for recent decreases in the time to initiation of breastfeeding (Holman & Grimes, 2001).

Nearly all Bangladeshi infants are breastfed (Holman & Grimes, 2001; Huffman, Chowdhury, Chakraborty, & Simpson, 1980; Khan, 1980). The duration of breastfeeding exceeds an average of two years (Ahmed, 1986; Huffman et al., 1980; Mannan & Islam, 1995; Mulder-Sibanda & Sibanda-Mulder, 1999). Urban women,

women with higher education, women whose husbands have higher education, and younger women tend to have a slightly shorter duration of breastfeeding (Huffman et al., 1980; Mannan & Islam, 1995). Most studies find no gender differences in breastfeeding patterns (Greiner, 1997). In Bangladesh, an important consequence of breastfeeding is reduced morbidity and mortality from diarrheal disease (Mulder-Sibanda & Sibanda-Mulder, 1999); this is particularly so for unsupplemented breastfeeding (Shahidullah, 1994).

Childhood

Childhood is recognized as a life stage that begins about the time a child starts school (age 6) and continues through about 10 years of age. Aside from a ceremony for the child's first school attendance, there are few rituals associated with the transition to childhood. Boys are usually circumcized during early childhood. The occasion is marked by a feast; the boy's status changes to that of a "senior" child. Most girls have an ear pierced between the ages of 1 and 9 (Aziz & Maloney, 1985).

The overall effects of poverty and malnutrition are reflected in 1996 statistics on stunting showing that 56% (rural) and 39% (urban) of children are more than two standard deviations below height-for-age standards. Likewise, 58% (rural) and 42% (urban) of children are more than two standard deviations below weight-for-age standards (Bangladesh Bureau of Statistics, 2001).

Adolescence

Adolescence encompasses three Bengali stages of life: *pre-adolescence, early adolescence,* and *late adolescence.* The pre-adolescence stage begins with the growth spurt around age 9–11. The stage ends around menarche in girls, but is less demarcated for boys. During this stage, a girl begins to adopt the dress of adult women. Both boys and girls start taking on gender-specific adult roles within the family. Individuals are expected to take on more responsibility for their behavior, especially in limiting their contact with individuals of the opposite sex. Pre-adolescent individuals are expected to sleep with same-sex kin (Aziz & Maloney, 1985).

The early adolescence stage begins around puberty or near the peak of the growth spurt. Boys remain in this stage until their facial hair begins to grow, and their voice changes. Menarche is considered private; it entails no rites or ceremony, although a girl will begin wearing a *sari* all the time. Girls will begin shaving pubic and underarm hair. Late adolescence begins around age 16 and continues through the early 20s or until marriage. During this time, males will either continue in their schooling or begin to work for the family. Females who are not in school will spend most of their time in the *bari*. Muslims believe that an individual has become morally mature at this stage (Aziz & Maloney, 1985).

Age at menarche was examined in a 1976 study that found a median age of just less than 16 years. The study was on the heels of the 1971 war and a famine in 1974, which may have delayed menarche on average. Heavier girls at a given age were more likely to have reached menarche (Chowdhury, Huffman, & Curlin, 1977). The girls were shorter and weighed less than same-aged U.S. girls, in part reflecting a later growth spurt (Riley, 1990).

Mortality is at its lowest during adolescence. Records from 1994 to 1999 in the Matlab intervention and comparison areas (ICDDR, B, 1996a, 1996b, 1998, 1999, 2000, 2001) show that only 2 per 1,000 individuals die between 10 and 20 years of age. For males, the leading causes of death are accidents (other than drowning, 25%), infections (other than tetanus or diarrheal disease, 9%), suicide (7%), gastro-intestinal disease, and cancer (5% each). For females, leading causes of death are suicide (14%), accidents (other than drowning, 13%), obstetric-related (10%), diarrheal disease (6%), and cancer (5%).

Adulthood

Bangladeshis recognize two stages of pre-senescent adult life: *young adulthood* and *middle age.* Young adulthood begins with marriage and lasts until the late 30s or early 40s, when one's children have grown. Following marriage, a woman moves into the household of her husband's family. Her role in young adulthood is to bear and raise children, and she helps maintain the household (Aziz & Maloney, 1985).

Middle age begins with maturation of one's children and continues into the 50s. At this stage, males may assume a leadership role in the community, and some women work outside the *bari*. Couples will avoid bearing children during this period, as it is considered embarrassing or shameful to reproduce when one's children are reproducing (Aziz & Maloney, 1985).

Adult mortality is highly gender-specific. About 24% of deaths to women aged 15–45 are from obstetric causes (Matlab comparison area, 1994–1999, ICDDR, B, 1996a, 1996b, 1998, 1999, 2000, 2001). The second leading cause of death is suicide (8%). Next are chronic obstructive pulmonary disease (6.5%), cancer (6%), TB (5.6%), and cardiovascular disease (5.6%). The leading cause of death in adult men is gastro-intestinal disease (14.0%), followed by accidents and drowning (12%), TB (8.5%), cancer (6.2%), and suicide (5.8%).

The ill health of adults can have significant consequences for their children. Deaths of rural adults put their dependent children at significantly increased risk of death, particularly girls (Strong, 1992). The effects of a mother dying are more severe than for a father dying. Older children are likely to have their education interrupted, girls are likely to leave the household earlier, and marry earlier (Roy, Kane, & Barkat-e-Khuda, 2001).

Women have traditionally had little say or control over reproductive decisions. Repeated cycles of reproduction followed by intensive breastfeeding combined with poor nutrition and bouts of diarrheal disease result in declining health with age for rural Bangladeshi women (Ahmed, Adams, Chowdhury, & Bhuiya, 1998).

A Bangladeshi woman usually gives birth in her husband's home or her natal home. Births are rarely attended by trained medical practitioners. Instead, they are attended by female family members, neighbors, or by traditional birth attendants (*dais* or *dhorunis*). Nearly one third of births are delivered by the mother alone (Croley et al., 1966). Social norms associated with *purdah*, shame, and family honor usually prevent male medical practitioners from performing gynecological examinations, even during obstetric emergencies (Rozario, 1995).

Maternal mortality, defined as deaths resulting from pregnancy or childbirth, occurs in about 5–6 out of every 1,000 live births. Increased risk of maternal mortality is associated with higher parity and mothers over 35 years (Alauddin, 1986; Khan, Jahan, & Begum, 1986; Koenig, Fauveau, Chowdhury, Chakraborty, & Khan, 1988; Rochat et al., 1981). About one quarter of all maternal deaths result from complications of induced abortion (Rochat et al., 1981).

Sexual violence occurs within and outside of marriage. In rural settings, women infrequently leave their homesteads, and ethnographic accounts suggest that rapists are likely to be known by the victim. Marital rape is tolerated in Bangladeshi society as a part of a husband's rights to his wife, even though the topic itself is taboo, making it difficult to study (Yasmin, 2000).

One study found that 27% of Bangladeshi women reported at least one instance of marital rape over a one-year period of recall. Marital rape most frequently occurred during taboo times (postpartum, third trimester of pregnancy, and menstruation). Prevalence was lower at older ages, for more educated women, and for women with an independent source of household income (Hadi, 2000).

Another form of sexual violence that has received widespread attention in the international media is the practice of "acid throwing," where men throw sulfuric acid on women. The crime is usually associated with refusals of sexual or romantic advances. Victims are permanently disfigured, and some are blinded, disabled, or die from their wounds. About 100 cases are believed to occur each year (Faga, Scevola, Mezzetti, & Scevola, 2000; Yasmin, 2000).

The Aged

Old age is recognized as a stage of life that begins in the mid to late 50s. Elderly parents are cared for by their sons, usually the oldest, although in urban settings elderly women may live with a daughter (Kabir et al., 1998). Care for the elderly is considered a beneficial responsibility, rather than a burden. The elderly are respected for their wisdom; they command a near religious reverence (Aziz & Maloney, 1985).

In the late 1990s, there were about 7 million elderly Bangladeshis; about 17.5 million elderly are projected for the year 2025 (Kabir et al., 1998). Men tend to live longer than women, but because of the age difference at marriage, women tend to outlive their husbands (Ellickson, 1988). Mortality risk for elderly men and women is lower while their spouse is alive. Mortality risk is also reduced while living with a son (Rahman, 1999, 2000). Older women report significantly more limitations in their daily activities, and have more physical performance limitations than do their male counterparts (Rahman & Liu, 2000).

Dying and Death

The final stage of life (*acal* or *marankāl*) begins as an elderly person becomes disabled and is reliant on caregivers for basic needs. Care-giving is intensified as an individual becomes progressively more disabled. All attempts are made to fulfill the wishes and needs of the individual. The oldest old are believed to recapture the mental disposition of children (Aziz & Maloney, 1985).

Funeral rites are important to both Muslim and Hindu Bangladeshis. A Muslim corpse is ritually bathed before burial and wrapped in a shroud. The body is placed in a plain wooden box and carried to the burial place by male family members. Women do not attend funerals (Gatrad, 1994). Hindu funerals involve elaborate ceremonies, including the cremation of the body (Maloney et al., 1981).

CHANGING HEALTH PATTERNS

Efforts to reduce diarrheal disease by establishing widespread tube-well sources has had unintended health consequences. In the mid-1990s, high concentrations of naturally occurring arsenic were discovered in the ground water in many regions of Bangladesh and West Bengal. Arsenic-related health conditions include skin lesions, skin cancers, cancers of internal organs, and neurological disorders, but the onset of many conditions has a 10-year latency period, so it is still too early to evaluate the long-term health effects. Because of the enormity of the problem—areas with high-concentration wells may include half of the population of Bangladesh—arsenic poisoning is being treated as a public health emergency (Smith, Lingas, & Rahman, 2000).

Another public health crisis is the re-emergence of malaria in recent years. Malaria had been almost eliminated from the region in the mid-1960s, primarily through heavy use of DDT. The practice was halted during the war of independence from Pakistan. Since the mid-1970s, cases of malaria have increased steadily, and reached a peak of nearly a million cases by 1997. Additionally, a higher fraction of cases involves the more severe and deadly falciparum malaria. The most heavily infected areas are difficult to access and lack surveillance infrastructure (Bangali, Mahmood, & Rahman, 2000).

HIV has not been a serious problem for Bangladeshis, although there is growing concern over the potential for a serious epidemic. As of 1999, about 13,000 individuals are believed to be HIV positive (UNAID, 2001); but only 17 cases of AIDS have been reported (Ministry of Health and Family Welfare, 2002). The infection has only recently made in-roads into vulnerable groups. Prevalence among sex workers is under 1%, but there is a low frequency of condom use among commercial sex workers. Female sex workers have an average of two to five clients a day and the number of clients is estimated to be half a million men daily. The high rate of other sexually transmitted diseases (STDs) in sex workers suggests that HIV could spread rapidly. The rate of HIV infection is about 2.5% among the relatively small injecting drug community. Needle-sharing is widespread in this group (World Bank, 2000). Blood screening in Bangladesh was initiated in 2000 (Ministry of Health and Family Welfare, 2002).

REFERENCES

Ahmed, M. M. (1986). Breastfeeding in Bangladesh. *Journal of Biosocial Science, 18*, 425–434.

Ahmed, S. M., Adams, A., Chowdhury, A. M. R., & Bhuiya A. (1998). Chronic energy deficiency in women from rural Bangladesh: Some socioeconomic determinants. *Journal of Biosocial Science, 30*, 349–358.

Alam, N. (2000). Teenage motherhood and infant mortality in Bangladesh: Maternal age-dependent effects of parity one. *Journal of Biosocial Science, 32*, 229–236.

Alauddin, M. (1986). Maternal mortality in rural Bangladesh: The Tangail District. *Studies in Family Planning, 17*, 13–21.

Ashraf, A., Chowdhury, S., & Streefland, P. (1982). Health, disease and healthcare in rural Bangladesh. *Social Science & Medicine, 16*, 2041–2054.

Aziz, K. M. A. (1977). Present trends in medical consultation prior to death in rural Bangladesh. *Bangladesh Medical Journal, 6*, 53–58.

Aziz, K. M. A., & Maloney, C. (1985). *Life stages, gender and fertility in Bangladesh*. Dhaka, Bangladesh: International Centre for Diarrhoeal Disease Research.

Bangali, A. M., Mahmood, M. A. H., & Rahman, M. (2000). The malaria situation in Bangladesh. *Mekong Malaria Forum, 6*, 20–24.

Basham, A. L. (1976). The practice of medicine in ancient and medieval India. In C. Leslie (Ed.), *Asian medical systems* (pp. 18–43). Berkeley: University of California Press.

Bangladesh Bureau of Statistics (2001). *2001 statistical yearbook of Bangladesh, Dhaka*. Retrieved February 28, 2001, from http://www.bbsgov.org/

Basu, A. M., & Amin, S. (2000). Conditioning factors for fertility decline in Bengal: History, language, identity, and openness to innovations. *Population Development and Review, 26*, 761–794.

Begum, H. (2001). Poverty and health ethics in developing countries. *Bioethics, 15*, 50–56.

Bertocci, P. J. (1984). Chittagong Hill tribes of Bangladesh. *Cultural Survival Quarterly, 8*, 86–87.

Bhardwaj, S. M., & Paul, B. K. (1986). Medical pluralism and infant mortality in a rural area of Bangladesh. *Social Science & Medicine, 23*, 1003–1010.

Bhuiya, A., Aziz, A., & Chowdhury, M. (2001). Ordeal of women for induced abortion in a rural area of Bangladesh. *Journal of Health, Population, and Nutrition, 19*, 281–290.

Bose, N. K. (1967). *Culture and society in India*. New York: Asia.

Chacko, A. (1991, May 31). Bangladesh convulsed by catastrophe. *India Today*, 70–75.

Chowdhury, A. K. M. A., Huffman, S. L., & Curlin, G. T. (1977). Malnutrition, menarche, and marriage in rural Bangladesh. *Social Biology, 24*, 316–325.

References

Chowdhury, M. K., & Bairagi, R. (1990). Son preference and fertility in Bangladesh. *Population and Development Review, 16*, 749–757.

Croley, H. T., Haider, S. Z., Begum, S., & Gustafson, H. C. (1966). Characteristics and utilization of midwives in a selected rural area of East Pakistan. *Demography, 3*, 578–580.

Dixon-Mueller, R. (1988). Innovations in reproductive health care: Menstrual regulation policies and programs in Bangladesh. *Studies in Family Planning, 19*, 129–140.

Ellickson, J. (1988). Never the twain shall meet: Aging men and women in Bangladesh. *Journal of Cross-Cultural Gerontology, 3*, 53–70.

Faga, A., Scevola, D., Mezzetti, M. G., & Scevola, S. (2001). Sulphuric acid burned women in Bangladesh: A social and medical problem. *Burns, 26*, 701–709.

Frank, N. L., & Hussain, S. A. (1971). The deadliest tropical cyclone in history. *Bulletin of the American Meterological Society, 52*, 444–483.

Gatrad, A. R. (1994). Muslim customs surrounding death, bereavement, postmortem examinations, and organ transplants. *British Medical Journal, 309*, 521–523.

Greaney, V., Khandker, S. R., & Alam, M. (1998). *Bangladesh: Assessing basic learning skills*. Dhaka, Bangladesh: World Bank.

Greiner, T. (1997). Breastfeeding in Bangladesh: A review of the literature. *Bangladesh Journal of Nutrition, 10*, 37–50.

Gupta, B. (1976). Indigenous medicine in nineteenth- and twentieth-century Bengal. In C. Leslie (Ed.), *Asian Medical Systems* (pp. 368–378). Berkeley: University of California Press.

Hadi, A. (2000). Prevalence and correlates of the risk of marital sexual violence in Bangladesh. *Journal of Interpersonal Violence, 15*, 787–805.

Haque, C. E. (1988). Human adjustments to river bank erosion hazard in the Jamuna floodplain, Bangladesh. *Human Ecology, 16*, 421–437.

Hartmann, B., & Boyce, J. (1979). *Needless hunger*. San Francisco: Institute for Food and Development Policy.

Holman, D. J., & Grimes, M. A. (2001). Colostrum feeding behaviour and initiation of breast-feeding in rural Bangladesh. *Journal of Biosocial Science, 33*, 139–154.

Huffman, S. L., Chowdhury, A. K. M. A., Chakraborty, J., & Simpson, N. K. (1980). Breast-feeding patterns in rural Bangladesh. *American Journal of Clinical Nutrition, 33*, 144–154.

International Centre for Diarrhoeal Disease Research, Bangladesh. (1996a). *Demographic surveillance system—Matlab: Registration of demographic events, 1994*, Vol. 25. Dhaka: Author.

International Centre for Diarrhoeal Disease Research, Bangladesh. (1996b). *Demographic surveillance system—Matlab: Registration of demographic events, 1995*, Vol. 27. Dhaka: Author.

International Centre for Diarrhoeal Disease Research, Bangladesh. (1998). *Demographic surveillance system—Matlab: Registration of demographic events, 1996*, Vol. 28. Dhaka: Author.

International Centre for Diarrhoeal Disease Research, Bangladesh. (1999). *Demographic surveillance system—Matlab: Registration of demographic events, 1997*, Vol. 30. Dhaka: Author.

International Centre for Diarrhoeal Disease Research, Bangladesh. (2000). *Health and demographic surveillance system—Matlab: Registration of demographic events and contraceptive use, 1998*, Vol. 31. Dhaka: Author.

International Centre for Diarrhoeal Disease Research, Bangladesh. (2001). *Health and demographic surveillance system—Matlab: Registration of demographic events, 1999*, Vol. 32. Dhaka: Author.

Islam, R. (1978). The Bengali language movement and emergence of Bangladesh. In C. Maloney (Ed.), *Contributions to Asian studies* (pp. 142–152), Vol. 11, Language and civilization change in South Asia. Leiden, The Netherlands: E. J. Brill.

Islam, S. (1982). Case studies of indigenous abortion practitioners in rural Bangladesh. *Studies in Family Planning, 13*, 86–93.

Kabir, M., Uddin, M. M., Chowdhury, S. R., & Ahmed, T. (1986). Characteristics of users of traditional contraceptive methods in Bangladesh. *Journal of Biosocial Science, 18*, 23–33.

Kabir, Z. N., Szebehely, M., Tishelman, C., Chowdhury, A. M. R., Höjer, B., & Winblad, B. (1998). Aging trends—making an invisible population visible: The elderly in Bangladesh. *Journal of Cross-Cultural Gerontology, 13*, 361–378.

Khan, A. R., Jahan, F. A., & Begum, S. F. (1986). Maternal mortality in rural Bangladesh: The Jamalpur District. *Studies in Family Planning, 17*, 7–12.

Khan, M. (1980). Infant feeding practices in rural Meheran, Comilla, Bangladesh. *American Journal of Clinical Nutrition, 33*, 2356–2364.

Khan, M. A., Smith, C., Akbar, J., & Koenig, M. A. (1989). Contraceptive use patterns in Matlab, Bangladesh: Insights from a 1984 survey. *Journal of Biosocial Science, 21*, 47–58.

Koenig, M. A., Fauveau, V., Chowdhury, A. I., Chakraborty, J., & Khan, M. A. (1988). Maternal mortality in Bangladesh. *Studies in Family Planning, 19*, 69–80.

Leslie, C. (1978). Pluralism and integration in the Indian and Chinese medical systems. In A. Kleinman (Ed.), *Culture and healing in Asian Societies: Anthropological, psychiatric, and public health studies* (pp. 235–251). Boston: G. K. Hall.

Maloney, C. (1974). *Peoples of South Asia*. New York: Holt, Rinehart and Winston.

Maloney, C. (1977). Bangladesh and its people in prehistory. *Journal of the Institute of Bangladesh Studies, 2*, 1–36.

Maloney, C., Aziz, K. M. A., & Sarker, P. C. (1981). *Beliefs and fertility in Bangladesh*. Dhaka, Bangladesh: International Centre for Diarrhoeal Disease Research.

Mannan, H. R., & Islam, M. N. (1995). Breast-feeding in Bangladesh: Patterns and impact on fertility. *Asia-Pacific Population Journal, 10*, 23–38.

Measham, A. R., Rosenberg, M. J., Khan, A. R., Obaidullah, M., Rochat, R. W., & Jabeen, S. (1981). Complications from induced abortion in Bangladesh related to types of practitioner and methods, and impact on mortality. *The Lancet, 24*, 199–202.

Ministry of Health and Family Welfare, Government of Bangladesh. (2002). *National AIDS/STD Programme*. Retrieved March from http://www.bangla-aids.org/document/organization.htm

Mostafa, G., Wojtyniak, B., Fauveau, V., & Bhuiyan, A. (1991). The relationship between sociodemographic variables and pregnancy loss in a rural area of Bangladesh. *Journal of Biosocial Science, 23*, 55–63.

Mputu, H. A. (2001). *Literacy and non-formal education in the E-9 countries*. Paris: United Nations Educational, Scientific, and Cultural Organization.

Mulder-Sibanda, M., & Sibanda-Mulder, F. S. (1999). Prolonged breastfeeding in Bangladesh: Indicators of inadequate feeding practices or mothers' response to children's poor health? *Public Health, 113,* 65–68.

Mushtaque, A., Chowdhury, R., & Kabir, Z. N. (1991). Folk terminology for diarrhea in rural Bangladesh. *Reviews of Infectious Diseases, 13,* S252–S254.

Nichter, M. (1989). *Anthropology and international health: South Asian case studies.* Dordrecht, The Netherlands: Kluwer Academic.

Pebley, A. R., Huffman, S. L., Chowdhury, A. K. M. A., & Stupp, P. W. (1985). Intra-uterine mortality and maternal nutritional status in rural Bangladesh. *Population Studies, 39,* 425–440.

Rahman, M. O. (1999). Family matters: The impact of kin on the mortality of the elderly in rural Bangladesh. *Population Studies, 53,* 227–235.

Rahman, M. O. (2000). The impact of co-resident spouses and sons on elderly mortality in rural Bangladesh. *Journal of Biosocial Science, 32,* 89–98.

Rahman, M., Laz, T. H., & Fukui, T. (1999). Health related research in Bangladesh: MEDLINE based analysis. *Journal of Epidemiology, 9,* 235–239.

Rahman, M. O., & Liu, J.-H. (2000). Gender differences in functioning for older adults in rural Bangladesh. The impact of differential reporting? *Journal of Gerontology, 55A,* M28–M33.

Rahman, M., Rahman, M. M., Wojtyniak, B., & Aziz, K. M. S. (1985). Impact of environmental sanitation and crowding on infant mortality in rural Bangladesh. *The Lancet, 8445,* 28–31.

Riley, A. P. (1990). Dynamic and static measures of growth among pre- and postmenarcheal females in rural Bangladesh. *American Journal of Human Biology, 2,* 255–264.

Rizvi, N. (1993). Issues surrounding the promotion of colostrum feeding in rural Bangladesh. *Ecology of Food and Nutrition, 30,* 27–38.

Rob, U., & Cernada, G. (1992). Fertility and family planning in Bangladesh. *The Journal of Family Planning, 38,* 53–64.

Rochat, R., Jabeen, S., Rosenberg, M. J., Measham, A. R., Khan, A. R., Obaidullah, M., & Gould, P. (1981). Maternal and abortion related deaths in Bangladesh, 1978–1979. *International Journal of Gynaecology and Obstetrics, 19,* 155–164.

Roy, N. C., Kane, T. T., & Barkat-e-Khuda. (2001). Socioeconomic and health implications of adult deaths in families of rural Bangladesh. *Journal of Health, Population and Nutrition, 19,* 291–300.

Rozario, S. (1995). Traditional birth attendants in Bangladeshi villages: Cultural and sociologic factors. *International Journal of Gynecology and Obstetrics, 50*(Suppl. 2), S145–S152.

Ruzicka, L. T., & Bhatia, S. (1982). Coital frequency and sexual abstinence in rural Bangladesh. *Journal of Biosocial Science, 14,* 397–420.

Ruzicka, L. T., & Chowdhury, A. K. A. M. (1987). *Demographic surveillance system—Matlab: Vital events, migration and marriages—1976,* Vol. 5. Dhaka, Bangladesh: International Centre for Diarrhoeal Research.

Sarder, A. M., & Chen, L. C. (1981). Distribution and characteristics of non-government health practitioners in a rural area of Bangladesh. *Social Science & Medicine, 15a,* 543–550.

Sarkar, P. C. (1982). Customs and beliefs associated with sexual behavior and human fertility in rural Bangladesh. *Eastern Anthropologist, 35,* 135–142.

Sarker, P. C., Rahman, A. P. M. S., Chowdhury, J. H., Nasrin, T., & Tariq, T. (1996). Infertility and practice of traditional methods of treatment in cross-cultural perspective in Bangladesh. *South Asian Anthropologist, 17,* 21–25.

Shahidullah, M. (1994). Breast-feeding and child survival in Matlab, Bangladesh. *Journal of Biosocial Science, 26,* 143–154.

Shaikh, K., Aziz, K. M. A., & Chowdhury, A. I. (1987). Differentials of fertility between polygynous and monogamous marriages in rural Bangladesh. *Journal of Biosocial Science, 19,* 49–56.

Siddique, A. K., Baqui, A. H., Eusof, A., & Zaman, K. (1991). 1988 floods in Bangladesh: Pattern of illness and causes of death. *Journal of Diarrhoeal Disease Research, 9,* 310–314.

Siddique, A. K., Zaman, K., Baqui, A. H., Akram, K., Mutsuddy, P., Eusof, A. et al. (1992). Cholera epidemics in Bangladesh: 1985–1991. *Journal of Diarrhoeal Disease Research, 10,* 79–86.

Sirageldin, I., Norris, D., & Ahmad, M. (1975). Fertility in Bangladesh: Facts and fancies. *Population Studies, 29,* 207–215.

Smith, A. H., Lingas, E. O., & Rahman, M. (2000). Contamination of drinking-water by arsenic in Bangladesh: A public health emergency. *Bulletin of the World Health Organization, 78,* 1093–1103.

Stewart, M. K., Parker, B., Chakraborty, J., & Begum, H. (1994). Acute respiratory infections (ARI) in rural Bangladesh: Perceptions and practices. *Medical Anthropology, 15,* 377–394.

Strong, M. A. (1992). The health of adults in the developing world: The view from Bangladesh. *Health Transition Review, 2,* 215–224.

Turner, B. L., & Ali, A. M. S. (1996). Induced intensification: Agriculture change in Bangladesh with implications for Malthus and Boserup. *Proceedings of the National Academy of Sciences of the United States of America, 93,* 14984–14991.

UNAID (2001). Retrieved December 2001 from http://www.usaid.gov/pop_health/aids/Countries/ane/bangladesh.html

United Nations Development Program (2001). *Human Development Report 2001.* New York: Author.

van Ginneken, J., Bairagi, R., de Francisco, A., Sarder, A. M., & Vaughn, P. (1998). *Health and demographic surveillance in Matlab: Past, present and future* (Special publication 72). Dhaka, Bangladesh: International Centre for Diarrhoeal Disease Research.

Vaughan, P. J., Karim, E., & Buse, K. (2000). Health care systems in transition III. Bangladesh, Part I. An overview of the health care system in Bangladesh. *Journal of Public Health Medicine, 22,* 5–9.

World Bank. (2000). *Bangladesh HIV/AIDS Prevention Project* (Report 21299-BD). Washington, DC: The World Bank.

World Health Organization. (1999). *The World Health Report 1999.* Geneva, Switzerland: Author.

World Health Organization. (2000). *The World Health Report 2000: Health systems: Improving performance.* Geneva, Switzerland: Author.

Yasmin, L. (2000). *Law and order situation and gender-based violence: Bangladeshi perspective* (Policy Studies 16). Colombo, Sri Lanka: Regional Centre for Strategic Studies. Retrieved May 26, 2002, from http://www.rcss.org/publications/POLICY/ps-16.html

Zeitlyn, S., & Rowshan, R. (1997). Privileged knowledge and mothers' "perceptions": The case of breast-feeding and insufficient milk in Bangladesh. *Medical Anthropology Quarterly, 11,* 56–68.

Baliem Valley Dani

Leslie Butt

ALTERNATIVE NAMES

The Baliem valley Dani are also known as the Dani and the Grand Valley Dani. The term Dani used here does not include the neighboring Western Dani, the Lani, or the Yali, who have been called "Dani" in the past.

LOCATION AND LINGUISTIC AFFILIATION

The Dani live in the 50 km-long Baliem valley, Jayawijaya district, Papua (also known as West Papua, or Irian Jaya), Indonesia. The valley is located 1,500 m above sea level in the middle of the mountain range which cuts through the center of the island of New Guinea. New Guinea was divided by colonizers in the 19th century; the Baliem valley sits on the western side and was part of the Dutch East Indies colony. Since 1969, the former Dutch colony has been incorporated into Indonesia. The Baliem valley Dani are speakers of the Papuan language, Dani, which falls into the Greater Dani language family.

OVERVIEW OF THE CULTURE

The Dani are well known because of the excitement that accompanied the discovery of their complex culture, terraced gardens, and densely populated communities in a mountainous region previously thought to be uninhabited. In 1938, an American pilot spotted from the air the valley's tracts of symmetrical gardens and circular dwellings. Excitement over this New Guinea discovery was intense, and the press dubbed the valley "Shangri-La." In the early 1960s, Harvard University organized a large expedition to the region. The well-known film *Dead Birds* (Gardner, 1963) and now-classic ethnography (Gardner & Heider, 1968; Heider, 1970, 1979; see also van Baal, Galis, & Koentjaraningrat, 1984) from this period have helped to fix the Dani as central to the global imagination about "primitive" populations and integral to the study of tribal cultures within the field of anthropology.

The outside world may not have known of the Baliem valley, but people have settled there and cultivated gardens for at least 7,000 years (Golson & Gardner, 1990). At present, the patrilineal Dani number some 60,000, and display the highest levels of cultural intensification and political integration of any group in the New Guinea highlands (Shankman, 1991). For their food staple, the Dani rely on root crops such as the sweet potato, introduced about 300 years ago, and the indigenous taro, which women cultivate in gardens on the valley floor and mountainsides. Women also raise pigs, which men strategically exchange to promote their status, and to strengthen their political alliances. People identify themselves by membership in a totemic clan. In the past, clans grouped into multi-layered political units, and large-scale pre-contact warfare dominated political activities (Heider, 1970). Even after pacification in the 1970s, clan groups still align to form large political alliances. Leadership is achieved through prowess in politics and exchange relations.

Dani men use their influence in the public arena to try to regulate the lives of women. The Dani are polygynous, and most men seek to acquire more than one wife, with between 36% (Aso-Lokobal, n.d.; Peters, 1975) and almost 50% (Butt, 1998; Heider, 1979) of men having at least two wives. Dani society is divided into two moieties, *weta* and *waya*. A person may not marry within his or her moiety. Marriage occurs ideally within a six- to seven-year cycle, culminating in a big pig feast, the *ebe akho*. In this feast, young girls either choose their marriage partner or have their partner arranged by their parents. The big pig feast is the climax of several ceremonial cycles which also include funerals and boys' initiation (Heider, 1972; Peters, 1975).

Dani gender roles are strongly demarcated. The strict division of labor appears to favor men, for women do most of the hard physical labor in gardens, and do not engage in the male-only rituals that give political power. Household compounds are divided into men's houses (*pilamo*) and family spaces. Women are forbidden to enter the *pilamo*, where sacred objects are stored. However, women expect support from husbands and male kin in domestic matters, and will run away or refuse to marry an unappealing suitor. About 30% of married women have ever run away from their husbands or former husbands (Butt, 2001c; O'Brien, 1969). As Peters (1975) summarizes, women are subject to men in this society, but they are by no means subservient.

Inroads into long-standing gender and social roles began in earnest when the first mission floatplanes landed in 1954. Although neighboring societies underwent mass conversion to Christianity, with many burning all their sacred ritual objects (O'Brien & Ploeg, 1964), the Dani remained uninterested. Instead, the Dani attacked police, resisted schooling, and refused to wear clothes (Peters, 1975). More rapid acculturation processes began in the late 1960s and early 1970s, as the former Dutch colony was handed over to Indonesia in 1969 in a contested vote. Under Indonesian rule, large-scale Dani warfare was stopped. Authorities further prohibited the custom of cutting off portions of the fingers of young Dani girls to commemorate a dead ancestor during a funeral. Government officials also implemented a process of "Indonesianization." A new lingua franca, a national education curriculum, and increased exposure to other Indonesians were meant to assimilate Papuans into the broader nation (Gietzelt, 1989). In 1971, "Operation Penis Gourd," for example, air-dropped clothing into the Baliem valley as part of an effort to "civilize" the Dani, who normally only wore penis gourds and grass skirts. Despite these efforts, the Dani were and remain largely self-sufficient, rejecting many aspects of modernization, Christianity, and Indonesian rule.

Since the 1960s, there has been overt, active resistance by the Dani and other indigenous groups to incorporation into Indonesia. However, up to 8,000 Indonesian migrants have moved to the Baliem valley, and most of them live in Wamena, a prosperous town built alongside the airstrip. Immigrants mostly run the government offices, businesses, and the military. The valley has many police posts and military barracks; these, along with racist and inequitable policies, have heightened resentment among the Dani. In 2000, in partial revenge for political deaths from the past four decades, and following police intervention into an independence rally, Dani activists massacred several families of Indonesian migrants. This tragic incident brings home the extent to which an imagined "Shangri-La" is in fact a contested territory enmeshed within the complex politics of a militarized state. Health and healing needs to be understood as much from the context of these political realities as from the cultural perspective of the inward-looking, self-interested Dani.

THE CONTEXT OF HEALTH: ENVIRONMENTAL, ECONOMIC, SOCIAL, AND POLITICAL FACTORS

Health challenges in the highlands of Papua are numerous. There are three major impediments to Dani well-being: environmental challenges; changing disease patterns; and unequal access to poor quality allopathic health services.

For most highland New Guinea populations, abundant rainfall, high altitude, and inhospitable terrain make horticulture a constant challenge. Populations living below 1,000 m must battle with endemic malaria. Above 1,800 m, misting from clouds limits agriculture. In contrast, the Dani live at 1,500 m, an altitude with more moderate rainfall, and in a region that favors dense cultivation. The Dani are self-sufficient in food production, with the sweet potato forming 90% of their diet. Pigs form an important supply of protein. Thus, the Dani are taller, more husky, and overall healthier than their neighbors, many of whom picture the Baliem as "a sort of paradise, where pigs abound" (Peters, 1975, p. 72).

Nevertheless, Dani health is affected by environmental constraints. Infant mortality rates are high. Ethnographic and local estimates range between 110/1,000 and 250/1,000 (Butt, 1999; WATCH, 1994), although government figures cite lower rates (cf. World Bank, 1991). The biggest health risks are pneumonia and other upper-respiratory infections, which cause over 50% of recorded infant deaths (WATCH, 1994). Smoke-filled traditional dwellings contribute to high rates of respiratory infection among children and adults.

Birth rates are low; the average number of children per family is around 1.5 children (Butt, 1998; Lokobal, 1992). Almost no mothers have more than three children. Women cite the risks associated with childbirth and the arduous work of caring for both garden and children as reasons for limiting family size. Women also lose autonomy if they have a second child, for it is harder to find a new husband if they have many children.

An overall low-protein diet takes its toll on growth patterns. Malnutrition rates increase as children grow. At 36 months, 7% of Dani children weigh less than 11 kg, and at 5 years, over 25% of children have poor nutrition

(WATCH, 1994). Fifty percent of children's growth patterns fall below established norms for Indonesia (Muslim, 1995).

Among adults, upper-respiratory infections remain a serious cause of illness. Smoky huts, combined with almost universal cigarette smoking of indigenous tobacco, make pneumonia and tuberculosis major health problems throughout the valley for adults of all ages.

Disease patterns are changing but most health problems remain acute. Malaria has gone from being absent in the valley in the 1950s to becoming a major killer of adults. Cerebral malaria, a fatal form, became a serious local concern after an outbreak in 1987 (Sudjito, 1987). Malaria Tropicana, in a 1998 outbreak, also claimed many deaths (Hanevik, 2000). Seven cases of HIV have been recorded from a random sample of 195 adults (Ingkokusumo, 2000). Actual rates of HIV infection are almost certainly higher. Within the valley, sexually-transmitted diseases have become a serious problem. In 1995, 50% of commercial sex workers in Wamena (including Dani and non-Dani respondents) tested positive for gonorrhea, 36% had chlamydia, and 25% had syphilis (Ingkokusumo, 2000). A survey in 1991 showed that 8% of adult hospital patients (including Dani and non-Dani respondents) had gonorrhea (Senis, 1995).

The last factor inhibiting higher standards of Dani health is poor-quality health services. There have been some successes. Since 1954, missionaries in some areas have carefully trained indigenous nurses in basic diagnosis and treatment, and still supply medicine and equipment. Two early campaigns successfully eradicated endemic health problems. The first controlled the incidence of endemic goiter by widely distributing iodized salt. The second campaign reduced the suffering caused by yaws (also known as framboesia), a chronic infectious skin disease. Among the Dani, yaws often progressed to the stage of open skin ulcers, at times eating away flesh, muscle, and joints (Gajdusek, 1961). Missionaries gained an early foothold in Dani communities by publicly curing known yaws cases with a single shot of penicillin.

Most other health interventions, however, have been markedly less successful. Indonesian health care programs put in place in the 1970s implement national health goals, often at the expense of local health needs. This format has been successful in reducing mortality rates in other parts of Indonesia (Yahya & Roesin, 1990). In Wamena, the government runs a hospital that is clean and comfortable, and the Dani will use it under certain conditions. At most smaller health centers and clinics run by trained Dani workers, however, medicines are unavailable, out of date, or improperly administered (Hartono, Romdiarti, & Djohan, 1999; Ingkokusumo, 2000). The health bureaucracy is controlled by Indonesian migrants who are perceived by the Dani as inaccessible and often racist (Butt, 1998; Hull & Hartono, 1999). Knowledgeable health advocates focus on increasing maternal education as a way to improve the health of women and children, yet they recognize that these aims often contradict Dani values, which do not privilege the education or autonomy of women (Hartono et al., 1999; Srini, 1999). Nonetheless, many women have been reluctant to bring their children for immunizations, in part because diseases prevented by the injections are not yet present in the highlands. Many women also reject Indonesia's family planning efforts, which promote the use of the long-term contraceptives Norplant and Depo-Provera, because they already have low birth rates.

MEDICAL PRACTITIONERS

There are three kinds of people with specific healing skills who contribute to the well-being of the Dani: part-time healers with specific practical skills, sometimes known as *hathale*; ritual participants, who seek to control the sacred realm; and trained health workers called *mantri*.

First, *hathale*, or part-time healers, span several realms. Many *hathale* are women. Some women have expertise in assisting with childbirth. Other women are experts at performing abortions. Others know herbal remedies or can perform small curing ceremonies. Certain individuals, mostly women (Heider, 1979), are known to be powerful practitioners of the arts of magic. They can cast spells and can sometimes cure the sick. Overall, there is no special status accorded to these healers.

An important healer role is played by participants in rituals designed to control sickness, food, and the threat of enemies. Political leaders have "to be able to involve the ancestral ghosts in the affairs of the group" (Ploeg, 2001, p. 34) and one way of demonstrating this is by bringing about good health through successful ritual practice. Most of these rituals involve killing a pig, cutting it up, cooking it, and eating it in a systematic manner (see Aso-Lokobal, n.d.; Lokobal, 1994; Peters, 1975). Pig sacrifices are intended to placate ancestors, thus ensuring fertile gardens, fecund women, and a healthy and prosperous population. Failure to carry out the rituals in their correct time and order can explain misfortune, sickness, and misery. The most important of these rituals is the *kaneke hakasin*, but Aso-Lokobal (n.d.) tallies 94 distinct rituals related to Dani health. In sum, political leaders produce Dani well-being, not by directly healing the sick, but by the successful execution of rituals.

Lastly, some Dani have undergone two years of nursing training, and are called *mantri*. For a small fee, *mantri* working in government clinics will dispense some medicines, treat wounds, and promote government health programs such as family planning. *Mantri* often find themselves in a position of conflict. On the one hand, all *mantri* believe in the power of Dani rituals to provide physical well-being through mediating social and spiritual relationships. On the other hand, they also dispense with confidence the medicines that they believe will heal their sick patients. Discretion in mediating two often competing belief systems is key to their success. One *mantri* summarizes a common strategy: "Never tell a patient you know he believes in spirits. Never tell them what disease they have. Just give them the medicine" (Butt, 1998, p. 199).

CLASSIFICATION OF ILLNESS, THEORIES OF ILLNESS, AND TREATMENT OF ILLNESS

Traditional beliefs about ancestor spirits or supernatural powers are intricately interwoven into Dani imaginings about why and how people become sick. The rich world of Dani *wesa* (the supernatural) includes ancestor spirits, magical powers, and forest and animal spirits. These intersect with the natural world, contemporary medical services, and individual illness in complex and often idiosyncratic ways. In the main, "illness and death are always considered to be more or less directly caused by ghosts" (Heider, 1970, p. 227).

One belief common to all Dani is the importance of "soul matter" or *etai-egen*. Each person who is born needs to grow an *etai-egen* over the course of his or her lifetime. The *etai-egen* is an immaterial substance that rests lodged just beneath the sternum. Throughout adult life the *etai-egen* is continually subject to potential disturbance or weakening, particularly through the work of malicious ghosts. Explanations for illness causation often refer to a weakened, shrunken, or disturbed *etai-egen*.

Another widespread belief is that ancestors, sorcery, and malicious spirits cause the majority of serious illnesses. Sometimes ancestors cause sickness directly (Lokobal, 1994). At other times, an individual's actions can cause another person to become sick. Theft, laziness, or disobedience on the part of one person can make a close relative sick. A man who has engaged in extramarital sexual relations, for example, might explain his wife's subsequent reproductive problems as punishment for his errant behavior.

Another form of illness can be brought about when an individual manipulates the spirit world. Vengeful sorcerers or jealous competitors can make another person sick. For example, *imak* sorcerers can cause small crab-like animals to leap into a man's penis if he urinates in a dangerous place. Sorcerers can poison food using a fine, white powder invisible to the target's eye (Gardner & Heider, 1968). Sorcerers often try to poison people from enemy alliances. Lastly, certain illnesses are always caused by angry spirits or sorcery. These include uncontrollable seizures, "inner heat" (*panas dalam*), mental disability, a high fever, and dry cough.

For the Dani, diagnosis of disease is an important art. Sickness will not leave the body, they believe, until a correct diagnosis has been made. Depending on the diagnosis, the sick person or his or her relative will choose from an array of treatment options. The hierarchy of resorts normally follows this pattern: (1) do nothing; (2) use herbal remedies or engage in bloodletting; (3) consult clan *hathale* who may use one of several diagnostic techniques, including blowing on the head of the sick person, dissecting a rat, eating a ritual pig, or preparing protection for the patient's *etai-egen*; (4) pray to the Christian God; or (5) consult a healer from outside the clan.

Most Dani look to the realm of the spirits to diagnose sickness and place clinics and Christian prayer a distant second. A "devout" Christian might immediately seek help from a healer to diagnose sickness, and only resort to prayer once the treatment begins to work. One informant summarized the trend of using allopathic medicine not as a treatment resort, but as a diagnostic tool for understanding the effects of ancestor spirits:

Government clinic medicine is just play medicine, not really important at all, just pills. Government clinic medicine is a test to see whether a sickness is really ancestors and magic or not, not a real treatment resource at all (L. Matuan cited in Butt, 1998, p. 193).

SEXUALITY AND REPRODUCTION

Dani sexuality has been a topic of active debate in anthropology, beginning with Karl Heider's (1976) argument that the Dani had a "low energy" sexual system. Heider had observed a seeming disinterest on the part of Dani men and women to engage in sexual relationships during his 1960s fieldwork. Their willing adherence to a four- to five-year post-partum taboo, low birth rates, and low rates of extra-marital sex, were interpreted by Heider as evidence of a general disinterest in heterosexual intercourse.

There is some data to support Heider's hypothesis. As with most other highland New Guinea societies, negative symbolic constructions about women and their reproductive capacities can make sexual intercourse an activity fraught with tension and uncertainty for men. Menstrual blood can poison men's bodies, food, and pigs, according to the Dani, and men consider vaginal sex toxic and potentially debilitating. Metaphors of poison and contagion continue to be employed to strengthen assumptions about male superiority. Women also fear men's bodily substances. Semen, in particular, can poison women's inner organs if it is ingested orally. Semen must never be scattered on the ground, for it can kill plants and poison the earth. A recent survey (Butt et al., 2002) suggested that Dani men and women continue to limit their location and kind of sexual activities in order to protect bodies and gardens from the debilitating effects of bodily fluids.

Another practice that lends support to Heider's argument is that Dani women continue to attempt to space their children so they are born every four to five years. One way they do this is by not having sexual relations with their husband. Parents of a newborn child are expected to refrain from having sex with each other until the woman has stopped breastfeeding, typically around the child's fourth year. Post-partum sexual abstinence is rooted in the conviction that semen is a highly dangerous fluid to an infant, whose body is weak and unfinished. If a couple has sex while the woman breastfeeds, semen can travel up through the woman's internal organs and come out in her breast milk; this can make the child sick or die.

However, other data substantially challenge Heider's claim of Dani "low sexual energy." Peters (1975) describes intense flirting and a high level of sexual knowledge among adolescents. Pre-marital sex was seen as commonplace in the 1960s (Hayward, 1980; Heider, 1976). Courting parties were common, and Hayward (1980) describes couples leaving parties together to spend a night in intimacy. Van der Pavert (1986) describes a ritual that cleanses people if they have engaged in sexual relations with members of their own moiety. Lastly, men say they expect to seek sex elsewhere while their wives are breastfeeding (Butt et al., 2002).

Explanations for low birth rates also challenge Heider's assertion that low rates of sexual relations result in small families. Pontius (1977) suggested that low birth rates were possibly due to tight scrotum strings holding the penis gourd in place, which reduced the transmission of spermatozoa. A further report describes the bark of a cinnamon tree being used as a contraceptive (Kostermans, 1969). Butt (1998; see also Peters, 1975, p. 31) records women obtaining abortions as standard practice.

There is, in sum, little evidence to support Heider's (1976, p. 189) claims of an "extraordinarily low level of sexuality" among the Dani. In a manner similar to most societies, the Dani do have a reasonably systematic set of beliefs about the negative effects of non-reproductive and extra-marital sex. Evidence suggests that stringent efforts were made in the past to regulate sexual practice. However, the Dani live out a complex esthetics of sexuality, based as much on ideas about beauty, desire, and entitlement, as on fears about poison and prohibitions.

HEALTH THROUGH THE LIFE CYCLE

Pregnancy and Birth

A number of Dani ideas about conception and birth draw on observed regularities in bodily growth and development. For conception to occur, the Dani believe that couples have to have sexual intercourse around 10 times. The fetus grows because semen from the man helps build the bones and blood from the woman helps build the flesh. When the fetus moves inside the womb, at approximately 20 weeks, couples abruptly stop having intercourse. Some women describe the fetus as roaming around the mother's body at will, drinking milk from the inside. The fetus is always protected by the placenta, termed *opase* (grandfather, or elder male kin), which looks after the baby in the mother's womb.

Most Dani women decrease food consumption in the final month of pregnancy to prevent their child from being too big at birth. The ideal newborn weight is around 2.5 kg. Women usually give birth alone or with a close relative or experienced assistant. Women like to hold themselves in a standing position, grasping onto the rafters of the cookhouse or the sleeping house roof. Giving birth inside the family compound is essential, as infants are perceived to have so little *etai-egen* that they cannot protect themselves from the spirits that constantly roam the forests and paths. Once the child is born, a relative or the birth assistant wraps the placenta carefully and take it down to the river where it can be disposed of safely. The mother and infant do not leave the site of birth until after the umbilical cord falls off, at around 10 days after giving birth.

Infancy

Dani women are skilled at transforming small, healthy newborns into large, plump infants. The newborn spends the first 4 months of life in a soft, smooth, and cool netbag. Women surround the infant with soft banana leaves or cloth bedding in the netbag, and carry the infant in the bag slung across their lower backs. The infant comes out of the bag only to change soiled bedding, or partially to breastfeed. Once the mother starts garden work, she takes special care to enclose her baby under 10 or 12 netbags to protect the child from the sun, and from being startled. The baby's vulnerable body, which is "still wet," with "closed eyes" and "soft skin," needs extensive protection from the spirit world (Butt, 1998). Infants grow very fast in these conditions. By 4 months, most babies have more than doubled their birth weight, and at 5 or 6 months almost all infants are plump and healthy.

Infant health deteriorates once mothers begin giving supplementary foods at around 6 or 7 months. Bananas, mashed sweet potatoes, and, increasingly, mashed noodles or crackers, are some of the first foods the baby eats. The protein levels of these foods are adequate, but many infants get sick because they are exposed to bacteria on the spoons that women increasingly use to feed their children. At 7–8 months, many infants experience their first bout of diarrhea. Growth rates slow over the first year, and by the age of 18 months, many children show signs of inadequate nutritional intake.

Infants and young children also suffer from numerous chronic skin problems, including infected insect bites and scabies. Scabies are endemic because many women have begun to use pieces of cloth instead of the more sanitary netbag to hold their babies. Intestinal worms plague significant numbers of young children. However, upper-respiratory infections are the biggest concern (see Figure 1) and many infants who get pneumonia die of it. The death of a child traumatizes parents, and blame usually falls on ancestor wrath brought about by poorly conducted rituals, or on errant behavior by the mother or the father.

Childhood

Dani childhood is carefree and pleasant. Children roam forests and fields, and do so in age sets distinguished along the lines of gender. Complex gender distinctions between boys and girls come into play as the child achieves more independence at around age 6 or 7. Bodies of boys are seen as more "soft" than those of girls, and as less likely to grow fast and well. Girls, in contrast, grow faster, mature faster, and are seen as hardier than boys. Despite claims to the contrary, the Dani exhibit preferential gender behavior. Boys are more likely to receive formal education than girls. Boys are also more likely to be recipients of medical care than girls. Boys under the age of 5 were taken to clinic one-third more often than girls (Butt, 1998). In ritual situations, boys over the age of 10 or so receive larger pieces of pork than do girls, and they are also fed before girls. Women compensate in part by slipping secret bits of food to their girl children.

Adolescence

There is little ritual importance attached to the transition from child to adolescent. Only young boys whose fathers are from the *waya* moiety undergo a ritual to be "made waya" (Heider, 1972). At this ritual, which occurs during the big pig feast and lasts about two weeks, boys have dangerous *wesa*, or sacred power, drawn from them. They are then purified through a series of rituals that involve some deprivation, hunting, eating ritual taro, and painting a ritual red stripe on their nose (Heider, 1972; Peters, 1975).

For girls, once a young woman has begun to menstruate, in most regions of the valley her clan holds a formal, public rite-of-passage event in which her reproductive capacities are extolled. This ceremony is known as the *hotalimo* (Peters, 1975). The girl cooks sweet potatoes and distributes them, and women dance non-stop to protect the young girl, for "the spirits of the dead want to kill people at moments of crisis, like menstruation or sickness" (Peters, 1975, p. 37). Young women also undergo a skirting ceremony, which usually happens during the big pig feast. Changing a grass skirt to the low-slung thread *yokal* worn by grown women signifies the girl's maturity and

Figure 1. Waiting at the clinic for the *mantri*, the 9-month-old infant on the right has attained the Dani ideal of plumpness. In contrast, the worried mother on the left holds her 2-year-old son, who is little bigger than the other baby. He has suffered from a respiratory infection for several weeks and is mildly malnourished. This was his first trip to the clinic to receive penicillin.

readiness for marriage. The skirting ceremony is still practiced, even though few women now wear the *yokal* skirt on a daily basis.

Adulthood

At the point where a girl begins to menstruate, she may already be married. Early Indonesian officials and missionaries were horrified to see girls aged 9 or 10 married off during the big pig festival, but they did not realize that the time between public, ritual marriage and consummation of marriage ties is often delayed by several years before the bridegroom's family fully pays off debts to the bride's family (Heider, 1972). Nonetheless, early age at marriage is common, and most young women give birth before the age of 20. This increases the chance of them contracting sexually transmitted diseases early in their reproductive years.

For men, adulthood is achieved later in life, at the time when assiduous attention to social and exchange relations with kin and alliance members begins to pay off.

Both men and women in the present run new health risks. The desire for money and material goods encourages young men to migrate to other parts of the province, in search of the success that cannot easily be attained through traditional routes of leadership. Mobility increases the risk of contracting sexually transmitted diseases. Wife abuse has increased as couples find themselves far from the regulating relations of village life. Youth regularly drink moonshine (made of bananas or pineapple), illegally imported whisky, and occasionally, distilled rubbing alcohol. Some young men and women in Wamena sniff glue every day.

Commercialized sexual relations are on the increase throughout the province, and in Wamena migrant Indonesian and indigenous women offer sexual services to both Indonesian and indigenous customers. Dani sex workers and their Dani clients are young, with most under the age of 21 (Yasukhogo, 2000). Women from the highlands find themselves at the bottom end of the sex work industry, where health risks are greatest. Dani women offer sexual services in insecure, often dirty, locations.

They are poorly paid, at around Rp. 25.000 (U.S. $2.50) per transaction. Dani women from across the valley have experienced coercive sexual relations with soldiers. Overall, their clients are more likely to be violent, and to refuse to wear condoms. Condom use among indigenous sex workers is almost nil (Butt, 2002).

For many Dani, health is an extension of politics. The use of metaphors of illness and sexuality to explain broader power relations recurs often in Dani political discourse. Many Dani believe that AIDS was deliberately brought in to the province by infected Javanese sex workers in order to render Papuans powerless and enable Indonesian takeover of the valley. Rape of indigenous women by members of the Indonesian military in other parts of the province is also understood as a political tool and contributes to negative Dani assessments of Indonesian rule (Butt, 2001a; Coomaraswamy, 1999; see also Kirsch, 2002). Promoting birth control is also easily understood as part of a broader political effort to control, or eradicate, indigenous Papuans (Butt, 2001b).

Dying and Death

Dani life expectancy is almost certainly lower than the Indonesian average. Contemporary observations suggest that women and men age rapidly, and often die after they become grandparents. Older people, whose *etai-egens* have become small, often refuse to eat or drink in their final days.

It has been said that the Dani are obsessed with death. The transition from the world of the living to the world of ancestor spirits is a critical time, for spirits are at their most vengeful right after death. The funeral of an important Dani man or woman can last for days. People bring their largest pigs as gifts, which are killed, cooked, and distributed in a highly ritualized fashion. After one to two days of mourning and feasting, the family lifts the corpse high onto a log pyre which has been set on fire. The corpse becomes engulfed in flames. Crying women now all raise their voices in a massive wail and men join in. As the smoke rises into the air and the crying voices cut through the night sky, the message carried to those ancestors, wood spirits, and humans lurking outside the compound is that this dead person has been fully honored, and has no cause to seek revenge after entering the world of ghosts.

Dani health is inextricably linked to broader cycles of life and death. Yet, while groups in other parts of the highlands abandon old ways of life, the Dani retain ritual as core to their ideas about healing. One reason for this might be because development and "Indonesianization" do not appear to have resulted in measurably improved standards of health. Infant mortality remains high, chronic illnesses such as scabies or sexually transmitted diseases have increased, and medical care is inadequate, both in basic quality and in cultural sensitivity (Hartono et al., 1999; Hull & Hartono 1999). For wider political reasons, as well as to protect their cultural strength as a self-interested society, Dani elders continue to insist that health and prosperity can best be achieved not through clinic medicines, but through properly executed rituals, through fecund women and fertile gardens, and through honoring the ancestors and spirits that roam the valley alongside the living.

REFERENCES

Aso-Lokobal, N. (n.d.). *Keterangan singkat mengenai beberapa unsur penting dalam kebudayaan masyarakat Balim*. Unpublished manuscript.

Butt, L. (1998). *The social and political life of infants among the Baliem Valley Dani, Irian Jaya*. Doctoral dissertation, McGill University, Montreal, Canada.

Butt, L. (1999). Measurements, morality, and the politics of "normal" infant growth. *Journal of Medical Humanities, 20*(2), 81–99.

Butt, L. (2001a). Women and the perils of reproductive "choice" in Irian Jaya, Indonesia. In H. Lansdowne & M. Dobell (Eds.), *Women, culture and development in the Pacific* (pp. 65–73). British Columbia: University of Victoria.

Butt, L. (2001b). "KB kills": Political violence, birth control and the Baliem Valley Dani. *The Asia Pacific Journal of Anthropology, 2*(1), 63–86.

Butt, L. (2001c). "An epidemic of runaway wives": Discourses of Dani Men on Sex and Marriage in Highlands Irian Jaya, Indonesia. *Crossroads: An Interdisciplinary Journal of Southeast Asian Studies, 15*(1), 55–87.

Butt, L., Numbery, G., & Morin, J. (2002). *Papuan Sexuality Project Research Report*. Jakarta: Aksi STOP AIDS (USAID-F.H.I.)

Coomaraswamy, R. (1999). *Integration of the human rights of women and the gender perspective: Violence against women*. United Nations Economic and Social Council.

Gajdusek, D. C. (1961). *West New Guinea Journal*: May 6, 1960 to July 10, 1960. Bethesda, MD: National Institute of Neurological Diseases and Blindness and National Institutes of Health.

Gardner, R. (1963). *Dead birds*. Boston, MA: Harvard University, Peabody Museum, Film Study Center.

Gardner, R., & Heider, K. (1968). *Gardens of war: Life and death in the New Guinea stone age*. New York: Random House.

Gietzelt, D. (1989). The Indonesianization of West Papua. *Oceania, 59*, 201–221.

Golson, J., & Gardner, D. (1990). Agricultural and sociopolitical organization in New Guinea highlands prehistory. *Annual Review of Anthropology, 19*, 395–417.

Hanevik, K. (2000). *Prayers, chants and drugs*. Medicins Sans Frontieres.

Hartono, D., Romdiarti, H., & Djohan, E. (1999). *Akses terhadap pelayanan kesehatan reproduksi: Studi kasus di kabupaten Jayawijaya, Irian Jaya*. Jakarta: PPT-LIPI.

Hayward, D. (1980). *The Dani of Irian Jaya before and after conversion*. Sentani: Regions Press.

Heider, K. G. (1970). *The Dugum Dani*. Chicago: Aldine.

Heider, K. G. (1972). The Grand Valley Dani pig feast: A ritual of passage and intensification. *Oceania, 42*(3), 169–197.

Heider, K. G. (1976). Dani sexuality: A low energy system. *Man, 11*(2), 188–201.

Heider, K. G. (1979). *Grand Valley Dani: Peaceful warriors*. New York: Holt, Rinehart & Winston.

Hull, T. H., & Hartono, D. (1999). Culture and reproductive health in Irian Jaya: An exploratory study. *Development Bulletin, 48*, 30–40.

Ingkokusumo, G. (2000). *Sexually transmitted illness: Perception and health seeking behaviour among the Dani men, in Wamena Jayawijaya District, Papua Province, Indonesia*. Master thesis, University of Amsterdam, The Netherlands, 2000.

Kirsch, S. (2002). Rumour and other narratives of political violence in West Papua. *Critique of Anthropology, 22*(1), 53–79.

Kostermans, A. J. G. H. (1969). A New Guinea cinnamon used as a contraceptive. *Reinwardtia, 7*(5), 539–541.

Lokobal, N. (1992). *Keberadaan dan peranan perempuan-laki-laki pada suku Dani di Irian Jaya*. Wamena, Irian Jaya.

Lokobal, N. (1994). *Catatan singkat antropologi kesehatan*. Wamena, Irian Jaya.

Muslim, Z. (1995). *Masalah kekurangan gizi di kabupaten Jayawijaya*. Wamena, Irian Jaya.

O'Brien, D. (1969). *The economics of Dani marriage: An analysis of marriage payments in a Highland New Guinea society*. Doctoral dissertation, Yale University, New Haven, CT.

O'Brien, D., & Ploeg, A. (1964). Acculturation movements among the Western Dani. *American Anthropologist, 4*(2), 281–292.

Pavert, J. van de (1986). I Ma Wusan, A purification ritual amon the Dani of West Irian. UNITAS, 59(1), 5–154.

Peters, H. L. (1975). Some observations of the social and religious life of a Dani-group. *Irian, 4*(2), i–vi, 2–199.

Ploeg, A. (2001). The other Western Highlands. *Social Anthropology, 9*(1), 25–43.

Pontius, R. (1977). Drawings of the body with phallocrypt contraptions of Baliem Valley Dani in West New Guinea: Clues for means of population control? *Journal of the American Medical Women's Association, 32*(6), 203–211.

Senis, J. (1995). Epidemiologi infeksi neisseria gonorrhoea dan STD lain di lembah Baliem, Irian Jaya. *Medika, 77–79*.

Shankman, P. (1991). Culture contact, cultural ecology, and Dani warfare. *Man (n.s.), 26*, 299–321.

Srini, S. (1999). *Analisis faktor-faktor yang mempengaruhi tingkat pemanfaatan pelayanan antenatal oleh suku Dani di kecamatan kurulu kabupaten Jayawijaya*. Master thesis, Universitas Gadjah Mada, Yogyakarta, 1999.

Sudjito, M. C. (1987). Management of cerebral malaria in the Karubaga primary health center, Jayawijaya District. *Irian, 15*, 11–17.

van Baal, J., Galis, K. W., & Koentjaraningrat, R. M. (1984). *West Irian: A bibliography*. Dordrecht, The Netherlands: Foris.

WATCH (1994). *1992 Project Report* Jayawijaya, Irian Jaya.

World Bank (1991). *Indonesia: Health planning and budgeting*. Washington, DC: The World Bank.

Yahya, S., & Roesin, R. (1990). Indonesia: Implementation of the health-for-all strategy. In E. Tarimo & A. Creeve (Eds.), *Achieving health for all: Midway reports of country experiences* (pp. 133–228). Geneva, Switzerland: World Health Organization.

Yasukhogo (2000). *Laporan hasil studi kwantitatif atas pengetahuan, sikap dan perilaku sehubungan dengan PMS dan HIV/AIDS dikalangan PSKJ dan klien, Wamena*. Jayapura: PATH (Program for Appropriate Technology in Health).

British

Ian Shaw and Louise Woodward

ALTERNATIVE NAMES

Great Britain, United Kingdom—component countries: England, Scotland, Wales, and Northern Ireland.

LOCATION AND LINGUISTIC AFFILIATION

Britain is an island nation geographically separated from France (and the rest of Europe) by the 22-mile stretch of sea called the English Channel. The dominant language is English, but Welsh is now enjoying a revival assisted by a recent devolution of government and the formation of a Welsh Assembly. Celtic is still spoken in a few remote parts of Scotland. Also there are some identifiable parts of major cities (linked to immigration) where the English may be a second language of use. Britain is a part of the European Union.

OVERVIEW OF THE CULTURE

The population of the United Kingdom is estimated at 58 million (the 2001 Census began reporting findings in

November 2002). Population density is toward the Midlands and the south-east of England with some areas in the north of Scotland very sparsely populated. However, there are some demographic drifts back to the countryside, particularly by the affluent middle class. Historically, England "annexed" Scotland, Wales, and Ireland into a United Kingdom by the 18th century, though much of Ireland gained independence in the 1960s. The British have been described as a "martial race," though this may be more a reflection upon Britain's history of Empire than an accurate description of current culture. Football (soccer) is the national sport.

Religion is predominantly protestant with the established Anglican Church of England and the Presbyterian churches of Scotland and Wales. However, Catholicism is a minority religion, and immigration has raised the profile of the religions of the Indian Subcontinent. Religious tolerance is high, with tolerance generally being cited as a cultural trait—though this does break down from time to time. For instance, the current influx of asylum seekers is seen as a major social and political problem. The influence of religion generally has weakened since the 1960s, which marked a rise of secularism, though there are links between Catholicism and nationalist politics in Northern Ireland. The economy has moved from predominantly manufacturing to predominantly service-based, with the main period of change and restructuring taking place during the 1980s. Globalization has had an impact on family and community ties, with particularly the middle class becoming more mobile, not just within Britain, but globally.

THE CONTEXT OF HEALTH: ENVIRONMENTAL, ECONOMIC, SOCIAL, AND POLITICAL FACTORS

There are four main demographic trends in Britain: a decline in fertility; an increase in dependency of the population; changes in family structure; and increased labor mobility.

A post-war "baby boom" occurred in the late 1940s and 1950s, which was followed by a "baby bust" in the late 1960s (Rimmer & Wicks, 1985). The present birth rate is less than half that of the "boom years." There has been a dramatic increase in the number of people over the age of 65 in the United Kingdom. Between 1901 and 1991 this number grew from 1.8 to 9 million (Shaw & Shaw, 19933). There is longer life expectancy for all age groups. Improvements in nutrition, medical services, housing, and work conditions have all contributed to this. The population of those over the age of 75 increased by a factor of 4 over the period 1901–1991 (Shaw & Shaw, 1993). The proportion of the population that is elderly (and the dependency ratio) is not set to "peak" until 2024. The elderly are the major consumers of health and social services and there is concern about how to fund state welfare, which is funded on a "pay as you go" basis, financed through general taxation. As an example, since 1983 state pensions have been linked to the cost of living rather than the average wage.

There have also been substantial changes within marriage. One in four will end in divorce on current trends. One third of all new marriages involve at least one partner who has been married before. One of the consequences is a rise in the population of single-parent families. The current estimate is 1.5 million single-parent families in Britain, and there are concerns about the issues arising from this for child poverty (poverty is used here as a relative rather than an absolute concept). The generally improving social status of women is recognized as important in determining their health outcomes. Also, women are more likely than men to invest the household's economic resources in their children's health and education.

Another key demographic effecting health within the United Kingdom is that of labor mobility. In 2000, approximately 84% of the British population lived in England. The years 1981–2000 have seen significant changes in geographical migration. Geographically, people have moved away from their homes for employment, subsequently reducing their available social capital. With the transition to the "nuclear family," people no longer have the traditional "family" as an institution of care. In contemporary society, people move geographically, which means it is unlikely that an individual will see the same doctor (general practitioner—GP) over a long period of time. The concept of the "family doctor" no longer exists, and unlike years gone by where the doctor was traditionally viewed as both a medical practitioner and also a friend, visiting a GP nowadays is strictly a consultation process. Accompanying these changing social and geographical conditions is a process of self-medicalization of individual problems. The role of medicine in society is now part of an ongoing social process in which health and illness are made relevant for aspects of everyday life. Personal problems are seen as becoming more medicalized with, for example, an increase in the prescriptions for anti-depressants.

Improvements in life expectancy in Britain have been attributed to access to clean water; effective sanitation systems; waste disposal services; safety precautions in food preparation and storage; increasing nutrient intake, especially following improvements in agricultural technology and productivity, for example; nutritionally fortified and higher-yielding crops; and environmental management. One issue affecting life expectancy is inequality between social classes. There is a significant relationship between social and economic position and health. People in unskilled occupations and their children are twice as likely to die prematurely in comparison with professionals. Similar gradients in mortality by social class are apparent for nearly all causes of death (Fox & Benzeval, 1995). Men in social class 5 lost 114 years of potential life per 1,000 of the population against 39 years in class 1. Such differences are not so marked amongst the elderly (Butler & Calnan, 2000). Data on social variation on patterns of morbidity show similar gradients. For example, age standardized long-standing illness rates for people who are unemployed are over 70% higher than for those who are employed. Also, various indicators such as body mass index, lung function, and blood pressure vary according to social class (Fox & Benzeval, 1995). This situation is mirrored across much of the Western world; however, economic inequality appears to be growing more quickly in Britain than in any other advanced industrial society (Butler & Calnan, 2000).

Britain has a National Health Service (NHS) that was formed on July 5, 1948, to provide a public health service "from the cradle to the grave." This service is organized and financed in ways that promote the values of equity and fairness and has been described as "the jewel in the crown" of Britain's welfare state. At the heart of this is the belief that good healthcare is a fundamental determinant of people's capacity to succeed in life and that as such, it should be free to those who need it. The NHS has been funded predominantly out of general taxation and based upon the principle that access to services should be determined primarily by medical need. A part of the role of the NHS is to combat health inequalities by improving access to those who most need it and by working with local authorities to improve environmental and other determinants of health. There is a strong commitment to the NHS by the Labour Government. The cost of the NHS is currently around 7.4% of gross domestic product (GDP). The target is to increase its budget to 10% of GDP over the next five years, to bring health spending in line with that of the United Kingdom's European neighbors.

Environmental issues within Britain also have an impact upon health. In the last 15 years, concerns have been raised regarding pollution, for example, hazardous waste, pollution in rivers and bathing waters. In more recent years, public concern about health has expanded to incorporate the importance of quality food production. Traffic congestion, noise, fumes, climate change, and air pollution have also increasingly become issues closely linked to health in the public mind. Economic conditions affecting health are significant in terms of access to formal healthcare and provision of services. An aging population is a key characteristic of the United Kingdom and has significant economic implications. Subsequently, as the population is living longer and birth rates are lower, there are fewer people in employment. The number of people in the tax base has consequently reduced, impacting on the economic resources available for health, though not public and political commitment to the NHS.

MEDICAL PRACTITIONERS

Medical practitioners in the United Kingdom are overwhelmingly employed and trained by the NHS. The service is organized into two sectors: primary and secondary care. Primary care includes health professionals such as GPs, who receive clients from within the community with undifferentiated and undiagnosed problems. The GP will, in many cases, be the first contact an individual has made regarding his or her condition. GPs serve the entire local community in their areas. Nurse practitioners serve the local community in a similar way to GPs. Primary care nurse practitioners decide upon the priority of patient needs by performing a full assessment of healthcare needs, thereby determining a patient's health status. Some patients may simply need to remain under the care of the nurse for treatment, whereas others will require consultation with the GP. Here, the patient has the opportunity to consult with either a GP, a nurse practitioner, or both. The role of the nurse practitioner is set to be given full consideration in extending his/her role to reflect that of the GP. However, while the nurse practitioner has a wide range of skills, knowledge base, and ability to deliver specific aspects of care at present, they may need to be supplemented by a specialist. This may be a district nurse, health visitor, another primary health care nurse, community psychiatric nurse, counsellor, or clinical nurse specialist working in an acute care setting. Essentially the expertise of the primary care nurse practitioner is becoming more akin to that of a GP. They have the ability to operate as a specialist generalist offering care to patients with a wide range of needs.

Secondary care is a specialized service. Patients are generally referred by their GP to a secondary service for continued more specialized care. The number of secondary services provided is vast and covers a plethora of specialities, ranging from adult intensive care, blood disorders and diagnostic hematology, to disability medicine, healthcare of the elderly, and mental health. This service is generally hospital-based, although not in all cases, such as the Community Outreach service that serves people suffering with long-term and enduring metal health difficulties. Essentially, this is a mobile service that offers treatment and service provision within the client's home.

Secondary care services are led by specialist consultants within given medical areas, but the largest group of practitioners within this service is nurses. In recent years, a blurring of professional roles has developed. For example, nurses now are seen to take on more clinical responsibility. This includes professionally autonomous decision-making; screening patients for disease risk factors; developing, with the patient, a care plan; and he/she has authority to admit or discharge a client from his/her own caseload to other healthcare providers. Nurse prescribing is under consideration.

CLASSIFICATION OF ILLNESS, THEORIES OF ILLNESS, AND TREATMENT OF ILLNESS

The biomedical paradigm and its classification of and theories about illness and treatments is dominant. However, this approach has been "modernized" with increasing acceptance amongst health practitioners of the more behaviorally oriented multiple-risk-factor model. This emphasizes the interaction between behavioral risk factors and biological/genetic factors in the causation of disease. The importance of structural and environmental factors is also becoming increasingly recognized in etiology and classification (Butler & Calnan, 2000).

There is strong medicalization in some areas, particularly childbirth and unhappiness. Births are increasingly taking place in a hospital environment, with the incidence of cesarean sections showing a sharp increase in recent years. The incidence of depression in the United Kingdom has also increased from virtually nil in 1950 to "epidemic proportions" today, with one in four people visiting a GP said to be suffering from depression (Shaw & Middleton, 2001). This also illustrates that "lay beliefs" in Britain are heavily influenced by professional (often biomedical) paradigms—namely, that people may seek medical responses to social problems. Common sense understandings are imbued with professional rationalizations. In British society, biomedicine not only has provided a basis for the scientific study of disease, it has also become Britain's own culturally specific perspective on disease; that is, its folk (common sense) model (Engel, 1977). "Lay beliefs" are tied up with the certainty of diagnosis and the legitimacy afforded by taking on medical rationality. It has even been argued that there are no indigenous cultural developments in Britain that are not informed by an expert (if not biomedical) conceptual framework (Shaw, 2002). This can lead to increasing demands for healthcare services as new improvements in technology and pharmacology lead to increasing public expectations of health services. The use of "complementary medicine," particularly herbal medicine, aromatherapy, and Chinese medicine (such as acupuncture), which is not available as a part of the NHS and has to be purchased privately, is also on the increase.

SEXUALITY AND REPRODUCTION

In the 1870s in Britain, two-thirds of women would have had five or more children. In the 1920s, two-thirds of women would have had two or fewer children. The proportion of families with two children has almost doubled over the last century. Now, less than one family in six has three or more children. The proportion of childless marriages has also continued to grow (Parker, 1990). This illustrates a long-term decline in fertility in Britain, despite a "baby boom" in the decade following World War II. Reasons for this decline include: women are having children later in life, many for career reasons; greater availability of birth control and a widening of the legal grounds for abortion since the 1960s has given women more control over conception; more women are working and may find difficulties in looking after children; marital breakdown is increasing; job security is decreasing (there may be a reluctance to have children in an insecure lifestyle); and women's attitudes have moved away from their traditional image of mother and housewife. An increasing number of live births are taking place outside of marriage (31% in 2001—about a half of these to cohabiting couples). Also, two groups of women have had an increase in fertility—middle-class women in their 30s, and teenagers of working-class parents. This latter group is the cause of considerable concern and the focus of current social policy.

HEALTH THROUGH THE LIFE CYCLE

Pregnancy and Birth

There are social class differences in attitudes toward pregnancy. For those in the "middle class", it is becoming increasingly the norm for a first pregnancy to take place in the woman's 30s. This is less the norm for lower socioeconomic groups, and teenage pregnancy is becoming less unusual in areas of the cities where there are high rates of unemployment. Abortion is legal up to 24 weeks unless the fetus is regarded as having a severe impairment, in which case it can occur at any time with a supportive medical judgment. People regard miscarriage as a "loss" and a time for expressions of sympathy and support. Pregnancy and childbirth are heavily medicalized in Britain. The development of in-vitro fertilization (IVF) and other fertility treatments has led to service demands upon the NHS. Such services are viewed as a right of citizenship (Shaw, 2002). There is still stigma associated with infertility, but this has reduced significantly in recent years. Similarly, pregnancy and childbirth are regarded as times of risk, both by the medical professions and by many women themselves. Pregnancy is monitored regularly by health professionals, both by midwifery services in the community and by regular ultrasound scanning of the developing fetus and examination in a hospital setting. Smoking and drinking alcohol are regarded as activities that put the fetus at risk and there is a strong stigma associated with any engagement in such activities during pregnancy. Friends and family can be particular agents of "social order" in this respect. Indeed, there has been some work suggesting that such restrictions on the activities of the pregnant woman lead to a sense of losing control over her own life (Bowen, 2001).

The vast majority of births, 98%, take place in hospital. Less than 8% of births in Britain are without any medical intervention and a higher percentage of these are associated with home births. There is government concern over the increasing number of cesarean births in recent years (Lane, 1995). Healthy newborn babies delivered in hospital are generally allowed home after 2–3 days, though mothers can opt to stay in hospital for longer if they wish.

Infancy

Infancy is not a clearly defined category in Britain. However, there is a general view that infants are those between the ages of birth and 5, at which point they start primary school. Infancy is generally broken down in the public mind into three categories: baby, toddler (from taking first steps to developing the ability to run, skip, and jump—the term is derived from the "toddling" walk of the child), and childhood. Health Visitors monitor the initial development of the baby. Health professionals heavily promote breastfeeding, and most mothers will breastfeed for the first 18 weeks (the period of paid maternity leave allowed from work, though many employers offer additional unpaid leave, some up to a year). Subsequent development is monitored by GPs, who carry out immunization. Infants are immunized at the ages of 2, 3, and 4 months against tetanus, diphtheria, whooping cough, hemophilus influenzae type B (HiB), and polio. At 12–18 months they receive vaccination against measles, mumps, and rubella (MMR), which they are also vaccinated against a second time at between 3 and 4 years when they will also receive additional vaccinations against diphtheria, tetanus, and polio. Vaccination take-up is high, over 80%, although the MMR take-up has dropped below this level in some areas in the last year over fears of links between the combined vaccination and autism. This has resulted in isolated outbreaks of measles during 2002. If taken ill, children will usually expect to receive separate specialist secondary care services.

There is little government support or provision of day care for children under the age of 3 in Britain. Childcare facilities do, of course, exist, and are run by private individuals, companies, and by local authorities. Extended families, where in existence locally, may also play a part. The cost of day-care varies across Britain, but £25 per day—2002 prices—(approximately US$36) is a good guide. This means that the low-paid or those with more than one child not living in an extended family relationship where there is support from those retired, may be financially better off if one parent remains at home. Labor mobility has reduced the number of available family members locally who could take on a childcare role. The main caretaker is usually the mother. The state does provide free preschool (during the school term only) for children aged 4 (or aged 3 if in an area defined as having poor educational performance). Children start full-time education from age 5.

In post-modern Britain, many of the social contacts that parents have may be other parents whose children attend the same day care.

Childhood

Childhood in Britain is a period of both protection and exclusion. Children's lack of competence in the highly specialized differentiated world of adult activity in this country provides an explanation for children's exclusion and the protectiveness that accompanies it (Mayall, 1996). It is true to say that the daily lives of children are structured through the organization of their parents' time (particularly when both parents work), although childhood is also a time when there is protection from the world by social and economic provision. Schooling from the age of 5 has the dual role of providing the education to enable them to function in the adult world and to protect them during the day while their parents are at work. Childhood is also seen as a time for developing interpersonal skills and relationships as well as providing creative space of their own. Particularly in early childhood, children's imagination is actively fostered by adults—Santa Claus really does come down the chimney and Mickey Mouse really does live in Disneyland. There is a desire amongst parents to extend this period of innocence as long as possible.

Adolescence

There is no legal definition of adolescence in Britain. The Children Act of 1989 covers everyone from birth to age 18, the legal age of consent. Legally, children are deemed to have the capacity to tell right from wrong from the age of 14. However, this age of legal responsibility was recently reviewed following concerns raised by the murder of the toddler Jamie Bulger by two children who were 10 and 11, and by other cases (including a rape committed by an 11 year old). There is still a presumption of incapacity for children between the ages of 10 and 14 in Britain. However, legally, children in this age bracket can be deemed to have, what is quaintly termed, "mischievous discretion." Here it is for the court to decide in individual cases whether the child knew the consequences of his/her actions and knew right from wrong.

In the mind of the general public the "lay distinction" between children and adolescents is in some sense distinguished by the direction of protection. Children need to be protected from dangers in society, whereas society needs to be protected against the dangers posed by adolescents. This is in part a perceptual, rather than a real issue as groups of adolescents playing or moving around the streets is not uncommon. However, in some, particularly inner city areas, school truants have been linked to crime and vandalism, particularly in the popular press. Mugging and theft for/of mobile phones, in particular, has strong links with adolescent age children (mainly adolescents stealing from other children), as does "joy riding" (fast driving in stolen cars). As a policy response the government has re-enforced the principle of parental responsibility. Parents are liable for fines imposed upon their children, and recent proposals have been made to impose reductions on the receipt of state benefits for parents whose children are truant from school.

Inoculation continues into later childhood, with children aged between 13 and 14 years receiving a Heaf test for tuberculosis and a BGC (Bacillus of Calmette and Guerin) vaccination if indicated. Also, children aged 15–18 receive a booster inoculation to protect against tetanus, diphtheria, and polio.

Adulthood

Health-related issues in adulthood are varied, and in recent years a shift in emphasis has taken place to a demand for the right to be healthy. As society has become adept at problem-solving within the medical realm, a discourse has developed in which health is seen as a basic human need and medical care is perceived as a basic human right. Related to this is the notion of individual responsibility, which is partly reflected nowadays in the attention given to self-help and complementary therapies; aromatherapy, massage, shiatsu, yoga and relaxation, and acupuncture, for example, have all become popular alternative forms of therapy. Within adult mental health creative therapies, such as art and horticultural therapies, are a developing area. In conjunction with this is the increasing number of fitness centres within the United Kingdom and increasing numbers of people attending them. Gender-related issues include concern about menopause and hormone replacement therapy (HRT), which has received mounting attention. Concerns have also been raised regarding the over "medicalisation" of this natural change within a woman's body, whereby women are seen more sexually and socially obsolete, and medically and emotionally out of control. Studies show that HRT prevents osteoporosis and may significantly reduce the risk of coronary artery disease, but they also indicate that HRT may increase a woman's risk for breast cancer.

Domestic abuse in the United Kingdom, either physical or verbal, is becoming recognized as a health issue for a significant minority of women and children. Most women experiencing domestic violence are not identified in medical records. Abusive relationships are still very much seen as a "private" affair, one where intervention has unclear boundaries. Women's services within mental health is a developing area, concentrating upon a service that identifies the unique needs and demands of women, including safety within in-patient and residential settings. In particular, motherhood and mental health is now a recognized area of need, including post-natal depression. Men's health has, in recent years, been receiving attention both in the media and in the academic literature. The demarcation between the health of a man and that of a woman, however, is that men's health is more closely associated with culture (and work) and women's, with nature (and health). Prostate cancer has become an area of concern for men; 22,000 men in the United Kingdom are newly diagnosed with this each year. The incidence of testicular cancer has doubled in the past 20 years. Depression also is widespread but more often goes unrecognized as a condition in men than it does in women. Furthermore, the effects of alcohol on the body are becoming an increasing medical concern within both genders; young women now drink more frequently than ever. Unequal access to medical care within society tends to be the experience of those individuals suffering from a mental health condition, rather than being unequal in terms of gender. Inequalities exist also in terms of social, economic, and environmental disadvantage. Despite overall improvements in health, the gap between socially disadvantaged and affluent sections of the population has widened. A significant movement within healthcare taking place at present is the recognition and need to involve users at all stages of their care. This spans the whole NHS to ensure the needs of clients are met, and a more pro-active role in their treatment is taken by clients.

The Aged

"The aged" are structurally defined in Britain with mandatory retirement at age 65, when the state retirement pension is also payable. However, those aged 65 in Britain often do not see themselves as elderly and generally reserve that phrase for those aged 75 and over. Life expectancy and general health have increased for all age groups and the numbers aged over 80 is increasing every year. Because of the good health of those in their late 60s, the government is currently considering introducing flexibility on retirement age. This is important because a longer retirement means that more resources are required to fund it. A more flexible retirement age would also provide more scope for individual choice in this respect.

Income in retirement is linked to income in working life. Although there is a state pension, it provides a very basic income and the expectation is that this will be used to supplement a private or employment-based pension scheme. The current focus on "welfare to work" itself tends to exclude those who have retired. Quality of life in retirement, and social status, is still linked strongly to income.

The elderly are the major consumers of Britain's NHS. There is an association between incapacity and increasing old age; those over 80 consume 27 times the health and social care resources as those between the ages of 14 and 44. Those over 75 are also heavy drug users, with an average of 12 prescriptions per year (Glennerster, 1997). The major medical problems of the elderly are related to their age: age-related healing and visual impairment, cognitive impairment, and cerebral vascular diseases and associated dementia. There are also skeletal problems resulting from arthritis and osteoporosis, which give rise to mobility issues. Some of these diseases and problems can be found in those younger than 65, but the incidence increases with age and concerns with mobility are particularly focused upon those over the age of 75. Labor mobility has meant that there is less likelihood of the children of the elderly living locally to them, and consequently, there is only limited practical support that the elderly can expect to receive from family members. This places increasing reliance upon primary health and social care services.

Dying and Death

Attitudes toward death within Britain have undergone something of a change over the years. Now viewed less as part of the natural birth–life–death cycle, death is increasingly viewed as the result of a pathological condition, caught up in medicine and disease. Although the dominant religion in Britain remains Protestant, the rise in secularization and the decline in religious influence and practice has led to an increased fear of death as an "end." This may be linked to the increased value placed on health in British society. Regardless of what condition a human is suffering from, attitudes toward death mean that it must be fought at all costs. The concept of euthanasia is an extremely political area. Although voluntary euthanasia is illegal in the United Kingdom, there are conflicting views concerning this. GPs are allowed to administer a potentially lethal dose of pain killing drugs to relieve pain and suffering, but with no intention to kill. This has been both a political and moral high ground in recent years. Public concerns sway between accepting voluntary euthanasia as a personal choice, and freedom to help people die in dignity and pain free. Other attitudes raise concerns that any system of legalized killing would be open to abuse. This sits very closely with treatment of those suffering from long-term, life-threatening illnesses, such as cancer and HIV/AIDS. Hospice or palliative care services are offered to the individual, which for some may mean living in a cared environment whilst "waiting to die." The type of care offered by these services is predominantly focused upon pain control to develop a quality of life for the person during the final stages of the illness. Included within hospice care are family support and bereavement services.

Bereavement within Britain is taken as an expected process a person will enter after losing someone to death, and is an acceptable process for an individual to experience. This is often a long and complex ordeal, especially with the loss of children or unexpected loss, and is often dealt with over a period of time, and for some, with the aid of supporting voluntary organizations. There are also now specialized services, such as the National Association of Widows, the Gay Bereavement Project, and Jewish Bereavement counseling. The attitudes toward the the loss of a child are seen as the most traumatic and profound of all losses. Parental loss of a child is seen as psychologically and biologically the same following the death of a child at any age, from miscarriage to adulthood.

Suicide within Britain is seen not as an act of will but rather as influenced by an individual's mental capacity. Someone who attempts suicide is in need of help, and there are services available to help address this.

The body is treated respectfully following death. It will be taken to a chapel of rest, either at the hospital (if that is where the person died) or to a funeral director. The body is then subsequently transported for purposes of cremation or burial. Cremation is becoming the more usual means of disposal of remains in Britain. A religious service will be held at the chapel and/or church prior to cremation or burial.

REFERENCES

Bowen, K. (2001). *Infant feeding: Maternal choice or social control?* Unpublished Master by research and thesis dissertation, University of Nottingham, U.K.

Butler, J., & Calnan, M. (2000). Health and health policy. In J. Baldock, N. Manning, S. Miller, & S. Vickerstaff (Eds.), *Social policy*. Oxford: Oxford University Press.

Engel, G. L. (1977). The need for a new medical model: A challenge for biomedicine. *Science, 196*(4286), 120–135.

Fox, J., & Benzeval, M. (1995). Perspectives on social variations in health. In M. Benzeval, K. Judge, & M. Whitehead (Eds.), *Tackling inequalities in health: An agenda for action* (Chap. 2). London: Kings Fund.

Glennerster, H. (1997). *Paying for welfare*. London: Harvester.

Lane, K. (1995). The medical model of the body as a site of risk: A case study of childbirth. In J. Gabe (Ed.), *Medicine health and risk*. Oxford: Blackwell.

Mayall, B. (1996). *Children, health and the social order*. Milton Keynes, U.K.: Open University Press.

Parker, G. (1990). *With due care and attention*. London: Family Policy Studies Centre.

Rimmer, L., & Wicks, M. (1985). The challenge of change: Demographic trends, the family and social policy. In H. Glennerster (Ed.), *The future of the welfare state*. Aldershot, U.K.: Gower.

Shaw, I. (2002). How lay are lay beliefs? *Health, 6*(3), 287–299.

Shaw, I., & Middleton, H. (2001). Recognising depression in primary care, *Journal of Primary Care Mental Health, 5*, 2, 24–27.

Shaw, I., & Shaw, G. (1993). Demography, nursing and community care: A review of the evidence. *Journal of Advanced Nursing, 18*, 1212–1218.

Burmese

Monique Skidmore

ALTERNATIVE NAMES

In 1989, Burma was renamed the "Union of Myanmar" by the newly constituted State Law and Order Restoration Council (SLORC). The "Burmese" were renamed the "Myanmarese" or "Myanmars", and the Burmese language was similarly changed to the Myanmarese language. Minority groups in Burma were renamed "National Races," with "Myanmars" used to designate Burmese citizens of all ethnicities residing within Burma. The United Nations recognized the name change but pro-democracy groups use the older terms, especially "Burma", "Burmans", and "Burmese" as a way of protesting at the undemocratic nature of the name changes.

From 1885 until 1937, Burma was designated a province of India as part of the British Raj, and literature of the period regarding Burma is found in Indian manuscripts and journals.

Several armed groups have claimed independence from the military government and insurgent-held areas often publish literature referring to their territory as an independent entity. The chief example is publications from Manerplaw, Kawthoolei, a self-designated independent homeland of the Karen people (and the Karen National Union rebel headquarters) from 1971 until its capture by Burmese military forces in 1995 (Hail, 1995).

LOCATION AND LINGUISTIC AFFILIATION

Shaped like a kite, Burma is the largest country in mainland Southeast Asia. Burma covers 671,000 km^2 and borders the Andaman Sea, the Bay of Bengal, Bangladesh, China, India, Laos, Thailand, and Tibet. Most of these borders are flanked by mountain ranges, including the Himalayas in the north. The central riverland plains are the agricultural heart of the nation with the major north–south river, the Ayeyarwady, flowing to the large delta area that spills into the Gulf of Martaban. The Thanlwin River forms much of the border with China, and the Mekong likewise forms the border between Laos, Thailand, and Burma (The "Golden Triangle"). The Shan Plateau is a rugged, mountainous area bordering China, and tropical jungle covers much of the area bordering Thailand. The eastern borders of the country are home to the Arakanese (Rohingyas), Chin and Naga groups, who are ethnically similar to neighboring populations in India and Bangladesh.

Over 135 minority groups reside in Burma and the range of languages spoken is correspondingly large with over 100 linguistic groups and sub-groups (WOB & NCGUB, 2000). Burmese is the lingua franca, and since the military coup of 1962, the language of internal colonization. It is the language spoken by the majority ethnic groups, the Bamars or Burmans. Karen, Shan, Kachin, Chin, Arakanese, Kayin, Palua, Hindi, Mandarin, and English are just a fraction of the languages one can hear spoken in contemporary Burma. Burmese is a member of the Tibeto–Burman family of languages. In addition, there are members of the Mon–Khmer, Austro–Thai, and Karennic language families. The many different minority groups in Burma are described by the State Peace and Development Council (SPDC) as falling into seven "National Races": Chin, Kachin, Kayah, Kayin (or Karen), Mon, Shan, and Rakhine. The Central Intelligence Agency (CIA) estimates the population to be broken down into the following percentages: Burman, 68%; Shan, 9%; Karen, 7%; Rakhine, 4%; Chinese, 3%; Mon, 2%; Indian, 2%; and "other", 5% (including Thais) (CIA, 2002).

OVERVIEW OF THE CULTURE

A primarily rice-based agricultural society practicing dry rice, wet rice, and swidden agriculture as well as aquaculture, Burmese society was monarchical until the Third Anglo-Burmese War in 1885. The deposing of King Thibaw ended governance by the monarchy and official patronage of the only other significant institution, the *Sangha* (monkhood). Education was in the hands of the monasteries, villages comprised approximately 500 people, and headmen in the central riverland plains were part of a complex political system of reportage, all the way up to the monarch. After British colonization, the complex political units of various groups within Burma (including the *gumlao* and *gumsa* systems of the Kachin, as documented by Edmund Leach) (Leach, 1970) was overlaid with a British administrative system. General Aung San, the founder of the Army, was the architect of Burmese independence. Independence was granted in 1948, one year after Aung San and his parliamentary cabinet was assassinated in a bomb blast. U Nu was elected Prime Minister. A devout Buddhist, U Nu reasserted traditional relationships between lay authorities and the *Sangha* and was seen by some members of the military as being too "traditional" and non-embracing of modernization. In 1962 General Ne Win launched a successful military coup and only resigned in 1988. This period, known as the "Burmese Road to Socialism," turned Burma from the largest rice exporter in Asia, to a UN-declared "Least Developed Country." Following Ne Win's resignation, country-wide pro-democracy demonstrations forced the new dictator, Sein Lwin, to resign. The bloody suppression of the democracy movement birthed the State Law and Order Restoration Council (SLORC), whose name was changed in November 1997 to the State Peace and Development Council (SPDC). Tertiary institutions have remained largely closed since 1988 due to student pro-democracy activism, including the December 1996 student demonstrations. Small technical and rural tertiary institutions opened in 2001. The pro-democracy leader, Aung San Suu Kyi (secretary of the National League for Democracy (NLD)), was released from house arrest in May 2002, but then arrested in May 2003.

Two-thirds of the population are agriculturalists, with 10% employed in industry and 25% in the service industry (CIA, 1999, estimates). Economically, the country is in a perilous condition, with inflation estimated at 50% per annum and a thriving black market. Sanctions by the European Union and the United States against investment in Burma killed a building and land speculation boom in the mid-1990s, a boom that screened the laundering of profits from the production and sale of heroin. External debt for 1999/2000 was estimated at US$6 billion, and Myanmar received US$99 billion in economic aid in 1998/99. The official rate of kyat per dollar in January 2001 was 6.6, as compared with the unofficial rate in 2002 of 1,000 kyat/dollar. The government does not publish economic data, but the CIA estimates the percentage of the population below the poverty line at 23% in 1997 (CIA, 2002). The Burma Freedom and Democracy Act, signed into law by the U.S. President in 2003, applied further sanctions and effectively closed down garment factories and other export industries, further imperiling the formal economy.

Burma vies with Afghanistan as the world's largest supplier of heroin and methamphetamines. The highest levels of the military regime are directly implicated in the sanction and control of the drug trade, as is the Wa State Army, and are, until recently, the Shan State Army and various other military groups operating in the Golden Triangle region. Profits from the heroin trade in part also fund the 40-year insurgency against the military regime as well as providing hard currency for some military officials.

Kin terminology emphasizes age groups rather than vertical linkages and there is no ancestor worship in Myanmar. Women do not take the names of their husbands when they marry, and anyone may change their name as often as they like without any formal procedure, although this is rapidly changing with the increased control local security offices have over the recording of significant life events. There is no division of Burmese names into first names and family names. A Burmese name usually consists of several syllables or words. "Tint Tint Khine," for example, means literally "strong but elegant."

Kin terms are self-referential. A man older than oneself is designated *U* (uncle) and someone younger than oneself is referred to as *Maung* (younger brother). A male of approximate or slightly older age is referred to as *Ko* (or elder brother). Older women are designated *Daw* and younger women *Ma*. Residence patterns are most commonly matrilocal, with extended families living in the same house or in separate houses within one compound. Inheritance is bilateral and kin ties are not of importance unless relatives wish to emphasize the bond (Henderson et al., 1971, pp. 67–68).

The Burmese are overwhelmingly Theravada Buddhist (89%), with small populations of Christians (4%) (Baptist 3%; Roman Catholic 1%), Muslims (4%), animists (1%), and others (2%). Astrologers are common, and Hindu temples, mosques and Jewish synagogues are also in evidence. There is an almost unanimous belief in animism in the *Nat* cult a vibrant system of spirit propitiation. *Nat* spirit wives have in recent years been largely replaced by transvestite men who find an acceptable social role in spirit mediumship and dramatic performance. The relationship between animist nature spirits, Nat spirits, and Buddhicized figures is complex and changes over time. Theravada Buddhism is strongest in Burma, where a mass lay meditation movement practicing *vissipana* meditation seeks to reform the monkhood and eventually social and political life, from below. Charismatic monks and various sects and cults (*gaings*) exist in Burma where a strong belief in the miraculous and millenarial beliefs hold significant sway in social life and strategies. Astrology, alchemy, magic runes (*in*), spirit propitiation, soul flight, reincarnation, karma, witchcraft, visions and portents, *samatha* and *vissipana*, are but some of the rich store of religious, cultural, and magical knowledge that inform Burmese people's worldviews.

Drama, dance, puppetry, music, painting, and poetry are central to Burmese life and to cultural, religious, and artistic expression. Comedy, political satire, romantic and nationalistic themes, and Buddhist and spirit universes and powers are all expressed through these media. In recent years popular magazines, DVD/video and karaoke huts, and movie theaters have fostered an indigenous movie and rock star scene.

Women's roles in society are limited due to the military hierarchy of the nation and their exclusion from monastic roles other than as donors, administrators, and nuns. Whilst many Burmese women will claim that they are equal because of shared property rules, lack of dowry, and the general ease with which women may enter and leave marital arrangements, human rights groups draw attention to the atrocities and suffering of Burmese women in war zones and in relocated areas where structural and domestic violence, sexual bartering, and prostitution are serious problems.

The Context of Health: Environmental, Economic, Social, and Political Factors

In the World Health Organization's (WHO) ranking of the health status of individual countries, Burma ranks almost last, 190 out of 191 nations (WHO, 2001). This statistic represents a spectacular fall in health standards (and living standards more generally) over the last four decades, the period of military rule.

The birth rate in 2001 is estimated at 20/1000 population (CIA, 2002), and the infant mortality rate in 1995 estimated by the WHO to be 50/1,000 live births and under-5 mortality at 101/1,000 live births (compared with the 2001 estimate of the CIA of infant mortality at 74/1,000). CIA health indicators are consistently worse than WHO and Burmese Government figures because they are based on estimates of deaths due to HIV and other future health estimates. The CIA estimates the total fertility rate to be 2.3 children per woman.

The 2002 health statistics issued by the Myanmar Ministry of Health put government expenditure on health at approximately US$278 million and capital investment in health at US$13.5 million. The reported number of cases of some childhood diseases has decreased in the past decade, with reported cases of diphtheria dropping from 204 in 1990 to 38 in 2000, and neonatal tetanus falling from 189 reported cases in 1991 compared with 61 in 2000. Similarly, since oral polio vaccine (OPV) coverage began in Yangon in 1982, the number of reported poliomyelitis cases has fallen from 390 to 50 in 2000 (Ministry of Health, 2002).

Burma has two major psychiatric hospitals, one in Rangoon [Yangon] and one in Mandalay. Qualified psychiatrists are rare in Burma. In 1994 there were 130 medical officers and 260 regional officers trained to identify five psychiatric conditions. There is very limited psychiatric training offered by Burmese medical institutes (Skidmore, 1998). There are four medical institutes in Burma, and since 1978 postgraduate training in psychiatry has been possible in Burma. From 1993, consultant psychiatric health services have operated in the 14 state and division hospitals (Sein Tu, 2002). The inadequate government wages mean that doctors working at the psychiatric hospitals operate their own private clinics in the afternoons or evenings. Medication consists almost solely of barbiturates and Thorazine (chlorpromazine). Differential diagnosis is very limited, with dramatic changes in the frequency of diagnostic categories occurring over the last 30 years (Skidmore, 1998).

Mandatory arrest, incarceration, and detoxification of people suspected of heroin and methamphetamine use occurs in Burma. Throughout the country a number of Buddhist detoxification centers exist. Withdrawal is managed with either no medical assistance or with a tincture of opium. Incarceration occurs for approximately 6 weeks. Two hospitals for drug addiction exist (in Putao and Myitkyina), six "major therapeutic centers," and 22 "rehabilitation" centers (Sein Tu, 2002).

Food shortages have been recorded across 10 of Burma's 12 provinces and inflationary pressures continue to push up the price of basic commodities such as rice, oil, and fish (People's Tribunal, 1999). Government quotas on rice production also assist in malnourishing

Burmese villagers. In peri-urban settlements ringing Rangoon and Mandalay, two meals per day consisting of rice with *ngapi* (fermented fish sauce) and a very small amount of protein (fish, meat, or legumes) is the norm. Malnourishment of children occurs when women do not have enough breast milk and feed their children the water in which the rice has been cooked. Many Burmese people in peri-urban and rural areas use rice-water and betel nut as appetite suppressants or as food substitutes (Skidmore & World Vision International, 1997a, 1997b).

The Ayurvedic humoral system of "hot" and "cold" foods has become irrevocably muddled in Burma. An enormous amount of indigenous medical knowledge was lost upon British colonization and in the subsequent decades of poverty and civil war. Indigenous medical practitioners trained through the apprenticeship system are now rare, with the new generation being trained in government traditional medicine institutes, hospitals, and clinics. Among the lay population, humoral medical concepts are used in non-consistent and often harmful ways. Pregnant women, for example, often refuse to eat foods that provide adequate dietary protein and fat because of beliefs related to harm of the fetus. In addition, the strong belief in *Nat* spirits means that foods favored by *Nat* are eaten only occasionally. These most ubiquitous of commodities, coconuts and bananas, thus provide a form of dietary proscription that further reduces the available foods to people only occasionally receiving adequate nutrition.

Medical hospitals provide excellent medical care for military officers and their families. Public hospitals are subject to routine theft of equipment and pharmaceuticals, and patients purchase their own medicines at pharmacies or market stalls for use in both biomedical and indigenous medical hospitals.

Medical Practitioners

Medical and medico-religious practitioners are utilized in Burma for a range of emotional, supernatural, mundane, and psychiatric problems. Biomedical physicians are trained at the government institutes of medicine, including training in many biomedical specialities. Overseas training is a common route to specialist training. Private diagnostic services such as X-ray and ultrasound exist in cities and large towns. Opticians and dentists are also common. Specialized nursing institutes also exist and government-trained midwives work in well-populated areas and also provide rural outreach programs.

Community healers (*yankus*) span the spectrum from charlatans to indigenous medical practitioners who practice a variety of traditional and biomedical procedures. This can often mean that medicine is given inappropriately and without regard to side-effects. Illegal trading of medicines from China, Thailand, and India means that Burma's markets are stocked with illegally packaged, misleading products, without cold storage, and often many years past their use-by date.

Indigenous medicine is the first recourse for most Burmese when self-diagnosis of the great majority of daily illnesses occurs in the home. Indigenous medicine is practiced privately by practitioners trained through apprenticeship, and officially, as managed by the traditional medicine department of the Ministry of Health. Utilizing the five tastes, Buddhist principles, knowledge of medicinal plant properties, and the preparation of dried and powdered compounds, traditional medicine is the primary form of medical care in the 40% of the country not covered by basic biomedical healthcare (IPPF, 2002). In addition to traditional medicine practitioners, *lethes*, or traditional midwives, practice delivering babies and the care of pregnant women, and care for women for 6 weeks after the birth. They also practice "massage" and a variety of other methods to abort unwanted fetuses. *Lethes* who specialize in abortion become known as abortionists, but the government will often arrest such women. Abortion is illegal in Burma but poverty and domestic violence lead many desperate women to abortionists.

Astrologers, white magicians, alchemists, wizards, *Nat* spirit mediums, and magic monks are among the medico-religious specialists (*daq sayas*) that Burmese people turn to for emotional, psychiatric, and economic aid. Among the rural population who present to psychiatric hospitals, a variety of these practitioners are sequentially utilized, with psychiatric personnel being used only as a last resort and usually without the patient's consent.

Classification of Illness, Theories of Illness, and Treatment of Illness

The most consistent finding arising from the author's fieldwork in peri-urban Burma is that the regularity and volume of blood at menstruation and childbirth is the fundamental indicator of health and well-being for women. Being in good health is contingent upon harmony in and between the body and the universe where the latter includes the socioeconomic and political dimensions of everyday life. Blood flow is the key symbol by which Burmese women's beliefs and practices concerning their well-being can be understood. This generic Burmese pathophysiology is used in the forcibly relocated peri-urban townships to experience and express the experience of illness and being unwell. Abortion, childbirth, madness, and "weakness" are the major domains in which understandings of the way that blood flows through the body are elaborated, and these lay diagnostic categories give coherence to women's illness, treatment, and local etiologies. Burmese women's health as articulated at the levels of pathophysiology, interpersonal relations, and the local peri-urban environment, necessarily include the wider political and moral economies of Burma. All of the women interviewed by the author are adamant that the body, mind, and soul cannot be well if the physical, political, and spiritual domains are not aligned in harmony. The moral economy of military dictatorship and the enforced economic impoverishment of the majority of peri-urban Burmese are understood by Burmese people to be the underlying causes of ill health in general, and of peri-urban "women's diseases" in particular (Skidmore, 2002a).

Well-being is divided into gender and age categories, with familial and sympathetic causes of illness understood as possible contributing factors to illness. A belief in the contagious spread of leprosy by family and friends of people with the virus leads, for example, to the stigmatization of entire families and social networks and their inability to find or keep employment if there is evidence of a member of the group with leprosy (Skidmore, 1997a).

SEXUALITY AND REPRODUCTION

Hepatitis, HIV, and other sexually transmitted diseases flourish for a variety of reasons. These include the prevalence of men using prostitution, the practice of polygamy, the rate of heroin addiction in Myanmar (4% for men, 2% for women) (UNDCP, 2001), and other unsafe practices such as re-use of syringes in medical clinics and penis-enlarging procedures common in Burmese jails and among fishing communities. Carrying condoms is widely believed by the police and military forces to constitute evidence that a woman is engaged in prostitution, and has been legal only since 1993 (Smith, 1996). In addition, men refuse to buy small- or medium-sized condoms, and a majority of both wives and women working in the sex industry argue that they cannot make their sexual partners use condoms. Most state that it is an inappropriate subject to discuss between the sexes.

HEALTH THROUGH THE LIFE CYCLE

Pregnancy and Birth

Women's health is divided into three life periods that correspond to three types of illness categories. *Miyet sa* covers all those illnesses believed to occur because of menstruation and up to the period of motherhood. *Miyet leh* refers to illnesses of the "middle years," specifically related to childbirth, breastfeeding, and women's health during their years of raising children. *Miyet so* refers to illnesses that occur around the time of menopause and continue on into old age for women. Each of these stages is concerned with blood flow, especially lack of adequate menstrual flow at puberty and leading up to menopause. "Weakness" is a major diagnostic category believed to result from too little or too much blood flow, but also arising from the use of contraceptives. Emmenogogues are used extensively to increase blood flow and also as an attempt at aborting unwanted fetuses. Many Burmese physicians translate weakness (*thwe aah neh*) as anemia but the category is broader and refers as much to social and material aspects of living as it does to physiological imbalance. The national nutritional program gives the statistic of 89% of pregnant women to be at risk from iron deficiency anemia (WHO, 2002), and the major reason for young women to present to traditional medicine hospital and clinics is amenorrhea (Skidmore, 2002a).

Too much bleeding during and immediately after childbirth and abortions is also a major cause of concern. Childbirth is seen as a particularly dangerous time, but a belief in potential harm to fetus and mother also marks pregnancy as a period when a large number of taboos concerning food and mobility come into play. Abortion is illegal in Burma, but it is not stigmatized in the community. In addition, marriage and divorce are easily acheived, and illegitimate children are not considered different from other children (Skidmore, 2002a).

Infancy

Burmese midwives deliver babies and perform abortions, primarily through abdominal massage. The postpartum period, *me-dwin*, is marked by sequestration of young mothers, dietary taboos, and attempts to stop the body from losing heat. This means that the body is not bathed in cool water (sometimes no bathing occurs for 6 weeks), and "cold" foods are not eaten. Burmese women usually breastfeed exclusively for at least 2 months, with the water that rice has been cooked in usually given as the baby's first supplemental food. Infants are carried and cuddled by mothers, and rocking cots, hammocks, and swings are used by mothers, sisters, daughters, and grandmothers to continually keep the child in motion when awake (Skidmore, 2002a). In urban, peri-urban, and regional centers, many mothers use antenatal care, but infant mortality remains high, as does maternal mortality, primarily from incomplete abortions. Maternal mortality is as high as 500–580 per 100,00 live births (WHO, 1997). Women present to regional hospitals on a daily basis with hemorrhage from incomplete abortions. Studies conducted throughout Burma indicate somewhere between 33% and 60% of maternal mortality is directly attributed to abortions (Ba Thike, 1997; Khin Than Tin & Khin Saw Hla, 1990; Ministry of Health & UNFPA, 1999). Two studies have found abortion to be the leading cause of maternal mortality in Myanmar (Krasu, 1992; UNICEF, 2000).

Vasectomy is illegal but there is a thriving trade by urban surgeons. Cesarian rates are very high in some areas and many women opt for a hospital birth, especially a cesarian birth so as to also have sterilization following the birth of their second child. Cesarian births are very expensive and there is no uniformity of charge across the urban and peri-urban areas (Skidmore & World Vision International, 1997a, 1997b). These difficult conditions for childbirth mean that infant mortality rates are almost 400% higher in the 40% of the country where no basic biomedical health services exist (IPPF, 2002).

Childhood

Twenty-nine percent of the population is aged 0–14 years (CIA, 2002). Thirty-eight percent of children aged 6–11 had, in 1994, recognizable signs of goiter (WHO, 2002). In 1994, 31% of children under three years of age had unacceptable weight-for-age ratios (WHO, 2002).

Children are traditionally allowed to run free until the age of 12, when they are sent to monastic centers for education and discipline. Many children attend government schools and are teased, mocked, and told parables to teach appropriate social behavior and norms. Children are regarded with great affection and as the embodiment of a family's happiness and wealth. Children have a keen sense of duty to their parents and teachers and may become monks or nuns for short periods of time to repay their parents' love by making merit for them in their coming lives.

Adolescence

The *shinbyu* (for boys) and ear-boring ceremony for girls traditionally mark, for Burmese Buddhists, the coming of age of adolescents and bestows upon them responsibilities of family and personal welfare. For impoverished families this means adolescents must earn money to contribute to family income. For wealthier families it means following Buddhist precepts and being respectful of monks, parents, and teachers, as well as trying to excel at school. In these ways, adolescents prove that they are worthy of the care that their parents have invested in their childhood. Many adolescents feel a fierce sense of duty toward their parents. As young adults, adolescents are required to dress in culturally appropriate ways and to be mindful of Burmese cultural norms and traditions, especially with regard to appropriate gender behavior. For young girls this includes having a chaperone in public.

Adulthood

Domestic violence, sexual bartering, sexual torture, and death from incomplete abortions are all facets of everyday life in the poorest of Burma's townships, especially those townships that have been forcibly relocated from central urban areas to paddy fields on the outskirts of the major cities. Alcohol use is very high in these areas as men's powerlessness becomes apparent in their lack of employment and social status that has resulted from their forcible removal from previous townships (Skidmore, 2002b).

Domestic violence, polygyny (due to the highly mobile nature of the young male workforce), and other forms of marital disharmony have led to a very high divorce rate, paralleled by an almost equal rate of remarriage. This means that families in the relocated townships frequently consist of many step-relations. Incest has been reported by girls and young women when step-brothers, step-fathers, and step-uncles take financial responsibility for the household. Fear of incest and marital rape is widely reported in these townships. Forcible relocation meant smaller land plots and splitting up of extended families which in turn meant less controls over male household leaders (Skidmore, 2002b).

Drug abuse is a significant problem in Burma. Nicotine, betelnut, codeine, and cough mixtures containing codeine, methamphetamine, opium, benzodiazepans, and, especially, heroin are all common substances easily purchased in urban and regional areas. It is relatively rare for women to be addicted to substances other than betel and opium, although alcoholic female patients do present to the psychiatric hospitals and the smoking of cheroot cigarettes is common among older women (Skidmore, 1998). The 1994 WHO smoking survey found that 47% of people aged over 15 are regular smokers (WHO, 2002).

Heroin is available in several grades in Burma, but the most common grade in known as No. 4 heroin and is highly refined. It is available on street corners, in schoolyards, in "shooting galleries," and via "courier" or "delivery" systems where regular deliveries are made to individuals' homes. Benzodiazepans are used as a form of self-medication when heroin is not available or as a form of slow withdrawal from heroin, although the recent availability of large numbers of inexpensive metamphetamines means that multiple drug use strategies are increasingly common (Skidmore, 1998).

The partial payment of government workers with palm oil has led to an enormous increase in heart attack, stroke, and hypertension as evidenced by both non-government organizations (NGOs) and by the presentation of patients at the traditional medicine hospitals (Skidmore, 1998).

The HIV/AIDS adult prevalence rate was estimated by the CIA in 2001 to be 1.99%, with 530,000 people living with HIV/AIDS and 48,000 deaths from HIV/AIDS (CIA, 2002). The explosion of prostitution in the forcibly relocated townships coincided with the narcotics-fueled building boom in the capital cities in the mid-1990s. Nightclubs, karaoke bars, and other venues for prostitution proliferated in this lawless atmosphere.

The 40 years of insurgency in the border areas means that trauma is a prevalent medical problem for patients presenting to biomedical hospitals, clinics, and psychiatric hospitals. Such trauma is not confined to adults and is prevalent among soldiers who have seen active service, insurgent fighters, and all adults and children caught up in war zones within the country's borders. In addition, landmines and unexploded ordinances pose grave risks to combatants and residents of contested territory. Finally, forced portering for the army, forced labor, the incarceration of political prisoners, and the existence of remote work camps all further endanger the health of Burmese people (Skidmore, 2002b).

The Aged

Life expectancy rates are low, there is no welfare system in Burma, and aged people have always been cared for within the extended family unit. Wealthier Buddhists may purchase a small home within the grounds of a monastery and so retire there. These Buddhist retirement communities are generally happy places with small garden patches, devotional and volunteer activities that revolve around the monastic cycle, and preparations for the next life. For destitute older people, cataracts and arthritis are major impediments to mobility and no help exists, outside of the charity of monasteries and neighbors.

Within family units, aged people command great respect and continue to wield power over the occupations and living arrangements of their children and grandchildren. They continue to control the family finances although they may spend increasing lengths of time at monasteries. Infrastructure for disabled people is non-existent and most roads are unpaved, making mobility difficult for aged people. They tend to stay home and "guard" the house against theft. When a family member becomes frail or incapacitated, other family members worry that they must stay on their own during the working day, afraid that unscrupulous people will take advantage of their aged relatives.

Dying and Death

As with all major life cycle events, death is managed for Burmese Buddhists by a series of religious proscriptions and practices that center upon the release of the soul, or butterfly spirit, from the corporeal body. Attachment to the body is not encouraged, but proper management of the enduring spirit is necessary to ensure an auspicious rebirth and avoid being trapped as a ghost.

Green deaths occur when a spirit is unable to leave surviving relatives. This can occur when women and babies die during childbirth, or when husbands die leaving their wives with several children to feed and raise alone. In the peri-urban townships ringing Rangoon and Mandalay, where infant and maternal mortality is high, several hospitals are known as "green hospitals." This term refers to the number of green deaths and resulting ghosts of babies who reside around the hospitals. Women, especially pregnant women, traveling home at dusk or at night, tie yellow ribbons around themselves or their bicycles to ward off these ghosts.

The CIA puts the death rate at 12 deaths per 1,000 population and life expectancy at 55 years. The WHO puts the death rate at 8.6/1,000 population and life expectancy at birth of 60 for men and 64 for women (WHO, 2002). Burmese people aged 65 and older constitute approximately 5% of the population (CIA, 2002).

REFERENCES

Ba Thike, K. (1997). Abortion: A public health problem in Myanmar. *Reproductive Health Matters, 9*, 94–100.
Central Intelligence Agency (2002). *CIA World Factbook: Burma.* Washington, DC: Central Intelligence Agency Government Publications.
Hail, J. (1995). Burmese attack could doom independence struggle. Bangkok, Thailand: UPI.
Henderson, J. W., Heimann, J. M., Martindale, K. W., Shinn, R. S., Weaver, J. O., & White, E. T. (1971). *Area handbook for Burma, foreign area studies.* Washington, DC: U.S. Government Printing Office.
International Planned Parenthood Foundation (IPPF) (2002). *Country profile: Myanmar.* Available from http://www.ippf.org/regions/countries/mmr/index.htm
Khin Than Tin & Khin Saw Hla (1990). *Causes of maternal deaths in affiliated teaching hospitals.* Yangon: Myanmar Medical Association.
Krasu, M. (1992). An Overview of Maternal Morbidity in Myanmar. *Proceedings of a Seminar on Maternal Morbidity Obstetric and Gynecological Section.* Yangon: Myanmar Medical Association.
Leach, E. R. (1970). *Political systems of Highland Burma: A study of Kachin social structure.* London: Athlone Press.
Ministry of Health (2002, May 20–26). *Ministry of Health Statistics.* Unpublished report cited by *Myanmar Times and Business Review, Health and Medicine Supplement, 6*(115).
Ministry of Health and United Nations Population Fund (1999). *A reproductive health needs assessment in Myanmar.* Yangon: Author.
Myanmar Times and Business Review (2002, May 20–26). *Health and Medicine Supplement, 6*(115).
People's Tribunal on Food Scarcity and Militarization in Burma (The) (1999, October). *Voice of the hungry nation.* Asian Human Rights Commission.
Sein Tu (2002, May 20–26). *New attitudes to mental health care. Myanmar Times and Business Review, Health and Medicine Supplement, 6*(115).
Skidmore, M. (1998). *Flying through a skyful of lies: Survival strategies and the politics of fear in urban Burma (Myanmar) UMI*, NQ50073. Doctoral dissertation, McGill University.
Skidmore, M. (2002a, June). Menstrual madness: Women's health and wellbeing in urban Burma (Myanmar). In A. Whittaker (Ed.), *Women and health* [Special ed.] *35*(4), 85–104.
Skidmore, M. (2002b). "Behind bamboo fences": Domestic violence against women and children in urban Burma (Myanmar). In L. Manderson & L. Bennett (Eds.), *Women and violence in Asia* (Chap. 5, pp. 90–106): Curzon Press.
Skidmore, M., & World Vision International (1997a). *A rapid assessment study of maternal and child health in Mandalay, Myanmar.* Yangon: World Vision.
Skidmore, M., & World Vision International (1997b). *A rapid assessment study of maternal and child health in Hlaingthayar, Myanmar.* Yangon: World Vision International.
Smith, M. (1996). *Fatal silence: Freedom of expression and the right to health in Burma.* London: Article 19.
United Nations Drug Control Programme (UNDCP) (2001). *World drug report.* Available from http://www.undcp.org/world_drug_report.html
United Nations International Children's Fund (UNICEF) (2000). *UNICEF Statistics.* Available from http://www.childinfo.org
Women's Organizations from Burma and Women's Affairs Department, NCGUB (WOB and NCGUB) (2000). *Burma: The current state of women—conflict area specific.* An unpublished shadow report to the 22nd Session of CEDAW.
World Health Organization (1997). *An assessment of the contraceptive mix in Myanmar.* Geneva, Switzerland: World Health Organization.
World Health Organization (2002). *Country health profile: Myanmar.* New Delhi, India: World Health Organization.

Cree

Naomi Adelson

ALTERNATIVE NAMES

Eeyou (singular); *Eeyouch* (plural); person, the people. The term Cree is only used when spoken in English and is not the term the people use for themselves.

LOCATION AND LINGUISTIC AFFILIATION

The entire population of Cree in Canada spreads from the province of Alberta to the province of Quebec. Historically ranging from plains-dwellers to woodland

and northern sub-Arctic hunters, the many Cree nations are the most widely distributed geographically of First Nations in Canada. All of the Cree dialects derive from the Algonquian language family. The eastern Cree of the James Bay region of Quebec, the population highlighted in this entry, speak Cree as their first language and English, French, or both after that (96% have Cree as their mother tongue, 90% speak it at home; 77% speak English, 29% speak French in addition to Cree, 20% speak neither French nor English [Schnarch, 2001]).

OVERVIEW OF THE CULTURE

The James Bay Cree or *Eeyouch* of *Eeyou astchee* (the people's homeland) have been living in, traversing through, and sustained by the sub-Arctic region for over 5000 years. Historically, small family bands navigated vast territories of *Eeyou astchee* over the course of a year, hunting, fishing, trapping, and harvesting the plant and animal resources required for food, clothing, shelter, and tools. These hunting territories remain a salient aspect of the lives of the Cree today, as people continue to hunt and consume the game of the region and define themselves through the lands that their ancestors marked as theirs by usufructuary right and on which they still hunt, trap, and fish. Animals of the sub-Arctic such as caribou, hare, beaver, and porcupine, as well as the fish and the fowl of the region, such as Canada goose, duck, grouse, and ptarmigan, were the principal diet of this hunting population and remain important foods today. Summer fruits such as blueberries and cranberries supplemented the meat diet. With contact, and especially the fur trade, came what are now considered staples of traditional fare: flour for bannock (soda bread), tea, and sugar.

The land and its resources are intrinsic to the rich animistic traditions of the *Eeyouch* whereby, most significantly, a successful hunt is the result of a system of mutual respect between the hunter and the animal and its spirit force. Indeed, the culmination of that alliance between animal and human is when the animal chooses to give itself to the hunter (Feit, 1986; Tanner, 1979). With animate spirit imbued in each and every element of the land, the nutritional or healing properties of plants and animals are enhanced. Life and spirituality are inseparable, and both are linked in innumerable ways to the places of *Eeyou astchee*. Thus, *Eeyou astchee* is not only a physical resource, but the source of spiritual and cultural heritage. For example, in the Cree language, specific sites are identified as often by their topographical features as by a significant experience that occurred there, such as burials, where individuals were born or where a particular spirit resides.

Christian influences on the *Eeyouch* have been both significant and complex ever since the earliest days of Christian missionary activity on *Eeyou astchee* and deserve mention here. Indeed, organized religion continues to demarcate certain boundaries of affiliation and identity among the people. More specifically, the Anglican church introduced Christianity to the Cree region in the mid-19th century. The early missionaries denounced all that they deemed to be heathen practices (especially drumming and the shaking tent ceremony) and those early resolutions resonate to this day. On the one hand, for many, Christianity was an extension of their familiar practices of humility on the land, respect for all living things, and deep love and commitment to their families. Thus, many were readily able to meld traditional practices with this new Christian faith, although strategically chose to practice indigenous ceremonies well outside of the purview of the missionaries. On the other hand, others were—and remain—less inclined to practice anything that was originally forbidden by the church. In some communities, too, evangelical Christianity has a growing and loyal following and it, too, separates indigenous spiritual practice from the more common activities of hunting and trapping. Moreover, there is a surge of interest among many *Eeyouch* in traditional spiritual beliefs and practices and a resurgence of ceremonies once thought lost to the younger generations. Thus, the church, as a symbol of organized religion, remains a complex influence on people's lives as they negotiate their contemporary cultural identities as sometimes part of and sometimes counter to its influences and authority.

Now numbering over 13,000 the *Eeyouch* live in nine communities, many of which originated first as summer gathering places and then as trading post sites. These sites became official villages when the Canadian federal government took over the management of the aboriginal communities in the early part of the 20th century. The nine *Eeyou* communities include four inland (Nemaska, Mistissini, Oujé-Bougoumou, Waswanipi) and five coastal (Waskaganish, Chisasibi, Wemindji, Eastmain on James Bay, and Whapmagoostui on Hudson Bay) settlements. The communities range in population size from about 500 to over 3,000. This is a young and growing population, with only 4% of the total population 65 and over and approximately 34% of the *Eeyouch* under 15 years of age. As well, with a crude birth rate of approximately 24%, the current trend suggests that the population will double between 1999 and 2027 (Schnarch, 2001). The income security program, set up under the James Bay and Northern Quebec Agreement, provides a cash income to those individuals who choose to spend the greater part of the year living in the bush and actively trapping and hunting. Recent figures indicate that for all nine communities, 37% of the population is registered in this program. Other forms of employment are largely community based and are divided almost exclusively between local administration, Cree entities, and community enterprises (approximately 12% employed in each of these sectors). The unemployment rate for those 15 and over in *Eeyou astchee* is approximately 17% (compared with 10% in Canada) and among those employed, only 39% had full-time, year-round jobs in 1995. The average total income for the *Eeyouch* in their territory is less than half that of non-Aboriginals living in *Eeyou astchee* (approximately $17,000 versus $40,000) (Schnarch, 2001).

The *Eeyouch* are represented nationally by the Grand Council of the Crees (GCC) of Quebec, a legal corporation of the Cree nation created in 1974 (GCC, 1996). The GCC originally arose out of political necessity in the early 1970s. In 1971, the *Eeyouch* found themselves in a dire struggle for their lands. The Quebec provincial government, without consultation with the *Eeyouch*, was planning to divert vast waterways on *Eeyou astchee* in order to construct one of the world's largest hydro-electric projects. After a long, gruelling, and bitter legal battle with both governments and Hydro-Quebec, the *Eeyouch* finally—but reluctantly—agreed to exchange the rights to over 600,000 km^2 of land for specific benefits, payments, and land rights. This modern treaty, known as the James Bay and Northern Quebec Agreement (JBNQA), was signed between the *Eeyouch*, the Inuit of Quebec, and the federal and provincial governments. The Agreement provides the context for a form of (delegated) self-government, control of wildlife management, input into environmental impact assessments, an innovative program of guaranteed income for men and women who trap on a full-time basis, as well as administrative control over education, health, local justice, and local and regional governments. The Cree Regional Authority was created to act as the regional government responsible for the administration of the JBNQA. Programs and services provided to the communities are co-ordinated by the Cree Regional Authority with the collaboration of the Cree Housing Corporation, the Cree Regional Boards of Health and Social Services, and the Cree School Board. The now self-governing *Eeyouch* are an established nation due to a large extent to the success of the assertive and, at times, uncompromising leadership of GCC of Quebec. The GCC rightly boasts consultative status on the UN Economic and Social Council and has had tremendous impact internationally, nationally, and regionally as an efficient, well-organized, and active government (Marvelle, 2001).

It is important to remember that prior to 1975 and really until about the early 1980s, the *Eeyouch* were organized into "village band societies," administered locally but according to the dictates of the federal Indian Act; their homes were substandard as were the facilities and services available to them (Salisbury, 1986). The JBNQA brought not only administrative control to the people and income security for full-time trappers but also improved housing and facilities, a voice in the administration of the lands and resources of *Eeyou Astchee*, and control over education and health services. The JBNQA, in other words, was viewed as progressive at the time of the signing. It was, however, never fully implemented, leaving the *Eeyou* leadership engaged in endless legal wranglings with the provincial and federal governments for alleged breaches of the Agreement. In 2001, the new leadership of the GCC agreed to drop all the outstanding lawsuits. In what must be seen as a pragmatic plan to accommodate future generations of the burgeoning population of *Eeyou astchee*, the current leadership of the GCC not only put a moratorium on the legal battles but, more importantly, has agreed to allow further extensive hydro-electric development on *Eeyou astchee*. In exchange, the new Rupert–Eastmain Accord will—if it passes environmental review—see the Quebec government paying out a large cash settlement over the next 50 years to be used for community and business development, job creation, and resource management (Dougherty, 2001; Macpherson, 2001). The signing of this new Accord did not come without dissent from some members of the *Eeyou* communities. The current leadership is attempting to plan for the future, however, and that future must include some form of guaranteed employment for the growing number of young people who have fewer and fewer opportunities to sustain themselves through traditional hunting or trapping activities.

THE CONTEXT OF HEALTH: ENVIRONMENTAL, ECONOMIC, SOCIAL, AND POLITICAL FACTORS

Beyond a simple dichotomy of health or ill-health, being healthy (*miyupimaatisiiun*) for the *Eeyouch* is the definitive expression of core cultural values. Being healthy has traditionally meant a balanced relationship between individuals and their social, spiritual, and natural worlds. In particular, it implies proper hunting practices and hence a respectful relationship between humans and the animals of the land: a successful hunt means healthy eating and appropriate social relations means that the hunted foods will be apportioned amongst oneself and one's kin. If, too, the hunted foods, the land, and the hunting traditions are all integral to *miyupimaatisiiun*, then the exploitation of the land would logically be the most profoundly felt impediment to being healthy for the *Eeyouch*. Thus each imposition upon the people and their lands—from the incursions of the earliest traders to internal colonization and present-day disputes to retain control over their territory—magnifies the meanings of *miyupimaatisiiun* as well as the direct and indirect impediments to health (Adelson, 2000).

The earliest days of contact brought an onslaught of infectious diseases to the indigenous peoples of the Americas. Diseases such as influenza, typhus, diphtheria, measles, smallpox, and whooping cough took a tremendous toll and wiped out generations directly, through death, or indirectly through the loss of hunting manpower or women's weakened ability to reproduce (Waldram, Herring, & Young, 1995). The more isolated Cree populations likely had comparatively fewer losses from these early infectious diseases because many simply were not in direct contact with the traders and other travelers to the region. As families began to live for longer periods of time near the trading posts, however, infectious diseases began to take a greater toll on the *Eeyou* population. Because of a decline in the fur-bearing animals vital to the trade economy, and subsequent real hunger, or job opportunities, or later to allow children access to the first elementary schools in the *Eeyou* north, by the early decades of the 20th century many more people were living in and around the trading post, swelling them into rudimentary villages. Living at the post, however, was no guarantee of health. Indeed, it was here that the second wave of infectious diseases, including influenza and tuberculosis, took their toll on this population. With large families living in small, inadequate housing facilities infectious diseases spread through these early settlements. Conditions had to reach an extreme before federally based health services were extended north to the people. In one *Eeyou* community, permanent health care personnel only arrived with the armed forces in the 1950s, as this one small, isolated post was an ideal site for the early distance radar warning systems that were integral to the Cold War. It was not until 1958, and after the federal government deemed the village to be a permanent settlement, that a clinic was established and a nurse available year-round.

Through the JBNQA the Cree Regional Board of Health and Social Services was established as an autonomous body overseeing the full range of health and social service needs of the *Eeyouch*. The Health Board administers all of the health care services in *Eeyou astchee*, including village nursing stations and one hospital (located in Chisasibi) and has worked hard to mediate between traditional therapeutics and biomedicine, always ultimately committed to offering locally meaningful, appropriate and efficacious care (see, for example, Bobbish-Rondeau et al., 1996). The contemporary health care and funding challenges faced by the Health Board, however, frustrate many of these efforts. A recent special assembly on health and social services in *Eeyou astchee* pointed to a host of issues that community members said need greater attention, including high staff turnover and burn-out, to language problems between English or French-speaking staff and Cree-speaking clients, the rise of chronic diseases such as diabetes and asthma, increased family violence, and the current housing crisis (Roslin, 1999). In the end, however, it is important to remember that the problems faced by the Cree Regional Board of Health and Social Services are a composite of the growing health care and social needs of the population, limited financial resources, and the constraints of a relatively new system of independent regional management that must contend with all these issues at once.

MEDICAL PRACTITIONERS

Historically, the *Eeyouch* traveled in small family groups and, hence did not develop any large, formal medical tradition. This does not mean, though, that there were no established healing practices or practitioners. There were those who had expert knowledge of the plant resources of the region, while others were gifted spiritual practitioners, employing sweat lodges or other healing ceremonies as required. As well, some women excelled as midwives, a practice that has all but disappeared with the hospital-based births now requisite in the north. See the next section for a summary of treatment practices.

CLASSIFICATION OF ILLNESS, THEORIES OF ILLNESS, AND TREATMENT OF ILLNESS

Illnesses are most broadly classified as either those that are historically known to the *Eeyouch* and those that came with the influx of non-Native peoples and influences, known colloquially, as "white man's" diseases.

Traditional illness theories are reflections of the land-based cultural knowledge and practice of the *Eeyouch*. Traditional therapies, in turn, draw directly from people's intimate knowledge of the land and its resources combined with a sense of the inherent spiritual force of all animate and inanimate beings as well as a healthy dose of pragmatic realism. Indeed, to this day, there is a general consensus that—regardless of treatment modality—there is a greater spiritual force, derived either from Christian or animistic roots, that ultimately decides one's fate. There were (and remain) individuals who were recognized for their healing abilities and it is this blending of healing power along with a sound knowledge of the indigenous pharmacopoeia which would lead to the greatest treatment success. Whether treatments would or would not work ultimately depended on the fate of a particular individual.

Indigenous therapies were used primarily for the kinds of injuries that one might sustain while traveling or working on the land, such as bone fractures, snow-blindness, burns, or lacerations. There are treatments, too, for common ailments such as toothache, boils, and stomach complaints as well as for post-partum bleeding. Treatments include therapeutic sweat lodges, either when the person is immersed entirely within the lodge or more localized heat treatments (such as around the abdomen or leg), to the specific use of the tamarack or spruce tree inner bark and gum, animal fats and especially rendered bear fat, gallbladder or lung, beaver castoreum, as well as breast milk, tobacco, and meat broth. The abundant bog plant, Labrador tea (*Ledum groenlandicum*), remains a vital part of the contemporary indigenous pharmacopoeia. Labrador tea is regularly harvested and used as a general tonic as well as for a range of pulmonary conditions. Not surprisingly, given the degree to which one depended in the past on mobility, there were a range of therapies devoted to walking disorders, sore muscles, and getting chilled or cold. In particular, feeling cold and lacking an appetite are viewed to this day as precursors to future illness and are treated immediately and with grave concern.

Bush food, and meat in particular, is a source of physical and symbolic strength, vitality, and health and hence not only nutritious but a therapeutic resource, too. There is a hierarchy of strength in bush foods so that the meat, fat, and broth of larger animals, such as bear or caribou, are considered to be especially "strong" whereas wild fowl (goose, ptarmigan, grouse) and fish are somewhat "weaker" foods. So, for example, whereas bear fat is a potent remedy for a number of ailments, the less potent fish broth is considered a sound treatment and an appropriate first food for those who have gone without or have lost their appetite. Most importantly, hunted foods are fundamental to well-being and thus, health, in the *Eeyou* sense of the term; good health is impossible without consuming a full range of (weaker and stronger) traditional foods.

White man's diseases are more readily classified as they cluster as all those that have come with contact, internal colonization, and its lingering and persistent influences and effects. White man's diseases follow, to a large extent, a trajectory common to many other indigenous peoples with similar histories of contact (see above). Today, white man's diseases are predominantly chronic (such as diabetes, cancer, and asthma) but also reflect some of the social ills of the communities that, one might argue, are part of the lingering effects of a system of internal colonization (see below).

White man's medicines are considered appropriate remedies to white man's diseases. Thus, even the most rudimentary biomedical services (such as treatment by 19th-century missionaries or their wives for infections or other ailments) was always accepted as part of the range of treatment options available to the *Eeyouch*. For the coastal dwelling Cree, the annual X-ray ship was oftentimes the only contact with biomedicine in the early half of the 20th century. Community members would line up for their annual exam when the ship arrived at the post site and anyone who was found to have tuberculosis was taken or later sent south to a sanatorium, sometimes for years at a time. By the 1950s, nursing stations, run by the federal government, were set up in the villages and were generally well-used by the *Eeyouch*. If a case required medical consultation, then a physician in an urban hospital center would be contacted by telephone and, if required, the patient would be evacuated by plane from the village. Today, the Cree Regional Board of Health and Social Services maintains all the nursing stations and the hospital in Chisasibi (see above). As throughout the rest of Canada, all medical care is provided free of charge; in addition, members of the communities also receive free all allied health services and required drugs. Despite any concerns about the quality or availability of care through these nursing stations, they are a vital aspect of each community, offering daily and emergency clinics, well-baby clinics, a host of medical information, and the only available pharmacy in the more remote communities.

HEALTH THROUGH THE LIFE CYCLE

Pregnancy and Birth

Each and every child born is a welcome member of the *Eeyou* community. By extension, abortion is considered anathema, whereas miscarriages and stillbirths are fully acknowledged. There is, for the most part, tremendous familial support extended to pregnant women and their infants. With a younger population having babies today, however, there are some basic constraints limiting the degree to which those who would, in the recent past, have cared for the young. Younger mothers, for example, are either still in school or employed and their support network shrinks as grandmothers, also still young, are themselves working outside of the home and unable to offer day-time care to their grandchildren. Child care facilities are now being created in communities where the need for them simply did not exist to this extent a decade ago. By extension, too, all children, regardless of any physical disadvantage, are incorporated into the community. Social stigma of any sort is rare and children are accommodated in whatever way necessary.

Just under 50 years ago all births took place in the family dwelling and, more often than not, this meant on the land. Traditional midwives attended these births and there remains a host of post-partum treatment practices that attest to the skill and knowledge base of these women. By the mid-20th century, most nursing stations were run by nurse–midwives who would regularly assist deliveries. More recently, standard practice is for pregnant women to leave their home community in order to deliver in a hospital. With a small hospital now permanently based in Chisasibi, many of the more northern dwelling *Eeyou* women at least remain in *Eeyou astchee* and do not have to travel late in their pregnancies to an urban center to have their children.

Approximately 70% of women breastfeed their children and while many still consider an extended duration of breastfeeding normal practice, one study has shown that between 1988 and 1993, the percentage of infants still being breastfed at 6 months had gone from 60% to 40% (although this is still higher than the general provincial target figure of 30%) (Schnarch, 2001). Again, the contingencies of work and family may be playing a large role in this overall reduction in breastfeeding, amongst the younger mothers in particular.

Infancy

Infancy begins at birth and ends with a formal "walking out" ceremony held at about the time when the child takes its first steps. From the time the child is born and right up until this ceremony, the infant is swaddled for warmth, protection, and comfort and is exceptionally well-bundled to protect it from the elements whenever outside. An additional protection used in the past, but cautioned against more recently, was a small necklace made of netting string that was used to "trap" any cold before it entered the infant's body and hence to shield it both physically and spiritually from the cold.

Many parents will prevent their infant's feet from touching the earth or ground until he or she has "walked out." The "walking out" ceremony, traditionally held at the break of dawn on a day soon after the child is just old enough to walk, is a celebration of the symbolic and social transformation of the infant into a future *Eeyou* man or woman. Little boys hold a carved wooden gun, symbolizing their future roles as hunters; little girls hold a carved axe, signifying their roles in maintaining the hunting camp. The young future *Eeyouch*, beautifully and ornately dressed, are assisted out of a ceremonial dwelling by a parent and, witnessed by family and community, can touch the ground for the first time. Symbolizing entry into the communal and natural worlds, the "walking out" ceremony remains an important part of contemporary cultural practice. The ceremony is complete only after a large feast is held, honoring the child as well as the community elders.

Unfortunately, infant mortality rates remain high—almost three times that of the province of Quebec (14.9 vs. 5.3 per 1,000 live births)—despite a significant reduction in infant deaths with improved pre-natal health services over the last 20 years (Schnarch, 2001). One infant disease is worthy of particular note. Cree leukoencephalopathy is a neurological disorder that occurs exclusively, albeit rarely, in Cree infants from northern Quebec (and Manitoba). This is a uniformly fatal disease of genetic origin that is triggered by a viral infection and has been found only in this population (Black et al., 1988).

Childhood

Children, and especially the youngest child in the family, are doted upon and given a fair degree of freedom outside of the age-appropriate tasks required of them. Children are not strictly disciplined and, traditionally, are taught primarily by example and by observing and listening to their parents and grandparents. This is especially so when learning about hunting-related activities such as the proper maintenance of firearms, hunting gear, and equipment, as well as weather patterns and other vital bush and camp skills. Children are always dressed warmly and their feet properly protected in order to safeguard them from the harsh sub-Arctic elements, as cold or a chill entering the body is viewed as a precursor to future illness.

There is some disjuncture for some parents between the more rigid processes of school-based learning and investing the child with the proper knowledge base and skills to live and work on the land. This is part and parcel of the contemporary tension between development and traditional practices that permeates so many aspects of the lives of the *Eeyouch*. In an attempt to balance these sometimes divergent objectives, the Cree School Board has implemented a formal Cree language program through to the high school years, instruction only in Cree in the first years of school, and a cultural program that ranges from craft production to land-based activities as part of the general curriculum. With control over education, the Cree School Board continues to work toward a reconciliation between the basic provincial educational requirements and the cultural needs of the children, and more and more children are not only completing high school locally but are traveling out of their communities to complete college and university degrees.

Adolescence

Adulthood symbolically begins for boys when they successfully kill their first large game animal. This has been known to happen when boys are between 8 and 10 years old. Thus there is, historically, no particular period of adolescence. And, given the number of pregnancies and young relationships and marriages today, there remains, for many, virtually no time between childhood and the responsibilities of adulthood (almost 25% of births in *Eeyou astchee* between 1985 and 1998 were to women under 20 years of age [Schnarch, 2001]). A growing concern is the rise in sexually transmitted diseases and, in particular, chlamydia, with a rate of 813/100,000 (compared with 87/100,000 for Quebec), occurring mostly in those between the ages of 15 and 30 (Schnarch, 2001). There has, to date, been no case of AIDS reported, although it is clear that unprotected sex is taking place and is a persistent concern for the public health community serving *Eeyou astchee*.

Adolescence, as something that has evolved with settlement living, is most clearly visible amongst the growing number of youth who are completing a secondary (high school) education, but are increasingly bored with what is for them the drudgery of village life, who see no particular future in their communities as jobs are at a premium and, perhaps through social networks, travel to urban cities or the media, such as satellite television or the internet, are discovering myriad, and oftentimes harmful, ways to alleviate their boredom. Many are smoking (a survey conducted in 1991 found that 77% of those between 15 and 24 years of age were either regular or occasional smokers [Schnarch, 2001]) and there are persistent concerns about substance abuse, vandalism, and accidental injuries and deaths in this young age group.

Adulthood

This stage of life, as mentioned in the previous section, often begins early as young men and women start having children and take on full familial duties while still in school or early on in their lives. The focus here, though, is on the older adults who have experienced dramatic transformations in the course of their lives. Many of the older adults today were born and raised to a young age on the land, then moved into sometimes squalid and certainly under-serviced settlement communities as children. They are the first generation to have had access to a formal educational system (some in more southern communities were placed in residential schools, which had profound effects, both negative and positive, on their future lives) complete with its formal structures and regimented classroom environment. This generation became, too, the first to lead the *Eeyouch* into their new relationship with government, moving out from under a paternalistic system of governance and into an historic, if uncharted, form of self-government. This generation, too, moved from predominantly land-based labor and resources to a substantially more sedentary and stable lifestyle as the village became a permanent home to so many. Foods, and especially those high in sugar and carbohydrates, have become the mainstay in many households whereas fresh fruits and vegetables, exorbitantly priced in the north and often blemished from long travel, remain both unpalatable and unaffordable to many. Many of these social and environmental changes in the adults' lives are directly reflected in their health statistics. Most telling is the shift in mortality and morbidity from infectious to chronic disease. Over the last decade, for example, mortality rates for circulatory disease, diabetes, and cancers overshadow all other causes of death except for those resulting from accidents and injuries. Obesity is a chronic problem for many and diabetes is taking a greater and greater toll on the population, with 11% of the total adult population and almost 30% of those over 50 now diagnosed with the disease (Bobbish-Rondeau et al., 1996; Schnarch, 2001).

Older adult women may now see, during their menopausal and post-menopausal years, the surfacing of a range of joint ailments that result, the elders say, from having been exposed to cold when young. Women are warned throughout their lives to keep warm, and especially so when menstruating. Any diversion from this practice in one's youth will now bear out in sore joints and, especially, sore legs. This is a time too, though, of positive transitions for women: from mothers to grandmothers and future elders. Once past menopause, for example, women may partake in certain symbolically powerful foods that they are prohibited from eating while still in their child-bearing years (e.g., fetal caribou).

The Aged

Older *Eeyouch*, or the elders, are the most highly respected members of the community. The elders were all born and raised on the land and their generation is viewed as the last ones who truly know the physical, spiritual, and symbolic contours of *Eeyou astchee*. They are referred to as grandparents to all of the children, which means, too, that they have a certain degree of authority in the social management of the community's children. The elders are provided with the most and the choicest parts of any hunted foods, are respected authorities on most issues, and are looked after as best as any community is able (given the constraints of the relative availability of adequate housing or other chronic care facility). The elders constitute a very small proportion of the total *Eeyou* population and are now living into their seventies.

The life expectancy for men is 74.8 (compared with 74.5 for Quebec) and 75.5 for women (81.1 for Quebec).

Dying and Death

While any death is always traumatic, the death of an elder is viewed as an especially significant loss given their status in the community. The rituals of death, for many *Eeyouch*, are an amalgamation of both Christian and traditional spiritual beliefs. Whereas prior to Christian influence the dead were buried near where they passed away, today burial takes place in the community cemetery immediately following a church service. With the range of spiritual and religious influences in communities today, there is a concomitant range of beliefs about life after death, including those that adhere most strictly to the Christian paradigm of a heavenly afterlife to a traditionalist belief in the transformation of spirit from this life to the next as one proceeds on one's sacred journey.

REFERENCES

Adelson, N. (2000). *Being alive well: Health and the politics of Cree well-being*. Toronto, Canada: University of Toronto Press.

Black, D. N., Watters, G. V., Andermann, E., Dumont, C., Kabay, M. E., Kaplan, P. et al. (1988). Encephalitis among Cree children in Northern Quebec. *Annals of Neurology, 24*(4), 483–489.

Bobbish-Rondeau, E., Boston, P., Iserhoff, H., Jordan, S., Kozolanka, K., & MacNamara, E. et al. (1996). *The Cree experience of diabetes: A qualitative study of the impact of diabetes among the James Bay Cree* (Final report). Chisasibi, Canada: Cree Board of Health and Social Services.

Dougherty, K. (2001, October 24). Cree get $3.5-billion deal. *The Gazette*, A1.

Feit, H. (1986) Hunting and the quest for power: The James Bay Cree and whitemen in the twentieth century. In R. B. Morrison & C. R. Williams (Eds.), *Native peoples: The Canadian experience* (pp. 171–207). Toronto, Canada: McClelland and Stewart.

Grand Council of the Crees (GCC) (1996). *Never without consent: James Bay Crees stand against forcible inclusion into an independent Quebec*. Toronto, Canada: ECW Press.

Macpherson, D. (2001, October 25). A new great peace. *The Gazette*, B3.

Marvelle, N. (2001). Retaliation and reconstruction: The James Bay Cree's success in the aftermath of development. In *Cultural survival*. Available from (http://www.cs.org/internships/cree.htm)

Roslin, A. (1999). Health concerns pour out at Assembly. *The Nation*, 6(8).

Schnarch, B. (2001). *Health and what affects it in the Cree communities of Eeyou Istchee*. Montreal, Canada: Cree Board of Health and Social Services of James Bay and the Public Health Module—Cree Region of James Bay.

Tanner, A. (1979). *Bringing home the animals: Religious ideology and the mode of production of the Mistissini Cree hunters* (No. 23). St. John's, Newfoundland: Memorial University, Social and Economic Studies.

Waldram, J., Herring, A. & Young, T. K. (1995). *Aboriginal health in Canada: Historical, cultural and epidemiological perspectives.* Toronto, Canada: University of Toronto Press.

Czechs

Zdenek Salzmann

ALTERNATIVE NAMES

The Czechs call their culture *česká kultura* (Czech culture). To call it "Bohemian culture" would be misleading because the term "Bohemian" has a historic and geographic rather than an ethnographic or linguistic reference. Furthermore, Bohemia (*Čechy*) is only the larger, western part of the Czech Republic; the smaller, eastern part is Moravia (*Morava*). In the north, Moravia includes a part of Silesia (*Slezsko*), a historical region that lies for the most part in southwestern Poland.

LOCATION AND LINGUISTIC AFFILIATION

The Czech Republic, located in central Europe, is bounded by Poland on the north, Germany on the west, Austria on the south, and Slovakia (the Slovak Republic) on the east. The area of the Republic is 78,866 km^2 (about the size of South Carolina).

The language spoken in Bohemia and Moravia is Czech. It is spoken in several regional dialects, but outside their homes most Czechs speak Common Czech, a supraregional variety of the spoken language. Czech, a Slavic language, belongs to the West Slavic subbranch of the Indo-European language family; it is mutually intelligible with the Slovak language.

OVERVIEW OF THE CULTURE

Population

The population of the Czech Republic as of December 31, 2000, was 10,266,546, with females exceeding males by 273,084. The ethnic composition of the republic is 94.9% Czech, with the remainder consisting of Slovak (3.1%), Polish (0.6%), German (0.5%), Ukrainian (0.4%), and Romany (Gypsy) minorities. Officially the Romany population amounts to 0.3%, but it may be as much as five to ten times larger. Because both the majority populations of Bohemia and Moravia speak Czech, they will be considered as belonging to the Czech ethnic group. When their practices differ, it will be indicated.

History

The ancestors of contemporary Czechs made the territory of what is now the Czech Republic their home in about the 6th century, replacing a Germanic people who migrated out of the area during the 5th century. In the early part of the 11th century what had become Bohemia and Moravia was brought under one control, becoming the kernel of the Czech state. The first Bohemian king was crowned in 1085, and the royal title became hereditary in 1198.

The peak of the kingdom's influence was reached under Charles IV, who reigned from 1346 until 1378, and in 1355 was also made emperor of the Holy Roman Empire. Prague became the seat of the archbishop and the site of the first university in central Europe (1348).

Czech national culture continued its development until 1620 when the Czech estates were defeated by the Hapsburg army in the Battle of White Mountain. Not only was the kingdom's independence lost, but its provinces were declared to be the hereditary property of the Hapsburgs. This loss led to a period referred to as "the darkness" (*temno*), which lasted until the end of the 18th century when, after the onset of national revival, a modern

Czech national consciousness began to form. Independence was regained at the end of World War I in 1918. From that year until the end of 1992 the Czechs and Slovaks shared a common state, Czechoslovakia, with the exception of the six years of World War II (1939–45). As of January 1, 1993, the Czechs and Slovaks peacefully separated to become two independent countries.

Economy and Occupations

The Czech Republic is and has long been a highly industrialized country. In 2000, the structure of employment in the civil sector broke down into services (54.9%), industry (40%), and agriculture (5.1%). The unemployment rate in 1999 stood at 8.7%, with the capital Prague experiencing the lowest unemployment, northwestern Bohemia and northeastern Moravia the highest.

Both imports and exports steadily rose over the final decade of the 20th century, with imports invariably exceeding exports, although not by a wide margin. In terms of monies expended, the main imports included (in descending order) machinery and equipment, chemicals and synthetic fibers, motor vehicles and trailers, communications equipment (including radio and television), basic metals, electrical machinery, crude petroleum and natural gas, rubber and plastic products, and fabricated metal products other than machinery.

The main exports included motor vehicles and trailers, machinery and equipment, electrical machinery, fabricated metal products other than machinery, chemicals and synthetic fibers, basic metals, communications equipment, certain nonmetallic mineral products, and rubber and plastic products.

The countries from which most goods are imported (in descending order) are Germany, the Russian Federation, Slovakia, Italy, France, Austria, the United States, the United Kingdom, and Poland; exports go to Germany, Slovakia, Austria, Poland, the United Kingdom, France, Italy, the United States, and the Netherlands.

The middle class is the largest socioeconomic class of the Czech Republic. Under the communist regime after World War II, the status of manual workers rose and that of the former upper middle class steeply declined. Since the velvet revolution of 1989, when democracy was re-established, some businesspeople have acquired luxury cars and expensive housing. By contrast, retired people must stretch their pensions to keep pace with the steadily, if slowly, rising cost of living.

Government

The Czech Republic is a parliamentary democracy with a Chamber of Deputies and a Senate. The President appoints certain high officials (for example, the prime minister) and can veto other than constitutional bills. During the post-communist times (from 1990 on), several dozen political parties have emerged, and the country has been governed by a coalition of some of those parties receiving the most votes. The results have not always led to an efficient government.

Family and Kinship

The effective kin group of urban Czechs is limited to the closest relatives. Those less close than grandparents, aunts, uncles, and first cousins usually meet only on such occasions as weddings and funerals, Christmas, or other holiday gatherings. Most villagers, however, especially in Moravia, maintain close contact with more distant relatives.

Religion

The first half of the 15th century was marked by a breach with the Roman Church as a consequence of the reform movement led by the Czech Jan Hus (John Huss). His death in 1415, when he was burned at the stake in Constance (Konstanz in southern Germany), initiated the ambivalent attitude of the Czechs toward Roman Catholicism; this ambivalence was later reinforced by the attempt at forcible re-Catholization begun during the 17th century by the Hapsburg rulers. More recently, the 41 years of communist rule (1948–89) undermined the observance of religious practices even further.

At present, about 40% of the population are Roman Catholic, about 4% Protestant, and the remaining 56% uncommitted, atheist, or agnostic. Compared with the people of Moravia, the Czechs of Bohemia tend to be lukewarm in their religious beliefs and practices.

THE CONTEXT OF HEALTH: ENVIRONMENTAL, ECONOMIC, SOCIAL, AND POLITICAL FACTORS

Demographic Data

At present, the population of the Czech Republic is becoming smaller due to natural decrease. The year 2000 was the seventh consecutive year in which the number of deaths exceeded the number of births. Because of the low number of births, the average age of the population has been slowly increasing. In 2000, 53% of the population was below the age of 40 and 47% above.

Life expectancy at birth in 1999 was 71.4 years for men and 78.1 for women. On reaching the age of 65, an additional 13.7 years is added for men and 17 for women. Among 24 European countries, life expectancy in the republic ranked seventh from the bottom.

The number of live births per 1,000 inhabitants stood at 19.1 in 1975. The number has been decreasing in recent years, from 135,881 in 1985 to 90,910 in 2000. During 1999, when there were only 8.7 live births per 1,000 inhabitants, the number was the third lowest among 24 European countries.

There were 49.24 abortions per 100 births in 2000; of these 12.03 were spontaneous, 35.68 legally induced, and 1.53 for other reasons (primarily the result of ectopic pregnancies). The number of abortions in 2000 was lower by 9.1% than in 1999 and less than half of what it was in 1990.

The rate of death for infants younger than one year per 1,000 live births was 4.1 in 2000. In a comparison with 24 other European countries, only four had a rate lower, nine had the same rate, and 11 a higher rate than in the Czech Republic.

The marriage rate in the Czech Republic dropped to a historical minimum during the late 1990s, and by 2000 only 5.4 marriages took place per 1,000 inhabitants. This low number of marriages has resulted in part from the steadily increasing age at first marriage, which at the beginning of the new millennium was approaching that recorded for Western Europe. And just as in many parts of the world today, a man and woman frequently live together for several years before deciding to legalize their cohabitation.

Since the early part of the 20th century, the choice of a spouse has been the decision of the young couple, but educational attainments and the likely socioeconomic status of a potential husband or wife are of some importance. Whenever possible, the young married couple establish a neolocal residence, that is, a residence independent of the family of either one. However, the help of the husband's or wife's mother is sought and appreciated at the arrival of a child, particularly if the mother expects to return to a job.

Environmental Pollution

Before 1990, protection of the environment was subordinated to the fulfillment of economic five-year plans. Lack of proper control over the sources of pollution caused increasingly serious problems for the population. Over 50% of forests were damaged, particularly spruce forests in the mountains along the Bohemian border. Of some 7,000 km (about 4,350 miles) of monitored streams, 60% were judged to be strongly or very strongly contaminated. Especially critical conditions existed in northwestern Bohemia around the towns of Most and Sokolov, both of which had large deposits of lignite and a variety of industrial enterprises. Altogether, about 2.5 million people were living in a substantially endangered environment.

Matters have greatly improved since 1990: for example, solid emissions decreased from 401,500 metric tons in 1990 to 16,100 in 1999; from 1,596,000 tons of sulfur dioxide in 1990 to 193,100 in 1999; from 493,900 tons of oxides of nitrogen in 1990 to 135,000 in 1999; and from 23,800 tons of hydrocarbons in 1990 to 17,700 in 1999. Only the amount of carbon monoxide has increased during that period, by about 50%; this increase parallels the increase in passenger and commercial automobiles.

Similar improvement is being achieved in the case of pollutants discharged into watercourses: for example, from 190,500 tons of undissolved substances in 1990 to 29,758 tons in 2000, and from 989,057 tons of dissolved inorganic salts in 1990 to 691,613 tons in 2000. The one remaining problem is the prevention of pollution from the disposal of the ever-growing amounts of toxic waste.

Diet

By American standards, traditional Czech food would be considered heavy and fat. Portions of meat served are not large, but potatoes or dumplings and substantial amounts of animal fats (lard, butter, and cream) are used both in gravies and in general cooking. Eaten as often as bread and butter is bread covered with lard rendered from pork fat (with cracklings) or with goose grease (and goose liver). Only since the 1990s have salad vegetables been available on a year-round basis; when out of season locally, fresh vegetables are now imported.

During 1999, consumption of the main types of food was as follows (amounts given are per person): pork 44.7 kg, poultry 20.5 kg, beef 13.8 kg, fish 5.2 kg, fats and oils 23.1 kg, lard and bacon 5 kg, butter 4 kg, fresh vegetables 85.3 kg, fresh fruit 75.6 kg, and potatoes 75.9 kg (1 kg = 2.2046 1b). However, during the final two decades of the last century vegetable shortenings, oils, and margarine began to replace animal fats. Despite some of the desirable changes in the basic diet, obesity is quite common because the current wide ownership and use of personal automobiles tends to offset some of the trends toward consumption of a more healthful diet.

Beer is the favorite beverage of Czech men (and many women) and its consumption in 1999 amounted to 160 l per person (1 l = 1.057 quarts). Wine and spirits are also drunk, but beer is the drink of choice. Breweries are plentiful and beer is relatively inexpensive.

Health Infrastructure

The physical resources required for health care stabilized during the last three years of the 20th century. The network of establishments for inpatient care included 211 hospitals with 67,457 beds (including 2,304 cots for newborns), 160 specialized therapeutic centers with 22,667 beds, and 63 health-spa centers with 22,179 beds. The total number of inpatient beds per 10,000 inhabitants was 109.4 and has been slowly decreasing despite the fact that the number of hospitalized patients in 2000 showed a 0.5% increase over 1999.

The number of pharmaceutical service establishments has been growing rapidly since 1990, replacing most of the earlier dispensaries (for example, an additional 80 pharmacies were established during 2000). While in 1990 all of the 917 pharmacies and 1,216 dispensaries were state-owned, in 2000 only slightly more than 4% of the 1,706 pharmacies and none of the 183 dispensaries was owned by the state.

Health Insurance and Medical Expenses

According to currently valid laws, all persons who permanently reside in the Czech Republic, as well as persons who are not permanent residents but whose employer is located in the republic, have health insurance. The insurance is paid either by the insured (one third) and the employer (two thirds) or by the state. Self-employed persons (and others who qualify as such) pay their own premiums. The state pays for all those who for one reason or another do not belong to either of the above two categories—for example, women on maternity leave, physically or mentally disadvantaged people, soldiers, and others.

Health insurance pays for all legitimate care extended to patients to maintain or improve their health. These services include ambulance transportation, emergency service, preventive care, hospital stay, balneological care, prescribed medicines, standard immunizations, and autopsy. Limitations and exclusions concerning the services provided by the health insurance plan are relatively few and reasonable.

Health care is financed partly from public resources and partly by individuals. Public expenditures for health in 2000 through the Ministry of Health amounted to 129.6 billion Czech crowns (Kč)—5% more than in 1999. The contribution from the state budget was 11%, while 89% was contributed by health insurance corporations, of which in 2000 there were nine. Amounts spent by individuals on health care have been steadily growing—from 5.282 billion Kč in 1994 to 12.245 billion Kč in 2000 (that is, from 511 Kč per person in 1994 to 1,192 Kč in 2000).

Significant among expenses for health care are medicines; such expenses almost tripled between 1993 and 2000. The most commonly dispensed medicines can be characterized as follows (the parenthetic percentages indicate the relative share of funds spent on them): for problems of the cardiovascular system (19.9%), of the digestive tract (13.2%), of the nervous system (11.8%), of the respiratory system (8.8%), to inhibit or prevent the growth and spread of neoplasms or malignant cells (8.6%), for antibiotics and chemotherapeutics (8.4%), for problems of the blood and blood-forming organs (7.3%), of the genitourinary system (6.1%), and of the musculoskeletal system (6%).

Compared with 18 other European countries the Czech Republic ranked sixth lowest in the percentage of gross national product spent on health services.

Domestic Abuse

According to the report of one among several agencies offering assistance to victims of domestic violence, out of 624 telephone calls received between September 4 and November 11, 2001, the victim was a woman in 538 calls, a child in 56 calls, and a man in 30 calls. Children were present when the violence took place in 239 cases. Both physical and psychological abuse occurred in 319 cases, psychological abuse in 127, physical abuse in 83, and sexual abuse in 16. According to a foundation concerned with children, during 2001 the most frequent forms of child abuse were infliction of pain, sexual abuse (both commonly under the influence of alcohol), forcing the child out of the family, and neglect.

Numbers of cases concerning child abuse and domestic abuse in general are never accurate because many instances of abuse are not reported. The data that follow must therefore be viewed only as illustrative. According to one particular public survey, 13% of those questioned stated that they had been victims of a partner's violence, while 3% of those surveyed admitted to taking the role of aggressor, yielding a total of 16% of instances of domestic abuse. According to specialists, however, the figure of 13–16% represents only the lower limit of incidents of abuse. The actual rate of domestic violence is thought to be much higher.

Another organization offering help to abused individuals provided the following data (which must also be considered tentative): female victims of abuse fall primarily in the age ranges (in descending order) of 40–44, 45–49, 30–34, 35–39, and 25–29; most of the males who commit abuse are in the age ranges of 45–49 and 35–39. As to educational background, most female victims and men who abuse them had completed secondary schooling; the least violence was reported for those with only basic education.

Medical Practitioners

Training of Medical Personnel

The training and education of future members of occupations related to the practice of medicine, pharmacy, and certain types of health services have been proceeding without problems. During the 2000–01 academic year, students of medicine numbered 8,251 (of whom 4,637 were women), of pharmacy 1,655 (1,280 women), and of types of health services requiring university training 3,383 (3,039 women). The numbers of graduates at the end of the 1999–2000 academic year were, respectively, 912, 254, and 665. At present, seven medical schools of three universities train physicians, two pharmaceutical schools at two universities train pharmacists, and two schools of social health at two universities train personnel for work in health and social care positions.

Paramedical schools in the academic year 2000–01 enrolled 26,301 students, of whom 3,405 attended evening courses.

Types of Medical Personnel

As of December 31, 2000, there were 39,342 (38,331 full-time equivalent) physicians serving the population of the Czech Republic. Of these, 22,212 were women. The number of physicians per 1,000 inhabitants was 3.8. All specializations are represented; most numerous were physicians in the following branches of medicine (listed here in descending order): dentistry and dental surgery, general medicine, internal medicine, pediatrics, surgery, gynecology and obstetrics, anesthesiology, neurology, X-ray diagnostics, psychiatry, and ophthalmology. As of the same date, pharmacists numbered 5,191 (4,726 full-time equivalent), of whom 4,256 were women.

According to the latest survey conducted by the Center for the Study of Public Opinion, physicians enjoy the highest prestige among the 27 occupations listed. Their average income, however, rates only twelfth.

The full-time equivalent of paramedical personnel (that is, nurses, children's nurses, midwives, dieticians, rehabilitation workers, assistants in hygienic services, medical and pharmaceutical laboratory assistants, radiological technicians, and dental technicians numbered 107,321 in 2000; of this total, 64,450 were nurses (about 60%). Of the paramedical personnel, 54.3% worked in hospitals (including hospital outpatient care) and 28.2% in independent establishments giving outpatient care.

The number of full-time equivalent lowest-tier health personnel (auxiliary nurses, disinfectors, dental assistants, and others) came to 6,537.

Alternative Medicine

Although the biomedical approach has been accepted and employed throughout the 20th century in the region, until World War II some individuals in villages and small towns preferred to consult herbalists (folk healers) and use the recommended infusions, decoctions, or ointments made from various herbs. Consulting herbalists was not approved of during the communist era (1948–89). During the last 10 years (since the early 1990s) folk medicine has again become popular. It is practiced nonprofessionally and usually involves the use of plant-derived remedies on an empirical basis. To prevent some illnesses or to cure them, herbalists as well as many physicians recommend carefully watching one's diet, exercising, and using herbs and other natural remedies, both domestic and imported. Many patients of physicians also consult herbalists.

Balneology

Balneological institutions—that is, health spas making use of thermal mineral waters and mud or peat baths—are numerous and popular. Some of these have been known for centuries and have become internationally renowned. Members of the European aristocracy and famous individuals (for example, Johann Wolfgang von Goethe) visited them on a regular basis. Among the best known are Karlovy Vary (Karlsbad) for gastroenterological and metabolic problems; Mariánské Lázně (Marienbad) for urological and orthopedic problems and for allergies; Františkovy Lázně for gynecological, circulatory, and orthopedic problems; and Luhačovice for gastrointestinal, metabolic, and endocrinological problems. These and other health spas are visited not only by Czech citizens but by thousands of foreigners.

CLASSIFICATION OF ILLNESS, THEORIES OF ILLNESS, AND TREATMENT OF ILLNESS

Medicinal Drugs

The biomedical approach to the diagnosis and treatment of illness has been accepted in the Czech Republic as long as it has been in other industrial countries of Europe. Furthermore, Czech medicine has always made every effort to keep up with modern advances in the treatment of illness.

During the first half of the 20th century extensive use of medicinal plants listed in Czech pharmacopeias was progressively replaced by the use of synthetic drugs. The pharmaceutical industry of the country has always been strong, but medications manufactured elsewhere are also prescribed and available in pharmacies.

The importation of medicinal drugs in 1999 and 2000 exceeded export by about 7,000 metric tons. The countries from which most of these drugs were imported are (in descending order) Germany, Slovakia, United Kingdom, Italy, Austria, and Slovenia; exports went primarily to Slovakia, Germany, Russian Federation, Poland, Austria, and Ukraine.

Diseases

Infectious diseases must be made known to the authorities. The most important infectious diseases reported during 2000 are listed in Table 1. Compared with 1990, increases in incidence occurred only for salmonellosis other than typhoid fever and paratyphoid fever B, whooping cough, viral encephalitis, syphilis, and AIDS.

Table 1. Infectious Diseases Reported to Authorities in 2000

Disease	Number of cases
Salmonellosis (other than typhoid fever and paratyphoid fever B)	40,233
Chicken pox	38,665
Scarlet fever	2,965
Viral hepatitis	1,979
Tuberculosis of the respiratory system	1,244
Syphilis	967
Gonorrhea	888
Rubella (German measles)	743
Viral encephalitis	719
Bacillary dysentery (shigellosis)	548
Whooping cough	395
Bacterial meningitis	227
Other types of tuberculosis	198
AIDS (of which 14 were newly reported cases)	148
Mumps	120
Measles	9
Paratyphoid fever B	1

Of the noninfectious diseases, diabetes is quite widespread and gaining. In 2000 about 650,000 diabetics were under treatment, with females exceeding males by about 60,000. To put it differently, there were well over 12,000 diabetics under treatment for every 100,000 inhabitants of the Czech Republic—more than one in every ten persons.

The incidence of malignant neoplasms (tumors) rose steadily over the last 40 years of the 20th century while the rate of mortality from this cause over the past thirty years has remained virtually the same in men and has risen only slightly in women. During 1999, the most common malignancies in men were (in descending order) those of the skin, bronchi and lungs, prostate, colon, bladder, kidney, rectum, and stomach; in women those of the skin, breast, colon, uterus, bronchi and lungs, ovaries, and cervix. Compared with 30 other European countries, the standardized mortality rate for malignant tumors in the Czech Republic was the fourth highest for men and the third highest for women.

During 2000, the most common causes of death for both males and females were (in descending order) diseases of the circulatory system, neoplasms, external causes (injuries and poisoning), diseases of the respiratory system, and diseases of the digestive system.

Mental Illness

Over two million (2,057,952) psychiatric examinations took place during 2000; they concerned 361,931 patients (or a total of 3,523 first examinations, that is, new patients, per 100,000 inhabitants)—not a small number. Neurotic disorders were most common; they were followed (in descending order) by affective (emotional) disorders, schizophrenia, organic mental disorders, personality disorders due to use of alcohol, developmental disorders during childhood and adolescence, and mental retardation.

Preventive Measures

The General Health Insurance Company of the Czech Republic pays for preventive examinations as follows: for men and women by a general practitioner once every two years; for women 15 and older by a gynecologist once a year; for pregnant women once a month by an obstetrician; for individuals under 18 by a dentist twice a year; for pregnant women by a dentist twice during pregnancy; and for persons older than 18 by a dentist once a year.

Required protection of children from transmissible diseases includes immunizations against tetanus, diphtheria, whooping cough, mumps, rubella, measles, polio, and tuberculosis; some of these immunizations are combined. Recommended immunizations are those against hepatitis B, hemophilia, tick-borne encephalitis, and other conditions.

SEXUALITY AND REPRODUCTION

Most young people feel relatively free to engage in sexual activity. While the fertility rate (the number of live-born children per 1,000 females of a given age) for women 15–19 was 44.7 in 1990, it fell to 13.2 in 2000. Because the number of induced abortions during the same period fell to only 31.1% of the induced abortions in 1990, it can be assumed that methods of contraception are being used more, and more effectively. Still, out of 90,910 live births in 2000, 19,792 (that is, 21.8%) were to single mothers.

The population of the republic has shown natural decrease over the last eight years. The size of a nuclear family averages between three and four members, but some couples have no children, and devoutly religious parents usually have three or four.

HEALTH THROUGH THE LIFE CYCLE

Pregnancy and Birth

Pregnancy is a condition that can be ended by legal abortion. It must therefore be assumed that pregnant women have chosen to bear a child. Whenever possible—in an overwhelming number of cases—children are born in hospitals. Under normal circumstances, mother and baby remain under hospital care for three days. Healthy infants are seen by a pediatrician once a month during the first six months, and then every other month until the end of the first year.

Infancy

Most babies are bottle-fed, but some mothers nurse their babies. Breast-feeding of children up to 3 months of age decreased between 1996 and 2000 from 32.85% to 24.15%. Over the same period of time, the rate of breast-feeding of children older than 3 months increased from 24.3% to 31.07%. Some mothers swaddle infants, continuing a practice very common two generations ago. They believe that swaddling will calm their children.

Much attention is given to children for the first two years. For example, it is customary to take small children outdoors every day in prams or strollers. At about age 2 or 3, some children are sent to day nurseries (crèches), and a year or two later to kindergarten. The number of kindergartens fell from 7,335 in 1990–91 to 5,776 in 2000–01 as did the number of children attending them, from 352,139 to 279,838.

Infants and children with birth defects or other problems receive appropriate care and, if necessary, are referred to special residential institutions or to day clinics and centers, all of which are serviced by physicians. Some of these institutions are state-run, others are private.

Czech parents, as a rule, expect their children to obey, and if they do not, spanking (for the most part symbolic) is likely to be administered. But children are loved and no expense is spared to supply them not only with all their needs but also such extras as attractive clothing and the popular toys.

Childhood

Special schools are available for children who are mentally handicapped, have impaired hearing, speech, or vision, are physically handicapped, or have a combination of such problems. Four types of schools serve these children. The numbers of students in each type are given parenthetically for the academic year 2000–01: nursery schools (boys 6,309; the number for girls is not available); basic schools (grades 1–10; 14,159); special education and reform schools (49,494); and secondary, vocational, and home economics schools (28,112). In the last two groups, the large majority of pupils are mentally handicapped, with the ratio of male to female 100 to 40.

The Aged

People in the republic seem to age faster and look older sooner than those of the same age in the United States. Because many have had a stressful life, they usually retire as soon as they become eligible. Widows and widowers sometimes live with one of their married children, but if their health and finances permit, are more likely to live an independent life in their own house or apartment, or in a retirement home. During 2000, 148 nursing homes with 12,129 beds were available for those pensioners who needed nursing services.

As for mortality age, in 2000 a full 7,118 men out of a total of 54,845 males died at 85 years of age or older; 16,734 out of a total of 54,119 females died at that age or older. The most common causes of death in 2000 were diseases of the circulatory system (53.39%), neoplasms (26.33%), external causes (injuries, poisoning, etc.; 6.49%), diseases of the respiratory system (4.55%), and diseases of the digestive system (3.89%).

Dying and Death

Death is accepted as the inevitable end of a person's life and is particularly mourned if the person has died prematurely. More than half of the republic's population consider death as final; those who are religious believe in some sort of an existence beyond earthly life. Most people hope for a sudden, painless death, but in actuality many die in hospitals while being treated for a serious illness, usually associated with advanced age.

Prior to World War II, burial in the ground was as common as, or more common than, cremation. As of 2002, three fourths of the deceased are cremated—four fifths of those dying in Prague and about two thirds in Moravia. No special treatment of the body is performed except for preparing the face if the body is to be viewed before the funeral. In very exceptional cases, at the request of the family, the body may be embalmed. An autopsy is usually required in cases of death by accident or under unusual circumstances.

REFERENCES

The information for the Czech Republic was obtained during the author's visit there during the spring of 2002 and from the following sources:

Statistická ročenka České republiky 2001 (2001). Praha: Scientia.
Všeobecná encyklopedie v osmi svazcích (1999). Praha: Encyklopedie Diderot.
Zdravotnická ročenka České republiky 2000 (2001). Praha: Ústav zdravotnických informací a statistiky České republiky.

Datoga

Astrid Blystad and Ole Bjørn Rekdal

ALTERNATIVE NAMES

Datoga, Tatog, Tatoga, Datoog, and Mangati.

LOCATION AND LINGUISTIC AFFILIATION

The core Datoga area has for several centuries been the Hanang and Mbulu Districts of Arusha Region in northern Tanzania, but Datoga groups are spread over much of Tanzania in small localized enclaves.

Datoga is a Southern Nilotic language. The various subsections have different dialects which are internally comprehended among all Datoga.

OVERVIEW OF THE CULTURE

The Size of the Population

The size of the Datoga population is rather uncertain owing to the lack of ethnic variables in recent censa and to the considerable dispersion of Datoga throughout Tanzania. The Datoga population of Hanang District has been estimated at 30,000 (Lane, 1996), but the total number of Datoga in Tanzania is probably several times that number.

Economy and Occupations

Datoga are pastoral in the sense that there is an immense cultural emphasis on cattle. The cultural elaboration of cattle does not mean, however, that Datoga have large herds. Except for a smaller number of rich cattle owners, the majority of Datoga today have merely a few head of cattle and increasing numbers have no cattle at all. With decimated herds, farming has become increasingly important for Datoga, and nearly every Datoga household in Hanang cultivates a smaller or larger plot of maize, beans, and/or sorghum. A large repertoire of myths with degrading descriptions of cultivation and cultivators nonetheless flourish.

Social and Political Conditions

Many Datoga have started to wear modern Tanzanian cotton dress, but the traditional and still commonly used dress is the bead-decorated leather cape and the leather skirt (*hanang'weanda*), a skirt which is vested with immense social and religious significance. Women carry heavy brass and bead decorations, while both sexes often have large facial scarification around the eyes and large extended earlobes with wooden or brass ornaments.

A husband and his wife or wives and their children and his old mother make up the prototypical inhabitants of a Datoga homestead. The traditional compound is characterized by a low flat-roofed men's house, several women's houses, and various animal dwellings placed in a half circle within a high thorn fence.

Datoga communities are linked together through rules and regulations that are enforced by an elaborate system of ad hoc meetings. The most common meeting is the "open meeting" (*geetabwaraku* or *girgweageeda emeeda*) which in practice is a male forum. The "clan meeting" (*girgweageeda doshta*), the "neighborhood meeting" (*girgweageeda gischeuuda*), the "youth meeting" (*girgweageeda gharemanga*), and the "married women's meeting" (*girgweageeda gademga*) are other examples.

Family and Kinship

Datoga have, like other East African pastoralists, commonly been described as polygynous, patrilineal with a strong "warrior" tradition and male domination of social and political life. With renewed focus on pastoral communities in recent writings a more nuanced picture has appeared. In writings on Datoga it has been pointed out that clan affiliation is retained throughout life for both men and women, socio-political responsibility include matri- as well as patrikin, a "dowry" institution facilitates that women as well as men receive gifts of livestock at the time of marriage, and that women are central actors in Datoga political and religious life (Blystad, 2000).

Religion

Aseeta, the Datoga deity, appears as an androgynous, powerful, and inherently good deity, invested with immense creative potential. Communication between human beings and the deity takes place with female and male spirits (*meang'ga*) as mediators. The spirits are more closely involved in people's lives than Aseeta. Aseeta and the spirits are particularly powerfully addressed by married Datoga women or "women who wear the leather skirt" during the frequently arranged *ghadoweeda* ritual ("people seeking blessing"), a ritual located at the heart of Datoga "tradition."

THE CONTEXT OF HEALTH: ENVIRONMENTAL, ECONOMIC, SOCIAL, AND POLITICAL FACTORS

During colonial times as well as in the post-independent Tanzania period, Datoga pastoralists were subject to attempts to do away with their way of life. Colonial discourse associated Datoga pastoralists with primitivism, barbarism, and savageness. The discourse was reflected in public executions, arbitrary imprisonment, forced conscription into the army, collective cattle fines, and discriminatory resource allocation. The colonial portrayal of Datoga influenced the attitudes and actions of neighboring populations, politicians, bureaucrats, journalists, and scientists in post-independent Tanzania.

The new state of Tanzania pursued these policies more rigorously with their swahilization, villagization, and education policies. The so-called "Operation Barabaig" in the 1970s developed into a dramatic encounter between Datoga and state employees, who forcefully removed the people from their homes and relocated them in villages.

The official discourse after independence in 1963 facilitated a new wave of agriculturalists moving into what was formerly Datoga grazing land, leading to serious conflicts and to ethnic clashes with large numbers of casualties. Mechanized farming was in particular encouraged by the new state. In 1970 the Tanzania Canada Wheat Project (TCWP) initiated the clearing of woodland above the Rift wall in what is now Hanang District. With the total allocation of some 100,000 acres of land for large-scale wheat farming during the 1970s and 1980s the problem of land shortage in the area was magnified. Adding to the impediment of land loss, the appropriation of Datoga land took place in a manner that caused reaction from both the national and the international community owing to its violations of human rights.

MEDICAL PRACTITIONERS

The landscape of indigenous Datoga medical practitioners consists of primarily three categories: Daremgajeega, Bajuta, and Gijoodiga. Most male Datoga healers are Daremgajeega. Daremgajeega is the name of a particular clan of healers as well as of a group of healing clans. The other major group of male healers is called Bajuta (Bajuta being both a healing clan and a Datoga subsection).

The practices of the Daremgajeega and the Bajuta are often quite similar. A common healing session implies that the healer spits on his client's body for extended periods of time, usually on the back, breast, and belly. Spitting, in the characteristic way of sending showers of spit through the teeth, is always perceived to be a blessing among Datoga, but gains in significance when carried out by a healer. The spitting is usually combined with the use of mixtures of herbs and animal fat which are rubbed over the patient's body with slow massaging movements to soothing accompaniment of chanted prayers for blessing. Patients say their healers make them "calm down."

Daremgajeega and Bajuta healers use a large number of techniques in addition to the somewhat standardized sessions of spitting, application of medicine, and prayers in their attempts to cure their patients. Creative use of product from domestic animals, particularly from cows and bulls, goats, and sheep is essential for Datoga healers. This particularly includes a practice linked to the consumption of fatty soups served with or without meat, the consumption or the smearing of milk, gee, butter, or blood onto the body, and the application of hides, manure, urine, or blood on parts or over the entire body. Incisions with or without the extraction of blood with a cup and the burning of parts of the body are other practices that commonly leave Datoga bodies with numerous little black scars. Quarantine or moving of particular homesteads or neighborhoods for shorter or longer periods of time are also common strategies used to prevent the spread of contagious conditions.

Although female healers are fewer in number than male healers, they make up a significant part of the landscape of Datoga healers. The most important category of female healers is *gijoodiga* (singular: *gijocheanda*), women whose mothers are of the Bajuta clan. *Gijocheanda* is a medium through which suffering individuals can communicate with the spirits, plead for help, and receive response and advice from the spirits. This communication takes place during night-time séances when the spirits "crawl onto the back" of the *gijocheanda*.

The abilities of the *gijoodiga* will commonly be revealed when they are still young due to peculiar conduct. Rigorous training is nonetheless required before a *gijocheanda* acquires proficiency in calling forth the voices of the living spirits. Some women of the Ghawooga, Hilbaghambowaida, or Bajuta clans may also in rare instances have special healing or spiritual abilities, but they are not as numerous as the *gijoodiga*.

Indeed, a prominent feature of the complex cultural environment of Mbulu and Hanang is that healing is often sought from individuals with foreign ethnic affiliation. Such cross-cultural therapeutic relationships are often established in spite of the fact that the ethnic groups involved have a long history of mutual hostility (Rekdal, 1999). Commonly consulted non-Datoga healers come from the Iraqw, Nyaturu, Rangi, Sukuma, and Tanga ethnic groups, as are health personnel at official or mission dispensaries, health stations, and hospitals.

CLASSIFICATION OF ILLNESS, THEORIES OF ILLNESS, AND TREATMENT OF ILLNESS

"Illness of God"/"Illness of Man"

There is a distinction made between "illness of God" (*Geyooda Aseeta*), and "illness of human beings" (*geating siida*). But Datoga illness classifications are ambiguous and internally inconsistent; most illnesses can be classified into both categories depending on the circumstances.

Witchcraft, Evil Eye, and Curse

The prototypical illnesses located in the category "illness caused by humans" are conditions caused by witchcraft, evil eyes, curses, and the use of poison or other dangerous remedies. Datoga consider themselves to have few witches, but witchcraft is nonetheless said to cause increasing harm to the health and prosperity of people, animals, and plant life due to the escalating proximity in relations between Datoga and their Iraqw, Iramba, and other neighbors. A related and not less frightening skill among some Datoga is "the evil eye." The evil eye can, with a mere glimpse, but more commonly with lengthy staring at its target, cause illness, death, and dying.

Curses commonly hit Datoga individuals who fail to adhere to fundamental Datoga norms of respect, and are commonly carried out by male or female meetings after the individual is given ample opportunities for redress. Curses cause social isolation of an individual as they are extended to persons who co-operate with the cursed individual.

Fever

Malaria and tuberculosis (TB) are conditions typically categorized as "illnesses of God." Living in a harsh and disease-ridden environment, malaria and TB are common conditions among Datoga of Hanang and Mbulu, and with relatively little variation, the theoretization and treatment linked to these conditions are similar to those of a whole range of common illnesses in the area. Normal fever conditions are distinguished from intermittent fever and fever in combination with bloody cough and weight loss as in TB.

These conditions receive similar treatment consisting of a combination of drinking of melted butter/gee, eating fatty soups cooked on beef meat, and drinking of herbal mixtures. The herbs utilized vary somewhat from one illness condition to the next. The aim of this treatment is to cause diarrhea and vomiting in order to expel the harmful substances from the body. For the pure fever conditions the drinking cow's, bull's, or goat's urine is added to make the expulsion more efficient. With worsening conditions patients suffering from any of these conditions may eventually be sent to hospital, but many Datoga choose hospitals as a last option, and their health may have severely declined by the time they arrive.

Physical Handicaps

Conditions that are particularly feared among Datoga are illnesses linked to burning, severe bleeding, acute diarrhea, and conditions affecting the bones, all conditions which may be attributed to either the will of God or to human agency. The last condition warrants a few remarks owing to its particular elaboration among Datoga.

The birth of malformed babies, that is, babies lacking fingers or toes or entire limbs, or accidents causing such losses, are particularly feared among Datoga. Such handicaps have caused expulsion and fear of infants as well as of adults throughout the known history of Datoga.

The fear of a person without a finger, "without feet," or without other "bony" parts of the body, must be understood with reference to the connection made between such conditions and a clan's poor semen. In Datoga semen is thought to be the substance that builds the bony structures of a body, and hence malformed bones imply poor contribution by the genitor and in extension by his clan. What is important in this context is that such an "incomplete" condition can be transmitted to others, particularly through the sharing of meals.

The fear of incomplete bones make Datoga shun amputations, which at times cause flights from hospitals. Severe and lengthy ritual activity carried out by Datoga healers is the only remedy that, to some extent, can cleanse individuals with such handicaps. Other categories of handicap, such as deafness, dumbness, or blindness are not feared. Individuals with such handicaps are rather often located at the heart of Datoga ritual life.

Illness of the Bones

Another and somewhat related condition to the one we just described is called "bones" (*geaka*) which strikes an individual who either purposely or accidentally kills a fellow Datoga. Such an act is said to be so grave that it enters into a man's bones—and in extension the semen which is to produce new "bony structures." The condition follows him for three generations, and in practice expels him from life in Datoga communities.

SEXUALITY AND REPRODUCTION

Datoga men and women intensely desire children. Both men and women need children to achieve adult status, and they gain prestige and influence with the birth of every additional child. A man or woman who dies without a son and a daughter, may never be given the honor of an "official" Datoga funeral (*bunged*), an ultimate goal among Datoga of this part of Tanzania. For fear of envy, great secrecy surrounds the fecundity of a compound, and the total number of children and cattle can only be traced with great difficulty. Any kind of population control in the married population is a highly foreign concept among Datoga who follow traditional religious concepts and ideals.

A woman's future is closely linked with her own ability to conceive and successfully give birth. Pressure to successfully conceive and give birth may follow a woman throughout her entire reproductive life if she fails to fulfill the goal of having at least one son and one daughter. A newly married woman will often have several sexual partners, classified as "brothers" of her husband, in order to ensure her first pregnancy. The younger "brothers" of her husband are usually particularly welcome in the young woman's house, as the "blood is often warm between them" and "the seeds of young men are good."

There is a strong belief that all women can conceive, although some may have problems carrying through a pregnancy or giving birth, due to the "will of *Aseeta*" or influences of "bad humans." If time goes by and there is no sign of a forthcoming offspring, a woman will set about seeking answers and cures. A husband is usually eager to cooperate in their wives' attempts to seek a cure against barrenness which often implies the spending of large amounts of wealth for treatment.

The status of a childless woman will remain ambiguous throughout her life, and her husband may demand a divorce. It is more common, however, that a woman who has difficulties conceiving will divorce her husband and remarry, in the hope that a new husband and his "brothers" may bring her better fortune. A childless couple will commonly receive a child or two from kin who are "rich in children."

HEALTH THROUGH THE LIFE CYCLE

Pregnancy and Birth

The Physical Growth of a Fetus. Semen and a woman's blood are considered to gradually thicken into a mass which forms the fetus. The blood produces the soft parts such as the flesh and blood, while the semen produces the white bony parts, namely the spine, bones, and skull.

For its growth, the child is said to be dependent upon a continuous supply of blood, as the fetus "nurses its mothers blood." The fetus is also said to benefit from the periodic addition of sperm for the growth of the bones and simultaneously to the development of hard, male qualities such as a tall and erect body posture, and a woman will therefore usually welcome sexual contact during a pregnancy.

Though fundamental in procreation, blood and sperm are both potentially destructive fluids that can harm the unborn child if contaminated by bad substances. A woman must watch her diet as "her blood becomes what she eats" and sperm may in some instances be too "strong" for a child in the womb and may cause the fetus to sicken and die. During the last couple of months of a pregnancy, the addition of sperm should cease completely since the sperm at this point unproductively merges with the milk in the mother's body and breasts.

Diet during Pregnancy. A pregnant woman strives to get the foods she craves for since these "go well together with the child inside." Raw blood products are desired by the pregnant woman as "they increase her blood"; whenever a slaughter has taken place, raw blood is served to the visibly pregnant. Generally speaking, the common diet of maize porridge, maize soup, and milk, with the periodic adding of millet and beans, is usually not altered dramatically during a pregnancy. In fact, far more attention appears to be given to the persons a pregnant woman eats with than with the actual food she consumes.

Precautions taken by the Pregnant. A pregnant Datoga woman considers herself to be surrounded by human beings, substances, and states that may harm her unborn child. There are certain sights that should be avoided by a pregnant woman at any cost; this applies to people, animals, or items that look peculiar or damaged, such as, for example, crawling snakes, broken pots or calabashes, and persons who are crippled. This fear of looking at people who do not have complete bodies—or, more seriously, of eating together with them—is not limited to pregnant women, but this fear takes on new proportions during pregnancy since the shape and character of the "soft" or "watery" child in the womb can easily be influenced by such abnormality. Foul smells, frightening sounds, or odd sensations are also shunned during pregnancy. When severe illness occurs during pregnancy, a sheep or a goat is commonly slaughtered in order to cook nourishing soup for the woman, and in due course the assistance of healers is sought.

Itself so vulnerable and dependant, the child in the womb may also be demanding and a source of potential peril. Indeed, a pregnant women's body is considered to be inhabited by a tiny strong-willed human being who has the ability to make her ill or even cause her death. Thus, a pregnant woman may experience herself as being vulnerable to both external and internal forces. Her leather skirt is tied tightly around her womb in a manner that pushes her growing belly downwards. "We close the womb with the leather skirt" informants would say, indicating the way they seek to prevent the unruly fetus from engaging in unproductive wild play. In addition, this technique prevents her condition from becoming visible until the final stages of pregnancy, thereby reducing the danger of evil eyes.

Stuck Pregnancy (*Muldeaneeda*). In Datoga experience, a fetus may suddenly cease to develop, in which case the child becomes stuck in the maternal womb (*muldeaneeda*). In such instances the woman experiences symptoms of pregnancy, but the fetus ceases to develop due to bleeding which leads to insufficient nourishment for the child. For years the child may "hide" in the back of a woman's womb "feeding on her body," leaving her thin and weak. Some women are bothered throughout their entire lives by recurring *muldeaneeda*, some of which are said to simply die off, while others are eventually born. Children that are born after years of *muldeaneeda*, are often given the names Gidamulda (male) or Udamulda (female), and these children are particularly adored, since their births relieve their mothers of years with pain and anxiety. The immense admiration of these children is moreover related to their alleged potency.

Miscarriage and Infant Death. Other dreaded outcomes of a pregnancy include miscarriages and the death of nursing infants. Such deaths are considered to be frightful, and constitute severe breaches in what is perceived to be the correct development: from conception through pregnancy and birth to convalescence and nursing. Such events produce "dirt" and those who are going to "eat the dirty condition"—usually the woman, her children, and sometimes her husband—are isolated in separate enclosures (*ghawiida*), in principle until a new pregnancy is ensured. Isolation is necessary because such events are considered to be contagious and can harm the potential fertility of others.

Ghawiida is part of the elaborate Datoga seclusion practices (*metiida*). *Metiida* involve the isolation of people and compounds for varying periods of time in order to prevent "dirt" from entering fertile processes. The most common conditions requiring a quarantine are deaths, abnormal births, conception outside of marriage, and conditions that require the purification or healing of people, places, or activities.

The most serious type of *metiida* is the *ghawiida jeapta* related to the death of a nursing infant, a stillbirth, or a miscarriage. It is particularly the milk dripping from the woman's breast that is feared. In contrast to the milk received by nursing infants, the unproductive mother's milk that drips from a woman's breasts after the death of a nursing infant is perceived as being entirely bad and infertile.

In principle, only a new pregnancy has the power to fully remove *ghawiida jeapta*, and a woman who does not become pregnant may remain in an ambiguous position, being neither completely clean nor completely dirty. A woman who experiences recurrent miscarriages or infant deaths will "have dirt" and substantial restrictions will be placed on her social conduct to the extent that years of her life may be spent isolated behind high thorn fences.

Giving Birth. There are commonly a number of older women ready to assist a woman in labor, and news about a woman who is about to give birth causes immediate moves by neighborhood women who move in to take part at this precarious time.

The woman in labor is placed in a half sitting, half hanging position, holding onto a rope which is hanging from the low ceiling. The rope the woman holds onto is fastened close to the fire "so that her blood can be heated and the delivery speeded up." Her feet are firmly planted on the large hide on her bed, with her leather skirt placed on top of it. With small modifications, a woman commonly remains in this position throughout the delivery. The midwife will fasten and unfasten a leather thong around the woman's abdomen—following the rhythms of labor—in order to increase the downward movement. She will also massage the woman's back and repeatedly change the position of her legs.

After the birth of the child, the leather strap is again carefully fastened around the womb to speed up the expulsion of the placenta. When the placenta appears it is closely inspected. The umbilical cord is then cut with a knife before the wound is covered with cow dung to heal. Later the placenta is placed in the cattle enclosure, while the remnants of the umbilical cord are tied to the foot of a heifer.

Both breech births and twin births are feared among Datoga. After such abnormal births there will be no *ghawiida* since no death has occurred, but the compound is isolated for 6–8 days. Despite their relatively light isolation, a breech birth is regarded as a most serious matter, and several of the practices that follow such a birth imply associations with death. The later consequences for the child are, moreover, serious, since the person may have difficulties finding a marital partner. Children born in breech births will also appear unpopular on the marital market. Twin births are commonly followed by the brewing of honey mead and diverse ritual activity which aims at "separating" the two children.

Increasing number of Datoga women will give birth at a health center or at a hospital.

Infancy

The sensation of having calmed down after birth is soon followed by another unsafe phase, post-natal convalescence (*ghereega*)—the months during which the woman is to regain her strength, and the infant is to face its initial harsh encounter with the world. *Ghereega* is a new liminal phase which is associated with almost total seclusion of the mother and infant in the low, dark, private room of the mother for a period of 2 months after the birth of a boy, and 3 months after the birth of a girl.

During the months the woman remains inside, she focuses on ensuring her own and her child's health. The midwife commonly stays on with the mother until the first peril is passed, and until a male goat has been selected, strangled, and butchered, and the "song of the child" initiated. A girl is then commonly selected to stay with her and assist her. Some women will move outside with care after the remnants of the umbilical cord have fallen off, but they usually remain close to the compound. In such cases she places her baby deep in a leather sling, and if the child is a boy she carries an arrow and a ritual stick for protection. If she meets a stranger she will commonly look at the ground or bend down to signal that she does not want contact.

A woman will generally let her infant suckle whenever it cries. However, breastfeeding does not begin immediately after the birth of a child. Water is substituted for mother's milk the first 2–5 days of an infant's life, or until every sign of the meconium, the first black feces of the child, has passed. Our informants would explain that water cleanses the infant's stomach. This practice cannot be understood without a fundamental recognition of water as a sacred life-giving source; rain water is referred to as "the spit of Aseeta" (*ng'usheanda Aseeta*). The feeding of the unboilt water causes diarrhea in many new-born Datoga babies.

Cow's milk, when available, is usually given a couple of days later in combination with mother's milk. When a mother starts to nurse her child, therefore, she also makes an effort to get some cow's milk. Mother's milk and cow's milk ideally provide the sole nourishment for the child the first year.

Milk has tremendous cultural significance among Datoga. Feeding the child cow's milk is crucial in order to gain the experience of sharing bodily substance with livestock, and is perceived as initiating a life-long pastoral education, implying the instilling of embodied, pastoral knowledge in the child. Later on ghee, butter, meat, and blood will be added to the diet. Despite the ideals of feeding Datoga infants cow's milk, Datoga mothers of Hanang and Mbulu are often forced to feed even tiny children maize-soup, soft maize-porridge, or bean stew since the shortage of cow's milk often becomes acute in the dry season.

Mother's milk and the flows that exist between the mother and her child during the period of nursing are considered to be inherently good, intimate, and nourishing. However, this milk is thought to be liable to be contaminated by other substances that enter the mother's body. Bad food, bad sperm, or other bad substances have the potential to negatively affect the quality of a woman's breast milk.

A Datoga woman commonly nurses her child for at least a year, and often up to two years. While the woman is nursing the child, the two are perceived to be "one," just as they were while the child was still in the womb. The intimate links between the two bodies during pregnancy are thus extended throughout the period of breastfeeding. The separation of the two bodies is said to start first after weaning. As we saw above, the death of a still-nursing infant requires that precautions parallel to those associated with a miscarriage be taken; and these events differ fundamentally from the way the death of a weaned child is handled. The notions that a mother and child "are one body" as long as the child is breastfeeding, and that death of a suckling child is like a delayed miscarriage have also been noted among neighboring peoples such as Iraqw and Maasai (Rekdal, 1996, p. 376; Spencer, 1988, p. 41).

The new mother will also spend a substantial amount of time studying every inch of the baby's body. She makes sure the painful but necessary burning of the infant's skull is carried out to "hasten the closing of the fontanels," and should the infant suffer from pains in the chest or stomach, she may allow the same treatment of burning to take place again, often leaving the little child with numerous small burn marks. She selects charms to hang on the baby's body to protect it from the first dangers of life. A weak baby girl may at this early point in her life be circumcized "in order to prevent further illness." While in seclusion a new mother spends cherished time on the composition of the "song of the child," which was initiated by the midwife. The mother holds her baby close while rocking her upper body slowly back and forth, thus providing it with soothing sounds and soft movements.

Childhood

The custom of mutating the physical body for beauty, socialization, or preventive health purposes takes place throughout childhood. The most common practices include the initiation of a process of extending the earlobes, the removal of the two lower incisors, and circumcision of both girls and boys.

The mandatory removal of the clitoris and labia minora is commonly carried out when a girl is between 2 and 5 years of age. This painful operation is carried out by a local elderly woman with only a handful of women present in the homestead. The incident is not talked about much and receives no cultural elaboration. This appears in stark contrast to male circumcision, which is a grand ritual occasion characterized by the brewing of large quantities of honey mead, sung prayer, and dancing. The operation takes place when boys are between 5 and 12 years old. Large numbers of boys, sometimes several hundred, are circumcized at the same occasion. The operation itself is commonly carried out with a knife or razor blade, and is followed by the feeding of milk and blood for the boys to regain their strength.

Adolescence

Youth is a time when Datoga pay substantial attention to their maturing bodies, and spend time on modifying it for beautifying purposes. For example, earlobes are extended with large wooden earrings and large facial scarifications are commonly inscribed around the eyes.

Particularly intricate and poetic youth talk takes place between potential boy/girlfriends. A man who fails to encourage the continuation of the relationship through dialogue may attempt to win a girl's heart through a spearing in the *lilicht* tradition, a traditional hunt for dangerous animals (usually a lion) linked up with youthful fertility.

Although sexual play is an integral part of youth activity, norms prohibit a couple from carrying out sexual penetration while the girl is unmarried. The norm is related to the absolute condemnation of children born without clan affiliation on the father's side. Indications of breaches of the sexual norm, and rumors that a particular girl "has become a woman," may lead to her being required to undergo a physical checkup by a group of older women, and they may conclude that "the girl has now become like us," and should be married off immediately. Such incidences severely reduces a girl's chances of influencing the choice of her marital partner.

This is not to say that premarital sexual intercourse does not occur. Preventive measures such as *coitus interruptus* are common and remedies that cause abortion are known. But girls were generally said to be reluctant to agree to sexual intercourse. Medical personnel confirmed that virginity among unmarried Datoga girls is the rule rather than the exception.

Adulthood

Alcohol Consumption. Datoga elders have customarily consumed the sacred honey mead primarily on ritual occasions and on grand occasions of festivity. Women and youth have according to Datoga custom not consumed alcohol. With the tremendous transformations taking place in Datoga communities, not the least with the increase in contact with outside peoples who brew a large variety of brews on which there is no customary restrictions, the consumption of alcohol has increased substantially. Youth and sometimes women may today be seen consuming non-indigenous brew, particularly on market days. The increase in alcohol consumption, the transformations of the people involved in the consumption, and of the timing and settings where the consumption takes place is readily addressed by the Datoga informant.

Domestic Abuse. Datoga have strong ideals of peaceful co-operation and complementarity between a husband and his wife, or wives, in order to ensure sound reproduction of a compound. Despite these ideals, however, conflicts between spouses and violent outbursts are common in many Datoga compounds. A generally accepted norm states that a man has the right to punish his wife or wives when they do not fulfill their obligations, and women will commonly defend their husband's abuse by referring to particular instances of lack of compliance or misconduct on their part.

There are however restrictions put on a husband's right to physically reprimand his wives, and men may be penalized by the female community when breaches occur where women experience their "procreative bodies," in the broad sense of the term, are infringed upon by male misbehavior. Such infringement must be instantly addressed lest all Datoga life suffer. The consequences of action taken by the married women's council (*girg-weageeda gademga*) against a man who has seriously infringed upon women's fertile domain may be quite severe. The offender will not only be fined a rare and cherished black bull, but will be humiliated by the women in front of the entire community.

Any male interference at critical moments of procreation—such as mistreating or quarreling with a pregnant or convalescent woman, or watching the birth of a child, if not specifically summoned to assist as a member of a healing clan—are regarded as particularly serious offences. Serious argument arising over funds for the treatment of barrenness, miscarriage, or illness related to pregnancy, birth, or nursing may require action, as may male violence inside a Datoga woman's private room. Men's behavior also comes under the scrutiny of women following incidences of misconduct in relation to what are perceived as sacred female activities in connection with ritual gatherings. Female animals, not only livestock, but also domestic animals such as cats and dogs, are protected in ways parallel to women when they have kittens or puppies.

Dying and Death

The treatment of dying persons who leave families and offspring behind them is colored by the notion that the person is soon to become a living guardian spirit; a condition requiring utmost respect and care.

Datoga fear corpses to the extent that individuals closely associated with the dead will be isolated for up to a year. The severe restrictions linked to miscarriages or deaths of still suckling infants were reviewed above. Deaths of spouses will experience a similar set of restrictions which imply that either the widow or widower experience months of isolation where they do not eat, work, or socialize with other than the potential person who might "eat the dirt" with them. Simultaneously they modify their dress, shave their hair, and remove all decoration except for a black bead necklace so their appearance becomes filthy.

The corpse is handled by female members of the household only. Dead bodies should ideally be placed in a horizontal position on a litter and carried to the east of the compound where it will be left to be "eaten" by the bush. The Government of Tanzania has prohibited the placing of corpses in the bush, and many Datoga will today bury the dead inside the compound even if a communal funeral is not held. Each year, however, a handful of deceased Datoga elders, both men and women, will be buried in grand communal funerals. In such instances the corpse is placed in a sitting, fetus-like position with the head bent toward the chest and the hands closed around the knees. The tomb that is to mark the grave for years to come is created in the image of a large pregnant female womb, which after 9 months "gives birth" to the new spirit. The funerary ritual is a ritual unique in scale and elaboration in Datoga culture.

The deceased who is to receive this honorary departure must have had many children, of whom at least one daughter and one son must be married and have children. The deceased must not have died "a bad death," which includes deaths caused by accidents, sudden diseases, severe bleeding, diarrhea, fire, or the breaking of bones. All limbs, including the fingers and toes, must be intact, without fractures or abnormalities such as the lack of a joint or even a fingernail. The most important criteria, however, are that the deceased is ritually clean, that is, is not suffering from "bones," which is related to homicide involving a fellow Datoga; and that the deceased comes from a wealthy family and clan. It is the male and female spirits born in such grand burials that are sought by "people seeking blessing" (*ghadoweeda*)—and whom Datoga informants consider themselves to be wholly dependent upon for health, wealth, and prosperity.

REFERENCES

Blystad, A. (2000). *Precarious procreation: Datoga pastoralists at the late 20th century*. Doctoral thesis, University of Bergen, Norway.

Lane, C. (1996). *Pastures lost: Barabaig economy, resource tenure, and the alienation of their land in Tanzania.* Nairobi, Kenya: Initiatives.

Rekdal, O. B. (1996). Money, milk, and sorghum beer: Change and continuity among the Iraqw of northern Tanzania. *Africa, 66*(3), 367–385.

Rekdal, O. B. (1999). Cross-cultural healing in East African ethnography. *Medical Anthropology Quarterly, 13*(4), 458–482.

Spencer, P. (1988). *The Maasai of Mataputo: A study of rituals of rebellion.* Manchester, UK: Manchester University Press.

Fore

David J. Boyd

ALTERNATIVE NAMES

None. The name "Fore" (pronounced FO-rey) was created by an Australian patrol officer during a patrol into the region in the early 1950s. Standing on the northern border of this different linguistic group, he asked local people for the name of those who lived further to the south. The answer was "*porekina*" (*kina* = people), meaning "people living downhill." "Pore" was transcribed as "Fore" (Lindenbaum, 1979, p. 39).

LOCATION AND LINGUISTIC AFFILIATION

The Fore are located in the southeastern region of the Central Highlands of the island of New Guinea. Their territory, centered on 6°35′ south latitude and 145°35′ east longitude, is a wedge of approximately 950 km^2, bounded on the north by the Kratke Mountains and on the west and the southeast by the Yani and the Lamari Rivers, respectively. In this mountainous montane zone, altitude varies from 400 to 2,500 m, although most people live within the altitudinal range of 1,000–2,200 m. Fore territory is divided into a northern and a southern region by the Wanevinti Mountains and the corresponding populations are referred to as North Fore and South Fore. Currently, the entire Fore region falls within the political boundaries of Okapa District, Eastern Highlands Province, Papua New Guinea.

The Fore language is the southernmost member of the East Central Family, East New Guinea Highlands Stock, Trans-New Guinea Phylum of Papuan languages. Three distinct dialects that reflect the geographical division are recognized: Ibusa dialect is spoken by the North Fore, and Atigina and Pamusa dialects are spoken by the South Fore. This is a region of considerable linguistic diversity and the Fore share common territorial boundaries with speakers of seven other mutually unintelligible languages.

OVERVIEW OF THE CULTURE[1]

Until the mid-20th century, the Fore, who currently number approximately 20,000 people, existed in relative isolation and had scant knowledge of the larger world beyond their territorial boundaries. They lived as subsistence horticulturalists practicing an extensive form of shifting swidden cultivation to produce their staple crop, sweet potato (*Ipomoea batatas*), and a variety of subsidiary foods. New gardens regularly were cleared in the surrounding forest and settlements were quite mobile as people moved their living sites to stay close to their gardens. They also tended domestic pigs, which were the most important form of local wealth and also essential objects of transaction in acquiring wives, settling disputes, rewarding allies, and compensating enemies. Although the Fore did share a common language, they did not consider themselves one people. They had no encompassing name for all speakers of Fore, no unifying political organization, and no collective ceremonies. In fact, social relations between local groups were generally antagonistic and intergroup warfare was common (Berndt, 1962). Groups would form impermanent alliances for intermarriage, small-scale ceremonial exchange, mutual defense, and trade, but these relationships were altered frequently as patterns of amity and enmity shifted.

Social and Political Organization

The primary Fore residential unit is the hamlet consisting of 70–120 people living in 12–20 houses. One or several adjacent hamlets combine to form a parish whose members share a corporate interest in the territory they occupy, including a sacred grove where spirits reside. Typically, parishes are subdivided into parish sections composed of several allied hamlets and these sections in the past were the effective political and military units. All sections and parishes were headed by a "big-man" leader. Since there were no permanent political positions or ranked statuses among the Fore, big-men won the respect and loyalty of fellow group members by showing superior skill in organizing activities that enhanced the well-being of the group. They directed most group activities, including warfare, managed exchange transactions, and recruited new members to enlarge and strengthen the group. Men competed with each other to achieve big-man status and those who succeeded were strong, dominant individuals, feared warriors, eloquent orators, and skilled negotiators. However, when a big-man faltered, the support of followers could be quickly transferred to a competitor who aspired to become a big-man.

Then, as now, the dominant organizing principles of Fore society were kinship and co-residence. All social groups are assumed to be founded on kin relatedness between members and patrilineal connections are stressed. But, genealogical memory is shallow, rarely extending beyond the grandparental generation. Also, in this very mobile society, local group membership changed often. In practice, then, unrelated newcomers were easily incorporated as new members of a group by adoption, affiliation, or mutual consent if they behaved as kin. People who lived together and demonstrated loyalty, cooperation, and mutual support could become kin, "one blood."

These putatively kin-based groups are still hierarchically organized. The smallest unit, called a *lounei*, consists mainly of co-residents of a single hamlet and is an exogamous group. Several *lounei* together form a subclan that may or may not be exogamous. And, several subclans form a non-exogamous clan and share a common territory.

Gender Relations

The relations between Fore men and women are relatively egalitarian, but important differences do exist in the social spaces men and women occupy. In the political arena, men dominate the public domain and vie with each other through oratory and subtle persuasion to influence the course of group activities. Also, in earlier times, men as warriors were responsible for defending their groups and for prosecuting offenses against them. In the domestic sphere, the gender division of labor still assigns relatively few tasks to only men or women. Men fell the trees for new food gardens and women assist in cutting the undergrowth and burning the debris. Men then build the garden fences and women prepare the soil and plant most of the crops. However, the bulk of labor required to cultivate, weed, harvest, and transport garden produce continues to be provided by women, although men do assist as they wish. The burden of pig rearing also falls largely to women who transport food and fodder from the gardens and feed the pigs each day. Most food for household members is prepared and cooked by women and men help with the gathering of firewood. Childcare again is largely the responsibility of women who enlist the willing aid of older siblings. Women, who once made all articles of clothing and net bags from pliable tree barks and other forest products, now purchase most clothing items or sew them on hand-crank sewing machines. Men, who used to craft weapons (bows, arrows and spears), no longer do so. In sum, although the division of labor assigns complementary tasks to men and women, women continue to provide a much larger portion of the labor required for essential household activities.

Female Pollution

The separation of male and female spheres is underpinned by an ideology of female pollution. Fore men, as do men in many other societies in New Guinea and elsewhere, fear contamination by wives and other actual or potential sexual partners. Especially debilitating would be any contact with vaginal blood associated with menstruation or childbirth. Should such contamination occur, either accidentally or as a deliberate act of assault, the male victim is expected to suffer severe respiratory symptoms and a gradual wasting away. Living arrangements in the past reduced the contact that men had with their wives. All adult men and older initiated boys lived together in large men's houses that were surrounded by the smaller houses of women and their children. Also, to safeguard the health of men, especially husbands, and to control the threat of contamination of public spaces and gardens, women were confined during menstruation and childbirth to small seclusion huts on the edge of residential settlements. Today, seclusion huts have largely disappeared, although a few families maintain them for birthing purposes. During their menstruation periods, women refrain from preparing food and have no contact with their husbands.

Male Initiation

This theme of male–female separation also is an important aspect of the Fore male initiation ritual. At the onset of male initiation, pre-pubescent boys about 10 years of age are taken from their mothers and sent to live with men in the men's houses. During the various stages of the ritual, which earlier lasted for several years, boys were initiated into manhood. In addition to being instructed in the proper behavior and responsibilities of men, they also were taught the techniques of nose bleeding, cane swallowing, and vomiting that were thought to promote their physical growth and maturation and to protect them from the polluting powers of women. Armed with these personal defenses, they eventually were deemed marriageable adult men capable of protecting themselves from the inherent dangers of marital relations. Male initiation ceremonies were halted in the late 1970s, but were resumed in the late 1980s in a truncated form. Fore boys now become men in a single week-long ritual.

Cannibalism

At some point in the early 1900s, the Fore adopted cannibalism, or anthropophagia. The consumption of human flesh was common among various neighboring groups to the north and gradually passed first to the North Fore and later to the South Fore. Among the South Fore, a selective form of endocannibalism was practiced: only the bodies of deceased group members were considered appropriate for consumption while those who had died of dysentery or leprosy, or had contracted yaws were avoided. Of the corpses selected for consumption, all body parts except the bitter gall bladder were eaten. As in the distribution of a cooked pig, specific body parts were given to designated relatives. But, not all Fore participated.

Most adult men and initiated boys living in the communal men's houses eschewed human flesh in favor of wild animals and domestic pigs. Women, children, and elderly men, then, were the principal consumers of human flesh.

Sorcery

Sorcery is an abiding concern in Fore society. All contacts between members of different parish sections, and even some contacts within one's own local group, are edged with the fear of sorcery. This appears to be a longstanding concern among the Fore. Genealogical evidence from c. 1900–62 for one South Fore parish attributes more than half of all deaths to sorcery (Lindenbaum, 1979, p. 65). In general, sorcery is thought to emanate from the intentional, malicious activities of political enemies and social inferiors who seek to harm rivals and those of higher social status or members of their families. Sorcery is judged to be a political act, a silent aggression, and big-man leaders are thought to be common targets. Although sorcery techniques are quite varied, most require some contact between the sorcerer and the victim or with items that have been in close contact with the victim. To protect against such contact, Fore go to extraordinary lengths to guard their hamlets and living spaces from unknown and suspected sorcerers. Typically, the sorcerer acquires bodily exuvia (hair, fingernail clippings, spittle, feces, etc.), personal items (clothing, tobacco, etc.), or food leavings of the intended victim. These are then treated by physical manipulation and verbal spells that transmit the sorcerer's evil intent to the victim. Alternatively, a sorcerer may cast an evil spell on some common item (food, tobacco) and place it in a location where the intended victim likely will come into contact with it. Angered wives, too, are thought to employ the latter technique against their husbands by contaminating their food, water, or tobacco with debilitating menstrual discharge. If the sorcery assault is successful, the victim will sicken and, in some instances, die. (See Lindenbaum, 1979, pp. 60–64, for a list of Fore sorcery practices and likely disease diagnoses.) As might be expected, Fore people carefully dispose of all excreta and food leavings and closely monitor personal possessions.

With the arrival in Fore territory of Australian colonial authorities in the late 1940s and as the increasing influence of various outsiders spread during the 1950s and 1960s, many of the above aspects of Fore life began to change. At the insistence of the colonial administration, warfare was quickly curtailed and cannibalism had ceased throughout Fore groups by 1960. The delineation of group boundaries by colonial agents reduced the mobility of settlements. The planting of coffee trees as a cash crop also restricted residential movement as people needed to remain near their coffee groves. The extensive trade networks that before had integrated large regions disintegrated as retail outlets sprung up in government outposts to offer alternative manufactured goods. Men gradually abandoned the communal men's houses and took up residence with their wives and families. Big-men remained important leaders at the local level, but their political influence was subordinated to new regional and national political authority. The introduction and adoption of many new vegetable crops expanded the array of garden foods. Many people also now benefit from the provision of modern health care offered by foreign Christian missions and government aid posts. The threat of sorcery remains a concern although many Fore now have converted to Christianity and deny the efficacy of sorcery. Indeed, the Fore proved to be quite receptive to adopting many new ways of living and sociocultural change was rapid. At present, local loyalties remain, but all Fore now recognize themselves as members of a common ethnic group within the nation of Papua New Guinea.

THE CONTEXT OF HEALTH: ENVIRONMENTAL, ECONOMIC, SOCIAL, AND POLITICAL FACTORS

Prior to the late 1950s, the Fore had no access to modern medical treatment. In that context, they suffered from a number of infectious ailments common to other peoples living in the montane environment who shared similar living conditions. Important among these were tropical ulcers and yaws followed by upper-respiratory infections, bronchitis, and pneumonia. These latter diseases likely are attributable to, or at least exacerbated by, the hearth fires in the houses. Houses often are filled with choking smoke and people routinely enter houses in a crouch position and quickly sit down to reduce exposure to the smoke. Also, most adult men and a few women smoked a harsh home-grown tobacco in bamboo pipes. Gastrointestinal and diarrheal diseases also were common, a likely consequence of poor village sanitation and casual approaches to food handling and personal hygiene.

People shared their living space with their domestic pigs, and Fore notions of personal hygiene did not include frequent bathing or washing of hands. Other diseases observed were meningitis, tetanus, scabies, and leprosy. Most of Fore territory is above the altitudinal limit for malaria (about 1,500 m), but it did occur in a few deep valleys at the lowest elevations. Today, malaria is a common affliction that has been brought home by labor migrants returning from coastal work sites.

The Fore experienced several epidemics of introduced diseases that swept through the area prior to the arrival of the Europeans who brought them. In the late 1930s and 1940s, mumps, measles, whooping cough, and dysentery ravaged the population. In the 1950s, visiting indigenous police officers introduced gonorrhea. In 1959 and 1963, influenza resulted in many deaths, hitting children under five years of age especially hard. Although Europeans and their associates did introduce new diseases, early government patrols also eliminated yaws and gonorrhea among the Fore with penicillin injections.

Pigbel

Another illness that afflicted the Fore people is *enteritis necroticans*, called "pigbel."[2] It was associated with ceremonial feasts during which pigs were butchered and cooked in earthovens, and large quantities of pig meat typically were exchanged and consumed. During the preparation of the carcasses for cooking, meat often was contaminated with a strain of the Clostridia bacteria present in the bowels and feces of the animals. This was transmitted by various means to people during and after the feast. Importantly, children were given raw entrails to play with and small pieces of raw meat to chew on. Also, undercooked meat was not consumed for several days as it passed from hand to hand in extensive exchange networks. Symptoms of severe abdominal pain, bloody diarrhea, and vomiting appeared within a day or two of consuming the contaminated meat. Most cases were mild, but if severe cases were left untreated, gangrene would progressively affect the small intestine and could cause death. Pigbel was endemic throughout the highlands region, but was largely unknown in the lowland areas of New Guinea. It appears that this restricted distribution of pigbel to the highlands was related to a diet that relied heavily on one staple, sweet potato, which commonly provided some 75% of calories. This low-protein daily diet resulted in low levels of protein-digesting enzymes.

Also, sweet potato contains an enzyme inhibitor that is not destroyed at the temperatures usually reached in earthoven cooking. This combination of factors meant that the toxic protein secreted by the Clostridia survived in the intestinal tract of infected humans. Pigbel most commonly affected children between the ages of 2–5. The blood of most adults contains an antibody to the toxin, which suggests that mild exposure as children provided some immunity for adults. Today, pigbel infection is extremely rare. Researchers attribute this decline to a public health immunization campaign and to somewhat increased protein intake in the diets of children in the highlands (M. Alpers, personal communication, 2002).

Kuru[3]

The Fore are most well-known in the medical and medical anthropology literature as the unfortunate subjects of the neurological disease, called *kuru*, which is a Fore word meaning to shiver, shake, or tremble. *Kuru* is a disease of the central nervous system that is progressive and always fatal. Major symptoms progress from a slight unsteadiness when standing and walking to severe tremors and the inability to walk without total support. Eventually, as the tremors become increasingly severe, the patient is unable to sit up or swallow and loses control of all motor functions, including speech. Death typically occurs about one year after the onset of symptoms.

Since *kuru* was initially noticed by outsiders in the early 1950s, it has become one of the great medical mysteries of the 20th century. Early epidemiological evidence showed that the epidemic was confined to the Fore and their close relatives in neighboring groups. Also, certain families and hamlets were more affected than others, and women and children of both sexes comprised the vast majority of cases. It initially was proposed that *kuru* might be an inherited genetic disorder, but the high incidence and lethal nature of the disease made this seem unlikely. Ethnographic investigations in the early 1960s by Robert Glasse and Shirley Glasse Lindenbaum (Glasse, 1967) found that *kuru* had first appeared during the lifetimes of older informants. Also, the first cases had occurred among the North Fore with subsequent outbreaks spreading south, the same spatial pattern reconstructed for the adoption of endocannibalism. This indicated a potential infectious agent. Also, it was known that in the distribution of body parts, the brain was consumed only by women and children. This conformed to the fact that the overwhelming majority of *kuru* victims were women and children. Transmission experiments finally proved that an infectious agent was involved as D. Carleton Gajdusek and co-workers at the U.S. National Institutes of Health succeeded in transmitting the disease to chimpanzees with symptoms appearing after a 20-month incubation period (Gajdusek, Gibbs, & Alpers, 1966). An unusual "slow" viral agent with an extremely long incubation period was suspected. Transmission, however, still occurred after all nucleic acids were destroyed, a result that seemingly would eliminate normal viruses from consideration. Finally, in 1982, S. B. Prusiner determined that the likely infectious agent for *kuru* and other related diseases is a small proteinaceous particle he termed a "prion" (Prusiner, 1989). If correct, *kuru*, along with Creutzfeldt–Jakob disease, Gerstmann–Straussler–Scheinker syndrome, and fatal familial insomnia in humans, scrapie in sheep and goats, bovine spongiform encephalopathy in cattle, transmissible mink encephalopathy in mink, and chronic wasting disease in mule deer and elk, comprise a new category of prion diseases.

Kuru is anthropologically significant for several reasons. First, the high mortality among women devastated the household economy, which is based on gender complementarity. This skewed mortality pattern created an overall male:female sex ratio of 2:1 in the total affected population, and was as high as 3:1 in some hamlets. Men of marriageable age were unable to find wives and many married men were made widowers. Without wives, men had to perform all domestic tasks themselves as well as provide childcare. Also, many infants lost their mothers. Second, much social disruption ensued as suspected sorcerers were hunted and pleas were made for them to cease their destructive activities. Suspicions and open accusations tore at the very core of communal cooperation as the Fore increasingly feared each other. Third, the Fore attributed the disease to a powerful new kind of sorcery. They were not certain of how this came about or who was responsible, but the fact that local curers were unable to counter its lethal effects proved that it was a previously unknown evil force. *Kuru* was a crisis that threatened the very survival of Fore society.

As news of the epidemic spread, the Fore became infamous throughout the region for possession of a powerful and deadly kind of sorcery. Neighbors lived in fear that this force might be turned against them and took special precautions whenever it was necessary to visit Fore territory. At the same time, some assumed Fore sorcerers were hired by members of other groups to attack their enemies. To this day, people from neighboring groups enter Fore territory with caution.

The *kuru* epidemic peaked in the early 1960s when the number of cases annually was about 200. With the cessation of cannibalism, the number of victims began to decline. The Fore, however, attribute this decline not to the end of cannibalism, but to the fact that elders stopped teaching the younger generation how to perform this powerful type of sorcery. By the early 1990s, about eight cases a year were recorded. Today, annual victims number two or less, all people in their 50s who likely were infected as children and survived an incubation period of more than 40 years. While confirmed cases now are few and occur only among the South Fore, people still fear *kuru* and often incorrectly assume that the feverish shaking associated with malaria is an initial indication of *kuru*.

Medical Practitioners

In the past, the Fore relied on local curers to treat their various maladies. Curers were individuals, both men and women, who possessed the knowledge, techniques, and spiritual power to cure the afflicted. The Fore recognize two categories of curers: Bark Men and Bark Women, and Dream Men. Bark Men and Bark Women usually reside in the same parish as their patients and use forest medicines to treat minor ailments. They also are capable of intervening with ghosts and forest spirits on behalf of the sick. In the 1950s, new, more powerful curers emerged who were thought to be able to counter illnesses caused by sorcery, including *kuru*. These Dream Men, or Smoke Men, always lived in distant parishes and in many cases were not Fore. Relying on dreams and trance states induced by the rapid inhalation of tobacco smoke and the ingestion of psychotropic plants, they were able to see beyond the realm of ordinary reality. They also were capable of identifying the offending sorcerers by using various divination techniques. To counter the sorcerers' evil acts, they prepared curative meals of pork laced with masticated ginger and special tree barks. They also would attempt to relieve the suffering of their patients by shooting small arrows into the skin at locations where the pain was most acute. In the context of the *kuru* epidemic, Dream Men enjoyed much influence and acclaim and many acquired substantial wealth. Today, the Fore still visit local curers, but also seek, when available, modern medical treatment for their ailments.

Classification of Illness, Theories of Illness, and Treatment of Illness

The Fore recognize two major categories of illness based on assumed causation. First, there are illnesses caused by the malicious actions of other people. Such diseases, including *kuru*, serious respiratory infections, and liver disease, are the result of sorcery and are life-threatening. These can only be treated by the intervention of powerful curers. Second, there are several kinds of less serious illnesses that are attributed to less dangerous forces. Forest spirits, called *masalai*, may become angered when people intrude on forested areas that are set aside by all hamlets as spirit abodes. Diagnosed retrospectively, their anger may result in minor illnesses or injuries for the violator or some close relative, or slight physical impairments of a child of the violator. The spirit can be assuaged with ritual offerings of pig fat and the victim consumes a medicinal meal of pork and appropriate herbs. Ghosts of the recently deceased also may punish individuals who remove produce from the deceased's gardens. Guilty parties can expect to suffer nausea, weakness, and fainting spells. The offended ghost can be placated only by an offering of pork and pig blood poured into the ground at the head of the deceased's grave. Infringement of social rules governing appropriate behavior among the living may cause minor ailments to the offender or a close relative. Relief can be gained by giving compensation to the aggrieved party, who in turn will provide a meal of pork and bush medicines to restore the health of the stricken person. The victims of all of these illnesses that are not caused by sorcery can be restored to good health by an indemnity payment to the aggrieved party, be they angered *masalai*, spirits of the recently dead, or neighbors.

Sexuality and Reproduction

Sexual relations among the Fore remain guarded encounters. Given the ideology of female pollution, partners rendezvous outside the residential living space and away from food gardens to avoid subjecting others to potential contamination. Young people are expected to refrain from engaging in sexual intercourse until marriage, and mature, unmarried women are carefully supervised by family members. Transgressions, however, are known to occur. Adultery also is considered a serious offense and, when discovered, causes intense social disruption and occasional violence. Marriage typically occurs when women are in their late teens or early twenties and men are in their mid- to late twenties. Marriage involves a public ceremony that lasts for several days that includes feasting and payment of bridewealth from the family of the groom to the parents and siblings of the bride. This effectively transfers her reproductive potential to the groom's social group and compensates the bride's family for the loss of her labor. Newly married couples usually take up residence in the same hamlet as the husband's close relatives. Children are highly valued, both for the vibrancy and the potential longevity they add to their natal group. Barrenness is thought to be extremely unfortunate. Fosterage and adoption commonly are used to provide childless couples with children and occasionally to make certain that sibling sets include both males and females. This is thought to ensure the viability of households and assistance for parents as they age.

HEALTH THROUGH THE LIFE CYCLE

Pregnancy and Birth

The beginning of life for the Fore is a time fraught with uncertainty and pregnancy is a time of considerable anxiety for the expectant mother. In the past, there were no means of surgical intervention, so complications during delivery occasionally resulted in the death of both mother and infant. Also, infant mortality was high. Pregnant women continue normal activities until the birth is imminent. When contractions begin, a woman goes to stay in a birthing hut at the edge of the hamlet. Some women prefer to give birth alone, but most are attended by close female relatives, who eat the placenta following the delivery. The birth of more than one infant was thought to be an aberration and in the past only one child survived the birthing hut. After a period of confinement in the birthing hut, during which the umbilical cord dries up and detaches from the infant, women return to their houses and gradually resume most normal activities. Post-partum sexual taboos, however, ideally remain in effect until the child is named during a celebratory feast approximately one year after birth. Today, most women go to local clinics to deliver and if serious complications occur, they are transported to urban hospitals by ambulance or helicopter. This has resulted in a decline of infant and post-partum mortality. Some women now seek contraception to limit the size of their families.

Infancy[4]

Infants are highly valued and the subjects of constant attention. They are nurtured most closely by their mothers and are rarely away from physical contact with their mothers for more than a short period of time. Other women and older siblings willingly assist when asked to do so. Breastfeeding is on demand at all times, with infants sleeping next to their mothers, and continues for a minimum of two years. Supplementary feeding of masticated sweet potato begins at about 6 months of age. Weaning is a gradual process. Mothers may ignore nursing requests, divert their toddlers' attention toward other activities, or rub bitter leaves on their nipples. If a subsequent pregnancy occurs during this period, weaning will be more abrupt. Toddlers are never scolded or disciplined by anyone and their requests for attention and physical contact with adults and older children invariably are met.

Childhood

Childhood is a time of relatively carefree play, but gradually children are given more responsibility. Girls are integrated into female work roles as soon as they are physically able to imitate their mothers. Boys, on the other hand, are allowed to roam hamlet territory with play groups, often away from adult supervision, and rarely are asked to help with household chores. However, this life of little responsibility comes to an abrupt end when they are taken to live in the men's house and begin the initiation process. Children gradually are expected to act responsibly and follow the requests of adults. They will be admonished for unacceptable behavior, but physical punishment is rare.

Adolescence

Fore adolescents experience increasing expectations of adopting adult work and social roles. Girls are instructed in proper adult behavior and provide important assistance to their mothers in gardening and household activities. A young woman's first menstruation is cause for a celebratory feast sponsored by her family to which the families of potential husbands are invited. Such women are then instructed by senior women of their group in the proper comportment of mature, but unmarried, women. If they prove themselves to be well behaved and strong and diligent workers, they reflect well on their families and gain reputations as desirable future wives. Boys, by the same token, are under increased supervision by adult men during the male initiation process. They are taught the important lore thought to be known only by men, are expected to learn the techniques required to protect themselves from female pollution, and are instructed in the proper demeanor and fighting skills necessary for defending group interests.

Adulthood

In the past, adult men frequently succumbed to injuries sustained in warfare. Adult women were most likely to die during childbirth or from contracting kuru. More recently, local health has been influenced by the movement of Fore people as labor migrants to coastal work sites. Beginning in the early 1960s, Fore men entered the labor flow and were joined by wives and other family members in the late 1970s. While this reallocation of labor to external cash-earning activities brought money into the local Fore economy, it also brought malaria and sexually transmitted illnesses. Gonorrhea, syphilis, and trichomoniasis are now common, and chlamydia and herpes also are present. HIV/AIDS has just been recognized among the Fore, but the details of incidence remain sketchy. Fore migrant workers did return home with money, but also with new diseases.

The Aged

As Fore people age and infirmities accumulate, they gradually and reluctantly withdraw from normal adult activities. They are respected by younger adults, but also acknowledged as a necessary burden. Close family relatives provide them with the necessities of life, but the influence of elderly men in the political arena and elderly women in the domestic sphere eventually is assumed by the next generation.

Death and Dying

Deaths of the very young and very old and of the physically or mentally disabled usually are mourned only by family members. Such deaths are not unexpected and are not politically significant, so only small groups of the living are directly affected. However, the deaths of older children and of all able-bodied adults are cause for serious concern and in most cases are attributed to sorcery. Mortuary payments for the death of an adult man are made by his brothers to his maternal relatives and to relatives of his wife; on the death of an adult woman, payments flow from her husband and his kin to her maternal relatives. If the deceased was especially prominent, such mortuary payments can involve large amounts of food and money. Burial usually occurs within three days of death, but mourners stay as guests for two weeks and eventually are sent home after a large feast and the distribution of expected payments.

Following death, the ghost of the deceased is thought to reside in the sacred spirit grove of his or her hamlet.

NOTES

1. This description of Fore culture relies on Lindenbaum (1979, and personal communication, 2002), and M. Alpers (personal communication, 2002).
2. Information on pigbel is taken from Lawrence (1992) with additions from M. Alpers (personal communication, 2002).
3. This account of *kuru* relies on Lindenbaum (1979) and Alpers (1992).
4. The sections on Infancy and Childhood draw on information in Sorenson (1976).

REFERENCES

Alpers, M. P. (1992). Kuru. In R. D. Attenborough & M. P. Alpers (Eds.), *Human biology in Papua New Guinea: The small cosmos* (pp. 313–334). Oxford, U.K.: Oxford University Press.

Berndt, R. M. (1962). *Excess and restraint: Social control among a New Guinea mountain people*. Chicago: University of Chicago Press.

Gajdusek, D. C., Gibbs, C. J., Jr., & Alpers, M. (1966). Experimental transmission of a kuru-like syndrome to chimpanzees. *Nature, 209*, 794–796.

Glasse, R. M. (1967). Cannibalism in the kuru region of New Guinea. *Transactions of the New York Academy of Sciences, 29*, 748–754.

Lawrence, G. (1992). Pigbel. In R. D. Attenborough & M. P. Alpers (Eds.), *Human biology in Papua New Guinea: The small cosmos* (pp. 314–344). Oxford, U.K.: Oxford University Press.

Lindenbaum, S. (1979). *Kuru sorcery: Disease and danger in the New Guinea Highlands*. Palo Alto, CA: Mayfield.

Lindenbaum, S. (2001). Kuru, prions, and human affairs: Thinking about epidemics. *Annual Review of Anthropology, 30*, 363–385.

Prusiner, S. B. (1989). Scrapie prions. *Annual Review of Microbiology, 43*, 345–374.

Sorenson, E. R. (1976). *The edge of the forest: Land, childhood, and change in a New Guinea protoagricultural society*. Washington, D.C.: Smithsonian Institution Press.

French

Michelle Lampl

ALTERNATIVE NAMES

France, French Republic, Republique Française.

LOCATION AND LINGUISTIC AFFILIATION

At slightly less than twice the size of the state of Colorado, France is the largest Western European nation. The French boundaries cover approximately 547,000 km^2 in Europe and its overseas administrative divisions add approximately another 600 km^2. With about 3,400 km of coastline on its western border, France is a boundary of Western Europe, from the Bay of Biscay in the Southwest to the English Channel in the Northwest. France is located between the present-day countries of Belgium and Spain, it is Southeast of the United Kingdom, and borders the Mediterranean Sea on the Southeast, situated between the borders of Italy and Spain.

French is the national language, with a rapid decline in numerous local dialects and languages (e.g., Provençal, Breton, Languedoc, Alsatian, Corsican, Catalan, Basque, Flemish).

OVERVIEW OF THE CULTURE

With a population estimated at 59,765,983 in July 2002 (ratio of 0.95 males/females), the demographic structure is comprised of 18% aged 0–14 years (1.05 male:female ratio), 65% 15–64 years (1.00 male:female ratio), and 16% 65 years and over (0.69 male:female ratio). The population growth rate is 0.35%, with a birth rate of 11.9 births/1,000 population, death rate of 9.0/1,000 population, and a net migration rate of 0.64 migrants per 1,000. Immigration has increased since unification of the European Union in 1992, creating increasing social tensions in a country already faced with significant unemployment at that time. The life expectancy at birth is 79 years, with females averaging 83 years and males 75 years. The fertility rate is 1.7 children born per woman.

The HIV prevalence rate was 0.44% in 1999 and estimated at 130,000 individuals living with AIDS.

Literacy, defined as reading and writing over 15 years of age, is said to be 99%.

Having suffered extensive losses in terms of manpower and economic well-being during World Wars I and II, France is one of the most modern countries in the world and a leader among the European Union. Since 1958, it has had a presidential democracy which has provided greater stability than its previous parliamentary democracy. It is a republic, divided into 22 regions, including the territorial collectivity of Corsica and is subdivided into 96 departments. France includes four overseas departments (French Guyana, Guadeloupe, Martinique, and Reunion) and three overseas territorial collectivities (Mayotte, Saint Pierre, and Miquelon). In addition, there are approximately 12 dependent areas over which France has transnational disputes of sovereignty. There are numerous political parties in France of opposing values, with a president elected by popular vote for a five-year term.

France is an ethnically heterogeneous population, representing peoples from North Africa, Indo-China, the Basque region, Celtic, Latin, and Teutonic populations, and nomadic Gypsies. France is presently predominantly Roman Catholic (85%), with Muslims comprising about 5–10% of the population and Protestant, Jewish, and unaffiliated sects equally apportioned.

France's economy is composed of historically extensive government ownership, much of which is presently undergoing privatization. Unemployment has been high over the past decade. Steps to improve this include mandatory retirement with an extensive and expensive pension system. France's economy suffers from a 35-hour work week, lengthy paid vacations, and national medical care. These expectations incur high taxes. The labor force is comprised of 70% in services, 25% industry, and 4% agriculture. Approximately 23% of the population are white-collar workers and 30% of the working population are manual workers.

France was the world's fourth largest economic power in the mid-1990s and the world's number one destination for tourists. Its industries include automobiles, machinery, metallurgy, textiles, food processing, and tourism. Seventy-seven percent of its electricity is nuclear generated. Its agricultural products include wheat, cereal, sugar-beet, potatoes, wine grapes, beef and dairy products, and fish. It exports 61% of its products to the European Union and 9% to the United States. France is a trans-shipment point for, and consumer of, South American cocaine, Southwest Asian heroin, and European synthetics.

THE CONTEXT OF HEALTH: ENVIRONMENTAL, ECONOMIC, SOCIAL, AND POLITICAL FACTORS

In its geographic position in the midst of industrial Europe, France faces major environmental issues with health consequences at the present time, including forest damage from acid rain, air pollution from industrial and vehicle emissions, water pollution from urban wastes, and agricultural runoff. Its beaches receive garbage from dumping grounds off the coast by numerous neighbors and it was within reach of air-borne radioactive waste from Chernobyl.

France has a state-subsidized medical system in which people are free to choose their own providers and are reimbursed up to 85% of most costs. Doctors are concentrated in the cities, leaving many rural areas underserved with facilities that are inadequate. The health profile in France follows its economic and social position in the industrial world in which the principal causes of death are cardiovascular disease and cancers.

The modern-day country of France consists of numerous small communities, each with long histories of self-identity. Historical geopolitical disputes notwithstanding, these are expressed in regional specialties in terms of consumable products from bread and chocolates to wines and cognacs. Salts from a particular coastal area, prunes from Agen, cheeses whose flavors reflect what sheep, goats, or cows find to eat in a particular region— all reflect the unique features of local natural environments. In 1986, a French doctor turned archeologist recorded his medical observations from 20 years of medical work in the Berry countryside (Allain, 1986). In these memoirs, he outlined the influence that social and geographic location had on patterns of health. He emphasized what an intimate relationship between man and animals, both wild and domestic, had on people's understandings of their own illness and health. Speaking from the viewpoint of a physician in need of understanding his patients, he wrote about his experiences learning to simply read his patients' anatomy, so that he could provide medical care. When they said they had a pain in the "name of the father," and two nerves, jumping one under the other, they meant that they had a pain in the forehead and a musculoskeletal joint problem. Some of them had "nerves stronger than the blood," referring to a disequilibrium between two antagonistic forces that referred to a polymorphic assortment of maladies ranging from psychologically based digestive troubles to menopause. Local diagnoses were based on the color and fluidity of the blood, and bread and wine were thought to generate the force of the blood—but not if the bread was toasted.

Local remedies included those for bloody noses and menstrual bleeding, a premier occupation in their health world. Some of the remedies were considered to be magic formulas. Oral ulcerations in infants prompted treatment by sympathetic magic, employing a herb with white spots on its leaves, visually similar to those evident in the mucosa, cooked and tied to the head of the infant's bed. While secret remedies have a long history, they were banned by decree in 1926 (Warolin, 2002). However, while no doubt this created a point of conflict between urban and rural settings as regulations were enforced in towns, the ancient remedies did not disappear by decree. While pharmacists took over the role of dispensing remedies in urban environments, rural peoples did not change their treatment plans to follow (Lafont, 2002). In the absence of alternative medical care—the alternative being biomedicine—traditional remedies shared a place with pharmacists for many years. Today, local pharmacists are themselves facing an increasing share of their work being taken over by international pharmaceuticals, in an increasing international arena since 1992.

MEDICAL PRACTITIONERS

French medical practice focuses primarily on physicians who undergo traditional biomedical training. In contrast to the United States, where four years of medical school follow four years of undergraduate work, in France, medical training is initiated immediately after secondary school education is completed. Competition is intense for positions in this field. Physicians may become either generalists or specialists, with foci on individual body functions (e.g., neurology, cardiac and renal specialties, rheumatology), age-related care (e.g., pediatrics), and surgical versus non-surgical approaches (e.g., internal medicine) in accord with international biomedical practice. Medical licensure examinations are given and extensive clinical training is the norm.

Chiropractic medicine is not regularly practiced in France, where historically homeopathic approaches became more acceptable in the early part of the 20th century. Some individuals choose to specialize in homeopathic treatment approaches and undergo further training. These physicians are not seen as practicing or offering alternative medicine: homeopathy is a moderately mainstream medical approach.

Individuals specializing in musculo-skeletal problems, known as kinesotherapy, are popular and are often sought by individuals with back problems, in particular. Somewhat the equivalent of physical therapists, they share with other non-medically trained practitioners a more narrow focus on functional aspects of the body. Pharmacists have historically had a wide range of abilities to prescribe treatments, and a number of medications that require prescriptions in the United States are available over the counter in France.

CLASSIFICATION OF ILLNESS, THEORIES OF ILLNESS, AND TREATMENT OF ILLNESS

Biomedicine provides the primary construct of illness and health in France. In the midst of the increasing internationalization of scientific knowledge and medical care, there remain differences in concepts of what constitutes a disease and its treatment. There are diseases whose symptom complex is uniquely culturally constructed, and there are symptoms and signs that are not yet culturally constructed in the medical domain. In the first category is the uniquely French disease, spasmophilia; in the second, osteoporosis among elderly French.

Spasmophilia has been called a polymorphic disease with numerous symptoms. But the overwhelming one is fatigue: there is never enough sleep and individuals awake in the morning exhausted. Difficulties in temperature regulation, ranging from feeling too cold or too hot; difficulties in eating from anorexia to overeating; sensations from nervousness to great anxiety; and being barely able to breathe are all typical complaints of spasmophilics. Muscle cramps from the throat to the extremities, headache, lightheadedness, and Reynaud's sensations are often reported, in addition to heart palpitations, digestive problems, hypoglycemia, and allergy. People who have spasmophilia take care to watch for the signs. The scientific question is: What could be biologically responsible for all of these signs and symptoms? Are they of psychogenic origin or do they reflect a metabolic disturbance? Some researchers suggested that it was a problem at the level of mineral salt exchange at the cell membrane and treatment included calcium supplements. More recently, it has been suggested that the symptom complex is a synonym for hyperventilation syndrome (Delvaux, Fontaine, Bartsch, & Fontaine, 1998) or, alternatively, similar to chronic fatigue syndrome or fibromyalgia (Maquet, Croisier, & Crielaard, 2000) and may reflect magnesium imbalance (Durlach, Bac, Durlach, Bara, & Guiet-Bara, 1997).

Alternatively, there are signs and symptoms that people exhibit that are not necessarily considered a disease or treatable entity in France, in contrast with the United States. An example is osteoporosis, which has been viewed merely as the normal process of aging—and untreatable. When do the processes of aging become treatable clinical categories and how is cross-cultural scientific information filtered? The high incidence of osteoporosis is evident from walking down any French street: numerous elderly women and men exhibit the so-called "dowager's hump," characteristic of vertebral compression resulting from bone loss. The statistical incidence in France is presently unreported. It is not a disease. It is not a condition in need of identification or treatment. In fact, one study identified that few general practitioners investigate the presence of osteoporosis (less than 6% in one study), and correct treatment was carried out by less than half (Laroche & Masieres, 1998). This is notable in a population where several studies have reported rather high prevalence of vitamin D deficiency in both the general adult population (Chapuy et al., 1997) and elderly women in particular (Cals et al., 1996). Aging of the skeletal system is to be expected and is not medicalized at this time.

Health is not something that comes merely as the result of a doctor's visit on the occasion that one does not feel well. It is not something to reflect upon because of a lab report indicating a high cholesterol level. Bodily well-being is a part of daily lifestyle. Taking care of oneself is not necessarily goal-oriented toward "health," but "well-being" is often a goal that has the side-effect of serving health. Teas, wines, and cognacs bolster and ameliorate; month-long vacations and rests at thermal baths rejuvenate. Social class has an important influence on daily life in France, and influences who has access to information, facilities, and cultural capital—but there are many aspects of lifestyle that cross these boundaries in modern-day France. National health insurance covers the greater part of payment for most culturally acceptable treatments.

Personal theories of health and well-being focus on all manner of daily practices. The importance of consumables set the background for a strong presence in France of a reliance on remedies in lieu of antibiotics. Too much to eat and struck by a "liver crisis" (crise de foie)? Try some champagne, it is also very good for a hangover. "Water? I have heard that some people drink it." Oysters are good for children weakened by disease, and are part of the sea cures. Indeed, oysters are filled with mineral salts—calcium, sodium, magnesium, iron, zinc, and copper—proteins, lipids, and vitamins A and B.

The body is constructed in a distinctly French manner, or habitus as Bourdieu put it (Bourdieu, 1990a, 1990b, 1999). Unlike the mechanically based approach of American medicine, where fixing the objectified body machine is the goal, whether through antibiotics, surgery, or therapy in the midst of a distinctive Cartesian dilemma, the French tend to their bodies outside of the biomedical domain on a daily basis as a lived experience. To be sure, the French access doctors and hospitals when called for, in a superb medical milieu where some of the great breakthroughs have occurred—from radiology and vaccines to HIV. But there is also a level of everyday bodily experience that involves responding to sensations in terms of culturally appropriate alternatives through which the French take care of their sense of well-being, their bodies, and in some cases, their "health."

There is an underlying sensibility that it is only natural for the body to become susceptible to discomfort as normal experience displaces the body's balance. From overeating and drinking, from stress to aging, people experience discomfort and require care. In an ancient society, food can poison people, weakness of the body is to be expected, and it is important that one knows what to do for these life experiences oneself.

Some people share in the sense of control over "health" as medically conceived and attend to physical exercise—there are parks filled with weekend runners and fitness centers that welcome people for work-outs. There is a strong influence of American lifestyle in these activities, and, thus, they are not shared by all people. It is entirely acceptable to claim that you are "allergic to exercise" and continue to enjoy a fine foie gras when it is available, regardless of one's last LDL blood level. C'est la vie.

What is called psychoneuroimmunology in the United States stands in stark contrast to a value system that involves life at a slower pace. One suspects that the reputed health benefits referred to by the "French paradox" (Perdue, 1992) and included in the "mediterranean diet" are more than merely drinking red wine and eating foods with olive oil. There are small reference texts that document the appropriate choice of wines—not to match the food, something every good wine purveyor can assist with—but to match bodily needs and contribute to health (Saint-Clair, 1993). Researchers have identified the potential chemical benefits of wine and epidemiological studies verify a correlation between daily wine consumption and decreased cardiovascular disease risk among middle-age and elderly men in France (Renaud, Gueguen, Schenker, & d'Houtaud, 1998). But, it is difficult to disentangle the direct effects of the wine from the lifestyle in which it is consumed. It is unlikely that these effects would be mimicked by extracting chemicals and taking them as a pill, transforming the French wine effects to an American lifestyle. Attention to details that keep people in touch with their senses and the natural world around them cross economic boundaries. Differences in social class notwithstanding, at the present time it is almost a cliché to say that for the French food consumption is more than eating. Much of health is grounded in gastronomy.

Depending on one's bodily state, a meal can be prepared to match sensations from head discomfort to stomach uneasiness based on the appropriate collection and sequence of herbs in the dish or liquid consumed. Infusions can be brewed to alleviate urinary troubles, joint pains, or lift the spirits, calm the nerves, and encourage sleep. Daily personal experience of the body can dictate subtle choices following the evening meal. There are a number of fermented fruit and herb-based drinks, each a specialty of a particular region, with a long history of local production and common consumption. Today, some are said to be best consumed before a meal, as they aid in relaxing the individual and assisting in the transition to a calm after work, and an enticement of the appetite. Others are known to aid in digestion. Some of these eau de vie (literally, water of life), or digestifs, may be so named not merely by chance. While the grand cognacs are enjoyed on special occasions, more modest eau de vie are consumed regularly. Some are reserved for specific needs—for example, when the meal may have contained a questionable food item. Numerous brandies and eau de vie are part of family heritage as being the *specific* amelioration for individual circumstances. To make an aperitif at home: macerate medicinal plants, add spices and aromatiques, ferment. During the harvest season, substitute fruit—each has a focal effect on the body's well-being. These practices are part of people's lived experiences.

At more than 100 natural hot springs, there are sites that serve as therapeutic centers, available to accommodate individuals with symptoms ranging from respiratory difficulties, skin diseases, gynecological disorders, arterial problems, and psychosomatic ailments. The water therapy may involve bathing, inhaling the vapors, drinking the waters, or application of the waters to the ailing body site. The daily consumption of mineral water from various locales, such as "Perrier," "Vichy," "Vittel," "Badoit"—each slightly different in terms of the minerals that it contains—speaks to the common acceptance of this approach. Each of the water sources is as distinctive as the result of its geological context. Containing different concentrations of local minerals, each of the spas is seen as benefiting distinctly different bodily conditions. For arthritis one might choose the conditions of the baths at Ax-les-Thermes, whereas problems of infant development might be better treated by the salts in the water found in Jura. These practices tie people to the earth and the correspondence of bodily sensations to regional distinctiveness in mineral baths conforms to a mind-set oriented toward choosing regional wines for the character each uniquely possesses due to the soil and climate conditions in the locale of origin.

French scientists have asked, "Do these water therapies really work?" One conclusion is that it is difficult to say. In their review of a number of research studies, Schilliger and Bardelay (1990) noted that very few have been prospective case–control studies with objective outcome criteria. This includes a large-scale prospective study conducted by a branch of the national medical service of more than 3,000 patients. The demographic profile of the people in this study is useful, providing insight into who seeks these treatments. Three groups were included in the study. The first had a mean age of 36 years (a group of children 5–15 years with their parents, 30–65 years), nearly evenly divided by sex, presented for treatment of chronic sinusitis and bronchitis, asthma, and recurrent ear infections. A second group of men, average age of 62 years, were treated for arterial problems, and a third group of women, mean age of 51 years, were treated primarily for urinary problems. The results of the study stated that more than 80% of these people reported being helped by the end of their first cure. However, no control groups were used and no quantifiable criteria of improvement were employed.

One of the few studies meeting objective research criteria was conducted at the Royat Spa, which receives 25,000 people annually. A prospective study of 140 people with a mean age of 57 years focused on a clinical versus control sample with vascular problems of the leg. They reported that more than half of the clinical sample experienced amelioration during the treatment by objective criteria (Fabry, Pochon, Trolese, & Duchêne-Marullaz, 1985).

Schilliger and Bardelay (1990) note that a number of negative effects have been reported from spa visits, ranging from exacerbation of symptoms to contagious respiratory outbreaks. Critics question their utility (Brockliss, 1987; Weisz, 2001). Nonetheless, national health insurance continues to support these options, and surveys report that many people say they feel better. Those who do not are not necessarily respondents.

In summary, the lived well-being of the body occurs largely outside of the medical domain for many French. Taking a cure and healing a discomfort is not merely an outcome, it is an ongoing process, a lifestyle of bodily experience. In a world filled with media regarding the health effects of smoking, drinking, and fat consumption, a study of more than 13,000 hypertensive French men in 1999 and 2000 found that fewer than one in four followed the recommended lifestyle changes of weight loss, decreased salt and alcohol consumption, and increased physical activity in order to lower their blood pressure (Tilly, Guilhot, Salanave, Fender, & Allemand, 2002). Physicians prescribed medications appropriately only two thirds of the time, and more than one fourth of these included a potentially contra-indicated drug. The study concluded that it was necessary to increase communication between healthcare administrators and clinicians to better educate patients to take responsibility for their health. It will be interesting to see if the French lifestyle will change for health, when so much of well-being is based on lifestyle. Will cultural values and ideas about the nature of human life alter, or are the outcomes of the well-lived life, well, simply, la vie?

Homeopathy

Homeopathic physicians are trained in the same way as all other physicians in France and then undergo further specialization. They learn the unique diagnostic and treatment approaches of homeopathy, first developed by Hahnemann in 1810 (Hahnemann, 1991). It is necessary for the homeopathic physician to get to know the patient, for even the same symptoms of disease may require different treatment depending on the personality of the patient. The diagnosis is fundamentally the treatment in homeopathic medicine. By identifying an individual's temperament, much can be ascertained about immune function and, with time, the appropriate chemicals will restore balance to an individual and, thus, one's sense of well-being. Each individual is treated in accord with his or her internal milieu. It is critical during the initial interview to identify the nature of this temperament. This requires a long conversation with the patient, directed toward the identification of his/her individual responses to temperature, stress, foods, and daily life. The homeopathic physician develops a profile of the patient based on a typological classification. While the reason for an individual seeking medical attention matters, it is not the core of the homeopathic interview. This is at first glance a singularly culture-bound diagnostic approach, and contrasts with the checklist approach posed by many doctors, where diagnosis follows a rote pattern, and treatment is focused on scientific prescription and crisis management. Homeopathic approaches follow a style that works quite well in France for a number of people.

The central tenet of homeopathic treatment is similarity, or like treats like: remedies are chosen to match the distinctive features of plants or minerals with the characteristics of the individual patient and his/her disease/discomfort. For example, herpes simplex virus of the lips (HSV II), results in chronic cold sores that are notoriously recalcitrant to treatments in biomedicine. This is treated homeopathically with success by small doses of *rhus toxicodendrum*—a derivative of poison ivy—a treatment chosen because the effects of the plant mimic the disease that is being treated. Additionally, several chemicals may also be prescribed to an individual apart from anything directed toward the presenting complaint. The goal of this further treatment is to improve the patient's mental and physical well-being and help him/her to avoid disorders for which he/she may be constitutionally and temperamentally predisposed (Vannier, 1992).

Homeopathic treatment involves small balls, approximately 3 mm in diameter, that are packaged in blue plastic vials (about 50 mm long by 15 mm in diameter). The vials have carefully constructed openings that permit only a single round pill to escape at one time. The pills must be deposited under the tongue, directly from the vial with no contact from the hands for maximum efficacy—and not taken in proximity to eating mints, drinking coffee, or brushing the teeth.

The principle of homeopathic treatment is microdoses, and traditionally is based on molecular-level ingredients. The packaging permits people to carry them in their pocket or purse and use them when they feel the need. The contents are chemicals, ranging from plant extracts to most of the elements of the chemical chart. Many people have a *Family Guide to Homeopathy* (such as Horvilleur, 1981) which they can consult for information on appropriate treatments for minor aches and pains of daily life. The blue homeopathic vials are available without prescription at homeopathic pharmacies. In consultation with a homeopathic physician, an individual can learn about his/her own temperamental type and self-medicate as life occurs.

There has been significant scientific debate regarding whether or not homeopathic remedies really *work* (reviewed in Poitevin, 1987), with numerous critiques matched by reports of efficacy in clinical trials (Kleihnen et al., 1991) with patients who suffer from symptoms as diverse as allergy (Reilly, Taylor, McSharry, & Aitchison, 1986) flu (Ferley, Zmirou, D'Adhemar, & Balducci, 1989), and anxiety (Benzecri, Maiti, Belon, & Questel, 1991). Like water therapy, there are people who simply feel better taking homeopathic treatments. Does this make homeopathy an example of the placebo effect? The people who take it say, "No, and anyway, does it matter?" Since often physicians do not know the underlying cause of people's symptoms and, therefore, how to treat them, why worry?

Homeopathic treatment often helps where allopathy fails. One woman reported that she had had recurrent sore throats prior to taking homeopathic remedies. After two years of seeking help from traditional physicians, she saw a homeopathist. She feels better: her discomfort "*was* in her throat, not in her head." This is a case where general practitioners did not know precisely what was wrong with the patient because a causative pathway has yet to be established to explain her experience.

There is an underlying sensibility that it is only natural for the body to be pushed off balance from overeating and drinking, from stress to aging; people expect to experience discomfort. In this sense, homeopathy is not an alternative medical system, a non-mainstream approach to cure. It is not likely that it can be exported and fill the gap where biomedicine leaves people's needs unmet more generally. It is tied into an essentially culturally distinctive lifestyle of individual and social experience. A culturally based experience of everyday life based in feelings and bodily senses inform our perceptions of health and ill-feeling, and our attempts at sensory alleviation or intensification. Small doses make sense in a society where local historical traditions design reactions to subtle sensibilities. It would be less likely to be effective in a society where "bigger is better" and "if a little is good, a lot must be better," which drives many people to take mega-doses of vitamins, for example, in the United States.

Homeopathic diagnosis gives voice to people's personal concerns and their theories about what is bothering them. They do not have to fit a biomedical conceptual category to be validated and receive treatment. The sensations experienced by the body are not reconstructed into a medical paradigm. Homeopathic treatments permit the individual—who knows himself or herself best—to maintain a sense of privacy regarding one's well-being and sensations. The over-the-counter availability of the blue vials further permits a person to respond to body fluctuations on a daily basis with an acceptance of individual strengths and weaknesses. The self remains private.

Perhaps, also, homeopathy finds agency because it continues a long tradition of home-based treatments, from foods to waters. It upholds the values of individual privacy in intimate matters, and the belief that health and well-being change with the natural order of life. The homeopathic interview fits into the social mores of polite interaction.

In terms of health-seeking, it has been reported that homeopathy is mainly used in mental, infectious, and rheumatological disorders where it is a useful alternative and avoids the abuse and adverse effects of sedatives, antibiotics, and anti-inflammatories (Colin, 2000). Some studies have shown that women use homeopathic treatments more than men (Bancarel et al., 1988), but the reasons are not clear. Perhaps this may be related to women seeking more healthcare than men in general.

In the year 2000, an investigative overview (Chaufferin, 2000) identified that the price of homeopathic products was only a quarter of the average cost of all health expenditures from insurance for reimbursable medicines. The total reimbursement for allopathic products, by contrast, was three times more than homeopathic. These figures were not explained by patient profile or diseases treated. Overall, homeopaths' annual financial outlays were one-half of general practitioners'. In terms of patient satisfaction, 87% of the patients whose physicians prescribed a homeopathic treatment did not seek a second opinion or further treatment. It is notable that since the patient is the decision-maker regarding treatment, issues of compliance are not significant in homeopathy.

SEXUALITY AND REPRODUCTION

Sexuality is a normal part of the life cycle and both advertising and social norms accept the body as a source of enjoyment. On average, sexual activity begins in adolescence and while a responsible and healthy emotional state of individuals is a stated concern, there is no requirement that love and marriage either precede or accompany sexual activity. While marriage is the expected social domain for child-bearing, a number of young French couples live together and have children prior to marriage. While many French would question women who never marry, but take occupations in lieu of child-bearing, this status is quite acceptable and the role of women in powerful positions depends greatly on the work domain and socioeconomic class.

The French have a strong sense of privacy in matters of their personal life. They are also quite tolerant of the right of individuals to choose a lifestyle that is best suited for themselves. Extra-marital affairs are not uncommon, although they are expected to be discretely managed.

The fundamental right to contraception was first affirmed in a public campaign in 1981–82. In 1992, a second national campaign was launched as the result of AIDS and focused on public education regarding condoms. In recognition of the 25th anniversary of the law establishing the fundamental right to contraception, a third national campaign was launched in 2000. With the slogan, "It is up to you to choose your contraception" appearing in advertisements on TV, radio, and in the press, the combined efforts of the Department of Health and the minister of education aimed to educate adolescents on sexuality. The goal was designed to inform teenage girls and women of the availability of state-funded contraception options, including the morning-after pill. The goal was to reduce the abortion rate among adolescents in France and in women in the French protectorates. For example, in the mid-1990s, in Guyana, 10% of births were to minors and 1% those were girls less than 15 years of age. The number of voluntary interruptions of pregnancy ranged from 73% in Guadeloupe to 30% in Guyana.

While great controversy from parents surrounded the reforms of 2000, 66% of the population supported them. A pocket guide to contraception was distributed to high school students and France lifted requirements for females under 18 to have parental consent for abortions or a morning-after pill. The timing of legal abortions was extended from 10 to 12 weeks to bring France in line with other countries in the European Union, and pills became available without prescription in pharmacies across France.

Presently, between the ages of 20 and 44, more than two out of three women use contraception. This is one of the highest rates in Europe. However, the Demographic Institute at the University of Bordeaux has identified that there is great regional and social inequality in access to contraception. For example, only 17% of public facilities propose RU 486 to women, compared with 60% of private facilities.

The effectiveness of these campaigns has been followed: while only 8% of women chose to use condoms in 1987, as a result of numerous advertising campaigns, 87% of young women and those not in permanent relationships chose this method in 1998. This is replaced by the pill in a stable relationship. However, about 15% of young women have no protection during their first sexual encounter. In fact, of women aged 20–24, 56% stated they used nothing if their first experience was at ages 14–15; and 23% if the ages were 18–19. The average age of first sexual experience is about 17 and it is estimated that about 10,000 young women each year have an unwanted pregnancy.

In 1994, two out of three women between the ages of 20 and 44 used contraception. The pill was the most common choice, and 58% of younger women (aged 20–24) were the primary users. Sterilization was the second choice overall (16%), chosen primarily by women between the ages of 35 and 44. The third choice was condom use (5%) and 31% used nothing (source: Demographic Institute of the University of Bordeaux).

According to a popular survey published in the magazine *Marie Claire* in 1999, while 8 out of 10 women believe that contraception is a liberty, one in three does not want to take the pill because she believes that the pill makes women fat. Only 4 in 10 young girls and 2 in 10 women were aware that the pill was available without prescription. Four out of 10 women responded that they believed withdrawal to be an effective mode of contraception, and 2 out of 10 thought that temperature taking was effective in preventing conception.

Health through the Life Cycle

Pregnancy and Birth

In the mid-1990s, there were 220,000 voluntary interruptions of pregnancy per year for 730,000 pregnancies. On average, each woman has had an accidental pregnancy in her life and one in two was interrupted (Demographic Institute, University of Bordeaux, 1994). These statistics prompted the increased efforts at public information regarding contraception, as noted.

For couples who want to have children but are having difficulties with infertility, France is one of the most advanced scientific communities and in vitro procedures are available. These, however, are confined to urban areas on central medical research teams.

Pregnancy is a normal event in a woman's life and is respected, supported, and culturally valued. Women do not hide their bodies and clothing that aims to follow fashion is available. Women follow the advice of their female relatives and friends regarding their health and well-being, and like other biomedically oriented countries, seek prenatal care for the best interests of their infants. With national health insurance, this care does not discriminate on economic grounds and is more readily available in urban areas than rural areas. Public health campaigns identify the risk to the fetus of AIDS, but fewer campaigns identify smoking or alcohol as potential risks.

Hospital births are the norm at the present time, with women speaking out against the medicalization of birth in recent years. Citing that medical interventions in France were the highest in Europe in 1998, with 85% of pregnancies employing the medical arsenal, feminists are now demanding greater consideration of midwifery and personal choice in birthing (perinatalite.chez.tiscali.fr).

Infancy

The birth rate in 1995 was about 12/1,000 individuals in France. As a modern European country, most births occur in hospital with medical attention and women are discharged approximately three days later, returning with their infants to their homes. For the predominantly Catholic population, this is an opportunity for family and friends to celebrate the arrival of a new family member and a christening follows. Godparents are chosen for the child from amongst family friends and relatives, and depending on the economic circumstances of the family, a party with various levels of decorum occurs. A popular mark is the distribution of candy-coated almonds—pink for girls and blue to announce a boy. Small shops exist that specialize in this confection for births, weddings, and other formal celebrations.

Infants are valued as small, dependent, and temporarily incompetent members of society who need to be cared for and who deserve patience while gaining the ability to learn how to become contributing members of society. They receive care from both parents, as well as doting grandparents, who often contribute significantly to caretaking. Infant feeding follows contemporary industrial countries' notions of what is best, with breastfeeding supplemented by infant foods within the first 6 months the most common pattern at the present time. Weaning normally occurs by one year of age and toilet training follows as soon as the child can manage. Social class differences, based on ethnic background and economic levels, affect the early patterns of care-taking and expectations, but appropriate behavior is valued in the larger society and children receive strong negative reinforcement for public displays of emotion beginning at early ages.

The primary health problems of infants follow other industrial countries, with ear infections and upper respiratory complaints at the top of the reasons for seeking medical care. Asthma and allergies have recently been noted to be on the rise.

Childhood

French children are expected to adopt appropriate behavior and customs from an early age. Moderate physical and negative verbal discipline are the norm as demands for children to conform to behavioral rules are clarified. A narrow range of individual liberties are tolerated as cultural values are inculcated.

Health problems of French children parallel industrial countries at the present time, with infectious diseases leading the common cause for pediatric consult. Childhood cancers of both tissue and blood are significant chronic diseases. French children seem to have poor eyesight diagnosed at an earlier age than their counterparts elsewhere, but it is unclear if this is due to diagnostic concerns or biological predisposition.

A cultural preference for males is sometimes reported, but is not universal. At times of economic/food shortage, meat, when available, was preserved for the boys and men who need it for strength, said one woman of her childhood. Girls received broths and vegetables while the heartier foods were saved for the males. No overt sex biases are evident in child-rearing practices and child abuse has not received any large measure of public health attention.

Adolescence

By the age of 12, French adolescents are mature young adults, with expectations of taking on fully adult roles in a short time. They are in training for their future jobs with a choice already being made in terms of the schools that they attend. Adolescents take responsibility in the home and family and are learning about the larger sociopolitical sphere in which they live. They are becoming sexually active and part of the school curriculum is designed to provide public health information about safe sex practices. Adolescents are expected to behave like adults, although parents are not surprised to find them having difficulties at times in this role.

The Aged

Aging is culturally reinforced by mandatory retirement at the age of 60. Seen as a socially responsible action to provide jobs for the younger generation, this practice moves an active individual into a category of retiree and forces them to adjust their lifestyle to new economic and social circumstances. For some people, this is a "new" time of life, a socially approved change of job and focus and an opportunity to explore new experiences. For others, this can be a serious economic hardship and the severing of ties with work-related social contacts can be isolating. While no formal epidemiological studies have been undertaken nationally to investigate the health consequences of taking retirement, it would be interesting to follow patterns of alcohol consumption and depression amongst the aging population.

There are a number of retirement homes that accept individuals as they become infirm. Many families, however, take care of their aging relatives themselves, with primary responsibility falling on eldest daughters as caretakers. Respect and consideration of aging family member's wishes is a priority in many families.

Dying and Death

Death is a natural part of the life process and the importance of religious faith in many people's lives helps them through the grieving process. A predominantly Catholic society, the church organizes death rituals and modern mortuary practices are followed. Each small community has its local cemetery, where people return to be buried with their families, and where their graves are visited by surviving family members. While appropriate grieving periods are socially expected, a surviving spouse is not expected to forfeit his or her remaining life in memory of the deceased. Re-marriage occurs frequently and is encouraged. With a history of a century of significant losses in war, each small village has a memorial to the dead from each battle in the center of town. These are carefully tended by local residents.

REFERENCES

Allain, J. (1986). *Anatomie, physiologie et pathologie du Berrichon vu par lui-meme.* (pp. 17–22, pp. 30–35).

Bancarel, Y., Blanc, M. P., Charrin, G., Chastan, E., Coldefy, J. F., Cottes, L. et al. (1988). Study of drug consumption in a working environment in France. *Fundamental Clinical Pharmacology, 2,* 37–46.

Benzecri, J.-P., Maiti, D.-D., Belon, P., & Questel, R. (1991). Comparaison entre quatre méthodes de sevrage après une thérapeutique anxiolytique. *Les Cahiers de l'Analyse des Données, 16*(4), 389–402.

Bourdieu, P. (1990a). *The logic of practice.* Stanford, CA: Stanford University Press.

Bourdieu, P. (1990b). *Distinction: A social critique of the judgement of taste.* Cambridge, MA: Harvard University Press.

Bourdieu, P. (1999). *A critical reader* (R. Shusterman, Ed.). London, UK: Blackwell.

Brockliss, L. S. (1987). Taking the waters in early modern France: Some thoughts on a commercial racket. *Soc Soc Hist Med Bull (Lond), 40,* 74–76.

Cals, M. J., Bories, P. N., Blonde-Cynober, F., Coudray-Lucas, C., Desveaus, N., Devanlay, M., et al. (1996). Reference intervals and biological profile in a group of healthy elderly population in the Paris region. *Annales de Biologie Clinique (Paris), 54,* 307–315.

Chapuy, M. C., Preziosi, P., Maamer, M., Arnaud, S., Gala, P., Hercberg, S. et al. (1997). Prevalence of vitamin D insufficiency in an adult normal population. *Osteoporosis International, 7,* 439–443.

Chaufferin, G. (2000). Improving the evaluation of homeopathy: Economic considerations and impact on health. *British Homeopathic Journal, 89*(Suppl. 1), S27–S30.

Colin, P. (2000). An epidemiological study of a homeopathic practice. *British Homeopathic Journal, 89,* 116–121.

Delvaux, M., Fontaine, P., Bartsch, P., & Fontaine, O. (1998). Tetany, spasmophilia, hyperventilation syndrome: Theoretical and therapeutic synthesis. *Revue Medicale de Liege, 53,* 610–618.

Durlach, J., Bac, P., Durlach, V., Bara, M., Guiet-Bara, A. (1997). Neurotic, neuromuscular and autonomic nervous form of magnesium imbalance. *Magnesium Research, 10,* 169–195.

Fabry, R., Pochon, P., Trolese, J. F., & Duchêne-Marullaz, P. (1985). Variations du périmètre de marche et des index de pression avant et après épreuves de march mesurés à un an d'intervalle chez 140 artériopathes traités á Royat. *Cahiers d'artériologie de Royat, 12,* 78–82.

Ferley, J. P., Zmirou, D., D'Adhemar, D., Balducci, F. (1989). A controlled evaluation of a homeopathic preparation in the treatment of influenza-like syndromes. *British Journal of Clinical Pharmacology, 27,* 329–335.

Hahnemann, S. (1991). *Organon of Medicine* (J. Künzli, A. Naudé & P. Pendleton, Trans.) London: Victor Gollancz.

Horvilleur, A. G. (1981). *Guide Familial de l'Homéopathie.* Paris: Hachette.

Kleijnen, J., Knipschild, P., & Riet, G. (1989). Clinical trials of homeopathy. *British Medical Journal, 302,* 316–323.

Lafont, O. (2002). Medicines for towns and medicines for countrysides. *Rev Hist Pharm (Paris), 50,* 211–220.

Laroche, M., & Masieres, B. (1998). Does the French general practitioner correctly investigate and treat osteoporosis? *Clinical Rheumatology, 17,* 139–143.

Maquet, D., Croisier, J. L., Crielaard, J. M. (2000). Fibromyalgia in the year 2000. *Revue Medicale de Liege, 55,* 991–997.

Perdue, L. (1992). *The French paradox and beyond.* Sonoma, CA: Renaissance.

Poitevan, B. (1987). Le devenir de l'homéopathie. Paris: DOIN Ed.

Reilly, D. T., Taylor, M. A., McSharry, C., & Aitchison, T. (1986). Is homeopathy a placebo response? Controlled trial of homeopathic potency, with Pollens in hayfever as a model. *The Lancet, 18,* 881–886.

Renaud, S. C., Gueguen, R., Schenker, J., & d'Houtaud, A. (1998). Alcohol and mortality in middle-aged men from eastern France. *Epidemiology, 9,* 184–188.

Saint-Clair, M. (1993). *Vins medicinaux et elixirs de santé.* Colmar: S.A.E.P.

Schilliger, P., & Bardelay, G. (1990). *La cure thermale.* Paris: Editions Frison-Roche.

Tilly, B., Guilhot, J., Salanave, B., Fender, P., & Allemand, H. (2002). Management of severe hypertension in France in 1999 and 2000: Intermediate results of a health insurance intervention program. *Archives des Maladies du Coeur et des Vaisseaux, 95,* 687–694.

Vannier, L. (1992). *Typology in homeopathy* (Marianne Harling, Trans.). Beaconsfield, U.K.

Warolin, C. (2002). Secret remedies in France until abolition in 1926. *Rev Hist Pharm (Paris), 50,* 229–238.

Weisz, G. (2001). Spas, mineral waters, and hydrological science in twentieth-century France. *Isis, 92,* 451–483.

Fulani

Kate R. Hampshire

ALTERNATIVE NAMES

Fulani is a Hausa term, which is also commonly used in English to describe this ethnic category. With the exception of Nigeria, most Fulani live in Francophone West Africa, where they are widely known by the French (and Wolof) terms *Peul* or *Peulh*. Fulani living in the westernmost parts of the Sahel (Senegal, Gambia, Sierra Leone, and Guinea) are usually referred to by the Manding word, *Fula*. The Fulfulde term *FulBe* (sing. *Pullo*) is often equated with Fulani, although it actually refers only to high status, pastoral Fulani.

LOCATION AND LINGUISTIC AFFILIATION

The Fulani are distributed over a very large geographical area, essentially spanning the West African Sahel, from Senegal and the Gambia, as far east as Chad and the Sudan. There are Fulani in every West African state, but in each country they constitute a minority of the population (Dupire, 1970; Riesman, 1992). Because of this wide geographical dispersion, it should be borne in mind that some beliefs and practice vary from place to place, and generalization is thus problematic.

Estimates of Fulani population size are problematic for several reasons. First, national censuses are often inadequate data sources because they frequently under-enumerate nomadic and semi-nomadic groups, and not all censuses include questions on ethnic identity. Second, and very importantly, the Fulani ethnic category is fluid and its boundaries are fuzzy (e.g., Burnham, 1996). That said, estimates range from 6 million in the early 1960s (Dupire, 1963) to 9 or 10 million today (Riesman, 1992). The largest national group of Fulani is in Nigeria (some 4.8 million) (Riesman, 1992).

The main Fulani language is *Fulfulde* (known in some parts of Senegal and surrounding areas as *Pular* or *Fula*). Because of the wide geographical dispersion of Fulani, the Fulfulde of different regions varies, often incorporating words of other local languages. However, Fulfulde is still more or less mutually comprehensible by speakers from nearly all areas, implying that geographical expansion was relatively recent. It should be noted that some non-Fulani groups have adopted the Fulfulde language (Burnham, 1996), and that some Fulani speak other languages, such as Hausa in Nigeria (Riesman, 1992).

OVERVIEW OF THE CULTURE

Pastoralism and the Importance of Cattle

The Fulani are the major cattle-herding group in West Africa. Much of the Fulani tradition is tied up with nomadic pastoralism, and the importance of cattle features prominently in most portrayals of Fulani society (Bonfiglioli, 1988; Hopen, 1958; Stenning, 1959). Cattle are seen not only as economically important, but as a social necessity, upon which much social organization is based (Hopen, 1958; Riesman, 1977; Stenning, 1958). Cattle are also considered to be inextricably bound up with FulBe identity, such that their loss represents something far more serious than economic or social insecurity.

However, by no means are all Fulani today are pastoralists. Particularly since the serious droughts of the 1970s and 1980s, many Fulani now combine pastoralism with other economic activities, and some have forsaken pastoralism altogether and live in cities (De Bruijn & Van Dijk, 1995; Maliki, White, Loutan, & Swift, 1984). Neither is this necessarily a new phenomenon: historically many Fulani may have moved in and out of pastoralism (Raynault, 1997).

Social Organization

Fulani society is hierarchically organized into "castes" associated with social status and occupation. The main categories are *FulBe*: free-born, pastoralist Fulani, and their erstwhile slaves or serfs: *RiimaaiBe* or *MaccuBe*. In most areas the system of slavery has been formally abolished, but the hierarchical relationships remain. In addition, there are other castes associated with particular trades, such as blacksmiths, merchants, bards, etc.

Fulani operate an essentially patrilineal descent system, with patrilocal residence. On marriage, a woman moves to her husband's family, and their children become part of his lineage. However, this is interpreted flexibly, and women typically retain important rights in their natal homes, and in their own patrilineages (*suudu baaba*) and matrilineages (*suudu yaaya*). Divorce is common (e.g., Hampshire & Randall, 2000; Stenning, 1959), and divorced women return to, and are generally accepted by, their *suudu baaba*.

Religion

Most Fulani are Muslim, and a strong identification between being Fulani and being Muslim has been noted in many areas (Burnham, 1996). Islam has been used as a vehicle of conquest, and Fulani became the rulers of large stretches of the Sahel in the 18th and 19th centuries through Holy War (*jihad*). However, the versions of Islam practiced by Fulani incorporate many aspects of pre-Islam cosmology. This is apparent in the way illnesses are conceptualized and treated. A minority of Fulani (e.g., the *WoDaaBe* groups in Niger) are not Muslim, but practice traditional religions.

Fulani Identity: Pulaaku

The sense of identity—what it means to be Fulani (specifically *FulBe*)—is strongly and explicitly expressed in the concept *Pulaaku*. *Pulaaku* is principally about control and restraint, linked with a strong sense of shame (*semteende*). This includes being in complete control of one's emotional and physical needs, and is often defined in opposition to assumed *RiimaaiBe* traits of lack of self-control. Another important aspect of *pulaaku* is cattle-ownership.

Despite the essentialist nature of this Fulani discourse on their culture, it would be wrong to see *pulaaku* as being uniform, unchanging, and representing all Fulani behavior. Individual Fulani construct and interpret *pulaaku* variously at different times and under different circumstances (Burnham, 1996; De Bruijn & Van Dijk, 1995). This is important in understanding people's responses to illness.

THE CONTEXT OF HEALTH: ENVIRONMENTAL, ECONOMIC, SOCIAL, AND POLITICAL FACTORS

Good quality demographic and epidemiological data for the Fulani are scarce, owing to the inadequacy of many national censuses and surveys. However, various indicators suggest that Fulani populations suffer as a whole from relatively poor health status and poor access to health services. The Fulani live in many of the world's poorest states, where health status is generally poor and health services under-developed (Table 1).

However, even *within* those countries, Fulani in rural areas may be particularly disadvantaged because of their nomadic or semi-nomadic status. Various pieces of research have pointed to barriers of access to health services experienced by mobile pastoralists (Foggin, Farhas, Shiirev-Adiya, & Chinabat, 1997; Loutan, 1989; Meir, 1987; Sandford, 1978; Swift, Toulmin, & Chatting, 1990; Zinsstag, Bidjeh, & Idriss, 1998). Pastoralists tend to live in sparsely populated, geographically marginal areas, while health services are typically concentrated in more densely populated areas of permanent settlement. Mobility in itself may restrict access, particularly where extended courses of treatment are indicated. Other barriers include political marginalization, cultural, ethnic, and linguistic differences with service providers, and very low levels of literacy.

Data collected among a sample of 9,000 Fulani in northern Burkina Faso suggest a high infant mortality rate, even compared with the rest of the country: 140 infant deaths per 1,000 live births (Hampshire & Randall, 1995). Maternal mortality is also extremely high among the Burkinabè Fulani, with a lifetime risk for women of

Table 1. Health Indicators and Services in Selected Countries Inhabited by Fulani

Country	Infant mortality rate (per 1,000 live births)	Life expectancy at birth (years)	Doctors per 100,000 people
Burkina Faso	106	46	3
Chad	92	47	3
Mali	134	47	5
Niger	191	48	2
Sources	*UNICEF 1998*	*UNDP 1997*	*UNDP 1997*

dying of maternal causes estimated to be around 1 in 17 (Hampshire, 2001, 2002b).

On the positive side, pastoralist Fulani may have better nutritional status than non-pastoralists, because of the availability of milk and meat. Again, while not shown specifically for the Fulani, this has been demonstrated for other nomadic pastoralists, such as the Turkana (Barkey, Campbell, & Leslie, 2001; Shell-Duncan & Obiero, 2000).

Medical Practitioners

For many common ailments, and for prevention of illness, everyone becomes a medical practitioner. For example, Riesman (1992) describes how all Fulani mothers in northern Burkina Faso prepare *basi*, an infusion of herbs and tree barks, which they give young babies to protect them from various ailments. Women learn the secrets of *basi* preparation from their mothers and grandmothers; these include both the ingredients and the method of preparation. In Chad, many common ailments are self-treated by Fulani using remedies made from ingredients found around the home, such as milk, butter, spices, and animals' urine (Hampshire, 2002a, in press).

For more serious conditions, the range of medical practitioners used by the Fulani reflects the variation in perceived cause of illnesses. Most illnesses are perceived to have their origins in malevolent external powers: spirits or djinns, witches, or non-human animals (De Bruijn & Van Dijk, 1995). Such illnesses can only be treated by confronting and defeating those forces, and are thus amenable only to intervention by those with power in the supernatural realm. *MooDiBaaBe* (sing. *mooDiBo*), Islamic learned men, are frequently called upon in these cases. *MooDiBaaBe* may use herbal medicine or Koranic texts in their practice, or a combination of both (De Bruijn & Van Dijk, 1995; Hampshire, 2002a; Riesman, 1992). An example of the use of Koranic texts is given by Riesman (1992, p. 92). As a cure for madness, a *mooDiBo* wrote Koranic verses on a wooden tablet in ink. The ink was then washed off, and the solution drunk by the afflicted person.

Today, the healing of *mooDiBaaBe* often goes hand in hand with the use of "modern" medicine. Because of the remoteness and poverty of many areas inhabited by Fulani, many of the modern health facilities to which they have access are fairly limited and basic health posts.

In addition to formal provision of modern medicines, in several areas, informal dealers of non-registered drugs provide another means of access to medicines in some areas (De Bruijn & Van Dijk, 1995; Hampshire, 2002a).

Classification of Illness, Theories of Illness, and Treatment of Illness

Most illness is thought by the Fulani to originate from external malevolent forces. It may be caused by non-human animals, particularly birds (*pooli*), spirits or djinns (*jinnaaji*), or by "witches" (*sukunyaaBe*) (De Bruijn & Van Dijk, 1995).

Jinnaaji may be anywhere in the bush, and the risk of encountering one makes many people fearful of going out at night. An encounter with malevolent spirits, *seedaani*, will almost certainly result in illness. The illnesses associated with *seedaani* are *henndu* (wind) and *haandi* (madness) (De Bruijn & Van Dijk, 1995). One who has come into contact with *henndu* is called *jom-henndu* (possessor of the wind). *Henndu* can attack the body, causing fever and wasting, or the mind, causing a person to lose their wits (*hakkile dillii*) (De Bruiju & Van Dijk, 1995). *Haandi*, or madness, is described by Riesman (1977) as being caused by the extreme fear a person experiences upon seeing a spirit. A crazy person (*kaanaDo*) is, by definition, someone who has encountered a djinn. The fear-induced craziness then drives the person to behave in ways opposite to that expected from normal people: tearing off his clothes, screaming, fleeing into the bush, etc.

SukunyaaBe ("witches") are another cause of illness. They are transformations of human beings that seek to suck the life-force/soul out of their victims. De Bruijn and Van Dijk (1995), from their work in Mali, describe them as being visible at night, appearing as white spots. Among some Fulani groups in northern Burkina Faso, *sukunyaaBe* are believed to be humans that transform themselves into birds of various kinds, who eat the souls of their victims by pecking at their chests. Unless a *mooDiBo* can effect a cure, the soulless victim is destined to die shortly afterwards (Hampshire, unpublished fieldnotes). *SukunyaaBe* are found mostly among outsiders and non-Fulani who live close to Fulani settlements.

A very different kin of witchcraft, which is also a common cause of illness, concerns the use of *dabare*.

SukunyaaBe, while evil, are generally believed to have no choice: as *sukunyaaBe*, they have to feed off human souls in order to survive. *Dabare*, by contrast, is the conscious practice of black magic against a particular person, and is usually related to jealousy (*haasidare*) (De Bruijn & Van Dijk, 1995). It is often used between co-wives, among whom jealousy might easily arise. Only *mooDiBaaBe* or non-Islamic practitioners of black magic (*bonngoBi*) can make *dabare*, which they do when commissioned and paid to do so by an aggrieved person. *Dabare* may be used to cause illness, or other evil (i.e., death), causing a person to act against his/her will, driving someone away, etc. (De Bruijn & Van Dijk, 1995; Hampshire, unpublished fieldnotes).

A third common set of causes of illness are non-human animals, particularly birds (*pooli*) (De Bruijn & Van Dijk, 1995; Hampshire, unpublished fieldnotes). *Pooli* are a particular danger for young children left unattended. Particularly feared is the owl, whose spirit may invade and attack a young child. Children taken by the *pooli* often have clenched fists, and they may have a fever and be restless (De Bruijn & Van Dijk, 1995). Mothers suspecting that their children are in danger may try to frighten the *pooli* away by making lots of noise (Hampshire, unpublished fieldnotes).

Finally, there are many illnesses that may be described as inconveniences rather than dangers, and which people expect to experience on and off, for much of the time. These include minor aches and pains, particularly headaches (*naoora hoore*), fever (*jonte*), stomach pains (*nyao reedu*), and diarrhea (*doggu reedu*) (De Bruijn & Van Dijk, 1995; Hampshire, unpublished fieldnotes). The causes of these generally are believed to be to do with relatively simple, physical causes: tiredness, lack of food, lack of tea. The remedies for such conditions are self-evident and straightforward, unlike the more serious conditions described above, where effective treatment can only come from a *mooDiBo*, skilled in interacting with external malevolent forces.

SEXUALITY AND REPRODUCTION

Having children is essential for Fulani men and women. Children are seen as conferring status in their own right, and are an important source of economic wealth and security (Hampshire & Randall, 1999, 2000). To have no children is seen as being a social disaster since, without children, it is very difficult for a Fulani to become a full social being and to set up his or her own household (Riesman, 1977; Stenning, 1959).

Child-bearing for the Fulani should ideally happen within marriage. Indeed, child-bearing is such an integral part of marriage that a marriage is often not seen as being formally completed until a child is born. Among the *WoDaaBe* of Niger, newly married couples do not initially have their own domestic space or hut in which to sleep. On becoming pregnant, the young wife returns to her own family. There she stays until the birth and often for an extended period afterwards (up to a year or more). Only when the woman returns to her husband's family with her year-old child is the marriage considered formally complete, and are the couple permitted to occupy their own domestic space (Dupire, 1970; Stenning, 1958, 1959).

Because of the central place of child-bearing as part of marriage, pre-marital fertility is generally regarded as extremely bad, and carries strong sanctions. In one case described by Hampshire (2001), the family of an unmarried young woman was driven out of their village when she became pregnant. Extra-marital fertility (for example, during the absence of a husband) and inter-marital fertility (following divorce or widowhood, and before re-marriage), though frowned upon, are more often tolerated, and are frequently the subject of joking. Hampshire (2001) describes how one young Fulani woman in Burkina Faso whose husband was working away said: "While the men are away, they go with other women. Why should we keep our legs closed?"

The wide geographical dispersion of the Fulani makes it impossible to generalize fully about factors influencing fertility and fecundity. A study carried out among Burkinabè Fulani in 1994–96 suggests a strong desire for very large families, for reasons of status, wealth, and security (Hampshire, 2001; Hampshire & Randall, 2000). For many Fulani informants, family size did not enter into the realm of conscious choice. It is, therefore, no surprise that contraceptive use in this population was virtually zero. Similarly, induced abortion was almost unheard of, although some old women claimed to know of very powerful herbs that could be used for this purpose. Demographic and Health Survey data of other countries with significant Fulani populations also indicate contraceptive prevalence rates among the lowest in the world, so it is likely that conscious birth control is not an important determinant of fertility in many Fulani societies.

Among the Burkinabè Fulani, the major proximate determinant of fertility is post-partum infecundity due to extended periods of breast-feeding—typically for around two years (Hampshire & Randall, 2000). Extended breast-feeding has been documented among other Fulani populations by Riesman (1992) and Dupire (1963). Other factors constraining fertility to a lesser extent among the Burkinabè Fulani were spousal separation following births and during migration and, more importantly, secondary sterility, probably acquired through sexually-transmitted diseases (Hampshire & Randall, 2000). Secondary sterility acquired through sexually-transmitted diseases was found by David and Voas (1981) to have a major impact on the fertility of Fulani living in Cameroon. Over half the women in their study were apparently sterile, due in particular to the high prevalence of gonorrhoea and syphilis.

HEALTH THROUGH THE LIFE CYCLE

Pregnancy and Childbirth

Pregnancy and childbirth among the Fulani occupy ambiguous positions. On the one hand, the birth of a child is greeted with joy and celebration. In particular, the birth of a first child is essential in cementing a marriage and giving the parents their new status in life (Dupire, 1963). On the other hand, pregnancy and childbirth are times of danger, and are surrounded by feelings of secrecy and shame, associated with *pulaaku*.

A pregnant woman can be the focus of much jealousy, particularly from any co-wives she may have, and other young wives in her husband's extended family. As such, she is at particular risk from *dabare* (black magic), which might be used to cause any number of serious misfortunes: miscarriage, death in childbirth, deformity of the child, etc. Pregnancy is, therefore, concealed for as long as possible. When it is no longer possible to continue concealment, a pregnant woman often returns to her natal home (*suudu baaba*), where fewer people will wish her harm.

Childbirth, therefore, particularly for younger women, usually takes place in the *suudu baaba* rather than the marital home. For first births, it is usual practice for the woman's mother to assist. For subsequent births, the ideal is for childbirth to be alone and unassisted (Hampshire, in press; Riesman, 1992). Adherence to the *pulaaku* ideal means that a woman should be able to face childbirth, and all its dangers, alone and without fear. It is important not to cry out, or to show any sign of succumbing to physical pain. Of course, it would be a mistake to assume that all women interpret *pulaaku* in the same way, and accounts of practice show considerable variation in childbirth experience (Hampshire, in preparation). That said, unassisted childbirth remains an important ideal.

Childbirth, therefore, represents a time of danger for women on many levels. Not least is the physical danger of death or serious incapacity. As stated above, lifetime risks of maternal death among the Burkinabè Fulani are in the order of 1 in 17 (Hampshire, 2001), and Demographic and Health Survey data from other regions inhabited by Fulani suggest that this risk may be similar elsewhere. Second, there is a risk of *dabare* (black magic) associated with jealousy, which may cause serious, long-term harm to mother and baby. Third, there is the fear of shame, through not being able to live up to the *pulaaku* ideal of stoicism in the face of pain and danger. It is little wonder, therefore, that a first pregnancy can be greeted with very mixed emotions by young Fulani women.

Among the DjelgoBe Fulani of Burkina Faso, Riesman (1992) describes how, when a Fulani baby is born, it is important that it comes out on the ground. This contact with the earth helps establish a connection between the child and the place of birth. The placenta, which is widely believed to have great power, is usually buried in the earth at the spot where the baby first touched it. In the following days and weeks, hot water is poured on the spot where the placenta is buried in order to prevent it from rotting (Reisman, 1992).

Infancy

Post-Partum Practices. From the day of the birth until the naming ceremony 8 days later, the mother is confined to her hut or tent (Riesman, 1992). She leaves only to urinate or defecate in the bush and, even for this, she is always accompanied by someone. A fire is lit when the baby is born, and this fire burns continuously for the first week of the child's life, until the naming ceremony. The fire is used to heat water for bathing and preparing infusions.

On the eighth day after birth, the naming ceremony (*indeeri*) takes place (Riesman, 1977, 1992). The infant is washed and its head is shaved. It is carried back and forth between the hut and the outside three times for a boy, four times for a girl. A goat is sacrificed and the name of the child pronounced. This represents the first stage of becoming a social being.

Protection of Newborn Infants. Newborn infants are thought to be vulnerable to many forms of harm and, as such, are protected in multiple ways. Every day for the first few days of life the infant is washed with hot water containing a medicinal herbal infusion. *Basi*, another medicinal infusion of herbs and tree barks, prepared by the mother, is given to the infant to drink. Various talismen or charms, often provided by *mooDiBaaBe*, are worn by the child from a very young age to offer protection against malevolent external powers (Riesman, 1992).

Because children are very highly valued, people are fearful that others may become jealous and wish harm on the child. To this end, it is regarded as being very dangerous to praise a young child by saying (s)he is good-looking, healthy, or fat (De Bruijn & Van Dijk, 1995; Hampshire, unpublished fieldnotes; Riesman, 1992). To make such a remark is interpreted as an intention to cause harm. Instead, visitors often make derogatory comments, remarking on the ugliness of the child. Another strategy used by some parents is to cover the child in dirt or dung, or to give it nicknames such as "cow dung" to make it appear unattractive.

Infant Feeding Practices. Fulani women typically breastfeed for extended periods of time. In northern Burkina Faso, this typically extends to the two years proscribed by the Koran (Hampshire & Randall, 2000). In cases where the child is weak or sickly, this may be extended further, and it is not unknown for mothers to nurse until 30 months or more (Hampshire & Randall, 2000). Early weaning occurs if the mother becomes pregnant again, because the milk in the breasts of a pregnant women is believed to "belong" to the new baby and is thought to be harmful to the breast-feeding child (Hampshire, 1998; Riesman, 1992). Accounts of other Fulani groups also point to extended periods of breast-feeding (Dupire, 1963; Riesman, 1992).

During the period of nursing, the breast is offered on demand, and is seen as a panacea to be offered whenever the child appears distressed. Co-sleeping of mothers and infants allows infants to feed on demand through the night too. Other women may also offer a breast to an infant, particularly grandmothers (Hampshire, 1998; Riesman, 1992) although, among the *WoBaaBe*, this may be extended to all nursing mothers in the camp (Dupire, 1963).

In addition to breast milk, Fulani infants are typically given supplementary foods from quite a young age. Butter and cow's or goat's milk may be given as early as the first week of life, and thin millet gruel may be tried after a few months (Dupire, 1963; Riesman, 1992). This is in addition to the medicinal *basi*, given from birth (Riesman, 1992).

Cessation of breast-feeding typically happens abruptly (Hampshire, 1998), although mothers may vary in how strictly this is enforced (Riesman, 1992). Because, until this point, feeding has been on demand, weaning is a difficult period for young children. To aid the process, the child may be physically removed from its mother by, for example, being taken by its grandmother until it has "forgotten" the breast (Hampshire, 1998).

Caretakers. While a mother is primarily responsible for the care of her infant, in practice this is often shared with many other women, since childcare happens as a part of women's ongoing life, while they cook, fetch water, wash, weave mats, and do each other's hair (Riesman, 1992). Consequently, all first-time mothers will have had years of experience watching and helping other mothers take care of their children. In turn, they will also be watched, criticized, and helped. Help may include anything from offering advice or carrying the baby to offering a breast. For the first few years of life, fathers are relatively uninvolved with the care of their children and in decisions relating to their health, welfare, and feeding.

Childhood

Between the ages of 5 and 7, children are thought to begin to develop *haYYillo* (a social sense) and are given their first responsibilities (De Bruijn & Van Dijk, 1995; Riesman, 1992). These responsibilities may include helping mothers with food preparation, or beginning to look after small animals. Until that time, children are not expected to know how to behave socially, or indeed to comply with the demands of their caretakers. From ages 5–7 onwards, children are thought to be more responsible for their actions, and are thus more likely to be reprimanded and corrected for behaving in inappropriate ways.

A particular danger to the health of children is *pooli* (described above), and mothers take precautions, such as making a loud noise, to try to prevent this (De Bruijn & Van Dijk, 1995). Other health risks to childhood include the belief that it is dangerous for a child to have contact with its mother, and particularly with her milk, for the last months of her pregnancy with her next child, and continuing after the birth of that child. To protect against this, an older child may be given an egg to eat, cooked directly in the coals, from a chicken that has not had any offspring (Riesman, 1992).

For very young children, the main responsibility for healthcare lies with the mother (De Bruijn & Van Dijk, 1995). It is she who decides when treatment for illness should be sought, and she who must find the means to pay for it, which are typically very limited. If very young children die, grief tends to be borne privately by mothers, rather than being a public matter. For older children (over age 5 or so), the responsibilities shift, with fathers beginning to take a more active role in the well-being of their offspring, and contributing financially to their healthcare (De Bruijn & Van Dijk, 1995).

Adolescence

Childhood ends quite early, for Fulani girls in particular. Marriage takes place typically at an early age—mean age at first marriage for Fulani women in Burkina Faso was found to be around 16 years, with a high proportion of marriages happening considerably earlier (Hampshire & Randall, 2000). For some time prior to marriage, girls are expected to take on roles appropriate to adult women, and much of the freedom associated with early childhood is lost. In Burkina Faso, some Fulani girls from the age of about 12 are sent to live with their future mothers-in-law, as part of preparation for married life (Hampshire, 1998).

Fulani men marry later than women: among the Burkinabè Fulani, mean age at first marriage for men was 24–25 years (Hampshire & Randall, 2000). Fulani men, therefore, enjoy a more extended period of adolescence during which they are free from many of the responsibilities of having their own family and herd. Among some Burkinabè Fulani men, it is common to begin to travel during this period, in particular to cities for temporary work (Hampshire, 2002b; Hampshire & Randall, 1999).

Adulthood

Until old age, Fulani are married for most of their adult lives. Marriage is effectively universal and re-marriage is typically rapid following divorce or widowhood (Hampshire & Randall, 2000). Polygamous marriage is practiced, although most marriages are monogamous: among the Burkinabè Fulani, for example, 7.6% of married men and 13.5% of married women were in polygamous unions (Hampshire & Randall, 2000).

For adult Fulani women, the most pressing health issues are pregnancy and reproductive health, and unequal access to health services. The health risks associated with pregnancy and childbirth have been discussed above. Maternal mortality ratios are typically high (e.g., 850 deaths per 100,000 live births among the Burkinabè Fulani, which corresponds to a 1 in 17 lifetime risk (Hampshire, 2001).

The reasons for such high risks of maternal death or injury relate in part to the lack of facilities for dealing with obstetric emergencies in particular (also ante-natal and post-natal care) in many of the areas inhabited by rural Fulani. They may also relate to a mismatch of services currently provided with the felt needs of Fulani women. Many of the beliefs and practices associated with pregnancy among the Fulani make use of standard maternity services problematic for many. Ante-natal care is difficult where women conceal pregnancy, and delivery in hospital or maternity units runs counter to the ideal of unassisted childbirth. Issues such as the disposal of the placenta in hospital are also areas of concern for many Fulani women (Hampshire, in preparation).

Gender inequalities in access to health services also exist in many Fulani populations, such as those in central Chad (Hampshire, 2002a). Because of both resource constraints and various cultural and institutional factors, Fulani women often have to rely on male affines and kin to gain access to particular health services and practitioners. Hampshire (2002a) found considerable variation in the ability of married women to mobilize their husbands' support in times of illness. Particularly vulnerable were childless wives, those least well established in the marital households, and those in polygamous marriages and/or in large, extended marital households. A woman's own kin network is important in terms of her access to health resources, particularly for unmarried women, or where husbands and other affines are unprepared to help (De Bruijn & Van Dijk, 1995; Hampshire, 2002a).

The Aged

Old people are generally treated with deference and respect in Fulani society (De Bruijn & Van Dijk, 1995; Riesman, 1992). According to De Bruijn and Van Dijk (1995), there are established ideas about help for old people in cases of illness or hardship. An old, widowed woman may receive help from her brothers, her own children, members of her *suudu yaaya* (mother's lineage), and neighbors. However, such support is not always forthcoming, and depends on the willingness and ability of such relatives to help out (De Bruijn & Van Dijk, 1995; Hampshire, 2002a).

The respect accorded to the elderly is not without ambiguity (Riesman, 1992). Important aspects of being a Fulani adult include being useful/able (*waawude*), and being in control of physical needs or wants—qualities that diminish in old age. Riesman (1992) discusses the sense of irony surrounding old age, in which the elderly are, on the one hand, respected, and yet, on the other hand, the objects of ridicule or joking, since they may no longer be capable of exercising that power in any meaningful way. Riesman even quotes elderly informants who describe themselves as being "dead," since they are no longer able to fulfil their social or productive roles.

Death and Dying

Reactions to death and dying in Fulani society are characterized by the restraint shown in other aspects of life. De Bruijn and Van Dijk (1995) describe how, for very young children (under about 5 years), death and mourning are private affairs, with no public expression of grief. For older children and adults, public mourning is typically very short. Burial happens very shortly after death, without an elaborate ceremony (De Bruijn & Van Dijk, 1995). Condolences are offered to the bereaved family but, thereafter, the grief becomes private, with no collective ceremony or remembrance at a later stage. Riesman (1977) points out that one should not mistake a lack of public mourning with lack of emotion or sensitivity. Fulani are expected to have strong emotional feelings surrounding death and other sadnesses. However, they are also expected to master those feelings, to control them, and not to display them publicly. This is particularly true for men; it is expected that women may shed tears over a death, particularly of a close family member, while men should be able to control and master their emotions.

CHANGING HEALTH PATTERNS

It is important to remember that the information presented here refers to particular groups of Fulani, documented by particular ethnographers at particular moments in time. It is now widely accepted within the field of anthropology that "culture" is neither static, nor rigid, nor is it interpreted by everyone in the same way. Within a Fulani camp or village, individuals will think about and interpret situations differently, with substantial variation in practice and action. This applies to beliefs and practice associated with health as much as any other element of life.

The variation is increased by the fact that the Fulani are a vast and fluid ethnic category, spanning the breadth of Sahelian Africa. In addition to the variation in health-related beliefs and practices within villages, there are also substantial variations between Fulani groups living in different areas. I have tried, throughout this entry, to draw the reader's attention to such differences and similarities. However, existing ethnographies of Fulani groups do not cover the whole range of Fulani life in every area. It is probable, therefore, that there is more regional variation in Fulani health patterns, beliefs, and practices than is documented here.

It must also be remembered that the various ethnographies of the Fulani drawn on in this entry are situated in particular moments in time. Health patterns, beliefs, and practices, like all other aspects of life, are not static, nor can they be properly understood ahistorically. Many Fulani societies have undergone substantial social and economic change over the period covered by the ethnographies, with important implications for health patterns, beliefs, and practices. Most of the states in which Fulani live achieved independence from colonial rule in the years around 1960, which precipitated, among other things, changing social and economic relations between the FulBe and slave groups. This, in combination with climatic and ecological changes, has meant that many Fulani have left behind agro-pastoral subsistence to pursue economic strategies that draw them increasingly into a national and global economic framework (Hampshire & Randall, 1999).

Modernity has brought about different changes for different groups of Fulani. Some of these changes can be illustrated by considering the case of Fulani groups in northern Burkina Faso. On the one hand, increased contact with health services has resulted in improving infant and child health for many, as indicated by falling infant and child mortality rates over recent decades (Hampshire & Randall, 1995). On the other hand, modernity brings with it new health challenges. Young Fulani men migrating temporarily to cities to find work are increasingly exposed to sexually-transmitted diseases, including HIV and conditions that result in sterility, such as gonorrhea and chlamydia (Hampshire & Randall, 2000). It is likely that these are being transferred back to Fulani villages, with possibly devastating future consequences.

References

Barkey, N. L., Campbell, B. C., & Leslie, P. W. (2001). A comparison of health complaints of settled and nomadic Turkana men. *Medical Anthropology Quarterly, 15*(3), 391–408.

Bonfiglioli, A. M. (1988) *DuDal: Histoire de famille et historique de troupeaux chez un groupe WoDaaBe du Niger.* Cambridge, U.K.: Cambridge University Press / Edition de la Maison des Sciences de l'Homme.

Burnham, P. (1996). *The politics of cultural difference in Northern Cameroon.* Edinburgh, U.K.: Edinburgh University Press.

David. N., & Voas, D. (1981). Societal causes of infertility and population decline among the settled Fulani of north Cameroon. *Man, 16*(4), 644–664.

De Bruijn, M., & Van Dijk, H. (1995). *Arid ways.* Amsterdam: Thesis.

Dupire, M. (1963). The position of women in a pastoral society. In D. Paulme (Ed.), *Women of tropical Africa* (pp. 47–92). London: Routledge.

Dupire, M. (1970). *Organisation sociale des Peul.* Paris: Edition Plon.

Foggin, P. M., Farhas, O., Shiirev-Adiya, S., & Chinabat, B. (1997). Health status and risk factors of seminomadic pastoralists in Mongolia: A geographical approach. *Social Science & Medicine, 44*(11), 1623–1647.

Hampshire, K. R. (1998). Fulani mobility: Causes, constraints and consequences of population movements in northern Burkina Faso. Unpublished PhD thesis, University College London.

Hampshire, K. R. (2001). The impact of male migration of fertility decisions and outcomes in northern burkina faso. In S. Tremayne (Ed.). *Managing Reproductive Life,* pp: 107–126. Oxford: Berghahn.

Hampshire, K. R. (2002a). Networks of nomads: negotiating access to health services among pastoralist women in chad. *Social Sciences and Medicine,* 54, 1025–37.

Hampshire, K. R. (2002b). Fulani on the move: Seasonal economic migration in the Sahel as a social process. *Journal of Development Studies, 38*(5), 15–36.

Hampshire, K. R. (2003, in press). What is "safe motherhood"? Insiders' and outsiders' preceptions of maternal health among chadian pastoralists. *Genus.*

Hampshire, K. R., & Randall, S. C. (1995). The single round demographic survey (Technical Report No.1, Land use, household viability and migration in the Sahel). Unpublished manuscript.

Hampshire, K. R., & Randall, S. C. (1999). Seasonal labour migration strategies in the Sahel: Coping with poverty or optimising security? *International Journal of Population Geography,* 5, 367–385.

Hampshire, K. R., & Randall, S. C. (2000). Pastoralists, agropastoralists and migrants: Interactions between fertility and mobility in northern Burkina Faso. *Population Studies, 54*(3), 247–262.

Hopen, C. (1958). *The pastoral FulBe family in Gwandu.* Oxford, U.K.: Oxford University Press.

Loutan, L. (1989). Les problèmes de santé dans les zones nomades. In A. Rougement & J. Brunet-Jailly (Eds.), *La santé en pays tropicaux* (pp. 219–253). Paris: Doin Editeurs.

Maliki, A. B., White, C., Loutan, L., & Swift, J. J. (1984) The Wodaabe. In J.J. Swift (Ed.), *Pastoral development in central Niger: Report of the Niger Range and Livestock Project.* Niamey, Niger: Ministère du Développement Rural & USAID.

Meir, A. (1987). Nomads, development and health: Delivering public health services to the Bedouin in Israel. *Geografiska Annaler, 69B*(2), 115–126.

Raynault, C. (1997). *Societies and nature in the Sahel.* London: Routledge.

Riesman, P. (1977). *Freedom in Fulani social life.* Chicago: University of Chicago Press.

Riesman, P. (1992). *First find your child a good mother: The construction of self in two African communities.* New Brunswick, NJ: Rutgers University Press.

Sandford, S. (1978). Welfare and wanderers: The organisation of social services for pastoralists. *ODI Review, 1,* 70–87.

Shell-Duncan, B., & Obiero, W. O. (2000). Child nutrition in the transition from nomadic pastoralism to settled lifestyles: Individual, household, and community-level factors. *American Journal of Physical Anthropology, 113*(2), 183–200.

Stenning, D. (1958). Household viability among the pastoral Fulani. In J. Goody (Ed.), *The development cycle of domestic groups.* Cambridge, U.K.: Cambridge University Press.

Stenning, D. (1959). *Savannah nomads.* Oxford, U.K.: Oxford University Press.

Swift, J., Toulmin, C., & Chatting, S. (1990). *Providing services for nomadic people.* (UNICEF Staff Working Papers No. 8.) New York: UNICEF.

Zinsstag, J., Bidjeh, K., & Idriss, A. (1998). L'interface entre la santé humaine et animale chez les nomades en Afrique: vers "une médicine unie." In M. Wiese & K. Wyss (Eds.), *Les populations nomades et la santé humaine et animale en Afrique et notamment au Tchad.* APT-Reports, *9,* 53–64.

Garhwali

Satish Kedia

Alternative Names

The current name of Garhwal is of relatively recent origin. Etymologically, *Garhwal* means a region formed by the integration of *Garhs* (fortresses) and their territories, an act that took place in the 16th century when King Ajaypal consolidated 52 small princely states and their forts to expand his empire over the whole Garhwal

(Rawat, 1989, p. 20). The original inhabitants of the Garhwal hills are known as *Garhwali*. They are also called *Pahari*, meaning "people of the mountain," by the neighboring communities in the plains.

LOCATION AND LINGUISTIC AFFILIATION

Garhwali live in the North Indian Himalayas in the western part of the State of Uttaranchal. Uttaranchal comprises 13 districts and was part of the State of Uttar Pradesh until 2000, when it was given autonomous status. The Garhwal region is in the middle zone of the Himalayas, which have a varying width of about 60–90 km and an elevation ranging between 1,000 m and 3,000 m above sea level. The region is full of peaks and valleys and is marked by rivers flowing through the deep gorges.

Most Garhwali people speak Garhwali and some Hindi, both in the Indo-Aryan linguistic group. Prior to the 19th century, Persian with Arabic script was used for administrative purposes, but Garhwali with Devanagari script remained the language of the commoners. Many inscriptions and state orders were written in Garhwali. In some regions of Garhwal, such as Rawain, Jaunpur, and Juansar-Bawar, dialects of Garhwali called Rawalti, Jaunpuri, and Jaunsari respectively are spoken.

It would not be prudent to generalize this commentary to all regions of Garhwal. This narrative is primarily based on the author's fieldwork among a peasant community near Tehri in the western part of Uttaranchal during the mid-1990s. However, a concerted effort was made to incorporate pertinent information from the writings of other researchers who have worked in various parts of Garhwal.

OVERVIEW OF THE CULTURE

Garhwali are predominantly peasants and have approximately 8.5 million people living in Uttaranchal. The initial archeological evidence in Garhwal includes inscriptions and artifacts from the time of King Ashoka (269–232 BC) and suggests that the region was inhabited long before this period. The mythical Hindu texts, *Puranas, Mahabharata*, and *Ramayana*, contain numerous indications that a society existed in this region. At that time, the people were variously known as *Kedarkhand, Uttarkuru, Kurujangal*, and *Uttarakhand*. However, it remains a matter of debate among historians as to who were the original inhabitants.

The economy in the Garhwal region is primarily agrarian with approximately 85% of the adult population practicing intensive agriculture. The other economic activities include horticulture, animal husbandry, forestry, and some cottage industries. The crop cycle and harvest determine the work and festivities in these villages. In addition, a sizable portion of adult males have either permanently or temporarily migrated to nearby cities to pursue wage-labor in factories, hotels, restaurants, and offices or to serve in the military.

While there are four major caste groups in India, Garhwali society is stratified in a tripartite caste system comprising priests (*brahmin*), warriors and landlords (*kshatriya/rajput*), and low occupation groups (*sudra*) (Berreman, 1997). Caste is a system of social hierarchy among Hindus where individuals are ascribed to a particular group by virtue of being born into it. The merchant caste is absent among Garhwali society because of the lack of trade at any significant level in the mountains. Social mobility has been largely upward among Garhwali. The non-caste local groups (*khas*) emulated the caste-like features to get into the fold of the caste system. Once in the caste system, attempts were made to move up to even higher categories of actual caste groups (*asal*) through marriage or by changing religious practices. Garhwali society is patrilineal with an emphasis on joint family structures consisting of parents, married sons and their families, and unmarried children.

The Garhwal region has a long tradition of Hinduism and is a hub of some of the most sacred pilgrimage sites. The Garhwali practice a syncretic form of orthodox Hinduism and animistic religion. The ritual practices differ among castes based on their past affiliation with primitive religion. Garhwali worship various forms of the Hindu trinity (*Brahma, Vishnu,* and *Shiva*) as well as different local deities such as *Narsing, Nagraj,* and *Bhairon*. Garhwali have ardent beliefs in the power of spirits and regularly worship them. Most villagers worship their village deity and each family has family deities (*kula devata*). The rituals and other religious practices involve magic, animistic beliefs, spirits, dancing of spirits and gods, and nature worship.

THE CONTEXT OF HEALTH: ENVIRONMENTAL, ECONOMIC, SOCIAL, AND POLITICAL FACTORS

Approximately 78% of the people in Uttaranchal live in the rural areas with fewer resources to maintain a balanced diet and healthy lifestyle. Fifty-three percent of the households have electricity, 44% have access to piped drinking water, and 39% have toilets in their houses (International Institute for Population Sciences & ORC Macro, 2000, p. 2). The environmental conditions limit the productivity of the land; even though the majority of the population work on the farm, there is always a food deficit. Local agricultural production fails to sustain the population, and the demand for additional food is met by importing food grains from the plains. The healthcare resources are abysmal in the region. Few biomedical doctors are available in urban areas, and those who use these doctors frequently complain about the poor quality of service. Government officials have found it difficult to provide healthcare services in the isolated villages of Garhwal. Many of these villages require hours of walking on steep, narrow, and winding mountain paths from the nearest road (Kumar, 1991). The state officials of the newly formed Uttaranchal are now paying closer attention to enhancing the healthcare services in this region.

Medical Practitioners

Garhwali use an eclectic mix of medical healers; biomedical practitioners are just one of the available options. Biomedical practitioners with a medical degree are rare in the Garhwal hills; most have only an associate's degree with a couple of years of medical training or apprenticeship and are in private practice. The district headquarters or towns have a local hospital with doctors that have a medical degree and, in some cases, other specialists. The fees and traveling distance to the biomedical healers make them unavailable to much of the population. There are health centers with a health worker and a nurse for every five or six villages.

Herbalists (*deshi vaidhya*) are available and found throughout the Garhwal region. They use herbs and medicinal plants indigenous to the Himalayas and are consulted for prolonged and serious illnesses. Along with the herbalists are the Ayurvedic doctors (*Ayurvedic vaidhya*) who have formal training in practicing medicine. An Ayurvedic doctor uses many natural substances and food in addition to herbs and medicinal plants to cure illnesses. In Garhwal villages, it is difficult to make distinctions between a herbalist and an Ayurvedic doctor; they are both synonymously called *vaidhya*.

A variety of religio-magical healers in this region specialize in warding off the malign effects of supernatural agencies. These healers are mostly part-time specialists and inherit or learn their skills from experienced healers. Garhwali have tremendous faith in these specialists. Priests (*pandit*) conduct worship, give sermons, perform rites of passage, and neutralize the effect of evil eye or evil spirits. They advise people to perform rituals, to worship deities, and to offer homage to their ancestors. They also perform *hawan* for clients and, on a larger scale, *yagya* can be arranged for the whole community. A trained priest chants sacred verses (*mantra*) and blows consecrated ash on their clients to get rid of some illnesses, particularly those believed to be caused by supernatural agencies.

Exorcists, another religio-magical specialist, chant *mantra* to cure illnesses. They make charms or talismans (*tabeez*) to prevent illnesses. They can give the patient consecrated water or sugar to consume or give a consecrated piece of stone for the patient to wear. In addition, special rituals are performed under the supervision of an experienced exorcist, such as placing consecrated vermilion, hair, salt, chili, and lemon on the road intersection (*chauraha*) for specific desired outcomes. Shamans (*baki*) are believed to have supernatural powers and mystical contacts with spirits. They invoke tutelary spirits to diagnose illness and prescribe treatment. The shaman's knowledge and power are mostly inherited or learned from someone who has had extraordinary life experiences. The traditional healers serve all segments of the Garhwali population, but people with less education, women, and the elderly appear to consult these healers more frequently.

Classification of Illness, Theories of Illness, and Treatment of Illness

The Garhwali explanatory scheme divides causes of illnesses into two sources: natural (*bhautik*) and supernatural (*daivik*). Natural sources include the weather, hygiene, food, pathogens, and physical and emotional imbalances. Supernatural sources incorporate spirit and ghost intrusion, ancestral neglect, wrath of deities, evil eye, witchcraft, sorcery, breach of a taboo, totem violations, community sin, fate, ominous sensations, and contagion. An illness can be caused either by one factor or a combination of factors. Among Garhwali, the confines of health beliefs are stretched to accommodate the demands of their emotional and psychosocial state. Generally, the more serious and chronic the illness, the more likely the perceived causes will include supernatural agents.

Natural Causes

Garhwali perceive factors in the surrounding environment as potential causes of illness, such as poor drinking water, air quality, and changes in the weather. Poor hygiene is an issue in some places of Garhwal, particularly due to overcrowding and poor sanitation. Hygiene problems cause and exacerbate the spread of diseases, especially those carried by mosquitoes and parasitic worms (Rizvi, 1991). While the lay Garhwali person has little knowledge of specific pathogens, there is an awareness that mosquitoes can spread malaria and that worms (*pillu*) cause stomach upsets and diarrhea. The balance of the body humors can be disturbed by rain, cold, wind, heat, and changes in weather conditions, such as from hot or cold. Garhwali believe that hot weather can cause fever, malaria, headache, joint pain, stomach ache, backache, and diarrhea, while cold weather can cause cold, cough, fever, pneumonia, and body and joint pain. The rainy season is said to cause fever and skin diseases, among other illnesses, while the cold east wind, called *purvia*, is believed to cause fever and other minor illnesses.

Garhwali consider food to be a major factor in the maintenance of good health. Most food has been classified according to its hot and cold properties. In the local humoral classification, most spicy and oily foods are considered hot and can imbalance the body fluids, resulting in indigestion and diarrhea. Heat-producing foods lead to headache, joint pain, backache, stomach ache, indigestion, diarrhea, and high blood pressure. Smoking *hukka* (a form of pipe tobacco) and cigarettes leads to coughing and other respiratory illnesses.

Supernatural Causes

Garhwali divide supernatural causes into three categories: animistic, magical, and mystic. Spirit intrusion (*opra, chhaya lagna, bhuta-pret lagna*) is the hostile or punitive act of a malevolent supernatural spirit possessing a person and causing dramatic symptoms of illness in the process. Garhwali believe that spirits and ghosts are invisible, intangible, and free-floating divine forces. However, they do make a distinction between spirits and ghosts: spirits belong to local deities, whereas ghosts belong to ancestors or other deceased human beings. The common feature of spirit or ghost intrusion is a general illness of the person. The illness caused by ghost intrusion has a number of symptoms, including the widening of the eyes and discoloration of the face, meaningless babbling, unconsciousness, dementia, excessive weeping or laughing, and convulsive movements or locking of the jaw (Rizvi, 1991). If someone suddenly gets sick with fever, headache, or backache and feels dizzy, then people diagnose the person with spirit intrusion, especially if the ill person is a young child, newlywed woman, or expectant mother. The symptoms tend to vary with the age of the afflicted person (Rizvi, 1991). Illnesses caused by the spirit influence are considered debilitating and are mostly chronic, prolonged, and recurrent.

Similar to spirit intrusion are illnesses believed to be caused by the wrath of deities (*devi-devata ka kop hona*). Deities are angered if they are not revered and cause illnesses and destruction. The most frequently mentioned disease believed to be caused by the wrath of a deity in this region is measles (*khasara*) among children. Leprosy is another disease that is supposed to be caused by the wrath of a particular deity, if she is propitiated improperly. The illnesses that are commonly attributed to the wrath of deities are insanity, babbling, extreme weakness, and any illness that is incurable by any other therapy.

Garhwali believe that all people have some supernatural power that can affect the well-being of others through the evil eye, witchcraft, and sorcery. They believe that compliments of any sort, especially those that have a sense of envy or jealousy, can trigger the evil eye (*najar lagna*). Extraneous objects such as fancy dress, jewelry, or food that evoke jealousy can make the person possessing them afflicted with illness. This means that anyone with anything admirable is susceptible to the evil eye. Young children and pregnant women are particularly vulnerable. Common symptoms of the evil eye include fever, loss of control, weeping, irritation, closed eye, sleeplessness, weakness, children's loss of appetite for milk, and loss of appetite in general (Kedia, 1997).

In addition to potentially casting an evil eye, Garhwali believe that certain individuals possess evil power or may control certain spirits and use their malevolent power to harm a person. Witchcraft can be defined as the intentional or unintentional act of a person possessing supernatural power that causes illness in a targeted person. The victims of witchcraft suffer losses and misfortune. Many different kinds of individuals are identified in the community as having these powers. Sometimes people in the community think a young widow cursed her husband and has the potential to harm other people. There is also a belief that if someone is born on the darkest midnight of *Bhando*, a specific month during the rainy season in the Hindi calendar, then he or she would possess this power naturally. Some people believe that the power is hereditary. For example, a woman may acquire evil power from her mother. Again, people believe that a female with malevolent power may be more dangerous than a male with similar power.

Sorcery differs from witchcraft in that it is a learned magical skill that is used primarily to harm others. A sorcerer employs an aggressive magical procedure to cast a spell that potentially can cause an illness. The practice of sorcery takes various forms such as spells, curses, magic performed over hairs, nails, excreta, or clothing, object intrusion, administration of presumptive poisons, or dispatch of an alien spirit to possess the victim's body. Many times Garhwali attribute sudden serious illness to sorcery, especially if they have a dispute with a family who believes in sorcery or has contact with a sorcerer.

Garhwali believe that illnesses can be caused by a violation of some taboo or moral injunction. These taboos consist of prohibitions on the consumption of certain foods, contact with people in certain ritual states, breach of appropriate behavior toward kinsmen, trespass or theft, use of forbidden words, urination on a religious site, incest, or pre- or extra-marital sex, and so forth. When a taboo is violated, the person may or may not be aware of it. For example, if someone urinates unknowingly over a site where a person was buried, it is an unintentional violation of a taboo. The consequences of unintentional taboo violations are milder than violations that are committed willfully, but illnesses still may occur. Intentional violations of a taboo result in guilt, which in turn can consume a person until he or she physically falls sick. Garhwali believe that violation of a taboo results in chronic and incurable illnesses such as skin infections or cancer. Illnesses are seen as a warning from punishing supernatural agents against breaking taboos.

One of the taboos prohibits killing certain animals, such as cows, peacocks, snakes, and monkeys, which are considered totemic. In the Hindu tradition, these animals are supposed to have mystic ties with the deities, and their death and consumption would incite supernatural anger and endanger entire communities with epidemics and other communicable diseases. In the event of a totem's death, it should be ceremonially cremated or buried to protect the person and the community from the wrath of a deity.

In the Hindu scheme of life, every human being has the responsibility to follow right conduct. Any act or even the mere thought of something that is detrimental to other living beings, including humans or divinities, is considered sinful (*pap karna*). Most sins accumulate until the death of the person and result in a divine justice in the "other world" or in the next life, which might include being born as an animal or other non-human life. Some sins, especially those committed against the deities, ancestors, or human beings, may receive a punishment in the current life in the form of illness. Most Garhwali believe that if one commits sins, he or she has to pay the price in this life only. Those who get dreaded diseases such as leprosy do so because of some sin they committed against mankind. Some of the commonly cited illnesses that result from committing sins are mental illness, barrenness, congenital malformation, and blindness. Garhwali attribute a number of illnesses to what can best be termed as fate (*bhagya*) and ascribe a number of illnesses to astrological influences or bad luck.

Treatment

Garhwali practice a pluralistic system of medicine with an emphasis on herbal and religio-magical healing with increasing use of biomedicine. In response to the onset of an illness that does not appear to be serious or life-threatening, Garhwali frequently opt for self-treatment (*gharelu ilaz*) at home. A number of therapeutic subcultures exist in the realm of self-treatment, such as herbal and food therapy, massage, and over-the-counter Ayurvedic and biomedicine. The strategies used in self-treatment are passed down within a family and are sometimes shared by the whole community. Self-treatment is less expensive and does not require consulting a biomedical practitioner or folk healer, saving time, money, and energy. Self-treatment also means either self-care or care provided by the family, who ensures that the patient gets appropriate food and proper rest. Restrictions may be imposed on the patient to avoid exposure to the wind or consumption of certain kinds of food. Another form of self-treatment involves doing simple rituals to ward off the effects of the evil eye or worshiping deities or ancestors for their blessings to cure an illness.

On many occasions, people buy over-the-counter medicines based on the recommendations of friends, relatives, and even strangers as most medicines can be bought from a pharmacy without a prescription from a doctor. People remember the name of the medicine recommended by a doctor for use in future illness. This practice saves them both the consultation time and the hassle of going to a doctor. Another mode of procuring medical advice is through a pharmacist. If a villager has an illness, then the person simply goes to the pharmacist and describes his/her condition, and the pharmacist prescribes and sells the medicine without any consultation fee.

Many Garhwali prefer to consult an indigenous herbalist or Ayurvedic doctor before going to a biomedical practitioner. However, for any illness presumed to be caused by a supernatural agent, they go to religio-magical healers. When Garhwali perceive an illness to be caused by a supernatural agent, but are uncertain as to the specific cause, they often try to find out the exact nature of the problem by consulting a shaman (*baki*). The shaman may identify the family with whom the patient might have had an altercation in connection with land, houses, cattle, or even chickens and hold their ancestral spirits or family deities responsible for the current situation. An exorcism is performed if evil eye, sorcery, or witchcraft is deemed the cause. If someone suddenly becomes very sick in the family, then people prefer to call a specialist who goes into a trance, a practice called *ghadiyala* (deities' dance ritual), to get rid of the invasive spirit. Family members and the people in the community gather and perform the ritual, and in the process these specialists or family members go into a trance. Once they achieve a trance, they challenge the spirit to leave the person. Depending on who the believed spirit is, the specialists sometimes severely beat the sick person in order to rid him or her of the spirit, frequently to the point of unconsciousness. The specialist also suggests remedies to cure mental ailments. However, not all Garhwali believe in these remedies. Those with more education say, "take the sick person to the hospital rather than wasting the time in making these deities dance."

SEXUALITY AND REPRODUCTION

Garhwali avoid any overt expression of sexuality. Adolescent girls and women are encouraged to cover their bodies with *duppatta* or *sari*. Chastity is highly valued in this society and one rarely indulges in premarital sexual relations. If a premarital relation results in pregnancy, the fetus is aborted in the most secretive way. Among Garhwali, premarital relations or pregnancy bring dishonor to the family and make it very difficult for a woman to find a husband. Sexual and gynecological issues are matters of shame and embarrassment; women rarely discuss gynecological or pregnancy concerns with the family or a doctor. This results in poor reproductive health and outcomes for many of these women. The median age of marriage for women in Uttaranchal is 18 years, four years younger than men (International Institute for Population Sciences & ORC Macro, 2000, p. 3). Homosexuality is not acknowledged as an option. Such practices are considered abnormal and immoral; people indulging in homosexuality are reprimanded by the family members.

The idea of maternity is stronger than the idea of sexuality. Fertility is important evidence of a woman's femininity. Infertile women are stigmatized and are not invited by the community on auspicious occasions. Women usually conceive within two years of marriage and continue to conceive until they reach a desired family size that, on average, in this society is three children. However, due to the joint family structure, most households have 9 to 11 members in the family. Families desire to have at least one son. Garhwali are patrilineal, not having a son endangers the continuity of the family and family heritage. Families with only daughters continue trying for a son, even if it means insufficient nutritional resources for the mother and children already born. Many women with one or two sons begin using some form of contraceptive. Men rarely participate in birth control measures; the use of condoms is resisted by men with the notion that it results in the reduction of pleasure. Some women use intra-uterine devices (IUDs), but it is discouraged because of a fear of tampering with the uterus and obstructing vaginal and menstrual fluid. According to a survey of 320 housewives in Garhwal, about 40% used some form of contraceptive. Those using contraceptives practiced the following: 59% tubectomy, 7% vasectomy (by their husbands), 7% IUDs, 10% pills, 10% condoms, and 7% other methods (Srivastava, 1997).

HEALTH THROUGH THE LIFE CYCLE

Pregnancy and Birth

Getting pregnant is a major transition for a married woman that dramatically enhances her status in the family. Now she is the bearer of potential sons who will continue the family lineage and provide economic security to their parents and grandparents. Pregnancy is a complex event that is influenced by cultural rules, supernatural beliefs, economic situations, and healthcare resources. Women work extremely hard (up to 16 hours a day) until very late in the pregnancy, causing health complications, poor pregnancy outcomes, and even miscarriage. Pregnancy does not alleviate the burden of the day-to-day household chores until the final weeks (Rao, 1995).

A pregnant woman is supposed to avoid any polluting agents and places, such as lonely, dark, haunted places, and cremation grounds. Women avoid heat- and wind-producing foods. Milk and milk products are avoided, especially in the second trimester, because they are believed to spoil the fallopian tubes and might increase the size of the fetus. A pregnant woman is not supposed to attend any rites of passage ceremonies, including birth, marriage, and death in the community. A pregnant woman is vulnerable to all kinds of evil forces. A pregnant woman avoids throwing clippings of her hair and nails because of the fear that a jealous person can perform sympathetic magic using these items.

The fetuses are rarely aborted. However, those who decide to abort use certain mechanical practices or local herbs to induce abortions for unwanted pregnancies. Garhwali believe that a pregnant woman can be invaded by the spirit of an unmarried woman, preventing a baby from settling in the womb. If a woman has multiple miscarriages, they perform a ritual called *aunja pujna*. This ritual is conducted on the darkest night of the month when the mother, two or three males, and a healer visit a water source in the natal village of the mother. The ceremony takes about an hour. The ritual paraphernalia include an animal for sacrifice (lamb, goat, or chicken), seven kinds of grains (*satmaju*), black glass bangles, a black quilt, and black thread. The animal is sacrificed and eaten raw with a pinch of salt and pepper by the members of the ritual team. The healer wraps black thread around the woman's wrist. She takes off all her clothes and discards them at the ritual site, covers herself with the black quilt brought by the group, and proceeds to her in-law's village. In her in-law's home, she remains confined in a completely dark room for two or three days.

Babies are usually delivered lying flat on the back, which can adversely affect the mother's blood pressure and pulmonary ventilation, decrease the normal intensity of contractions, inhibit efforts to push the baby with contractions, increasing the need for forceps, and inhibit the spontaneous expulsion of the placenta. Access to healthcare services is fairly restricted. According to one survey of 320 housewives, 88.9% of deliveries took place at home assisted by family members or midwives. The majority of deliveries were performed by midwives (Srivastava, 1997). The usual practice of applying force on the abdomen has led to traumatic rupture of the uterus in some cases. Sometimes more than one person applies force, even using their feet. A razor, household knife, or sickle is used to cut the umbilical cord. None of the instruments are sterilized, and boiled water is not used in all cases. The outer tip of the umbilical cord is burned, turmeric and castor oil are applied, and the placenta is buried in an isolated place. Roughly one or two women out of every 100 die from birth-related complications. A majority of these maternal deaths are due to anemia, hemorrhage, toxemia, sepsis, induced abortion, and miscarriage.

Infancy

The birth of a baby in the family brings mixed emotions among people. While it is a time of celebration, the process of the birth itself is considered unclean. Special care is taken to ensure that all ancestral worship is performed, especially on the patrilineal side. Spirits and other people can be a threat to both mother and baby, who are vulnerable to spirit intrusion and evil eye. Visitors are regulated and need to wait outside before they can get in. Any kind of visible birth defect is considered an act of a deity or spirit or an outcome of sins previously committed by one of the parents. After the baby's birth, the mother is not brought into the sun for five days and is not exposed to even an oil lamp lit in the same room. On the fifth day, the village priest mixes cow urine with cow milk, puts a copper coin in it, and chants verses. This liquid is then given to the mother and all family members to consume as ambrosia, which is supposed to have purifying effects for the whole family.

Midwives or a sister-in-law usually takes over the responsibility of maintaining the physical well-being of the new mother. Sometimes women return to their natal home for deliveries to be taken care of and to avoid daily chores in their homes. Women are given a semi-solid diet for 21 days after delivery. These 21 days, called *sujan*, are considered polluting for the entire family; the family limits their participation in community celebrations and exchange of food with others. On the twenty-first day, the family performs rituals, prepares special food, and thereafter returns to mainstream life in the community. According to the NFHS-2 survey, only 44% of the mothers with children born in the preceding three years received an antenatal check-up (International Institute for Population Sciences & ORC Macro, 2000, p. 7).

In Garhwal, children are gently rubbed with warm oil twice a day until they are two years old. Most mothers squeeze their first milk before feeding the baby and may not breast-feed the new baby until the second or third day after delivery. On average, children are breast-fed for two years, which also helps prevent conception for these women. Since many infants are taken care of by grandmothers during the day, some infants suckle on their grandmothers' breasts as well, even though there is no milk. Children co-sleep with their mothers until 8 to 10 years of age. Sometimes sons get preferential treatment in terms of food and healthcare because they are considered essential for continuation of family life and potential support for parents and grandparents.

Childhood

Childhood is considered to last until 12 years of age. Since mothers are busy with household chores, children are taken care of by grandparents or older siblings. Only about 69% of children get immunizations. The children in Garhwal suffer from a variety of communicable diseases. Common diseases among children are pneumonia, upper respiratory infections, diarrhea, and fever. Approximately 4% of the children experience disability, with hearing loss being the most common. Child mortality in Uttaranchal is 56 per 1,000 (International Institute for Population Sciences & ORC Macro, 2000, p. 6). Malnutrition is commonplace due to inadequate and imbalanced nutrition because of a higher dependence on the carbohydrate-based diet as opposed to the traditionally diverse diet. According to NFHS-2, more than half of the children under the age of three are underweight and, between 6 and 35 months, 77% are anemic.

Adolescence

When girls reach the age of puberty, they are expected to take the role of adult women—increasing participation in the household chores, restricting lifestyle, and avoiding mixing with boys. Menstruation for a teenage girl is a traumatic event and is considered unclean and a matter of shame. Menstruating girls are not allowed in the kitchen or in any religious place. Family members will not even drink water brought by her. She is considered untouchable and is not supposed to touch green and fruiting trees. After the fifth day, she can be reintegrated with the family and community after a purifying bath. The boys also reach adolescence at the age of 12 and begin to notice changes in their physical appearance. They are expected to help their fathers in the field with the agricultural work and take care of the animals. Neither girls nor boys go through any specific rites of passage during this transition, and there is no cultural practice of genital operation either for the girls or boys. Many of the health issues of childhood continue through adolescence.

Adulthood

Women in Garhwal have to work extraordinarily hard compared with men. Many men work in the cities or are in the military. Even those who are home do not assist the women with the household work. Women complain of general weakness, body aches, and anemia and are prone to accidents in the fields and forests. Women report respiratory infections, chronic bronchitis, gall bladder and urinary stones, and gynecological problems. Many married women have limited access to healthcare and suffer from reproductive health problems. According to one survey, 67.7% women reported abnormal vaginal discharge but 87.6% had not received treatment for it. Many others suffer from urinary tract infections, pain, and bleeding. Sometimes men bring sexually transmitted diseases. Adult men suffer the health consequences of drinking country liquor, which many Garhwali men do. As described in the disease causation section, mental illnesses, presumably caused by the animistic, magical, and mystical agents, are common among Garhwali adults.

The Aged

The life stage of the elderly starts at roughly 60 years of age. The elderly continue to live in joint family households and contribute significantly by watching fields and taking care of children. Women sometimes help prepare meals, and both men and women help feed the animals. The elderly hold considerable power within the family and community and play important roles in village and regional politics.

With age and inadequate and imbalanced nutrition, the majority of the elderly suffer from chronic diseases, including body aches, stomach ailments, respiratory illnesses (due to prolonged use of cigarettes or pipe tobacco), diabetes, cardiovascular diseases, blood pressure, and hearing and vision loss. Women also suffer from back problems stemming from lifting heavy loads and gynecological problems. Owing to poor dental hygiene, both older men and women have very poor oral health. As much as half of the elderly are chronically ill. The elderly are taken care of by family members. Their generation relies primarily on religio-medical healing and herbal treatment.

Dying and Death

Death is distressing for most Garhwali. The entire lineage, and sometimes the whole village, is in a state of mourning. High levels of emotion are expressed in private and in public. The body is taken to the cremation site in a procession. Family members take the clothes off the body and apply purified butter and sandalwood paste to the body. The body is then cremated in the presence of male members of the family, relatives, and community. The bone remains and the ashes are thrown into the river. After the cremation, an elaborate ritual ceremony lasts for more than 11 days. During these days, the family and the lineage of the dead are considered polluted and do not attend any celebrations. On the last day, relatives and members of the community are invited for a feast. Because Garhwali believe in ancestral worship, the dead are treated with special reverence. They also believe that the dead could be reincarnated in human or animal form.

Changing Health Patterns

Like anywhere else in the world, changes are taking place in the Garhwal region. While there is an increased use of biomedicine, there is a gradual loss of traditional healing practices and resources. Owing to exposure to the outside world, the Garhwali beliefs about disease causation are consistently changing from supernatural to natural causes. Women are becoming more aware of their health problems and seeking more healthcare services. With increasing contact with outsiders, Garhwali are also contracting new infectious diseases including sexually transmitted diseases. At this time, there is not much knowledge and awareness among Garhwali about the threats of HIV/AIDS.

References

Berreman, G. D. (1997). *Hindus of the Himalayas: Ethnography and change*. Delhi, India: Oxford India Paperbacks.
International Institute for Population Sciences & ORC Macro (2000). *National Family Health Survey (NFHS-2), 1998–99: India*. Mumbai: International Institute for Population Sciences (IIPS).
Kedia, S. K. (1997). *Impacts of involuntary resettlement on ethnomedical systems: A case study from North India*. Unpublished doctoral dissertation, University of Kentucky.
Kumar, K. T. (1991). Health and culture in rural Garhwal. *American Journal of Preventive Medicine, 7*(10), 63–64.
Rao, B. (1995). *"Is she ill or is she not": Female sexuality, gender ideologies and women's health in Tehri Garhwal, North India*. Unpublished doctoral dissertation, Syracuse University.
Rawat, A. S. (1989). *History of Garhwal, 1358–1947: An erstwhile kingdom in the Himalayas*. New Delhi: Indus.
Rizvi, S. N. H. (1991). *Medical anthropology of the Jaunsaris*. New Delhi: Northern Book Centre.
Srivastava, S. K. (1997). *Women and health, as reported from a community health need assessment survey in 1997*. Retrieved August 4, 2002, from http://www.education.vsnl.com/phalguni/health.html

Garifuna

Nancie L. González and Gloria Castillo

Alternative Names

Garifuna, which is actually an adjective, is the term now preferred by the people themselves, and most widely used, but until the 1970s they were more often known in anthropological and travel literature as *Black Caribs*, and as *Morenos* (*Dark Ones*) to their compatriots of other ethnicities. When speaking their own language, however, the correct usage is "(we) the *Garinagu*," which derives from Arawakan *Kalinago*.

LOCATION AND LINGUISTIC AFFILIATION

The Caribbean coastline of Central America, extending from Belize to Nicaragua, has been home to the Garifuna since 1797 when the British forcibly transported them there from the island of St Vincent in the Lesser Antilles (González, 1988). Their phenotype reveals their African heritage, but their language, culture, and genotype demonstrate that they are a hybrid people, with both African and (South) Amerindian roots (Crawford & González, 1981; Taylor, 1955). The people encountered by the first Europeans in the Caribbean became known as Caribs, and the language family widely spoken there and in the tropical forests of South America, as *Carib*. The term is probably a corruption of *Galibi*. By the middle of the 18th century, after numbers of enslaved Africans had arrived—by capture or through flight—on St Vincent, the term Black Carib was coined to refer to the population darkened by intermixture.

Ironically, the language, still spoken today by perhaps 75% of all Garifuna, is not of the Cariban, but of the Arawakan family—also originating in South America. Despite some allegations to the contrary, it does not contain more than a few possible words of African origin (Taylor, 1952; Taylor & Hoff, 1980). It seems that it was important for this new group to hide and deny their African roots, at least at first, in order to escape recapture by Europeans. In time, the Amerindian culture that had received and nurtured them became theirs. However, both the people and their culture were further "Africanized" through continued contacts with other African people, especially after 1763, when the British seized the island and began major sugar cultivation, importing large numbers of African slaves as laborers.

OVERVIEW OF THE CULTURE

Subsistence fishing and the cultivation of cassava, yams, rice, plantains, and various other crops have been the basis of the economy and the diet for centuries. However, even on St Vincent there is evidence that the Garifuna learned to desire some European foods and goods, and were willing to do migrant wage labor in order to obtain them. In Central America this practice escalated into what became a full-time endeavor for many men, and a part-time experience for most. Increasingly the men remained away for longer periods during their mature years, and although their sense of belonging to their home villages remained strong, their women, by default, took over the roles of parent/provider/protector/disciplinarian/guardian of ritual/ and more (Kerns, 1983).

Since the 1960s, however, more and more women have also chosen to emigrate—primarily to the United States, where they either join their men, or make their way independently. As more Garifuna are born in the U.S. and benefit from the educational opportunities there, the diaspora has enlarged, become more dispersed, and changed its character. Many Garifuna are now well educated, reasonably affluent, and are truly transnational in their outlooks. Some, having achieved a comfortable seniority, have chosen to retire to their original villages, where they live in such relative affluence, that it sometimes makes them targets of unkind gossip, and a certain alienation. Others, having experienced such, give up and return to the United States.

Among these expatriates are some who rue the loss of what they still define as "their" culture, and they have organized several mechanisms for retaining, rehabilitating, and reeducating themselves and the world at large as to their origins and their ethnicity. Belizean Garifuna, both at home and abroad, have been especially active in this, and one of their successes has been the May 2001 United Nations proclamation of the Garifuna culture as "Masterpiece of the Oral and Intangible Heritage of Humanity."

The growth of international tourism has directly affected the Garifuna, whose culture is seen as deliciously exotic by nearly everyone. This has, no doubt, helped to slow out-migration as individuals and villages struggle to organize themselves to receive and entertain visitors from abroad. This has had both beneficial and negative impacts on health. Although providing some cash income, it has also brought new diseases, drugs, and prostitution.

The Sources of Garifuna Ethnic and Political Identity

The symbols of ethnicity for the Garifuna, wherever they happen to live, are language, indigenous spiritual beliefs and practices, and increasingly, what they believe to have been a forgotten and buried African past. Although they see themselves as different from other African-derived peoples whom they encounter in the United States, there are good reasons for them to identify their culture as a variant of African, rather than North Amerindian, culture. It is difficult to have it both ways, so today's Garifuna increasingly cling to the African component of their heritage. They are witness to the success of the civil rights movement in the United States, as well as to the developing awareness on the part of Latin American "Blacks" of the discrimination against people of their color.

Spiritual Life

Traditional Garifuna religion focused attention of the living on the deceased—both ancestors and the spirits of the many children who died young. The former were believed to be either benevolent or malevolent, depending upon the relations one had with them in life, and on how well they are attended to ritually after death. Spiritual guides or shamans, called *buwiyes*, are capable of communicating with these spirits, often using as helpers the spirits of dead children, many of whom seek to be born again to their mothers. Although conversion to Roman Catholicism occurred for many on St Vincent, and for most others during the early years in Central America, it is rare to find a Garifuna today who does not harbor beliefs in the reality and the power of the spirits of their own relatives (Foster, 1982; Howland, 1984). Rituals to honor and placate the dead occur apart from, but in conjunction with Catholic masses, and enlightened Catholic priests close their eyes to what some might think of as blasphemy.

THE CONTEXT OF HEALTH: ENVIRONMENTAL, ECONOMIC, SOCIAL, AND POLITICAL FACTORS

Overview

There exist no precise figures on how many Garifuna there are, either in the world as a whole, or in Central America. Davidson (1979) estimated that there were about 60,000–70,000 in Central America in the 1970s; estimates given by the senior author for Honduras alone suggested a figure of 100,000 in 2000. Guatemala has less than 10,000 today, Belize perhaps 20,000, and Nicaragua less than 5,000. Due to permanent out-migration and developing transnationalism, the population is difficult to estimate, and is probably diminishing in Central America, while it grows in the United States. No one doubts that there are more Garifunas living full or part-time in the United States than in all of Central America today. Because of this wide dispersion, with people living in any one of five different countries, it is impossible to do more than make guesses about this and about mortality and morbidity rates.

Probably the most significant health problem today is AIDS, but other venereal diseases, cardiovascular problems, viral respiratory illnesses, and diabetes have plagued those in Central America for generations. Because of the relatively high frequency of the sickle cell gene, homozygotes suffer sicklemia and its consequences, while heterozygotes enjoy resistance to malaria. (Crawford, 1984; González et al., 1965).

Environment

The Caribbean coastline of Central America is hot, humid, and has long been considered unhealthy by Europeans, primarily because of their susceptibility to the fevers that were endemic in the area until the middle of the 20th century. The Garifuna flourished there, both because of their resistance to malaria, and because they were pre-adapted to the tropics. They knew how to build their houses so as to catch the sea breeze, and to stay out of the sun during the height of the day, and they knew how to harvest the natural resources. Most importantly, they were excellent seafarers (McKusick, 1960; Nicholson, 1976).

Wild palms and other fibrous plants provided raw materials for their traditional baskets, fish traps, mats, hammocks, cassava presses and sifters, and roof thatching. An abundance of hardwoods was available for manufacturing canoes and other wooden implements, such as cassava-graters. Houses were made of reeds, wattle and daub, or palm boards, although cement blocks and lumber are increasingly used today.

Citrus, avocado, mango, and other tropical fruit trees adorn every village and most home sites. The oil-bearing "milk" extracted from the fruit of the coconut palm is a major ingredient in their cuisine, a primary source of cholesterol. Fish and shellfish such as shrimps, lobsters, and crabs, were easily caught using nets and traps. At one time the men also did some hunting in the interior, but this was never so important as the harvests of seafood. Chickens and pigs were introduced while still on St Vincent, and their meat continues to be appreciated, although fish is still preferred. Cows and goats are rare. Upland (dry) rice, beans, and plantains became important after they reached Central America. Unlike the indigenous people of the interior, they prefer bread and tortillas made of refined wheat flour rather than of maize.

In recent decades their healthy traditional diet has deteriorated, due to several interactive factors, including the out-migration of both women and men during their most productive years, the loss of lands suitable for horticulture, the rise of commercial fishing enterprises that seriously affect the availability of fish and shellfish in local waters, the influence of television and other advertising for processed, often "junk" foods, and the scarcity of cash to buy foods that were once locally produced.

Economic Factors

The coastal location was also important for the opportunities it afforded the men to seek employment among the increasing numbers of foreign explorers, casual travelers, missionaries, and would-be settlers arriving in the area after the 1821 independence. Even before that time they earned a reputation for being skilled, valiant mercenary soldiers, serving both government and rebel leaders. Concomitantly, the women, who were the chief agricultural producers, found markets among the same people.

At first the men traveled only on occasion, providing transport for both people and goods in their large seaworthy canoes, and as stevedores loading and unloading ships in the several ports. In Belize and Honduras they cut mahogany and dyewood, and in Guatemala they helped with the construction of buildings in the ill-fated colony of Santo Tomás (Blondeel Van Cuelebrouk, 1846). The migration gradually became more intense, some men traveling daily or weekly to and from their villages, others working seasonally, and still others virtually disappearing for months or years at a time. In the early 20th century they began working on ships carrying fruit—especially bananas—to the United States, and many elected to stay on there, working first on the docks in ports such as New Orleans, Mobile, and New York. Later, many signed on as seamen, cooks, or waiters on ocean liners (Sherar, 1973).

Women began to emigrate in large numbers in the 1960s. Small children were left with their grandmothers, and the households, which formerly contained three generations of related women and their children, were increasingly composed primarily of only two—the middle generation living abroad (González, 1984). Emigrants originally sent money to care for both the children and the elderly, but too often the amounts were insufficient to provide what is needed for optimum health and well-being. Many Garifuna leaders have long recognized this problem, but there is no simple solution. The United States has not always offered the easy and inexpensive life that many expected, and as many emigrants are undocumented visitors, they must accept the lowest paying jobs and cannot take advantage of government sponsored assistance programs. Thus, they often do not have sufficient resources to share with those at home.

Social Factors

Garifuna society has long been characterized as "matrifocal," meaning that women are prevalent and influential, especially in household and village affairs, and in maintaining ritual and other traditions (Kerns, 1983). Households are consanguineally based, centered on a group of related women and their children (González, 1970). This stems from two circumstances. First, there were only 496 men, 547 women, and 422 children upon arrival in Trujillo, Honduras in 1797 (González, 1988). None of the children were under three years of age, which meant that virtually every woman of childbearing age was at risk of pregnancy. High fertility would have been expected and desired under these demographic circumstances. Second, as described above, the out-migration of men became more frequent, and because their return and length of stay were always uncertain, mother–daughter households became the norm. Polygyny had been customary on St Vincent (Helms, 1981), and in Central America this developed into a system of serial monogamy for the women as various men came and went in their lives, while the men had temporary or lasting common-law wives in more than one village. The resulting pattern, which condoned multiple partners for the men, brittle unions, and the right of the woman to seek new partners as old ones disappeared, was beneficial in increasing fertility and in maximizing a woman's ability to seek men who would be economically successful and supportive of her household (Brown, 1975; McCommon, 1982). It also contributed to widespread venereal diseases, and to difficulties in gender or psychological differentiation for the men (González, 1979; Mertz, 1977).

Political Factors

The Garifuna have until recently lived virtually outside the framework of national politics in Central America, although the circumstances have varied from country to country. In part, this was because they lived in relatively isolated villages on a sparsely settled coastline that others eschewed. They are readily identifiable by their color, their accents when speaking the dominant language, and by other cultural characteristics. Despite the fact that the general populations of both Honduras and Belize include a strong component of genes from Africa, the Garifuna are looked down upon, even by others with similar racial characteristics. In Guatemala, which has a much smaller "Black" population, they are feared and despised by the indigenous people, and either ignored by the others or appreciated only for their singing, dancing, and sexual prowess. Ladinos have nearly always held the important local offices, and as a class, hold a higher social position than Garifuna of comparable economic and educational status.

The Garifuna did not participate actively in Guatemala's civil violence (1964–98), but the peace agreement of 1998, in promising to better the human and civil rights of minority peoples, specifically mentions the Garifuna as one of the beneficiary groups. This has helped politicize those still living in the country, and many have left the coast to attend university level courses in the capital city. Formerly this had been a career path open to and chosen by only a handful of others. Although basic literacy and fluency in a second language are nearly universal today, many cannot do advanced academic work owing to poor preparation—itself the result of poor schools, inattention to studying at lower levels, lack of discipline, and expectations of emigration.

Politicization of Garifuna in Belize and Honduras began earlier, and although they do not vote as a minority block in either country, they are taking various actions to make their concerns known and addressed by the national governments. In this, they have been assisted by the United Nations and other organizations that bring together indigenous peoples of the world to share their concerns, ideas, and agendas.

Above the level of the extended family women wield considerable influence, but little political power. Only in Belize have women achieved public office at the national level, and that occurred more often a generation ago than now. New organizations aimed at strengthening ethnic solidarity and pride are dominated by men, and women are now more often relegated to serving as secretaries and as spokespersons for the leaders.

Medical Practitioners

In the villages now and previously, formal biomedical assistance is rare. The elders, as always, are a source of information on home cures, some of them based upon purchased remedies such as paregoric, aspirin, cough medicine, and even antibiotics, the latter, like most medicines, being available without medical prescriptions. In choosing these they may be advised by pharmacists, if there are such. Some villages have government-run clinics, which are consulted as a last resort. Traditional curers, who may be either men or women, include herbalists and diviners, as well as buwiyes. The latter are thought to have innate abilities and to be "called" to their profession, after which they undergo lengthy training with a practicing buwiye. As discussed below, they are usually consulted after home remedies are exhausted, or if the illness is suspected of having a supernatural cause (Cohen, 1984; Staiano, 1981).

A few Garifuna, including the second author, have taken university degrees in medicine, but they rarely practice in their home villages. Some have studied pharmacy, medicine or nursing at different levels in the United States, but they also have, in effect, left their traditional culture and people, and practice in cities. It is not known what percentage of their patients, if any, is Garifuna.

Classification of Illness, Theories of Illness, and Treatment of Illness

Western (Hippocratic) medical ideas were first introduced by French missionaries in St Vincent in the 17th century, but these were of a pre-biomedical character, and became part of what is thought of today as traditional, or pre-scientific medicine. This lore included many or most of the medicinal plants still in use (Cosminsky, 1976), as well as concepts concerning "hot" and "cold" conditions of food and the body, "humors" and so on (Sanford, 1976).

Diseases and illnesses are classified according to whether they result from natural or supernatural causes. The former are often susceptible to treatment with home remedies, including both herbal and chemical agents. Some of these are traditional, others recent introductions through biomedicine. If, however, the condition is not relieved, or if it returns frequently, a biomedical doctor may be consulted. If the symptoms persist, suspicion grows that either some living person has used obeah, or "black magic" against the patient, or that a deceased ancestor is feeling neglected and seeking to punish the sick one and to goad him or her into sponsoring one of several different rituals. All of these begin with a mass in the local Catholic church, and continue, either in the patient's home, or in the case of the more elaborate ceremonies, in a special sanctuary reserved for such occasions.

But before a ceremony is prescribed, either a diviner or a buwiye may be consulted merely to advise as to the cause and the cure. As such, the buwiye is both priest and physician. In his latter role he or she may use a variety of techniques, some of ancient Arawakan origin, such as blowing tobacco smoke or aguardiente over the patient, or using ventriloquism to suggest the voices of spirits talking from the rafters (Glazier, 1980; Taylor, 1951). The buwiye's knowledge and rituals have undergone modification over the past generation, largely through interactions with other Caribbean peoples in the United States, and probably stem from one or another African culture.

If a ceremony to placate an ancestor is prescribed and the patient does not comply, psychological consequences may ensue. The sick one may experience dreams in which the ancestor first beseeches, and later threatens to do even more harm if he or she is not bathed, or fed. The duration of a ceremony varies, but the activities always include drumming, singing, and dancing, and the offering of favorite foods to the deceased. It is expected that one or another participant will experience "possession," a state of trance in which the spirit of an ancestor speaks to the congregation, offering advice, and information (González, 1995; Goodman et al., 1974; Howland, 1984).

In addition to seeking supernatural remedies for those illnesses caused by spirits, Garifuna may also seek help from their ancestors through the buwiye to address diseases such as cancer, for symptoms of mental distress, or for any other condition of unknown or misunderstood etiology. This compares with the use of prayers to various gods in other societies (González, 1966). Garifuna see no conflict between their primarily Catholic faith and the ritual honoring of their ancestors. Indeed, over the past generation more and more Christian elements have been included in the latter (González, 1963, 1966, 1988).

SEXUALITY AND REPRODUCTION

Garifuna have a reputation in Central America for what is perceived by others as promiscuous, guiltless, sexual freedom for both sexes. This has led, over the years, to the higher fertility mentioned above—a necessary element for survival of the group. Their reputation for sexual prowess has brought many European and American women tourists to the area seeking new experiences with the younger men. Some of these have taken their lovers home with them—a possibility that has made such encounters attractive for many men. Male tourists also seek partners—among both men and women, and sometimes these have led to lengthy liasions.

Among themselves, heterosexual unions, whether permanent or fleeting, have long been the norm, but modernity and gay tourism have brought out what may have long been an unconscious homosexuality in some men. Psychologists may attribute this in part to matrifocality, which has penetrated the culture, as well as the household structure, in many ways (González, 1979; McCommon, 1982; Mertz, 1977; Munroe & Munroe, 1971; Sanford, 1976).

Children are desired by both women and men, the latter because they offer proof of their virility, even when they are otherwise disconnected from both the mother and the child. Women desire children to validate their femininity and maturity, and in an effort to tie the father(s) to them and their household, even though this is not always the result. Having children by more than one man increases the number of possible sources of income and other assistance, not only from the various fathers, but from the grandparents as well (González, 1969, 1988; McCommon, 1982). Beliefs about the afterlife also give impetus to reproduction, for to die without issue means that there will be no one to remember and care for one in the spirit world.

HEALTH THROUGH THE LIFE CYCLE

Pregnancy and Birth

Birthing in the villages usually occurs at home, assisted by a midwife, but in cities hospital or clinic births are preferred. Latin American food taboos related to hot and cold apply during pregnancy and the postpartum period, during which the mother is urged to rest, preferably in bed, for 40 days. The placenta is buried, in earlier times in the floor of the house, but the amniotic sac, after careful examination for what it may reveal of the child's character, is dried, then wrapped in a red or black cloth to protect it and the child from evil spirits or living enemies (Idiáquez, 1982). The stump of the cord is also dried, preserved, and may be used for medicinal purposes during the child's lifetime.

The couvade appears to be an ancient, probably Arawakan, custom (Riviere, 1974), of which there are only traces today (Coehlo, 1949; Kerns, 1976; Munroe & Munroe, 1971; Munroe et al., 1973; Taylor, 1950). Fathers are still said to suffer some of the symptoms of pregnancy, such as nausea and labor pains, and they are advised to refrain from certain activities and foods, as well as sexual relations with other women, so as not to endanger the fetus.

Abortions are usually termed miscarriages, and both occur with some frequency. Many of them are never reported to authorities, so there is little information concerning them, but recent observations by the second author suggest they are now increasing, especially among younger women. Most women will admit to having lost children, either before or after birth. Stillbirths also occur with some frequency, for unknown reasons.

Infancy

Children are breast-fed, supplemented almost from birth with a pap made of manioc starch. Unfortunately, although this is thought to be especially nourishing, it provides only calories and tends to suppress the amount of breast-milk, leading to early weaning—often after only a few months, although many continue up to 2 years. The broths from boiling either meat or black beans, with bread, are offered after about 6 months, and the mother will premasticate other foods before offering them to a child without many teeth (González, 1963; Jenkins, 1984).

Respiratory and intestinal illnesses, exacerbated by undernutrition, are two of the most common ailments of childhood. Vaccinations against the common childhood diseases are available through government clinics, and most children are so protected. Still, infant mortality is relatively high in the villages, where biomedical assistance is not always available. Sickle-cell disease may also contribute to many early infant deaths. Protection against the evil eye and other malevolent dangers includes painting blue crosses upon the forehead, and the wearing of amulets containing various charms.

Childhood

Children are cared for primarily by women—the mother, grandmother, older sisters, and aunts—but men who may be resident in the household (e.g., mother's brothers) will be called upon to help discipline little boys when the others fail to control their behavior.

From the time children are able to walk without assistance and to understand directions, they will be given chores, such as carrying small pails of water and other burdens, running errands in the neighborhood, or selling homemade candy in the streets.

Adolescence

In early adolescence, when they can be really useful in household and productive activities, girls and boys may be passed around within the larger kin group wherever help is needed. A child learns to feel comfortable sleeping and eating in several different households (Sanford, 1971).

Adulthood

Despite the relatively high status of women, many of them suffer physical abuse by the men with whom they consort, but since sexual dimorphism tends to be slight, the women are not adverse to striking back—but usually not enough to risk losing the man. Due to the migrant labor customs, men are in short supply between late teenage and retirement. Women often vie with one another for a man's favor, sometimes coming to blows, although shouting matches, including name-calling, insults and denunciations, are more common.

Most people have several different mates during their lifetimes, but as late middle age approaches, many of them settle down into a monogamous relationship, that may or may not include co-residence. It is not uncommon for such couples to go through a legal marriage ceremony, often just days or even hours before the death of one of them.

The Aged

The elders are respected, but often neglected or ignored by their adult offspring or grandchildren. In return, or in order to secure a favor, an individual may threaten to haunt their children after death.

Dying and Death

With advancing age, references to "taking a trip"—a euphemism for death—will be heard. Roman Catholic beliefs concerning what happens later, are blended with those from other traditions, some very old and probably of Amerindian origins, and others clearly showing the influence of more recent associations with other Caribbean peoples of African descent (González, 1995).

Deaths of infants and of the elderly are considered "normal," but persons in the prime of life are not expected to die, and when they do some evil is suspected—either from living enemies, a variety of evil spirits, or from jealous or angry dead ancestors. Even when the death is recognized as the result of an incurable illness or an accident, it is thought that these misfortunes may have been brought about by one or another of the above agents.

Immediately after death the women begin to keen; in the villages the news may be further announced by drumming, and relatives and neighbors begin to arrive for help. If there is a resident Catholic priest, he may be called to administer the last rites, but this is not always done. The body is washed and dressed, placed in a wooden coffin, and set upon a table for viewing. An allnight wake is held, during which paid mourners join the closest family members, offering prayers at intervals (Cayetano, 1977). Coffee or tea and sweet bread is offered at midnight to visitors. Burial occurs early the next morning in the community cemetery. In Belize in the 19th century, missionaries reported that there were rituals on the beach that first night, suggesting that the deceased would be leaving by sea, but this is no longer done.

On the ninth night after death, which today is usually held on a Saturday night, another wake is held. This is considered to be so important that it is attended by nearly the whole community, as well as by friends and relatives from other villages, and even from overseas. It is believed that the soul wanders the earth, and may present a danger to the living, until this final ceremony. The atmosphere is not one of grief, but of celebration, including drinking and feasting, traditional drumming and dancing, gambling, and formal story telling (McCauley, 1981). The custom has been reported from societies throughout the Caribbean area, as well as in Ireland, and its origin among the Garifuna is not well understood. In this way, the community shows the deceased that she/he is remembered, but that it is now time to leave the living in peace (González, 1995; Howland, 1984).

CHANGING HEALTH PATTERNS

The Garifuna have been in contact with Western civilization for some 300 years. Although some traditional beliefs and practices remain, most people today are aware of and respect biomedicine. Most communities have running water, electricity, government schools, television, telephones, fax machines, and computers. Thus, everyone knows something about bacterial and viral infections, diabetes, intestinal parasites, cancer, high blood pressure, and the like. They do not, however, always connect these or their own symptoms with poor hygiene or nutrition, or with potentially damaging behavior patterns involving substance abuse or irresponsible sexual activity.

Many young Garifuna believe themselves to be immune to AIDS, even when they prove to be positive for HIV. Ironically, despite this belief, there is considerable stigma attached to having the disease, so it is probably under-reported. Once diagnosed, many believe it to have been a punishment from God or from the ancestors. Castillo (2002) believes it to be the major cause of death among adult Garifuna in Livingston today. It is also ravaging the communities in Belize and Honduras. It is also no doubt responsible for a good bit of morbidity among adults, including probable psychological problems that go undiagnosed.

Genetically related diseases such as diabetes, arthritis, and sickle cell anemia are similarly less well understood, and are sometimes not recognized for what they are. Recent studies in both Honduras and in Guatemala indicate that drug use and alcoholism are increasing, as is obesity. All are undoubtedly related to aspects of the modern lifestyle, including relative sedentariness, transnationalism, migration, poverty, tourism, and a general decline in the quality of life.

Those Garifuna who do make these connections in general and in their own lives, live in the cities of both Central America and the United States, and at the most visit the villages only occasionally and for short periods. In their effort to learn more of their own roots and traditions, in what anthropologists call a "nativistic" or "revivalistic" movement (Sanford, 1974), they concentrate on the more romantic symbols of their past, and fail to recognize the sometimes-serious problems that beset the people left behind. Local leaders have become more sophisticated in political matters, and in Honduras and Belize there are efforts underway to improve economic opportunities so as to reduce emigration. However, the remaining Central American communities need all the help they can get if the people and culture are to survive.

REFERENCES

Blondeel Van Cuelebrouk, M. (1846). *Colonie de SantoTomas*. Brussels: Le Ministre des Affaires Etrangeres.

Brown, S. E. (1975). Low economic sector female mating patterns in the Dominican Republic. In R. Rohrich-Leavitt (Ed.), *Women cross culturally: Change and challenge* (pp. 149–152). The Hague: Mouton.

Castillo, G. (2002). *Factores sociales y culturales asociados a la prevalencia de VIH/SIDA en la población Garifuna de Guatemala*. Master's Thesis, Public Health, Universidad de San Carlos, Guatemala.

Cayetano, E. R. (1977). Garifuna songs of mourning. *Belizean Studies, 5*, 17–26.

Coehlo, R. (1949). The significance of the couvade among the Black Caribs. *Man* (n.s.), *49*, 64–71.

Cohen, M. (1984). The ethnomedicine of the Garifuna (Black Caribs) of Rio Tinto, Honduras. *Anthropological Quarterly, 57*, 16–27.

Cosminsky, S. (1976). Medicinal plants of the Black Caribs. *Actes of the Forty-second International Congress of Americanists, 6*, 535–552.

Crawford, M. H., & González, N. L. (1981). The Black Caribs (Garifuna) of Livingston, Guatemala: Genetic markers and admixture estimates. *Human Biology, 53*, 87–103.

Crawford, M. H. (Ed.). (1983). *Population biology and culture history of the Black Caribs (Garifuna) of Central America*. New York: Plenum Press.

Crawford, M. H. (Ed.). (1984). *Current developments in anthropological genetics. Black Caribs: A case study in biocultural adaptation* (Vol. 3). New York: Plenum Publishing Corporation.

Davidson, W. V. (1979). Dispersal of the Garifuna in the Western Caribbean. *Actes of the Forty-second International Congress of Americanists, 6*, 467–474.

Foster, B. (1982). An interpretation of spirit possession in southern coastal Belize. *Belizean Studies, 10*, 18–23.

Glazier, S. D. (1980). A note on shamanism in the Lesser Antilles. *Proceedings of the international congress for the study of pre-Columbian cultures in the Lesser Antilles, 8*, 447–455.

González, N. L. (1961). Family organization in five types of migratory wage labor. *American Anthropologist, 63*, 1264–1280.

González, N. L. (1963). Patterns of diet, health and sickness in a Black Carib community. *Tropical and Geographical Medicine, 15*, 422–430.

González, N. L. (1966). Health behavior in cross-cultural perspective. *Human Organization, 25*, 122–125.

González, N. L. (1969). *Black Carib household organization: A study of migration and modernization*. Seattle: University of Washington Press.

González, N. L. (1970). Toward a definition of matrifocality. In Norman Whitten, Jr. & John Szwed (Eds.), *Afro-American anthropology: Problems in theory and method* (pp. 231–243). New York: Free Press.

Gonzalez, N. L. (1984). Rethinking the consanguineal household. *Ethnology, 23*, 1–12

González, N. L. (1988). *Sojourners of the Caribbean: Ethnogenesis and ethnohistory of the Garifuna*. Urbana: University of Illinois Press.

González, N. L. (1995). African-derived religious behavior in the Caribbean and New York. In Michael W. Coy & Leonard Plotnicov (Eds.), *African and African-American sensibility* (pp. 21–34). Pittsburgh: University of Pittsburgh Ethnology Monographs, *15*.

González, N. L. et al. (1965). El factor Diego y el gene de células falciformes entre los Caribes de raza negra de Livingston, Guatemala. *Revista del Colegio Médico de Guatemala, 16*, 83–86.

González, N. L., & González, Ian. (1979). Five generations of Garifuna migration. *Migration Today, 7*, 18–20.

Goodman, F. D., Henney, Jeannette H., & Pressel, Esther. (1974). *Trance, healing, and hallucination*. New York: Wiley and Sons.

Helms, M. W. (1981). Black Carib domestic organization in historical perspective: Traditional origins of contemporary patterns, *Ethnology, 20*, 77–86.

Howland, L. G. (1984). Spirit communication at the Carib dugu. *Language and Communication, 4*, 89–103.

Idiáquez, J. (1982). *El culto a los ancestros en la cosmovisión religiosa de los Garífuna de Nicaragua*. Managua: Universidad Centroamericana, Managua.

Jenkins, C. L. (1984). Nutrition and growth in early childhood among the Garifuna and Creole of Belize. In Michael H. Crawford (Ed.), *Current developments in anthropological genetics. Black Caribs: A case study in biocultural adaptation* (Vol. 3). (pp. 135–147).

Kerns, V. (1976). Black Carib (Garifuna) paternity rituals. *Actes of the Forty-second International Congress of Americanists, 6*, 513–524.

Kerns, V. (1983). *Women and the ancestors: Black Carib kinship and ritual*. Urbana: University of Illinois Press.

McCauley, E. (1981). *No me hables de muerte . . . sino de parranda*. Tegucigalpa: ASEPADE (Asociación para el Desarrollo).

McCommon, C. S. (1982). Mating as a reproductive strategy: A Black Carib example. Doctoral dissertation, Pennsylvania State University.

McKusick, M. (1960). Aboriginal canoes in the West Indies. *Yale University Publications in Anthropology, 1*, 159–178.

Mertz, R. (1977). Psychological differentiation among Garifuna male students. *Belizean Studies, 5*, 17–22.

Munroe, R. L., & Munroe, Ruth H. (1971). Male pregnancy symptoms and cross-sex identity in three societies. *Journal of Social Psychology, 84*, 11–25.

Munroe, R. L., Munroe, Ruth H., & Whiting, J. (1973). The couvade: A psychological analysis. *Ethos, 6*, 30–74.

Nicholson, D. V. (1976). Precolumbian seafaring capabilities in the Lesser Antilles. *Proceedings of the International Congress for the Study of Pre-Columbian Cultures of the Lesser Antilles, 6*, 98–105.

Palacio, J. O. (1983). Food and body in Garifuna belief systems. *Cajanus, 16*, 149–160.

Riviére, P. G. (1974). The couvade: A problem reborn. *Man* (n.s.), *9*, 423–435.

Sanford, M. (1971). Disruption of the mother-child relationship in conjunction with matrifocality: A study of child-keeping among the Carib and Creole of British Honduras (Doctoral dissertation, Catholic University, 1971).

Sanford, M. (1974). Revitalization movements as indicators of completed acculturation. *Comparative Studies in Society and History, 16*, 504–518.

References

Sanford, M. (1976). Disease and folk-curing among the Garifuna of Belize. *Actes of the Forty-second International Congress of Americanists, 6*, 553–560.

Sherar, M. G. (1973). *Shipping Out.* Cambridge, MD: Cornell Maritime Press.

Staiano, K. V. (1981). Alternative therapeutic systems in Belize: A semiotic framework. *Social Science & Medicine, 15*, 317–332.

Taylor, D. M. (1950). The meaning of dietary and occupational restrictions among the Island Carib. *American Anthropologist, 52*, 343–349.

Taylor, D. M. (1951). *The Black Carib of British Honduras.* Viking Fund Publication No. 17, New York: Wenner-Gren Foundation.

Taylor, D. M. (1952). Tales and Legends of the Dominica caribs, *Journal of American Folkore 65*: 267–79.

Taylor, D. M., & Hoff, B. J. (1980). The linguistic repertory of the Island Carib in the seventeenth century: The men's language—a Carib Pidgin? *International Journal of American Linguistics, 46*, 3001–3312.

Greeks

Eugenia Georges

ALTERNATIVE NAMES

None.

LOCATION AND LINGUISTIC AFFILIATION

Greece is located in the southeastern corner of Europe and occupies the southernmost tip of the Balkan peninsula. Its territory of 131,990 km^2 also includes some two thousand islands, only a few hundred of which are inhabited. Greek, the official language, is a member of the Indo-European family.

OVERVIEW OF THE CULTURE

Greece has a population of approximately 10.6 million. With over four million of its inhabitants residing in Athens, the nation's cultural and political center, and another million in Thessaloniki, Greece is a highly urbanized country. The rural population has been steadily declining since at least World War II and comprises less than 20% of the total population. Today, only about a quarter of Greeks are employed in agriculture. Another quarter work in industry and construction, and nearly half of the population works in the service sector. Tourism, both internal and international, is one of the most dynamic areas of the economy. In 2000, per capita GNP had reached $15,400. Unemployment, which officially hovers at about 10% of the work force, is a serious and persistent problem (United Nations Development Program, 2001).

Greece was established as a nation-state in 1832, following a protracted war of independence against the Ottoman Empire. The new kingdom of Greece embraced only a fraction of the Greek-speaking population under Ottoman rule however, and irredentism was a recurrent theme in the century that followed. By the early 20th century, expansion of national territory through war, as well as an exceptionally high rate of natural increase, resulted in tremendous population growth. This growth was not easily absorbed by Greece's relatively underdeveloped economy, which consisted of large plantations producing export crops such as currants and small, fragmented peasant holdings growing wheat, olives, and grapes largely for subsistence.

Throughout its history, Greek society has been characterized by sizable communities of Greek-speakers scattered far beyond the lands of its circum-Aegean core. In the 19th century, diaspora Greeks, many of whom were wealthy and well educated, played an important role in the struggle for Greek independence. Beginning around 1890, a massive wave of emigration sent 500,000 Greeks, 90% of them men, abroad, principally to the United States. A second significant wave of emigration began around 1950, when the economic hardship of World War II and the Civil War that followed prompted many to seek work in Canada, Australia, and Western Europe. By the 1990s, however, economic growth had transformed

Greece into an attractive destination for thousands of migrants from Albania and the former Eastern bloc countries.

Political instability characterized much of the 19th and 20th centuries. The Greek population suffered terribly from the occupations of German, Italian, and Bulgarian troops during World War II. In 1941–42, close to 200,000 people died of famine. The hardship of the war was followed closely by a devastating Civil War that finally ended in 1950. Hopes of political stability were shattered when a military junta seized power in 1967. Its demise in 1974 signaled a new era in Greek society and politics, as peaceful democratic elections became routine, and new social movements, such as feminism, flourished. Greece's full membership in the European Union in 1981 has been a major impetus for further social and political changes (Clogg, 1979; Gallant, 2001).

The family is the central institution of Greek culture. Individuals derive their primary identity from their family, and ideally, family members provide lifelong support, love, and care to one another. Marriage is regarded as the destiny of men and women, and children are highly desired as sources of happiness in themselves and because they perpetuate the family. Most families are nuclear, but residence patterns vary by region. On the mainland, and especially among shepherds, couples live virilocally, sometimes in the same house as the husband's parents. On the islands of the Aegean, the practice of providing a dowry house to daughters commonly results in uxorilocality and close-knit networks of related women. In urban areas, residence tends to be neolocal (Loizos & Papataxiarchis, 1991).

Nearly all Greeks belong to the Greek Orthodox Church. Although everyday practices do not necessarily correspond to the teachings of the Church, the sensibility of Orthodoxy permeates most aspects of Greek life. Under Ottoman rule, religion was an important means of identifying and governing subject populations. With the founding of the Greek nation, Orthodoxy was appropriated by the state to help forge a sense of national identity (Just, 1989).

Greek cultural identity is premised on the often competing pillars of Greek Orthodoxy, which traces its heritage through Ottoman rule to Byzantium, and the heritage of Classical Greece, as idealized by Western European thinkers (Herzfeld, 1982). The tension between these two legacies has led to what some observers have called a "perennial crisis of national identity." In recent years, integration into the European Union has fostered an increasingly cosmopolitan orientation that favors modernization and development.

THE CONTEXT OF HEALTH: ENVIRONMENTAL, SOCIAL AND POLITICAL FACTORS

Greece's health indicators are among the best in the world. Life expectancy at birth in 2000 was 78 years, higher than in the United States and several Western European countries. In 1999, the infant mortality rate was 6 per 1000 live births, and the under-five mortality rate was 7 per 1000. The probability of a female born between 1995 and 2000 surviving to age 65 years was 91.4%; for a man, it was 81.6% (United Nations, 2001).

The economic development of the second half of the 20th century is closely related to the nation's positive health profile, but it cannot fully explain Greece's advantage over many far richer nations. Among the reasons for Greece's favorable indicators are: (1) The continued social significance of strong familial ties, which provide economic and psychological support across the life cycle. Family members commonly rally around pregnant women and new mothers, provide childcare for children when both parents work, and absorb and support the unemployed and the aged. (2) A diet that, despite considerable increase in meat consumption with the growing prosperity still includes abundant consumption of fresh fruits, vegetables, nuts, and legumes. In recent years, the positive publicity surrounding the health benefits of the so-called "Mediterranean Diet" has helped reverse a local trend away from some traditional foods, such as olive oil. (3) The low rate of divorce and the almost complete absence of births outside of marriage, which elsewhere have been associated with poor health outcomes for infants and children. Abortion is an integral part of Greek contraceptive culture, and pregnancies either lead to marriage or end with abortion. (4) Universal access to health care through the National Health System (ESY, established in 1983), as well as to subsidized prescription drugs.

Beginning in the early decades of the 20th century, and accelerating rapidly after World War II, Greece has witnessed a marked decline in the birth rate. At 1.3 births per woman, the total fertility rate is among the world's lowest. In conjunction with the related "graying" of the population, the low birth rate, known in Greece as "the demographic problem," has become the focus of much official concern and policy debate (Georges, 1996).

Greeks have high frequencies of the genes for the blood disorders beta thalassemia (8% of the population) and G6PD deficiency, both of which have been linked to the endemic malaria that plagued the region until fairly recently. G6PD deficiency is a red blood cell deficiency that can result in acute hemolytic crisis, often provoked by eating fava beans, a food commonly consumed in rural areas, especially during the Lenten period. To avoid the stigma attached to this inherited disorder, hemolytic episodes were popularly attributed to "fava bean poisoning" (Trakas, 1981). Children who inherit the thalassemia gene from both of their parents develop a serious disease characterized by severe anemia starting at a few months of age, distinctive deformities of the facial bones and enlarged spleens. They can only survive with the aid of frequent blood transfusions. Heterozygote individuals manifest only mild anemia and can live normal lives. In some regions of Greece, such as the Dodecanese Islands, up to one third of the local population are carriers of the genes for G6PD deficiency or thalassemia. The Greek government has mounted highly successful public health campaigns to screen for the trait and educate the public about thalassemia, known popularly as the "stigma" in Greek. As a consequence of aggressive prenatal screening, fewer than ten infants a year are born with thalassemia major.

MEDICAL PRACTITIONERS

Over 90% of the Greek population are covered by some form of health insurance to which both employers and employees contribute. Of the approximately 35,000 physicians working in Greece, two thirds are employed by the National Health System and one third work in full-time or part-time private practices. There is an ongoing oversupply of medical doctors in Greece, many of whom may wait years for a job opening in the public sector (Colombotos & Fakiolas, 1993).

Lay practitioners such as bonesetters, practical midwives and herbalists were important providers of health care in rural areas through the middle of the 20th century. Today, however, they are completely extinct. In recent years, complementary healing traditions such as homeopathy have become available in the major cities as alternatives to biomedicine.

CLASSIFICATION OF ILLNESS, THEORIES OF ILLNESS, AND TREATMENT OF ILLNESS

The biomedical model of disease is unquestionably hegemonic in contemporary Greece. Many older beliefs about demonic forces and spirits as the causes of serious illness have almost completely disappeared. Nevertheless, classification, etiology, and treatment of illness are characterized by a situation of medical pluralism, as biomedicine coexists with longstanding and persistent humoral models, religious healing, and belief in the evil eye and witchcraft. Furthermore, aspects of the biomedical model display specifically Greek inflections.

Demonic Forces

Up until the middle of the 20th century, many Greeks living in rural areas maintained belief in a variety of demonic forces known collectively as the *xotika*, forces existing "outside" (*exo*) the bounds of Greek Orthodox Christianity. Haunting the woods, streams, and crossroads beyond the protected enclosure of the community, these forces were prone to attack people during liminal periods in the life cycle, particularly pregnancy, birth, and the postpartum, and at the time of one's wedding. Unless protected by symbols and objects associated with the Greek Orthodox Church, such as crosses, icons, amulets, phylacteries, holy water, or olive oil, a person encountering these demonic forces could fall seriously ill or become paralyzed, lose their sanity, and eventually die. These beliefs today are chiefly found among the oldest Greeks, who only reluctantly divulge them to avoid the derision of the younger generations (Stewart, 1991).

Humoral Pathology

Humoral understanding of health and illness underpinned many of the everyday practices of Greeks until relatively recently and still guides aspects of popular health care. Humoral theory was premised on an essentially Galeno-Hippocratic approach to the body that understood health and illness prevention in terms of maintaining an equilibrium between the opposed qualities of "hot" and "cold." These qualities could be literally thermal, as in cold currents of air or water, or metaphorically so, as in the symbolically "cold" valence of particular foods, such as lemons. Illness was caused by a two-step process: first, a person became vulnerable when their body's thermal balance was somehow disturbed; then, exposure to an assault of excessive "hot" or "cold" precipitated an episode of illness. Diagnosis was normally a retrospective process that reconstructs a person's recent history of vulnerability and exposure to specific environmental insults. Deducing the etiology of an illness and deciding on a diagnostic label, in turn, pointed the way to appropriate therapy through the application of the "principle of opposites": illness caused by "cold" is treated with "hot" therapies, and vice versa. This humoral logic persists today, and is implicit in many daily practices whose aim is to protect the body from exposure to cold assaults.

Evil Eye

Another common cause of illness is the evil eye, *to kako mati*. The evil eye is the conscious or unconscious product of human envy. Its vector is a psychic force that emanates from the eyes. Ultimately, the evil eye derives from the devil: it is the devil's work if people are envious and covet the good fortune or possessions of others (Campbell, 1964; Herzfeld, 1986). Once struck by this malignant force, the person who is the object of envy may take ill, or if the coveted object is inanimate, a car, for instance, it may begin to malfunction (Stewart, 1991). The evil eye may also be inadvertently caused by more benign admiration, even on the part of those closest to a person and who otherwise sincerely wish them well. Whatever the origins and intentions, the harmful consequences are the same: headaches, body aches, depression, and severe illness (Blue, 1991). To forestall the undesirable effects of the evil eye, and to avoid being blamed for causing another's misfortune, one should either avoid making compliments, especially of babies and children, who are thought to be especially vulnerable to the evil eye, or be thoughtful enough to bracket one's admiration with an apotropaic gesture: formulaic spitting, or making the sound of spitting ("ftou") three times, or uttering a ritual phrase.

Religious Healing

Religious healing in Greece encompasses a wide variety of practices, ranging from supplications and promises in the form of vows made privately to a saint or to the Panayia, or All Holy One, as the Madonna is most commonly addressed in Greek, to dramatic public displays of self-sacrifice performed during pilgrimages to specialized shrines (Dubisch, 1995). Although women predominate, both genders participate in healing rituals for the gamut of illnesses, from mouth sores (Stewart, 1991) to cancer. Some holy personages specialize in particular ailments or body parts, while others, such as the Panayia, have more general capabilities and may be called upon for all sorts of ills. In addition, the Panayia has her local manifestations, and some of these may be specialists in particular illnesses.

Mental Health

Mental illness, popularly associated with violent behavior, is highly stigmatized and infused with strong sentiments of shame for both the patient and his or her family. Severe mental illness is believed to be inherited through the blood. Thus, the mental illness of one family member may affect the chances of marriage of others in the family. For this reason, efforts may be made to conceal mental illness or to isolate the afflicted family member in a mental asylum.

Mental illness is commonly expressed through somatic complaints, such as headaches and chest pains, and through culturally specific idioms of distress, such as "nerves" *nevra*, and lack of *kefi*, roughly glossed as a positive mood that is associated with high spirits and zest for life. To avoid the stigma associated with the profession of psychiatry, sufferers usually seek care from internists, who will often prescribe tranquilizers. They may also prefer to visit a neurologist who will treat their "nerves." Mental illness may also be attributed to the evil eye, possession by the devil, or to black magic, in which case, religious and magical practitioners may be consulted in addition to internists and psychiatrists. Pilgrimages may be undertaken to holy sites to pray and make votive offerings to saints and to the Panayia for the patient's recovery (Blue, 1991).

Greek psychiatrists, aware of the intense stigma attached to mental illness, often avoid precise diagnoses, such as schizophrenia, in favor of more vague, but less socially damaging labels (Blue, 1991).

SEXUALITY AND REPRODUCTION

Sexual norms and attitudes have changed dramatically over the last few decades. In the past, chastity was a highly valued component of an unmarried woman's identity and reflected positively on the reputation and honor of her entire family. Men, in contrast, were believed to be physically incapable of enduring abstinence for any length of time, and their sexuality was not confined to marriage in the same way as was a woman's (Hirschon, 1989).

In contemporary Greece, the historical value placed on virginity is often described as a "taboo," a relic of the past that has been transcended. Sexual intercourse is now a routine element in courtship, and many brides are pregnant when they marry (Loizos & Papataxiarchis, 1991). However, Greeks have an almost negligible rate (approximately 1%) of births outside of marriage, as unwanted pregnancies are almost always aborted (Agrafiotis & Mandi, 1997).

The one to two child family is today the norm, if not precisely the ideal. Family size is limited primarily through the use of condoms, coitus interruptus and abortion. Abortion has been legal since 1986, but safe medical abortions were available and widely resorted to decades before legalization. Greek women and couples have a marked aversion to medical means of contraception: only 2% of Greek women use the pill and another 7% use the IUD. Abortion is used as a backup when the less reliable, but vastly more popular methods of birth control fail, or fail to be used. An estimated 200,000–300,000 abortions occur each year, in a country with roughly 100,000 live births. Across the political spectrum, the low birth rate and high abortion rate have caused alarm, and concerns for the continuity of the Greek nation and "race" are commonly voiced in the media and other public arenas (Georges, 1996).

Homosexuality, especially among men, and in particular for the passive partner, was highly stigmatized in the past. In recent years, the loan word "gay" has begun to replace the older, pejorative terms and new sexual identities are being articulated. The islands of Mykonos and Mytilene have emerged as popular destinations for both internal and international gay and lesbian tourism.

HEALTH THROUGH THE LIFE CYCLE

Pregnancy and Birth

In the prewar period, pregnancy was not regarded as a state that required the care, advice, and esoteric knowledge of specialists. Pregnancy was nonetheless a condition marked by special behaviors that were carefully observed by the pregnant woman and those around her. Because children were regarded as sacred "gifts from God," the pregnant woman, as bearer of this gift, was herself imbued with a sort of sacred aura. As a consequence, not only her immediate family and wider circle of kin, but her co-villagers generally, treated her with special consideration, forbearance, and respect. Pregnant women, for example, could demand and expect special foods to satisfy their cravings. After their fifth month or so, they could sit comfortably with their legs apart, an immodest posture that would otherwise certainly have provoked criticism and gossip (Chryssanthopoulou, 1984).

If carrying a child conferred a sacred aura and a degree of behavioral latitude on a woman, it also increased her vulnerability to a panoply of dangerous forces. Pregnancy was among those desirable states and qualities, such as youth and beauty, which acted like magnets for demons eager to attack and destroy both the woman and the new life within her (Stewart, 1991). Thus, by virtue of her pregnancy, a woman entered into an endangered state from which she would emerge only at the completion of the 40-day postpartum period.

By far the most significant dietary rule during pregnancy concerned the aroma of cooking food. If a pregnant woman smelled a food as it was being prepared, she was obligated to taste at least a bite. Failure to do so could cause something to go badly awry with the pregnancy. In the worst case, a miscarriage could result; in the best of outcomes, the baby would be born with a birthmark in the shape of the food not eaten on the precise spot that its mother happened to scratch after catching the scent of the food.

In the past, each village had at least one practical midwife, known as the *mammi*. The midwife assisted and supported the woman throughout her labor and delivery, and visited her daily during the week following the birth to examine her and the infant, wash the infant's clothes, and help clean the house. When called to assist a birth, the midwife examined the woman internally to determine her dilation and to ascertain the baby's position. If the baby did not present head first, the midwife attempted to turn it internally by inserting her fingers and manipulating the baby's limbs. External version was also performed by having the woman lie on a blanket held by several people who then tossed her in the air. To prevent tearing and to help women achieve an optimal state of "openness" essential for a successful birth, midwives massaged the perineum with olive oil or soap. The auspicious state of openness was also sympathetically invoked by always leaving the windows and doors of the house slightly open during the birth. In many parts of Greece, husbands removed their clothes, unbuttoned their shirts, and loosened their belts for the same purpose (Chryssanthopoulou, 1984).

Women usually gave birth in a sitting or semireclining position, often astride special birthing stools. To help the woman to push more effectively, in the Dodecanese Islands, a rope was tied to the iron window bars for women to pull down on. In Northern Greece, midwives put some of the exhausted woman's hair in her mouth, causing her to gag and automatically contract her abdominal muscles.

Postpartum. The postpartum period was the most richly elaborated dimension of a woman's procreative experience. The 40-day seclusion period represented a kind of symbolic death and entombment after the act of physiological birth. During this period, the woman who had just given birth, known as the *lehona*, was said to live with "one foot in the grave." Ideally, she should not leave her house for the full 40 days. Confinement protected both her and her newborn from the many dangers, both spiritual and physical, that literally threatened their lives. Safely tucked away within the house, whose every means of entry was guarded by layers of protective devices, woman and infant were buffered from the spiritual attack of the demons whose malevolent designs, activated by her pregnancy, were now driven into high gear with the arrival of new life. At the same time, the *lehona* herself was considered polluting and her isolation and confinement were necessary to prevent the endangerment of others as well. Only at the conclusion of the postpartum period, marked by the special Orthodox ritual of "churching," would both woman and child be safely incorporated into the social life of the community.

Today, pregnancy and birth have become fully medicalized. In distinct contrast to the prewar period, elaborate attention is now focused on the prenatal period as opposed to the postpartum. Women are intensively monitored throughout their pregnancies by a battery of procedures. Fetal ultrasound imaging is universal, and on average, Greek women are scanned four times per pregnancy. The rate of birth by cesarean section has risen rapidly in recent years, from 16% in 1988 to about 40% today. There is no "natural childbirth" movement in Greece, and all births now occur in the hospital (Georges, 1997).

Infancy

In the past, infants were not immediately nursed after birth. The colostrum, the clear, nutritious fluid that precedes the first flow of milk, was always expressed and discarded. In contrast to the colostrum, which, perhaps because of its transitional status, had an ambiguous valence that led to its avoidance, breast milk was regarded as a pure and even quasi-sacred substance. It was used to cure eye ailments of all sorts (Blum & Blum, 1965). Until its mother's milk came in, the infant was spoon-fed chamomile tea. Chamomile also helped the baby expel the meconium, its first, dark sticky stools, and clear the phlegm from its throat. Although the infant was a highly desirable, even quasi-sacred presence, these effluvia were considered "dirty," and had to be eliminated from its body before nursing could begin. Chamomile tea was also occasionally fed to the baby afterward as well, especially if it had a stomach problem. Mothers nursed their infants exclusively for the first 5 or 6 months. At that time, weaning foods, often made of toasted wheat flour, were introduced.

On the third, fifth or seventh day after birth, depending on the region of Greece, a "salting" ritual was performed. The infant's body was rubbed with salt and then bathed. Salt, which has positive connotations and protective powers in Greek culture, helped the baby develop physical toughness, good sense and a logical mind, and "cured" its skin, just as meats were cured for preservation, to prevent rashes.

Infants were wrapped in three layers of inner cloths and then swaddled from head to toe. Swaddling helped the baby's legs to grow straight (although more than one woman I interviewed during my research observed that her child had turned out bow-legged, nonetheless). Swaddling also effectively immobilized the infant and made it easier for women to mind as they went about their chores. When mothers removed the swaddling cloth to change or bathe their babies, they vigorously stretched the infants' limbs to help them develop properly. Stretching was done according to a standard formula, with each leg or arm first gently tugged and then crossed over to touch the opposite leg or arm.

From its ears to its feet, the infant's body was the focus of intense attention and concern, and mothers followed a number of practices to encourage its proper physical and esthetic development. In the weeks following the birth, the new mother was fed plenty of soup made from roosters so that the baby's neck would become strong. To insure that the baby developed a well-shaped head and to prevent colds, a special head covering was worn continuously for at least the first 2 months. To prevent one side of the head from becoming flatter than the other, mothers took care to turn their swaddled infants at least twice a day and after each nighttime feeding.

Childhood and Adolescence

Today, as in the past, considerable vigilance is exercised to prevent children from catching a chill. As with infants, young children may be bundled in layers of clothing that from an American perspective would seem excessive (Sutton, 1998). In the past, girls in particular, were admonished not to stand in pools of cold water or walk barefoot or sit on a cold tile floor, lest the currents of cold air or water attack their womb and cause health problems or even infertility later on. In the past, young girls generally were not told about menstruation, and menarche was often a frightening experience.

Improvements in nutrition and sanitation have resulted in a secular increase in height among Greeks. Average height over the last 70 years has increased 9 cm for boys and 7 cm for girls. Mean age at menarche has decreased to 12.1 years. In recent years, childhood obesity has become increasingly common (Papademetriou, 1999).

In contemporary Greece, older children and adolescents are subjected to considerable stress from the demands of the highly competitive educational system. To pass the grueling exams that until very recently were required for entry into the prestigious national university system, students had to study long hours and attend private cram schools in the afternoons and evenings.

Adulthood

As in the past, marriage continues to mark the passage to full adulthood for Greek men and women. Even in contemporary Greece, marriage is usually the first time that individuals will establish an independent household.

Patterns of parenting have undergone massive changes in the postwar period. Older gendered idioms of maternal suffering and sacrifice have been largely replaced by the discourse of "stress" and "anxiety," as likely to be heard from fathers as from mothers. Both motherhood and fatherhood have been profoundly rewritten, each in their own ways, by the child-centered, emotionally-intensified, financially taxing, consumerist model of contemporary parenting. The Greek context stands out for the intensity with which children's education is culturally valued as a means for confronting an increasingly uncertain future (Tsoukalas, 1977). The cultural expectation that parents will make every effort to assist children in obtaining formal schooling as well as the "shadow education" provided by private cram schools has generated considerable stress and anxiety for parents. To facilitate their children's educational advancement, mothers may forego their daughters' help with housework and fathers may work extra hours or take a second job to pay for their children's private classes. In many parts of Greece, an added burden is the expectation that parents will provide their daughters with a dowry, often in the form of an apartment or house.

Menopause. Middle-aged Greek women regard menopause with some ambivalence. Because the regular flow of blood in the form of menses is an important mechanism for cleansing a woman's body and renewing her health, the cessation of menses is seen by many older women as a harbinger of health problems. On the other hand, menopause puts an end to the constant worry about unwanted pregnancy and the need for abortions, especially acute among Greek women because of the lower reliability of the forms of the most popular birth control methods. Menopause remains largely unmedicalized in contemporary Greece (Beyene, 1989).

The Aged

Institutionalization of the infirm elderly is rare in Greece. The aged are usually cared for by their families, and increasingly, by immigrant home caretakers. Grandmothers in particular often provide valuable household labor and assistance with childcare and housework that enables their daughters to work full-time.

Dying and Death

In rural areas of Greece, a priest is summoned to pray over the dying person and offer communion. At the moment of death, the soul leaves the body through the mouth. The body of the deceased is washed, dressed in new clothes, and laid out in the house. The closest female relatives express their grief by donning black clothes, and kerchiefs, crying and singing heart-wrenching laments known as *miroloyia*. The deceased must be buried within

24 hours, in a loosely made wooden coffin that will facilitate decomposition. After a period of 3, 5 or 7 years, a rite of exhumation is performed, after which the bones of the deceased will usually be deposited in the community ossuary. The condition of the exhumed bones reflects the moral and spiritual status of the dead: clean white bones indicate that sins have been forgiven and that the soul of the deceased has entered paradise (Danforth, 1982; Panourgiá, 1995).

CHANGING HEALTH PATTERNS

Into the early decades of the 20th century, the major killers were infectious diseases such as malaria, which affected an estimated two thirds of the Greek population, tuberculosis, and diarrhea. Today, the top causes of mortality in Greece are the well-known "diseases of civilization": heart disease, cancer, and stroke. To date, Greece has among the lowest rates of HIV/AIDS infection in the European Union. Automobile accidents are a major cause of death. Smoking rates are also very high.

REFERENCES

Amy Victoria. B. (1991). *Culture, Nevra, and Institution: Greek Professional Ethnopsychiatry*. Doctoral Dissertation, Case Western Reserve University.
Anastasios, P. (1999). The increase in height as a mirror of socio-economic change in Greece. *Paidiatriki, 62*, 100–103 (in Greek).
Campbell, J. K. (1964). *Honour, family and patronage: A study of institutions and moral values in a Greek mountain community*. Oxford: Clarendon Press.
Charles, S. (1991). *Demons and the devil: Moral imagination in modern Greek culture*. Princeton: Princeton University Press.
Constantinos, T. (1977). *Dependency and development*. Athens: Themelio (in Greek).
David, S. (1998). "He's too cold!" Children and the limits of culture on a Greek island. *Anthropology and Humanism, 23*(2), 127–138.
Deanna, T. (1981). *Favism and G6PD deficiency in Rhodes, Greece*. Doctoral Dissertation, Michigan State University.
Dimosthenis, A., & Mandi, P. (1997). Greece. *The international encyclopedia of sexuality*. New York: Continuum Press.
Eugenia, G. (1996). Abortion policy and practice in Greece. *Social Science and Medicine 42*, 509–519.
Eugenia, G. (1997). Fetal ultrasound and the production of authoritative knowledge in Greece. In R. Davis-Floyd & C. Sargent (Eds.), *Childbirth and authoritative knowledge* (pp. 91–112). Berkeley: University of California Press.
Jill, D. (1995). *In a different place: Pilgrimage, gender, and politics at a Greek island shrine*. Princeton: Princeton University Press.
John, C., & Fakiolas, N. (1993). The power of organized medicine in Greece. In F. Hafferty & L. McKinlay (Eds.), *The changing medical professions: An international perspective* (pp. 138–149). New York: Oxford University Press.
Loring, D. (1982). *The death rituals of rural Greece*. Princeton: Princeton University Press.
Michael, H. (1982). *Ours once more: Folklore, ideology and the making of modern Greece*. Austin: University of Texas Press.
Michael, H. (1986). Closure as cure: Tropes in the exploration of bodily and social disorder. *Current Anthropology, 27*, 107–120.
Neni, P. (1995). *Fragments of death, fables of identity*. Madison: University of Wisconsin Press.
Peter, L., & Evthymios, P. (1991). *Contested identities: Gender and kinship in modern Greece*. Princeton: Princeton University Press.
Rene, H. (1989). *Heirs of the Greek catastrophe*. Oxford: Clarendon Press.
Richard, B., & Eva, (1965). *Health and Healing in Rural Greece*. Stanford: Stanford University Press.
Richard, C. (1979). *A Short history of modern Greece*. Cambridge: Cambridge University Press.
Roger, J. (1989). Triumph of the ethnos. In Elizabeth Tonkin, Maryon McDonald, & Malcolm Chapman (Eds.), *History and ethnicity* (pp. 71–88). London: Routledge.
Thomas, G. (2001). *Modern Greece*. London: Arnold Publishers.
United Nations Development Program. (2001). *Human development report*. New York: Oxford University Press.
Vassiliki. C. (1984). An Analysis of rituals surrounding birth in modern Greece. Masters Thesis, University of Oxford.
Yewoubdar, B. (1989). *From menarche to menopause: Reproductive lives of peasant women in two cultures*. New York: State University of New York Press.

The Hadza

Frank Marlowe

ALTERNATIVE NAMES

Hadzabe, Hadzapi, Hatsa, Tindiga, Watindiga, Kangeju, Wakindiga.

LOCATION AND LINGUISTIC AFFILIATION

The Hadza are located at approximately 3° south, 35° east, around Lake Eyasi, North Tanzania, Africa. Their language, Hadzane, has clicks, and for that reason has often been classified with the San languages of southern Africa, but may be only very distantly related (Sands, 1995).

CULTURAL OVERVIEW

The Hadza are nomadic hunter–gatherers who live in a savanna–woodland habitat around Lake Eyasi in northern Tanzania (Woodburn, 1968). They number about 1,000 (Blurton-Jones, O'Connell, Hawkes, Kamuzora, & Smith, 1992), of whom many are still full-time foragers and almost none of whom practice any kind of agriculture. Men collect honey and use bows and arrows to hunt mammals and birds. Women dig wild tubers, gather baobab fruit, and berries. Camps usually have about 30 people and move about every month or so in response to the availability of water and berries and a variety of other reasons, such as a death.

The Hadza are very egalitarian and have no political structure, indeed they have no specialists of any sort (Woodburn, 1979). There is a slightly greater respect afforded to older people but not very marked compared to that in other East African societies. One manifestation of this respect is the fact that camps are usually referred to by the name of some senior man, usually in his 50s or 60s. The core of a camp, however, tends to be a group of sisters, one of whom the man has long been married to. There is no higher level of organization than the camp and people move into and out of camps with ease. Post-marital residence is best described as multi-local. Of those marriages where one spouse had parents living in the same camp, in about 60% it was the wife, 40% the husband (Woodburn, 1968).

There are no clans, or unilineal kin groups of any kind. Descent is traced bilaterally with overlapping kin ties so that any Hadza can usually decipher some kin connection to any other. Generation and gender are distinguished. For example, gender is distinguished among grandparents but matrilineal and patrilineal grandparents are not distinguished (though a suffix can be added to distinguish them). Cousins are distinguished by gender but matrilineal and patrilineal are not distinguished, nor are parallel distinguished from cross-cousins. The term for a female cousin is the same as for sister and male cousin the same as brother, though in both cases, they can be distinguished from siblings with a prefix. A distinction is made between maternal and paternal aunts and uncles. Father's brother is called by the same term as father, which may be related to the fairly common practice of the levirate in which a man marries his dead brother's widow. Mother's brother is called by a different term than father. Maternal and paternal aunts on the other hand, are both called by the same term as mother. When personal names are used, there is only a given name. In recent times, when asked to give a surname by government officials, missionaries, or researchers, Hadza use the first name of the father as the child's second name.

There is no organized religion and no belief in an afterlife. There is a cosmology with the sun and moon in the role of mother and father of all the stars. There is a creation myth that explains how people came to be, and how there came to be different tribes. Religious symbolism is associated with epeme meat (the heart, kidney, back, and genitals) of larger game animals. There is a ritual epeme dance performed at night. Men perform one at a time, stomping and singing and whistling to the women who sit and return their calls. The man attempts to rouse the women into getting up and twirling around him. There must be no moonlight, nor firelight, but must be pitch dark. The women try to guess who the man is through the call and shout interchange and his anonymity allows them to interact with him in a way they would not otherwise do, suggesting sexual overtones.

There are several different neighboring tribes of farmers and herders, the Nilotic-speaking Datoga and Maasai, the Cushitic-speaking Iraqw, the Bantu-speaking Isanzu, Iramba, and Sukuma. Since Hadzane is in a completely separate linguistic phylum, this means there are four different language phyla represented, which is a high degree of linguistic diversity for such a small area. Some of these neighboring tribes have been in the area for a long time, the longest being the Iraqw, who moved down from Ethiopia 2,000–3,000 years ago. Relations between the Hadza and their neighbors are somewhat hostile but do involve some trading. For example, the Hadza give the Datoga honey which is made into beer and the Hadza in return get some beer or meat. The Hadza also trade meat and snakebite medicine for iron, cloth, and food. The Hadza resent the encroachment of the pastoralists, especially during the dry season when their herds can drink up all the water and eat up the plants needed to support the wildlife the Hadza hunt. In days past, Hadza would occasionally hunt a cow belonging to the pastoralists but if caught, would be hunted down and killed by a posse of pastoralists. When the first European explorers traveled in Hadza country, the Hadza would hide, which was probably their response to many outsiders (Marlowe, 2002b). Obst, the first person to write about the Hadza who actually met them, was told that around the turn of the 19th century the Maasai would hunt and kill Hadza (Obst, 1912).

There is no written Hadza language and until very recently few Hadza had had any schooling. Today, about half of the Hadza have been to school for a year or two. Only some of those who have attended school know their ages so we have to guess at their ages. In the late 1950s, James Woodburn collected genealogies, in the 70s, Lars Smith began collecting demographic data and this was continued from 1982 on by Nicholas Blurton Jones (Blurton-Jones et al., 1992). Thus, for those who were born from the early 60s or 70s, we know their ages well. In the late 1960s a team of researchers did an anthropometric study including measurement of color blindness, blood pressure, cholesterol, levels of certain diseases, among other things (Barnicot, Bennett, & Woodburn, 1972; Barnicot & Woodburn, 1975). Only recently has any genetic research been conducted and results are yet to be published.

The Context of Health: Environmental, Economic, Social, and Political Factors

The most important fact about Hadza health is that they are hunter–gatherers who live wholly outside during the half of the year when it is dry and only sleep in minimal grass huts during the rainy season. They occupy an area of about 2,500 km^2 at a population density of about 0.24/km^2 (Blurton-Jones et al., 1992). They live in camps that average 29 individuals and move about 10 times per year, though the number of moves is slightly decreasing these days (Marlowe, 2002b). Because they live in the open at low densities and move frequently, they are less vulnerable to many of the contagious diseases that spread among their farming and herding neighbors, who live indoors. In 1964, soon after independence, most Hadza were rounded up by the army and forced to settle at Yaeda Chini where a school and clinic were built in order to settle and modernize them. Within a few months however, many Hadza caught contagious diseases and many died with, "respiratory and diarrheal infections" (McDowell, 1981, p. 7). This caused the Hadza to return to the bush. Today, there are no Hadza children in that school and the clinic is used mostly by the other ethnic groups who were attracted to Yaeda Chini by the school and clinic.

The Hadza have a much less monotonous diet than their agricultural neighbors, who eat maize or rice almost every day with only the occasional bit of meat. The Hadza eat a variety of berries, tubers, honey, baobab fruit, and a wide variety of game from birds to mammals. Of course, there is more fluctuation in the quantity of food consumed by the Hadza than among their agricultural neighbors. However, while the Hadza are often hungry, they do not recall any Hadza ever starving to death. When some big game animal is killed, they gorge themselves for days. During the berry season however, they may sometimes eat almost nothing but one type of berry for 2 months. A variety of wild tubers, three species in particular, are the staple of the diet since they can be found all year round. Medium-sized to large game is shared pretty equally among all those in an average size camp and this sharing helps minimize the variance in daily consumption. There is probably less equitable sharing of other types of food but still some sharing occurs, which also minimizes variation in daily consumption and variation in the amount consumed by each household. This extensive sharing of food that is taken back to a central place, which is typical of foraging populations, must have had a significant impact on human life history since it subsidizes young even after weaning. However, children begin foraging for themselves quite early and by age 10 acquire about half their own needs (Blurton-Jones, Hawkes, & O'Connell, 1989).

Even though the Hadza are sometimes spared epidemics that hit their neighbors, they have an appreciable infant and juvenile mortality rate. Infant mortality in the first year is 21%, and juvenile mortality by age 15 is 46% (Blurton-Jones et al., 1992), both of which are close to the mean for foraging populations (Marlowe, 2001). By the time women have completed their child-bearing years, they have given birth to an average of 6.2 children (also about the mean Total Fertility Rate for foragers). The total population of Hadza is slightly increasing, perhaps partly because it is rebounding from past declines caused for example by the Maasai expansion in the late 1800s, and the deaths during the 1960s settlement attempt (Blurton-Jones et al., 1992). The life expectancy at birth is about 31 years but this is greatly driven by the infant mortality rate, and does not mean there are few old Hadza around. A woman who survives to age 45 has a life expectancy of about 21 more years (Blurton-Jones, Hawkes, & O'Connell, 2002). The mean inter-birth interval regardless of whether children live or die is about 3.4 years.

The overall sex ratio is very close to equal, as is the operational sex ratio (OSR), the ratio of reproductive-aged men to reproductive-aged women. The sex ratio of those under 5 however, is quite skewed toward males, who die at a higher rate. This is despite the fact that male infants nurse at a higher frequency than females. Fathers spend more time holding and interacting with male infants and toddlers but this is almost balanced by mothers spending slightly more time with female children, so that overall care received by children is not significantly different for males or females (Marlowe, 2002a).

Adult male weight was 53.1 kg in the late 1960s and 46 kg for adult females (Barnicot et al., 1972). In 2000, I found male weight had not changed (53.6 kg), but female weight had increased slightly to 47.2 kg. Height was 1.625 m for adult males and 1.513 m for adult females, up from 148.6 cm in the late 60s. I found the percent of body fat was 20.4% for adult females and 11% for adult males. Body Mass Index (BMI) is 20.2 for adult males and 20.6 for adult females. These statistics show that the Hadza are not malnourished, and in fact are in quite good shape for a subsistence population, with men having plenty of muscle and women plenty of fat.

Despite being in good health generally, the Hadza have a hard life and many have had broken bones or serious wounds. This is evident in their fluctuating asymmetry (FA). FA is a measure of the deviation from perfect symmetry in bilaterally symmetrical traits, which is assumed to reflect the degree of environmental stress experienced by organisms. Measuring 10 body traits, Hadza FA is significantly greater than FA in the U.S. (Gray & Marlowe, 2002). The Hadza sleep on the ground on an Impala skin, with little covering, sometimes a thin shawl to keep off the cold before dawn. Both men and women get lots of exercise since women go foraging about 4 hr a day and men about 6 hr. While foraging, women dig with sticks in hard ground to get under big boulders, which is very tough work. Men often use axes, they make, to chop into trees and get the honey in bee hives, which is jarring to the body.

The Hadza smoke as much tobacco as they can get their hands on (women chew it), but it seems their vigorous activity keeps them from suffering from emphysema because they are not short of breath compared to non-Hadza who try to keep up with them when walking. They have been making stone pipes for many centuries, suggesting that they have been able to get tobacco through some trading with others for a long time, and probably marijuana or another plant before that (Fosbrooke, 1956; Sutton, 1990). The Hadza also drink as much alcohol as they can get, which is very little. They do not make alcohol themselves but trade honey to their agro-pastoralist neighbors who use it to make beer and give them some in return.

Normally, the Hadza receive little or no standard medical treatment. When injuries occur or someone is seriously ill, unless someone like a researcher is around to dispense medicine or take them to the nearest clinic or hospital, they simply endure (though they do have certain medical practices as described below). There is one hospital, which is quite good by Tanzanian standards, only a day's walk from part of Hadza country but since it is still a long walk up steep hills, few Hadza are treated there. There are three small clinics with very limited facilities and medicines a bit closer but unless someone pays for them, Hadza are rarely treated. For the most part, the Hadza continue to exhibit natural mortality and morbidity, only slightly influenced by medical attention. Due to their foraging lifestyle however, they have extremely good eyesight, hearing, teeth, no obesity, and apparently little cancer.

Medical Practitioners

With the exception of the few older women who know how to perform a clitoridectomy, there are no medical specialists or specialists of any kind among the Hadza. Every adult knows about the various medicinal plants and practices that the Hadza use. Any adult present may treat someone with an ailment. Women often sit and groom children, removing lice, washing them, and blowing their noses. When anyone is injured and cannot forage for a while, their close kin usually attend to them and bring food back for them until they recover.

Classification of Illness, Theories of Illness, and Treatment of Illness

Illnesses and causes of death tend to be explained in four ways. (1) If someone falls to their death, the Hadza say the cause of death is simply an accident. (2) If someone is killed by a lion or a snakebite, this is just part of the dangers of the natural world. Likewise, malaria is understood to be caused by mosquito bites and sleeping sickness (African trypanosomiasis) the bite of the tsetse fly. (3) When the cause is less obvious however, and especially when death is sudden, a heart attack or poisoning, and the person had been healthy before, then the witchcraft of their non-Hadza neighbors is often said to be the cause of death. The Hadza do not themselves practice witchcraft but fear that of their neighbors. Even when a death is clearly caused by some disease, the Hadza may say this was due to witchcraft if that person had had some quarrel with someone earlier. The percentage of deaths attributed to such witchcraft is on the rise, and these days even a few Hadza are suspected of having learned witchcraft from their neighbors. (4) Finally, there is the supernatural cause not involving witchcraft. There are few rules or taboos in Hadza society, but one such taboo relates to the eating of the epeme meat. Only the adult epeme men can eat these parts of larger game (heart, kidneys, genitals, and back). Sub-adult males and all females cannot eat this meat nor can they even see the men eat it. If they do, it is said they can get ill or die.

There are several types of medical treatment (Woodburn, 1959). For example, there is a certain type of plant that is boiled and then drunk to relieve the symptoms of malaria. Several plants are used to cause one to vomit after being bitten by a poisonous snake. With the most poisonous snakes, there is little hope since they can kill a person in a minute or seconds, but these plants can apparently work with less poisonous but still deadly snakes, as many people say they have been saved this way. Bark from a certain tree is boiled and consumed like tea to treat syphilis and gonorrhea. A certain plant is given to someone who falls down with a seizure, presumably epilepsy, which must be rare, given that there are only 1,000 Hadza. Some say there are plants that can induce miscarriage in the early stages of pregnancy, though it seems they are rarely used, or that they are ineffective, since women sometimes complain that they just keep having babies and cannot stop. There is also a plant that is supposed to help men overcome impotence.

When one has general pain in the body, a horn is used to create suction to suck "the poison" out. A knife is also used to make cuts and let blood run for general pain. Some Hadza have several scars on their arms and backs as a result. When one is badly cut, a tourniquet is applied after boiled animal fat or honey is applied to the wound. The Hadza do not like to wear bandages and believe it is better to let wounds have fresh air, which is probably true most of the time, but they also have a problem keeping wounds clean and free of infection since hygiene is unavoidably poor.

Sexuality and Reproduction

Boys and girls begin playing house around the age of 6 or 7 and probably begin having sex for real in their mid teens. First marriages follow courtship that is carried on clandestinely and if the couple like each other enough, they begin living together, hopefully with parents' approval though it is not required. Because a man might kill his wife if he catches her having an affair, female marital infidelity is probably fairly rare, though many marriages end when the husband is away for so long (usually pursuing another woman) that his wife begins an open relationship with another man, saying that her husband has left her. Most extramarital sex occurs between a married man and a single woman. Polygyny is rare and polyandry, at least of an overt kind, does not exist. Serial monogamy is the rule. Apparently syphilis and gonorrhea have been present for some time since the Hadza have a treatment which they say cures it. So far there have been few cases of AIDS, even though the frequency is high throughout Tanzania.

Men sometimes experience impotence, and try to treat it with a plant. Infertile women feel sad about not having any children and others feel sorry for her. On the occasions when men admit to having left their wife (they usually say they were left by her) they most often say the reason is that she bore no children. There appears to be no ideal family size. Women sometimes say they wish they could stop having babies. Men seem to think more is better but both men and women think having no children is not good.

HEALTH THROUGH THE LIFE CYCLE

Pregnancy and Birth

The Hadza understand much about conception. A Hadza woman will wash on the last day of her period and then the couple has sex. They think that if she gets pregnant, it is on that first day after her period, though whether she does they would not know until she misses her period a month later. Young women, up until their early 30s perhaps, often carry around a thin, 2–3 foot long stick with designs carved by their husbands, which they say is a fertility stick that will make them get pregnant.

Women give birth at home, squatting. They are attended by their mothers, sisters, and/or friends, or their husband if there is no other woman to help. The attendant cleans the baby and cuts and ties the umbilical chord. Women apparently die at fairly high rates from giving birth, given that I was able to elicit the names of several women who had died in childbirth from many different men and women. Men say women are clitoridectomized because if they were not, the baby would have trouble coming out, since the opening would be obstructed by the protruding clitoris.

There is no evidence of infanticide and the Hadza say only in the case of severe deformity, might a baby be killed. Even when twins are born, they do not kill or neglect one, though the risk is higher that one will die since it is difficult to rear two at the same time. Even though some women say they know of ways to induce miscarriage, those same women complain of not being able to stop having babies, so these methods are either ineffective, or they do not really want to use them.

Infancy

Infants are carried at almost all times during the first 6 months of life and nurse on demand. Women carry infants on their back wrapped in a skin, or in these days a shawl. When the baby cries, they swing it around to the front or side so that the infant can nurse while the mother continues her work. Infants are carried on their mother's back when she goes foraging and can sleep right through the vigorous movements of digging for tubers. Children are usually completely weaned by 3 and nurse at low frequency by age 2 years (Marlowe, n.d.). Mothers chew some weaning foods before giving them to the infant. Most mortality occurs in the first year of life. These deaths are apparently due to various respiratory and diarrheal infections, malaria, and at a later age measles, over which parents have little influence.

Childhood

By age 3, children have small slits made on their cheeks by their mother, uncle, or grandparent, which leaves them with small scars to identify them as Hadza. Toddlers are not easy to take foraging because they are too young to walk very far and too old and heavy to carry. They are therefore, usually left in camp and that means someone must be in camp to watch them. This can be almost anyone, but is often a grandmother, and during the dry season when men hunt at night, it is often the father who is dozing in camp all day. Men interact most with their young children when they are 1–3 or 4 years of age (Marlowe, 2002a).

There is very little disciplining among the Hadza. When children are in their "terrible two's" and throw violent tantrums, they pick up sticks and beat adults, who merely fend off the blows, rather than take the stick away. Children tend to learn how to behave from older children because when a 2-year-old hits a 4-year-old, the 4-year-old does take revenge. The most disciplining adults do is to simply make a noise of displeasure. Their leniency extends to letting young children play with whatever they want. A baby can often be seen with a sharp knife or other dangerous object in its mouth with adults not bothering to take it away. Nor do they try to keep little ones out of the fire. Children get burned or get cut and learn on their own. By the standards of modern America therefore, Hadza adults would appear to be guilty of child neglect, if not abuse. Actually Hadza parents are very loving and very rarely hit or abuse their children, and children never feel unloved. They grow up to be extremely well-behaved by the age of 4, without parental disciplining. They do not disobey their parents, nor argue with them, and by 4 or 5 will wait on adults or run errands often without even being asked.

Boys usually get their first bow when they are about 4 and spend hours each day in target practice. By the time they are 7 or 8 they are already very good and kill small birds and rodents. Girls begin accompanying their mothers on foraging forays by around 8–10 and also look after their younger siblings.

Adolescence

By age 10–12, boys and girls may begin having some sex but only begin courting when girls are about 15 or 16 and boys 17 or 18. There is no disapproval of premarital sex but it is kept secret anyway. There is no public display of affection even between married couples. Age at menarche is about 16 or 17. At this time, or close to it, there is a puberty ritual for girls attended only by females, called Mai-to-ko, during which the tip of the clitoris is cut off with a knife by one of the few older women who know how to perform this. The Hadza may have acquired this practice from their neighbors, the Iraqw, Datoga, or Maasai, since all clitoridectomize women, though several Hadza claim it is not a borrowed custom. Males are not circumcised and there is no puberty ritual for boys. There is a ritual eating of the epeme meat when a teenager has killed his first big animal.

By the age of 10, girls are not only supplying about half their own food needs but bringing food back to share with others. They also tend younger siblings and other children much of the time. Boys by the age of 10 or 12 go hunting in groups of 2 or 3 and kill birds and small mammals. There is also no generation gap. Teenagers look up to adults and get along well with the elders. This is at least partly due to the fact that adults do not try to control them and rarely express a strong opinion about whom they should marry. The fact that there is little polygyny means the young males are not in such intense competition for females as they are in many cultures. In addition, since there is no wealth, men do not have the same kind of leverage over their sons as they do in societies where inheritance is important (Marlowe, 2003).

Adulthood

Age at first marriage for females is about 17–18 and median age at first reproduction is 19 (Nicholas Blurton-Jones, personal communication). Age at first marriage for men is around 20. Marriage is not arranged and there is no ceremony, it consists of a couple that has been secretly courting for a while, beginning to live together. Men are expected to kill a large animal to become epeme men and eat epeme meat with the other men. This often happens when a male reaches about 18 or 19. Even men who have not killed such an animal, however, would be considered epeme men after reaching the age of 25 or 30. Men feel the need to bring in enough meat to keep their mother-in-law from counseling her daughter to look for someone better.

There is a fairly high rate of divorce, especially in the early years of a marriage (Blurton-Jones, Marlowe, Hawkes, & O'Connell, 2000). Divorce consists of a couple simply ceasing to live together. Because women usually have their kin around them, there is not a heavy price to pay for being divorced. While a child has increased risk of mortality if its mother dies, there is no evidence of increased child mortality resulting from a child not having its father present (Blurton-Jones et al., 2000). There is also little or no stigma associated with out-of-wedlock births. Normally, there is little domestic abuse, though a man may sometimes hit his wife and be forgiven by others if there is considered to be good cause (infidelity or laziness). However, wife beating is now on the rise (see changing patterns below).

Being a natural fertility population with frequent nursing, strenuous exercise, and mere subsistence level intake of calories, women do not cycle frequently. They probably have less than $\frac{1}{4}$ of the number of menses of American women, perhaps 80 in a lifetime. When they finally reach menopause, they are usually nursing a child and are not sure whether they will resume cycling or not. This may mean they experience fewer side-effects, such as hot flashes, upon reaching menopause, since none of these symptoms has been reported. No cancer has been observed among the Hadza but with such a small population it would be difficult to know whether it exists. A similar absence among other foragers as well has led some to speculate that reproductive cancers may be absent in natural fertility forager populations (Boyd Eaton et al., 1994).

The Aged

Even though the life expectancy at birth is only 31.5 years, there are plenty of very old Hadza. A person who makes it to age 18 is likely to live to be 60 and one who makes it to 45 is likely to live to be 66. Most women over 60 are single, either because their husband has died or left them for a younger woman. These women usually live with one or more of their daughters and tend their grandchildren. They certainly do not feel alienated, no Hadza does, and they are an integral part of the camp and family life, but they express bitterness over the fact that men leave them once they get too old. Old men are shown extra respect until they reach the age of perhaps 70 or 75, when their status drops a bit. Old men are the most likely to fall out of tall baobab trees and to their deaths since they continue to try to collect honey into old age. They are somewhat grudgingly fed and are expected to watch children in camp when they are not out foraging.

Dying and Death

The Hadza leave corpses out for hyenas to eat, or if the deceased is an older person, they may push his or her hut down on top of them and set it ablaze, then move away (Woodburn, 1982). Increasingly, they are under some pressure from the government to bury dead. There is no belief in an afterlife. While most Hadza say they have never heard of anyone committing suicide, there are two recent cases of people attempting to hang themselves. One was a man whose wife left him for another man, another a woman who thought her husband was pursuing other women. In both cases the rope broke and they fell from the tree.

Common illnesses and injuries include: scabies, backache, malaria, eye infections from hearth smoke, broken bones, and wounds from accidents. Causes of death include tuberculosis, malaria when young, sleeping sickness, viral diarrhea, falling from baobab trees when collecting honey, murder by another Hadza, snakebite, and being charged by a buffalo after hitting them with an arrow. Hadza often scavenge meat from the kills made by lions, leopards, and hyenas and this sometimes gets them killed by one of these predators. Finally, childbirth apparently results in the death of the child and the mother at a fairly high rate.

CHANGING HEALTH PATTERNS

The Hadza habitat is being damaged by their pastoralist neighbors' overgrazing and their horticultural neighbors' felling of trees for firewood and planting of crops. They have almost lost the best spot with a large underground spring to the burgeoning number of villagers. More Hadza are attending school where a steady, if monotonous diet, means they grow faster and larger and more are becoming literate. However, future development and a more sedentary existence will probably prove deleterious. While the consumption of tobacco might pose the greater direct threat to health, the ramifications of drinking are much more severe. Promiscuous sex near the village is bound to result in sexually transmitted diseases and death from AIDS. In the village, begging and money from tourists (for whom they perform a song and dance) leads to drunkenness and quasi-prostitution, injuries from fights, and murder. Murder is becoming more common, mainly as a result of increasing alcohol consumption. Other Hadza do not approve of a man beating his wife and will reluctantly intervene to stop it, but when drunk, men may seriously injure or kill their wives with little or no provocation before anyone can prevent it.

REFERENCES

Barnicot, N. A., Bennett, F. J., & Woodburn, J. C. (1972). Blood pressure and serum cholesterol in the Hadza of Tanzania. *Human Biology, 44*.

Barnicot, N. A., & Woodburn, J. C. (1975). Colour-blindness and sensitivity to PTC in Hadza. *Annals of Human Biology, 2*.

Blurton-Jones, N., Hawkes, K., & O'Connell, J. (2002). Antiquity of postreproductive life: Are there modern impacts on hunter-gatherer postreproductive life spans? *American Journal of Human Biology, 14*(2), 184–205.

Blurton-Jones, N., Hawkes, K., & O'Connell, J. (1989). Modelling and measuring costs of children in two foraging societies. In V. Standen & R. Foley (Eds.), *Comparative socioecology: The behavioural ecology of humans and other mammals* (pp. 367–390). London: Basil Blackwell.

Blurton-Jones, N., Marlowe, F., Hawkes, K., & O'Connell, J. (2000). Paternal investment and hunter-gatherer divorce rates. In L. Cronk, N. Chagnon, & W. Irons (Eds.), *Adaptation and human behavior: An anthropological perspective* (pp. 69–90). New York: Elsevier.

Blurton-Jones, N., O'Connell, J., Hawkes, K., Kamuzora, C. L., & Smith, L. C. (1992). Demography of the Hadza, an increasing and high density population of savanna foragers. *American Journal of Physical Anthropology, 89*, 159–181.

Boyd Eaton, S., Pike, M. C., Short, R. V., Lee, N. C., Trussell, J., Hatcher, R. A., et al. (1994). Women's reproductive cancers in evolutionary context. *Quarterly Review of Biology, 69*, 353–367.

Fosbrooke, H. A. (1956). A stone age tribe in Tanganyika. *The South African Archeological Bulletin, 11*(41), 3–8.

Gray, P., & Marlowe, F. (2002). Fluctuating asymmetry of a foraging population: The Hadza of Tanzania. *Annals of Human Biology, 29*(5), 495–501.

Marlowe, F. (2001). Male contribution to diet and female reproductive success among foragers. *Current Anthropology, 42*(5), 755–760.

Marlowe, F. (n.d.). *Who tends Hadza children?* In B. Hewlett & M. Lamb (Eds.), Culture and ecology of Hunter-Gatherer children.

Marlowe, F. (2002b). Why the Hadza are still hunter-gatherers. In S. Kent (Ed.), *Ethnicity, hunter-gatherers, and the "other": Association or assimilation in Africa*. Washington DC: Smithsonian University Press, 247–275.

Marlowe, F. W. (2003). A critical period for provisioning by Hadza men. *Evolution and Human Behavior, 24*(3), 217–229.

McDowell, W. (1981). *A brief history of Mangola Hadza*. Mbulu, Arusha Region: Mbulu District Development Directorate.

Obst, E. (1912). Von Mkalama ins land der Wakindiga. *Mitteilungen der Geographischen Gesellschaft in Hamburg, 26*, 2–27.

Sands, B. (1995). *Evaluating claims of distant linguistic relationships: The case of Khoisan*. Unpublished doctoral dissertation, University of California, Los Angeles.

Sutton, J. E. G. (1990). *A thousand years of East Africa*. Nairobi: British Institute in East Africa.

Woodburn, J. (1959). *Hadza conceptions of health and disease*. Paper presented at the one day symposium on Attitudes to Health and Disease Among Some East African Tribes, East Africa.

Woodburn, J. (1968). An introduction to Hadza ecology. In R. B. Lee & I. DeVore (Eds.), *Man the hunter* (pp. 49–55). Chicago: Aldine.

Woodburn, J. (1968). Stability and flexibility in Hadza residential groupings. In R. B. Lee & I. DeVore (Eds.), *Man the hunter* (pp. 103–110). Chicago: Aldine.

Woodburn, J. (1979). Minimal politics: The political organization of the Hadza of North Tanzania. In W. A. Shack & P. S. Cohen (Eds.), *Politics in leadership* (pp. 244–266). Oxford: Clarendon Press.

Woodburn, J. (1982). Social dimensions of death in four African hunting and gathering societies. In M. Bloch & J. Barry (Eds.), *Death and the regeneration of life* (pp. 187–210). Cambridge: Cambridge University Press.

Haitians

Robert Lawless

ALTERNATIVE NAMES

The ethnonyms for Haitian include the Creole word *ayisyen* and the French word *haïtien*. In English language sources the word is also sometimes spelled *Haytian*. Haiti in Creole is *Ayiti*; and in French, *Haïti*.

LOCATION AND LINGUISTIC AFFILIATION

Occupying the western third of the Caribbean island of Hispaniola, which it shares with the Dominican Republic, Haiti covers 27,750 km^2 and is located 90 km southeast of Cuba, 187 km northeast of Jamaica, and about 1,000 km from Florida. Although Haiti has some flat, semiarid valleys, much of the country consists of rugged and sharply dissected mountains with about two thirds of the almost 28,000 km^2 divided into three mountain ranges. La Selle Peak, the highest elevation, tops off at about 2,680 m. The average temperature falls within a typical Caribbean range with the capital city of Port-au-Prince having a mean annual temperature of 26.3° Celsius.

Although difficult and chronically misunderstood, the language situation is not complex. All Haitians speak Creole, sometimes referred to as Haitian Creole. For most of modern history, however, the official language of government, business, and education has been French, though only about 3% of the population speaks French with any recognizable fluency. Traditionally the educated elites have used the requirement of French to exclude the masses from competing for positions in government and business. Creole has, nevertheless, come into its own in recent years and is gaining prestige as the natural tongue of Haitians and of Haiti. The current constitution states, "All Haitians are united through one common language: Creole. Creole and French are the official language of the Republic." Due to the recent flood of Haitian migrants to Florida, the international decline of the French language, and various economic and cultural trends in the Caribbean, there has been a considerable expansion in the use of English.

OVERVIEW OF THE CULTURE

Demography

Although demographic information is highly undependable, an estimation of the total population would be 8,500,000 million, and the capital of Port-au-Prince would be estimated at 1,500,000 million people. In addition, Haiti has a tradition of emigration, and many Haitians live in the neighboring Dominican Republic, on other Caribbean islands, in the countries of Central America, northern South America, and in North America, especially New York City, which hosts the second largest Haitian community after Port-au-Prince.

History

The Republic of Haiti is the second oldest independent nation in the Western Hemisphere, and it is the only one with an overwhelmingly African culture. The people who occupied the island of Hispaniola at the beginning of the 16th century when Europeans first arrived in significant numbers rapidly succumbed to imported diseases, died in battle, or were killed off by slavery in the first 50 years of Spanish occupation. The Europeans then brought slaves from Africa.

These slaves primarily worked the sugar cane plantations that made the French colony of Saint Domingue an economic success. This success, however, was based on unimaginable brutality and cruelty, and in 1789 the slaves began their 5-year struggle for freedom. On January 1, 1804, the ex-slaves of Saint Domingue renamed their country Haiti and proclaimed its independence.

From 1915 until 1934 the United States occupied Haiti and suppressed peasant movements, revamped the army, and concentrated sociopolitical power in Port-au-Prince. Until 1946 the Haitian administrations were pale reflections of U.S. political interests in the Caribbean. Then from 1946 to 1950 President Dumarsais Estimé ushered in a progressive era that saw an interest in African heritage, cooperation with other Caribbean nations, the development of peasant economic cooperation, the introduction of progressive income tax, expanded education and economic opportunities for the poor, and the rise of a middle class.

In 1957 President François Duvalier emerged as the proclaimed heir to Estimé. The Duvalier regime was marked by brutal oppression against opponents and isolation from the international community. Many professionals fled into exile, and the economy descended into a serious slump. With the 1971 transition from Duvalier to his 19-year-old son Jean-Claude, the United States guided Haiti to a new economic program that featured private investments from the United States featuring no custom taxes, a very low minimum wage, the suppression of labor unions, and the right of U.S. companies to repatriate their profits. For most of the population living conditions continued to decline.

With little to show after 14 years of rule by a second Duvalier, Haitians began protest demonstrations in 1984. In February 1986 Jean-Claude Duvalier fled to France, and an era ended. After several provisional governments the popular priest Jean-Bertrand Aristide was elected president and installed in office in February 1991—5 years to the day after the end of the Duvalier dynasty.

Despite the widespread popularity of Aristide and the heightened expectations of the masses, the military ousted Aristide after only 7 months. No government or state except the Vatican recognized the de facto military regime. This brutal and illegal 3-year occupation of Haiti by its own army saw a rapid downward spiral in the economic and health conditions of Haitians.

Under U.S. sponsorship Aristide was returned to power in October 1994. Since Aristide could not succeed himself, his protege René Préval was elected and took office in February 1996. Four years later Aristide was again elected president and took office in February 2001.

Economy and Occupations

After defeating the colonial government of the French slave owners the newly independent nation faced the threat of a French army returning to re-enslave them. To counter this perceived threat, the new government at first confiscated private land and imposed forced labor in an attempt to develop an export agriculture leading to the importation of war material. Such a plan proved impractical, and eventually the confiscated plantations were distributed to the ex-slaves and the Haitian elite retreated to the provincial cities. The result was the fragmentation of land holdings, peasantization, and the alienation of the masses from the government and the ruling elites. Currently an estimated 80% of the rural population owns its own land, though the plots are fragmented and small, and about 65% of the labor force is in agriculture. Despite the importance of agriculture and the peasantization of Haiti, the government traditionally expends little effort or money on agricultural research or on integrating the rural population into the politics of the country.

Haiti's primary export products have traditionally been coffee, sugar, rice, and cocoa, though the political uncertainties of recent decades have meant a low rate of export. Haiti has some light manufacturing along with a few cotton mills. Before the so-called de facto regime, when offshore industries were in operation, Haiti was a major source of garments, toys, baseballs, and electronic goods for the United States. Many people engage in the manufacture of tools and small items.

Market women from the rural areas bargain their produce in open-air markets, and most of this produce moves by foot as these women often carry heavy loads. All sorts of merchandise may be found in the city markets, including black-market items. Much of the trading is done in kind. The annual per capita monetary income is estimated at only US$480.

Social and Political Conditions

The early distribution of land to the rural population created a large class of peasants who generally regard the government as a nuisance. Haiti does, nevertheless, have a political structure. The nation is traditionally divided into several *départements*, each of which is further divided in several *arrondissements*, and each *arrondissement* consists of several *communes* that usually coincide with church parishes. Finally, each *commune* is divided into several *sections rurales*, each of which is headed by an appointed *chef de section*, who reports to the *commandant* of the *commune*, who reports to the *préfet* of the *arrondissement*. The government official that rural Haitians deal with is generally the *chef de section*. In urban areas the most important government official is usually the *préfet*. For the most part, however, the rural population polices itself under the watchful eyes of village elders. A rather subfunctional police force and a largely corrupt court system exist in the urban areas.

International Relations

Internationally Haiti is closely tied to the United States with a sizeable majority of its exports coming to North America and a goodly portion of its economy dependent on government and nongovernment aid from the United States. Although twice-elected Aristide remains popular among Haitians, his socialistic and anti-American rhetoric means that influential sectors in the U.S. government will continue to oppose his administration.

Since May 2000 when the party of President Aristide won approximately 80% of the seats in a parliamentary election and the U.S.-backed opposition front Democratic Convergence alleged that the election was rigged, Haiti has been in a political and economic crisis. Largely under pressure from the United States, more than US$500 million in international aid has been frozen until the government and the opposition reach an agreement to hold new elections.

Family and Kinship

The family and kinship pattern most prevalent in rural Haiti is the somewhat patrilineal extended family and patrilocality resulting in a cluster of consanguineously and affinely linked joint households headed by the oldest male member. Traditionally the rural population has tended to avoid involvement both with the Roman Catholic church and with the government, which means that this population also avoids legal and church marriages. The result is a wide variety of mating and parenting patterns, including completely informal unions, non-conjugal couples, fathers who do not participate in rearing their children, and non-nuclear family households, as well as conventional church weddings, long-term monogamous unions, and neolocal nuclear family households. Also, both men and women may simultaneously, or in succession, enter several different kinds of union with the same or with different partners. Since all children from all the varieties of conjugal unions have equal rights of inheritance, the complexities of the domestic unit often lead to significant inheritance problems. In addition, the specification of the adult responsible for the care and health of a child can be easily disputed.

Religion

All religions have a close connection with health and illness, and an understanding of the religions of Haitians is essential for developing an appreciation of their health care system. Although some of the population is nominally Christian, the major religion of Haiti is Voodoo. Between 50 and 75% of the population of rural Haiti actively practice Voodoo, and 90% believe in it to some degree. Voodoo is also very popular among the urban working class, and to some degree among people in all classes, including the educated elite. Many important ceremonies revolve around celebrations of milestones in the life cycle, such as birth, maturity, marriage, and death. Other important ceremonies focus on agriculture, planting, harvesting, and insuring a sufficient crop yield. An ancient, affirmative, and legitimate religion that focuses on contacting and appeasing ancestral spirits, Voodoo provides a folk medical system that attributes illnesses to angry ancestors. The performance of ceremonies that appropriately appease these ancestors is, then, extremely important in curing illness. Such ceremonies include divination rites, used to find the cause of illnesses; healing rites, used to interact directly with sick people; propitiatory rites, used for offerings to specific spirits; and preventive rites, used to offer sacrifices to prevent trouble.

THE CONTEXT OF HEALTH: ENVIRONMENTAL, ECONOMIC, SOCIAL, AND POLITICAL FACTORS

Demography

According to various studies, the birth rate is probably around 35.5 per 100,000, and the annual growth rate somewhere just less than 2% per annum; the mortality per 1,000 is approximately 13; the infant mortality per 1,000 is estimated to be about 140; and life expectancy at birth is about 54 years.

Colonial Era

No reliable information exists on the conditions for slaves during the colonial era. The treatment of the slaves was most certainly deplorable, and their diet was no doubt substandard. Various estimates suggest that the life expectancy of slaves after their arrival in Saint Domingue was 7 years. As inhuman as it sounds, the economies of slavery dictated that it was cheaper to export labor and quite literally work the slaves to death rather than waste time and energy on reproducing and maintaining a domestic work force.

Postcolonial Era

Currently tuberculosis is Haiti's most serious disease. The country is also infected with malaria, influenza, dysentery, tetanus, whooping cough, and measles. Eye problems are endemic, and blindness is not uncommon and is usually caused by cataracts, scarring of the cornea, and glaucoma.

Many Haitians, especially the poorer masses, suffer many health problems associated with malnutrition. The daily per capita food consumption is estimated at 1600 calories, and measles, diarrhea, and tetanus claim the lives of many children before they reach 10 years of age.

A 92-page study released in 1992 by the Permanent Commission on Emergency Aid, which represents more than 60 non-governmental development and democracy organizations in Haiti, reported that the death rate has been rising and the health of the population dropping since the 1991 military takeover. Pointing out that there has been a deterioration in state services amounting to a descent into chaos, the report stated, "The situation is extremely critical and just waiting for cholera to strike" (Staff, 1992). Other problems that were reported included an increase in garbage in the streets, a rise in the number of preventable illnesses, and a deterioration in mental health. This report is the most recent one covering a general overview of health in Haiti. There is every reason to believe that the situation has deteriorated further as Haiti enters the 21st century. The most recent estimate, for example, is that only 40% of the population has easy access to potable water (Farmer, 2002).

Current Era

An obvious current difficulty is the U.S.-sponsored embargo. In an electronic message sent in March 2002 to the Haitian Discussion Group the anthropologist and physician Paul Farmer spells out the health problems resulting from this embargo (Farmer, 2002). Farmer's 80-bed charity hospital has delivered health services in the central plateau for the last 18 years, and he writes that the situation there has gravely deteriorated because of a serious lack of resources, medical personnel, and an increase in diseases. He draws a direct connection between this suffering and the aid embargo.

Farmer points out that while the brutality of the 1991–1994 de facto military regime is well known with thousands killed outright and hundreds of thousands made homeless, the current catastrophic decline in overall health is much less publicized. There have been measles epidemics, outbreaks of dengue fever and polio, as well as the appearance of other vaccine-preventable diseases. Most of these diseases are related to the increasing prevalence of malnutrition. According to Farmer, "The nationwide network of public clinics and hospitals was left to fend for itself, and many health professionals left Haiti as this network foundered" (2002).

Although democratic rule was restored in 1994, and a coalition of international donors promised more than

$500 million of aid, no monies have yet been disbursed. During 2001 the United States declared a formal aid embargo that blocks all loans and grants from the Inter-American Development Bank, the World Bank, and the International Monetary Fund. As a result 40,000 or so inhabitants of the commune where Farmer's hospital is located (Thomonde in the Departement du Centre) are without a single doctor or nurse, except for Farmer's hospital, which is now overwhelmed with patients. In the nation at large there is only one physician for every 11,000 patients.

Health Infrastructure

Although Western medicine has been available to the urban elite for several decades and is, indeed, available from a few rural clinics, Voodoo healers are a major part of the medical system of Haiti. Interestingly Haitians place an enormous emphasis on physical cleanliness. It is very important to be clean-looking and clean-smelling in your body, your clothing, and your home. This value may be important in helping Haitians to accept modern medical practices. Bathing is explicitly health-oriented. According to Haitian folk model, regular morning baths are necessary to maintain the balance of the body. Irregular bathing causes heating up of the body and may result in stomach boils. Incidentally bathing, which is often done in a nearby river, is usually done with the clothes on, and it is thought that keeping the clothes on helps cool the body and prevents loose bowels. For the most part, however, health and healing for most Haitians are handled by herbal medicine, bonesetters, injectionists, Voodoo rituals, and by a rich body of folk knowledge.

MEDICAL PRACTITIONERS

Voodoo is a particularly egalitarian religion with both men and women serving as intermediaries, and this rule follows for most of the indigenous therapists. There are at least four types of indigenous therapists who can treat illnesses caused by natural phenomena and two who treat illnesses caused by unnatural phenomena: (1) herbalists, (2) bonesetters, (3) injectionists, (4) midwives, (5) Voodoo priests and priestesses, and (6) sorcerers.

> (1) Among the most common therapists are herbalists (leaf doctors, or in Creole *dòktè fèy* or *medsin fèy*). As with the other indigenous therapists, herbalists have no manuals or organized schools for training. They can be either men or women, and they treat natural illnesses through administering teas and with the use of various baths and compresses, which are usually accompanied by incantations and rituals. They seem to come in families, and every village has several people who specialize in herbal treatments, most of whom work only part-time as herbalists. Many herbalists are also Voodoo priests and priestesses. The herbalists may also use injections and pharmaceutical preparations.
>
> (2) Bonesetters (*dòktè zo* or *medsin zo*) are not very numerous, and frequently those in rural areas with broken bones may have to hike for several hours to the home of a bonesetter. The bonesetters are generally recruited through spiritual revelation and serve a fairly long apprenticeship. They realign broken bones and fix obstructed blood vessels. Most use a flour-based plaster cast that is applied and left on for several weeks.
>
> (3) Injectionists (*pikirist*) usually learn their trade through affiliation with a medical center and are usually found in urban areas. They often are patronized by tuberculosis patients, and they administer the daily streptomycin in the patient's home. Although some injectionists use pharmaceutical preparations, they may also be found in the streets pushing carts holding jars of medicine that often contain only cough syrup. The injectionists diagnose the patient's illness on the spot and select a syrup to inject, which they may do with a previously used syringe. Due to the obvious implications for the spread of disease, especially AIDS, some of the several recent administrations have cracked down on these itinerant injectionists, and nowadays few are found roaming the streets.
>
> (4) Midwives (*fanmsay*) are always women, and it is estimated that they deliver at least 85% of all babies born in Haiti. Some learn their trade from an apprenticeship and others from their mothers, though some have in the past been trained and paid by the state. They all use herbal treatments for all neonatal complaints, and they sometimes consult with herbalists.

The most typically Haitian medical practitioners are the Voodoo priests (*ougan*), priestesses (*manbo*), and sorcerers (*boko*). Their role will be treated in more detail at the end of the next section.

CLASSIFICATION OF ILLNESS, THEORIES OF ILLNESS, AND TREATMENT OF ILLNESS

The traditional healing system of Haiti is complex, but it can profitably be discussed in terms of (1) humoral systems, (2) the concept of personhood, (3) the anatomy of the body, (4) the etiology of disease, and (5) mechanisms of diagnosis.

> (1) Haitian concepts of health incorporate the familiar hot-cold dichotomies imported from ancient Europe and used throughout Latin America. Haitian concepts may also incorporate

humoral-related ideas from West Africa, but little information exists on related African indigenous health systems, which themselves may have been influenced by Islamic sources. In its simplest form humoral systems depend on controlling health through the monitoring of intakes of hot and cold items, primarily food.

Good health requires people to maintain their individual balance of the four humors. Sickness results from a humoral imbalance and from extremes of hot and cold, dry and wet. One's humoral equilibrium can be influenced by age, climate, style of living, and the seasons, and especially by one's diet.

Over the generations in Latin America the humoral system of the Spaniards was altered and simplified with the wet and dry categories disappearing. Although not well studied, the humoral system is widespread in Haiti, and "is a guiding principle in rural Haitian behavior" (Wiese, 1976, p. 198). Interestingly H.J.C. Wiese lists Haitian foods' hot and cold ratings and gives the vitamin and mineral content of Haitian food (Wiese, 1976, pp. 196–197).

P. Minn supplies a recent account of an illness rooted in humoral pathology that is called simply heat (*chalè*) (Minn, 2001). When Haitians say, "*M gen chalè*" (I have heat), they mean that they have been exposed to excessive amounts of heat and are therefore ill, with the most common symptoms being dizziness and headaches, though almost every part of the body, especially the skin, can be affected in various ways.

(2) Religious beliefs in general and religious practices in particular involve some degree of loss of control over the natural world, and all, therefore, leave the individual vulnerable. The defense against being damaged when in a vulnerable state is the establishment of good relations with the supra-natural world. Such good relations come primarily through being possessed by entities from the supra-natural world. Such possession is easier when the person is composed of various tenuously related parts. Such a characteristic is, indeed, found in the Haitian concept of personhood. The head is, for example, seen as the seat of the good big angel and the good little angel (*gwo bon anj e ti bon anj*). The little angel is the one who goes to heaven and is the guardian of the person. The big angel feels emotion and can be displaced through possession. The usual struggle before a person is possessed, which is sometime quite violent, is the attempt of the little angel to prevent the possession. At the death of the natural person the big angel goes back to Africa (*Ginen*).

(3) The head is also the seat of the spirits. It must be washed, fed, and, at death, the master of the head (*mèt tèt*), who controls thinking, must be removed. When people are worried, their head is "loaded." In excitement the head heats up. The medium of heating and cooling is the blood. As might be expected in a system largely based on the humoral system, a major part of both causing and curing illnesses involves manipulating heat and cold, and influencing the state of the blood, and the condition of the head. The tools of manipulation usually focus on the intake of food; food as a sacrifice to the ancestors, food as charity, food as poison, and also oral and injected medicines.

(4) In terms of etiology illnesses have various causes, but the basic distinction is between natural and supra-natural phenomena. Natural illnesses include everything from infectious to chronic degenerative diseases and may be treated by priests and priestesses or physicians or other types of indigenous therapists, according to varying criteria such as severity of illness, customary local practices, availability of competent practitioners, and ability to pay. Supra-natural illnesses are attributed either to sorcery or to the supra-natural spirits and can be treated only by priests and priestesses.

(5) The first step in diagnosis is to determine the complexion of the disease and the natural humoral balance of the patient. Following the principle of opposites, a disease thought to be hot will be treated with cold herbal remedies and the patient will be advised to eat mainly cold foods. A person complaining, for example, of *chalè* in the head will usually be treated with cold water and with medicinal leaves on the head. In general, treatment consists of attempts to restore the normal humoral balance. Actual methods of diagnosis and treatment vary from one priest and priestess to the next, but the pattern is generally something like this: After receiving the initial complaint, which is often very general—a headache or a stomachache—the priests and priestess call for the aid of their spirit helper. They most likely sit before a domestic altar, recite the rosary, light candles, and purify the floor with holy water, often chanting in a language that is supposed to be African. They then become possessed by their spirit helper, who gives them advice about how to proceed. Usually they then collect a detailed medical history from the patient and the patient's relatives. They may also try to get more specific details through the use of various divination techniques.

If the problem is caused by black magic, the priests and priestesses may have to employ counter-magic techniques, often calling again on their spirit helper. Mischievous or neglected spirits can cause illnesses. Sometimes spirits are purchased and told to make trouble for others. Indeed, much of the Haitian population firmly believes in the existence of sorcery and the effectiveness of sorcerers, though they are rare and difficult to find. A sorcerer may also use bad spirits to cause various diseases, such as tuberculosis and epilepsy.

SEXUALITY AND REPRODUCTION

The health focus in terms of sexuality is clearly on AIDS. One of the poorest countries in the world, Haiti also has the highest seroprevalence rate in the Caribbean and the highest rate outside sub-Saharan Africa. In May 2002 representatives of the Haitian health community met with international AIDS experts to discuss the AIDS pandemic in Haiti. Again the focus was on the harm being perpetuated by the U.S.-sponsored embargo. The problem was no longer labeled simply a health crisis but a national development crisis.

Throughout the meeting the specialists, both American and Haitian, recited the overwhelming statistics of poverty and the high rate of HIV/AIDS in Haiti. It is estimated that Haiti accounts for more than 90% of HIV/AIDS cases in the Caribbean, and the number of children orphaned by the epidemic is estimated as approaching 200,000.

HEALTH THROUGH THE LIFE CYCLE

A bibliographic account of works on Haiti in English and Creole lists only 228 items (Lawless, 1992, pp. 194–170). This rather meager literature does not allow for any detailed accounts of the health beliefs, attitudes, and practices as they might change through the life cycle.

REFERENCES

Alleyne G. A. O. (2001). Health in Haiti. *Pan American Journal of Public Health, 10*(3), 149–151.
Arthur, C. (2002). *Haiti: A guide to the people, politics and culture.* New York: Interlink.
Barnes-Josiah, D. L., Myntti, C., & Augustin, A. (1998). The "three delays" as a framework for examining maternal mortality in Haiti. *Social Science and Medicine, 46*, 981–993.
Bentivegna, J. F. (1991). *The neglected and abused: A physician's year in Haiti.* Rocky Hill, Connecticut: Michele.
Bordes, A., & Couture, A. (1978). *For the people, for a change: Bringing health to the families of Haiti.* Boston: Beacon.
Burgess, A. L., Fitzgerald, D. W., Severe, P., Joseph, P., Noel, E., Rastogi, N., et al. (2001). Integration of tuberculosis screening at an HIV voluntary counselling and testing center in Haiti. *AIDS, 15,* 1875–1880.
Coreil, J. (1983). Parallel structures in professional and folk health care: A model applied to rural Haiti. *Culture, Medicine and Psychiatry, 7,* 131–151.
Daniels, J. (2000). U.S. funded AIDS research in Haiti: Does geography dictate how closely the United States government scrutinizes human research testing? *Albany Law Journal of Science and Technology, 11,* 203–224.
Dash, J. M. (2001). *Culture and customs of Haiti.* Westport, Connecticut: Greenwood.
Deschamps, M.-M., Fitzgerald, D. W., Pape, J. W., & Johnson, W. D. (2000). HIV infection in Haiti: Natural history and disease progression. *AIDS, 14,* 2515–2522.
Direksyon Edikasyon Sanite ak Direksyon Ijyen Familial ak Nitrisyon (1986). An nou aprann byen manje pou nou an sante. [Port-au-Prince]: Ministè Sante Piblik ak Popilasyon.
Farmer, P. (1988). Bad Blood, spoiled milk: Bodily Fluids as moral barometers in rural Haiti. *American Ethnologist, 15,* 62–83.
Farmer, P. (1992). *AIDS and accusation: Haiti and the geography of blame.* Berkeley, CA: University of California Press.
Farmer, P. (2002). Dr. Paul Farmer outlines connection between suffering and aid embargo. Haitian Discussion Group. Http://www.webster.edu/~corbetre/haiti-archive/msg11016.html.
Farmer, P., Walton, D., & Carter, L. (2000). Infections and inequalities. *Global Change and Human Health, 1*(2), 94–109.
Fitzgerald, D. W., Behets, F., Caliendo, A., Roberfroid, D., Lucet, C., Fitzgerald, J. W., et al. (2000). Economic hardship and sexually transmitted diseases in Haiti's rural Artibonite Valley. *American Journal of Tropical Medicine and Hygiene, 62,* 496–501.
Freeman, B. C. (1992). *Medical dictionary: Haitian Creole–English; English–Haitian Creole.* Port-au-Prince: Evangélique.
King, K. W., Fougere, W., Webb, R. E., Berggren, G., Berggren, W. L., & Hilaire, A. (1978). Preventive and therapeutic benefits in relation to cost: Performance over 10 years of Mothercraft Centers in Haiti. *American Journal of Clinical Nutrition, 31,* 679–690.
Laguerre, M. S. (1987). *Afro-Caribbean folk medicine: The reproduction and practice of healing.* South Hadley, MA: Bergin and Garvey.
Lawless, R. (1990). *Haiti: A research handbook.* New York: Garland.
Lawless, R. (1991). Haitians, AIDS, anthropology, and the popular media. *Florida Journal of Anthropology, 16*(7), 1–7.
Lawless, R. (1992). *Haiti's bad press: Origins, development, and consequences.* Rochester, Vermont: Schenkman.
Lawless, R. (2002). Haiti: Voodoo, Christianity, and politics. In R. G. Mainuddin (Ed.), *Religion and politics in the developing world: Explosive interactions* (pp. 39–49). Aldershot, Hampshire, England: Ashgate.
Miller, N. L. (2000). Haitian ethnomedical systems and biomedical practitioners: Directions for clinicians. *Journal of Transcultural Nursing, 11,* 204–211.
Minn, P. (2001). Water in their eyes, dust on their land: Heat and illness in a Haitian town. *Journal of Haitian Studies, 7*(1), 4–25.
Staff (1992). *Report.* [Port-au-Prince]: CPAU.
Wiedeman, J. E., Zierold, D., & Klink, B. K. (2001). Machete injuries in Haiti. *Military Medicine, 166,* 1023–1025.
Wiese, H. J. C. (1976). Maternal nutrition and traditional food behavior in Haiti. *Human Organization, 35,* 193–200.

Han

Xingwu Liu

ALTERNATIVE NAMES

Chinese, *Zhongguo-ren, Han-ren, Hua-ren, Tang-ren*. Because of their predominant numbers in China, the Han are commonly referred to as Chinese in the English language, though in China (including Taiwan) they have always called themselves "*Han-ren*" (Han people). Outside China they tend to call themselves "Chinese" (*Zhongguo-ren*) and many overseas Han Chinese also use their provincial names or other local origins in China (such as Guangdong, Fujian, Hakha, Taishan, etc.) to refer to themselves since their Han identity is virtually unknown outside China.

The name "Han" was derived from the Han River, an upper tributary of the Yangtze River. It was further strengthened by the famous Han Empire (206 BC–220 AD) which lasted for several hundred years when the people began active interactions with the outside world. It is understood the Han are the result of amalgamation of numerous ancient ethnic groups.

While the term "*Han-ren*" is used in China to differentiate the Han from other ethnic groups, "*Hua-ren*" now mostly refers to Overseas Chinese, such as *Mei-ji Hua-ren* (American Chinese), *Yindunixiya Huaren* (Indonesian Chinese), etc. "*Hua*" is an ancient name for China, and it is still used when referring to China in combination with the word "*Zhong*" (middle), such as *Zhong Hua Ren Min Gong He Guo* (People's Republic of China) and *Zhong Hua Min Guo* (Republic of China). Chinatowns in other countries are referred to as *Tang-ren-jie*, meaning "Town of the Tang People." *Tang* refers to the ancient Tang dynasty (618–907) in China.

LOCATION AND LINGUISTIC AFFILIATION

The Han are the majority ethnic group in China in East Asia, accounting for 91.59% of the whole Chinese population. The Han language (*Han Yu*), outside China known as "Chinese," belongs to the Tibetan-Han Language Family. It consists of seven major dialects, namely the Northern dialect, the Wu dialect, the Xiang dialect, the Gan dialect, the Hahka dialect, the Min dialect, and the Yue dialect. The modern common speech (*pu-tong-hua*) is based upon the Northern dialect with Beijing pronunciation.

These dialects are mutually intelligible in most cases, especially between the Northern dialect and others. However, since 221 BC, there has been an unified idiographic (actually it is an idiographic script with phonetic elements) script which has survived the long history of China and which has been one of the most important factors in holding the people and the country together.

OVERVIEW OF CULTURE

The population of Han in China mainland was 1,167 million, and about 22 million in Taiwan in 2000. As of 1999, the Han Overseas Chinese population was about 35 million (about 27 million in other Asian countries; 6 million in the Americas; and about 1 million in Europe).

Confucianism, which made its appearance between the 6th and 5th century BC began its glorious era during the Han Dynasty not too long after the First Emperor's ruthless persecution of Confucian scholars. While Daoism shaped the accepting and yielding, joyful, and carefree side of the Han character, Confucianism molded the Han people with a moral and duty-conscious, austere, and purposeful group mentality. As a set of social ethics governing interpersonal relations, Confucianism is largely responsible for the secular nature of Han social thought. Its social philosophy advocates filial piety in the family and benevolence in the administration of the government. Confucianism virtually played the role of a state religion together with ancestor worship, and has been a strong binding force that has kept Chinese society a super stable agrarian one in spite of more than two thousand years of dynastic changes.

In Confucianism an ideal society is a peaceful and harmonious one in which everyone plays by the rules and everyone loves the other. It is the enlargement of the family model in which responsibilities and obligations to the group reign supreme and personal freedom and individual rights are unheard of.

The Han family is patriarchal, patrilineal, and patrilocal in nature. The key link that binds everyone together in the family is *xiao*, filial piety. It means obligations of descendants to their progenitors, especially one's father and mother. The family is an unending chronic continuum linking the dead, the living, and the unborn. The principles underlying the kinship system are lineage, generation, sex, and seniority, and the most important relationship in the family is between the father and the son (Hsu, 1948). In this closely-knit network everyone is taken care of and nobody is left alone. The family clan may exert enormous influence over individual families and their members.

Extensive use of kinship terms outside the family and family clan, and even with strangers is a strong indication that among the Han the ideal society is one constructed on the family model.

While Confucianism provides the guidelines for interpersonal relations in this world, ancestor worship provides the solution for the next. As a religion Daoism plays an auxiliary role. Mahayana Buddhism which originated in India also has considerable influence. Christianity, both Roman Catholic and Protestant, is present and the followers are a small minority.

The Context of Health: Environmental, Economic, Social, and Political Factors

Environment and Population Distribution

The Han people live mainly in the most developed agricultural regions and large metropolitan areas in the Song-Liao plain, drainage areas of the Yellow River, Huai River, Yangtze River and the Pearl River, and in Taiwan. Most of the areas where the Han live belong to temperate and subtropical zones with four distinct seasons. The north has a long winter when things are covered with ice and snow. The south is warmer and has more rain. 36.09% of the population live in urban areas and 63.91% in rural areas. Illiteracy and semi-illiteracy rate was 6.72% (2000).

Sedentary Agriculture in History and Recent Industrial Development

The Han people have practiced sedentary agriculture for thousands of years and developed sophisticated irrigation systems that played an important role in the economy. While in the north one crop is raised, in the south two crops are planted. As majority of the population live in the countryside and practice agriculture, most of the activities have evolved according to a lunar calendar, that is agricultural routine. But recent changes have accelerated the country's industrial development. Now China ranks seventh in the world in GDP following the United States, Japan, Germany, France, Great Britain, and Italy. The way of life is changing fast.

Diet and Nutrition

Han diets vary with geographic location. The main staple foods in the north are wheat, corn, and millet and in the south rice and sweet potatoes. Vegetables form an important part of the Han diet. In the south, fresh vegetables are available all year round, while in the north, preserved cabbage, potatoes, and radishes are mostly relied on. With economic development, northern areas have more fresh vegetables in the winter. Red meats, poultry and fish are also important, but in the past they were consumed by the common people only during major festivals or on the occasions of entertaining guests. A Han meal is divided into two parts: *fan* and *cai*. *Fan* includes food grains and potatoes, and *cai* may consist of vegetables, meats, poultry, fish, and pickles. Dofu (soybean curd) made of soy is a popular food item which provides good nutrition and is a major protein source.

According to a 1990s survey, an average of 66.8% of dietary energy originated from cereals, 57.4% for urban residents and 71.7% for rural people. The proportion of protein coming from cereals was 50–60%, beans 5–6%, and from animals 20–30%; dietary fat provides 22% of energy intake (Ge, 1996).

As there is no clear demarcation between food and herbal medicine, Han people often mixed food with herbs and call them "Yaoshan" (medicinal foods). Herbs used are either tonic or are effective cures for some health problems.

Han people generally do not drink milk and dairy products were not very common, though milk and milk products are on the increase. Many of them do not have the tolerance for milk. Protein is obtained mainly from soy products like tofu and soy milk.

With the new economic policy and affluence, McDonalds, Hamburgers, Coca-Cola, Pepsi, etc., have made their way into the Chinese cities and are quickly becoming favorites of the children. Some nutritionists are critical of the new changes and predict serious health consequences.

Nutrition has improved substantially since the turn of the century. According to *The Dietary and Nutritional Status of Chinese Population, 1992 National Nutrition Survey*, average energy intake per person per day was 2,328 kcal, accounting for 97.1% (99.8% for urban residents and 95.7% for rural) of the Recommended Dietary Allowance (RDA). This shows adequate food has been consumed and hunger is no longer a problem. However, there are still serious problems with regard to nutrition.

Average protein intake was 68 g, accounting for 90.3% of the RDA. The intake of nicotinic acid (15.7 mg) and ascorbic acid (100.2 mg) was adequate, the intake of thiamine (1.2 mg) was fair, and that of retinal equivalent (Vitamin A, 156.5 mg) and riboflavin (vitamin B_2, 0.8 mg) was low.

Deficiency of calcium is rather common, and the intake (405.4 mg) accounted for only 50% of the RDA. As a result rickets is relatively common among children. The high incidence of osteoporosis among elderly women is also related to calcium inadequacy.

The apparent iron intake (23.4 mg) is adequate. However, iron deficiency and iron deficiency anemia are the most common nutritional deficiency problems, particularly among women and children. This is due to poor absorption of iron from plant food and specialists suggest enriching food with iron (Ge, 1992). In 1998, more than 50,000 people were found to suffer from Keshan disease due to lack of selenium, more than 1 million had Keschin Beck disease, while more than 9 million had diseases related to lack of iodine (China Yearbook of Health, 2000).

According to another report, inadequacy of zinc among children and early youth, and of iron among women is worrisome. Fifteen thousand people die every day (70% of total deaths) due to chronic diseases caused by malnutrition and nutrition imbalance, 21.6 million children are below their normal body weight, and 42 million children suffer retarded growth (China News Service, 2002).

Major Diseases

The ten leading causes of death in rural areas in order of incidence are; (1) respiratory disease, (2) malignant tumors, (3) cerebral vascular disease, (4) heart disease, (5) traumatic injuries and toxicosis, (6) digestive diseases, (7) urinary and reproductive diseases, (8) new born baby diseases, (9) TB, and (10) endocrine, nutritional, metabolic and immune diseases. In metropolitan areas they are (1) malignant tumors, (2) cerebral vascular disease, (3) heart disease, (4) respiratory disease, (5) traumatic injuries and toxicosis, (6) digestive diseases, (7) endocrine, nutritional, metabolic and immune diseases, (8) urinary and reproductive diseases, (9) mental disease, and (10) neuropathy. Malignant tumors have risen 69% for the last 20 years. Specialists attribute this to more intake of red meats and heavy use of tobacco. China now has 3 million smokers. Environmental pollution is the third major factor contributing to increased rate of cancer (New China News Service, 2001).

Leading infectious diseases are viral hepatitis (incidence rate per 100,000 is 68.93), dysentery (45.91), gonorrhea (20.63), and newborn baby tetanus (16.55).

Since the Han people use nightsoil for crops, fecal-borne diseases are very common. Typhoid fever, dysentery and cholera were once causing the highest mortality in rural areas (Simoons, 1991). Parasitic infections and other major health hazard diseases that used to be prevalent have been put under control after the 1950s. In 1999, out of 1919 rural counties, 409 found snail fever (schistosomiasis) infecting 366,784 people (China Yearbook of Health, 2000).

Venereal diseases that were once rampant in the country were almost eliminated after the 1950s. But following the opening up and reforms that began in 1970s and the issuing economic development, they have come back at an alarming rate.

AIDS has become an urgent issue in recent years. According to UNAID, an agency of the United Nations, estimated persons infected with HIV might be as high as 1.5 million at the end of 2001. Chinese official reports estimated the figure of HIV carriers at one million during the first half of 2002, a 16.7% increase over 2001 (China Daily, October 16, 2002). Predictions are, that if no effective measures are taken, AIDS cases may reach 10 million before the year 2010. Drug use, illegal blood transfusion practices, and unsafe sex are the main problems. In a culture where the talk of sex is almost a taboo, dissemination of the right information concerning AIDS is very difficult.

According to an official report, in 1999 0.054% of the total population were drug addicts, which means that about 681,000 people were drug users. Of the drug addicts, those taking heroin made up 71.5%, and those under the age of 35 were 79.2% (CPIRC). The suicide rate is rising, especially among rural women. According to an official report 250,000 people commit suicide every year and the rate is 22 per ten thousand. Women are three times more likely to commit suicide when faced with unsolved problems, and suicide has become the leading cause of death among people between 15–34 years of age (People's Daily Online, 2002).

Alcohol and Tobacco Use

Drinking strong liquor brewed from rice and other food grains is viewed as entertainment. There is no legal age for drinking; some may begin to drink from childhood. However, alcoholism is not a serious threat. People begin to smoke very early in life. No legal age limit is set and almost everyone smokes in the countryside. To offer a cigarette is considered a very friendly gesture few can decline. This is one of the reasons why respiratory diseases are so common.

Infrastructure

Among the Han people there are two medical systems, namely biomedicine that they call "Western medicine," and traditional medicine which they call "Chinese medicine" (*Zhong Yao*). In both Japan and Korea, Chinese medicine is called the "Han medicine." For clarity this article will use the terms "Western medicine" and "Chinese medicine." Though Western medicine came to China during the early 20th century, it has never been able to replace Chinese medicine which has long become part of daily life.

Today Chinese medicine and Western medicine enjoy equal status. All medical doctors must be trained in both Chinese medicine and Western medicine although they may have either as their major area of training. All hospitals provide both services and when a patient comes to see a doctor, he or she may have a choice with the help of the doctor.

Health care in China was notorious before the 1950s. Steady progress has been made in public health since then. At the end of 2000, there were 325,000 health institutions (including clinics) throughout the country, with a total of 3.18 million beds, 2.21 million of which were in hospitals and public health centers. There were 4.49 million health workers, including 2.08 million doctors and 1.27 million senior and junior nurses (China Yearbook of Health, 2000).

By mid-1960s, an effective healthcare network was established and the whole population was provided with basic preventive and curative health care. Strict control by the government on drug production and sales ensured low prices for quality drugs. Since early 1980s however, things have changed drastically. Market oriented reforms brought about an economic boom in the country, but at the same time, medical costs have risen sharply as public health sector is put to the mercy of profit-making markets. Corruption is rampant in the country; some reports reveal that many doctors are on commissions from drug suppliers. Major operations cost dearly. In addition, families of patients who need surgery give to the surgeons and other personnel they consider important good amounts of money and expensive gifts called "red envelopes" (*hongbao*).

While higher medical costs do not affect urban residents too much because they are mostly covered by medical insurance and special aid, higher costs have caused enormous difficulties for rural populations. The three-tier (county, town, village) health network that covered most of China's rural area with active participation from what were called "barefoot doctors" is no longer effective because of lack of public funding, privatization of village clinics, and profiteering of medical institutions. A 1998 survey conducted by the Ministry of Health revealed that due to financial difficulties, about 36% of sick farmers did not seek medical treatment and 61% of inpatients had to leave the hospital halfway through getting proper care. The fact that only one fifth of government total public health spending in the late 1990s went to rural areas, where more than 70% of the total population live, has caused controversy among government circles (Zhu, 2002.)

Some pilot programs with village-level public clinics which pool funds from villages, local government, and businesses are underway and collective medical insurance or social medical insurance systems are being tried, but the results are yet to be seen.

Status of Chinese medicine is very different among Overseas Chinese (Han). In some countries, they are still using it in their health care. In the United States and Western Europe, stringent policies against traditional practices favor biomedicine. The Dietary Supplement Health and Education Act of 1994 (DSHEA) and the issuing policy by the U.S. Food and Drug Administration created another category that is neither regular food nor drug: dietary supplements. Since Chinese medicine is not recognized as medicine, many Chinese medicinal patent formulas came to the U.S. under this category.

Some methods of treatment in Chinese medicine may be considered abusive by Westerners outside the culture. Skin-scraping, a procedure of scraping the surface of the body until it is purple (near bleeding), and even acupuncture and moxibustion occasionally cause legal problems.

MEDICAL PRACTITIONERS

Four Thousand Years of Unbroken Tradition

The beginning of Chinese medicine was lost in antiquity. Legend of the Han attributed it to Shennong, "The Divine Peasant," one of the earliest emperors who was said to have tasted hundreds of herbs. Later *Shennong Ben Cao Jing* (Shennong's Herbal) was compiled in his name. It is perhaps the worlds first pharmacopoeia. Thereafter the investigation of herbal medicine has never been interrupted. A great amount of information has been recorded. Methods of preparing herbal formulas have been successfully refined, and the number and variety of clinical applications of herbal medicine have grown in proportion (Hsu, 1986).

In the 3rd century BC, *Huangdi Nei Jing* (*The Yellow Emperor's Classics of Internal Medicine*) was compiled showing how advanced practical medical knowledge was at that time. Many of the formulas described in it are still being used today. Hua Tuo, a historically famous medical specialist in the beginning of the 3rd century, was able to perform major surgery including opening the stomach cavity and repairing the intestines, under anesthesia.

As a tradition, Chinese medicine is shaped by the influence of Confucianism in many ways. Its secular nature came from the Confucian attitude "to respect gods and ghosts but hold them at a distance." In history, many Confucian scholars were also famous doctors, and their strong sense of history and responsibility to society made them instrumental in making the medical system a strong and unbroken tradition (Li, 1990). Landmark medical writings including materia medica and treatment experiences are countless including such works as Zhang Zhongjing's *Shang Han Lun* (*Discussion of Cold Diseases*, about 200 BC), Sun Shimiao's *Invaluable Prescriptions for Ready Reference* (652 AD), Li Shizhen's *A Compendium of Materia Medica* of the Ming Dynasty (1368–1644), *A Grand Dictionary of Chinese Medicine* (1979), *Encyclopedia Sinica, Volume of Traditional Chinese Medicine (Zhongguo Dabaike Quanshu Zhongyi* 2000*)* and the *China Pharmacopoeia (2001).*

Shamanism

Shamanism may have been practiced since the beginning of the Han civilization. The famous bone-shell script that evolved into the present Han writing (Chinese script) and was used to predict the future may also have been used to diagnose diseases. Since people also believed in supernatural causes of diseases, including gods, spirits of both dead people (ancestor or dead relatives) and some mysterious animals such as the fox and weasels, shamans were much needed to solve the problems acting as a go-between or a messenger between these supernatural beings and the human world. These shamans played an important relieving role in the countryside. Even during the time after 1949 when the Chinese government waged extensive campaigns against superstition and religion, shamanism has never disappeared entirely. With softening of the policy toward traditional beliefs, shamanism seems to have come back to the countryside especially in economically poor areas and remote regions.

Religious Personnel

In cosmology, disease and destiny (in the Chinese language destiny and life share the same word) are closely connected. Therefore it is understandable that religious personnel played a part in Chinese medicine. Most of the Daoist (Taoist) priests and Buddhist monks had very good knowledge of traditional medicine and used them in their congregations. People also looked to these religious premises for ultimate cures, especially in case of tenacious diseases or in case of barrenness.

Traditional Herbal Practitioners

Chinese medicine as a strong empirical tradition has always kept its secular nature, and to diagnose, treat, and cure diseases has always been an important profession. Some of these professionals people call "doctors" were full time herbalists while others, depending upon the circumstances in the community, might be part time practitioners and part time peasants.

Doctors of Chinese medicine were trained mainly in master–apprentice relationships. More often than not this master–apprentice relationship was just between the father and son. The founding of the People's Republic of China brought about many changes in medical practice. Chinese medicine is officially recognized and enjoys the same status as Western medicine. Training of doctors of Chinese medicine became formalized and institutionalized and research institutions have been set up with the China Academy of Traditional Chinese Medicine as the nation's highest authority. Universities and colleges have been set up to train traditional doctors and other specialists. They major in Chinese medicine, but at the same time they must have basic training in Western medicine as well. After graduation they would work in metropolitan hospitals or rural clinics. While they mainly use herbs to treat and cure, they also can legally use Western medicine, integrating the two systems as they see fit. Patients would also have a choice to elect for either Western drugs or herb formulas.

View of Western Medicine

Doctors in Chinese medicine tend to be more personable than their counterparts in Western medicine due to their training. They inquire into many aspects of the patients' way of life and tend to form closer interpersonal relationships with the patients and their families.

Many Han people feel that Western medicine is particularistic and analytic, and focuses on the ailing part of the body, "seeing only trees, but not the forest." On the other hand, Chinese medicine, though slow in efficacy, tackles the root cause of the diseases. So people would tend to go to a western medical doctor when the illness is acute, and go to a doctor of Chinese medicine when it is not.

Since the 50s, a process of integration took place in China and combination of the two systems has shown promising results. A popular statement in this connection is that $1 + 1 > 2$, and the experience has enriched both Western medicine and Chinese medicine.

CLASSIFICATION OF ILLNESS, THEORIES OF ILLNESS, AND TREATMENT OF ILLNESS

Theories of the Universe and of the Human Body

The Universe and Nature. The "Universe" or "Cosmos", as expressed in Han language, is *Yu Zhou*, designating space and time. "*Yu*" is the collocation of three-dimensional spaces and time. What they call "*Zhou*" is constituted by the one-dimensional series of changes in succession—the past continuing itself into the present and the present, into the future. *Yu* and *Zhou* taken together represent the primordial unity of the system of space with the system of time. Han Chinese consider the world, man, and history in terms of comprehensive harmony that permeates anything and everything (Fang, 1986). Everything in the Universe including the Universe itself, is changing all the time, and each has two opposite aspects: *Yin* and *Yang*, which are in conflict and at the same time interdependent; any change is the result of the *Yin-Yang* change within it. Everything is related to everything else in the Universe in the way they should be: the *Dao* or *Tao*.

The Human Body and Five Elements. Humans are part of the Universe, or nature, and the ideal relationship between humans and nature is harmony. Adaptation to nature and to be one with it is the way (*Dao*). The human body, like the larger Universe, is an organic whole. The various organs and tissues are all connected to one another. The connected system includes meridians (or channels) and collateral which are the lines along which blood, vital energy, and the impetus to functional processes move. While meridians are the cardinal lines, the collaterals are the branches at various levels, which form a structural network (Xie, 1995).

Due to Confucian norms of filial piety which forbid hurting the human body, anatomy is not very developed in Chinese medicine (Li, 1990). That is why it is quite different from Western medicine with regard to the structure and functions of the human body. What are called the visceral organs are referred to rather as comprehensive systems of physiological functions than as anatomical entities. Among the most important ones are the heart, liver, spleen, lung, and kidney. But they are not the exact organs as understood in the English language. They should be read when used in the Chinese medical context, as the heart and/or higher nervous system (heart); the system that controls emotional activities, muscle action, bile secretion, and blood storage (liver); the system responsible for digestion, absorption, assimilation, and energy metabolism (spleen); respiratory system (lung); and the system which works to secrete urine and provide vital essence for heredity, reproduction, development as well as replenishing the brain, nourishing the bone, and producing marrow (kidney) (Xie,1995).

The relationship among the five systems is represented with the relationship among the five elements: fire,

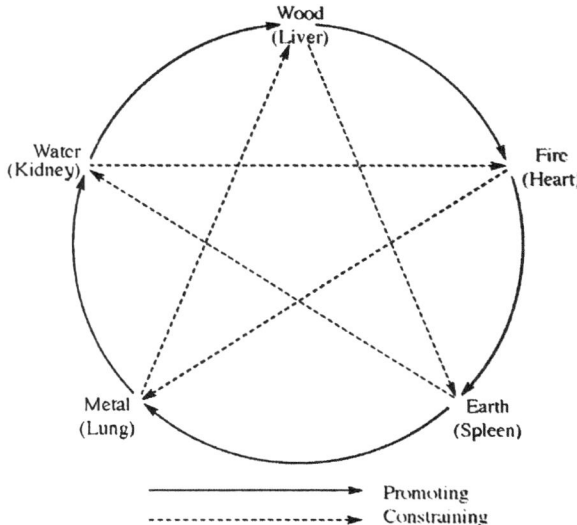

Figure 1. Relationship among the five elements. (Translated from Li, 1989.)

earth, metal, water, and wood in a complete circle as shown in Figure 1.

Here one can see the attempt of the ancient Han people trying to illustrate the most important human organs or systems with things they considered basic in nature. Properties and nature of the five elements are being used to explain the relationships of the five visceral organs. In the scheme, things go clockwise in a circle with one promoting the one next to it, but constraining the one next but one to it as is shown in the five point star. This must be considered in case of diagnosis and treatment as everything must be taken into consideration in both processes.

In order to carry out their normal functions, these visceral organs and other organs and tissues need Qi, blood, vital essence, body fluid, and nutrients. Here Qi are the basic particles which constitute the cosmos and produce everything in the world through their movement and changes. In Chinese medicine it refers to the force or energy required for various functional processes. It consists of two sources, the inborn and the acquired.

Concept of Illness

In Chinese medicine disease is defined in terms of breaking up of the relative balance in the human body. According to this theory, there is an endless process of adaptation among all parts of the human body and between the human body and the environment. This process maintains a relative dynamic balance that in turn supports normal biological activities. When the human body is in harmony with the environment, pathogens will not harm it.

Diseases happen under two circumstances, one is when functional disorders occur in the human body and the "right Qi" is relatively weak, and the other is when strong "evil Qi" has affected the body. Right Qi means normal functional activities and the body's ability to resist diseases, and the evil Qi means pathogenic factors. Onset and changes of the disease reflect the course of the struggle between the right Qi and the evil Qi.

The Han people attach primary importance to maintaining the right Qi which largely means "immunity" in terms of Western medicine. Diseases strike only when the right Qi is relatively weak. Outcome of the struggle between the right Qi and the evil Qi depends on the constitution, mental state, environment, habits, nutrition and exercise.

Pathogenic factors include atmospheric factors, epidemic pestilence, personal emotions, improper foods and physical fatigue, traumatic injuries, parasites, phlegm-rheum, and blood stagnancy.

There are six excessive atmospheric factors (*liu Yin*) that are wind, cold, heat, damp, dryness, and fire. Normally these natural changes will not cause diseases. They do only when the body's resistance is low and adapts poorly to natural changes, or abnormal natural changes become too much for the body. Sometimes in-coordination among different organs may also produce the same symptoms. These are called "internal wind," "internal cold," "internal heat," "internal damp," "internal dryness," and "internal fire."

There are seven emotions that are the normal reactions to the outside world, but when subjected to repeated long term stimuli, especially drastic ones, each may cause imbalance between the *Yin* and *Yang*, between the Qi and blood, and malfunction of the internal organs, thus leading to diseases.

Classification of Illness

Chinese medicine classifies diseases according to their major patterns in terms of *Yin* and *Yang*, internal and external, heat and cold, vacuity and excess. In connection to the state of Qi, blood, body fluids, and different body functions, it identifies the nature of the major problems as external cold, external heat, internal cold, internal heat, excess, vacuity, *Yin* problems, *Yang* problems, *Qi* inadequacy, blood inadequacy, etc.

Methods of Diagnosis

Doctors of Chinese medicine diagnose in four major ways including inquiry (asking questions concerning body temperature and how the patient feels, perspiration, general body feeling, urination and stool, diet, feelings of the lung and stomach, monthly discharges in case of woman, etc.), visual observation (patient's mood, face color, shape of body, and color, shape of the tongue, etc.), smelling and listening (patient's body odor and voice), and pulsation. Generally the information collected and the observed symptoms are enough to help the doctor determine the nature of the disease. Today doctors of Chinese medicine are also using modern equipment and methods such as stereoscope, X-ray, CT, etc., to verify conclusions.

Treatment Principles

Prevention as Priority. The overall guiding rule in treating diseases in Chinese medicine is determining the root cause. It includes two aspects: one is taking precautions before disease strikes; the other is prevention after a disease strikes. There are many formulas and methods to help enhance the immune system. As early as the 16th century, the Han people invented inoculation for smallpox. After a disease strikes, medical practice tries to intercept possible pathological changes. For instance, when the liver is out of order, it is important to protect and strengthen the spleen.

Balance as the Ideal State of Health. Because everything is connected to everything else in a relationship of cooperation and coordination it is the mission of Chinese medicine to redress any imbalance. What is more, in reestablishing this balance, efforts must be taken to avoid creating any new imbalance. That is why treatment with Chinese medicine is expected to be slow and noninvasive (Liu, 2002).

Holistic Considerations and Individualized Treatment. In prevention, diagnosis and treatment, everything, including the season and local conditions, physical and psychological environment, the patient's age, gender and general physical conditions, family history, etc., all have to be considered. A patient from a Western culture may feel uncomfortable with a doctor of Chinese medicine when the doctor asks questions that are too personal.

Chinese medicine is an art of healing, and a good doctor of Chinese medicine treats the patient rather than the disease. Since everyone is a complete organism himself subject to different physical and psychological environments and every organism is changing all the time, every patient is necessarily different in Chinese medical concept, and therefore everyone must necessarily be treated differently. Every patient needs to be examined individually and individualized diagnosis must be made, and the treatment is unique. In high plateaus that are cold and dry, exterior pathological factors are mainly related to cold and dryness and treatment should use herbs that have the property of dissipating aridity, enriching, and moistening; and in an environment of heat and damp such as lowland area, herbs that can dispel heat and transform damp should be given the priority. Even in the same location, patients are different. For an instance, two persons are both diagnosed by Western medical doctors as having essential hypertension. But patient one is robust with a red face, red eyes, constipation, irritability, a thick yellow coated red tongue, and a wiry full pulse. A doctor of Chinese medicine would provide a treatment to calm down his or her liver fire (to cleanse the liver). Since patient two has pale and frail appearance, loose stools, low energy, pale flabby tongue, and a weak pulse, the same doctor would formulate a treatment plan to invigorate the patient's kidney *Yang* (to tonify the kidney).

Strengthen the Immune System and Eliminate Pathogenic Factors. In case of weak defenses, the emphasis has to be on enhancing the immune system in order to eliminate pathogens. On the other hand, when pathogenic factors are too strong, the focus has to be the elimination of pathogenic factors in order to strengthen the body's defenses. Sometimes, it is also necessary to do both at the same time.

Differentiation of Root Cause from Symptoms and Determination of Acuteness of the Disease. To treat the root cause instead of the symptoms is always the principal aim of Chinese medicine, but in treatment considerations are given to the actual circumstances. For instance when it is a mild disease and not acute, it is right to treat the root cause. In case of acute diseases however, it may be necessary to treat the symptoms first. When the patient is running a high fever, bleeding, or suffering from severe pain, the symptoms themselves may lead to death if no drastic measure is taken to treat the symptoms. Eliminating the high fever, stopping bleeding and pain are imperative before going for the root cause. In some cases it may be better to treat both the root cause and the symptoms.

Straight Treatment and Paradoxical Treatment. Straight treatment is to meet cold with heat and heat with cold, supplement in case of inadequacy (vacuity), and discharge in case of excess. These are the normal ways of treatment. However, when symptoms do not reflect the root cause, the opposite is called for. For instance cold symptoms may be the result of extreme heat, and in such case the right way is to use cold method. This paradoxical treatment is a hallmark of a good doctor.

Methods of Treatment

Since Chinese medicine is nutrition oriented, it treats various health problems in terms of excess syndromes, deficiency syndromes, and deficiency with excess syndromes. Deficiency includes deficiency in *Qi*, in blood, in *Yin* and *Yang*; excess includes excess in wind, cold, heat, damp, dryness, fire, phlegm, and *Qi*.

There are numerous methods of treatment in Chinese medicine, but basically there are eight approaches that are used singularly or in combination as the situation calls for: (1) The diaphoretic approach induces sweat to expel pathogenic factors. It is used for external problems such as unripe pox, sores, and boils. It may be due to external cold or external heat, and therefore there are two basic ways which are resolution of exterior with coolness and acridity, or resolution of exterior with warmth and acridity; (2) the emetic approach induces vomiting to expel pathogenic factors or toxins. It is used in case of thick phlegm in the throat or stagnant or poisonous food in the stomach; (3) the purgation approach is used in case of serious constipation, bloating, stagnant phlegm and blood, or in case of parasites. It is comprised of cold purgation, warm purgation, expelling of water, dissolution of stagnation, or expelling parasites; (4) the harmonization approach is used to resolve poor co-ordination between internal organs in their functions. Malaria and irregular menses are also treated this way; (5) the warming approach is taken mainly to dispel cold pathogens, etc., as in case of internal cold pattern of diseases; (6) the febrifugal approach is to remove internal heat to protect body fluids; (7) the tonification approach is employed in case of deficiency or vacuity including tonification of the *Yin*, tonification of the *Yang*, tonification of blood and tonification of the *Qi*, etc.; and (8) the dispersion approach is used for stasis and accumulation in blood, *Qi*, phlegm, rheum and foods, etc.

It is obvious that the objective of all these is to achieve balance and harmony of the various organs and their functions while dynamic biological changes and processes are maintained.

Classification of Medicine

Sources of Chinese Medicine. Chinese medicine is characterized by natural, low cost, and nutrition oriented sources. So far there are 12,807 medicinal materials out of which 11,146 are herbs, 1,581 are from animal sources, and 80 minerals (*Encyclopedia Sinica*, Volume of Chinese Medicine, 2000 Edition). Because of the predominance of herbal sources, people refer to Chinese materia medica as herbal medicine. Based upon principles of Chinese medicine in formulation, there are more than one million patent formulas that can be adjusted with addition or reduction of some ingredients to suit particular needs of the patient. They represent a great rich treasure house for health care and fitness.

Properties of Herbs. Herbs are said to have four properties: cold, cool, warm, and hot, and five flavors: sour, bitter, sweet, pungent, and salty. The four properties have nothing to do with the temperature of the herbs. They are the resulting effects produced by the herbs. Coptis, phellodendron bark, gardenia fruit are classified as cold medicine because they eliminate heat, dryness, remove toxins, and they are normally used to tackle heat patterns of diseases, while aconite and dry ginger are classified as hot medicine because they are normally employed to warm the center and dispel cold.

From the beginning, Chinese medical practitioners found that herbs of certain taste possessed certain medical properties. With long historical development, the five flavors actually come to represent the properties rather than the actual flavor or taste. Herbs with sour flavor possess the ability of astriction and are used to treat seminal emission, night sweating, enuresis, enduring diarrhea, anal desertion, etc. Those with bitter flavor including such herbs as coptis, phellodendron bark, gardenia fruit, etc., have the properties of dispelling heat, dryness, and toxins. Sweet herbs generally tonify and supplement, relax tension (acuteness), and harmonize the functions of different herbs. For instance, codynopsis, astragalus, and cooked rehmannia tonify and supplement in case of vacuity. Licorice and sugar relax tension, and harmonize different herbs. Pungent herbs disperse external pathogens, move the *Qi* and promote blood circulation. Mahuang treats wind cold and external problems, Cnidium (Chuangxiong) activates the blood, and carthamus disperses accumulation. Salty herbs soften hardness and drain precipitation and are used to treat scrofula, phlegm nodes, lump glomus, and dry and hard stool. Oyster shell, for instance, may help disperse hard lumps, and mirabilitum can ease constipation.

Flavor and properties of an herb must be considered together. Herbs of same properties may be different in their flavor, and vice versa.

Herbs are also classified according to their tendency to reach certain parts of the body or channels. This selectivity of the herbs in their functions is called *gui jing* (channel entry). For instance, both phragmite and gentian root belong to cold herbs used to clear away heat. However, the former is particularly effective in clearing lung heat and the latter heat of the liver.

According to their functions herbs are classified into more than 20 categories, including diaphoretics or exterior resolution herbs, either for dispelling wind-cold or wind-heat; Antipyretis or Ferifuges; Antitussive, Expectorants and Anti-asthmatics; Digestives; Tonics including *Qi* replenishing herbs, blood replenishing herbs, *Yin* replenishing herbs and *Yang* replenishing herbs; Carminatives or *Qi* flow herbs; Blood circulators to remove blood stagnation; Hemostatics; Laxatives; Diuretics; Fragrant herbs for resolving damp; Antirheumatics including herbs for arthritic pain, for muscles and collaterals and for strengthening the bone and tendons; Warming herbs; Anticonvulsants; Sedatives; Aromatic stimulants; Astringents; Anti-malarial herbs; Anthelmintics; Analgesics, and Topical agents. These herbs are used as soldiers for specific missions whether singly or in groups to achieve prevention or curative effects.

Herb Preparation. Doctors of Chinese medicine are very sensitive to authenticity of the herbs depending upon the areas where they are produced. Through longtime observation and composition they know that some herbs produced in certain areas are best in quality and therefore in curative effects. Ginseng, deer antler, and schisandra fruit in northeast China, rehemannia root, Chinese Yam of Henan, coptis root and fritillary bulb in Sichuan, wolfberry in Ningxia, and notoginseng in Yunnan are the most famous. They are more sought after than others.

Herbs must be carefully prepared and processed before they are used. Preparations include stir baking with or without auxiliary fluids (such as vinegar, wine, honey, salt water or ginger juice), calcination, roasting, steaming, boiling, water purification, fermentation, germination, and frosting. Different preparations may enhance the curative effects, reduce toxicity, remove undesired ingredients and taste, and make them easy to use and store.

Prescription Formulation Principles. After diagnosis and determination of the treatment principles, the doctor of Chinese medicine decides the principal herb and the auxiliary herbs in the formulation to achieve the curative effects wanted. Sometimes a single herb is used, but most often it is a compound preparation that may consist of anything from four to twenty herbs.

A compound prescription normally includes four different component parts. They are called Monarch (principal), Minister (adjuvant), Assistant (auxiliary), and Guide (conductor) respectively.

The principal ingredient provides the main curative action. The adjuvant helps strengthen the principal action; the auxiliary is a corrector ingredient to relieve secondary symptoms or to temper the action of the principal when it is too potent, and the conductor is to direct the actions of the principal and adjuvant herbs to the affected area or site or acts as a minor ingredient.

In a compound prescription, drug interactions must be considered. According to Chinese medicine, herbs may be either mutually reinforcing, mutually restraining (to weaken or neutralize each other's actions), mutually counteracting (one ingredient reduces the potency or toxicity of another), mutually neutralizing (one counteracts the toxic reaction of another), or mutually incompatible. Modern research has found that some of these relationships are valid and some are not. However a good herbal doctor is one who knows really well the properties of the herbs and uses them correctly and innovatively to treat his patients.

"Food Therapy"

A very important medical tradition is what is called *shi liao*, that is, food therapy. The origin of this can be traced back in history to several thousand years. *The Shennong's Herbal* (*shen nong ben cao jing*) carries 365 herbs classified into three categories including superior, medium, and lower grades. Most of those listed as superior are food grains, vegetables, fruits, meats, and herbs with a friendly nature. As herbal drugs are strong and taste awful, and long-term use may hurt the stomach, the best way is to use food to do the work. This is considered an ideal since foods may not only cure in the long run, but may also be made into something enjoyable. Food therapy follows the same principle as herbal treatment, that is, to warm up when cold is present, cool off when heat is the problem, to supplement in case of deficiency (vacuity) and discharge in case of excess. For this, food items are classified into different categories according to their properties and diet is planned in such a way as to achieve therapeutic result in different situations. Food therapy is often employed to supplement medical treatments, especially for chronic diseases.

Health through Proper Diet

Another tradition based on Chinese medicine is called *shi yang*, health through proper diet. It is to select certain diet to regulate various biological functions of the related organs of the body, to nurture the *Qi*, the blood, body fluids, to build up the body's resistance to diseases according to different needs in terms of constitution, age, gender, the season or local conditions. For this, people are classified into several types with different constitutions. For instance, those who tend to suffer *Qi* deficiency should include yam, lotus seeds, pork, and eel in their diet, and those who have blood deficiency problem should include longan, wolfberry, mulberry leaf, chicken, carrots, etc. Those who have *Yin* deficiency should eat white fungus, honey, sesame, black bean, etc. And those who are identified to have *Yang* deficiency should add mutton, shrimp, chives, etc. Foods are different in different seasons, and local conditions differ. People in different regions have different needs in foods.

"Yangsheng" (Life Preservation)

Shi liao and *Shi Yang* are included in another wider approach called "Yangsheng," that is, life preservation. This includes a variety of ways to prolong life.

A very important part of *Yangsheng* is *Qigong*. It is to achieve good health through breathing exercises, meditation, mental channeling, etc. As a component part of Chinese medicine it has a history of over 4,000 years originating from a kind of dance to maintain good health and to prevent diseases. Along the way it absorbed many elements from Daoism, Confucianism, and Buddhism and in a way it also acquired some religious characteristics to certain extent (He, 1998). The key concept is energy cultivation. It is said to be very effective in eliminating fatigue and achieving relaxation, enhancing the immune system for prevention of diseases, and treating and curing diseases connected with certain systems such as the nervous system, the circulatory system, the respiratory system, the digestive system, the endocrine system, and the immune system.

Many schools have developed over the years. Most of the Chinese medical writings cover *Qigong* exercises. The best known is the *Taiji* exercise, a set of slow movements which have spread far and wide outside of China. During the so-called "Great Proletarian Cultural Revolution" from mid 1960s to mid 1970s, *Qigong* was suppressed until the chaos was over. In 1986, the Degree Committee under the State Council made a decision to adopt *Qigong* in Chinese medicine as a new science discipline, and special funds were set up for further research and it became part of clinical treatment. Doing *Qigong* became part of daily life, especially in the morning in parks.

Currently the enthusiasm for *Qigong* seems to have had a setback with the government suppressing the *Falun Gong*, a quasi-religious group officially named an evil cult. It started as a *Qigong* organization and attracted many people. The government was taken by surprise when suddenly it organized a mammoth sit-in demonstration surrounding the government seat in Beijing.

Another area of *Yangsheng* is to control desire and avoid indulgence, especially sexual indulgence. Self-emotional control is also a very important part of these exercises. All these are aimed at achieving balance between *Yin* and *Yang*, promoting normal flow of the *Qi* and blood, increasing the essence and preserving the semen.

Tuina

Another treatment method in Chinese medicine is *tuina*, a special type of therapeutic massage. The origin can be traced back to pre-historic times. Its basic therapeutic orientation is the Chinese medical theory of the flow of *Qi* through the meridians. *Tuina* has many unique techniques including one finger pushing, rolling, *neigong* (internal *Qigong*), pointing, bone setting, etc. It can be done alone or in combination with *Qigong*.

Acupuncture and Moxibustion

Originally a component part of Chinese medicine, acupuncture and moxibustion have developed into a medical system in themselves. Two things are involved, one is to stimulate locals on the human body to achieve therapeutic effects, and the other is to use heat to achieve therapeutic results. They are used either separately or in combination. The origin can be traced to Neolithic times, 8,000 to 4,000 years ago. Needles were made of stones, wood, bamboo and finally metal, especially silver or stainless steel. Moxibustion uses leaves of mugwort (artemesia vulgaris) made into a cone and burned on oilment or ginger slice. The basis of both acupuncture and moxibustion is the concept of channels (meridians and collaterals) in the human body through which *Qi* travels. Channels (meridians and collaterals) are distributed all over the body and coordination or various functional processes are realized through the *Qi*. Acupuncture and moxibustion rely on the relationship between the *Qi* and the meridians for successful therapy (Xie, 1995).

Now electricity, magnets, laser, infra-red, and microwave are added to needle stimulation. Clinically it is comprised of four aspects: treatment with acupuncture and moxibustion, maintenance of good health with acupuncture and moxibustion (mainly to enhance the immune system), anesthesia with acupuncture and diagnosis through channel-local manipulation. Acupuncture has been accepted as legal and valid method of treatment in many countries in the world.

SEXUALITY AND REPRODUCTION

Sexual Attitudes

As noted above, Chinese medicine holds that congenital essence is responsible for reproduction. It is stored in the kidney and serves as the origin of life. Congenital essence can be transformed into acquired essence and vice versa. Sexual over-indulgence is regarded as one of the major causes of disease. Exhaustion of the reproductive essence stored in the kidney impairs other organs and has serious health consequences. For this serious efforts were made, especially among the leisure class, to preserve semen. Many writings, mostly Daoist in nature, discussed ways of sex without ejaculation. Occasionally some "masters" would advocate sex with young girls as it was thought to be tonifying for the man. In a way, women are considered dangerous for a man since she could deplete him of semen.

People generally avoid talking of sex. The purpose of marriage is understood to be to produce children to continue the family line. Sex, except for producing children, is generally considered immoral. So there has been great secrecy surrounding sex.

Seclusion of Women

Historically, seclusion of women was practiced and premarital sex was strictly forbidden. Marriages were arranged by parents and often the nuptial night was the first time the couple met. It was a great scandal if the bride was found to not be a virgin. Even now it is still common to have a relative or a friend to "introduce" a girl or a boy since it may be considered improper to approach the opposite sex directly for purpose of marriage.

Traditionally, only men could initiate divorce and a divorced woman carried stigma and little hope of a happy remarriage.

Fertility

The average lifetime fertility rate of Chinese women was reported to be 1.81 (1.98 in rural and 1.22 in urban areas) in 2001 (National Family Planning & Reproductive Health Survey, 2001). As the Han are a huge majority, this may well represent the fertility of Han women.

Barren women are considered unlucky and for a long time it was blamed on the woman if a couple failed to produce children or failed to produce a boy. It was generally hoped that a married couple should have as many children as possible and begin to have children as early as possible. It was common to find a woman having eight or even ten children. After 1949, there was a time when having more children was encouraged by the Chinese government until there was a population explosion. Since then there has been a policy encouraging late marriage and fewer children. During the seventies and eighties, the policy was to allow only one child per couple among the Han people while minorities may have more. The policy was quite successful in urban areas while in the countryside it was not so successful. As only the male child would be able to continue the family line, people generally prefer a boy, and the government One Child Policy met with great resistance. Forced abortion took place frequently, but rural people could evade it in several ways. Abandoning of unwanted children and infanticide due to poverty were quite common before 1949. After 1949, they were virtually eliminated. But the One Child Policy led to frequent female infanticide and abortion of female fetuses. The Fifth National Census (2000) reported the sex ratio at 116.9:100, alarmingly different from the international ratio of 105:100. Specialists attribute this to social factors and biological factors, but to what extent infanticide of female children, selective abortion or other factors each contributed to this ratio remains unexplained.

Ideal Household

Large families comprising three or four generations are generally considered ideal. Historical records often highlight large families, sometimes with a few hundred people living under the same roof. But these are exceptions rather than the rule. Statistics show that the average family size through different times in history was 4.84 persons (Qiao, 1990), though the clan, mostly under the same surname and therefore having the same ancestry, could be quite strong organizationally. The 2000 census indicated that the current average family size is 3.44 per family. There are many joint families and extended families in the countryside, but most of the families in the urban areas are nuclear families.

HEALTH THROUGH THE LIFE CYCLE

Pregnancy and Birth

Since the purpose of marriage is to have children, especially male children, the news of pregnancy is always a good one to the family and it is also the beginning of good expectations, particularly for a boy. People begin to be protective of the pregnant woman and she may be excused from heavy labor. Habits and practices around pregnancy and birth differ from place to place. The husband's mother may even go to a temple to pray for a healthy grandson. Generally the expectant mother should not go to unlikely places such as the burial ground or participate in a funeral. There are taboos as to what she may or may not eat. For instance, in some places she is not to eat rabbit for fear of giving birth to a child with cleft palate.

Han people had the practice of what is called "fetus education" in order to have a child healthy both physically and mentally. The mother should avoid sex, noise, bad thoughts, and bad images, and adhere to correct and nice language and demeanors.

A birth of a boy was called "great happiness" which was an occasion of elaborate celebrations and the mother's position in the family could be said as established, while a birth of a girl was called "small happiness." Normally there was not a problem if the first child was a girl as she may help take care of her younger siblings. It would become worrisome if the couple continued to have girls. The One Child Policy puts great pressure on women to have a boy.

Traditionally babies were born at home with help from a midwife and other experienced women. No man or child should be present, though the husband was normally kept close by for emergencies. His urine was sometimes given to the wife to drink in case of difficult labor. Children are generally kept from any knowledge about sex and birth. The answer to the question of where the baby is from is answered evasively. Often it is said to have been picked up from a dung-heap outside by accident.

As of 1999, 70% of deliveries were in hospitals. Hospital births are much higher in urban areas (83.3%) than in the countryside (61.5%). Births with new delivery methods (both in and out of hospitals) account for 96.8% of all births (China Yearbook of Health, 2000). Formerly the greatest risk for the new born baby was tetanus due to use of dirty scissors and un-sterilized cloth ropes to tie the umbilical cord, and for the mother it was puerperal or childbed fevers. Infant mortality in China was rated the highest in the world, 275 per 1,000 live births in 1927, and maternal death was estimated to be about 200 or more per 10,000 live births, at least three times that of western countries (Simoons, 1992). Since then things have changed and infant mortality in 1990 was 33 per 1,000 and maternal death was 95 per 100,000 (China Yearbook of Health, 2000).

For the first month after birth, both mother and child are kept indoors for fear of severe environmental conditions such as drastic temperature changes and evil spirits. For the mother, this is called "doing the month" (*zuo-yue-zi*) when she is given nutritious food to recuperate and to generate enough milk for the baby. This was perhaps the only occasion for her to enjoy extra treatment since in daily life women were expected to eat last and poorest.

Traditionally, a new mother had to avoid any exercise and washing of her body and hair for fear of future chronic diseases such as arthritis and headache. For government employees these mothers are given a paid leave. If no one in the family can take care of the baby, she would put the child in a nursery during the day when she returns to work.

Infancy

A newborn baby is taken care of by everybody in the family. A preferred caretaker both of the new mother and the baby is the husband's mother. In some regions the baby is tied with three cloth-made ropes, one just under the shoulders, the second around its waist with both hands on the sides, and the third just above the ankles. From time to time the grandmother would turn the baby to a different direction and a well-cared baby would have a round head and long limbs. Breast-feeding is generally the rule and may continue for up to 3 or even 5 years; 90–95% of infants were reported nursed by their mothers (Simoons, 1992). Newborn babies are kept from people considered unlucky, such as widows and those who have serious diseases. When traveling with young children, it was important to come back the same way so that their souls may not get lost. When the child was thought to have lost its soul the mother or grandmother would knock the door upper lintel with a ladle to call it back. In some places children are shaved with a pigtail behind called "*gui-jian-chou*," meaning "devil's fear" to ward evils away. It is quite common to have a boy with a name of a girl to appear cheap so he may be left alone by evil spirits. Children whom the family considered vulnerable and had fear of losing may even be given a name meaning, a dog, or a monk to ensure their safety. A boy may be called "*ya-tou*" (girl) for the same reason. To ensure safety of the child, the family might also make clothes for it from oddments of cloth, or a meal of rice collected from a hundred families.

Childhood

Children were taken care of by the elders, preferably the grandmother on the father's side, (though it is quite common, and even more preferably by the grandmother on the mother's side in urban areas), and older siblings. Corporal punishment is common and people believe it is good for the child. Traditionally Han Chinese thought children belonged to and were at disposal of their parents, grandparents, etc., and theoretically these elders could do anything with them. It was not even a crime to kill them if they so wished. Generally there was no concept of child abuse as such. How to treat children was a moral issue rather then a legal one until recently.

With older siblings as caretakers and playmates, it helped to build up the pecking order according to age and form close sibling relations called *Ti* (fraternal duty) among them. But the One Child Policy of the government in which a couple is allowed to have only one child has changed this enormously. Social scientists are worried that a future generation without siblings will lead to social problems. A popular name for these children is "small emperors" characterized as spoiled, selfish, overweight, etc.

Adolescence

Children participate in adult activities early and learn life skills by direct observation or by assisting adults in chores. Anthropologists say that traditional Chinese (Han) children did not have adolescence. Girls were taught skills they would need after marriage. There was virtually no sex education especially in the countryside where young people had to pick up scraps of sexual knowledge from older friends. Now children begin to go to school at 6 or 7, mostly in their own neighborhoods in urban areas and villages in the countryside. They are encouraged to help their parents in their work, especially during vacations.

Adulthood

Traditional Han society considered men and women adults at marriage. The basic objective in life was to take care of the man's parents and other elders in the family and to have children to continue the family line. Marriage was arranged by parents and other elders in the family with help from go-betweens. Ideal age of marriage was 16 for the bride and 18 for the bridegroom, but often a girl child was taken to her husband's house to be raised by her parents-in-law first. After 1950s, free choice has been encouraged though in most cases people still use friends or relatives as go-betweens to introduce the couple to avoid embarrassment. Menopause starts between 45 and 49.

The Aged

Age carried great respect and authority in traditional Han society. The eldest male was generally the family head with decision-making powers. Normally they had light work and would be given better food and tonics to eat. They were the center of everybody's attention and it was the obligation for everyone to take good care of them. On important occasions they were worshiped by everybody in the family. Even today, elderly people continue to live with family until they die, and only those who do not have sons or daughters are living in what is called "*jing-lao-yuan*" (home to respect the elderly), the Chinese version of a nursing home paid by the community.

China is an aging society. In 2000, one in every 10 people was 60 years of age or older. Those beyond 65 made up more than 7% of the population. The sharp decrease in birth rate since the mid 1970s has resulted in a huge number of one-child families. By around 2010, parents of these single children will be getting frail or senile. In the future, an adult couple will have to support four parents and one child. With rising cost for child rearing, including education and medical care, the financial burden on the couple will be formidable. Besides, senile dementia, common for senior citizens, will rank the third highest health problem in incidence by the middle of the 21st century, next only to cardiovascular and cerebrovascular diseases. Caring for these people will become a big problem for families psychologically and economically. There is an urgent need to build a social security system (Zhang, 2002).

Dying and Death

Traditional Han people believed in a next life after death and it was imperative for the living to send money and necessities for their comforts. This is the reason why one should not die sonless. Only those who had a grandson (son's son only) could avoid becoming hungry ghosts. The dead would be washed and dressed in their best, normally black suits or coats. Mourning would continue for days depending upon family means and what the astrologer would say. The occasion was important for the large clan members and relatives to revisit the notion how closely related they were with each other either through the dead or other connections.

Han people used to bury their dead and every clan or family would have their own burial ground where people went to burn paper money and other things, and cremation was limited to abnormal death such as death in traumatic accidents and painful diseases. Since the late 1960s, the Chinese government began to encourage cremation and forbid burial in order to save land for agriculture. Now urban residents have accepted cremation, but many rural people still bury their dead.

A rich family may invite Buddhist or Daoist monks or both to recite sutras to add to the merits of the dead so that they would be treated well in their next life by being sent to heaven.

REFERENCES

China *People's Daily Online*, April 4, 2002.

China Population Information and Research Center (COIRC), Available online at www.cpirc.org.cn

China Yearbook of Chinese Medicine (Zhongguo Zhongyi Nianjian) 2000 (in Chinese). Beijing: China TCM.

China Yearbook of Health (Zhongguo Weisheng Nian Jian) 2000 (in Chinese). Beijing: People's Medical.

Encyclopedia Sinica, Volume of Nationalities (Zhongguo Dabaike Quanshu Minzu, 1986) (in Chinese). Encyclopedia Sinica.

Encyclopedia Sinica, Volume of Traditional Chinese Medicine (Zhongguo Dabaike Quanshu Zhongyi 2000) (in Chinese). Encyclopedia Sinica.

Fang, Thome. (1986). *The Chinese view if life*. Taipei, Taiwan: Linking.

Ge, Keyou. (1996). *The dietary and nutritional status of Chinese population (1992 National Nutrition Survey)*. People's Medical.

He, Shaochu. (1998). *An overview of Chinese medicine and culture (Zhong Yi Yao Wen Hua Tong Lan)* (in Chinese). World Books Publishing Company: Beijing, Guangzhou, Shanghai & Xi'an.

Hsu, Francis L. K. (1948). *Under the ancestors' shadow*. New York, NY: Columbia University Press.

Hsu, Hong-yen et al. (1986). *Oriental Materia Medica: A concise guide* (pp. 3–42). Keats.

Li, Jingwei et al. (1990). *Ancient Chinese Culture and Medical Science (Zhongguo Gudai Wenhua Yu Yixue)* (in Chinese). China: Hubei Science and Technology.

Li, Xiangzhong. (1989). *Elementary Traditional Chinese Medicine (Zhongyixue Jichu)* (in Chinese). Beijing, China: People's Health.

Liu, Xingwu. (2002). *Good Health the Chinese Way*. In Llewellyn's 2003 Herbal Almanac (pp. 143–164). Los Angeles, CA: Llewellyns.

National Family Planning & Reproductive Health Survey. (2001). March 4, 2002.

Qiao, Jitang (1990) Chinese Life Rituals (Zhongguo Rensheng Liyi Daquan). Tianjin People's Publishing.

Simoons, Frederick J. (1991). *Food in China: A cultural and historical inquiry*, CRC Press, Boca Taton.

Wu, Gangzi et al. (1993). *The treasure house of greater Chinese culture*. China: Hubei People's.

Xie, Zhu-Fan. (1995). *Best of traditional Chinese medicine*. Beijing, China: New World.

Zhang, Benbo. (2002). *Aging Population Requires New Action*. China Population Information and Research Center (CPIRC), Available online at www.cpirc.org.cn

Hausa

Murray Last

ALTERNATIVE NAMES

"Hausa" is the most common term for these people. Earlier terms include "Habasha" (from which "Hausa" probably derives) and "Afnu" (used by Kanuri-speakers of Borno), but political scientists are apt to use "Hausa-Fulani" to indicate the fact that the old ruling elite is often labeled "Fulani" for historical reasons (they no longer speak the language of the Fulani, fulfulde) (Paden, 1975).

LOCATION AND LINGUISTIC AFFILIATION

Hausa-speaking people number at least some forty million and live mainly in the most northerly states of Nigeria and in the Republic of Niger. Hausa-speakers are also found in the Republic of the Sudan as well as in Eritrea, Chad, Cameroon, and Ghana. And there is a diaspora to North Africa and the Middle East (Tunis, Tripoli, Cairo, the Saudi cities), to Europe (especially France, Britain, with immigrants to Spain and Germany coming from the Hausa communities in Ghana), and to the USA. In the past Hausa traders traveled widely, and could be found living in India and southern Russia for example; there were also Hausa-speaking communities in 18th and 19th century Brazil due to the slave trade. Hausa was a written language using Arabic script (*a'jami*) for poetry and personal correspondence; prose works were later, dating from the late 19th century, and in the 20th century Hausa has been printed using a slightly modified roman script for textbooks (e.g., on subjects like health), religious works, and novels. The Hausa language is one of the "Chadic" languages, within the Afro-Asiatic language group (Newman, 2000). It became the language of cities and long-distance trade, and has spread as a lingua franca. It was used as the language of government by the British colonial administration in northern Nigeria. There is also an elaborate traditional sign-language used by the Hausa deaf and their neighbors (Schmaling, 2000).

OVERVIEW OF THE CULTURE

In the 19th century, Hausaland constituted the political and economic core of the Sokoto Caliphate, then the largest autonomous state in pre-colonial Africa (Last, 1967). The caliphate, governed from Sokoto and from Gwandu, consisted of a large number of emirates, in Hausaland the most important of which were Kano, Katsina, and Zaria. Under colonial rule (1903–60), the emirates formed the core of separate Provinces, and these became states during the Nigerian civil war (1967–69). The original states have since been sub-divided to create more states; each state is in turn divided into Local Government Areas.

The economic base of Hausaland was its farming (Hill, 1972; Mortimore, 1989). Cotton growing and weaving was a major activity, and still is today. Under colonial rule, groundnuts became an important export crop. Otherwise, trading and transporting was the staple of the urban economy, along with a variety of craft production using metals, wood, leather as well as cotton; the production and trading of books in classical Arabic was also significant, requiring a large learned class (cf. Koki, 1977). Since the 1980s, oil has supplied over 95% of Nigeria's national income. The rural economy is focused on food production, while the urban economy is heavily dependent on revenue funneled to the states and to the local governments from the Federal Government and its oil account.

Hausa society is Muslim (Sunni, using the Maliki school of Islamic law). In the countryside, there exist small non-Muslim communities (*Maguzawa*; *Arna* or *Anna*), and there are converts to Christianity, many of whom speak Hausa as the language of the church (Crampton, 1979; Last, 1993b; Nicolas, 1975). Inheritance is patrilineal, and polygyny is permitted; though the limit is four wives (and, in the past, any number of concubines), two wives are not uncommon for a man in his 30s or 40s. Divorce is common and easy; a woman might get through four husbands in a lifetime, even taking her very young children with her. Slavery was made illegal by the British administration in the 20th century.

The relatively brief period of British administration (1903–60) transformed Hausaland, not only by making it part of the new confederation, Nigeria, and eliminating local warfare, but also by slowly introducing new schools, a new medical system (including public health), new methods of transport (railways, roads, later airlines), and an economy of export crops (groundnuts, cotton); a market in hired labor replaced the previous use of slave labor, while dry-season migrant labor traveled far south, beyond Hausaland, for work. Existing Quranic schools and the tradition of Islamic learning were maintained, as were the courts administering the sharia law (with modifications). Hausa medical and surgical practices continued, as did the methods for controlling and curing mental illness; healers were neither illegal (as elsewhere under colonial regimes) nor persecuted. Christian missionaries were not allowed until the 1930s, and then restricted to non-Muslim areas; apart from leprosaria (and an eye hospital in Kano since the 1940s) their medical work was minimal (Schram, 1971).

The sixth largest oil-producer in the world, and with a population of some 120 million people, Nigeria has several major medical schools (among its 36 universities). There are some 3000 Nigeria-trained doctors working abroad, so that there is a shortage of high-caliber medical staff in Nigeria itself. Currently the shortcomings of most biomedical facilities have meant that Hausa people are still using traditional medicines, self-medicate, or opt to take no medicines at all.

THE CONTEXT OF HEALTH: ENVIRONMENTAL, ECONOMIC, SOCIAL, AND POLITICAL FACTORS

Oil-derived wealth has transformed young people's awareness of health and medicine. Traders in village markets opened booths selling a variety of biomedicines, with some having needles to offer injections of antibiotics as well as capsules. Medicines have become a profitable business; little expertise is needed, as diagnoses and prescriptions can be had from hospital staff even when the hospital pharmacy has no drugs on its shelves. Cut-price but fake drugs are commonplace and undermine confidence. By contrast, leprosy clinics handing out a weekly supply of dapsone (thought useful for a range of disorders) have all but disappeared. And cigarette smoking has plummeted as women, once the greatest purchasers of cigarettes, have turned against it, persuaded by Muslim preachers in the marketplaces.

The elimination of smallpox and the control of sleeping sickness have made a significant difference in rural areas. A more common yet major hazard for country-dwellers used to be guinea-worm (often acquired when washing after a day's work). New hazards include the widespread use of amphetamines (and now a cheaper local tonic known as *gadagi*), to make one work faster or for longer hours, and some take overdoses and have to be taken to a mental hospital. Another new hazard is road accidents; emergency services are poor, with neither phones nor ambulances and stocks of blood rarely ready for transfusion. Most recently (in the late 1990s) HIV has been diagnosed in country villages, with occasional deaths noted as being from AIDS. In the big cities of Hausaland it has been more commonly reported, but the stigma (and the lack of diagnostic facilities) has kept figures down. Less than 5% are infected (according to the government), with much higher levels among the sex workers surveyed; the military are also concerned about higher-than-average infection rates in army barracks. Epidemics of measles and cerebro-spinal meningitis cause higher numbers of deaths, as does cholera. Malaria, too, is a seasonal cause of death.

In rural areas meat is rarely eaten (dogs are vegetarian too), except at the annual festivals. Milk products have become rare as reduced grazing areas have reduced cattle herds; soured milk and butter were once regularly available (raw milk, being indigestible by adults, was not drunk). Only in far northern areas of Hausaland might camel's milk be available for making cheese. Hens' eggs are not eaten (except in cities), while those of guinea fowl are sent for sale to southern Nigeria as they have stronger shells and last three weeks unrefrigerated. No game remains; even birds are rare as children shoot them with catapults to eat roasted. Some insects (*gara*) are also roasted seasonally by rural children. In recent years bread has been widely marketed, with sugar but not salt added; biscuits, boiled sweets and mints are widely available, as is Coca-Cola and similar drinks in village market shops. Sugar consumption has therefore increased, even though the once-common supplies of semi-refined brown sugar blocks (*mazarkwaila*) have all but disappeared. Honey (*zuma*) from wild bees is comparatively rare now too. In general, foods that are sweet (and "sweet" includes salt) are thought to be bad for you, producing a mucus-like substance (*majina*) in the body; many snack foods are indeed sweet but are taken in small quantities. Late-onset diabetes is quite commonly diagnosed among the well-to-do (who have access to the routine tests of clinical medicine), and may be a response to an unusually high consumption of imported sugars taken with tea and breakfast cereals, foods that are not commonly found in the village diet.

Hausaland thus offers two distinct dietary regimes: the rural areas which would normally be self-sufficient in foodstuffs and export surplus grains to the big towns; and urban economies which import their food supplies not only from surrounding areas but also from abroad. Imported rice and maize as well as processed "fast" foods such as semolina and pasta, along with a variety of ready-made sauces and spices, have altered the "traditional" diets of many wage earners whose staples in the past were sorghum and millet, with gravies including products of the baobab or locust bean tree, or with local groundnut- or palm-oil. Cowpeas added variety, as did spinach leaves mixed with processed groundnut; so too did cassava, grown on special plots and sold, cooked as snacks; in the cities especially, fried fish of various kinds are available on the street, cheaper now that dams have been built for irrigation and rivers are no longer the main site for fishing. Goats, sheep, hens, and ducks are reared within houses, especially by women, but their role is not as food for the household but as items for sale: they serve as a woman's "savings bank," especially in cities.

Water from wells (and not from open pits) is commonly drunk in villages. Tap water is relatively scarce even in cities, where water-carriers distribute water from boreholes. Wells within houses in old cities such as Kano are now often contaminated, and their water at best can be used for washing; its taste precludes use in cooking (Last, 1999). Rarely is water boiled before use, nor is it filtered. Hence soft drinks like Coke and Pepsi are "pure," as is bottled beer, which, being alcoholic, is not allowed for Muslims. Traditionally, non-Muslim Hausa brewed beer, which was drunk once or twice a week as a social drink, especially by elders; it was cloudy, nutritious, and (when very fresh) not very alcoholic. It was the only boiled liquid available. Today, it is more rarely found, and those who drink it drink to get drunk (if there's enough for sale); alternative sources of alcohol are distilled spirits, imported from southern Nigeria. But alcoholism is unusual in Hausaland, or is invisible; hard drinking is done in private.

Wells have other uses, too. Deep wells are the most common site for suicide and attempted suicide (the fire brigade is called upon to rescue people from wells). They are also places where the bodies of victims of violence can be hidden. Contamination of wells and water supply is a standard source of panic, and scare stories against a pariah group may accuse them of this crime. Water, then, is dangerous; only children swim for pleasure—no one is taught to swim. Bilharzia is common where children have been tested for it. Defecation in rural areas is traditionally done on waste ground or on fields close by the farmstead. Only in towns are there pit latrines (*salga*) which are emptied periodically by hand by specialists hired by the residents. The night soil is taken to the edge of town where it dries out and solidifies and is eventually sold to farmers for use as manure. While the latrine is partly full, ash from the cooking place is sprinkled on the surface to reduce the smell. The hole, being infested with cockroaches, is often kept covered with a plate. A rising water table (as in Kano city today) causes latrines to fill more rapidly with water than was usual in the past. Where possible, the preferred direction to face when defecating is south (one prays, in contrast, facing east). Most people defecate very early in morning or at night; and if they are defecating outside they often choose a high place or a slope too steep for housing.

Houses are regularly swept clean by the women of the house, though not in the middle of the day when spirits might be playing (you normally say "excuse me" as you sweep, in case a spirit is about to be disturbed). Inside a house the ground is kept very tidy; but outside the house, rubbish tends to accumulate as the public authorities have generally ceased to clean the streets even in relatively wealthy parts of town. There is a good side to this: as spirits like rubbish to play in, heaps in the street keep them occupied there rather than coming over the wall and inside houses. Spirits are coming into towns, cities, and even villages in ever larger numbers from the "bush" because their "homes" are disturbed by people and especially by motor traffic; hence modernity brings mental illness.

MEDICAL PRACTITIONERS

Healers

There is a range of terms for healers. The most traditional are *maye* ("witch," implying he deals with spirits) and *boka* who is more aligned with herbal medicine. *Maita*, the ability to be a witch, is inherited; there are lineages of *mayu*. Itinerant sellers of medicine may call themselves (or be labeled by villagers) *mayu*, and advertise their wares by a particular roar that can be heard from the street. Their witchcraft substance, ideally *kankara* (hailstone or ice), is in my experience a glass marble flecked with red paint ("blood"); this is elaborately coughed up from the stomach and an object of real panic to villagers. A *maye* in serious medical practice does not travel, but runs an asylum for the mentally ill, often deep in the countryside. Patients are brought to him from afar (sometimes from outside Hausaland), and they may stay with him for a year before they are discharged. He gives patients herbal medicines to drink, and shackles them if they are prone to running away. He initiates them eventually into the *bori* cult, but the primary treatment is herbs, incense, and time. A *boka* may do much the same work (Last, 1976, 1981; Wall, 1988).

Other healers include the *masu magani* (skilled in medicines, mainly herbal) and *masu bori* (skilled in the *bori* spirit cult). The *wanzami* is the barber who may also do some surgical procedures such as circumcision on boys (occasionally on baby girls too, when he merely nicks the hypertrophied hymeneal tags) and a uvulectomy; he is likely too to incise the identity marks on babies and the decorative patterns on girls' necks or upper chest at puberty. He usually tours his district, often on horseback, and is quite distinctive with his hat and equipment. The *madori* sets bones that have been broken (now more common with football being such a popular sport), using bandages, herbs, and prayer (plus a chicken bone), with repeated visits to his patient; no anaesthetic is used, and manipulation is minimal. Finally there is the *ungozoma*, who is any woman serving as a midwife at childbirth in a home delivery. With obstructed labor or especially difficult deliveries, women try to get to a hospital; the traditional midwives are not really experienced in turning a fetus and do not use instruments such as forceps. Cases of retained placenta are treated at home, using a log of wood pressed on the abdomen, but again husbands try and take their wives to the nearest hospital. Pregnant women fearing an obstructed labor may take a razorblade and do a kind of episiotomy on the anterior wall of the vagina. Since the area is particularly vascular, considerable blood loss is possible; worse still, the urethra may be cut, resulting in incontinence that is very difficult to repair in hospital. What women are worried about here is *gishiri*, an infection that "causes" the anterior wall to swell and "block" the birth canal. It is transmitted from woman to woman, via cooking stools (Last, 1979). The cutting of the umbilical cord and using herbs to seal the cut sometimes results in tetanus in the newborn, whose death is attributed to spirits or to witchcraft within the house. Accusations of witchcraft can split the household, leading to divorces or to physical violence against the "witch" (*mai dodo*).

Other specialists include men expert at pulling teeth without an anesthetic; and there are those who specialize in treating livestock. Oxen are castrated, by smashing the testicles with a stone; wounds from fights involving horns have to be sewn up; cattle need de-worming. But small livestock are treated by their owners, using the common wisdom of the house which may include opening up capsules containing antibiotics and applying the powder to wounds. Snakes are relatively rare, and snakebites even rarer. Lancing (*sakiya*) a boil or an infected site is done with a red-hot arrowhead (*kibiya*). It is a procedure done at home, as is the procedure for removing "dead blood" from the back of a man by applying suction through a cow's horn. Cesarean sections were apparently not done traditionally.

The main modes of healing involve (1) spirits and/or ghosts of the dead; and (2) remedies made from plant materials, parts of dead animals and minerals ground down into powders (Etkin, 1979, 1981, 1996; Etkin & Ross, 1991; Ross, Etkin, & Muazzamu, 1991, 1996). Some healers have their own brews, unguents, or powders, whose composition is a secret; others combine these with verses of the Quran, with numerological "squares" or with special prayers and words written on scraps of paper folded into a "charm" (*laya*). "Islamic medicines," sold in bottles, are now gaining popularity in towns. Farmers often bring plants in very early on market day in order to sell them to the medicine seller (*mai magani*). Although herbs are collected from forested areas, in practice most useful plants grow on farmland or around settlements. In demand are the various incense woods that are burnt at night in people's rooms; Muslim spirits especially are associated with particular scents.

The cult of spirits known as *bori* has been the subject of much study (Besmer, 1983; Last, 1991a; Lewis, 1991; Nicolas, 1972). *Bori* divides into two kinds, one public and the other private. The first, accessible to visiting Europeans and others, is run by public healers competing in the bigger towns for the prestige of being the leading traditional practitioner. Many of these healers are from out of town; in Kano they are often men and have come from the Niger Republic. Equally public used to be the *bori* performances put on occasionally by the women in brothels (*gidan karuwai*; *gidan mata*), where they danced late into the night to the specific tunes of their spirits. The second kind of *bori* is usually private (if not secret), performed within a purdah'd house with women guests invited; it is usually run by women and is focused on a specific problem, such as illness or a misfortune in the family. In a crisis, a mother may go into possession-trance in the middle of the night to identify what is killing her daughter; desperate to find a remedy, she'll have awakened a co-wife or a neighboring woman to help in posing the questions and hearing the answers.

A dead person's soul (*kurwa*) may possess a kinswoman or descendant, and speak through her; curing requires acceding to the soul's demands and then exorcising the patient who might then go on to be initiated (*girka*) into the spirit possession cult (*bori*). A spirit (*iska*) can trouble a person (usually a woman), either through illness or dreams or loss of children, and "call" her, so that her kin then finance her initiation into the spirit cult (cf. Erlman & Magagi, 1989). At its end, one or more spirits have been identified as linked to the initiate, and they are escorted to her room in her husband's house, where they will act as guardians or helpers. A particular kind of hen, goat, or sheep will then be kept by the initiate, as symbol of the spirit's readiness to help. That spirit may on occasions possess her and, speaking through her, express the initiate's complaints and her wants. These wants may include perfume, new clothes, or some luxury, or more prosaically the rightful share of some property or a special ritual (mobilizing her kin to carry out a sacrifice of an animal on her behalf). If a marriage is in difficulty, spirit possession may be a means of negotiating a better relationship with the husband or a co-wife.

The work of healers is to mediate between spirits and patients. The healer supervises initiations (which can be done for more than one initiate at a time, thus reducing the costs of hiring a musician, his assistant, and the healer), and helps to identify the particular spirit that is to act as long-term guardian of the patient. But the healer, who has a spirit of her own, can herself become possessed by that spirit and offer answers to questions posed by the women coming to seek her advice. This consultation is normally done in the healer's room in her house, and is more or less private. Questions are usually about the underlying causes of illnesses, or how to prevent a child's ill health or some other misfortune, but may also be about a pending move to another house, or re-building a house, or even about the consequences of converting to Islam. As spirits do not always speak "normal" Hausa but disguise it by switching sounds, their pronouncements are not always clear, and can lead to discussion afterward and reinterpretation.

While women are common as *masu bori*, the major asylums where the mentally ill live (in the healer's compound) are run by men, with the help of wives. Such asylums are usually found in the midst of farmland, not far from villages or even towns where inmates on the mend can find odd jobs or extra food. Inmates may come from non-Hausa backgrounds and include both men and women, adolescents and adults. They may stay up to a year in an asylum, being treated with medicines to drink (which may include pills acquired by the healer from a government mental hospital).

Hospitals

Government hospitals, whether run by university medical schools or by state ministries of health, exist in the major cities, but in recent years their services have been marred by lack of drugs and by basic equipment being out of operation. Electricity has also been intermittent. Shortages of running water have made hygiene difficult for in-patients, whose relations are expected to bring in food and drinking water and do the washing of bed linen and patients' clothes. Private clinics have provided good services for the rich. Mentally ill patients are treated in special units (Last, 1991c). In rural areas medical services vary enormously, from a well-staffed and managed hospital to dispensaries or "cottage hospitals" where the few staff can offer little or nothing beyond advice; private clinics now compete for patients too. Drugs may be out-of-date, dumped by first-world suppliers; or they may even be "fake," sold by unscrupulous local "manufacturers" in plausible packaging. Where blood transfusions are possible, blood may be supplied by a relative; if not, a person waiting by the hospital may be hired to supply blood. This is screened, and if all right, added to the blood bank; if not, it is disposed of.

Ambulance services are almost non-existent, in towns or in rural areas, so in an emergency adult patients are carried to the roadside and put in a bus. But cases of obstructed labor, for example, pose a real problem, as they cannot be carried using a bed as a stretcher; that is taken to be a bier and to presage death. Since hospitals are often seen as places where the sick die, there is a reluctance to be taken to hospital. Taking a body back means specially hiring a vehicle (given the reluctance of public vehicles to carry corpses). Furthermore, a post-mortem in the hospital can leave a body mutilated, giving rise to fears that hospital staff have robbed the body of significant parts for magical use (or for sale). In short, those considered near death are not necessarily rushed to hospital; a decent death at home is preferable. A hospital can ameliorate pain or cure illness but it cannot of itself prevent death, though some of its staff may believe they can delay it.

CLASSIFICATION OF ILLNESS, THEORIES OF ILLNESS, AND TREATMENT OF ILLNESS

Biomedical ideas about disease and about human anatomy and physiology are quite widespread, being taught in schools. But the older generation as well as many young also know the "traditional" ideas, especially about mental illness. Some illnesses are caused by spirits or sorcery: that is, illness originates outside the individual and involves malice. The idiom is that of someone "shooting" you as with an arrow; the "wound" (and the pain) is in the outer layer of the body. In Hausa thought, the body is composed, broadly speaking, of two layers: an outer layer (*jiki*) and an inner core (*ciki*). The outer layer is essentially fluid, with breath, blood, sweat, saliva, and semen, as well as sounds, smells, and sights. Entry points into the outer layer include the throat, ears, nose, eyes; the penis and the vagina are exits. The inner layer contains solids, brought inside via a separate throat and expelled through the anus.

Post-mortems on humans are not part of traditional Hausa mortuary culture, so that there is no formal anatomical knowledge of what is inside the human body. Parts of the body are, of course, given names—heart, lungs, kidney, liver, stomach—and as butchery of livestock is commonly witnessed, the innards of goats and sheep are well known. But people are reluctant to equate human with animal anatomy; to do so would presume acquaintance with the inside of a human, and that is knowledge that only witches have.

The outer layer of the body is not just the site for illness-as-injury, but for all social interactions. Popularity and pleasures occur on the outer layer, whereas innate characteristics are "solid" and located "inside." So too is witchcraft substance and a propensity for evil. Life is there too; in death, the solids within liquefy and pour out. Ageing is the gradual drying out of the outer layer, leaving the elderly almost all "solids" and scarcely mobile. The young are predominantly "fluid" and mobile. One symptom of illness is for the fluids of the outer layer to get blocked, and form a swelling that may need to be lanced. Liquids, like semen or urine or "dead" blood ("dead" because of exertion in a hot sun), must be eliminated, or else they back up and go bad (for example, in the small of the back). Too much sweat is a bad sign; liquids should not pour out through the skin.

Medicines enter the outer layer as fluids, whether drunk or (nowadays) through injections. The outer layer can also be massaged with unguents and scarified; it is accessible through incense as well as music and words. It can be terrified by a horrible sight or soothed by what's pleasing. Mental illness is an affliction of the outer layer; spirits, ghosts can enter that layer, just as a person's self can go out in dreams, without depriving the person of life.

Many illnesses afflicting people are seasonal and caused not by spirits or sorcery; they are God-given. Seasonal variations affect the outer layer, too: cold and rain, mosquitoes, biting flies and insects, guinea-worms (in the past), as well as the hot sun (noontime is especially dangerous). Night, too, has potential sources of illness, in particular, dew (*raba*) and the full moon, and no one should sleep outside. The countryside is criss-crossed by spirit "highways," and it is wise not to rest under one of these highways, let alone site a house there. Similarly, paths or cross-roads where people deposit rubbish, old medicines, and any other contaminated article, are sources of "infection": such items are placed there specifically in order that their "infectiveness" can be carried away by a passerby. It is possible to map sites of danger in a community, and these are usually also sites where extra protection against harm is required (Last, 1988, 1991b). Each adult would expect to have sufficient protection for everyday life. But chance encounters, especially with strangers, carry risks: the marketplace is one such site.

Protective medicine in the form of charms (*laya*; commonly a combination of text in Arabic script plus a few dried leaves and a twig) can be carried on the body. Some special underwear (*bante*) can give magical protection. But the main source of preventive medicine is based in the house, where the ancestors and domesticated spirits are located, where there are medicines built into the walls of each room and the homestead as a whole, as well as in the structures of the granaries. The regular five prayers a day and the strict practice of Islam ensure well-being, too; the annual rituals are carried out to cleanse the house and sanctify its residents and their kinsfolk. Almost daily the elder men of the household eat together and resolve any conflicts or preempt tensions; senior wives are within earshot, and relay news of problems. The regular succession of illnesses and deaths shows how important prevention is.

SEXUALITY AND REPRODUCTION

Both men and women have "semen" (*maniyi*) that is necessary for reproduction. Without regular expression, the semen can back up, perhaps go bad, and cause illness especially in the lower back. Marriage and regular coition for both men and women are therefore necessary for health. In a polygynous household, wives take turns to cook for their husbands two days at a time, and by implication the two nights are theirs too. (Concubines in the past had one night only; irregular unions occur in the afternoon.) Whereas marriage may take place at a young age (even before puberty) a very young bride is not usually sexually active until she is "strong enough," that is large enough to bear children safely; this may be a year or two after menarche. Her first child may die in infancy, particularly if the newborn loses maternal immunity at the height of the malaria season (August/September). Given that weddings often occur in March/April, a baby conceived very early in the marriage and born in December/January is at extra risk.

Pregnancies may go to "sleep" (*kwanta*), and not come to term in nine months (Kleiner-Bosaller, 1993). In Maliki (Islamic) law, such sleeping pregnancies can last up to 7 years, so that a divorced woman who conceives out of wedlock can within that period claim her former husband as the father of the fetus; otherwise the formal penalty for fornication outside marriage is death by stoning. But such "sleeping" pregnancies also occur when a wife is having difficulty becoming pregnant and feels threatened by divorce; it can occur too when there is an ectopic pregnancy or a cyst that mimics an early pregnancy. There is no traditional treatment for it. Sleeping pregnancies are not, of course, confined to Hausa culture, but are potentially found wherever the Maliki School of Islamic law is in force, as for example in Morocco.

A further complication of pregnancy occurs when a woman's uterus has been entered by a spirit, and a changeling child (*dan ruwa*) is later born to her. Spirits are apt to enter women when they bathe or are near water. There are no clear signs of such a compromised pregnancy; and the child only shows his or her changeling status later, for example by sickle-cell anemia or by dying prematurely. A child born to a mother after earlier children have died is known as a *dan wabi*. It is not uncommon for a woman to lose all her children while a co-wife has all hers alive; one assumption is that the repeated deaths are a single child or spirit repeatedly reentering the mother and dying early (Last, 1992). Another assumption is that her breast milk may be bad, or even contain some poison.

A wife who is losing all her children may well decide to leave her husband and "drink other water" in the hope of bearing a child who will survive. While divorce is common and easy, I have known women who have remained with their husband despite having borne him eight or more children all of whom died young. The explanation is that wife and husband both love each other deeply (Last, 1992). Romantic love is not uncommon and may lead to a couple, who were childhood sweethearts, marrying finally in their old age after both have been through other marriages. Some wives concerned over their husband's apparent infertility may discreetly become pregnant outside the marital home. Similarly, where the husband is from a lineage that only produces daughters, the pressure on a wife to provide him with his only son may result in her taking special measures. There are special sites and rituals (including *bori*) that are supposed to restore fertility, but the most commonly tried cure is a change of husband.

Apparently infertile women may often remarry, and be the third or fourth wife in a large household with many children, one or more of whom she may help to bring up. Such junior wives can be hugely popular, with their stories, songs, and humor bringing fun to the children of the house; they can devote themselves to making snack foods or to petty trading or a craft to earn cash and give presents. They cannot live on their own, nor can they readily return permanently to their father's house (at least not until old age). Modern conditions may make women-headed households more socially acceptable: for example, where the husband has died of an illness that has been labeled as AIDS (*kanjemau*), his widow may have to live alone, as people fear contact with her and her cooking. Otherwise, every woman has the right to a husband, no matter how disabled she might be (Last, 2000a).

HEALTH THROUGH THE LIFE CYCLE

Pregnancy and Birth

A pregnancy is not considered established until 6 months, when the pregnant mother (*mai ciki*) will cover herself up (e.g., with a cloth over her breasts) and be teased (e.g., asked if she's been eating beans). When spontaneous abortions (*bari*) occur, the fetal matter is disposed of without ceremony. Miscarriages (also known as *bari*) are similarly treated; the dead child is not formally buried in the graveyard. Until recently, women might commonly smoke cigarettes, resulting in babies of low birth-weight. Six pounds was not an uncommon weight, and obstructed labor was relatively rare. Birth is done kneeling down (*nakuda*); another woman (who is then known as *ungozoma*) picks up the child and cuts the umbilical cord. It is she who bathes the cord stump, and washes the baby each morning and evening, in principle until the day for naming. But in practice mothers may take over the care of their babies straightaway, and breast-feed them, despite taboos on letting a neonate suck the colostrum. The placenta (*mahaifa*) and cord (*cibiya*) are both buried behind the mother's room by the *ungozoma*. First births are given this full attention; later births can happen so easily and unexpectedly that the mother may find herself giving birth in a field or in a latrine.

Once delivered, the mother has to be purified and "heated" to remove the damp-cold of birth; the full procedure (*wankan jego*) lasts for 40 days and thereafter continues once a day for 3 or 4 months. If she fails to do this, she will swell and die, or at least become pale and chill. Twice a day she has to wash herself by sprinkling boiling water over herself with special leafed branches (*kunfu* or *cediya*). In Zaria (since at least the 16th century) the mother also lies on a special clay bed under which a fire is kept burning; she is gradually "roasted." Elsewhere she is "steamed" over a pot of boiling water. Finally, she is given a special food made from the feet and head of an ox and medicines (especially potash, but also spices). A consequence of this excessive heat and salt intake is the above-average incidence of peripartal heart failure (*daita*) from which women can die. The local remedy is a plant found in marshy areas similar to foxglove, but it is not always given in sufficient quantity to prevent death. A baby whose mother dies is usually not nursed by a wet-nurse, though if some kinswoman is lactating and is willing to take on the orphan, that is acceptable (but the baby already on the breast and the orphan cannot later marry; sharing the same breast milk makes them siblings).

In theory, on day three a barber will come and cut the uvula, incise marks on the baby's chest, and also cut marks on the baby's face. But I have known these to be done later, on the occasion when a girl baby has her hypertrophied hymeneal tags nicked as if it was the clitoris that was being cut. The baby's first haircut is on day seven, when it is named in a big ceremony.

Infancy

The baby is solely breast-fed until about 6 or 7 months old. Weaning occurs after 2 years and some months, usually by being sent away on a Friday to the mother's mother; sometimes a bitter substance has earlier been smeared on the nipple to discourage the infant from suckling. A baby may react badly to weaning. If an infant is weakening before it is weaned, it may be a sign that the mother's husband has started touching her, or even that she is pregnant with another baby. Such a woman feels shame. Her infant, in its distress, typically has flies hovering over its face. More routine are the diarrheas associated with teething. A mother watches the fontanel for signs of stress. Another hazard is the fire within the mother's room; babies have occasionally been known to fall into the fire as they sleep with their mother. The baby by day is carried on her mother's back (or by an elder sister), and is therefore very early taught not to urinate or defecate except when told (the baby is held between the mother's ankles, and a "ssss" sound is made). The twinning rate in Hausaland is much lower than among the Yoruba to the south; twins may be unusually bad-tempered.

Childhood

Children work as well as attend school. Punishment is given to those who fail to do their work properly; a child absconding from school may find himself kept in shackles (*mari*). But willingness to punish varies between Hausa communities: Muslims are readier to punish the young than are non-Muslim Hausa (Last, 2000b). Child sexual abuse does occur, if rarely: there are stereotypical stories about sadistic or sexually active teachers who abuse their pupils. So, too, young girls may be sexually abused in cities by adult men on weekends (and need urgent hospital repairs the next day). But sometimes children use their time on the street to experiment sexually with each other.

Circumcision of Muslim boys occurs about 7 to 9 years old, and a group of boys is operated on by the barber at a major ritual in the cold dry season; non-Muslim Hausa boys may be circumcised just before puberty. The operation is done with a knife with the boy sitting on the ground over a small hole; the fraenum, though, is cut not with a blade but with the barber's fingernail (*akaifa*). The wound is then smeared with juice from acacia pods, and bound up. It can take a month before he is fully recovered. Newly circumcised boys usually do not sleep on a mat or wear clothes; and there is traditionally an adult man to keep an eye on them at night, so as to stop anyone from rubbing his penis. Nowadays, bandages are used.

Hausa girls are not excised (nor were they in the 19th century), apart from the nicking of the hypertrophied hymeneal tags shortly after birth, at the time when the uvula is also cut. The object in both cases is to keep open the two orifices (the throat, the vagina), on the assumption that both might otherwise become blocked at maturity; it is not about controlling sexuality or closing the vagina. Confusion arises because the term used, *beli* (cf. *belu* for uvula), implies that it is the clitoris, seen as a parallel organ to the uvula, that is being cut, when observation of the operation clearly shows that it is not cut, and could not be at that age. I have known people worry that their uvula has grown (or re-grown) and is causing them trouble in their throat; I have never heard of the clitoris growing to cause a blockage. Vaginal "blockage" occurs only in late pregnancy and is caused (if at all) by a temporary swelling of the anterior wall of the vagina. This swelling is relieved by using a razorblade to remove the liquid, not by cutting the clitoris.

Adolescence

While menarche may mark the start of adolescence, courtship may precede menarche, with the girl receiving presents from older suitors. The period of courtship in villages may start with girls having decorations (depicting, e.g., drums or even a boxer) incised on the neck and upper chest; pot black darkens the drawings. Ideally in the middle-class urban context, the would-be husband should be 10 or more years older than his bride. In rural areas, the age difference is less pronounced, and adolescents go courting in marketplaces and in the evenings once the harvest is brought in (October to January). Courting involves petting but not coition. The process from betrothal to wedding takes some 3 months, and is a dry-season activity involving elders as well as the couple's friends and kinswomen. Male adolescents may go to a village brothel if they have the funds after the harvest, but today, with AIDS and the reintroduction of Islamic law, many brothels have been closed down and the women dispersed.

Young men have to work hard to accumulate cash to buy presents for their girlfriend. Many go on dry-season migration (*ci rani*), to farms in the south of Nigeria and even Ghana, or to cities where laboring jobs are to be found. They gain experience by seeing the world and experimenting with different foods, drinks, and social habits; by being away, they also save their parents the costs of feeding them from the family's granary (cf. Last, 1993a). Adolescent girls stay at home, possibly going to school or trading on their mother's behalf. Stricter households, particularly in the towns, may insist on the girl being in purdah by day, and only allowed out with an escort after dark. Ideally, she will be betrothed or married soon after menarche, thus transferring responsibility for her well-being (and her further education) from her parents to her new husband.

Adulthood

An adolescent male becomes an adult when he takes a wife; he is responsible for another person. A woman becomes an adult, not when she marries, but when she has her first child, when she too is responsible for another person. The couple are housed by the husband's parents, with the bride having a room of her own, filled with her bridal goods, and she has her own waterpot as well as food bowls there; she cooks at her own fire in front of her room (unless there is a single family kitchen). She is likely to get pregnant within 2 years; if not, there may be pressure on her to leave her husband, or on him to seek a second wife (if he can afford one). If she does get pregnant and bears a child, it is not unusual for the first child to die young. Although there is pressure on her to "avoid" her baby (and not over-coddle it), in practice mothers are very solicitous of their firstborn. If all goes well, she will be pregnant again some 2 years after the first child is born. And the process will continue for some six to ten births. Women do seek means of restricting the number of conceptions, but there are no formal means of contraception. Were she to divorce her husband, she would conventionally have to leave behind all her weaned children; alternatively, she could declare herself menopausal, and stop sleeping with her husband. If he had not already done so, he would seek another wife or wives. The first wife may well encourage him to do so, to save her some of the hard labor of being his sole wife; a good polygynous household, run by a competent senior wife and a just husband and filled with children, is more desirable than a small domestic unit. An unhappy polygynous household can be riven by conflict and jealousies, with fears of competitive love-magic and sorcery resulting in mental or other illness; a husband's favoritism adds to the tension. Violence between wives, symbolic or actually physical, may result, and the husband may resort to beating a wife. Divorce is always a possibility, more often initiated by an unhappy wife, but she may just go away awhile to cool off; she would return to her parents' home or seek out a sympathetic brother.

As a divorcee a woman can expect to re-marry as a junior wife. Although a husband has to treat all his wives with strict equality, her status is still subordinate to the senior wife. A woman may move on from husband to husband until she finds a congenial household; some in the period between marriages set themselves up as courtesans, keeping a salon where men come to chat and one or more of these men court her with a view to sleeping with her regularly or marrying her. This is not the same as prostitution based in a brothel; relations of courtesanship are built up over time, and sexuality is controlled and only part of the scenario. Finally a divorcee may decide to marry a farmer, in which case he may give her fields to farm on her own account, and she re-builds a life around her own earnings and new friendships. Although adultery (*zina*) occurs in towns and villages alike, only pregnancy outside marriage is really serious; as noted, it can now result in the woman being sentenced to stoning.

As a married man's household expands, he may section off part of his father's compound, dividing it from his brothers' area; but where a house is crowded (as in an old city), a wash-room may have to be converted into another bedroom, in which case it may still retain the "wrong" aura and give rise to illness. Alternatively he may (especially in rural areas) build himself a new house of his own, choosing a propitious site with the help of an Islamic scholar. The well-being of his household may depend on how carefully the sacred rituals have been carried out at the house building. Repeated illness may force him to move house again, or his wives to leave him. Spiritual well-being can be enhanced by regular religious practice, for example by being part of a Sufi brotherhood and joining in the extra recitations after the Friday prayer; extra fasting, twice a week, may be done too, as well as regular reading of the Holy Quran. Drinking alcohol or smoking or intemperate behavior generally calls into question not just his own health but that of his whole household.

The Aged

A man starts becoming an elder (*dattijo*) around 40 years old, but can cease to be regarded as an elder once he becomes senile. An elderly woman too, after menopause, acquires a special status, staying on with her grown-up sons who have taken over the house and its land after the death of her husband. Such old women are no longer subject to purdah, but they may have little reason to go out except to the life cycle rituals of their children's families and other kin. Elders seldom leave the house and its environs, spending long hours in their dark room with enough of a fire to keep them warm and may be fed by their sons' wives. Often toothless, they need to grind such luxuries as kolanuts or groundnuts on a small tin grinder; other foods require no chewing. Now they may have a radio with its songs to listen to.

Dying and Death

In the 19th century, March/April and October/November saw more deaths than other seasons. Malaria is particularly lethal in August/September, and was attributed traditionally to the newly harvested grains. A study I organized of burials in the old city of Kano showed that more women and children died on a Friday (and Thursday). To die on a Friday is a blessing given by Allah, and it might seem that people "choose," if they can, to die on that day. But the timing of death is Allah's alone, people say. Yet people also report mothers (and fathers) summoning their children to come to them and awaiting their arrival; after speaking to them in front of others (a sure sign of impending death), the dying mother closes her eyes. Foreknowledge of death does not provoke attempts to avoid it, but rather to ensure that what needs to be done right is done. Near-death experiences occur on occasions of very high fever, which may provoke visions of ancestors. The corpse is washed in the ritual Islamic manner. No post-mortem is done. A death in the house is marked by a deep, almost audible silence; no wailing is heard (as it implies impatience with Allah). No work is done that day (two days if the dead person was important), and visitors come to express condolences. Grief is meant to be restrained.

Burial is done within a few hours; or if the death occurred at night, early the following day. Graves in city cemeteries are dug in advance each day by a gravedigger and on Fridays by volunteers. A head of a house is likely to be buried in the yard behind his room; others are carried out on a bier and buried in a cemetery, with graves unnamed and all aligned toward Mecca (head to the south, facing east). It can soon be difficult to locate a specific grave in a city cemetery where 10 to 15 new graves are dug daily; the body is meant to have anonymity in death. But in recent years gravestones with names have become fashionable. The body is never disinterred, and though graves can in theory be reused after seven years, they seldom are. There is a persistent fear of grave robbers digging up recent corpses at night for body parts to be used in magic or for sale; anyone loitering in a graveyard therefore attracts suspicion.

Some bodies have to be treated differently. Wives brought back dead from the hospital to their husbands' house are buried in a field near where they lived, and not in the family or village graveyard. In the past, slaves who died in town were not necessarily buried; their bodies were thrown into ponds (where crocodiles ate them). War dead were quickly buried, unwashed and in their clothes, on the battlefield, but many such bodies were in fact eaten by hyenas. Nowadays hyenas, crocodiles, and vultures are all so rare that carrion (*mushe*) rots uneaten. Hausa, being Muslim, have no concept of reincarnation.

REFERENCES

Besmer, F. E. (1983). *Horses, musicians & gods: The Hausa cult of possession-trance.* South Hadley, MA: Bergin & Garvey.

Crampton, E. P. T. (1979). *Christianity in Northern Nigeria* (3rd ed.). London: Geoffrey Chapman.

Erlman, V., & Magagi, H. (1989). *Girka: Une cérémonie d'initiation au culte de possession boorii des Hausa de la région de Maradi.* Berlin: Dietrich Reimer Verlag.

Etkin, N. L. (1979). Indigenous medicine among the Hausa of Northern Nigeria: Laboratory evaluation for potential therapeutic efficacy of antimalarial plant medicinals. *Medical Anthropology, 3.4,* 401–429.

Etkin, N. L. (1981). A Hausa herbal pharmacopoeia: Biomedical evaluation of commonly used plant medicines. *Journal of Ethnopharmacology, 4,* 75–98.

Etkin, N. L. (1996). Ethnopharmacologic perspectives on diet and medicine in northern Nigeria. In E. Schroder, G. Balansard, P. Cabalion, J. Fleurentin, & G. Mazars, (Eds.), *Médicaments et aliments: Approche ethnopharmacologique* (pp. 8–62). Metz: Société Française d'Ethnopharmacologie.

Etkin, N. L., & Ross, P. (1991). Recasting malaria, medicine and meals: A perspective on disease adaptation. In L. Romanucci-Ross, D. E. Moerman, & L. R. Tancredi (Eds.), *The anthropology of medicine: From culture to method* (2nd ed., pp. 130–258). New York: Bergin & Garvey.

Hill, P. (1972). *Rural Hausa: A village and its setting.* Cambridge: Cambridge University Press.

Kleiner-Bosaller, A. (1993). Kwantacce, the "sleeping pregnancy": A Hausa concept. In Gudrun Ludwar-Ene, & Mechthild Reh (Eds.), *Gros-plan sur les femmes en Africa; Afrikanische Frauen im Blick; Focus on women in Africa* (pp. 17–30). Bayreuth African Studies Series 26. Bayreuth: Bayreuth University.

Koki, A. M. (1977). *Alhaji Mahmudu Koki: Kano malam.* (Neil Skinner, Trans. Ed.). Zaria: Ahmadu Bello University Press.

Last, M. (1967). *The Sokoto caliphate.* London: Longmans Green.

Last, M. (1976). Presentation of sickness in a community of non-Muslim Hausa. In J. B. Loudon (Ed.), *Social anthropology and medicine* (pp. 104–149). ASA monographs 13. London: Academic Press.

Last, M. (1979). Strategies against time. *Sociology of Health and Sickness, 1.3,* 306–317.

Last, M. (1981). The importance of knowing about not knowing. *Social Science and Medicine, 15B,* 387–392.

Last, M. (1988). Charisma and medicine in northern Nigeria. In D. B. Cruise O'Brien & C. Coulon (Eds.), *Charisma and brotherhood in African Islam* (pp. 176–224). Oxford: Clarendon Press.

Last, M. (1991a). Spirit possession as therapy: Bori among non-Muslims in Nigeria. In I. M. Lewis (Ed.), *Women's medicine: The zar-bori cult in Africa & beyond* (pp. 49–63). Edinburgh: Edinburgh University Press.

Last, M. (Ed.) (1991b). Adolescents in a Muslim city: The cultural context of danger and risk. In Youth & Health in Kano Today. *Kano Studies* (special issue, pp. 3–17). Kano: Bayero University.

Last, M. (1991c). The presentation of mental illness at the Gorondutse Psychiatric Hospital, Kano (pp. 33–54) (with G. Ilyasu). In M. Last (Ed.), Youth & Health in Kano Today. *Kano Studies* (special issue). Kano: Bayero University.

Last, M. (1992). The importance of extremes: The social implications of intra-household variation in child mortality. *Social Science and Medicine, 35.6,* 799–810.

Last, M. (1993a). The power of youth, youth of power: Notes on the religions of the young in northern Nigeria. In H. d'Almeida-Topor, C. Coquery-Vidrovitch, O. Goerg, & F. Guitart (Eds.), *Les Jeunes en Afrique* (pp. 375–399). Paris: L'Harmattan.

Last, M. (1993b). History as religion: Deconstructing the Magians ("Maguzawa") of Nigerian Hausaland. In J-P. Chrétien (Ed.), *L'invention religieuse en Afrique: histoire et religion en Afrique noire* (pp. 267–296). Paris: Karthala.

Last, M. (1999). Rubbish: Public health in northern Nigeria. In Jean-Pierre Olivier de Sardan (Ed.), *Second rapport intermédiaire (1999)* (pp. 37–43). Marseilles: CNRS/EHESS.

Last, M. (2000a). Social exclusion in northern Nigeria. In Jane Hubert (Ed.), *Madness, disability and social exclusion: The archaeology and anthropology of "difference"* (pp. 217–239). London: Routledge.

Last, M. (2000b). Children and the experience of violence: Contrasting cultures of punishment in northern Nigeria. *Africa, 70.3*, 359–393.

Lewis, I. M. (1991). *Women's medicine: The zar-bori cult in Africa & beyond*. Edinburgh: Edinburgh University Press for the International African Institute.

Mortimore, M. (1989). *Adapting to drought: Farmers, famines and desertification in West Africa*. Cambridge: Cambridge University Press.

Newman, P. (2000). *The Hausa language: An encyclopedic reference grammar*. New Haven: Yale University Press.

Nicolas, G. (1975). *Dynamique sociale et apprehension du monde au sein d'une société Hausa*. Paris: Institut d'ethnologie.

Nicolas, J. (1972). *Ambivalence et culte de possession: Contribution à l'étude de bori Hausa*. Paris: Editions Anthropos.

Paden, J. N. (1975). *Religion and political culture in Kano*. Berkeley: University of California Press.

Ross, P. J., Etkin, N. L., & Muazzamu, I. (1991). The greater risk of fewer deaths: An ethnodemographic approach to child mortality in Hausaland. *Africa, 61.iv*, 502–512.

Ross, P. J., Etkin, N. L., & Muazzamu, I. (1996). A changing Hausa diet. *Medical Anthropology, 17*, 143–163.

Schmaling, C. (2000). *Maganar hannu: Language of the hands. A descriptive analysis of Hausa sign language*. Hamburg: Signum Verlag.

Schram, R. (1971). *A History of the Nigerian Health Services*. Ibadan: Ibadan University Press.

Wall, L. (1988). *Hausa medicine: Illness and well-being in a West African culture*. Durham NC: Duke University Press.

Hmong in Laos and the United States

Kathleen A. Culhane-Pera, Dia Cha, and Peter Kunstadter

ALTERNATIVE NAMES

Miao or Meo (considered derogatory by Hmong).

LOCATION AND LINGUISTIC AFFILIATION

Hmong are a distinct ethno-linguistic group who originated in China and who migrated into northern Southeast Asia during the 19th century. After the end of the "Secret War in Laos" in 1975, many Hmong fled to Thailand and then were resettled around the world, including the United States. Most Hmong in the United States speak two closely related dialects of the Miao–Yao language: White Hmong (*Hmoob Dawb* or *Moob Dlawb*) and Green Hmong or Blue Hmong (*Hmoob Ntsuab* or *Moob Lees*). These two groups were traditionally distinguished by their dialect, dress, housing style, and other cultural differences.

The Hmong diaspora is extensive, with over seven million in China, one million in Southeast Asia (Schein, 2000), and almost 16,000 in France, Canada, French Guyana, Australia, and Argentina (Rice, 2000). According to the 2002 Census, 169,428 Hmong live in the United States, with most in California, Minnesota, and Wisconsin (Pfeifer, 2002). However, Hmong community leaders feel the U.S. Census 2000 undercounted the Hmong and estimate that 227,217–268,747 live in the United States (Hmong National Development Links, 2001).

OVERVIEW OF THE CULTURE

History

Hmong myths suggest Hmong lived in northern China and migrated into southern China. Early Chinese historical records indicate that Hmong were lowland irrigated rice farmers. Resisting the political control and population pressures of the Han Chinese, many ended up in the mountainous areas of central and southern China. Some Hmong migrated into the highlands of northern Southeast Asia in the mid-1800s. In the first half of the 20th century, French administrators in Laos granted opium-growing concessions to some but not all of the Hmong in Xieng Khouang province. This was one basis for the split between Hmong who allied themselves with the French and the Americans after 1954 and those who allied themselves with the Pathet Lao and northern Vietnamese.

From the 1960s through 1975, many Hmong were recruited by the United States Central Intelligence Agency

(CIA) to fight on the side of the Royal Lao government against the communist Pathet Lao and Vietnamese in the "Secret War in Laos" (Warner, 1999; Weldon, 1999). Hmong suffered many casualties and great economic disruption during the war. After the collapse of the U.S. side, more than 100,000 Hmong fled to Thailand (Robinson, 1998; Yang D., 1991). Some emigrated to the United States starting in 1975, but some were not resettled until the end of the 1990s. Tens of thousands of Lao Hmong remain in Thailand with unresolved immigration status, and approximately 10,000 Hmong were repatriated to Laos (Cha & Chagnon, 1993; Cha & Small, 1994; Kirton, 2002).

Economy

The 20th century economy of Hmong in Laos was based on subsistence swidden (slash and burn) cultivation of upland dry rice, vegetables, and maize, with opium as a medicine and as a cash crop in some areas. They also raised pigs and chickens for food and horses for transport. Large multi-generational patrilineal extended families cooperated economically in subsistence production. Other than the activities of the CIA and the United States Agency for International Development (USAID), there was no wage labor and no Hmong village-based commerce. Very few Hmong received any formal education or became fluent or literate in Lao, French, or English (Yang D., 1991). Thus few Hmong refugees had skills or basic knowledge that was appropriate for employment in modern Western industrial, market, and service economies and only a small proportion of middle-aged and older adults have been employed in wage work outside the house. In the 1980s to 1990s, some Hmong were self-employed farmers or gardeners, especially in California, but many have depended on welfare programs for income and for health services. Many adults work in menial jobs, such as janitorial and factory work, often working two jobs or interchanging day and night shifts so parents can care for their children. However, many of the younger generation have become educated and have entered the mainstream U.S. economy (Lo, 2001; Mills & Yang, 1997; Yang & Murphy, 1994).

Social and Political Organization

Membership in exogamous patrilineal clans is the major organizing principle of Hmong society. Traditionally Hmong lived in large multigenerational patrilineally extended family households that were economic and ritual units. Elder or middle-aged men led households and groups of closely-related extended families. Clan members had some obligations to provide each other with support, such as shelter (Cooper, 1984; Tapp, 1986), and had a strong influence in ethnic identity (Leepreecha, 2001). Political organization was limited to the village level. Village leaders were usually the heads of the largest and economically most influential families. French colonial authorities recognized some of these leaders and gave them positions in the local administration, but Hmong villages retained their autonomy. During the war, some Hmong men became high-ranking military officers in the Royal Lao Army (Quincy, 1995).

In the United States, housing and economic conditions preclude forming large household units, but relatives often live in close proximity. Most Hmong still follow rules of clan exogamy, but some do break the traditional taboos. Some clans have expanded the assistance functions of clan membership, to other functions, such as encouraging higher education (Hones & Cha, 1999). Elders still tend to occupy leadership positions, leading important rituals and influencing important decisions including medical care (Cha, 2000; Hones & Cha, 1999; Pfeifer, 2002). Former military leaders often continue to lead, acting as heads of social welfare organizations that are often based on clan or Laotian region affiliation and attempting to serve as spokesmen for an increasingly heterogeneous "Hmong community." As the younger generation become educated and acquire professional qualifications, they are replacing the elders as political leaders. Some Hmong organizations, including Christian church groups, cut across the traditional kinship or regional lines and serve a variety of social welfare and community development purposes. At a national level, Hmong have several organizations for political lobbying, and for staging conferences to discuss various issues (Hmong National Development Links, 2001).

Religion

Traditional Hmong religion is animistic, with beliefs in spirits (*dab*), ancestor worship (*dab niam txiv pog yawg*), multiple souls (*ntsuj plig*), reincarnation, incantations for blessing, curing, or cursing, and selection of spiritually appropriate locations for houses and graves. The most powerful spirits, including the Master of the Universe, are believed to live in the sky; lesser spirits dwell on earth, or underground (Lemoine, 1986; Morechand, 1968). In Laos, Hmong learned about Buddhism from lowland ethnic

Laotians and learned about Christianity from Catholic and Protestant missionaries (Barney, 1957; Capps, 1994). About 25% of the Hmong in the United States are Christian, having converted in refugee camps or having accepted the religious affiliation of their immigration sponsors (Capps, 1994).

THE CONTEXT OF HEALTH: HMONG HEALTH CONDITIONS IN LAOS AND THE UNITED STATES

There is no systematic record of Hmong health conditions in Laos, but data collected from Hmong refugees in the United States suggests that infections and parasitic diseases predominated in Laos, and that wartime trauma was also a leading cause of injury and death. Reproductive histories collected from Hmong women in Merced, California in 1987 suggest that infant mortality of children born in Laos was around 120 per 1,000 live births, declined to about 90 per 1,000 for children born in refugee camps in Thailand, and fell to about 9 per 1,000 among children born in the United States (approximately the same as the U.S. infant mortality rate) (Kunstadter & Kunstadter, 1990).

Historically, Hmong in Laos had access mostly to their traditional system of healing. Modern preventive or curative health services were not available until USAID established a few hospitals in Laos and trained some Hmong military medics and nurses in the 1960s (Weldon, 1999). Modern health services were provided in refugee camps in Thailand. There were conflicts between Hmong refugees and a fundamentalist Christian organization that provided medical care in Ban Vinai, the largest Hmong refugee camp in Thailand (Bouvier, 1994; Wright, 1986). Thus some Hmong refugees associated modern health care with coercion, an idea that has been reinforced by several notorious cases of court-ordered treatments (Culhane-Pera, 1989; Fadiman, 1998). Occasional major conflicts between Hmong and health care providers have received wide attention in the media, especially related to surgery, chemotherapy, and radiation therapy, and sustained medical treatments for tuberculosis (Arax, 1994; New York Times, 1994; Snyder & Kunstadter, 2001).

Changes in environment, risk behavior, and the availability of modern health services in the United States have resulted in changes in Hmong health and health-seeking behavior. While child mortality has declined dramatically, health of adults appears to be deteriorating. Comparisons of self-reported survey data from Hmong in Fresno CA with non-refugee Hmong in Thailand (Kunstadter P., unpublished data 1997) indicate the U.S. Hmong adults age 40 and above have three times as much illness including about 20 times as much hypertension, 20 times as much diabetes, 13 times as much depression or mental distress (all differences significant at $p < 0.001$); and similar rates of gastrointestinal ailments. Fresno Hmong parents reported twice as many respiratory ailments, and approximately the same rate of gastrointestinal ailments for their 0–4-year-old children than did Thailand Hmong parents. Analysis of Hmong death certificates in Fresno County between 1980 and 2001 (Kunstadter & Vang, 2002) revealed significant reductions in deaths of all ages from infectious diseases while deaths of adults 40 years and older from degenerative diseases have increased. There were marked reductions in proportions of deaths of young children with increased proportions of deaths in adults 40 years of age and older over this period. Deaths associated with external injuries and violence (motor vehicle accidents, drowning, suicide, and homicide) increased, especially for those aged 10 to 39. There were no significant gender differences in proportions or causes of death.

Improvement of child survival among infants born to Hmong refugee mothers, who are generally poor, have no education, have high fertility, and short birth intervals, is similar to reports of an "epidemiological paradox" among Hispanic migrants to the United States (Markides & Coreil, 1986). The apparent deterioration of the health of older Hmong in the United States is similar to reports of increases in cardiovascular diseases and diabetes among Japanese and Samoan migrants to the United States and some Native American groups (Baker et al., 1986). In all of these populations the increase in degenerative diseases appears to be associated with a decline in physical activity and changes in diet (increases in consumption of fats, sugars, and total calories). The similarity in rates of gastrointestinal ailments reported for all ages among Fresno and Thailand Hmong suggests that despite the availability of protected water supplies and sanitary waste disposal in the United States, hygienic conditions related to spread of these diseases might not have improved. The decline in infectious diseases as causes of death, especially among children, suggests that public health measures (immunizations and environmental sanitation) and modern curative medicine for acute illnesses have had important effects on the U.S. Hmong population.

Unlike some other groups, Hmong in the United States do not seem to have increased their consumption of tobacco greatly as compared with Hmong in Southeast Asia, but they may have increased frequency and amount of alcohol consumption. There is an apparent increase in the amount of violence directed at self or other Hmong (Hmoob Thaj Yeeb, 1998) as well as an increase in gambling addiction (Zander & Xiong, 1996). These socially dysfunctional behaviors are indicators of mental distress probably related to stresses of social adjustment. Depression and post-traumatic stress disorder have impaired people's adjustment to the U.S. (Westermeyer, 1988; Westermeyer, Lyfoung, & Neider, 1989; Westermeyer, Neider, & Vang, 1984). Mental distress and depression contribute to poor physical health, and compound the difficulty in following modern treatments for chronic diseases—along with the lack of traditional knowledge about chronic diseases requiring long-term therapies.

MEDICAL PRACTITIONERS

The Hmong have several traditional healers who continue to influence the community's health beliefs, values, and practices in Southeast Asia and the United States (Cha, 2000; Culhane-Pera & Xiong, 2003; Kirton, 1985; Lemoine, 1986; Spring, 1989; Thao X., 1986).

Shamans

Shamans (*tus ua neeb*) are generally men, but women shamans are also known. There are two main types of shamans: *muag dub* (covered face) and *muag dawb* (uncovered face). The *muag dub* shamans are chosen by the shaman's helper spirits (*dab neeb*), go into trance while wearing a black or red cloth over their face, and enter the spirit world in order to battle the offending spirits, aided by their helping spirits. In contrast, the *muag dawb* shamans are not chosen by spirits, and do not heal while in a trance, so do not cover their faces with cloths (Chindarsi, 1976; Cooper et al., 1996; Lemoine, 1986).

Tus ua neeb perform a wide range of ceremonies for people in need of spiritual healing, as determined by themselves, other healers, or family members. They perform preventive rituals such as soul calling (*hu plig*) at New Year celebration and tying strings on wrists (*khi tes*); diagnostic rituals and therapeutic ceremonies (Bliatout, 1986; Cha, 2000; Cooper, 1996). To ward off spirits in healing and preventive ceremonies, shamans may attach strings on sick people's wrists (*khi tes*) or ankles (*khi hlua*), or place necklaces, bracelets, or anklets on sick people made of three twisted metals (or three twisted strips of red, white, and black cloth that represent three metals). The shaman's power varies depending on how much they have studied or learned, their innate abilities, and the power of their helping spirits. Additionally, their success is influenced by the match of the shaman's inherent power (*hwj huam*) and the patient. If the shaman's power is stronger than the patient's power, the shaman's healing efforts are more likely to be successful (Cha, 2000).

Herbalists

The *kws tshuaj* are herbalists who diagnose illness and dispense herbal medicines. They are usually women who acquire their knowledge apprenticed to an older female relative. The *kws tshuaj* are guided in diagnosing disease and prescribing medicines by their helping spirits (*dab tshuaj*) whose altar is beside the household altar to a main house spirit (*dab xwm kab*). Sick people or their family members bring spirit paper money and incense for the *dab tshuaj*. Herbalists burn the offerings to inform the spirits that people have come for assistance and to ask for their spiritual guidance when gathering the appropriate healing herbs (Cha, 2000; Cooper et al., 1996; Rice, 2000; Thao X., 1986).

In the United States, herbalists dispense dried herbs (*tshuaj qhuav*) they receive from Southeast Asia or China, fresh herbs (*tshuaj ntsuab*) they cultivate from Asian plants, and medicines they collect in the wild or obtain from botanical gardens. Herbalists also distinguish between "wild" herbs (*tshuaj qus*), which are forest plants and "tame" herbs (*tshuaj nyeg*) which are domestic plants. Additionally, the *kws tshuaj* may know other diagnostic or healing techniques, such as *hu plig, zaws hno, kav, nqus, hno koob* that are described below. Some *kws tshuaj* specialize in specific areas, such as fertility or childbirth (Cha, 2000; Cooper et al., 1996; Rice, 2000; Spring, 1989; Thao X., 1986).

"Magical" Healers

The *kws khawv koob* are "magical" healers. They are usually men who learn their skills and obtain their connections with spirits (*dab khawv koob*) as an apprentice to an experienced healer. There are various types of *khawv koob*, each with specific rituals that use metal, water, incense, and chanting. The *kws khawv koob* diagnose and treat people with ailments such as burns, broken bones, foreign bodies in eyes, vomiting, babies' chronic nocturnal crying, children's febrile illnesses with rash (*ua qoob*), children's fright (*ceeb*), bleeding, and other illnesses caused by evil or wild spirits. Their powers and abilities vary widely and some have extraordinary powers (*muaj leej*) (Cha, 2000; Cooper et al., 1996).

Other Healers

The *tus hu plig* are ordinary men and women (not necessarily healing specialists such as shamans) who have gained the knowledge and skills to return people's wayward souls. They can return souls that have left people's bodies when the person was frightened or when the souls ware abducted by spirits and they can secure souls to bodies during preventive ceremonies, such as during New Year's celebrations or prior to a long journey. They help people with soul loss as identified by themselves, other traditional healers, or by a household diagnostician. They have varying power and abilities. Some perform basic ceremonies where they chant, entice the soul with eggs and chickens, tie strings to wrists that secure the soul (*khi tes*), and interpret the soul's return. Others perform additional, more elaborate ceremonies (Cha, 2000; Chindarsi, 1976; Cooper et al., 1996).

Ordinary people have healing knowledge that is not recognized as specialist knowledge. Many women know about herbal medicines, without being an herbalist. Some people know how to divine the presence of spiritual problems (*tsawv qe* and *nchuav qe*). Some people know how to massage the abdomen and extremities, and then to gently pierce the skin with a needle to release illness, built-up wind, or bad blood (*zaws hno*). Others apply ointment on the skin, then rub with a silver coin (coining or dermabrasion *kav*), or apply suction to the skin with a cup (cupping or moxibustion *nqus*) to bring the illness to the skin's surface in order to release toxins, wind, and stress. Some people massage specific neural pressure points to stimulate blood flow and relieve muscle strain (*xais ceeb*) and others pierce the skin with needles, akin to acupuncture (*hno koob*) (Cha, 2000).

Changes in the United States

Hmong in the United States continue to conceive of health in a traditional holistic way, which integrates the body and its souls and perceives natural, supernatural, social, and personal causes of illness. Science classes, advanced studies in nursing and medicine, health care providers, and conversion to Christianity are influencing ideas about sickness such that people have varied concepts of illness and treatment. People continue to seek the assistance of traditional therapies, as well as other healers. However, there are several important factors that impede traditional healers. There are constraints that limit the use of traditional methods, such as difficulty in obtaining medicines; restrictions on fires in hospitals and communities; injunctions against butchering of animals in cities (Arax, 1995); limits on noise; and conflicts between the divergent nature of traditional and modern medical practices. Also, fewer people are learning traditional healing methods. In addition to these constraints, fewer people believe that traditional healing methods are efficacious, due to the influences of formal education, modern medicine, and Christianity. Hmong families also seek assistance from non-traditional healers, such as Christian ministers and priests; traditional healers of Cambodian, Lao, Vietnamese, Chinese, and Thai ancestry; as well as Hmong and non-Hmong licensed health care professionals (Cha, 2000; Culhane-Pera & Xiong, 2003).

CLASSIFICATION OF ILLNESS, THEORIES OF ILLNESS, AND TREATMENT OF ILLNESS

Traditional Hmong classification of illness can be divided into four classes: natural, supernatural, personal, and social (Culhane-Pera & Xiong, 2003; Helman, 2000; Thao, X., 1986).

Natural Etiologies: Illnesses and Treatments

Hmong traditional ideas about natural etiologies include imbalance of metaphysical forces, germs, genetics, behaviors, constitution, and accidents (Culhane-Pera & Xiong, 2003). Metaphysical ideas similar to the Chinese concept of *yin/yang* indicate that the balance of natural elements is essential to health and that an imbalance causes disease. People get sick from hot or cold, dry or wet, windy or calm weather, and particularly from weather changes with increased wind or air pressure. Foods or water that are thermally hot or cold or that are traditionally classified as hot or cold, can cause illness when people's bodies are thermally or metaphorically out of balance. Small creatures that are observable (lice or parasites) or unobservable (microorganisms) can cause infectious diseases. New "American" germs that cause diseases that Hmong were unaware of in Laos can be particularly worrisome. Hmong traditionally believed that some diseases run in biologically related families and may also affect women who marry into a family. Pregnant women's behaviors can cause congenital birth defects. Susceptibility to disease is related to bodily constitution: people with bad fat and blood, weak immunity, heavy and weak bones rather than light and strong bones, flabby muscles rather than firm muscles, people who are skinny and/or emaciated rather than fat, are more likely to get sick (Xiong & Culhane-Pera, 1995). Hmong also believe that chemicals, including pesticides, fertilizers, horticultural medicines, and Yellow Rain (chemical weapons to which they believe they were exposed to in Laos) can cause a wide range of diseases.

Naturally caused illnesses are diagnosed by physical appearance and symptoms and by history. They are treated by massage, cupping, coining, poking with needles, herbal medicines, and, in America, by physical therapy, chiropractic, or osteopathic manipulation and the whole range of modern medical care, including surgical operations and pharmaceutical preparations (Culhane-Pera & Xiong, 2003).

Supernatural Etiologies: Illnesses and Treatments

Hmong distinguish between five types of supernatural problems (Cha, 2000; Cooper et al., 1996; Culhane-Pera & Xiong P., 2003; Lemoine, 1986).

Souls (*ntsuj plig*). Souls (*ntsuj plig*) can make people sick in many ways, with soul loss (*poob plig*) being the most common. Wandering souls may go off by themselves and not be able to find their ways back; may be caught in dreams; or may leave after an emotional trauma, such as being frightened from a fall, an attack, or seeing someone die. Reincarnated souls may leave if they are enticed by seductive spirits, stolen by evil spirits, or being reincarnated. Other types of problems with souls include a child's soul being unhappy with its parents, clan, or name; or a reincarnated soul wanting a grievance from a previous life remedied or a debt from a previous life settled. The soul who guards the grave may make family members ill if the grave is disturbed or if non-disintegrating elements are in the body, coffin, or grave.

Shaman's helper spirits (*dab neeg*). People become sick when shaman's helper spirits (*dab neeg*) chose them to become a shaman, and they recover when they accept the responsibility of becoming a shaman. Shamans must maintain good relationships with their helping spirits; if they do not thank their helping spirits or offend the spirits by performing rituals not consistent with spirits wishes, they may get sick.

Tame and wild spirits (*raug dab*). Even though they usually protect people, many types of tame or domestic spirits (*dab nyeg*) and wild spirits (*dab qus*) cause illness. Tame spirits include the ancestral spirits and the seven household spirits that reside in houses. Families honor and feed them with spirit money and food during special ceremonies. In turn, the spirits protect families from evil spirits or warn them of impending doom. However, if families neglect their responsibilities, or if spirits need something from families, then spirits communicate their need by making someone sick.

Wild spirits live outside the house. They can cause illness, seizures, or death by disturbing or stealing people's souls. People may inadvertently disturb the land or water where spirits reside, thus inviting wrath. Evil spirits may purposefully seek out people to cause death and destruction. Some spirits sit atop sleeping people and squeeze the breath out of them. Also, souls of dead who were not properly buried and cannot find their ancestors in the afterworld may cause illness, as a plea for assistance. Souls or ghosts that have not achieved reincarnation because of the suicide or violent death of their previous owners, may entice other people to die in similar ways. In the United States, interactions with wild spirits seem to occur less frequently than in Laos. People speculate that spirits are scared away by electricity. Still, spirits in lakes cause drownings, roaming ghosts cause car accidents, and evil spirits cause suffocation (diagnosed as Sudden Unexpected Nocturnal Death Syndrome).

Sorcery (*raug pob zeb, nyuj ciab*). People who are motivated by hate or revenge to harm others can hire black magic specialists to send stones, bones, or other objects to lodge in others. Sorcery was rare in Laos, and seems to be even less common in the United States.

God and sin (among Christian Hmong). Christian concepts of supernatural etiologies vary by denomination, but include sinful thoughts, words, and actions, as well as God's displeasure.

Treatments for the above supernatural etiologies include soul-calling ceremonies (*hu plig*), shaman rituals (*ua neeb*), releasing black magic, Christian prayer, as well as modern medicines or operations in addition to specific supernatural treatments. Generally, Christians pray or ministers lay on hands instead of performing traditional ceremonies, although Catholics may be more likely than evangelical Christians to integrate traditional spiritual healing with Christians prayer healing.

Social Etiologies: Illnesses and Treatments

Stress and anxiety of day-to-day life as farmers in the highlands of Laos or as refugees in American society can cause illness. Human conflicts, such as between generations and genders, can cause illness. Mocking a sick person may result in having that same illness, and teasing a handicapped person can result in children being born with that same disability. When people curse each other, the gods can hear the dispute and cause the guilty person to be harmed. Women who do not revere their husband or parents-in-law can have a difficult time at childbirth.

Treatment requires resolving intra-personal stresses, interpersonal conflicts, and ritually removing the strength of the words said in curses. Medication may also ease pain and disability, but the illness may not be completely cured without repairing the rift in social relationships.

Personal Etiologies: Illnesses and Treatments

Personal behaviors can affect one's health. Using tobacco, opium, or alcohol can cause weakness and a range of illness. Accidental injury can cause impairment immediately after the accident, as well as years later. Not following health promotion proscriptions can make people sick (but too much medicine, even if it has been medically prescribed, can also cause illness, and some modern medicines are known to have side-effects). Frying spicy odiferous food when children have fevers may cause children to become sicker. Not following specific restrictions in eating, sex, and physical activity during their postpartum month may cause women to suffer from headaches, arthritis, infertility, or prolapsed uterus. Treatments are often aimed at the physical ailment, with massage, poultices, Hmong medicines, or pharmaceutical medications bringing relief. Changing underlying behaviors may also be important, such as stopping tobacco or opium consumption.

Medical Decision Making

Traditionally, the sick person is passive, staying in bed and letting family members provide care—feeding, clothing, bathing, and making treatment decisions (Culhane-Pera & Xiong P., 2002). Family members use many sources of information to identify the illness, its cause, the best healer, and best therapy. They may consider the sick person's symptoms, bodily signs, and prior events (such as trauma or conflicts); they may generate ideas about potential causes; they may perform diagnostic divination procedures; and they may consult clan and community members before deciding on which healers to consult or which therapies to employ. The vast majority of illnesses are initially considered to be caused by natural etiologies and are treated accordingly, but if people are chronically ill, seriously ill, or have some significant historical event, then supernatural etiologies are considered, and supernatural prevention or treatments are pursued. Any illness event can have multiple explanations for causation.

While this traditional sick role and family based decision-making persists, in the U.S. there are changes. Some individuals will tell the family what they want to do; some will make medical decisions without the family approval; and some will even take actions without informing the family. Some individuals shun while others embrace modern modalities, including operations, medications, chemotherapy, and radiation therapy. People may still consider the above cultural information, but other issues are also relevant: religious orientation, insurance availability, language services, previous experiences with invasive therapies, and relationships with and reputation of the healer/ provider (Cha, 2000; Culhane-Pera et al., 2003).

When considering treatments, family members evaluate both risks and benefits. Often times, people who follow the traditional animist perspective are concerned about the spiritual risks of invasive procedures as well as the physical risks. Souls can be frightened and leave the body during an operation and during anesthesia. People can be reincarnated with physical disabilities related to loss or destruction of tissues from operations performed on their spiritual ancestor. Also, metal staples or metal prosthesis placed in the body during operations can cause the soul that stays with the grave to make living family members sick. These dangers may influence people to refuse life-saving or reparative surgical operations (Mouacheupao, 1999; Westermeyer & Thao, 1986).

SEXUALITY AND REPRODUCTION

Sexual Identity

In traditional Hmong society, gender was interwoven with the division of labor and considerations of social worth. Both men and women worked in the fields and at home. Women were responsible for child rearing, cooking, weaving, embroidering, and animal husbandry. Men were responsible for clearing land, creating tools, building houses, making decisions, maintaining and overseeing clan rituals, and supervising family matters. Men also took an important part in child rearing. Generally, men had higher social status and more power than women although women gained power as they grew older and had larger households to manage (Cooper, 1984; Donnelly, 1994; Tapp, 1986; Thao C. T., 1986). Also, there was cultural variation, such that some wives had more power than their husbands, related to their recognized intelligence, ability to manage money, treat illnesses, or make decisions. Generally, at marriage, women left their natal family and lived with their husband's family. Marriage had both brideprice and bridewealth components. The groom's family gave money (*nqi mis nqi hno*) to the bride's family, compensating for the family's economic loss and emotional hardship, and illustrating their promise to love and take good care of the bride. The bride's family sent a dowry with the bride, consisting of clothes and silver, for her contribution to her new household. Women were generally considered to be worth less than men (*tsis muaj nqis*), as they were raised always knowing that they would one day leave the family. Changing gender relations and women's increasing power in the household is causing social conflict (Her & Heu, 2003; Lyfoung, 2003).

Women were at a disadvantage with regard to courtship and marriage. Traditionally, girls were expected not to initiate courtship, and they were enjoined not to openly express their preference for a husband. This is changing in the United States, as some women and teenage girls initiate courtship and express their desires about potential husbands. Newly married daughters-in-law (*tus nyab*) had little power in their husbands' families, performed many chores, and had to adjust to customs and manners of their new families (Cooper, 1984; Donnelly, 1994). Men could discipline their wives physically, in order to teach them and help them conform to familial expectations. If women were mistreated, they could appeal to their male family members for assistance, and if abused, women could ask their male family members to bring civil charges (*ua plaub*) against their husband and his family (Donnelly, 1994; Ovesen, 1995). In the United States, domestic violence continues to occur, and is possibly even escalating, as recently publicized cases of murders and murder–suicides illustrate (Haga & Her, 2001).

Hmong consider marriage a way of life. Traditionally, every individual was expected to marry and raise a family, as Heaven created women and men to love, care for, and help each other (Thao C. T., 1986). Traditional forms of betrothal included marriage by parental arrangement and bride capture, but agreement between groom- and bride-to-be and elopement were more common ways of marrying. Levirate was practiced in a few cases if an older brother died leaving a widow with young children. Most marriages were monogamous, but polygyny occurred (Kunstadter, 2003). In the United States, people are choosing their mates more often, and arranged marriages are becoming rare.

Traditional sanctions against divorce were strong for women (Cooper, 1984; Donnelly, 1994; Lyfoung, 2003). Young couples and couples with problems were repeatedly counseled against divorce. Divorced women brought shame upon their families and were characterized as having moral defects. Divorced women were almost always separated from their children, who belonged to their former husband's clan and stayed with their father. Usually there were strong economic and religious constraints on divorce for women, because they had limited means of independent livelihood. They had already been separated at the time of marriage from their parents' ancestral and household spirits, and the divorce separated them from their husbands' ancestral and household spirits. Those who did not re-marry often returned to live with their natal families, but they could not sleep in the same room with their father's household and ancestor spirits, and could not be included in the clan rituals. If they were seriously ill, they were not allowed to die inside their father's house, and could not receive funeral rituals that connected them with ancestral spirits. In the United States, divorced women may become Christians, or their families may ask Christians to conduct the funeral. While the divorce rate is low in traditional society (Thao C. T., 1986), current literature indicates an increase in the United States (Strohl, 2000).

Reproduction

In traditional society, the purpose of sex was for reproduction, and reproduction is for both economic reasons and for continuation of the patrilineal links with ancestral spirits. Hmong traditionally prefer large numbers of children for economic and social reasons (Kunstadter, 2002). Additional hands in agricultural fields result in enhanced family income and needed assistance in times of trouble and old age. In Thailand, the total fertility rate was 8 in 1987, which may have been similar to the number in Laos. Fertility rates remained high in California in the 1990s (Kunstadter et al., 1993).

Contraception

Traditionally, most couples did not use deliberate methods of contraception but accepted children as fate (*hmoov*) determined. It was believed that if a woman did not deliver the pre-determined number of babies, she would be reborn as a woman to deliver these infants. Birth spacing was increased by near universal and prolonged breast-feeding. If contraception was used, women secretly obtained herbal medicines that could cause sterility from herbalists. Herbalists were familiar with abortifacients, although the extent of their use in Southeast Aia and the United States is not known. It seems that abortifacients were used sparingly, given the danger of hemorrhage and death; women and their husbands had to agree about taking the medicine.

In the United States, some people are choosing to limit the number of children although many people are concerned about side-effects of Western methods of contraception, from changing normal hormonal cycles to increased weight to cancer. Most people prefer *caiv*, which has multiple meanings: natural family planning, or withdrawal, or abstinence. Decision-making about contraception varies, influenced by people's acculturation and education levels; most men decide alone, many men and women decide together, and some single or married women make the decision alone (Kunstadter, 2003; Spring, 2001; Spring & Luchongvu, 2003). Over the past 25 years in the United States, surgical abortions have increased, as people have found the procedure more acceptable over time. Some people voice concerns about consequences of denying the infant soul's desire for life, and of performing multiple abortions on women's health.

Sexual Pleasure

In traditional society only heterosexuality was approved, as reproduction was the purpose of sexual intimacy (Thao C. T., 1986). Homosexual relationships were generally unknown, although in the United States some homosexual relations are being recognized. However they are apparently still very strongly disapproved of, as witnessed by the recent suicide of a lesbian couple in Fresno California. Generally, intercourse in the "missionary position" took place with clothes on, in a bed that contained sleeping children. No kissing, foreplay, oral sex, or verbal sexual expression occurred during lovemaking. Men initiated sexual advances and took the lead in sexual intercourse, and women were not to display any sexual eagerness. Men and women did not discuss sexual pleasure with each other, but men would discuss with men, and women with women (Symonds, 2003).

Sexual modesty was the traditional cultural norm especially for women. Sanctions against public displays of affection—such as extended conversation and any physical contact, even shaking hands—were implicit. While premarital sex was strongly discouraged especially for young women, it happened often, with ensuing pressures for marriage. Extra-marital affairs were permissible for men, but prohibited for women (Donnelly, 1994; Trueba, Jacobs, & Kirton, 1990). In the United States, social sanctions are more relaxed; couples can hold hands and otherwise openly express their feelings in public. Premarital sex and ensuing marriage is still common. Men's accusations of their spouses' sexual affairs are a common reason cited for marital discord, divorce, and domestic violence (Haga & Her, 2001; Strohl, 2000).

HEALTH THROUGH THE LIFE CYCLE

Pregnancy and Birth

Traditionally, pregnant women followed their usual work activities. Modest, they did not inform anyone about the pregnancy until people noticed. They were instructed to avoid some activities (*caiv*) for the well-being of their pregnancy (reaching above their heads could cause a miscarriage), health of their baby (cutting cloth in their bedrooms could cause a cleft palate), and easy deliveries (being disrespectful to in-laws could cause a difficult labor). There were no traditional midwives who followed women through their pregnancies. If abnormal events occurred—such as vaginal bleeding or premature labor—then herbalists, message therapists, or shamans were consulted (Spring et al., 1995; Rice, 1997, 2000).

In Southeast Asia, women gave birth at home, attended by their mothers-in-law and husbands. Women pushed while squatting, with their husbands supporting them. If problems occurred, families sought assistance from elderly experienced women or men for medication to ease the delivery, massage to turn the baby, chanting to call the baby's soul, or appealing for spiritual assistance. After delivery, the placenta (or baby's shirt *lub tsho menyuam*) was buried under the earthen floor (by the ancestral post for a boy and under the bed for a girl). If the couple preferred a specific gender for the next baby, a ritual could be performed on the placenta before burial (Symonds, 2003).

Traditional proscriptions (*caiv*) during the postpartum month assured the mother's immediate and long-term health as well as her breastmilk supply. Women ate only thermally and metaphysically "hot" foods (rice, eggs, and chicken cooked with herbs). For the first three days after delivery, women sat by fires on beds of grass wearing hemp skirts. For the first month, women did not perform household duties; were not sexually active; did not visit other people's houses; and covered their heads and bodies so the wind would not enter their joints. These healthy proscriptions ensured the flow of lochia, restored their metaphysical "balance," and protected them from spirits (Symonds, 2003). Since women had lost "hot" blood, they must only eat hot foods; cold foods could make their blood clot inside and make them infertile; and cold wind could enter their joints and cause old-age pains, such as arthritis. Also, if they ventured outside, evil spirits could visit them and cause hemorrhage and death, or if they visited other people's houses, the spirits accompanying them could cause other people to hemorrhage.

In the United States, women usually obtain prenatal care from physicians or midwives in the second trimester. Women's motivation may be more for ease of entering the hospital for deliveries in order to obtain needed birth certificates, or for assistance after birth such as from the WIC program than for desire for a healthy pregnancy. Some women refuse to have blood drawn, have pelvic exams, or take prenatal vitamins as a result of modesty and traditional prohibitions on anyone except a husband touching a woman's genitals, for fear of harming themselves, their fetus, or having large babies, and difficult deliveries. Many women want fetal ultrasounds for reassurance about the baby's well-being, and for knowledge of the baby's sex (Bruce & Xiong, 2003; Spring et al., 1995).

Most women deliver in hospitals in the United States, but some women deliver at home either because the labor and delivery occurred too quickly to get to the hospital or because they stayed home to avoid conflicts with hospital personnel. While most women lie down, some may squat if labor is difficult. For decades, many women have resisted obstetrical interventions, such as rupture of membranes, internal monitors, medicine for induction, medicines for pain relief, or Cesarean sections. Recently use of pain medications is on the increase, as is acceptance of interventions. Placentas have been taken home for burial, but this practice is virtually abandoned, due to difficulty of burying placentas under floors in modern dwellings (Symonds, 2003).

Most women follow some of the traditional postpartum proscriptions. Women eat "hot" foods, wear hats, cover their bodies, and refrain from sexual intercourse in order to ensure that the blood flows out of the uterus so they will not become infertile or crippled with arthritic pains. Families object to nurses and doctors vigorously massaging the postpartum uterus, as women fear resulting problems will occur, such as infertility, pelvic pains, abdominal pains, and cancer (Culhane-Pera, 2003; Spring, 2001; Spring et al., 1995). Other proscriptions are not followed strictly. As breast-feeding is no longer the norm, proscriptions to insure breastmilk are no longer important. And nearly universal bottle-feeding has ended the biological influence of breast-feeding on birth spacing and fertility rates.

Infancy

Traditionally, at three days of life family members welcomed a newborn infant by calling the soul (*hu plig*) and bestowing a name (*tis npe*) (Symonds, 2003). Today, animists continue these practices while Christian families have altered or abandoned this ceremony, depending upon their denomination. Traditionally, breast-feeding mothers cared for their infants, with their families' assistance. In the United States, the vast majority of infants bottle-feed so family members can care for babies while mothers return to school or work (Culhane-Pera, Naftali, Jacobsen, & Xiong, 2002; Tuttle & Dewey, 1994, 1996).

Multiple births are a blessing. Birth defects are caused by fate (related to events in infants' or parents' previous lives) or by mothers' actions while pregnant, or by a curse. While the ideal is to have equal numbers of boys and girls, people prefer boys, since boys stay with the family and girls marry outside of the clan. Mortality risk is equal for girls and boys, indicating that boys are not given preference over girls such that girls are at risk of increased mortality (Kunstadter et al., 1993).

There were many traditional health promotion and disease prevention activities. Infants fed on-demand; they wore silver necklaces, ornately decorated hats, and bright bracelets with bells to please their souls; and wore amulets to ward off frightening spirits. Parents refrained from unpleasant words that could upset the infant souls and from praising children, a practice believed to attract evil spirits. Parents did not wash fontanels, the location of a soul; and they cut infants' hair to protect from fright of thunder (Culhane-Pera, unpublished ethnographic research). Many of these activities persist today, except some Christians who refrain from engaging in all animist practices. Some adults are concerned about childhood vaccinations causing pain, fever, illness, and disability, and refuse vaccines until children are older and less vulnerable to side-effects (Xiong & Culhane-Pera, 1995). Several Hmong children died in measles epidemics in California and Minnesota in 1990 from inadequate vaccinations (Henry R., 1999).

Traditional Hmong practices for sick children included *khawv koob* for fevers with rashes; *ua ceeb* for fright or startle; *nqus qe nyiaj* to reduce fevers; herbal medicines (solutions and poultices) for various illnesses; and pharmaceutical preparations from the United States and Asia. In the United States, parents also take children to see physicians. While desiring medications, and accepting intravenous fluids, families have concerns about adverse effects of venipunctures, lumbar punctures, and operations on their children's health. Conflicts between families and physicians have sometimes resulted in court-ordered treatments for diagnostic and therapeutic procedures (Brunnquell & Kuracheck, 2003; Culhane-Pera, 1989; Culhane-Pera & Thao, 2003; Fadiman, 1998; New York Times, 1994; Plotnikoff, 2003; Snyder & Kunstadter, 2001).

Childhood

A general philosophical orientation toward childrearing is characterized by *hlub*, a loving permissive attitude (Culhane-Pera, Naftali, Jacobsen, & Xiong, 2002; Xiong Z. B., 2000). As children become older and gain more responsibilities, parents are more likely to discipline them, which can include physical punishment. The perception is that sexual abuse was rare in traditional society and that child abuse occurred more often for orphans and step-children. Social workers in the United States note that probably more child protection orders have been brought against parents for neglect (allowing children to have too much independence) than for abuse.

Some diseases are especially significant in Hmong children: (1) Hepatitis B: all infants need Hepatitis B vaccines, and infants of carrier mothers need to be tested for immunity (Franks et al., 1989; Gjerdingen & Lor, 1997; Poss, 1989). (2) Measles: a 1990 measles epidemic in Minnesota hit Hmong children, as they had a low immunization rate (Henry R. R., 1999). (3) Milk anemia: a high rate of iron-deficiency anemia in toddlers is related to a diet heavy in cow's milk and low in solid foods (Culhane-Pera, Naftali, Jacobsen, & Xiong, 2002). (4) Thalassemia: a genetic disorder of hemoglobin, thalassemia can be fatal (Choy et al., 2000; Yang P., 2000). (5) Baby bottle tooth decay: prolonged and frequent sucking of milk in a bottle leads to tooth decay as well as ear infections (Tuttle & Dewey, 1994, 1996). (6) Lead toxicity: Exposure to lead results from ingesting lead paint chips in older homes, and when given a Chinese red powder as medicine (MMWR, 1993). (7) Obesity: obesity in Hmong children is increasing, probably due to over-bottlefeeding, over-eating, and under-exercising (Gjerdingen et al., 1996).

Adolescence

Traditionally, teenagers had much work responsibility and little freedom beyond the household and extended family (Xiong Z. B., 2000). While there were no traditional "rites of passage," teenagers were encouraged to marry when they had accomplished agricultural and domestic life skills. The expectation to assume adult responsibilities was high, and discipline tended to be strict. While premarital intercourse was tolerated, sexually active teenagers were expected to marry. Conflicts in the United States are resulting from tensions between parental demands for conforming to traditional values and youth's desires for increased freedom. Parents feel their traditional values are being disrespected, and teenagers feel parents are not realistic in U.S. society (Wheeler, 1998). Parental concerns are compounded by the high rates of pre-marital intercourse, sexually transmitted diseases, gangs, and use of tobacco and drugs. External injuries and violence are leading causes of death for adolescents and young adults in Fresno California, and the rate is increasing (Kunstadter & Vang, 2002).

Adulthood

In Laos and the United States, middle-aged adults have the most responsibility for children, aging parents, and society at large. Men are active in leading society, and increasingly gain respect as they become elders. Women also have increased status, with increased influence over family matters, and may have some community influence especially if their husbands are clan leaders (Donnelly, 1994).

In Southeast Asia, adults suffered from infectious diseases, accidents, and other ailments. In the United States, the "modern" diseases of diabetes, hypertension, strokes, heart attacks, and cancers are on the rise. Many suffer from these ailments, and are challenged to respond to these chronic and life-threatening conditions.

The Aged

Traditionally, elders—people older than 50 years of age—were respected for their opinions and life experiences, and enjoyed a reduced work load. Usually youngest sons and their wives were responsible for the aged, including providing care as they became infirm. This continues in the United States, although elders' opinions and life experiences may be less relevant to their children and grandchildren than in Southeast Asia, and more elders live alone or in nursing homes. Their health issues include the modern diseases, as well as degenerative diseases (arthritis, osteoporosis, blindness, and deafness).

Dying and Death

Death is understood as a transition from the world of the living to the world of the spirits and as a preparation for the next reincarnation (Symonds, 2003). Living family members continue to revere, remember, and appease ancestral spirits with offerings. Nonetheless, family members will engage in many actions to heal sick family members, including traditional and modern healing options. A "good death" occurs without pain and at home, surrounded by the loving family who attend to physical needs and visited by friends and relatives who give them encouraging words about their long life. In addition, a good death occurs in the company of ancestral spirits while wearing traditional clothes, and with the chance to impart their last words to the family (Vawter & Babbit, 1997).

Traditionally, at death family members wailed their grief and washed the body before dressing it in multiple layers of ancestral clothes. Funerals of un-embalmed bodies lasted up to 9 days, depending on the deceased's social importance, with many rituals. Burial occurred in a place chosen by geomancy rules for good luck. Rituals continued at the gravesite every day for 3 days, with a final ritual to release the soul at 13 days. In the United States, animists still wail and conduct traditional funeral ceremonies, whose details have changed to adapt to current social, economic, and political realities. Christians conduct prayer services with songs and sermons, and do not wail (Bliatout, 1993).

REFERENCES

Arax, M. (1994). Cancer case ignites culture clash: Hmong parents refuse to agree to court-ordered chemotherapy for teen-age daughter. *Los Angeles Times* Nov. 21, v113, pA3, Col 1.

Arax, M. (1995). Hmong's sacrifice of puppy reopens cultural wounds: Immigrant shaman's act stirs outrage in Fresno, but he believes it was only way to cure his ill wife. *Los Angeles Times* Dec. 16, v114, pA1.

Baker, P. T., Hanna, J. M., & Baker, T. S. (1986). *The changing Samoans: Behavior and health in transition*. New York, Oxford: Oxford University Press.

Barney, G. L. (1957). Christianity and innovation in Meo culture: A case study in missionization. Unpublished MA thesis, University of MN.

Bliatout, B. T. (1986). Guidelines for mental health professionals to help Hmong clients seek traditional healing treatment. In Glenn L. Hendricks, et al. (Eds.), *The Hmong in transition*. New York: Center for Migration Studies.

Bliatout, B. T. (1993). Hmong death customs: Traditional and acculturated. In D. P. Irish, K. F., Lundquist, & V. Jenkins-Nelson (Eds.), *Ethnic variations in dying, death and grief: Diversity in universality*. Washington, DC: Taylor & Francis.

Bouvier, B. (1994). Hmong need more respect. (Letter). *Bangkok Post*, August 19.

Bruce, H., & Xiong, C. (2003). Pregnancy complications. In K. A. Culhane-Pera, D. E. Vawter, P. Xiong, B. Babbitt, & M. Solberg (Eds.), *Healing by heart*. Nashville, TN: Vanderbilt University Press.

Brunnquell, D., & Kuracheck, S. (2003). Children with high fevers. In K. A. Culhane-Pera, D. E. Vawter, P. Xiong, B. Babbitt, & M. Solberg (Eds.), *Healing by heart*. Nashville, TN: Vanderbilt University Press.

Capps, L. L. (1994). Change and continuity in the medical culture of the Hmong in Kansas City. *Medical Anthropology Quarterly* (new series), *8*(2), 161–177.

References

Cha, Dia. (2000). *Hmong American concepts of health, healing, and illness and their experience with conventional medicine* (Doctoral Ph.D. dissertation). University of Colorado at Boulder: Boulder, CO.

Cha, Dia, & Chagnon, J. (1993). *Farmer, war-wife, refugee, repatriate: A needs assessment of women repatriating to Laos.* Washington, DC: Asia Resource Center.

Cha, Dia, & Small, C. A. (1994). Policy lessons from Lao and Hmong women in Thai refugee camps. *World Development, 22,* 1045–1059.

Chindarsi, N. (1976). *The religion of the Hmong Njua.* Bangkok, Thailand: The Siam Society.

Choy, J., Foote, D., Bojanowski, J., Yamashita R., & Vichinsky, E. (2000). Outreach strategies for Southeast Asian communities: Experience, practice, and suggestions for approaching Southeast Asian immigrant and refugee communities to provide thalassemia education and trait testing. *Journal of Pediatric Hematology Oncology, 22*(6), 588–592.

Cooper, R. (1984). *Resource scarcity and the Hmong response: Patterns of settlement and economy in transition.* National University of Singapore: Singapore University Press.

Cooper, R., Tapp, N., Lee, G. Y., & Schworer-Kohl, G. (1996). *The Hmong.* Bangkok, Thailand: Artasia Press.

Culhane-Pera, K. A. (1989). *Analysis of cultural beliefs and power dynamics in disagreements about health care of Hmong children.* Master's thesis. Mpls, MN: University of MN.

Culhane-Pera, K. A. (2003). Hospice patient with gallbladder cancer. In K. A. Culhane-Pera, D. E. Vawter, P. Xiong, B. Babbitt, & M. Solberg (Eds.), *Healing by heart.* Nashville, TN: Vanderbilt University Press.

Culhane-Pera, K. A., Nafali, E. D., Jacobson, C., & Xiong, Z. B. (2002). Cultural feeding practices and child-raising philosophy contribute to iron deficiency anemia in refugee Hmong children. *Ethnicity and Disease, 12*(2).

Culhane-Pera, K. A., & Thao, V. (2003). Children with high fevers. In K. A. Culhane-Pera, D. E. Vawter, P. Xiong, B. Babbitt, & M. Solberg (Eds.), *Healing by heart,* Nashville, TN: Vanderbilt University Press.

Culhane-Pera, K. A., Vawter, D. E., Xiong, P., Babbitt B., & Solberg, M. (Eds.), *Healing by heart.* Nashville, TN: Vanderbilt University Press.

Culhane-Pera, K. A., & Xiong, Phua. (2003). Hmong culture: Tradition and change. In K. A. Culhane-Pera, D. E. Vawter, P. Xiong, B. Babbitt, & M. Solberg (Eds.), *Healing by heart.* Nashville, TN: Vanderbilt University Press.

Donnelly, N. D. (1994). *The changing lives of refugee Hmong women.* Seattle: University of Washington Press.

Fadiman, A. (1998). *The spirit catches you and you fall down: A Hmong child, her American doctors and the collision of two cultures.* New York: Farrar, Straus and Giroux.

Franks, A. L., Berg, C. J., Kane, M. A., Browne, B. B., Sikes, R. K., Elsea, W. R., & Burton, A. H. (1989). Hepatitis B virus infection among children born in the United States to Southeast Asian refugees. *New England Journal of Medicine, 321*(19), 1301–1305.

Gjerdingen, D. K., & Lor, V. (1997). Hepatitis B status of Hmong patients. *Journal of the American Board of Family Practice, 10*(5), 322–328.

Gjerdingen, D. K., Ireland, M., & Chaloner, K. M. (1996). Growth of Hmong children. *Archives of Pediatrics & Adolescent Medicine, 150*(12), 1295–1298.

Haga, Chuck, & Her, Lucy, Y. (2001). At 18, she steps up to preserve a family of 13. *The Star Tribune,* December 9, 2001.

Helman, C. G. (2000). *Culture, health and illness: An introduction for health professionals* (4th ed.). Woburn, MA: Butterworth-Heinemann.

Henry, R. R. (1999). Measles, Hmong and metaphor: Culture change and illness management under conditions of immigration. *Medical Anthropology Quarterly* (new series), *13*(1), 32–50.

Her, Mymee, & Heu, C. P. (2002). Domestic violence. In K. A. Culhane-Pera, D. E. Vawter, P. Xiong, B. Babbitt, & M. Solberg (Eds.), *Healing by heart.* In press.

Hmong National Development, Inc. (2001, Fall). HND Links: A Quarterly Newsletter. Washington, DC.

Hmoob Thaj Yeeb. (1998). *Taking a Public Stand: Completing the journey from war to peace through the ending of violence. Initiative for violence-free families and communities.* St. Paul, MN: Hmoob Thaj Yeeb.

Hones, D. F., & Cha, C. S. (1999). *Educating New Americans: Immigrant lives and learning.* Mahwah, NJ: Lawrence Erlbaum Associates.

Kirton, E. S. (1985). *The locked medicine cabinet: Hmong health care in America.* Doctoral dissertation, University of Santa Barbara, CA.

Kunstadter, P. (2000). Hmong marriage patterns in relation to social change. In G. Y. Lee, J. Michaud, C. Culas, & N. Tapp (Eds.), *The Hmong in Southeast Asia: Current issues.* Chiang Mai: Silkworm.

Kunstadter, P. (2003). Controlling fertility. In K. A. Culhane-Pera, D. E. Vawter, P. Xiong, B. Babbitt, & M. Solberg (Eds.), *Healing by heart.* Paper presented at Am. Public Health Assoc. meeting, Philadelphia, PA.

Kunstadter, P., & Kunstadter, S. L. (1990). Health transitions in Thailand. In J. C. Caldwell, S. Findley, P. Caldwell, G. Santow, W. Cosford, J. Braid, & D. Broers-Freeman (Eds.). *What we know about health transition* (Vol. 1). Canberra: Health Transition Centre, The Australian National University.

Kunstadter, P., Kunstadter, S. L., Podhisita, C., & Leepreecha, P. (1993). Demographic variables in fetal and child mortality: Hmong in Thailand. *Social Science and Medicine, 36*(9), 1109–1120.

Kunstadter, P., & Vang, VaKue. (2002). *Mortality transition among Hmong refugees in Fresno County, California, 1980–2001.* Unpublished manuscript.

Leepreecha, P. (2001). *Kinship and identity among Hmong in Thailand.* Unpublished Doctoral dissertation. Seattle, WA: University of Washington.

Lemoine, Jacques. (1986). Shamanism in the context of Hmong resettlement. In Glenn L. Hendricks et al. (Eds.), *The Hmong In transition.* New York: Center for Migration Studies.

Lo, F. T. (2001). The promised land: Socioeconomic reality of the Hmong people in urban America 1976–2000. Lima, Ohio: Wyndham Hall Press

Long, Lynellyn D. (1993). *Ban Vinai: The refugee camp.* New York: Columbia University Press.

Lyfoung, P. (2003). Domestic violence. In K. A. Culhane-Pera, D. E. Vawter, P. Xiong, B. Babbitt, & M. Solberg (Eds.), *Healing by heart.* Nashville, TN: Vanderbilt University Press.

Markides, K. S., & Coreil, J. (1986). The health of Hispanics in the Southwestern United States: An epidemiological paradox. *Public Health Reports, 101,* 253–265.

Mills, P. K., & Yang, R. (1997). Cancer incidence in the Hmong of Central California, United States, 1987–94. *Cancer Causes Control, 8*(5), 705–12.

Morbidity and Mortality Weekly Report. (1993). Lead poisoning associated with use of traditional ethnic remedies—California 1991–92. *42*(27), 521.

Morechand, G. (1968). Le Chamanisme des Hmong. *Bulletin de l'Ecole Francaise d'Extreme Orient LXIV*.

Mouacheuapo, S. (1999). Attitudes of Hmong patients to surgery. Paper presented at Minnesota Academy of Family Physicians Annual Spring Research Forum. Minneapolis, MN.

New York Times. (1994). Girl flees after clash of cultures on illness: Should Hmong refugees have to accept Western medicine? November 12, 1994, p. 6.

Ovesen, J. (1995). *A minority enters the nation state: A case study of a Hmong community in Vietiane Province, Laos*. Sweden: Uppsala University.

Pfeifer, M. E. (2002). U.S. census 2000: Trends in Hmong population distribution across the regions of the United States. St. Paul, MN: Hmong Cultural Center.

Plotnikoff, G. (2002). Child with Down syndrome and a heart defect. In K. A. Culhane-Pera, D. E. Vawter, P. Xiong, B. Babbitt, & M. Solberg (Eds.), *Healing by heart*. In press.

Poss, J. E. (1989). Hepatitis B virus infection in Southeast Asian children. *Journal of Pediatric Health Care, 3*(6), 311–315.

Quincy, K. (1995). *Hmong: History of a people*. Cheney, WA: Washington University Press.

Rice, P. L. (1997). Giving birth in a new home: Childbirth traditions and the experience of motherhood among Hmong women from Laos. *Asian Studies Review, 20*(3), 133–148.

Rice, Pranee L. (2000). *The Hmong way: Hmong women and reproduction*. Westport, CT: Bergin & Garvey.

Robinson, W. C. (1998). *Terms of refuge: The Indochinese exodus and the international response*. New York, NY: United Nations High Commissioner for Refugees.

Schein, L. (2000). *Minority rules: The Miao and the feminine in China's culture politics*. Durham & London: Duke University Press.

Snyder, D. M., & Kunstadter, P. (2001). Providing health care in a multicultural community. In D. Wedding (Ed.), *Behavior and Medicine* (3rd ed., pp. 49–59). Seattle, Toronto, Göttingen, Bern: Hogrefe and Huber.

Spring, M. A. (1989). Ethnopharmacologic analysis of medicinal plants used by laotian Hmong refugees in Minnesota. *Journal of Ethnopharmacology, 26*, 65–91.

Spring, M. A. (2001). *Reproductive health and fertility of Hmong immigrants in Minnesota*. Doctoral dissertation. Minneapolis, MN: University of Minnesota.

Spring, M. A., Ross, P. J., Etkin, N. L., & Deinard, A. S. (1995). Sociocultural factors in the use of prenatal care by Hmong women in Minneapolis. *American Journal of Public Health, 85*(7), 1015–1017.

Spring, M. A., & Lochungvu, M. (2003). Family planning. In K. A. Culhane-Pera, D. E. Vawter, P. Xiong, B. Babbitt, & M. Solberg (Eds.), *Healing by heart*. Nashville, TN: Vanderbilt University Press.

Strohl, L. (2000, May). Asian ascending. *The Horizon Magazine*.

Symonds, P. V. (2003). *Calling in the soul: Gender and cycle of life in a Hmong village*. Seattle: University of Washington Press.

Tapp, N. (1986). *The Hmong of Thailand: Opium people of the Golden Triangle*. London, UK: Anti-Slavery Society.

Thao, Christopher T. (1986). Hmong Customs on marriage, divorce and the rights of married women. In Johns, Brenda, & David Strecker. (Eds.), *The Hmong World*. New Haven, CT: Yale Southeast Asia Studies.

Thao, X. (1986). Hmong perception of illness and traditional ways of healing. In Glenn L. Hendricks et al. (Eds.), *The Hmong in transition*. New York: Center for Migration Studies.

Trueba, H. T., Jacobs L., & Kirton, E. S. (1990). *Cultural conflict and adaptation: The case of Hmong children in American society*. New York: The Falmer Press.

Tuttle, C. R., & Dewey, K. G. (1996). Potential cost savings for Medi-Cal, AFDC, food stamps, and WIS programs associated with increasing breast-feeding among low-income Hmong women in California. *Journal of American Dietetic Association, 96*(9), 885–890.

Tuttle, C. R., & Dewey, K. G. (1994). Determinants of infant feeding choices among Southeast Asian immigrants in Northern California. *Journal of American Dietetic Association, 94*(3), 282–286.

Vawter, D. E., & Babbitt, B. (1997). Hospice care for terminally ill Hmong patients: A good cultural fit? *Minnesota Medicine, 80*(11), 42–44.

Warner, R. (1999). *Shooting at the moon: The story of American clandestine war in Laos*. South Royalton Vermont: Steerforth Press.

Weldon, C. (1999). *Tragedy in paradise: A country doctor at war in Laos*. Bangkok: Asia Books.

Westermeyer, J. (1988). A matched pairs study of depression among Hmong refugees with particular reference to predisposing factors and treatment outcome. *Social Psychiatry and Psychiatric Epidemiology, 23*, 64–71.

Westermeyer, J., Neider, J., & Vang, T. F. (1984). Acculturation and mental health: A study of Hmong refugees at 1.5 and 3.5 years postmigration. *Social Science and Medicine, 18*(1), 87–93.

Westermeyer, J., & Thao, X. (1986). Cultural beliefs and surgical procedures. *Journal of American Medical Association, 255*(23), 3301–3302.

Westermeyer, J., Lyfoung, T., & Neider, J. (1989). An epidemic of opium dependence among Asian refugees in Minnesota: Characteristics and causes. *British Journal of Addiction, 84*, 785–789.

Wheeler, S. R. (1998). Hmong parents strive to connect: Cultural rift divide adults from children. *The Denver Post*. Nov. 15, 1998: B1, B5.

Wright, A. (1986). *A never ending refugee camp: The explosive birthrate in Ban Vinai*. Unpublished paper, Bangkok: Thailand

Xiong, P., & Culhane-Pera, K. A. (1995). Hmong perceptions and attitudes about immunizations. Paper presented at Hmong National Education Conference, St. Paul, MN.

Xiong Z. B. (2000). *Hmong American family problem-solving interactions: An analytic induction analysis*. Doctoral dissertation. St. Paul, MN: University of Minnesota.

Yang, Dao. (1991). The Hmong: Enduring traditions. In Judy Lewis (Ed.), *Minority cultures of Laos: Kammu, Lua', Lahu, Hmong, and Iu-Mien*. Sacramento, CA: Southeast Asia Community Resource Center.

Yang, Dao, with Blake, J. (1993). *Hmong at the turning point*. Minneapolis, MN: Yang Dao/WorldBridge Associates.

Yang, P. (2000). A case of thalassemia in a Fresno Hmong child repeatedly misdiagnosed as a respiratory disorder. Department of Family

Medicine, UCSF-Fresno Medical Education Program. Personal communication.

Yang, P., & Murphy, N. (1994). *Hmong in the '90s: Stepping toward the future.* St. Paul: Hmong American Partnership.

Zander, D. B., & Xiong, L. P. (1996). *The effects of problem gambling on Southeast Asian families and their adjustment to life in Minnesota.* St. Paul, MN: The Council on Asian-Pacific Minnesotans.

Iroquois

Barbara W. Lex and Thomas S. Abler

ALTERNATIVE NAMES

The Iroquois were and are a confederacy of several Native North American nations. The original five members of the confederacy were the Mohawk, Oneida, Onondaga, Cayuga, and Seneca. Accordingly, they were known to English colonial officials as the Five Nations. When the Tuscarora joined them early in the 18th century, they became the Six Nations. The Iroquois saw their confederacy as a metaphorical longhouse, the multifamily dwelling which housed them in settlements at the time of contact with Europeans. Hence they referred to themselves as the Hodénosaunee, meaning, roughly, People of the Longhouse. French colonists in Canada used the term Iroquois, a name they probably learned from a 16th century Basque–Algonquian pidgin used in the St. Lawrence valley (see Bakker, 1990; Goddard, 1978).

LOCATION AND LINGUISTIC AFFILIATION

Iroquois territory stretched through what is now upstate New York from the Mohawk River valley to that of the Genesee. The Mohawks lived in the Mohawk valley; the Oneidas lived near Oneida Lake; Onondaga territory was lands surrounding present-day Syracuse, New York; the Cayugas lived near Lake Cayuga; and the Senecas occupied lands to the west. In historic times there was an expansion westward and the Iroquois claimed as hunting territories lands now in Pennsylvania, Ohio, and Ontario. In the late 17th century Mohawks who had been converted to Christianity by French Jesuits moved north to the St. Lawrence valley where they remain in communities in Quebec, New York, and Ontario.

Following the American Revolution, a large portion of those Iroquois who had fought as allies to the Crown moved to lands in Ontario. Others remained in New York, the Onondagas with a reservation near Syracuse and the Senecas currently living on three reservations in western New York. The Oneidas, who had fought as allies to the rebellious Americans, initially remained in their homeland, but in the early 19th century emigrated in large numbers to both Wisconsin (near Green Bay) and Ontario (near London). The Tuscaroras moved northward from South Carolina to reside in the Iroquois country about 1720, and now occupy a reservation near Niagara Falls, New York. Some Iroquois who had settled in Ohio eventually were established on a reservation in Oklahoma (Campisi, 1978; Fenton & Tooker, 1978; Sturtevant, 1978; Tooker, 1978a). Table 1 lists the many contemporary Iroquois reservations and reserves and their enrolled memberships. However, a large portion of those enrolled live off-reservation or off-reserve and the residents of a reservation or reserve include many not officially enrolled as tribal or band members.

The nations of the Iroquois Confederacy, as well as neighboring, politically independent groups such as the Huron, spoke languages of the northern branch of the Iroquoian language family. A single language, Cherokee, survives in the Iroquoian language family's southern branch. Of northern Iroquoian languages, Tuscarora is the most divergent. Mohawk and Oneida are the most closely

Table 1. Contemporary Iroquois Reservations and Reserves

Reservation or reserve	Enrolled population
New York	
Seneca Nation (Allegany and Cattaraugus)	6,241
Tonawanda Band of the Senecas	1,050
Oneida	1,100
Onondaga	1,600
Tuscarora	1,200
Akwesasne (St. Regis)	5,638
Ontario	
Akwesasne (St. Regis)	9,500
Tyendinaga	7,046
Six Nations	20,876
Wahta Mohawk (Gibson)	659
Oneida of the Thames	4,776
Quebec	
Kahnawake (Caughnawaga)	8,888
Kanesatake (Oka)	1,943
Wisconsin	
Oneida	11,000
Oklahoma	
Seneca-Cayuga	2,460

These figures reflect the enrolled or registered membership of the above communities in or about 1990 for the United States and 2000 for Canada. Many enrolled members live off-reservation or off-reserve. The reserves and reservations also are home to many non-enrolled individuals, Indian and non-Indian, who have married into or otherwise have a right to reside on the reserve or reservation.
Sources: Abrams, 1994; Campisi, 1994; Canada, 2001; Hauptman, 1994; Oneida Indian Nation, 2000; Patterson, 1994; Starna, 1994; Wells, 1994.

related of northern Iroquoian languages still spoken while Mohawk itself has distinct dialects (Bonvillain, 1984; Lounsbury, 1978).

OVERVIEW OF THE CULTURE

The Iroquois lived in compact villages usually located in defensible positions near water sources, ranging in size from approximately 500 persons up to 3,000 or possibly even more. Typically surrounded by palisades, a village consisted of parallel rows of longhouses, each elm-bark and pole structure housing a number of matrilineally related women with their spouses and unmarried children. Open hearths were spaced about 8 m apart in the central aisle running the length of the house. Each was shared by two families, and the usual longhouse at time of contact had three or four hearths, hence housing six to eight families (Abler, 1970).

Surrounding the village were its agricultural fields that provided the larger portion of the Iroquois diet. Here grew what the Iroquois referred to as the "life-supporters" or the "three sisters"—maize (corn), beans, and squash. Men cleared the fields that the women planted, cultivated, and harvested. Nearby, women gathered firewood, necessary for cooking and heating, and collected edible wild green plants, nuts, and fruits to supplement cultigens. When local sources of firewood were exhausted and fields drained of their fertility by repeated crops of maize, a new village would be established at some distance from the old settlement. Males were frequently away, hunting, trading, or waging war. Deer was the most important mammal that was hunted, both for its meat and its hide. Large quantities of fish were also harvested at semipermanent fishing stations. With the coming of Europeans, the harvesting of beaver pelts, to exchange for goods of European manufacture, became important. Wallace (1952) has contrasted the "village" and the "forest" in Iroquois life with the former being the women's domain while the latter was the sphere of male activities.

The dog was the only native domesticated animal, although captured bear cubs were raised in pens until of sufficient size to warrant slaughter for food. From their European neighbors the Iroquois obtained pigs that thrived in the temperate forest of North America and were found in large numbers in Seneca villages by the last quarter of the 17th century. Other European animal and plant domesticates were adopted, and through the 18th century the multi-family longhouse was used with lessening frequency, being replaced by single-family dwellings. As threats of outside invaders subsided, large villages fragmented into smaller hamlets that did not exhaust local resources.

Iroquois communities were divided into exogamous matriclans. The number varied from nation to nation, with the Mohawk and Oneida having only three clans each while the other nations had from eight to ten matriclans. A smaller clan might have just one lineage, but larger clans were divided into two or more lineages, each lineage headed by a senior woman or matron often referred to as a "clan mother." This woman had considerable responsibility in organizing the activities of her sisters, daughters, and nieces (sisters' daughters) in the lineage and in choosing from among her brothers, sons, and nephews (sisters' sons) in the lineage one qualified to hold a position as a political leader in the community (Fenton, 1978; Tooker, 1978b).

Marriage was monogamous. Men upon marriage left the home of their own clan and lineage to reside in houses belonging to their wives' lineage. As is often the case in matrilineal societies, divorce was relatively easy and common, with the children remaining in the home of their mother. Bonvillain (1980, pp. 52–53) notes the role of women in initiating pre-marital and extramarital sexual encounters which "occurred easily." She also notes that in the "ideal pattern" the lineage matron of an eligible male would arrange a marriage with the senior women of the lineage of a prospective bride. Extramarital affairs were among the behaviors condemned by the prophet Handsome Lake and he also condemned those who gossiped about such activities leading to the breakup of a marriage (Parker, 1913, pp. 32–33; Wallace, 1971, pp. 370–371). Fenton (1941b) reports infidelity of a husband as a common cause of female suicides.

Extensive documentation of Iroquois religious practices postdates both the revitalization of Iroquois religion by the prophet Handsome Lake, who experienced his first vision in 1799, and the conversion of substantial numbers of Iroquois to Christianity (Wallace, 1970). The yearly cycle of ceremonies was first outlined in Morgan's pioneering ethnography (Morgan, 1851). The major ceremonies, Midwinter and Green Corn, as well as many lesser ceremonies, relate to the agricultural year, but others such as the Strawberry ceremony and Maple ceremony celebrate the wild foods which had been important in the Iroquois diet (Fenton, 1936, 1941a). Another important aspect of religion among the Iroquois involved societies involved in the curing of illness.

THE CONTEXT OF HEALTH: ENVIRONMENTAL, ECONOMIC, SOCIAL, AND POLITICAL FACTORS

Skeletal remains excavated from early historic (late 16th century) sites yield some clues about general health status in several Seneca villages (Wray, Sempowski, & Saunders, 1991; Wray, Sempowski, Saunders, & Cervone, 1987). Lacking information to be gained only from analyses of soft tissue or body fluids, the fragmentary nature of this record and possible collection bias dictates cautious interpretation. Given their close spatial and temporal proximity, variations among sites and even between cemeteries at a single site are puzzling. Pathologies at the Adams and Culbertson sites suggest many suffered from anemia or iron deficiencies (Wray et al., 1987, pp. 28–29, 188). Over half of the analyzable remains from Cemetery 2 at the Tram site exhibit pathologies, and the greater portion of these exhibit nutritional deficits or growth-disrupting illness (Wray et al., 1991, p. 390). The pathology rate at Cameron was low, but interment of immature individuals was very high, indicating contagious disease either of European origin or the result of poor sanitary conditions resulting from village size increases in the 16th century which could have promoted indigenous dysentery epidemics (Wray et al., 1991, p. 397).

Dobyns (1983, pp. 313–318) argued that European diseases introduced elsewhere in the Americas in the 16th century also swept through Iroquoia. A number of scholars (Henige, 1986; Snow & Lanphear, 1988, 1989) strongly reject that assertion, seeing no evidence of European disease epidemics among Iroquois people before 1634. Ramenofsky (1987), who concurs with Dobyns about the early impact of European disease on Native American populations, found insufficient archaeological evidence indicating that European diseases led to a 16th-century Iroquois population collapse.

None deny, however, the impact of European diseases such as smallpox, measles, and mumps on the Iroquois population during the 17th century. Snow (1994, p. 98) estimates half of the Mohawk population died over fewer than 100 days in 1634. Mortality from periodic epidemics of European diseases continued among Iroquois from that date until well into the 19th century. To maintain their population, the Iroquois adopted large numbers of refugee populations and war captives.

Estimates reporting the number of Iroquois fighting men from the latter half of the 17th century suggest 2,000 to 2,500 warriors (see Table 2), hence a total population of 10,000 to 12,500. A half-century of contact with European diseases had considerably reduced the population. Archaeological and ethnohistoric work with the Mohawk suggests a population in 1633, prior to epidemics of European diseases, of 8,000 to 10,000 for the Mohawk alone (Guldenzopf, 1984; Snow, 1992; Snow & Lanphear, 1988; Snow & Starna, 1989; Starna, 1980). Snow (1992, 1994, p. 100) notes the Mohawk population

Table 2. Estimates of Numbers of Iroquois Warriors

Nation	1660	1665	1677	1689	1698
Mohawk	500	300 to 400	300	270	110
Oneida	100		140	200	180	70
Onondaga	300	300	300	350	500	250
Cayuga	300	300	300	300	200
Seneca	1,000	1,200	1,000	1,300	600
Total	2,200	2,240 to 2,340	2,150	2,550	1,230

Source: Abler, 1970, p. 21.

had expanded greatly in the previous 50 years with an influx of refugees and/or war captives and speculates that the total Iroquois population was 22,000 in 1633, prior to the first epidemics of European diseases to strike Iroquoia.

Warfare was intensified in historic times (Abler, 1992), and skill in treating wounds was praised by a contemporary observer (Lafitau, 1974–77 (2), pp. 204–206). Accidents were also a danger (traditional Iroquois still pray that dead tree limbs not fall on children playing underneath them) and fire was a constant hazard in villages composed of densely packed bark-covered dwellings.

Suicide has been a continuing pattern in Iroquois communities (Fenton, 1941b, 1986). Abuse or mistreatment can lead one to suicide. Both children abused by parents and middle-aged women abandoned by husbands are prominent among those who commit suicide. Also, a political leader who had lost the support of his followers might take his own life. A typical method for committing suicide was ingesting the root of the water-hemlock, *Cicuta maculata*.

MEDICAL PRACTITIONERS

Recent ethnographic studies of conservative portions of Iroquois communities report that diagnosis of illness occurs separate from its treatment. There are a small number of clairvoyants or traditional diagnosticians who are regularly consulted about the cause of a patient's problem. These practitioners then direct patients to appropriate cures. Advice may be to consult a herbalist for treatment, to have a rite performed by one of the medicine societies, or to consult specialists in western medicine (Blau, 1969, p. 7; Shimony, 1994, pp. 270–274; Isaacs, 1973, p. 77).

Herbalists are specialists who have acquired from an earlier generation knowledge of local flora and the efficacy of their use (Herrick, 1995). Fenton (1940, p. 793) notes a herbalist of either sex may pass his or her knowledge on to either a son or daughter or even, skipping a generation, to a grandchild. The knowledge of herbalists is often extensive; Fenton reports that knowledgeable herbalists with whom he worked could identify from 200 to 300 species of plants, perhaps a third of the plants available locally (Fenton, 1942, p. 504; see also Isaacs, 1973, pp. 76–79). However, plants in the ethnopharmacopoeia have changed over time (Isaacs, 1976–77, pp. 272–281).

Religious practices and health practices overlap. Specific illnesses may require performance of specific rites by medicine societies with a restricted membership. Some of their rites occur in public ceremonies; others can be witnessed only by the patient and society members. Generally, having been cured by a specific medicine conferred membership, with its obligations, in the society. The nature of the illnesses treated by these societies and the sort of rites performed are briefly discussed below, although it should be recognized many participants are uneasy about public discussion of these matters. Here we summarize only material which has appeared in print.

Fenton reports a ritual pattern followed by the medicine societies in conducting their rites. Invitations are sent out by a headman. A thanksgiving and tobacco invocation begins the meeting. These are followed by the ritual with its cycle of songs. The participants are then thanked and appropriate food is provided for the participants (Fenton, 1979, p. 1607). Medicine societies are the means by which individual illness becomes a group concern because participation involves mutual aid, allays anxieties about health, and increases involvement in traditional Iroquois culture (Lex, 1977, p. 284).

Certainly the best documented among the medicine societies is the Society of Faces, often named the False Face Society. Participants wear wooden masks, usually painted either red or black, with perforated brass plates for eyes and hair made from tails of horses. They carried large rattles made of the shell of the snapping turtle, its long neck reinforced with splints to form the handle. The Society of Faces journeyed through the community in the spring and fall visiting houses to drive out disease (the Traveling Rite). Fenton (1941c, p. 425) described his encounter with the Faces on the Allegany Reservation in the 1930s: "The company afforded a wild spectacle as they sped up the valley road in open Fords with their hair whipping in the chill winds; they grated their rattles on the car body and uttered their terrifying cries whenever they swerved to pass a stranger." They also perform publicly at the Midwinter ceremony. Individuals suffering from illnesses which the Faces have the power to cure can also have private rites performed for them in their homes and subsequently become members of the society (Fenton, 1987). Both private and public rites manipulate fire as a source of power for preventing or curing illness (Isaacs & Lex, 1980).

A second masked medicine society is the Husk Faces or Bushy Heads. They wear masks made of braided corn husks. They serve as heralds to announce the arrival of the Society of Faces during the Traveling Rite and at Midwinter. Among the ailments cured by Husk Faces is backache (Fenton, 1987, p. 400).

The third medicine society to use masks has been named by Parker (1913, pp. 122–123) the Society of Mystic Animals (it is also known as the Medicine Company or Shake the Pumpkin). The "blind masks" once used by this society, with which the wearer demonstrated his power to find objects even though the wooden mask had no eye holes, were out of use by 1900 (Fenton, 1987, p. 48). Pig masks carved of wood are also used by this society which performs its rituals in a darkened room, and Speck (1949, p. 104) reports the medicine used by this society to be "extraordinarily powerful."

The most powerful of the medicines is held by the Little Water Society. Strictly speaking members of this society did not cure, since all their rituals served to maintain the potency of the Little Water medicine. The rites are conducted in the dark, between 11:00 pm and dawn and both gourd rattles and a flute are used to accompany the cycle of songs. Visitors may listen in the next room but only the initiated should witness the renewal of the medicine (Parker, 1913, pp. 116–118). As the most potent of available medicines, the Little Water medicine is also the most dangerous, and one who has custody of such medicine must exhibit exemplary behavior (Shimony, 1994, p. 284).

Other important medicine societies include the Pygmy Society or Dark Dance, the Eagle Society, the Bear Society, the Otter Society, and the Buffalo Society. Speck (1949, pp. 59–60) describes several dissociative illnesses which one or another of these societies were able to cure, such as convulsions in an Oneida woman treated by the Buffalo Society and howling hysteria in a brother and sister treated by the Bear Society.

Not all illnesses require interventions by medicine societies. In some instances a cure necessitates performance of a particular rite or dance with general participation, for example the war dance. Outcomes of games, such as lacrosse or snow snake or the Bowl Game, could have curative as well as predictive effects (see Blau, 1969, p. 144; Speck, 1949, pp. 115–126).

CLASSIFICATION OF ILLNESS, THEORIES OF ILLNESS, AND TREATMENT OF ILLNESS

Iroquois theory asserted that misfortune and ill health could be caused by several sources. These could be simply physical, as with injury or war wounds. Another source was witchcraft in which illness is caused by foreign objects that have been magically projected into one's body. A failure to perform obligatory rites or rituals was another. Spirits, including those of the dead, could attack. An additional and significant cause of illness was unfulfilled desires of the soul that were expressed among the Iroquois through dreams. In his evaluation of this last category, Wallace, a psychological anthropologist, considers the Iroquois dream theory as "basically psychoanalytic" and phrased "in language which might have been used by Freud himself." The Iroquois recognized that the mind possessed both conscious and unconscious desires and "were aware that the frustration of these desires could cause mental and physical ('psychosomatic') illness." They were also aware that the dreams often had latent content which required considerable interpretation to uncover their symbolism (Wallace, 1958). A frequent desire of the soul as expressed in dreams was for the performance of a particular ceremony by one of the medicine societies.

Shimony (1970) reports that witchcraft is universally attributed to "jealousy" but notes this is most often envy of unusual achievement on the part of the bewitched or a past grievance. She reports cases in which auto accidents and the death of a child were attributed to witchcraft. Witches can transform themselves into animals, but they are also sometimes seen at night as flying lights. Shimony points out that those who prove themselves able to cure illnesses caused by witchcraft sometimes come under suspicion of being witches themselves (on witchcraft, see also Herrick, 1995, pp. 37–38, 42–44).

Isaacs (1973, pp. 72–72a) notes external causes in addition to witchcraft which could cause physical distress. These include being attacked by a spirit of a dead individual or because one has offended an animal spirit.

The former is often revealed by the fact that one dreams of the deceased. Affliction by an animal spirit could call for the performance of the rites of one of the medicine societies. Speck (1949, pp. 65–67) cites the case of a woman suffering from St Vitus's dance whose father had mistreated his catch while fishing. Her illness called for a performance of the rites of the Otter society.

Herrick (1995, p. 37) reports "offensive behaviorial acts or taboo violations" as a primary cause of disease and misfortune. This includes contact with a menstruating woman or even eating rich foods. Possession of powerful medicines and charms can pose a danger to one and one's family if one fails to treat and "feed" (e.g., perform rituals, burn tobacco) the medicine in the required manner (see Fenton, 1987, pp. 143–144; Shimony, 1970, pp. 250–254, 1994, p. 285). Also, witnessing rites of a medicine society by the uninitiated can lead to hysteria or other mental or physical harm. Initiation into the society is often the cure (Shimony, 1994, p. 282). Herrick (1995, pp. 50–63) presents a list of 287 conditions (with some repetition or overlapping testimony) of illnesses with their causes.

There are specific illnesses that demand the performance of a medicine society for their cure. Fenton (1987, p. 143) lists various symptoms of False Face sickness which can be cured by the Society of Faces. These center on the head, shoulders, and joints and include inflamation and swelling of face, nose bleeding, earache, toothache, other facial pain, and facial paralysis (see also Isaacs & Lex, 1980, p. 8).

SEXUALITY AND REPRODUCTION

Women typically bore four or fewer children and spaced their children by 5- or 6-year intervals. There was a herbal remedy which was believed to induce abortions, but its use was forbidden in the teachings of the Seneca prophet, Handsome Lake (Parker, 1913, p. 30). Contraception was also forbidden by Handsome Lake, but it is reported that it is practiced by his contemporary followers, that boiled sassafras shoots is thought to be an abortifacient, and that they are also convinced that prolonged breast-feeding reduces fertility in women (Shimony, 1994, pp. 208–209). Shimony also observed a gender difference in average duration of breast-feeding, with two years usual for girls but a range of from one year to 1 year and 9 months for boys. Engelbrecht (1987) concludes from evidence from historic sources discussing the Iroquois and their neighbors and from archaeological sites that the average Iroquois family size was small. Those who have investigated Iroquois culture have ignored the issue of homosexuality. A recent study reports the belief that masturbation leads to insanity (Herrick, 1995, p. 70).

HEALTH THROUGH THE LIFE CYCLE

Pregnancy and Birth

Conception is thought to take place during a new moon. Pregnant women are expected to refrain from certain activities, such as associating with men engaged in hunting or making medicines (Shimony, 1994, p. 207). Pregnant women had to bring their own cup to use when strawberries were distributed as part of the annual Strawberry Ceremony in June (Shimony, 1994, p. 159). Skinning mink was avoided since the odor was thought to cause abortions and certain behaviors (such as sitting in a doorway or sitting upon one's foot) are to be avoided as detrimental to the birth or the fetus. Contact with a menstruating woman would also cause abortion (Shimony, 1994, pp. 208, 217). While geophagy was known among the general population, it "is quite common among pregnant women" (Shimony, 1994, p. 208). Kneeling was the traditional position for delivery, a practice still "found occasionally" a half-century ago (Shimony, 1994, p. 207). Maidenhair fern was prescribed by midwives for labor pains and sassafras was brewed into a tonic for use by women after childbirth (Fenton, 1942, p. 517).

Infancy

Public recognition of the place of an infant in a community came at the Midwinter Ceremony or the Green Corn Ceremony that followed its birth. Today the exact day on which the names are announced as one element within these lengthy ceremonies varies. A speaker announces the name of each new member of the community. Each infant receives a name held or owned by the matriclan into which she or he was born (Sturtevant, 1984). Only one living person bears a specific name, and infants thought destined to fulfill important ritual roles in adulthood are given names of deceased ritualists (Blau, 1969, p. 27).

Infants are reported to have been nursed for lengthy periods, 3 years being common with observations nearly two centuries apart reporting cases of children over 5 years of age still nursing (Engelbrecht, 1987, p. 19). Children were weaned by painting the nipple black to frighten the child, placing a chicken feather on the breast, or coating the nipple with a noxious, non-toxic substance (Shimony, 1994, p. 209). A recent study among the Mohawks of Kahnawake, Quebec, reported that breast-feeding increased from 45% of infants in 1978 to 64% in 1986 (Macaulay, Hanusaik, & Beauvais, 1989).

Traditionally infants would be strapped to a cradle board, a plank approximately 0.5 m long and 0.25 m wide. At the bottom there was a foot-board and the head was protected by a hoop or bow. Morgan (1851, p. 390) reported "the patience and quiet of the Indian child in this close confinement are quite remarkable."

High infant mortality is indicated by the Iroquois proverb, "an infant's life is as the thinness of a maple leaf" (Fenton, 1978, p. 314). Engelbrecht (1987, pp. 21–22) argues that the fact that Handsome Lake found it necessary to preach against infanticide indicates its earlier practice. Scattered references found in the historic literature suggest that infants may have met death through conscious actions or neglect in cases in which their mother died or their mother was distraught because of the death of a husband or other social circumstances. Handsome Lake specifically stated that childless women should adopt infants born to their sisters (Parker, 1913, p. 35).

Childhood

Restraint on the actions of children was limited. The prefered form of punishment for bad behavior was to splash the child with cold water. A thorough dousing or the threat of a dunking in a cold stream were used on children who do not reform. A red willow whip, to which tobacco had been burned, was used to strike particularly disobedient children. Traditional values disapproved slapping. Indeed, severely punishing children was thought to bring disease, hysteria, or the vomiting of worms upon the disciplinarian (Shimony, 1994, pp. 209–210). Parents also refrained from disciplining children lest they commit suicide by eating the poisonous root, *Cicuta*, or that they might mature to abuse their elderly parents (Fenton, 1941b, p. 125). Field workers have noted considerable respect for the rights of children to make their own decisions. Shimony (1994, p. 269) records the case of parents accepting the refusal of a 9-year-old to undergo a heart operation deemed necessary by the local hospital.

Adolescence

Oral traditions suggest that vision quests by males at puberty were once a significant element in Iroquois culture, but this does not appear to have been widely practiced in historic times. Similarly, oral tradition suggests the isolation of a girl for up to a year at the time of her first menses, but again this seems to have disappeared as a practice by the time observers were recording Iroquois culture and behavior. The girl did have to observe the prohibitions of behavior observed by any menstruating woman for the first three days of her period (Shimony, 1994, pp. 215–216).

As a mark of maturity, the "baby name" previously held was replaced, each person receiving an adult name from the roster of names of his or her matriclan. As was the case with the "baby name," these names were publicly announced at either the Green Corn Ceremony or the Midwinter Ceremony. One of the boys who had received a new name then led a Great Feather Dance as part of the worship (Fenton, 1963).

Adulthood

Since the time of Morgan (1851, p. 83), clan exogamy has been recognized among the Iroquois. However, this rule has been frequently breached, even among conservative Iroquois. Lineage exogamy is more frequently observed. In addition to these matrilineal relatives, all persons related to one through either parent are forbidden as a spouse or sexual partner (see Shimony, 1994, pp. 30–32).

Adult males, particularly those with ritual obligations, undergo a regimen of purging for three days in the spring. In the past this included baths in a sweat lodge. Various herbal concoctions are used as emetics and laxatives to cleanse the body. An 18th-century observer noted the use for purging of "very strong medicines which clear them out to excess and might well kill a horse" (Lafitau, 1974–77 (2), p. 206). One fasts during the purge but then takes tonics to regain one's strength. Sassafras is thought to restore the blood following the period of the purge. In addition to this period of "spring cleaning," men may also purge themselves in preparation for an important ritual (Parker, 1913, p. 77; Shimony, 1994, pp. 265–266). A phobia apparently rare or absent among Iroquois males is fear of heights (Wallace, 1951, p. 64).

Onset of menses was believed to occur at the new moon. Shimony (1994, pp. 216–218) reports that women were considered "poisonous" or "dangerous" during the first 3 days. They should not attend curing rituals or be in the presence of medicines, nor should they come into close contact with males. They even constitute a danger to themselves, since if they comb their hair it might fall out. They may attend longhouse events, but should drink from their own cup rather than the common dipper. These restrictions are no longer followed, and the timing of the Midwinter Ceremony no longer allows 5 days for completion of mestruation (Blau, 1969, pp. 59–60; Shimony, 1994, p. 174). Contact with a menstruating woman causes bloody diarrhea and bleeding piles. Menstrual blood is also an ingredient in love medicines that can lead to bad luck or even insanity.

The Aged

Morgan (1851, p. 171) recorded the teaching of the Longhouse religion speaker from Tonawanda, Jimmy Johnson: "It is the will of the Great Spirit that you reverence the aged, even though they be as helpless as infants." The prophet Handsome Lake preached that it was "ordained that people should live to an old age" and that "an old woman should be as a child again and when she becomes so the Creator wishes the grandchildren to help her" (Parker, 1913, p. 35). Despite this, Fenton (1941b, p. 125) feels "that the number of cases where Iroquois adults have maltreated their aged parents is great enough to warrant investigation" and notes that two abused elderly Onondaga males committed suicide. In one of these cases there were factors involved other than simply abuse by adult children.

Randle (1951, p. 171) presents an idyllic view of the role of the aged woman: "Honored as heads of clans and household, the old age of women could be rewarding, surrounded by her offspring." Although unmarked by a rite of passage, after menopause women can partake in activities involving curing and medicines that previously had been barred to them (Shimony, 1994, p. 218). Wallace (1951, p. 64) notes that the absence of fear of heights continues into old age—"even old men of 60 and 70 will take, and efficiently perform, such jobs as pruning high trees, painting the roofs of buildings, and carpentry work on scaffolds."

Dying and Death

Several authors have commented upon the excessive grief and intensity of mourning found among the Iroquois. Wallace (1970, p. 77) noted: "Descriptions of Seneca mourning behavior read like psychoanalytic essays on the dynamics of depressive states, and the paranoia of bereavement, which generated blood feud and fear of witchcraft, was regarded by the Iroquois themselves as a continuing threat to the solidarity of the community."

Unusual behaviors of birds often foretold deaths (Shimony, 1994, p. 234). People preferred to die at home, and those believed terminally ill had the clothes in which they will be dressed in their coffin placed within their view. Although an innovation of the modern New York Onondaga involved a ritual of lying in state in their Longhouse to show respect for a Confederacy chief (Blau, 1969, p. ix), traditionally the corpse was on view in a coffin in the home. This was a time of much unease, since the dead attract other dead who may take offence at some action among the living— especially inattention to ritual details—and cause illness or bad luck. Children under the age of five were considered particularly vulnerable, and typically a deerskin thong tied around the wrist was used to protect them from the spirit of the dead. Shimony reports that a wake involving special songs and a wake game was conducted on the Six Nations reserve, but this was not practiced by the conservative element in other Iroquois communities. The grave was dug by a member of the moiety opposite to that of the deceased.

The whole community mourned for 10 days, culminating in a tenth-day feast organized by matrilineal kin of the deceased and marked by distribution of his or her property. The spirit of the dead then may depart the community, but would be addressed again during a feast on the first anniversary of death. It is believed that Handsome Lake forbade mourning for an entire year, considering that practice too disruptive to daily life (Shimony, 1994, pp. 228–250). An all-night Feast of the Dead (*Ohgiwe*) is held at least once annually to honor and propitiate potentially restive spirits (Blau, 1969, pp. 245–252; Fenton & Kurath, 1951, pp. 139–166; Lex, 1977, p. 294; Shimony, 1994, pp. 229–233). Certain songs are sung falsetto, and dancers move clockwise as does distribution of special feast foods. At dawn a procession of the living was intermingled with spirits of the dead.

CHANGING HEALTH PATTERNS

In the mid-20th century no strict dichotomy of beliefs and behaviors associated with health and illness distinguished Longhouse adherents from those who eschew traditional beliefs (Weaver, 1972, p. 33), and acceptance of Western medical services by some Six Nations reserve inhabitants is documented before 1850 (Weaver, 1972, p. 39). Contagious diseases such as smallpox (Weaver, 1971, pp. 361–378) diminished as preventive measures such as vaccination and sanitation improvements became available on reservations and reserves; most prevalent currently are chronic diseases stemming from behavioral risk factors (Lex & Norris, 1994, pp. 193–196).

Among Iroquois in both the United States and Canada, alienation of land, breach of treaty rights, and monetary exploitation—as well as discriminatory attitudes and actions of the dominant societies—have promoted resentment and distrust of governmental authority, and concomitantly reinforced desire to be treated as sovereign nations (Abler, 1997, pp. 27–28; Weaver, 1971, pp. 361–378; Weaver, 1972, pp. 32–37). Prevention and treatment in Canada are provided by Health Canada, and in New York by the American Indian Health Program of the state Department of Health (not the Federal Indian Health Service). Governing bodies on each reserve or reservation would need to grant permission for health-related surveys (Lex & Norris, 1994, p. 195), and some reject enumeration (Department of Health, 1999, p. 32). Given intragroup factionalism, participation could be neither required nor guaranteed despite needs for knowledge to plan appropriate programs.

Some chronic disease data are available from selected samples. Indian populations have high rates of Diabetes mellitus (Type 2 diabetes) linked with obesity, hypertension, anemia, and nutrient deficiencies, as well as complications of pregnancy (Lex & Norris, 1994, pp. 198–199). For 1980 to 1986 in upstate New York, birth certificate data were compared for Indian (predominantly Iroquois), white, black, and other race infants (Buck et al., 1992, pp. 569–575). Mothers of Indian infants not only were younger, had less education, had more children, and took longer to obtain prenatal care than mothers of white infants, but also had more post-term births and excessive-sized (>4,000 g) babies—factors associated with gestational diabetes or diabetes mellitus. Death certificate data for the same interval (Mahoney, Ellrott, & Michalek, 1989, pp. 403–412) showed that Indians died at an average age 9 years younger than others (women 65 versus 74 years; men 58 versus 67 years), with ages 50 to 59 years and 20 to 29 years showing higher rates for females and males, respectively. Diabetes mellitus was the leading cause of death for women, followed by liver cirrhosis, nephritis, and homicide, and for men major causes were tuberculosis, diabetes mellitus, cirrhosis, and pneumonia.

A tribal survey of the Seneca Nation of Indians during the mid-1970s reported 26% had diabetes mellitus. Epidemiologists assessed mortality causes by analyzing death certificate data for 914 deaths among 3,262 Seneca (391 of 1,690 females and 523 of 1,572 males) listed on the tribal roll January 1, 1955, resident in New York State, and followed until December 31, 1984, in comparison with deaths of other upstate New York inhabitants during that interval (Mahoney, Michalek, Cummings, Nasca, & Emrich, 1989, pp. 816–826). Tuberculosis was the leading cause for Seneca women, followed by diabetes mellitus, pneumonia, liver cirrhosis, nephritis, accidents, and homicide; for Seneca men ranked causes were tuberculosis, diabetes mellitus, atherosclerosis, hernia/intestinal obstruction, cirrhosis, and accidents. Median age at death increased from 55.9 years during the first decade studied to 64.6 years during the last. A similar study analyzed 74 deaths among 3,033 Seneca children (47 of 1,483 females and 27 of 1,550 males) ages 0 to 24 years born between January 1, 1955, and December 31, 1989 (Michalek, Mahoney, Buck, & Snyder, 1993, pp. 403–407). Most deaths before age 5 were from infectious diseases. Between ages 15 and 24 years, accidents, particularly motor vehicle accidents, were predominant causes for both males and females, with males also exhibiting elevated suicide rates.

In a 1985 study, Kahnawake Mohawk adults with and without diabetes mellitus were chosen randomly and matched for age and sex. Data from clinical records, interviews, and body measurements showed male and female diabetics to have 5.51 times more peripheral vascular disease, with ratios of 4.57 for cerebrovascular disease and 3.56 for ischemic heart disease. Among diabetics, 48% had ischemic heart disease (versus 22% of non-diabetics), the highest known rates for North American Indians. Moreover, 86% and 74%, respectively, were obese. Persons with diabetes also had high rates of hypertension, hypercholesterolemia, and diabetic complications. These factors in combination indicated need for community-wide interventions (Macaulay, Montour, & Adelson, 1988, pp. 221–224; Montour, Macaulay, & Adelson, 1989, pp. 549–552).

Children between ages 9 and 10 years exhibited increased weight, height, body mass index, and subscapular skinfold thickness (SSF) associated with increased television viewing and decreased physical fitness. Findings led to a pioneering community-based primary prevention program, the Kahnawake Schools Diabetes

Prevention Project, to change diets and promote physical activity (Macaulay, et al., 1997, pp. 779–790). In 1994, 103 girls and 95 boys attending elementary schools in two Mohawk communities were surveyed for demographic and lifestyle variables, height, weight, and SSF were measured, and children performed a run/walk fitness test, with a follow-up assessment of SSF in 1996.

Despite improvements for some children, risk factors, especially television viewing, were confirmed for Kahnawake girls (Horn, Paradis, Potvin, Macaulay, & Desrosiers, 2001, pp. 274–281). Also, among asthmatics ages 4 through 12 years, Seneca girls were over-represented when compared with pupils from other ethnic groups; their "triggers" were associated with adverse housing conditions (Lwebuga-Mukasa & Dunn-Georgiou, 2000, pp. 745–761). Accordingly, lifestyle factors associated with risk for development of chronic diseases and associated complications need to be discerned and addressed on reserves and reservations, with special emphasis on understanding gender differences

REFERENCES

Abler, T. S. (1970). Longhouse and palisade: Northern Iroquoian villages of the seventeenth century. *Ontario History, 52*, 17–40.

Abler, T. S. (1992). Beavers and muskets: Iroquois military fortunes in the face of European colonization. In R. B. Ferguson & N. L. Whitehead (Eds.), *War in the tribal zone: Expanding states and indigenous warfare* (pp. 151–174). Santa Fe: School of American Research Press.

Abler, T. S. (1997). Iroquois: The tree of peace and the war kettle. In M. Ember, C. R. Ember, & D. Levinson (Eds.), *Portraits of culture: Ethnographic originals, Vol. 1, North America* (pp. 1–34). Englewood Cliffs, New Jersey: Prentice Hall.

Abrams, G. H. J. (1994). Seneca. In M. B. Davis (Ed.), *Native America in the twentieth century: An encyclopedia* (pp. 580–582). New York: Garland.

Bakker, P. (1990). A Basque etymology for the word "Iroquois." *Man in the northeast, 40*, 89–93.

Blau, H. (1969). Calendric ceremonies of the New York Onondaga. Doctoral dissertation, New School for Social Research, 1969.

Bonvillain, N. (1980). Iroquoian women. In N. Bonvillain (Ed.), Studies in Iroquois culture (pp. 47–58). *Occasional publications in northeastern anthropology* 6. Rindge, NH: Franklin Pierce College.

Bonvillain, N. (1984). Mohawk dialects: Akwesasne, Caughnawaga, Oka. In M. K. Foster, J. Campisi, & M. Mithun (Eds.), *Extending the rafters: Interdisciplinary approaches to Iroquoian studies* (pp. 313–323). Albany: State University of New York Press.

Buck, G. M., Mahoney, M. C., Michalek, A. M., Powell, E. J., & Shelton, J. A. (1992). Comparison of Native American birth in upstate New York with other race births, 1980–86. *Public Health Reports, 107*, 569–575.

Campisi, J. (1978). Oneida. In B. Trigger (Ed.), *Handbook of North American Indians, Vol. 15, Northeast* (pp. 481–490). Washington: Smithsonian.

Campisi, J. (1994). Oneida. In M. B. Davis (Ed.), *Native America in the twentieth century: An encyclopedia* (pp. 407–408). New York: Garland.

Canada Department of Indian Affairs and Northern Development (2001). *Registered Indian population by sex and residence 2000*. Ottawa: Minister of Public Works and Government Services Canada.

Dobyns, H. F. (1983). *Their number become thinned*. Knoxville: University of Tennessee Press.

Engelbrecht, W. (1987). Factors maintaining low population density among the prehistoric New York Iroquois. *American Antiquity, 52*, 13–27.

Fenton, W. N. (1936). An outline of Seneca ceremonies at Coldspring Longhouse. *Publications in Anthropology 9*. New Haven: Yale University.

Fenton, W. N. (1940). An herbarium from the Allegany Senecas. In W. J. Doty, C. E. Congdon, & L. H. Thorton (Eds.), *The historic annals of southwestern New York* (pp. 787–796). New York: Lewis Historical Publishing.

Fenton, W. N. (1941a). Tonawanda longhouse ceremonies: Ninety years after Lewis Henry Morgan. *Smithsonian Institution. Bureau of American Ethnology Bulletin, 128*, 140–156. Washington: U.S. Government Printing Office.

Fenton, W. N. (1941b). Iroquois suicide: a study in the stability of a culture pattern. *Smithsonian Institution. Bureau of American Ethnology Bulletin, 128*, 79–137. Washington: U.S. Government Printing Office.

Fenton, W. N. (1941c). Masked medicine societies of the Iroquois. *Annual report of the Smithsonian Institution for 1940*, pp. 397–430. Washington: U.S. Government Printing Office.

Fenton, W. N. (1942). Contacts between Iroquois herbalism and colonial medicine. *Annual report of the Smithsonian Institution for 1941*, pp. 503–526. Washington: U.S. Government Printing Office.

Fenton, W. N. (1963). The Seneca Green Corn Ceremony. *The New York conservationist, 18 (October–November)*, 20–22, 27–28.

Fenton, W. N. (1978). Northern Iroquoian culture patterns. In B. Trigger (Ed.), *Handbook of North American Indians, Vol. 15, Northeast* (pp. 296–321). Washington: Smithsonian.

Fenton, W. N. (1979). The "great good medicine". *New York State Journal of Medicine, 79*, 1603–1609.

Fenton, W. N. (1986). A further note on Iroquois suicide. *Ethnohistory, 33*, 448–457.

Fenton, W. N. (1987). *The False Faces of the Iroquois*. Norman: University of Oklahoma Press.

Fenton, W. N., & Kurath, G. P. (1951). The feast of the dead or ghost dance at Six Nations Reserve, Canada. *Smithsonian Institution. Bureau of American Ethnology Bulletin, 149*, 139–166. Washington: U.S. Government Printing Office.

Fenton, W. N., & Tooker, E. (1978). Mohawk. In B. Trigger (Ed.), *Handbook of North American Indians, Vol. 15, Northeast* (pp. 466–480). Washington: Smithsonian.

Goddard, I. (1978). Synonymy. In B. Trigger (Ed.), *Handbook of North American Indians, Vol. 15, Northeast* (pp. 319–321). Washington: Smithsonian.

Guldenzopf, D. (1984). Frontier demography and settlement patterns of the Mohawk Iroquois. *Man in the Northeast, 27*, 79–94.

References

Hauptman, L. M. (1994). Seneca-Cayuga. In M. B. Davis (Ed.), *Native America in the twentieth century: An encyclopedia* (p. 582). New York: Garland.

Henige, D. (1986). Primary source by primary source? On the role of epidemics in New World depopulation. *Ethnohistory, 33*, 293–313.

Herrick, J. W. (1995). *Iroquois medical botany*. Syracuse: Syracuse University Press.

Horn, O. K., Paradis, G., Potvin, L., Macaulay, A. C., & Desrosiers, S. (2001). Correlates and predictors of adiposity among Mohawk children. *Preventive medicine, 33*, 274–281.

Isaacs, H. L. (1973). Orenda: An ethnographic cognitive study of Seneca medicine and politics Doctoral dissertation, State University of New York at Buffalo, 1973.

Isaacs, H. L. (1976–77). Iroquois herbalism: The past 100 years. *International Journal of Social Psychiatry, 22*, 272–281.

Isaacs, H. L., & Lex, B. W. (1980). Handling fire: Treatment of illness by the Iroquois false-face medicine society. In N. Bonvillain (Ed.), *Studies in Iroquois culture* (pp. 5–13). *Occasional publications in northeastern anthropology 6*. Rindge, NH: Franklin Pierce College.

Lafitau, J. F. (1974–77). *Customs of the American Indians compared with the customs of primitive times*. W. N. Fenton & E. L. Moore (Eds. and Trans). Toronto: Champlain Society.

Lex, B. W. (1977). Altered states of consciousness in Northern Iroquoian rituals. In A. Bharati (Ed.), *The realm of the extrahuman: Agents and audiences* (pp. 277–300). The Hague: Mouton.

Lex, B. W., & Norris, J. R. (1994). Health status of American Indian and Alaska Native women. In A. C. Mastroianni, R. Faden, & D. Federman (Eds.), *Women and health research. Vol. II. Ethical and legal issues relating to the inclusion of women in clinical studies* (pp. 192–215). Washington: Institute of Medicine.

Lounsbury, F. G. (1978). Iroquoian languages. In B. Trigger (Ed.), *Handbook of North American Indians, Vol. 15, Northeast* (pp. 334–343). Washington: Smithsonian.

Lwebuga-Mukasa, J. S., & Dunn-Georgiou, E. (2000). The prevalence of asthma in children of elementary school age in western New York. *Journal of Urban Health, 77*, 745–761.

Macaulay, A. C., Hanusaik, N., & Beauvais, J. E. (1989). Breastfeeding in the Mohawk community of Kahnawake. *Canadian Journal of Public Health, 80*(3), 177–181.

Macaulay, A. C., Montour, L. T., & Adelson, N. (1988). Prevalence of diabetic and atherosclerotic complications among Mohawk Indians of Kahnawake, PQ. *Canadian Medical Association Journal, 139*, 221–224.

Macaulay, A. C., Paradis, G., Potvin, L., Cross, E. J., Saad-Haddad, C., McComber, A., et al. (1997). The Kahnawake schools diabetes prevention project: Intervention, evaluation, and baseline results of a diabetes primary prevention program with a native community in Canada. *Preventive Medicine, 26*, 779–790.

Mahoney, M. C., Ellrott, M. A., & Michalek, A. M. (1989). A mortality analysis of Native Americans in New York State, 1980–1986. *International Journal of Epidemiology, 18*, 403–412.

Mahoney, M. C., Michalek, A. M., Cummings, K. M., Nasca, P. C., & Emrich, L. J. (1989). Mortality in a northeastern Native American cohort, 1955–1984. *Amerian Journal of Epidemiology, 129*, 816–826.

Michalek, A. M., Mahoney, M. C., Buck, G., & Snyder, R. (1993). Mortality patterns among the youth of a northeastern American Indian cohort. *Public Health Reports, 108*, 403–407.

Montour, L. T., Macaulay, A. C., & Adelson, N. (1989). Diabetes mellitus in Mohawks of Kahnawake, PQ: A clinical and epidemiologic description. *Canadian Medical Association Journal, 141*, 549–552.

Morgan, L. H. (1851). *League of the Ho-dé-no-sau-nee or Iroquois*. Rochester: Sage.

Oneida Indian Nation. (2000). Available online at, http://oneidanation.net/ Oneida, New York.

Parker, A. C. (1909). Secret medicine societies of the Seneca. *American Anthropologist, ns, 11*, 161–185.

Parker, A. C. (1913). The code of Handsome Lake, the Seneca prophet. *New York State Museum Bulletin 163*. Albany: New York State Museum.

Patterson, K. (1994). Tuscarora. In M. B. Davis (Ed.), *Native America in the twentieth century: An encyclopedia* (pp. 663–664). New York: Garland.

Ramenofsky, A. (1987). *Vectors of death: The archaeology of European contact*. Albuquerque: University of New Mexico Press.

Randle, M. C. (1951). Iroquois women, then and now. *Smithsonian Institution. Bureau of American Ethnology Bulletin, 149*, 167–180. Washington: U.S. Government Printing Office.

Shimony, A. A. (1970). Iroquois witchcraft at Six Nations. In D. Walker, (Ed.), *Systems of North American witchcraft and sorcery* (pp. 239–265). Moscow, Idaho: University of Idaho Press.

Shimony, A. A. (1994). *Conservatism among the Iroquois at the Six Nations Reserve*. Syracuse: Syracuse University Press.

Snow, D. R. (1992). Disease and population decline in the northeast. In J. W. Verano & D. H. Ubelaker (Eds.), *Disease and demography in the Americas* (pp. 177–186). Washington: Smithsonian Institution.

Snow, D. R. (1994). *The Iroquois*, Oxford: Blackwell.

Snow, D. R., & Lanphear, K. M. (1988). European contact and Indian depopulation in the northeast: The timing of the first epidemics. *Ethnohistory, 35*, 15–33.

Snow, D. R., & Lanphear, K. M. (1989). "More methodological perspectives": A rejoinder to Dobyns. *Ethnohistory, 36*, 299–304.

Snow, D. R. & Starna, W. A. (1989). Sixteenth-century depopulation: A view from the Mohawk valley. *American Anthropologist, 91*, 142–149.

Speck, F. G. (1949). *Midwinter rites of the Cayuga long house*. Philadelphia: University of Pennsylvania Press.

Starna, W. A. (1980). Mohawk Iroquois populations: A revision. *Ethnohistory, 27*, 371–382.

Starna, W. A. (1994). Onondaga. In M. B. Davis (Ed.), *Native America in the twentieth century: An encyclopedia* (pp. 408–409). New York: Garland.

Sturtevant, W. C. (1978). Oklahoma Seneca-Cayuga. In B. Trigger (Ed.), *Handbook of North American Indians, Vol. 15, Northeast* (pp. 537–543). Washington: Smithsonian.

Sturtevant, W. C. (1984). A structural sketch of Iroquois ritual. In M. K. Foster, J. Campisi, & M. Mithun (Eds.), *Extending the rafters: Interdisciplinary approaches to Iroquoian studies* (pp. 133–152). Albany: State University of New York Press.

Tooker, E. (1978a). Iroquois since 1820. In B. Trigger (Ed.), *Handbook of North American Indians, Vol. 15, Northeast* (pp. 449–465). Washington: Smithsonian.

Tooker, E. (1978b). The league of the Iroquois: Its history, politics, and ritual. In B. Trigger (Ed.), *Handbook of North American Indians, Vol. 15, Northeast* (pp. 418–441). Washington: Smithsonian.

Wallace, A. F. C. (1951). Some psychological determinants of culture change in an Iroquoian community. *Smithsonian Institution. Bureau of American Ethnology Bulletin, 149*, 55–76. Washington, U.S. Government Printing Office.

Wallace, A. F. C. (1952). The modal personality structure of the Tuscarora Indians as revealed by the Rorschach test. *Smithsonian Institution. Bureau of American Ethnology Bulletin, 150*. Washington: U.S. Government Printing Office.

Wallace, A. F. C. (1958). Dreams and the wishes of the soul: A type of psychoanalytic theory among the seventeenth century Iroquois. *American Anthropologist, 60*, 234–248.

Wallace, A. F. C. (1970). *The death and rebirth of the Seneca.* New York: Knopf.

Wallace, A. F. C. (1971). Handsome Lake and the decline of the Iroquois matriarchate. In F. L. K. Hsu (Ed.), *Kinship and culture* (pp. 367–376). Chicago: Aldine

Weaver, S. M. (1971). Smallpox or chickenpox: An Iroquoian community's reaction to crisis. *Ethnohistory, 18*, 361–378.

Weaver, S. M. (1972). Medicine and politics among the Grand River Iroquois: A study of the non-conservatives. *National Museum of Man publications in ethnology 4.* Ottawa: National Museums of Canada.

Wells, R. N. (1994). Mohawk. In M. B. Davis (Ed.), *Native America in the twentieth century: An encyclopedia* (pp. 353–354). New York: Garland.

Wray, C. F., Sempowski, M. L., & Saunders, L. P. (1991). Tram and Cameron: Two early contact era Seneca sites. *Rochester Museum & Science Center research records, no. 21.*

Wray, C. F., Sempowski, M. L., Saunders, L. P., & Cervone, G. C. (1987). The Adams and Culbertson sites. *Rochester Museum & Science Center research records, no. 19.*

Jamaican Maroons

George Brandon

ALTERNATIVE NAMES

Windward Maroons, Leeward Maroons. Also call themselves Nyankimpong Pickibo ("children of the Creator" in Twi).

LOCATION AND LINGUISTIC AFFILIATION

There are two major centers of Maroon life in Jamaica. The Leeward Maroons are centered in the mountainous Cockpit country of the Western half of Jamaica in the parishes of St. James, St. Elizabeth, and Trelawney. The spiritual and physical center of the Maroons in this area is the village of Accompong with significant Maroon populations in Aberdeen, Maroon Town, and Whitehall. On the eastern half of the island, in the Blue Mountains, is the other center of Maroon culture, the village of Moore town. Other Windward Maroon settlements in this area include Scots Hall and Charlestown. Most of the time Maroons speak a Jamaican patois that derives most of its vocabulary from English but often has syntactical and grammatical features more akin to those common in West African languages. Both Maroon groups also possess an archaic language they call Kromanti, an African-based tongue that nowadays has no consistent native speakers but survives in ritual songs and in old folktales. In Accompong there is a small but significant Rastafarian community that has added its own distinctive form of Rasta-talk to an already complex linguistic situation.

OVERVIEW OF THE CULTURE

Maroon identity is based more on history, land and sacred charter, and the primacy of certain social values than on cultural or linguistic distinctiveness. Jamaica's Maroons trace their origins to explorers, livestock managers, and militia brought to Jamaica from Africa and the Iberian peninsula by the Spanish in the 1550s when various European colonizers contended for possession of the island. The Africans had been commissioned by Spain to raise livestock, intercede with the remaining Arawak Indians, and protect the island from other European powers should they attack. When slavery became the operative principle of colonization and the British contested Spanish possession of the island, this growing group—who had already intermingled with the remaining

Arawaks to some extent—refused to align themselves with either the Spanish or the British; instead they grabbed their weapons and took to the hills to forge a new life for themselves on their own terms. It was because of this that they came to be referred to as Maroon, a term deriving from the Spanish *cimmaron*, meaning "wild" or "untamed" and referring especially to domesticated animals that escape and return to the wild.

By no means were the Maroons a homogeneous group. Their numbers included people from various ethnic groups of West and Central Africa, Spanish deserters, runaways who had found their way to Jamaica from Barbados, Africans mixed with Arawak, and refugees from shipwrecked slave ships including a few who were native to Madagascar. Out of this heterogeneous collection of peoples they forged a new group that fanned out across the island to form the Leeward Maroons centered in Accompong in the western mountains of Jamaica's Cockpit country, and the Windward Maroons in the Blue Mountain region on the eastern side of Jamaica. Despite the distance that separates the Leeward and Windward Maroons, the two groups are in contact and have long maintained relations with each other.

Jamaican Maroons possess an abstract conception of history focused on the military and political events of the Maroon Wars and on the 1738 and 1739 treaties that the Leeward and Windward Maroons concluded with the British government. In this conception of history the "first time Maroons"—that is, the original Maroons and those who waged the warfare that led to the treaty—achieve a mythic and heroic status. The "first time Maroons" and the next generation of "Old People" were venerated in rituals after they died and the most important war leaders such as Cudjoe, Nanny, and Accompong were promoted into a spiritual pantheon. Origin myths grew up around them and the oldest Maroon religion was essentially an ancestor cult devoted to the spirits of these people. The treaties ending more than 80 years of intermittent warfare with the British have also achieved a sacred status. The 1738 and 1739 treaties ceded land to each of the Maroon groups as a whole rather than to individual Maroons and the communal ownership of these lands has remained an important anchor of Maroon identity among both Leeward and Windward Maroons up to the present day.

Today Maroons are a part of Jamaica's rural peasantry and share much of its culture. Most rural Jamaican communities have populations of less than 4,000 and Accompong and Mooretown are probably at the midrange with populations of about 2,500 to 3,000 each if you include their environs. Men and women both farm but housework is performed mainly by women and children. The influence of orthodox Protestant religious denominations, various heterodox Afro-Jamaican forms such as Revival and Rastafarianism, and survivals of Myal, an early creolized version of African ancestor veneration, are as evident in Maroon communities as they are throughout the rest of rural Jamaica. Maroon family and household structure exhibit the same extreme variability of form, frequent dispersal of children across households, multiple forms of mating, late marriage, and female-headed households that we find all across the rural areas. Still Maroons' history as rebels against slavery and their self-image as fighters are keys to many of their social and communal values.

Independence, communalism, and self-sufficiency are important Maroon values stemming from their warrior heritage. The peasant's attachment to land has been given a particular valence by the sacred status of the Maroon treaties and because Maroons anchor their conception of history and some important social and cultural values to specific geographical sites. Maroons have a reputation throughout Jamaica as skillful practitioners of the Jamaican magic tradition of Obeah and as makers of powerful traditional medicines. While Maroons are apt to protest to outsiders that they do not practice Obeah, in fact some of them do. No such stigma or denial attaches to Maroon traditional medicine, though. Traditional medicine is now being seen as an anchor of contemporary Maroon identity and non-Maroon entrepreneurs have appropriated this connection between Maroons and associate it with their own commercial preparations to increase customer appeal. A ceremonial and ritual figure unique to the Jamaican Maroons is the Maroon abeng player. The abeng, a side-blown animal horn producing a narrow range of pitches, once communicated messages across the distances separating Maroon groups. When the first time Maroons were alive the abeng conveyed important military information between camps in coded messages that reflected the tonal features of some African languages. Nowadays, the abeng summons the community to town meetings, announces difficulties and deaths, and plays an important symbolic role in funerals and holidays.

CONTEXT OF HEALTH

Maroons are politically, socially, and economically marginalized within Jamaica. The 1738/39 treaties gave the colonial government the right to appoint Maroon chiefs (called "Colonels") and also created a position for "Whites" who fulfilled the twin functions of representing the Maroons to the central government and also representing the interests of the central government to the Maroons. In effect the Colonel became a middleman to a middleman and the colonial government refused to meet with or recognize Maroons. The political situation did not improve substantially after Jamaica gained independence from England in 1962. Both the Peoples' National Party and the Jamaica Labor Party have attempted to cultivate a broad base of support among Jamaican Maroons from time to time because of the party loyalties of their large extended families. Partisan politics, however, has brought few tangible rewards to Maroons and has even fomented serious divisions within the communities.

Because of the 1738/39 treaties Maroons pay no taxes to the national government and so get little in the way of services in return. Their roads are badly maintained—even by Jamaican standards—and the government contributes meagerly to Maroon educational and health services. Maroon involvement in government-sponsored agricultural initiatives has been a roller coaster affair subject to dramatic market fluctuations, expensive chemical inputs to agriculture, and high transportation costs. At the same time as traditional farming methods are waning and cooperative labor arrangements supplanted by wages, Maroons find themselves vulnerable and dependent upon an external economic system over which they have no control.

All in all, an estimated 100,000 people in Jamaica regard themselves as Maroons or Maroon descendants. Not all of them live in the traditional Maroon areas. Many Maroons now live in nearby cities such as Montego Bay or Port Antonio, or they move to Kingston in search of work and shuttle back and forth between urban life and the rural environment of the Maroon settlements. In the late 1950s, just before Jamaica's independence, a significant number of Maroons left for England and a steady stream of migration there has continued ever since. This overseas Maroon population sends remittances home and continues to have rights to land in the settlements. More recently a new stream of migration has begun, this time it is elderly and retired Maroons who are returning from England to the communities in which they were born.

MEDICAL PRACTITIONERS

There are no full-time Maroon healers. Every healer practices their particular form of medicine part-time, alongside farming and sometimes even a third occupation or trade. There are five types of Maroon healers: herbalists, midwives, bonesetters, dancers, and science men.

Herbalists. The training of Maroon herbalists frequently combines apprenticeship and family tradition with self-teaching. Everyone begins learning about the healing properties of plants from an early age but only a few pursue it conscientiously. There is often a religious component in the herbalist's practice and orientation; many of them use the Christian Bible and see no opposition between prayer and healing by faith, and healing through herbal medicine. Some are community workers with high standing and live in the towns; a few spend much of their time in the forest and are marginal to the community but bring in plants that are difficult to find in the settlements or the fields. A few herbalists have established reputations wide enough to bring in clients from the towns and coastal cities or even overseas. In recent years some herbalists have formed groups to share knowledge, promote their craft, and mount small-scale economic ventures.

Midwives. Midwives are trained by older women through observing and assisting them in births. Midwives form a loose network of cooperation and work as teams with shifting membership assisting the delivery of Maroon babies under the direction of an elder woman, the nana.

Bonesetters. Bonesetters train primarily by observation and the empirical experience of seeing bones set. There is sometimes a family tradition in this work but the role can be assumed by anyone with the requisite experience and skill. The bonesetter needs to be able to fashion splints, do some elementary massage, and administer herbs that help reduce swelling and pain and assist healing. Bonesetters usually work as individuals who gradually establish a reputation within the Maroon community.

Dancers. The dancer was a distinctive ritual and religious figure that was last seen in Mooretown in the 1940s. Dancers specialized in ailments caused by spirits and in severe cases that had not been resolved by other medical means. They conducted public and private rituals in which the dancer, assisted by prayers and drumming, went into spirit possession and then healed from the possessed state.

"Science Men". The distinctive characteristics of this kind of Maroon healer relate to their role as seers and diviners, and their use of European magic and occult sciences. Invocations, fashioning charms and talismans, and using oils, candles, and holy water in their rituals are part of their expertise. They may specialize in fortune telling, sexual and love magic, magic related to farming or treating particular health problems with a combination of rituals and herbal medicine. "Science men" or "scientists" are frequently associated or identified with obeah (the Afro-Jamaican sorcery tradition) and herbalism, and may or may not perform any of the other healer roles already mentioned. They are often ambivalently valued figures inspiring both fear and respect. A few science men occasionally have clients coming in from abroad or from distant regions within Jamaica.

CLASSIFICATION OF ILLNESS, THEORIES OF ILLNESS, AND TREATMENT OF ILLNESS

Maroon Disease Theory: Etiology

Cold and Blood. Cold and blood are important concepts in Maroon explanations of the causes of sickness. "Cold" is a fundamental element of Maroon etiology, an explanation for a general type of disease process with wide ramifications. When Maroons speak of "contracting a cold" or "contracting cold" they are referring not to the sickness itself as we might in speaking about a head cold, but to "cold" as a causal factor underlying the symptoms that have developed (Cohen, 1973, p. 69). According to them, "cold" is responsible for the majority of the sicknesses afflicting Maroons. "Cold" is a range on a continuum of judgments that Maroons use in describing bodily sensations related to temperature. While this scheme of explanation tends to focus on the effect of cold, the other extreme is not neglected and a few ailments are traced to the effect of too much heat, the most prominent of these being "belly hot."

"Cold" is a pervasive generalized sensory quality or force that emanates from certain objects or places in the environment. The earth or ground is a source of cold; so is immersion in cold water or getting soaked in the torrential downpours that mark the lives of both Leeward and Windward Maroons day after day for weeks at a time. But encounters with these cold-emanating elements of the environment are common everyday occurrences.

Under certain conditions, however, contact with cold can adversely affect people's health. This contact is even more dangerous if cold is actually able to enter the body and reside there. Cold permeates the whole body once it has made its way inside.

Blood is the most important part of the body, its vital force. Maroons evaluate and describe blood by referring to its ability to flow, its purity, and its strength. Break down and impurity of blood cause the organs to malfunction and fail. For Maroons a warm body is a healthy body. It is the blood and its circulation through the body that keep the body warm. When cold intrudes itself into the body the thickness and stubbornness of cold mix with the blood, rendering it stagnant, obstructing perspiration (itself a cleansing process). If the blood is pure, strong, and flowing it will force the cold back out of the body otherwise the cold mixes with blood and causes it to "sleep" and not flow, and the body becomes vulnerable to disease.

What causes and facilitates the intrusion of cold into the body is the sudden or rapid alternation of extreme states of heat and cold. This is especially true if the rapid/sudden alternation of heat and cold is repetitive and reoccurs over a few months or goes on over years. Cold accumulates in the body because of poor defenses (i.e., impure, stagnant, or weak blood). Cold becomes "rooted" in the body as subsequent intrusions occur, only emerging as an identifiable disease years later. The accumulation and repeated intrusions of cold trigger a gradual development from transient discomfort to chronic disease and pain (arthritis, asthma) or even permanent disability (blindness, deafness) (Cohen, 1973, p. 75.) Some health problems ultimately traced to cold intrusion include: loss of appetite, malaise, asthma, pneumonia, diarrhea, boils, deafness, blindness, arthritis, catarrh, some fevers, blockage of the urine, sore throat, tuberculosis, and earaches.

Germs. Maroons attribute some diseases to entities they call "germs." Germs are little organisms living in the ground (Some people say that they can see them). Germs breed in dust and in dirt and it is from there that they venture out to attack human beings. Once they attack, germs survive by "feeding on the flesh" of their victims. "Cold" in the body provides germs with an environment where they can thrive, but germs will not abide long in "heat" (Cohen, 1973, pp. 79–80). While it can be argued that their conception is an inadequately understood mixture of the antique and the contemporary, nevertheless it does overlap with biomedical understandings in so far as

Maroons think of germs as very small living things capable of causing disease.

Germs provide Maroon etiology with an agent to use in explanations of disease causation that draw on the idea of contagion. This contagion can be between one human being and another, and also between human and nonhuman life forms as in the case of flies and mosquitoes. According to them, germs may be transmitted by food. (The necessity of boiling pork before cooking it any further comes from the need to kill germs as much as anything else.) Coughing in someone's face, putting one's foot in another's boot, and drinking from a sick person's glass are other means of transmitting germs and sickness. Malaria, gonorrhea, "night fever," urinary blockage, tooth decay, grunitch, ringworm, other body itches, and sores are all problems Maroons attribute to germs. Urinary blockage comes from drinking unclear water, that is water contaminated with dirt and therefore, with germs.

Maroons have no direct action they can take against germs and the preventive strategies they employ against germs are mainly prompted by the idea of contagion between human beings. They do not use their germ ideas to explain the upper respiratory infections that are so common among them and which the visiting medical teams treat with antibiotics. Although Maroons appreciate the efficacy of the antibiotics they get from physicians, they do not understand the basis of that efficacy, and they do not relate it to their own indigenous germ theory.

Things People Do or Do Not Do. Maroon's etiological system also implicates behaviors related to food and physical exertion as causes of sickness.

Diet. Mooretown Maroons trace problems such as constipation, diarrhea, stomach pains, chest and gas pains, nausea, vomiting, and high blood pressure (and the headaches and nose bleeds they connect with it) back to an imbalance between "hard" and "soft" foods in the diet. Mooretown Maroons consider yellow and Negro yams to be "hard" foods. (Pumpkin, dasheen, and coco are hard foods, too.) Other yams, though, such as renta and St. Vincent yams, are "soft," as are boiled banana, boiled rice, chocho, and mango (Cohen, 1973, p. 80). The amount of meat a person consumes is also considered important. Too much beef, mutton, or "rich" food injures the body.

There is a small minority of Maroons in Accompong who are Rastafarians and follow that religion's dietary practices. For them even following the dietary strictures noted above does not ensure good health, either physically or spiritually. The most orthodox of the Rastas are vegetarians, eschewing all meat especially pork. For the most part they either grow their own food or trade for it with other Rastas, and promote a doctrine of "ital" or "natural" eating that bans salt, sugar, alcohol, and the store bought foods that so often contain these substances. Their views have begun to spread but they still constitute a marginal minority group among Maroons. One point where the Rastas and more general Accompong Maroon ideas converge, however, is on the issue of pesticides and chemical fertilizers. Quite a few Accompong Maroons oppose the use of chemical fertilizers and pesticides for growing food crops. They would not use them and say that eating foods grown this way affects people's health and causes disease.

Work, Exercise, and Physical Exertion. We noted the importance Maroons attach to blood and blood circulation when we described their ideas about what happens when "cold" enters and lodges itself within the body. Contrary to the stagnating and contaminating effects that cold has on the blood; work, exercise, and physical exertion cause blood to flow freely hence directly and positively affecting health. Physical work also cleanses the body of internal poisons that adversely affect blood quality. The poisons leave the body through perspiration. Maroons view perspiration itself as a cleansing process.

Hard work has a social value for Maroons and is an important component of how men assess each other as farmers. They also believe that hard work has health consequences. When people are too sedentary, do not work enough, and do not get enough exercise, sometimes they get sick. On the other hand, a person can exert themselves too much or inappropriately and end up with health problems too, especially if they fall or get hit by something. Excessive or inappropriate exertion underlies back pain and pains in the heart, biliousness, stiff neck, cataracts and eye inflammation, hernias, and some instances of urinary bleeding (i.e., those not caused by falls or gonorrhea). Physical exertion also includes sexual intercourse. "Running around with too many women" leads to impotence. Sometimes Maroons use exercise, particularly walking, as a form of self-treatment as when Accompong Maroons with chest colds attempt to treat the cold by "walking it out."

Emotional Reactions. Although emotions are a minor component of their disease theory system, Maroons do believe that emotional reactions can cause health problems. Too much excitement, for example, or a startling sudden fright can cause pains in the heart (Cohen, 1973, p. 82). If a pregnant woman has a strong emotional response to a particular person or animal, the attributes that ignited the pregnant mother's emotional reaction—(a man's deformity or hirsuteness, an animal's crippled limb)—get transferred to her unborn child. Later on, though it may stop short of tears, a mother's disappointment or anger at some aspect of her child's behavior can seek vengeance and harm the child on its own without her conscious intention. In the last two instances the mother–child bond provides the social and biological context in which the mother's emotional reaction creates an unusual relationship between the child and an external event.

Spiritual Intervention: Duppies, the Devil, and God. Religious ideology thoroughly permeates the lives of rural Jamaicans and spirit concepts supply a pervasive idiom for explaining misfortune. Although Maroon etiology charts the actions of duppies, the Devil, and God as they relate to the domain of human sickness; the full domain of spiritual interventions in Maroon life is much wider and the true domain being dealt with is not disease and/or sickness but misfortune: road and farming accidents, problems with neighbors, marital tensions, sexual difficulties, economic woes, and personal anxieties. The actions of spirits may underlie all of these problems.

Duppies are ancestral spirits, also called "shadows," "shades," and "wandering spirits." The duppy differs from the soul in that it remains in the grave after death; the soul does not. With the proper rites the shadow reaches the land of the dead where the fact that it has made the transition from being a live person to being a dead one is confirmed. Without the proper funeral rites the "shadow" becomes a duppy, a dispossessed and dissatisfied ghost that brings misfortune and terrorizes the community. When a duppy's attack comes through illness, it is called a spirit illness. The Devil also afflicts humans with sickness and, though the range of sicknesses is no different from those associated with duppies, the Devil is more selective in his choice of victims. These are primarily people who are sinful as opposed to those who are righteous and of strong Christian faith. God can also visit sickness upon a person as a punishment for wrong doing, for someone who has malevolently, intentionally, and knowingly harmed other people.

Spirits can cause any sickness. Whether the sickness results from spiritual intervention is not something Maroons can recognize by the fact that the sickness has occurred or by the symptoms it presents. In addition, using spiritual intervention to explain the cause of a particular disease does not automatically cancel out other possible explanations for it. Although Maroons may attribute "madness," "fits," leprosy, and stroke to duppies and spiritual interventions, they will acknowledge other causes as well. If the sickness or misfortune has struck suddenly and for no apparent reason, Maroons may suspect that spirits have been involved.

Some Maroons have an idea of how the sicknesses they encounter are supposed to progress and this colors how they explain them. Often this concept of disease progression resembles an S-curve with a gradual mounting of disease symptoms and discomfort that reaches a peak of maximum intensity and is then followed by a less gradual falloff. When the sickness does not follow the progression healers and lay people recognize as normal for that ailment, they may be forced to consider whether there has been some intervention by spirits. In other cases, they are forced to resort to spiritual intervention as an explanation because they have exhausted all the other possibilities their etiological system offers and because the sickness has proved refractory to both folk and biomedical treatments.

Treatments

All of the healer types make use of a wide variety of locally available plants in the forms of infusions of leaf teas, decoctions made from the roots of plants to create "roots tonics" or "bitters," and tinctures using white rum, wine or some other alcohol as a solvent medium. The Maroon herbal pharmacy supplies them vermifuges, cathartics, sedatives, diuretics, emmenagogues, antihypertensives, and medicines serving many other uses. The uses often overlap and are emphasized variously in the practices of the different healer types. For example, many of the same herbs that are used to modulate menstruation, ease, and speed up labor in childbirth, and perform purifying "washouts" of the body also induce abortion, so midwifery requires a sophisticated knowledge of these herbs and their dosage effects. "Bush baths" are another means of using herbs medicinally. The healer ties up bundles of leaves and dips them in a tub or small basin of hot bath water to steep. The patient either submerges himself in the water or sits over the water's steam. Treatment with poultices involves crushing the plants into a pulpy mucilaginous mass, which is then positioned over and wrapped around the affected area. Poultices can be made from either fresh or dried herbs. If dried herbs are being used a common practice is to mix them with cornmeal or some other starchy substance to bind the dry leaves together and wet them in an effective way. Another use of medicinal plants in healing involves striking or lightly hitting the patient's body with the plants themselves.

Religious ritual is an integral part of the recognition and treatment of some medical problems. Some Maroon healers claim a kind of psychic ability that allows them to diagnose a patient's ailments with no information from them. They may do this through divination with devices such as a chunk of crystal or the Bible, or simply by looking at the patient. Bush baths are thought to have both medicinal and spiritual properties and may be accompanied by songs. Striking the patient's body with healing plants is regarded as a cleansing act and may be accompanied by songs and associated with other rituals. And the midwife augments her knowledge of herbal medicine not only with skills in physically manipulating the fetus and the pregnant woman's body, but also with the ability to perform the ritual procedures needed after the birth.

All three spiritual causes of sickness and misfortune—duppies, the Devil, and God—are resorts when other explanations within the etiological system fail, or the sickness is anomalous, and common therapies do not suffice. As Maroons' ability to predict, explain, and treat specific sicknesses diminishes, spirits' explanatory role increases. Furthermore, as the spiritual actor and the mechanisms involved become more remote from the human and less alterable by human action, they become more and more powerful. There are defenses against sickness-causing duppies (among them obeahmen, science men, and good duppies). If God did not defeat the Devil in the battle for the sinner's heart, in the Mooretown of olden days at least, it was possible to call on the dancer, a healer who could succeed where ordinary healers could not and was able with his spirits' aid to stand and confront the Devil himself. But before the wrath of God even extraordinary healers are helpless. Any sickness that God causes can only be healed by God. The utter helplessness of humans before God testifies to the ultimate power of God over all. This kind of sickness brings Maroons into direct confrontation with that ultimate power and imparts the knowledge that in this situation they have reached the absolute limit of their human and spiritual capabilities.

The national government has provided biomedical training to a small number of Maroons as health aides and these health aides render first aid for minor problems to people in the community and assist visiting nurses when they come through. Church-sponsored medical teams including physicians are an infrequent presence in the Maroon settlements and may only appear for a few days two or three times a year. Sometimes antibiotics obtained through a doctor's prescription become items of trade, are sold for a few dollars, and get into the hands of people who use them inappropriately and ineffectively. Given the sporadic availability of biomedical care, most Maroons rely on self-treatment with over-the-counter items or the teas they know as a first resort. Vicks Vapo-Rub, the bathing soap sometimes incorporated into poultices, kerosene, and headache and cold remedies such as Comtrex and aspirin are all commercial items available locally and used to self-treat health problems. White rum also figures prominently in self-treatment and in the materia medica of herbalists.

SEXUALITY AND REPRODUCTION

Pollution, Purity, and Sexual Experience

In rural Jamaica generally sexuality is closely connected with ideas of pollution, waste, and impurity. The male's semen is viewed as a kind of waste product that women's bodies are only able to tolerate and fully absorb when it fertilizes an egg. Otherwise women's absorption of semen fosters decay, pressure and sickness, things they avoid only because of menstruation, a process that cleanses the body of semen, excess blood and a host of other impurities. Men's bodies do not accumulate impurities as rapidly as women's do; and men also tend to cleanse themselves more rapidly of these impurities through sweating. Maroons do not consider sexual intercourse to be either good or bad in itself. Instead they consider it to be inevitable, but also healthy, for most people. It is denial of the sexual impulse and of sexual gratification that they regard as unhealthy, as it leads, in their thought, to mental instability and "madness" in women and back problems and accumulation of toxic substances in men due to their failure to ejaculate.

Sexual experiences generally begin at mid-adolescence so that 14- and 15-year-olds are often sexually active. Both females and males are likely to have a succession of sexual partners. But males are more likely to have several sexual partners at once. Women are more likely to have a series of short-lived relationships before settling into a more steady sexually exclusive relationship with one male; over her life there may be a series of these longer lasting unions, which may or may not eventuate in legal marriage. This mating pattern is reflected, albeit negatively in certain ideas about the health effects of multiple mating and multiple sexual partners. These operate differently for females and males. Each man's semen is supposed to be unique, differing from that of all other males. If a woman has more than one sexual partner at a time their semen will mix inside her body after she absorbs them. Mixing different men's semen is both taxing and dangerous to the woman's body and accelerates the process of decay and sickness we described earlier. This belief tends to decrease women's attempts to obtain sex outside of a current relationship. The corresponding belief for men is one that regards excessive intercourse or having too many women as a cause of impotence or erectile dysfunction. In this case men may seek out any of a number of roots tonics that claim to deal with these problems (Cohen, 1973; Sobo 1993, p. 221).

Having a baby automatically sweeps out the impurities in a woman's reproductive system. The new mother becomes clean while the childless woman's level of purity ebbs and flows with her menstrual cycle. Post-menopausal women, however, escape this purity bind altogether. After menopause their wombs are thought to "close up," and a man's discharge cannot lodge within them. They neither generate impurities relevant to sexuality nor can they absorb them from men; nothing accumulates within them. The woman who is childless is pitied rather than stigmatized or derided. In part this is because the cause of childlessness is not thought to lie within her but with her mate or their relationship. Mooretown Maroon males believe men are the most important and powerful element in reproduction; women are just containers. If a couple is infertile, then, the blame falls on the man because his "seed" (semen) must have become sick or weak (Cohen, 1973). Maroons also believe that disharmony between sexual partners may cause infertility.

Contraception and Abortion

There seems to be a high degree of ambivalence about both contraception and abortion among Maroons. Reproduction forges links between males and females that imply a regime of reciprocity and kin connections which—fragile, conflicted, and unreliable as they often are—still have to be dealt with and often channel decisions around contraceptive practices and abortion. According to some Maroon women indigenous birth control practices have been lost in Accompong, leaving no local alternatives for contraception. Couples in Accompong have become almost totally dependent on outside sources and non-indigenous methods for birth control. Indeed contraception goes against the implicit ideology surrounding both sexual intercourse and social intercourse. Contraception places the power of continuity within the hands of women as opposed to men. Contraception also removes the woman from the round of reciprocity with the network of the partner. On the other hand contraception also allows people to limit the size of their families and strike the balance between what they can invest in the care and rearing of children and what the children can return to the parents in terms of love, care, and labor.

A survey of clinic records of women utilizing family planning in Accompong gives a tentative picture of contraceptive use that differs from the conclusions of the National Survey for the region. Depo-Provera injection is clearly the most common contraceptive method used by women in Accompong. Younger women favor condoms. Condom use peaks between ages 18 and 24 and drops dramatically after 30 years of age. Intrauterine device use is uncommon and is only found in women under the age of 30.

Just under 23% of the women in the Accompong clinic sample had had one or more abortions. Undoubtedly this figure understates the abortion rate for women because it only accounts for the surgical abortions of women referred by clinic physicians. While contraceptive knowledge has been lost in some Maroon communities, knowledge of herbal abortifacients has not. Furthermore, as we noted earlier, herbal abortifacients form part of a continuum of treatments aimed at purifying the blood, regulating menstruation, and managing childbirth. Early abortions, whether spontaneous or accidentally or deliberately induced by natural medicines, might not even be classified as abortions by those experiencing them, nor reported as such to anyone at all.

The abortion rate reflected in clinic records appears to be related to the length of time a woman has been involved in family planning on one hand, and the number of times she has been pregnant on the other. The number of abortions per woman decreases the longer they have been on birth control and is highest for women who have used contraception for a year and a half or less. Elisa Sobo's rural Jamaican informants supported a suggestion put forward by Brody that Jamaican women's use of abortion increased with successive pregnancies (Brody, 1981, p. 51). Even my very limited quantitative data from Accompong also tend toward this conclusion and implies that Maroons share this correlation of abortion and pregnancy with the wider Jamaican population.

HEALTH THROUGH THE LIFE CYCLE

Pregnancy and Birth

The traditional Maroon birth process involved a team of women headed by an elder midwife called a nana. There are few nanas today. Contemporary Maroon women frequently voice the need to train midwives and revive the practice of midwifery but the remaining midwives are not called on very often except in emergencies. Nowadays most women trek considerable distances to hospitals in neighboring towns to give birth. First I will describe what was the traditional practice until fairly recently.

The mother in labor is taken to a birthing room or some other area where she is prepared to give birth. Typically the nana administers the woman a variety of teas specialized for different purposes: teas to cleanse the bowels; a tea made from fresh cut cerassee to cleanse her uterus and womb. Other teas will have already been given to her—in advance of the labor pains—to prepare her to dilate. The nana "bands" or places a cloth around the mother's stomach, anoints it with leaf herbs and oil, and massages her throughout the delivery. If the nana comes to believe that the labor is taking too long she will give the mother castor oil, penny royal tea, or a tea made from piabah to speed it up (Crellin et al., 1998, pp. 41–42, 63, 65).

After the baby is born, the nana and birth team have to do a number of things to protect the baby from sickness, and establish the baby's relationship to the land where it will reside, as well as its relationship to the community of ancestral dead. They lightly wash the newborn child with white rum and massage the bottoms of the baby's feet, its head, and navel with this liquor to ward off cold. Newborns attract spirits, so the nana uses bluing derived from plant dyes to mark an "X" on the baby's forehead and also rubs some onto the infant's eyebrows to prevent spirits from troubling or harming it. Putting asafetida on the baby's navel discourages playful spirits from even coming around.

The nana and the birthing team burn the placenta under a tree that is dedicated to the baby and bury it there in a brief ceremony that includes making a nonalcoholic libation to the ancestors and reading psalms from the Bible. The baby's navel they tie off with a cotton string and burn it and bury it at the same spot as the afterbirth or, failing this, under another tree dedicated to, and thereafter symbolizing, the baby's link to its ancestors. After the delivery the nana and birthing team also give the new mother a special bush bath. (Some midwives have the mother drink some of this bath mixture before they throw it out.) Mother and baby then spend the next 9 days together in the birthing room where the women will look after them during the subsequent days of "special care and welcoming."

Infancy

Depending on the number of other children a mother has and the nature of her other responsibilities in the household, babies usually feed at the breast until their first teeth appear. Sometimes women use a commercially prepared formula along with breast-milk. Goat milk is available if there is a problem with feeding but it is used only in emergencies. When infants cease breast-feeding they are weaned onto soft foods, often a preparation composed of seasonal fruits and vegetables in a pulverized form. Mothers may also spoon-feed infants a thick porridge, fruit juice, and mashed fruits. It is not unusual to see a 3-year-old child receiving porridge from a bottle. Infants' diets may lack variety but malnutrition is uncommon. Sometimes, though, there are problems: When infants' teeth come in: their gums hurt and they may develop diarrhea. A Maroon remedy for the gum pain is to rub the gums with a young tomato. The diarrhea can be helped with a "rehydration salt" of lime juice, sugar, and soda given by the spoonful.

A not uncommon malady of infants is the "mole cold." According to Maroons, this infantile malady develops either because the mother takes her infant outside while the rain is falling and the baby's fontanel becomes wet; or because the mother has not done a good job of drying the baby's scalp after bathing it. Mole colds are fateful for later life. If the mole cold takes root, Maroons predict the child will become a sickly adult (Cohen, 1973, p. 75).

Childhood and Adolescence

Mothers or grandmothers care for small children by using home remedies and herbs. Within the household there is a value placed upon listening to parents and elders. While a small percentage of parents beat "bad" children, parental discipline is usually light, consisting of a spank or yelling as well as reasoning. Children assist in household labor and, when they get older and strong enough, they help with farm work in the fields. Children are often dispersed or moved between households. These shifts in residence and between caretakers sometimes make a dramatic difference in the health status of infants and young children.

Children and adolescents quite early take on the dietary patterns of adults: yams, rice and kidney beans, breadfruit, dumplings, bananas, and plantain predominate at meals. Pineapple and papaya, oranges, and apples, neesberry, and ackee go in an out of season. Salt-preserved codfish is a mainstay because fresh fish is expensive and has to be brought in. Maroon diet makes protein deficiency a real problem for adolescents and both young and mature adults. Beef and pork rarely appear at meals and the diet is high in sugar, starches, and salt.

Immunization for common infectious diseases affecting children depends upon the visits of medical teams from non-governmental organizations or the existence of a clinic. Often sponsored by religious groups which dispatch the teams 2 or 3 times a year, the medical teams are conscientious in giving, recording, and maintaining a program of childhood immunizations, and doubtless the disease picture for early childhood in Maroon communities would be very different without them. But immunizations do not take care of everything and there are number of very common health issues for Maroon children.

The most prominent children's health problems include worms (round worms, pinworms, tapeworms, long worms, etc.), upper respiratory infections, sores, asthma, rashes, "runny belly," scabies, eczema, "loose bowels," ear infections, iron and zinc deficiencies (often a result of worm infestation), and pica. Accompong Maroons associate children's worm infestations with excessive sugar consumption.

A significant part of Maroon childhood is spent in school. These are usually Protestant mission schools that emphasize Christian religious teachings while also giving instruction in secular subjects, mete out harsh discipline, and have corporal punishment as a standard practice. The poorer Maroon children may only wear their shoes—if they possess them at all—to school or church and go barefoot the rest of the time, a situation that predisposes them to pedal fungal infections. About a quarter of the students drop out at the primary school level, especially boys. Whether they have remained in school or not, by age 12 Maroon children are often able to take care of simple medical problems on their own using traditional medicines, especially teas, and fresh picked leaves.

Adulthood

The two largest age groups in Mooretown and Accompong are the youth and elders. Adults and young adults, particularly the males, are highly mobile and are often away from the villages for long periods. Working away from home forces them to leave their children in the care of grandparents or other relatives. Of the middle aged as many have migrated as have remained in the countryside, gone to England or the coastal cities. Hypertension, back pain, insomnia, arthritis, and epigastric pain bedevil the adult men, as do the injuries and wounds that result from farm work. The health problems of Maroon women are very similar to the males' except that the women are more likely to suffer from headaches and psychiatric disorders.

Adults use both traditional medicines and the non-prescription commercial medications available at small local stores for self-treatment. Maroons treat themselves for common mild problems or may seek assistance from another adult or a relative. Mild problems that linger may occasion a trip to a town where there is a biomedical practitioner or an older mature Maroon woman who has had much experience with diagnosing or treating similar problems. Severe, painful, or disturbing problems of long duration call for either a biomedical doctor or a more specialized Maroon healer. Sometimes a failure of biomedical diagnosis and treatment confirms the patient's own self-diagnosis and directs them back to the Maroon folk system where herbal therapies predominate, sometimes supplemented by divination and other rituals.

Menopause is dealt with through dietary prescriptions (such as increase in eating green, leafy vegetables) and a variety of herbs that are said to reduce the symptoms and strengthen the body. Domestic violence (men beating women) flares up because of rumors of female infidelity or men finding out that their women have been with other men. If attacked by men, women are encouraged to fight back. Men's drunkenness is an important cause of arguments and violence between men and women. Alcohol use also figures prominently in fights between men. Knives and machetes are the weapons of choice but injuries from using these weapons in fights are much less common than injuries from using the same tools in farm labor. No one I talked to could recall a local incident of murder or rape.

The Aged

Prominent health problems of Maroon elders include chronic pain (hands, feet, stomach, and "all over"), high blood pressure, arthritis, stomach problems, nerves, weakness, and fatigue, also diabetes. Quite a few elderly Maroons continue doing light farm work into their sixties and seventies. Subsistence requires it. With the old cooperative work, labor exchange, and barter arrangements long laid low, elders must either farm or hire wage workers—not a viable option for most elderly Maroons. Instead the elders depend on their adult children who have remained in the Maroon communities to assist and look after them. Often, however, their children have left and gone to the cities or overseas. These children may send money back to elders but they are not there to look after them. In these cases other relatives take care of them.

Death and Burial

Although women usually outlive men, Maroons generally die in their 60s and 70s. The abeng player is the first to be notified after a Maroon dies; his job is to blow the abeng to announce the death of a Maroon to all the Maroon community. Nowadays the body of the deceased is given over to a mortician in a nearby city for 2 or 3 weeks for embalming and to allow relatives across Jamaica or from overseas time to get to the Maroon community for the funeral. In Accompong they used to preserve the body within the village for 3 or 4 days before the funeral. This was done by sinking a zinc lined shaft into the ground, putting the body in it and keeping the shaft continuously filled with ice while draining off the water from the melting ice. The body orifices of the deceased (their ears, anus, nose, and eyes) would be stuffed with native coffee and cotton before placing it in the shaft.

Maroons have always buried their dead in coffins. People used to pitch in spontaneously to dig the grave but now at least some of the diggers have to be paid. Anyone can dig a grave and there is no stigma attached to being a gravedigger. At the graveside libations of white rum bring the ancestors near and give energy to the gravediggers who sometimes sing Kromanti songs while they work. It takes several days to dig the grave and construct the concrete burial vault. The deceased's family supplies the diggers with food at the gravesite and hosts a wake the night before the funeral.

After the abeng player has announced the arrival of the body, the funeral begins at a Christian church with a service usually called a "Thanksgiving." After the service the abeng player leads a procession from the church through the town to the cemetery blowing the abeng to make continuing musical announcement of the key events of the burial up to its final moments. The process of interment in the burial vault is accompanied by the blowing of the abeng and by hymn singing at the graveside.

This burial sequence is the same for men and women. Children who die after reaching school age receive the same kind of funeral and burial as adults, including the blowing of the abeng. Children who die before school age do not receive a funeral. Their mothers bury them without ceremony in a shallow vaultless grave dug behind the house.

The modern public cemeteries near churches supplement much older burial ground, which may or may not have the concrete slabs or raised headstones that are in common use to mark graves now. Some Maroons bury their dead relatives near their homes as opposed to the communal burial sites. They say they do this simply because they want their dead relatives near them.

REFERENCES

Barker, D., & Spence, B. (1988). Afro-Caribbean agriculture: A Jamaican Maroon community in transition. *Geographical Journal, 154*, 198–208.

Barrett, L. (1976). Healing in a balmyard: The practice of folk healing in Jamaica, W. I. In W. D. Hand (Ed.), *American folk medicine* (pp. 285–300). Berkeley: University of California Press.

Besson, J. (1979). Symbolic aspects of land in the Caribbean: The tenure and transmission of land rights among Caribbean peasantries. In Malcolm Cross & Arnaud Marks (Eds.), *Peasants, plantations and rural communities in the Caribbean* (pp. 86–116). Location/ Guildford Press.

Besson, J. (1997, Fall). Caribbean common tenures and capitalism: The Accompong Maroons of Jamaica. *Plantation Societies in the Americas, 4*(2 & 3), 201–232.

Bilby, K. (1994). Maroon Culture as a distinct variant of Jamaican culture. In Kofi E. Agorsah (Ed.), *Maroon heritage: Archaeological, ethnographic and historical perspectives* (pp. 72–85). Kingston: Canoe Press.

Brody, E. B. (1981). *Sex, contraception and motherhood in Jamaica.* Cambridge: Harvard University Press.

Campbell, C. (1984). Missionaries and Maroons: Conflict and resistance in Accompong, Charles Town and Moore Town (Jamaica) 1837–1838. *Jamaican Historical Review, 14*, 42–58.

Campbell, M. (1988). *The Maroons of Jamaica, 1655–1796: A history of resistance, collaboration and betrayal.* South Hadley: Bergin and Garvey.

Center for Natural and Traditional Medicines & the Accompong Traditional Medicine Group. (n.d.). *Welcome to the World of Maroon Traditional Medicine.* Washington, DC: Center for Natural and Traditional Medicines.

Cohen, M. (1973). Medical beliefs and practices of the Maroons of Mooretown: A study in acculturation. Doctoral dissertation, New York University.

Comitas, L., & Lowenthal, D. (1964). Occupational multiplicity in rural Jamaica. In E. Garfield and E. Friedl (Eds.), *Proceedings of the American Ethnological Society* (pp. 41–50). Seattle: University of Washington Press.

Crellin, J., & the Accompong Traditional Medicine Group (1998). Traditional Maroon Medicine in Jamaica. *Memorial University Occasional Paper in the History of Medicine, No. 15.* St. Johns, Newfoundland: Memorial University.

Kopytoff, B. (1973). The Maroons of Jamaica: An ethnological study of incomplete polities 1655–1905. Doctoral dissertation, University of Pennsylvania.

Kopytoff, B. (1976). Jamaican Maroon political organization: The effects of the treaties. *Social and Economic Studies, 25*(2), 87–105.

Kopytoff, B. (1978). The early political development of Jamaican Maroon societies. *William and Mary Quarterly, 35*, 287–307.

Laguerre, M. (1987). *Afro-Caribbean folk medicine.* South Hadley: Bergin and Garvey.

Lowe, H., Payne-Jackson, A., & Duke, J. (2001). *Jamaica's ethnomedicine: Its potential in the health care system.* Kingston, Jamaica: Pelican Publishers.

Patterson, O. (1970). Slavery and revolt: A socio-historical analysis of the First Maroon War 1655–1740. *Social and Economic Studies, 19*, 289–325.

Price, R. (1979). *Maroon societies: Rebel slave communities in the Americas.* Baltimore: Johns Hopkins University Press.

Robertson, D. (1988). *Jamaican Herbs.* Montego Bay: De Sola Pinto Distributors.

Sheridan, R. (1986). The Maroons of Jamaica, 1730–1830: Livelihood, demography and health. In Gad Heuman (Ed.), *Out of bondage: Runaways, resistance and marronage in Africa and the Americas* (pp. 152–172). New York: Frank Cass Publications.

Sobo, E. (1993). *One blood: The Jamaican body.* Albany: State University of New York Press.

Sobo, E. (1996). Abortion traditions in rural Jamaica. *Social Science and Medicine, 42*, 495–508.

Smith, M. G. (1965). Community organization in rural Jamaica. In. M. G. Smith (Ed.), *The plural society in the West Indies* (pp. 176–195). Berkeley, Los Angeles: University of California Press.

Watts, D. (1987). *The West Indies: Patterns of development, culture and environmental change since 1492.* Cambridge Studies in Historical Geography, Cambridge: Cambridge University Press.

Wendenojo, W. (1989). Mothering and the practice of "balm" in Jamaica. In C. S. McClain (Ed.), *Women as healers: Cross-cultural perspectives* (pp. 76–97.) New Brunswick: Rutgers University Press.

Japanese

Denise Saint Arnault

ALTERNATIVE NAMES

China referred to the Japanese in ancient writings as "Children of *Wa*," using the characters of the sun and the tree in script. This name acknowledged the indigenous mythology of the Japanese as ancestors of the sun goddess, and the cultural focus of harmony. From around AD 500, the Japanese referred to themselves as *Yamato*, which was also the name of their central island, created in myth by the eighth son of the sun goddess; they currently pronounce these characters as *Nippon* or *Nihon* (de Bary & Dykstra, 2001).

LOCATION AND LINGUISTIC AFFILIATION

Japan consists of several thousand islands, however most of the population inhabits four major islands, located between the Pacific Ocean and the Sea of Japan, near Korea to the northeast, China to the east, and Russia to the north. Despite the massive urbanization over the last 50 years, the mountainous topography of the islands has helped the people retain regional characteristics, including dialects in at least 21 regions and regional family names. Japanese administration is by prefecture (*ken* or *to*). Within the 47 prefectures, there are cities (*gun* or *shi*), each of which are subdivided into wards (*ku*), and blocks, or neighborhoods (*chome*).

The linguistic origin of the Japanese language is debated, however the leading theories are either that it is from the Ural-Altaic family (Miller, 1972) or that it is related to Korean (Ohno, 1970). The Japanese written language includes a complex collection of characters derived from Chinese, and two native Japanese syllabary, or *kana*, each of which include 39 syllables. The Japanese use *Hiragana* for Japanese words, and certain grammatical elements. The *Katakana* includes the same sounds of the *hiragana* syllabary, but is used to write foreign loan words and words written for emphasis, such as in advertising. Japanese language is flexible, accepting loan words for new and evolving social ideas. Recent trends in Japan include not only the adoption of loan words, but also the adoption of Roman characters in advertising. Despite the liberal sprinkling of Western concepts and Western text, their meanings and pronunciation retain a distinct Japanese flavor.

OVERVIEW OF THE CULTURE

Demographics

There were 127 million Japanese in Japan in 2000 (World Health Organization, 2002a), making it the eighth largest population in the world (see Table 1). The annual growth rate in 2000 was 0.3%, much lower than the 1.1% in the United States. Twenty-three percent of the Japanese population is over 65, higher than the 16.6% in the United States. The fertility rate is below replacement level in Japan, at 1.4 children per woman, while, in the United States, the fertility rate is 2.0. Seventy-five percent of the Japanese live in urban areas. In the 1930s, the Japanese household was primarily multigenerational, with around 70% of the elderly living with their children. In 1955, about half of Japanese households were still three generations, and by 1990, 60% were nuclear family households and only 15% were three generations (Feeney & Mason, 2001). The Japanese citizenry are educated and literate, with 100% of the population completing primary school since 1955, and 95% completing secondary school during the same period. In 1996, estimates were that 31% of the population was enrolled in tertiary education. However, a gender gap persists in higher education, with 27.2 per 1,000 women in third level education, compared with 35.8 per 1,000 men

Table 1. Population Statistics (World Health Organization, 2002a)

WHO member states	Total population (000) 2000	Annual growth rate (%) 1990–2000	Percentage of population aged 60+ years		Total fertility rate	
			1990	2000	1990	2000
Germany	82,017	0.3	20.4	23.2	1.4	1.3
Japan	127,096	0.3	17.4	23.2	1.6	1.4
Sweden	8,842	0.3	22.8	22.4	2.0	1.4
United Kingdom	59,415	0.3	20.9	20.6	1.8	1.7
United States of America	283,230	1.1	16.6	16.1	2.0	2.0

enrolled at the same level in 1996. In addition, only 22% of third level teachers are women (United Nations, 2000).

History

The Japanese migrated to the islands in AD 470, where they were primarily a tribal people, with heterogeneous, regional, cultural, and social patterns. The Empress Suiko reigned from 592–628, and institutionalized several elements of Chinese civilization into Japanese social and political life, including the adoption of the Chinese writing system, rules of rank and etiquette, the Chinese calendar, diplomatic relations with China, a highway system, and a constitution. The Nara period (709–784) institutionalized Buddhist religion and with it, Chinese images, books, ritual devises, astronomy, and architecture. This period also gave rise to the public bath, cremation, maps, and an irrigation system. The Heian period (8th to 12th century) was a period of cultural evolution, including the development of an indigenous Japanese *kana* writing system, poetry, and scroll paintings. Feudal institutions, a militaristic government, and political power struggles resulted in the social disorder and feudal warfare that characterized the Medieval period (12th to 16th century). However, this period also saw the rise of Zen Buddhism, and the development of an indigenous healing system based on Chinese medicine. The Tokugawa period (1603–1868) marked military reunification, and the development of the *bushido*, or the way of the Samurai warrior. Confucian rules of civil order, including rules of loyalty, social hierarchy, and filial piety, were institutionalized into the society, as well as an aversion to outside cultural influences.

The 700 years of cultural isolation during the Medieval and Tokugawa periods (1185–1853) allowed not only a richly developed cultural identity, but also an intense nationalism. The 1847 Meiji restoration was the official end of this international isolation, as the government centralized, deposing the feudal landholders, as well as modernizing education, the judiciary, the military, and their economic system (de Bary & Dykstra, 2001; Reischauer, 1981; Vogel, 1967; Yoshino, 1992). Japan's late but rapid rise in industry and urbanization has set it apart from other East Asian countries, and its capitalistic economy was based on a Confucian social and political order, creating an industrial organization very different from most other industrialized countries.

Economy and Occupations

Japan had an agricultural economy until the turn of the 20th century, and the percentage of the total labor force in agriculture shifted from 33% in 1960 to 7% in 1990. As of 1990, the percentage of employment for all men and all women over age 15 was 77% and 50%, respectively. A statistical "M" shaped curve characterizes Japanese women's employment since 1965, with the highest rates of employment from 20–24 and 40–49, and lower rates during child-rearing ages (Okunishi, 2001). In 1998, 36% of women were employed part-time, compared with 12% of employed men in the same year (United Nations, 2000). Most women (54%) work in clerical, sales, and service fields, and the wage differential for women was 60% of male wages in 1990 (Bauer, 2001).

Family and Kinship

Japan had a patrilineal kinship system organized around the household, or *ie*. Within this system, the eldest son succeeded the father, and inherited the family property and the responsibilities of the householder. Other sons established stem families (Befu, 1980, 1993; DeNoon, Hudson, McCormack & Morris-Suzuki, 2001). The household was a cooperative unit, and included the parents, eldest son and his wife, and their children, and sometimes workers for the family, as well as adopted sons-in-laws (when there was no oldest son). The *ie* is also a concept, and it refers to the entire lineage of ancestors. Stem families then were essentially equivalent to nuclear families, as they were established outside the *ie* or household structure. These stem families had a much less important role in the economics of the family. Therefore, this kinship pattern allowed for the mobility of nuclear families, which were free to move to the cities with the coming of urbanization and industrialization, perhaps facilitating the rapid trend toward urbanization of Japan after World War II (Vogel, 1967).

Religious Traditions and Philosophical Underpinnings

Shrines, temples, and doorways draped with folded paper punctuate the seemingly endless urban landscape of Japan. These sights belie the sustained importance of Japanese traditional, philosophical, and religious underpinnings. Japan's first religion, Shinto, continues to influence day to day cultural patterns. Based in earlier animistic and shamanistic spiritual traditions, Shinto shrines honor the gods and goddesses of ancient Japanese mythology. Shinto concepts that are important to Japanese daily life include cleanliness of the body and the dwelling structure, symbolic cleansing required before entering Shinto structures, at important seasonal events, and funerals. Ritual devices include salt for purification, rice and *sake* for offerings, and folded paper, bells and rope as communication with the deities.

Buddhism arrived from China in the 5th century. While Shinto was the people's religion, the elite adopted Buddhism. Japanese Buddhism had numerous sects, which rose and fell with the important families that adopted them. Zen Buddhism is especially important in Japanese culture, with an emphasis on personal effort, personal sacrifice, dedication, exertion, and attunement to the body, as well as meditative activities involving daily activities such as tea, flower arrangement, or gardening.

Neo-Confucian philosophical traditions are embedded in religious and secular institutions today, including loyalty and piety in the five essential human relationships, which are father–son, ruler–son, husband–wife, older brother– younger brother, and between friends (de Bary & Dykstra, 2001).

Sociocultural Norms

Research shows that two opposing dynamics are central in Japanese social interactions. One dynamic is the perception of relative intimacy between oneself and another. Shared group membership and a commitment to group harmony and solidarity foster intimacy among group members. The opposing dynamic is the hierarchial organization within any given group. This hierarchy arises from differences in age, social status, and social roles. Appropriate behavioral styles and norms about social exchange in a given situation depend on the accurate perception of relative social distance between any given two participants (*kejime*). Interaction between intimates (*uchi*) includes relatively free expression of emotions and needs (*honne*) and by non-verbal, unrestrained exchange of support (*amae*). Social distance decreases intimacy; prompting people to communicate using polite deference (*enryo*) in order to avoid offending the higher-status person. Strict rules of reciprocity between non-intimates include the edict that each favor incurs a reciprocal exchange. Therefore, the social exchange norm in Japanese culture is to ask for help only within one's intimate social group (Bachnik, 1994; Hendry, 1992; Lebra, 1976; Saint Arnault, 2002). Smooth functioning within the group requires a person to sensitively assess each context and one's role within it. One important task for a Japanese person involves appropriate role behavior. Behaviors that foster conflict or indicate deviance are frowned upon, and may result in ostracism (Bestor, 1996; Johnson, 1995; Smith, 1961).

THE CONTEXT OF HEALTH: ENVIRONMENTAL, ECONOMIC, SOCIAL, AND POLITICAL FACTORS

The Japanese Health Care System

In the 1999 fiscal year, the Japanese health government social security expenditures totaling 75.0 trillion yen (Ministry of Public Management, Home Affairs, Posts, and Telecommunications, 2002). According to the WHO, in 1998, Japan's health care expenditure was 7.5% of its GDP (World Health Organization, 2002a) (see Table 2). Like other nations with national health care coverage, Japan pays over 78% of its health care expenditures with public monies; of these monies, over 89% come from social security revenues. Private expenditures are 77% out-of-pocket, as Japan uses virtually no private insurance.

Japan maintains a westernized, technological health care system, and has as many or more physicians and nurses than the industrialized nations selected for this chapter (see Figure 1). Another interesting difference is the higher number of pharmacists per 100,000 in Japan, which may be related to factors such as the heavy use of pharmaceuticals in Japan, the health care payment structure, governmental support of medical technology research

Table 2. Selected Industrialized Countries' Health Care Expenditures for 1998 (World Health Organization, 2002a)

WHO member state	Percentage of GDP Total	Percentage of total expenditures		Percentage of public expenditures		Percentage of private expenditures	
		Public	Private	Social security	Tax funded	Private insurance	Out-of-pocket
Germany	10.3	75.8	24.2	91.6	8.3	29.5	52.8
Japan	7.5	78.1	21.9	89.2	10.8	1.3	77.8
Sweden	7.9	83.8	16.2	0.0	100	—	100
United Kingdom	6.8	83.3	16.7	11.8	88.2	20.8	66.8
United States of America	12.9	44.8	55.2	33.2	66.8	60.7	28.3

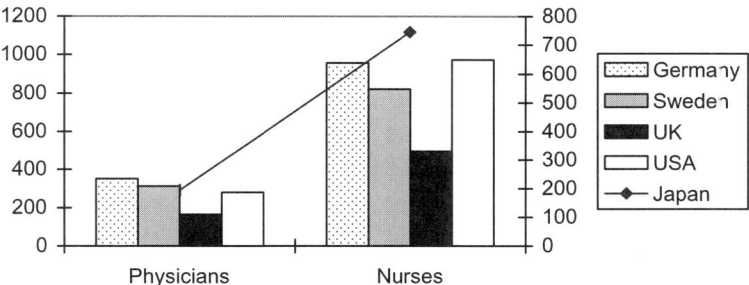

Figure 1. Selected WHO members physicians and nurses per 100,000 around 1988 (Ministry of Health Labour and Welfare, 2002; World Health Organization, 2002a).

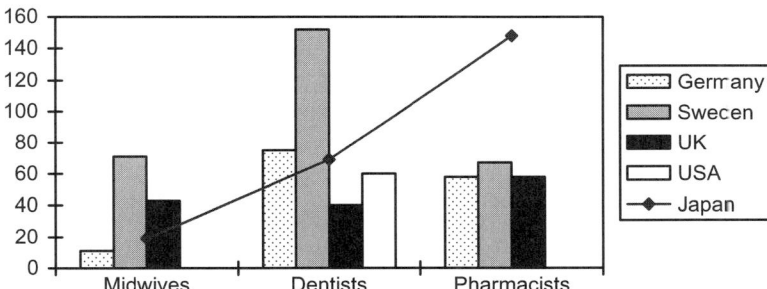

Figure 2. Selected WHO members midwives, dentists and pharmacists per 100,000 around 1988 (Ministry of Health Labour and Welfare, 2002; World Health Organization, 2002a). (Note: USA midwife and pharmacist data unavailable.)

and development and limited cost control monitoring (Steslicke, 1987) (see Figure 2).

Morbidity and Mortality

In 1999, the average life expectancy was 77.10 years for men and 83.99 years for women, the highest level in the world. Infant mortality rate in 1990 fell to 3.4 per 1,000. The three leading causes of mortality in 1999 were malignant neoplasms (cancer), heart disease, and cerebrovascular disease, with 29.6% of all deaths due to cancer (Ministry of Health Labour and Welfare, 2002; World Health Organization, 2002a) (see Figure 3). The rise in cancer, heart disease, and cardiovascular diseases since the 1950 is probably related to an increased exposure to the risks of poor diet, pollution, work related threats, and stress. A comparison of international disease trends reveals that Japan ranks high among these nations in cardiovascular accidents (strokes), digestive diseases, chronic liver diseases, external causes of death, and suicide. However, Japan ranks lower in diseases of the circulatory system and heart attacks (see Figure 4). The symbolic prominence of the stomach and abdomen as the seat of the self for Japanese plays a role in the perception of somatic distress (Ohnuki-Tierney, 1984). This symbolic seat is different from that seen in the West, where heart and the blood are symbolic centers. For example, interview data reveal a concept of "swallowing sadness," noting that concealing the true self for the sake of social harmony may create a symbolic, but somatic, abdominal form. In the West, the cultural prominence of syndromes such as type A personality and hostility may be similarly related to cardiovascular disease and heart attacks.

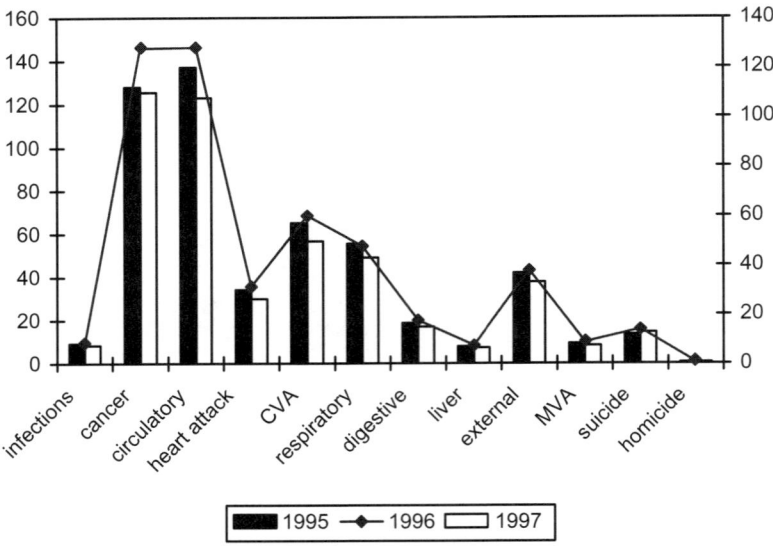

Figure 3. Leading causes of death in Japan 1995–97 (World Health Organization, 2002a).

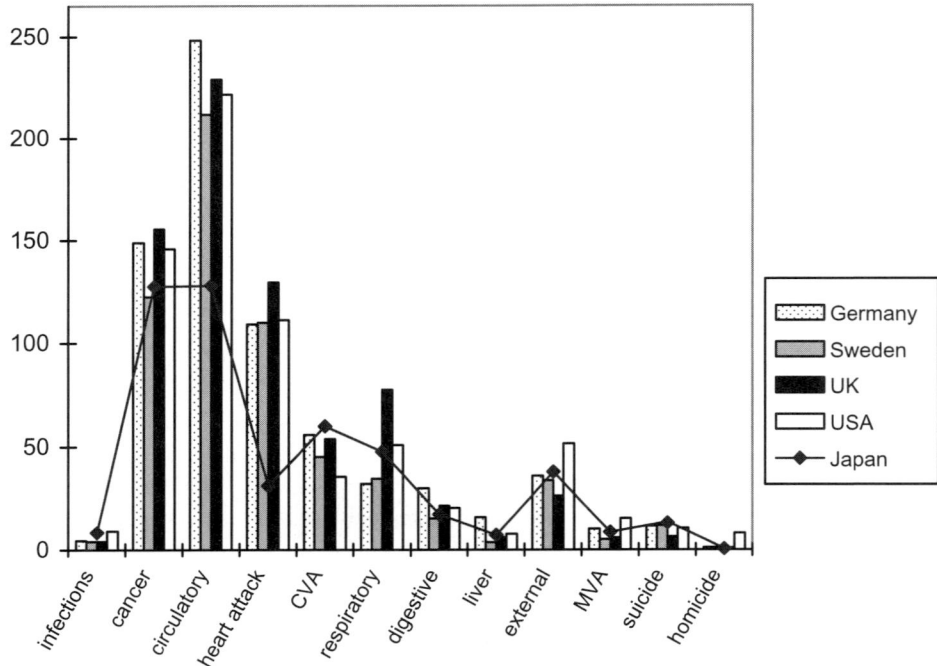

Figure 4. Selected WHO members death rates (per 100,000): 1996 (World Health Organization, 2002a).

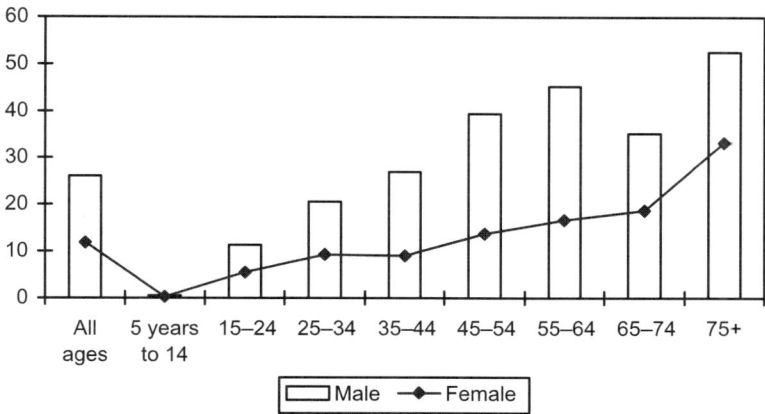

Figure 5. Japanese 1997 suicide rates per 100,000 by sex (World Health Organization, 2002a).

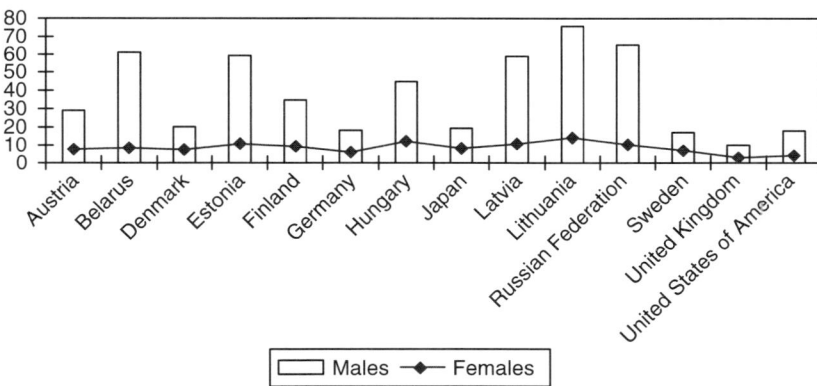

Figure 6. Selected WHO members suicide rates per 100,000 by sex: 1996 (World Health Organization, 2002b).

Mental Illness and Suicide

In 1999, the largest total number of patients in hospitals per 100,000 were tied between mental illnesses and circulatory diseases (259 each) (Health Statistics Division, 1999). However, the Japanese tend not to use outpatient mental health services. Mental illnesses rank second to the lowest number of patients using outpatient services (at 124 per 100,000) compared with digestive disorders (at 1198 per 100,000). Most of the mental hospital beds are occupied by persons diagnosed with schizophrenias (68.6%), then dementia (6.4%), mood disorders (5.5%), and alcohol dependence (5.0%). The incidence of neurosis, depression, psychosomatic disorders, and other mental illnesses can be attributed to urbanization and the related stress (Ministry of Health, 2002). Suicide is a major source of concern for Japanese men (see Figure 5). The rise in the rates for men aged 35–55 are probably related to the economic downturn in the 1990. While Japan's suicide rates for men are higher than some nations, they compare favorably with those of the former Soviet Republic and many Eastern block countries (see Figure 6).

MEDICAL PRACTITIONERS

Japan's primary medical system is western biomedicine. The government sponsored medical scholars to study Dutch medicine and translate German medical texts in the 1700s. In the late 1800s, the Japanese government formalized the adoption of the German medical system of medical training and medical care delivery (Long, 1987).

The second system, *Kampo*, was formalized in Japan during the Heian period (794–1192). Rooted in Traditional Chinese Medicine, *Kampo* is a uniquely Japanese healing system. Despite periods of decline, in the 1950s the government granted *Kampo* official status and governmental regulation. One aspect of the regulation is that practitioners must have a medical degree before becoming a certified *Kampo* doctor (Ohnuki-Tierney, 1984; Rister, 1999). Almost 77% of Japanese physicians have used *Kampo* medicine, and nearly 60% of physicians surveyed considered *Kampo* to be the best choice for certain diseases (Long, 1987; Rister, 1999). Currently, *Kampo* has educational institutions, journals, a scientific presence and development corporations, as well as national insurance coverage for some types of *Kampo* treatment, including *Kampo* hospitals and clinics.

CLASSIFICATION OF ILLNESS, THEORIES OF ILLNESS, AND TREATMENT OF ILLNESS

The primary concept in *Kampo* is balance in the entire physical, mental, social, and emotional state. *Kampo* practitioners gather information about subtle aspects of personal experience, such as cravings, warm or cold sensations, sweating, eye, nose, and mouth states such as dryness, redness or congestion, digestive changes, energy level, quality of sleep and the like. Causes of imbalances include both internal and external pathogens, such as wind, cold, heat, dampness, improper diet, or overwork. The state or condition is then understood according to three sets of complementary sets of indicators—hot: cold, excess: deficiency, and internal: external. Hot and cold refers to the experience of temperature throughout the body, as well as under or over activity of physiological processes. Excesses and deficiencies refer to the constitutional state of the person, including the degree of vitality, resistance to stress and imbalance, and energy level. Finally, internal: external refers to the location of the pathogenic processes, and attunes the practitioners to the "depth" of the problem. Internal problems indicate that pathogenic situation has existed longer and the problem is more entrenched. In addition to these complementary states of balance, *Kampo* is concerned with blood flow, bodily fluids, and vital energy (*qi*). Bodily fluids and *qi* need to circulate freely and can become blocked or stagnant, causing states of illness.

Kampo treatment is indicated when multiple organs are affected, such as in immune disorders or in the failing health of the elderly, and when the patient's condition does not have a ready biomedical diagnosis or effective treatment, such as is the case in psychosomatic illness or idiosyncratic illness expressions. Finally, they are widely used as alternative and complementary treatments to biomedicine (Long, 1987; Ohnuki-Tierney, 1984; Rister, 1999; Tsumura, 1991; Yamamura, 1987).

Sexuality and Reproduction

Japanese women are socially expected to be modest, polite, and deferent to men, consistent with sexual hierarchy and social propriety. The social prescriptions to avoid open display of emotion or private feelings creates a climate that discourages public discussion of sexuality, and schools do not teach family planning or sex education. Despite the ethos of modesty and propriety, sex hotels, prostitution and pornography, including provocative, explicit, and often violent *manga* (comic books) are widely available. As Japanese women have become more assertive about their rights and legal protection, they have complained about sexual advances at work and in public areas, such as trains. During field experiences in the mid 1990s, sexually explicit advertisements were common on trains; in 2002, they were much less present. However, sexually explicit *manga* and advertisements in newspapers are still common. The molestation of women on trains is such a problem that some train lines offer women-only cars during rush hour.

The concept of the "good wife and wise mother" (ryosai kempo) historically saw women as productive members of society, with important roles in the family and the community. In the last 40 years, though, the mothering role has become increasingly important in self-definition, and as a foci of resistance among women (Lock, 1987; Rosenberger, 2001). Women in the last few decades are marrying later and having fewer children. Still, most women do have one or two children, and devote substantial amounts of patience, energy, devotion, and personal sacrifice to them.

Methods of birth control have been primarily male condoms in Japan. The availability of the birth control pill was an issue of rancor and concern for the Japanese people and their government. The birth control pill was finally legal for sale in Japan in 1999, however, a survey of over 1,500 women found that only 4% of Japanese women use

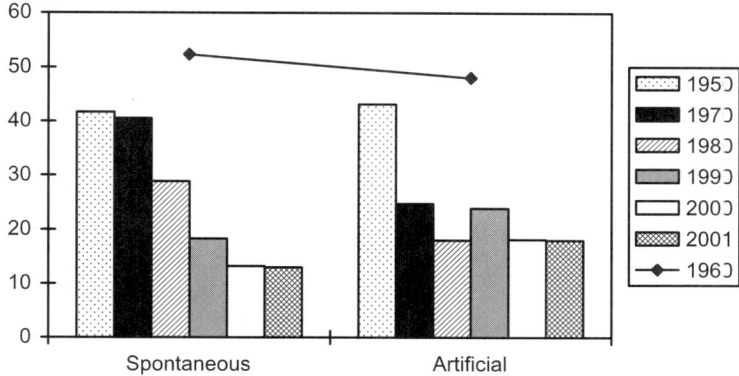

Figure 7. Japan fetal death rates per 100,000 by year and cause (Ministry of Health Labour and Welfare, 2002).

this method, compared with 87% in the United States and 93% in Germany (Pharmacia, 2001). A physician must monitor the use of the pill regularly, and it is not covered by the national health insurance. Other forms of birth control have included infanticide and abortion in historical Japan (Cornell, 1996). Figure 7 depicts the prevalence of abortion in Japan since 1940. The highest periods of the practice were in the 1960s, and the practice has decreased somewhat, but remains a means of birth control in modern Japan. Abortion is not taken lightly, but is considered an extreme and painful necessity to provide for the security and health of surviving members, and is often accompanied by ritual and lifelong grief.

The death of an unborn child, whether spontaneous or artificial, is marked by ritual in contemporary Japanese society (Klass & Heath, 1996–97; Namihara, 1987). The *Mizuko Kuyo* is a Buddhist ritual in which parents express regret and gratitude to their aborted children for their sacrifice for the family. Aborted fetuses are believed to have no connection with the living or ancestors. An unconnected spirit remains bound to the world, wandering about and feeling vengeful. These vengeful spirits can cause illness for the parents and the living children. One aspect of the ritual is to form connections with the spirit of the fetus and the *Jizu*, who is the guardian bodhisattva.

HEALTH THROUGH THE LIFE CYCLE

Pregnancy and Birth

Women tend to give birth around 26–32 years, and 99% of women giving birth in 1992 were married (Miyaji & Lock, 1994). Pregnancy is both a natural act and a technological event in Japan. Pregnancy is considered a natural state in which daily events such as mood changes of the mother, changes in diet, and even the weather have important effects on the health of the child (Ohnuki-Tierney, 1984, 1989). With the advent of modern, westernized medicine, new mothers are encouraged to monitor the pregnancy carefully using the Maternal and Child Health Handbook (*boshitecho*). Women have monthly visits to the physician to monitor the baby's growth, usually with ultrasound. Mothers prefer to either return to their natal home to have the baby delivered, or have their mother or sister come to be with them for the birth and postnatal period. Almost all births happen within a medical facility, and 98% occur under physician supervision, with a full range of medical technology available. The birth tends to be without pain control, and episiotomies are routinely practiced (Miyaji & Lock, 1994; Yeo, Fetters & Maeda, 2000).

Infancy

Mothers tend to stay in the hospital for a week, but may remain confined to the house for as long as a month. About half of all Japanese mothers breast-feed, and women experience a great deal of anxiety about producing enough milk for the infant, fostering the use of therapeutic activities such as breast massage, heat applications, and supportive nursing care. The nursing event is considered a natural, intensely personal time for the mother and infant to connect in a nonverbal, mutually dependent way. Fathers, and Japanese society believe that childcare is within the purview of women, and that the only one who can truly love and nurture an infant is its mother (Miyaji & Lock, 1994).

Maternal–infant bonding in Japan has been the subject of numerous international psychological studies (Shwalb & Shwalb, 1996). Japanese mothers quietly and physically reassure and interact with their baby, anticipating needs rather than waiting for the infant to signal needs. The intense bond formed during this period is part of the interdependent relationship pattern that will continue throughout life. This socialization pattern fosters individuation within an interdependent social frame. Throughout their lives, Japanese individuals enjoy periods of "refueling" into the indulgence of the intimate relationship, referred to as *amae* (Freeman, 1998).

Childhood

As the child enters into toddlerhood, gendered differences in the maternal–child relationship have been noted (Freeman, 1998). Boys are rough, active, and vocal, receiving minimal interference from the mother. When the male child is disciplined, it tends to be with a gentle encouragement to be concerned about his role as the older brother (if appropriate) and to recognize that his behavior reflects on his family. Girls are expected to help the mother at a young age, and are encouraged to recognize their role in the home, and their role as a sweet, loving, polite daughter in public.

Children are rarely cared for outside of the home during their first few years. When mothers return to work, it tends to be after their children enter school. The child's schooldays are similar to that of adolescents, wearing uniforms, and commuting to school by train. Visitors to Japan are often enchanted by groups of well-groomed and well-behaved, eager, and boisterous children on trains and buses. Mothers are expected to prepare balanced, fresh foods for children, be involved in school activities and to devote their evenings to the children's studies.

Adolescence

The Japanese school year begins in April. The average high school day begins around 8:30 am. Since most people in Japan travel by rail, a teenager in Japan may leave home as early as 6:30 am. At the end of the day students are often involved in club activities. Consistent with the tendency for people in Japanese culture to affiliate with one primary group, Japanese students may join only one club. These clubs may be sports or culture clubs. The Confucian hierarchical relationship pattern of seniors and juniors is created and maintained during the club activities. It is the responsibility of the seniors (*senpai*) to teach, initiate, and take care of the juniors (*kohai*). It is the duty of the *kohai* to serve and defer to the *senpai*. Most of the interaction the typical adolescent has is with peers at school, on the long rides to and from school, and through phone and e-mail contact.

Japanese students spend as many as 240 days a year at school. In addition to this long school year, it is highly likely that students attend after school "cram schools" where approximately 60% of Japanese high school students go for supplemental lessons. These schools operate after school and extend into the evening, and require extra homework, leaving the Japanese adolescent little free time during the week. Tuition for these schools is expensive and parents may carefully select and pay for the highest level of schools affordable. Parents and adolescents accept this burden on their time, effort and finances because admission to a good university can be a primary factor in the success of a Japanese adult, determining employment opportunities, and future income. Students who fail an entrance examination may feel as though they have failed their family and experience intense worries about their future, making them a highly vulnerable sector of Japanese youth (Johnson & Johnson, 1996; Rohlen, 1983; White, 1993).

The behavioral problems of Japanese adolescents have risen throughout the 1980s and 1990s at an alarming rate. Many of these problems, including truancy, rebelliousness in school and school refusal, can be seen as social reactions to the incredible stress of life in a conformist society, where academic success is held to be the most important social goal. The school-refusal syndrome is characterized by somatic complaints of headaches and stomachaches, followed by refusal to go to school, moodiness, and sometime verbal and physical abuse of parents. The child may then become listless, depressed, withdrawn, and remain secluded in their rooms. These somatic symptoms, passive resistance, and retreatism are consistent with the Japanese forms of resistance in the face of social pressure which cannot be directly changed (Lock, 1987).

Some of these behavioral changes may also be related to the modern ambivalence toward Japanese cultural rules, and the westernization of Japanese cultural expectations. One example is the problem of bullying (*ijime*). One survey found that 31% of third-graders, 25% of fourth-graders, 13% of fifth-graders, 10% of sixth-graders, 17% of seventh-graders, and 8% of eighth-graders are victims of *ijime* (U.S. Department of Education, 2002). One aspect of *ijime* is that students who participate in it tend not to feel guilty about cruel acts in which they engage. This may be because *ijime* may be a modern distortion of the *senpai–kohai* relationship.

The Aged

The rapidly aging population has created a burden on the health care infrastructure of Japan. Much of the distresses related to the chronic illnesses described above are age related. The most controversial element of the burgeoning heath care needs of the elderly is about who should provide their care. Japan has only recently begun to create an institutional infrastructure to accommodate these needs (Health Statistics Division, 1999). Instead, the Japanese government has called upon a reawakening of the traditional structure of the Japanese family and the Confucian ideal of filial piety among the citizenry. This expectation leaves the burden of long term care of the elderly to the women in Japanese society.

Another facet of the burden of care is the cultural expectation of dependency. Japanese patients, especially the chronically ill or elderly, are expected to remain dependent. Western models of rehabilitation for the purpose of self-care are considered inappropriate. The concept that an elder may become bedridden, and be cared for by family, is alive in the Japanese consciousness. Even when an infirmed elderly person is in a facility, or when they live with family members, all present hold no expectation that they will resume self-care (Keifer, 1987; Shibusawa & Mui, 2001). One example of this phenomenon is the average length of stay in Japan's hospitals is 38.3 days, compared with 8 days in the United States (Keifer, 1987; Sonoda, 1988), suggesting an expectation of dependency.

Another aspect of the plight of the elderly and the chronically ill in Japan, as in much of the west, is that these conditions do not respond to technological advances. In Japan, fortunately, there is an alternative health care system, *Kampo*, which specializes in chronicity and quality of life. *Kampo* medicines are often paid for by the national insurance, and it is common for the elderly to seek *Kampo* (Keifer, 1987; Sonoda, 1988).

Dying and Death

The concept of pollution is at the heart of most funeral rituals. Japanese culture has well articulated notions of pollution and rituals devoted to the eventual restoration of purity, particularly within the Shinto religion. Contact with death, especially in the material realms, such as utensils and clothing, are important places for ritual intervention. One way that people have historically provided protection from pollution is through reversal—that is, object, patterns, furniture, dishes—are placed on the opposite side or in reverse during the funeral period. Spirits of the dead are believed to wander about the earth until they are appeased. Eating and drinking among the living provide protection during this vulnerable, transitional state, and food and drink are available as offerings to the spirits of the newly deceased. There is usually a wake the day of the death, with burial of cremated remains in a family area of the cemetery the day after the death. Rituals may occur as often as weekly until the 49th day. The family member is remembered with a blessed placard bearing their name placed with the other family names in the family gravesite. Families often have a Buddhist altar in the home, with photos of the deceased. Remembrances yearly and at special days throughout the year are marked by rituals including the offering of incense and prayers. The altar also bears other symbolic elements, including salt, *sake*, water, and rice. Some families also practice a Shinto purification ritual of "pouring water," (misogi) at the family gravesite. This might be the responsibility of the wife of the eldest son, but, more generally, the wife of the household nearest to the grave does it.

REFERENCES

Bachnik, J. (1994). Uchi/Soto: Challenging our conceptualizations of self, social order, and language. In J. M. Bachnik & C. J. Quinn (Eds.), *Situated meaning: Inside and outside in Japanese self, society, and language* (pp. 3–37). Princeton: Princeton University Press.

Bauer, J. (2001). Demographic change, development, and economic status of women in East Asia. In A. Mason (Ed.), *Population change and economic development in East Asia* (pp. 359–384). Stanford: Stanford University Press.

Befu, H. (1980). The group model of Japanese society and it's alternative. *Rice University Studies, 66*(1), 169–187.

Befu, H. (1993). *Cultural nationalism in East Asia* (Vol. Berkeley: University of California Press.

Bestor, T. (1996). Forging tradition: Social life and identity in a Tokyo neighborhood. In W. P. Zenner (Ed.), *Urban life: Reading in urban anthropology* (2nd ed., pp. 524–547). Prospect Heights: Waveland Press.

Cornell, L. L. (1996). Infanticide in early modern Japan? Demography, culture, and population growth. *Journal of Asian Studies, 55*(1), 22–50.

de Bary, W. T., & Dykstra, Y. (2001). *Sources of Japanese tradition* (2nd ed.). New York: Columbia University Press.

DeNoon, D., Hudson, M., McCormack, G., & Morris-Suzuki (Eds.) (2001). *Multicultural Japan: Paleolithic to postmodern*. New York, NY: Cambridge University Press.

Feeney, G., & Mason, A. (2001). Population of East Asia. In A. Mason (Ed.), *Population change and economic development in East Asia* (pp. 61–95). Stanford: Stanford University Press.

Freeman, D. M. A. (1998). Emotional refueling in development, mythology, and cosmology: The Japanese separation-individuation experience. In S. Akhtar & S. Kramer (Eds.), *The colors of childhood: Separation-individuation across cultural, racial and ethnic differences* (pp. 17–60). Northvale: Jason Aronson.

Health Statistics Division. (1999). *Patients and medical institutions in Japan: Graphical review of health statistics*. Tokyo: Health and Welfare Statistics Association.

Hendry, J. (1992). Individualism and individuation: Entry into a social world. In R. Goodman & K. Refsing (Eds.), *Ideology and practice in modern Japan* (pp. 55–71). New York: Routledge.

Johnson, M. L., & Johnson, J. R. (1996). *Daily life in Japanese high schools*. Japan Digest, U.S.–Japan Database, Indiana University. Retrieved September 12, 2002, from http://www.indiana.edu/~japan/digest9.html

Johnson, T. (1995). *Dependency and Japanese socialization*. New York: New York University Press.

Keifer, C. W. (1987). Care for the aged in Japan. In E. Norbeck & M. Lock (Eds.), *Health, illness, and medical care in Japan: Cultural and social dimensions* (Vol. pp. 89–110). Honolulu: University of Hawaii Press.

Klass, D., & Heath, A. O. (1996–97). Grief and abortion: Mizuko Kuyo, the Japanese ritual resolution. *Omega, 34*(1), 1–14.

Lebra, T. (1976). *Japanese pattern of behavior*. Honolulu: University of Hawaii Press.

Lock, M. (1987). Protests of a good wife and wise mother: The medicalization of distress in Japan. In E. Norbeck & M. Lock (Eds.), *Health, illness, and medical care in Japan: Cultural and social dimensions* (pp. 130–157). Honolulu: University of Hawaii Press.

Long, S. O. (1987). Health care providers: Technology, policy, and professional dominance. In E. Norbeck & M. Lock (Eds.), *Health, illness, and medical care in Japan: Cultural and social dimensions* (pp. 66–88). Honolulu: University of Hawaii Press.

Miller, R. A. (1972). *The Japanese language*. Tokyo: Zohoban.

Ministry of Health Labour and Welfare. (2002). *Statistics and other data*. Retrieved September 10, 2002, from http://www.mhlw.go.jp/english/

Ministry of Public Management, Home Affairs, Posts, and Telecommunications. (2002). *Statistical handbook of Japan: 2002*. Retrieved September 10, 2002, from http://www.stat.go.jp/english/data/handbook/c13cont.htm#cha13_2

Miyaji, N. T., & Lock, M. M. (1994). Monitoring motherhood: Sociocultural and historical aspects of maternal and child health in Japan. *Daedalus, 123*(4), 87–112.

Namihara, E. (1987). Pollution in the folk belief system. *Current Anthropology, 28*(4), S65–S74.

Ohno, S. (1970). *The origin of the Japanese language*. Tokyo: Kokusai Bunka Shinkokai.

Ohnuki-Tierney, E. (1984). *Illness and culture in contemporary Japan: An anthropological view*. New York: Cambridge University Press.

Ohnuki-Tierney, E. (1989). Health care in contemporary Japanese religions. In L. E. Sullivan (Ed.), *Healing and restoring: Health and medicine in the world's religious traditions* (pp. 59–88). New York: Macmillan.

Okunishi, Y. (2001). Changing labor forces and labor markets. In A. Mason (Ed.), *Population change and economic development in East Asia* (pp. 300–331). Stanford: Stanford University Press.

Pharmacia. (2001). *What women want*. Pharmacia and Upjohn. Retrieved September 10, 2002, from http://www.birthcontrolresources.com/results.htm

Reischauer, E. L. (1981). *The Japanese*. Cambridge, MA: Harvard University Press.

Rister, R. (1999). *Japanese herbal medicine: The healing art of Kampo*. Garden City Park: Avery.

Rohlen, T. P. (1983). *Japan's high schools*. Berkeley: University of California Press.

Rosenberger, N. R. (2001). *Gambling with virtue: Japanese women and the search for self in a changing nation*. Honolulu: University of Hawaii Press.

Saint Arnault, D. M. (2002). Help seeking and social support in Japanese company wives. *Western Journal of Nursing Research, 24*(3), 295–306.

Shibusawa, T., & Mui, A. (2001). Stress, coping, and depression among Japanese American elderly. *Journal of Gerontological Social Work, 36*(1/2), 63–82.

Shwalb, D. W., & Shwalb, B. J. (Eds.). (1996). *Japanese childrearing: Two generations of scholarship*. New York: Guilford Press.

Smith, R. (1961). The Japanese rural community: Norms, sanctions and ostracism. *American Anthropologist, 61*, 522–533.

Sonoda, K. (1988). *Health and illness in changing Japanese society*. Tokyo: University of Tokyo Press.

Steslicke, W. E. (1987). The Japanese state of health: a political-economic perspective. In E. Norbeck & M. Lock (Eds.), *Health, illness, and medical care in Japan: Cultural and social dimensions* (pp. 24–65). Honolulu: University of Hawaii Press.

Tsumura, A. (1991). *Kampo: How the Japanese updated traditional herbal medicine*. San Francisco: Kodansha.

U.S. Department of Education. (2002, November 3, 2001). *Secondary education in the life of Japanese adolescents: Adolescents' problem behaviors*. Contemporary Research in the United States, Germany, and Japan: Japan. Retrieved September 12, 2002, from http://www.ed.gov/pubs/Research5/Japan/secondary_j3.html

United Nations. (2000). *The world's women: Trends and statistics* (Vol. 16). New York: Author.

Vogel, E. (1967). Kinship structure, migration to the city and modernization. In R. P. Dore (Ed.), *Aspects of social change in modern Japan* (pp. 91–111). Princeton: Princeton University Press.

White, M. (1993). *The material child: Coming of age in Japan and America*. New York: The Free Press.

World Health Organization. (2002a). *WHO Statistical Information System (WHOSIS)*. WHO. Retrieved September 10, 2002, from http://www3.who.int/whosis/menu.cfm

World Health Organization. (2002b, September 4). *Mental health: Global suicide rates by country, age and gender*. World Health Organization. Retrieved September 12, 2002, from http://www5.who.int/mental_health/main.cfm?p=0000000149

Yamamura, Y. (1987, August 19–21). *Recent advances in the pharmacology of Kampo (Japanese Herbal) medicines*. Paper presented at the 10th International conference of Pharmacology, Auckland.

Yeo, S., Felters, M., & Maeda, Y. (2000). Japanese couples' childbirth experiences in Michigan: Implications for care. *Birth, 27*(3), 191–198.

Yoshino, K. (1992). *Cultural nationalism in contemporary Japan: A sociological inquiry*. New York: Routledge.

Jat

Sunil K. Khanna

ALTERNATIVE NAMES

The Jats constitutes one of the largest and diverse communities living in northwestern India and Pakistan. According to Westphal-Hellbusche and Wesphal (1964), the Arabic equivalent of Jat is *Zutt*, a generic term used for "men from India." The word *Jat* also means "bunch of hair" and the Jats themselves claim that they have descended from the hair of lord Shiv. According to Ibbetson (1916), Jats are of Indo-Aryan (or Indo-Scythian) descent. Bowles (1977) argues that the word *Jat* in the Punjabi language means a "grazer" or "herdsman," but notes Ibbetson's (1916) suggestion that a shift from the Punjabi soft "t" to a hard "t" in some Muslim areas means an agriculturalist.

LOCATION AND LINGUISTIC AFFILIATION

The Jats are distributed over a wide and diverse geographic area—from the hot and humid regions in northwestern India to the hills and plains in southern Pakistan—presenting extensive cultural, linguistic, and religious diversity. Some Jat groups have also been identified in the Maldives, Russia, and Ukraine. Different Jat groups living in India and Pakistan speak different dialects of Hindi, Urdu, Sindhi, and Punjabi.

OVERVIEW OF THE CULTURE

It is important to note that each Jat community presents certain unique cultural characteristics, which makes it difficult to generalize about the Jats as a culturally homogenous group. Any attempt to generalize about Jat culture and/or its characteristics is therefore problematic. This account does not attempt to represent the vast cultural diversity and regional variation among the Jats. Instead, it presents some of the cultural characteristics shared by the Jat communities inhabiting the northwestern provinces of India and southern Pakistan.

There is considerable disagreement among scholars over the caste ranking of the Hindu Jats. Throughout the 1900s, several scholars and Jat politicians and activists have used three very different labels for Jat identity. While some have identified the Jats as members of the warrior group (*Kshatriya*) (Qanungo, 1982), others have argued that they belong to the "backward castes." Freed and Freed (1993) argue that until 1958, the Jats were not considered members of the three twice-born *varnas*. Instead, the Jats were ranked as "clean menial workers" (*Shudras*). Some suggest that the Jats rank below the *Rajputs* in the warrior group primarily because of the practice of widow remarriage (Lewis, 1965). Recently, the Jat community has been added to the list of "Other Backward Communities" primarily based on poor economic and education status of the Jats in India.

The Hindu Jats in India generally follow *Arya Samaj*—a reform sect of Hinduism, which originated in the mid 1800s. Generally, the Jats follow the teachings of Swami Dayananda Saraswati, the founder of the *Arya Samaj*. According to Datta (1999), the Jats living in the northwestern plains of India belong to the *Shudra* group. Primarily because of the influence of the Arya Samaj, the Jats claimed a *Rajput* descent and a *Kshatriya* status.

Fuchs (1974) defines the Jats as a Central Asian nomadic group that immigrated into northwestern India. Serological and anthropometric studies of the Jats in Haryana—a north Indian state, suggest a close association between the *Rajputs* and the Jats (Khanna, 1995). The Jats in India primarily practice Hindu religion, however, some Jat groups in Punjab embraced Sikh religion around the 17th century. These Jat groups are called Jat Sikhs or Sikh Jats. In western Punjab, now in Pakistan, the Jat community adopted Islam between 8th and 10th centuries AD.

Notwithstanding social, linguistic, and religious and diversity, the Jats are one of the major landowning agriculturalist communities in South Asia. Generally, Jat communities in India primarily engage in agriculture and live in permanent village settlements in rural and urbanizing areas. However, some groups are nomadic herdsmen.

As agriculturalists, the Jats grow cereals such as wheat, maize, and millet, and cash crops such as sugarcane, fruits, and vegetables. Typically, Jats live in villages in which their community is in numerical majority and dominates the economic and sociopolitical aspects of the village life. In Haryana as well as in Delhi, Jats are locally referred to as chiefs (*chowdharies*). The title symbolizes their ancestral control over the village land and their socioeconomic dominance in the village (Pradhan, 1966).

The Jats practice subcaste (*gotra*) exogamy in arranging marital alliances. Ideally, subcastes of the father, mother, paternal grandmother, and maternal grandmother are to be avoided. Residence after marriage is patrilocal and the inheritance of property patrilineal. Agrarian needs also forced the practice of widow remarriage among the Jats, especially levirate (Chowdhry, 1994). The practice of levirate is called *Karewa, karao*, or *chaddar andazi* among the Jat communities in northwestern India. It involves marriage by the simple ritual of a man throwing a white sheet (or chaddar) over the widow's head, signifying his acceptance of her as his wife (Chowdhry, 1994). Such marriages are described as "wearing bangles in the name of her husband" (*chura pahenana*). Sometimes levirate alliances are primarily symbolic in order to protect the right over property and to avoid a sexual indiscretion on the part of the widow.

The Context of Health: Environmental, Economic, Social, and Political Factors

Throughout northwestern India and southern Pakistan, state-sponsored health care delivery programs and services as well as efforts by numerous non-governmental organizations have led to significant improvements in health care access for all rural and urbanizing communities. Members of the Jat community in these regions now receive health care from allopathic and non-allopathic physicians. In India, the state sponsored Maternal and Child Health (MCH) programs and the Integrated Child Development Scheme (ICDS) have expanded the public health care system to urbanizing and rural areas. These programs involve recruitment of village-level workers (*Anganvadi*) and their training in prenatal, perinatal, and neonatal care. Most commonly, local midwives are selected as village-level workers and are trained at state sponsored instructional workshops. These trained health care workforce then serve as crucial links between the community and the government trained health care practitioners.

Most Jat villages in northwestern India are part of one of the Maternal Child and Family Welfare (MCFW) target zones of the regional Primary Health Center (PHC). A recent family planning survey suggests high acceptance of family planning methods among the Jats of New Delhi (Khanna, 1995). Child immunization records in this region indicate that Jat children are immunized in a timely manner. The success of the state sponsored programs among the Jats living in urbanizing villages can be attributed primarily to the easy availability and utilization of state sponsored health care services. Furthermore, the social position and networks of the *Anganvadi* workers in these villages play an important role in this process. They regularly visit and pursue the parents to utilize the available health care services. Khanna (2001) reports that in Shahargaon—an urbanizing Jat village in New Delhi—increasing awareness of health care and close proximity to state-sponsored and privately run health care services, especially those for mothers and children, have led to an overall increase in the utilization of these services and a corresponding decrease in family size among members of the Jat community.

Medical Practitioners

Traditionally, Jats received health care from traditional doctors (*vaidys*) practicing *Ayurvedic* and/or Greek (*Unani*) medicine, *homeopathic* physicians, midwives, and local self-trained village healers who practice popular medicine. With increasing government support and sponsorship for western medicine, a large number of Jats are now using the services of allopathic doctors at the local dispensaries, primary health centers, hospitals, and private clinics.

For childbirth, Jat women prefer at-home deliveries with the help of a local village midwife. In Shahargaon, just like in many other Jat villages, the village midwife belongs to the low caste of Untouchables (*Balmiki Harijans*). It is important to note that due to increasing urban contact and improvement in state-sponsored health care delivery, most Jat communities now have access to biomedical services (Khanna, 2001). This has brought significant changes in birth practices among the Jats; gradually leading to increased medicalization of pregnancy and childbirth and a corresponding decrease in the roles of the village midwife. Nowadays, Jats have access to trained midwives and/or biomedical doctors for prenatal, perinatal, and postnatal health care. Notwithstanding these changes, health care during prenatal and perinatal periods is primarily a responsibility of Jat women and Jat men play little or no role in this arena.

CLASSIFICATION OF ILLNESS, THEORIES OF ILLNESS, AND TREATMENT OF ILLNESS

In spite of the increasing use of western medicine and technology among the Jats, traditional ideas of health, illness, and healing, generally associated with *Ayurvedic*, *Homeopathic*, and *Unani* healing systems constitute the core theories of illness. Generally, the Jats use non-biomedical systems of healing for chronic health conditions. The non-biomedical treatments prescribed generally involve herbal and plant medicines, seasonal do's and dont's, dietary changes, etc. Among the Jats, non-biomedical categories "hot" and "cold" foods are generally associated with seasons and physiologic effects (Freed & Freed, 1993).

The *Ayurvedic* system of medicine adopts a holistic approach through dietary and lifestyle changes, herbal medicine, and exercise to cure chronic disease and maintain individual health. The *Ayurvedic* system of medicine is based on the ancient knowledge contained in *Atharvaveda*. It deals with the totality of individual and social health including preventive and curative aspects. In fact, *Ayurveda* is a way of life based on certain emphasis on diet, lifestyle, and Yoga practices suitable for an individual according to his/her constitution. The constitution, in turn, is determined on the basis of the predominance of or loss of equilibrium in one or more of the humor, viz. Gas (*Vata*), bile (*Pitta*), or phlegm (*Kapha*). Based on the symptoms produced due to excess or deficiency of particular fault (*dosha*), the *Ayurvedic* practitioner selects remedial measures in the form of herbs, plant medicine, and metal salts.

The present health and healing culture among the Jats of northwestern India and Pakistan involves an eclectic use of treatments and therapies from several distinct healing traditions, namely, *Ayurveda*, *Homeopathy*, *Unani*, and biomedicine. In addition, local healers use a combination of ideas related to disease causation and treatment modalities borrowed from a variety of healing traditions and practices.

The Jats in north India continue to believe in numerous mother goddesses, many of whom serve as the names of diseases, especially those of childhood. Most Jat communities identify two mother goddesses namely, *Sitala* and *Khasra* for smallpox and skin rashes, respectively. In order to protect children and family members from illness, evil eyes, and harmful spirits, the Jats tie protective amulets made of iron, gold, silver, copper, beads, cloths, and strings on children's wrists, ankles, and around their necks (Freed & Freed, 1993). Beliefs associated with evil spirits are more common among Jat women than among Jat men. More commonly in rural Jat communities, diseases like tetanus, diarrhea, and measles are often associated with spirit possession and evil eye. Jat women are expected to worship local deities and observe charms (*totkas*) in order to ward off evil spirits and cure diseases among family members, especially children (Chowdhry, 1994).

SEXUALITY AND REPRODUCTION

Before discussing sexuality and reproduction among the Jats, it is important to note that members of this community no longer live in isolated communities. Most Jat communities are currently experiencing rapid social change brought about by urbanization and technological diffusion, especially in terms of increasing access to education, health care, and the mass media. Such changes have perceptibly influenced everyday life among the Jats. Amidst rapidly changing social and economic environments, the Jats are experiencing significant shifts in their health and economic well-being as well as in their traditional cultural ethos.

Among the Jats, culture-bound beliefs and practices associated with gender, conception, prenatal period, childbirth, postpartum care, childhood, puberty, adulthood, and old age are central to perceptions of health, illness, and well-being. Such beliefs and associated practices significantly influence the utilization of health care. Like other agriculturalist communities in South Asia, the Jat patriarchal system constitutes core cultural values, customs, laws, social roles, and metaphors that clearly subscribe to the larger Hindu ideas and ideals of male dominance. Under the Jat peasant patriarchal system, there is little equality or symmetry between men and women. As compared with men, Jat women are generally given little choices, especially with regard to marriage, mobility, education, and employment. Extensive ethnographic evidence suggests gender-specific neglect against women among the Jats (Das Gupta, 1987; Khanna, 1995). The customs and traditions of a patriarchal society are suggested as underlying causes of the overall neglect of women leading to their poor overall physical health (Jeffery et al., 1989; Khanna, 1997). These studies conclusively demonstrate the adverse effect of the culturally prescribed subordination of Jat women on their reproductive health and survival. Culturally prescribed practices associated with a biased allocation of intrahousehold resources for health care and food have been argued to be responsible for the poor overall health and high morbidity rates among Jat women.

Among the Jats, marriage marks the socially acceptable initiation of reproduction. However, it involves two distinct events—the actual social event associated with marriage and the consummation of marriage. Traditionally, marital alliances are arranged with the help of the village barber who assists Jat parents in finding a suitable match for their children and in finalizing the timing, exchange, and other important details associated with marriage. After a suitable match has been arranged by Jat parents, an elaborate religious ceremony marks the marital bond. The consummation of marriage (*gauna*), however, takes place usually within one or two years of marriage. Before a marriage has been consummated the bride stays at her natal home. The term *liviayo* refers to the time in the *gauna* ceremony when the bridegroom, along with his other relatives, comes to take the bride to her affinal village. Recently, there has been a considerable decrease in the time gap between marriage and its consummation among the Jats. On both occasions, however, the bride's family is expected to arrange for feasts for the marriage party and to provide gifts for the groom's relatives. Some Jat communities have traditionally preferred "double marriages," which involve marital alliances between sets of sisters and sets of brothers. Such marriages are described by Jats as more realistic and suitable to an agricultural way of life and three-generation patrilocal extended family system (Khanna, 1995; Kolenda, 1987).

HEALTH THROUGH THE LIFE CYCLE

Anthropologists have often used events marking life transitions as age categories signifying major divisions of life. Earlier research among the Jat communities in north India provides detailed accounts of the rites of passage, highlighting the importance of rituals and ceremonies associated with birth, marriage, and death (Lewis, 1958). The Jats recognize an individual's life cycle into the three major stages (*ashram*) as prescribed by the traditional Hindu views. In *brahmacharaya-ashram*, the individual as a student follows a strict code of chastity (*brahmacharaya*) and learns from his teacher (*guru*). In the next stage or *grahastrha-ashram*, the individual is expected to take household responsibilities (*grahastha*) and fulfill his duties toward his family and his *dharma*. *Vanprastha-ashram* marks the third state of life in which the individual begins the hermitage phase of life and returns to contemplation and for guiding family and society. The fourth and final stage, *sannyas-ashram*, marks the beginning of the renunciation phase of life. The individual is expected to renounce all the "outer" goals of life and to begin learning about spirituality away from all social or political concerns (Lewis, 1958).

Among the Jats, specific ceremonies or events are often associated with the transition of an individual through different stages of life and that in each life stage, individuals are expected to perform the social roles prescribed for that life stage. It is, however, important to note that the traditional Hindu views do not separate life stages based on gender. Peasant patriarchal system among the Jats and the corresponding gender stratification place severe constraints upon the activities and roles of Jat women throughout the lifecycle. These constraints directly affect Jat women's ability to access health care.

Pregnancy and Birth

Both pregnancy and childbirth are considered auspicious events among the Jats. Childbearing constitutes an important part of an adult Jat women's life. During the time of a woman's labor, family members gather in the house and women perform important roles associated with childbirth. If the child is a boy, the relatives bang a metal place (*thali*) or fire in the air to announce his birth. Because of a strong son preference among the Jats, the birth of a son is considered an auspicious occasion of happiness and rejoicing. No such celebrations take place on the birth of a girl. The happiness on the birth of a son is shared by members of the affinal and natal families of the mother as the birth of a son improves the social status of the mother in the affinal patrilineage. The affinal family of the mother receives gifts, special food items, and money from the natal family. These contributions are comparatively smaller in case of the birth of a daughter. This practice, however, is limited to the first-born daughter. For subsequent daughters the natal family generally does not send gifts and money to the affinal family.

Infancy

A great deal of demographic, ethnographic, and health research among northwestern peasant communities in India, especially among the Jats, indicates sex differentials in health and mortality during infancy and early childhood. While several factors have been identified as causing excessive female mortality during infancy and early childhood, culturally prescribed patterns of son preference and daughter neglect are considered important proximate determinants of gender differentials in morbidity and mortality among the Jats. The mortality patterns among the Jats correspond with the South Asian pattern of sex differentials: males generally have equal or slightly higher mortality rates than females during the first month of life and lower mortality rates than females after the first month of life.

Ethnographic research suggests that Jat parents tend to show preferential treatment for sons, generally investing more household resources toward ensuring their survival. Infant girls are considered relatively more resistant to disease than an infant boy. A daughter is often equated with *kikkar*—a thorny bush that grows wild and does not require much nurturing.

Childhood

Gender role socialization begins very early in the life of a Jat individual, often before children are even aware of their sexual identity. This may happen even before the development of an internal motive for conforming to sex role standards. Jat parents and the community play important roles in reinforcing norms of expected behavior. Jat girls, generally socialized under strict patriarchal control, came to understand their limited sex role options and the rigid patterning of these options.

For the young girl discrimination in terms of nutrition and health care allocation occurs in conjunction with a constant reinforcement of her gender identity.

The individual and additive effects of cultural and biological factors invariably lead to lower survival rates for Jat girls than boys, and negatively influence their overall growth and development patterns, reproductive health, and fertility (Khanna, 1997).

Adolescence

Son preference and daughter neglect has continued among the Jats in spite of increased urban contact as well as access to and utilization of education and health care services. In the Shahargaon—an urbanizing Jat community located in the outskirts of New Delhi, India, recent shifts in the economy and occupation based on agriculture to urban job-based economy, and improvements in education and health care facilities have not significantly changed traditional patterns of son preference and daughter neglect. Jat sons are viewed as economic and political assets, adding to the strength and prestige of the family, but daughters are perceived as an economic and moral burden. While Jat parents expect their teenage sons to enroll in schools and colleges, for a majority of Jat daughters, adolescence invariably marks the end of their education experience. Rarely are they allowed to study beyond the middle school level available at the village school. Most Jat parents in Shahargaon feel that the daughter should be educated up to a level so that she can write letters to her parents and take care of her children's health and education.

As the development of secondary sexual characteristics and the onset of menarche signal a girl's emerging sexuality, Jat girls are expected to spend more time in the house and take responsibility of the domestic workload. Jat girls are expected to help prepare and serve food and take care of their younger siblings. Culturally mediated concerns for virginity, marriage, and workload expectations lead Jat parents to enforce seclusion and strict parental control of mobility, education, and occupation of Jat daughters. The imposition of a strict conduct of behavior invariably leads to agitation, apprehension, and emotional anxiety among young Jat girls. Although the onset of menstruation is regarded as a natural event, women rarely discuss this issue amongst themselves. Delayed menarche is often considered an indicator of infertility. Irregular menstrual cycles are believed to be associated with a sexual indiscretion on the part of the girl and a source of great anxiety for parents. One of the major parental concerns is the fear of a daughter's emerging sexuality and of any dishonor which sexual activity could bring to the family. Anxiety over menstruation and lack of knowledge adds emotional distress to the psychosocial development of an adolescent girl.

Health care behavior among the Jats also indicates a strong bias favoring sons over daughters. Generally, Jat sons receive "modern" health care at an earlier stage of their illness than Jat girls. In comparison to Jat girls, Jat boys are more likely to be taken to biomedical "specialist" doctors and clinics. An ill son is a matter of great concern and anxiety for the family members while an ill daughter soon becomes the target of insults and parental frustration.

Girls during this stage of their life have little or no control over their health and are dependent on decisions of the elder family members.

An adolescent Jat girl's life in her natal home is often emically described as life in a state of transition, at the threshold, or living inbetween two worlds. She is considered a "commodity" (*paraya dhan*) that belongs to some one else. Life stages like puberty and development of secondary sexual characteristics reinforce the idea of her temporary existence in her natal home.

Adulthood

Jats prefer early marriage, especially in the case of girls. Among the Shahargaon Jats, the average age at marriage is 16 years (Khanna, 1995). Jat parents observe strict patterns of subcaste (*gotra*), village, and at times, regional exogamy while arranging marital alliances for their children. Jats practice patrilocality—a cultural practice that requires the bride to leave her natal home after marriage in order to live with her husband's family. The "adjustment" process for the young Jat bride in her affinal home involves meeting conjugal responsibilities, often from a position of complete ignorance, and dealing with the daily demands of the household chores. She is expected to maintain veil (*purdah* or *ghunghat*) from senior men in the family and the community.

Expectations to produce a son usually result in early pregnancy among the Jats. Early marriage followed by early childbearing increases the risk of obstructed labor and reproductive morbidity. Due to increase availability of health care in most rural and urbanizing areas of northwestern India, the Jats generally seek prenatal and perinatal health care from state-sponsored health clinics. Jat women in the household are primarily responsible for health care during and after pregnancy.

Among the Jats, sex and gender are contrasting concepts. Jat parents do not think of their children only in terms of a child's sex because what concerns them most is whether the child is a son (*beta*) or daughter (*beti*). These terms invoke an entire set of cultural values and behavioral norms associated with the sex of the child. While the birth of a son is considered a good gift or an indication of the family's good fortune, that of a daughter is considered a sign of distress and anxiety. Some of the commonly used proverbs expressing strong son preference among the Jats include:

"The number of sons is equal to the number of sticks and the number of sticks determines the amount of land controlled by a family" (*jitney ladke utney lath, jitney lath utna kabza*); and "selling milk is equivalent to selling a son" (*jisney gharka dudh baech diya usne apna puut baech diya*).

On the other hand, common household names for daughters include:

Rambatheri (God, this one is enough) or *Rambheji* (this daughter belongs to God and he will take care of her).

Among the Shahargaon Jats, increasing urban contact and access to health services and reproductive technology, especially prenatal diagnostic technology, have provided Jat parents with a reliable means for reducing family size and limiting the number of daughters in the family. Shahargaon Jats are using temporary methods of contraception for reducing the number of children per family and achieving the desired family composition by using ultrasonography to identify the sex of the fetus and, in some case, aborting female fetuses. In a survey of 127 Shahargaon Jat women in 45 or less age category, 28.3% women used ultrasonography to identify the sex of the fetus and 13.4% women opted for sex-selective abortions between 1989 and 1994. The majority of Shahargaon Jat parents (68%) expressed desire for a small family size, and considered two sons and a daughter to be the ideal family composition (Khanna, 2001).

It is important to note that despite considerable improvements in economic and educational status brought about by urbanization, the use of prenatal diagnostic technology and sex-selective abortion of female fetuses suggests the continuation of the son preference and daughter neglect reminiscent of the Jat agrarian ethos. Notwithstanding the legality of abortion in India, well-equipped and professionally operated screening and abortion facilities are not readily accessible to all rural and urbanizing Jat communities. By seeking repeated ultrasonographic examinations and, at times, abortion services, Jat women are at increasing risk of complications resulting from unsafe and unhygienic abortion procedures, including reproductive tract infection. In the context of the generalized lack of knowledge and social stigma associated with reproductive health problems in the Jat community, Jat women are less likely to seek treatment for reproductive health problems.

The Aged

Older members in the Jat community command considerable social prestige and respect. The Jats prefer to live in joint families with older men as heads of the household. While men enjoy high status and respect in the family primarily by virtue of their gender, Jat women have to contribute male heirs to the husband's patrilineage in order to gain authority and power in the family. Often older Jat women's primary responsibilities include direct supervision of their children's socialization. Among the Shahargaon Jats, the elderly women lack physiological knowledge associated with menopause and experience a wide range of menopausal symptoms such as tension, headaches, swelling, and loss of appetite. Some informants also reported a feeling of resentment and anger. Although the elderly Jats rarely seek health care for minor health problems, the joint family system among the Jats provides financial security and facilitates timely health care. Even at this stage in life, Jat men have a privileged social status over Jat women.

Dying and Death

The Jats generally observe 13 days of pollution after death. Jat men are expected to show considerable behavioral restraints during the days of pollution, while Jat women are expected to express their grief in public. The dead are cremated by following the traditional Hindu ritual of cremation in which the eldest son plays an important role. At the death of a married woman, rituals and practices reinforce the notion that she does not belong to her husband's patrilineage. Her sister(s) and natal women relatives prepare the body for cremation and funeral rites.

REFERENCES

Bowles, G. T. (1977). *The people of Asia*. London: Weidenfeld and Nicolson.
Chowdhry, P. (1994). *The veiled women: Shifting gender equations in rural Haryana 1880–1990*. New Delhi: Oxford University Press.
Datta, N. (1999). *Forming an identity: A social history of the Jats*. New Delhi: Oxford University Press.
Freed, R. S., & Freed, S. A. (1993). *Ghosts: Life and death in north India*. Washington, DC: The American Museum of Natural History.
Fuchs, S. (1974). *The aboriginal tribes of India*. Delhi: Macmillan India.
Ibbetson, D. (1916). *Punjab castes*. Lahore: India Press.
Khanna, S. K. (1995). *Gender discrimination, maternal health, and pregnancy outcomes in north India*. Doctoral dissertation, Syracuse University.
Khanna, S. K. (1997). Traditions and reproductive technology in an urbanizing north Indian village. *Social Science & Medicine, 44*(2), 171–180.
Khanna, S. K. (2001). *Shahri* Jat and *dehati Jatni*: The Indian peasant community in transition. *Contemporary South Asia, 10*(1), 37–53.
Kolenda, P. (1987). *Regional differences in family structure in India*. Jaipur: Rawat Publications.
Lewis, O. (1958). *Village life in northern India*. Urbana: University of Illinois Press.
Pradhan, M. C. (1966). *The political system of the Jats of northern India*. Bombay: Oxford University Press.
Qanungo, K. (1982). *History of the Jats: Contribution to the history of northern India*. New Delhi: Surajmal Memorial Education Society.
Westphal-Hellbusch, S., & Westphal, H. (1964). *The Jat of Pakistan*. Berlin: Duncker & Humbolt.

Lijiang Naxi

Sydney Davant White

ALTERNATIVE NAMES

Currently, the "Lijiang Naxi" and the "Yongning Naxi" (or Mosuo) are officially classified by the People's Republic of China government as members of the same Naxi "minority nationality," with the Mosuo designated as a "branch" of the Naxi; however, both groups see themselves as significantly different peoples. This entry focuses on the Naxi of the Lijiang area, specifically the Naxi of the Lijiang basin. During the Republican Period (1912–49 CE), the contemporary Naxi of the Lijiang area of the People's Republic of China (PRC—1949–present) were referred to by Joseph Rock (1947) as the Na-khi; the contemporary Mosuo of the Yongning area (officially referred to as the Yongning Naxi, and sometimes also as the Naru, Naze, or Na by various scholars) were referred to by Rock as the Hli-khin.

LOCATION AND LINGUISTIC AFFILIATION

The Lijiang basin has historically been the geographical, political/administrative, economic, and cultural center of the Naxi. The town of Dayanzhen (referred to by outsiders as Lijiang) is located at the center of the 2,400-meter Lijiang basin at the base of the spectacular 6,300-meter Jade Dragon Snow Mountain. Numerous villages surround Dayanzhen, and the Naxi area extends to the north, northwest, and south of the Lijiang basin. Dayanzhen serves as both the county seat for the Lijiang Naxi Nationality's Autonomous County and as the seat for the larger Lijiang Prefecture (which encompasses a total of four counties). The Lijiang area is located in the northwestern corner of Yunnan Province, adjacent to ethnically Tibetan areas in the province's Himalayan sweep toward Tibet. Yunnan is located in the southwesternmost corner of the contemporary PRC. The first language of most Lijiang basin Naxi residents is *Naxi-hua*, a Tibeto-Burman language, though most basin Naxi who have come of age since 1949 have acquired the western Yunnan dialect of Mandarin through the schools as they have grown up.

Overview of the Culture

Naxi are the predominant "nationality" of the Lijiang basin and of the larger county, constituting approximately 250,000 of the county's 300,000 residents. As of 1990, Dayanzhen had a population of approximately 60,000. In order to understand Naxi identities and therapeutic practices both prior to and since 1949, however, it is critical to note that not only has the Lijiang basin been the center of Naxi culture for centuries, but it has also been an arena of longstanding engagement between basin Naxi and the Chinese state—the latter encompassing both state policies and Chinese popular culture practices. Thus, while "Naxi culture" is the hegemonic culture of the Lijiang basin, at the same time it has been shaped in many arenas—including the arena of therapeutic practices—by several centuries of influence from Chinese society.

Historically, Naxi are believed to have originally been a Qiang people from the Qinghai Plateau, who migrated to the Lijiang area approximately 1,400 years ago. Beginning in the Yuan Dynasty (1206–1368 CE), the Naxi "kingdom" entered into a tribute relationship with the Chinese state under the *tusi* system; this marked the formalization of a two-tiered structure of elites and commoners within Naxi society. During the Ming Dynasty (1368–1644 CE) and the Qing Dynasty (1644–1911 CE), basin Naxi sociocultural identities were influenced by the inmigration of Han Chinese soldiers and other Han settlers from areas such as Nanjing, with whom basin Naxi apparently frequently intermarried. The Lijiang area was formally incorporated into the Chinese state under the system of "regular government" in 1723. To a degree under the Republican state, but of special note under the Socialist (i.e., post-1949) state, basin Naxi have been strong adherents to the official/Chinese state status quo.

With respect to economic practices, historically most Naxi were agriculturalists and pastoralists. Wheat, corn, and various legumes have long served as the primary basis of basin Naxi diets, and the raising of horses, oxen, and pigs has also long been central to the village-based Naxi economy. For contemporary basin residents, the primary salient social distinction is along lines of urban versus rural/town versus village residence (as is the case throughout the PRC). Dayanzhen has for centuries been an important stop on the trade route between Lhasa (in Tibet) and Kunming (the provincial capital of Yunnan)—many basin Naxi men historically took part in the muletrain-based, long-distance transport trade as well as in a number of other itinerant trades such as leatherworking, and many Dayanzhen Naxi women controlled the local markets with formidable marketing and business skills.

In the post-Mao period (i.e., post-1979), most Dayanzhen residents have had relatively secure employment in state or collective work units (though this is changing). Additionally, many of the town-based enterprises that are controlled by Dayanzhen residents (as opposed to outsiders) are run by Naxi women. Over the past decade, however, the ethnic tourism that has emerged as a booming industry in the Lijiang area has drawn (primarily male) Han entrepreneurs from a number of different provinces to the area. In the context of post-Mao decollectivization, basin village residents for the most part rely upon their own family labor for their livelihood—primarily in terms of agricultural labor, but also in terms of wage-labor opportunities in village industries and via migration to Dayanzhen and other urban centers throughout Yunnan, Sichuan, and the PRC in general.

Graduates of Dayanzhen high schools compete with top students throughout the PRC for admission to provincial and national level institutions of higher education. Basin Naxi males in particular have historically sought to associate themselves with the Chinese state; since 1949, Communist Party membership has been an important marker of social status for Naxi men.

Patrilineal descent, patrilocal postmarital residence, and (patrilineal) ancestor worship continue to be practiced by both village and town Lijiang basin Naxi residents (in contrast to the system of matrilineal descent and "walking marriage" of the Mosuo). Prior to 1949, arranged marriages were also the norm, and older Naxi women describe their pre-1949 lives as "terribly bitter." Dayanzhen Naxi women have an advantage over village Naxi women in that they do not "marry out" of their natal villages, and they are able to utilize this advantage to maintain their childhood social networks and enhance their entrepreneurial talents as small business managers.

In addition to the neo-Confucian practices that basin Naxi incorporated in their formal and informal encounters with the Chinese state, basin Naxi culture has historically been influenced (at various historical moments) by Buddhism in both Chinese and Tibetan (Nyingmapa) forms, by Daoism, by Tibetan Bon practices, and by a variety of Chinese popular cultural practices. For ethnographically-based research on the Lijiang Naxi, refer to Chao, 1995, 1996, 1999; Goullart, 1955; Hansen, 1999; McKhann, 1992, 1995; Mueggler, 1991; Rees, 2000; Rock, 1947, 1963, and White, 1993, 1997, 1998a, 1998b, 1999, 2001. For ethnographically based research on the Mosuo, see He, 1994, 1999; Mathieu, 1996, 1998, 2000; Rock, 1947; Shih, 1993, 1998, 2001; Walsh, 2001a, 2001b; Weng, 1993; and Yan, 1982.

THE CONTEXT OF HEALTH: ENVIRONMENTAL, ECONOMIC, SOCIAL, AND POLITICAL FACTORS

Epidemiological Profile

According to official PRC Public Health Bureau representations, infectious diseases of all varieties were of dire proportions in the pre-1949 basin, and poverty insured that much of the population was particularly susceptible to them. The relatively dramatic transformations in the epidemiological profile of the basin over the past several decades are undoubtedly due to the vigilant vaccination system that has been put into place in both urban and rural contexts since 1949, along with the overall increase in the standard of living for the majority of basin residents since 1949. These shifts in the overall epidemiological profile occurred in spite of the large-scale starvation and malnutrition experienced primarily by basin villagers during the Great Leap Forward (GLF—1958–61). In basin villages, virtually every family lost at least one or two members during the GLF due to starvation and malnutrition—a famine (provoked primarily by Maoist policies) that was far worse than any in villagers' memories. (See Mueggler, 2001.)

In the contemporary basin, the most common afflictions vary between town and village contexts, but colds, influenza, and stomach problems are common among all ages in both contexts; coughs, pulmonary infections, and both gallstones and kidneystones are common among older basin residents; rheumatoid arthritis, gynecological disorders, high blood pressure, chronic headaches, and "neurasthenia" are particularly common among middle-aged and older Naxi women (see below). During the summer months in the villages, dysentery is common among villagers of all ages, though especially among young children; intestinal worms are also common among young village children during the summer.

Overall, while infectious diseases such as hepatitis A and tuberculosis are common (as elsewhere in the PRC), the epidemiological profile in the Lijiang basin, and throughout most of the PRC, is closer to that of a "developed" rather than a "developing" country. The widespread PRC practice of using only boiled water for drinking and cooking (in addition to improved standard of living and vaccination factors) can be credited for this. However, in basin villages the use of human waste for fertilizer, the lack of physical containment of poultry and livestock, and the lack of screens are important factors influencing the spread of infectious disease by flies and other vectors. Limited access to bathing opportunities in basin villages is also a factor that influences a number of afflictions (including vaginal infections). See below for discussion of risk factors for HIV/AIDS and hepatitis B.

Environment/Global and Local Factors Influencing Health

Environmentally, the Lijiang basin does not suffer from the dramatic increases in air pollution, water pollution, and forms of environmental pollution that plague most contemporary PRC cities. However, as noted above, over the past decade, the Lijiang basin has become a prime national and international tourist destination for an estimated million plus tourists annually, and the considerable economic development in the basin has resulted in increased environmental degradation. The town/village divide and the gendered division of labor are the key local factors influencing health (see below).

Public Health Infrastructure

Beginning in the 1950s and the 1960s, the Lijiang basin has benefited from being the location of both county-level and prefectural-level public health bureaus, clinics, and hospitals established by the Chinese socialist state. In addition to the County Hospital and the Prefectural Hospital, there are also county-level and prefectural-level Hygiene and Prevention Health Stations, Women and Child Health Protection Stations, and Pharmaceutical Inspection Stations. At the county level there is a specialized Chinese Medicine Hospital and a Schistosomiasis Station; at the prefectural level, there is a Hygiene School and a Pharmaceutical Company (the Lijiang area and Yunnan in general being prime areas engaged in the production of herbal Chinese medicine). Public health bureaus that are under the joint administration of the public health system and other government administrative units include the Patriotic Hygiene Movement Committee, the Birth Planning Committee, and the Leprosy (Hansen's Disease) Hospital.

During the Maoist period, particularly during the Cultural Revolution (1966–76), brigade-based clinics served as the first of three tiers of resort, the second tier being commune-based health centers, and the third tier being town-based hospitals. The average brigade consisted of approximately 500 families or 2,000 individuals. These cooperative medicine clinics were (with a few exceptions) dismantled during decollectivization in the post-Mao period. "Integrated Chinese and Western medicine" (see below) was originally implemented as the lynchpin therapeutic practice for cooperative health care.

Therapeutic Practices

Prior to 1949, the primary therapeutic epistemology informing basin Naxi understandings of affliction was the "medicine of systematic correspondence" following Paul Unschuld's (1985) term—the classical text-based humoral therapeutic practice associated with the Confucian state that serves as the theoretical underpinning of the post-1949 socialist Chinese state's formulation of "traditional Chinese medicine" (TCM). The hegemony of this epistemology reflects both the degree of influence of the Chinese state in the Lijiang basin and the significance of this epistemology for the "civilizing project" (Harrell, 1995) of the Chinese state. Some other therapeutic practices existed in the basin as well. These practices for the most part were like popular therapeutic practices in other (Han) parts of Yunnan and China in general. They encompassed fortune telling, the exorcism of ghosts or demons, the retrieval of lost souls (Goullart, 1955), and *gu* witchcraft. Rituals associated with Buddhist and Daoist voluntary associations also factored into basin Naxi therapeutic practices. Neither Tibetan medicine, "Western medicine," nor indigenous Naxi *dongba* or *sanyi* practices played a significant role in pre-1949 basin Naxi therapeutic practices (though *dongba* and *sanyi* practices were influential in other parts of the Naxi area).

In the early 1950s the Chinese Communist Party (CCP) established a distinction between "official" and "unofficial" therapeutic/religious practices. Therapeutic practices designated as official during the Maoist period—in the basin as well as throughout the PRC—included four practices, which were identified as "Chinese medicine," "Western medicine," "integrated Chinese and Western medicine," and "folk medicine." All other types of therapeutic practices were officially labeled as "unscientific," "backward," and reflective of the "feudal superstition" associated with the "old society" prior to 1949.

For basin Naxi, this meant that fortune-telling, exorcism of ghosts and demons, retrieval of lost souls, and *gu* witchcraft were no longer etiologies that could be openly acknowledged or strategies of therapeutic resort in which citizens could be openly engaged. It also meant that other ritual practices of various members of basin Naxi society, such as those associated with Tibetan Buddhism, Chinese Buddhism, Daoism, Confucianism, and/or other aspects of Chinese/Naxi popular culture were banned.

"(Traditional) Chinese medicine" (hereafter referred to as TCM), the official PRC practice of Chinese medicine, reflected a reformulated medical practice whereby the medicine of systematic correspondence, the pharmaceutical canon, and a variety of techniques such as acupuncture, moxibustion, and acupressure were standardized and "scientized" (see Croizier, 1968, 1976; Farquhar, 1998; Hsu, 1992; Kleinman, 1980; Scheid, 1995; Sivin, 1987; Unschuld, 1985; White, 1993, 1999, 2001).

The official PRC practice of biomedicine, designated as "Western medicine" (hereafter referred to as WM), was also a Maoist period creation. In the 1950s,

Mao broke with physicians who had been trained during the Republican era in an elite-focussed, urban-based, hospital-centered practice of biomedicine. It was at this juncture that Mao declared TCM a "national treasure-house," and that biomedicine was re-envisioned, reformulated, and reorganized to enable the new, more pragmatic, broad-based, public health oriented practice of WM.

The aforementioned "integrated Chinese and Western medicine" (hereafter referred to as integrated medicine) emerged as a consciously formulated hybrid medical practice that was introduced by Mao during the Cultural Revolution as the cornerstone of national health policy (see RHCHP, 1977). Ideally, it was to represent a synthesis of the best of TCM and of WM. And although it was initially implemented in both urban and rural areas, it was in fact geared toward the health care needs of the rural areas in particular.

"Folk medicine" was the official category established during the Maoist period to acknowledge medical practices other than TCM and WM. It came to encompass aspects of folk and popular Chinese therapeutic practices (particularly local herbal medicines) that were perceived as relatively "scientific" and therefore as relatively legitimate. In the Lijiang basin, "Naxi folk herbal medicine" was recognized as a form of folk medicine, albeit one described in basin Naxi and state public health discourses alike as "without (its own) theory" (i.e., no theory distinguishable from TCM theory) informing the pharmaceutical usage of its materia medica.

The Maoist period establishment of these four official practices through public health policies and institutions reshaped the contours of the previously existing Chinese medicine and other therapeutic practices in the basin. In Dayanzhen, where most of the public health bureaus, hospitals, and clinics were established beginning in the 1950s, the newly reformulated practices of TCM and WM were the ones with which town residents engaged when they sought health care. As was the case elsewhere during the 1950s and early 1960s in the PRC, Dayanzhen medical institutions operated according to an institutional division of labor between TCM and WM. The Prefectural Hospital and the County Hospital each maintained separate WM and TCM departments that mutually engaged in cross-referrals, but maintained no other coordination. There was additionally the County Chinese Medicine Hospital that engaged solely in the practice of TCM. For the duration of the Cultural Revolution, all three institutions shifted to the revolutionary practice of integrated medicine (indeed, it was counterrevolutionary to do otherwise), but returned to their former institutional division of labor arrangement as soon as the Cultural Revolution began to wind down. In contrast, in basin brigades (now "administrative villages") integrated medicine became thoroughly institutionalized as the prevailing therapeutic practice starting from its implementation in the late 1960s. Rural residents engaged in this new practice that endeavored to combine TCM and WM when they sought out treatment from brigade cooperative health care practitioners.

In the post-Mao period (as was the case during the Maoist period), the four officially recognized therapeutic practices—TCM, WM, integrated medicine, and folk medicine—have continued to be the only therapeutic practices technically regarded as "medicine" in the PRC. While medicine, "hygiene," and public health have remained centerpieces of ongoing PRC discourses of socialist modernity and "culture," however, all four official practices have become more professionalized, "scientized" (in a manner different than the Maoist period), and commoditized in the post-Mao period. Additionally, there has been a post-Mao "reemergence" of certain practices formerly banned as unofficial into the officially tolerated (if not sanctioned) arena of "popular culture." Most prominent among these practices have been fortune-telling and, in basin villages, *gu* witchcraft—both of which reflect long-standing Chinese popular culture legacies.

MEDICAL PRACTITIONERS

Prior to 1949, basin Naxi practitioners of Chinese medicine (namely, the medicine of systematic correspondence combined with Chinese herbal pharmaceuticals, acupuncture, and massage techniques) were virtually all male, and had either acquired their knowledge as apprentices of their fathers or other older male relatives, or through an apprenticeship with a non-related practitioner usually arranged by their parents. They started as young boys and gradually developed expertise over the years through learning from their masters, reading the classic texts of Chinese medicine, and learning through personal experience.

During the Cultural Revolution, the practitioners in the basin who were recruited to staff the brigade clinics and to carry out the practice of integrated medicine were generally from one of three backgrounds. Middle-aged and younger doctors and medical technicians trained in

"Western medicine" (sometimes via state medical training schools or universities and sometimes via People's Liberation Army training programs) were "sent down" to the countryside from town and city hospitals and clinics for at least the early years of the Cultural Revolution. Older Chinese medicine practitioners were selected from the brigade itself. They usually came from families that had a multigenerational legacy of practicing Chinese medicine or that at least had a repertoire of secret medical (usually herbal) remedies. The "barefoot doctors," the majority of whom were male, in their twenties, and of "good" political background, were selected at the brigade level as well. It was the barefoot doctors in particular who were targeted to learn the newly minted practice of integrated medicine, but practitioners from all three backgrounds agreed that what they collectively strove to practice was integrated medicine.

By the early 1980s, most of the "Western medicine" practitioners who had been sent down to the villages of the basin at the beginning of the Cultural Revolution had returned to their original town- or city-based clinics and hospitals, and the dissolution of cooperative health care did have an impact on the lives of rural practitioners. Former cooperative health care practitioners were officially re-designated as "village doctors," and they became independent, private practitioners. A national examination system for village doctors was implemented in 1982 and 1983, whereby doctors were to choose and ideally pass either the TCM exam or the WM exam. In the Lijiang basin, while all village doctors eventually took one of these exams, not all of them passed their respective exams. Of the village doctors I interviewed, approximately half had passed one of the exams (basic literacy usually being a problem for many of those who did not). While passing one of the exams generally added to the prestige of a given doctor among villagers, those doctors who did not pass an exam were not stigmatized or prevented from practicing.

As a result of the exams, most basin village doctors presently categorize themselves as either TCM or WM specialists. However, they virtually all refer to the type of medicine that they practice as integrated medicine, just as was the case during cooperative health care. While there have been changes in the rural health care infrastructure, the practitioners for the most part are still the same individuals. And they identify integrated medicine as the prevailing epistemology that informs their therapeutic practice.

In the hospital contexts of Dayanzhen (and other cities in China), there is more of a clear-cut division of labor carved out between TCM and WM. Both the Lijiang Prefectural Hospital and the Lijiang County Hospital concentrate most of their space and staff on the practice of WM, and have a relatively small space and staff focused on the practice of TCM (TCM is one department among the rest). In this context, the practice of integrated medicine is restricted to those patients who choose to visit both WM and TCM doctors of their own volition, and to pursue both kinds of diagnoses and treatments simultaneously; although there is a mutual referral system, there is no formal consultation between the concerned TCM and WM doctors who share the same patients.

CLASSIFICATION OF ILLNESS, THEORIES OF ILLNESS, AND TREATMENT OF ILLNESS

In the contemporary basin, the options open to residents with respect to strategies of therapeutic resort are for the most part determined by whether or not they have state jobs. This division breaks down primarily along urban and rural lines: the vast majority of village residents in the Lijiang basin (as throughout the PRC) are not state employees and consequently have no health insurance. Since most villagers also have limited access to cash income, this socioeconomic distinction means that they tend to avoid visits to the clinics and hospitals located in Dayanzhen. Villagers who have average or higher than average incomes will attend the latter clinics only in the event that they have a major illness, or for the birth of a first child. Villagers who have lower than average incomes just let the disease take its course, or have their babies in the village.

For most afflictions in both town and village contexts, depending on the illness, the first resort is usually either to use herbal medicine which one collects oneself, to purchase some medicine (either WM, TCM, or integrated medicine) at the local pharmacy, or to go to the nearest local clinic for an injection (usually of antibiotics) depending on one's resources. While visits to village clinics are on average at least 50% less expensive than visits to town clinics, villagers must use their own hard-to-obtain cash, whereas town residents generally pay very little of the cost themselves.

Additionally, there is a pecking order among the therapeutic institutions that villagers tend to frequent as opposed to those that members of work units tend to frequent. Most villagers in Lijiang County go to the somewhat less expensive county level institutions, whereas it is primarily members of state work units or those who can afford to pay out of their own pockets who tend to frequent the prefectural hospital. Most villagers complain bitterly about the fact that, since decollectivization, city hospitals and clinics demand payments up front before admitting anyone without health insurance. For most villagers, then, the primary form of medical care available (aside from self-treatment) is from village doctors.

The most common afflictions in basin villages, by the accounts of village practitioners and non-practitioner village residents alike, included colds, influenza, dysentery, parasites, stomach afflictions, gallstones and kidneystones, arthritis, heart disease, tracheitis, and gynecological disorders. The etiologies of these afflictions—according to basin village practitioners and lay villagers—are rooted in a variety of factors, including inattention to properties of heat and cold, dryness and wetness, and "wind," and/or inattention to "water–earth relationships," seasonal weather cycles, basin microclimates, one's diet, one's body type, one's emotions, one's gender, and/or one's position in the life cycle. For example, colds are caused by one's inappropriate response to excessive heat or cold in the weather in terms of one's choice of clothing (especially failure to wear enough clothing), or the "hot" or "cold" properties of the food one chooses to eat without consideration for the weather. Influenza is an especially potent form of a cold, which is "infectious" (i.e., defined as being more widespread) during the spring and fall seasons when the weather is particularly prone to changeability. Stomachaches are caused by eating irregularly, exposing oneself to excess cold through work, eating cold food, and drinking cold water, and/or by harboring unhappy feelings. Dysentery is primarily caused by exposure to too much heat, which is the explanation for its prevalence during the period of the Chinese solar calendar known as *san fu tian* (the hottest period of summer); dysentery is also caused by eating too many chili peppers, eating too many kinds of food with oppositional "qualities" (e.g., sour versus sweet, etc.), eating unpeeled fruit, or by eating cold food and drinking cold water. Gallstones and kidneystones are frequently also attributed to excess cold, and/or seen as resulting from many years of stomach problems. Heart disease is caused by chronic illnesses in other parts of the body that spread to the heart. Arthritis is a "wind" affliction caused by exposure to too much dampness, cold, and wind, as well as by exposure to sudden shifts in temperature and by working to exhaustion. Tracheitis and bronchitis are caused initially by exposure to cold and then continual exhaustion from work. Gynecological disorders are generally due to the intrinsically depleted nature ascribed to women's (reproductive) bodies (see below).

The etiologies of these afflictions overwhelmingly reflect a view of disease and illness causality that has been called the "medicine of systematic correspondence" (Unschuld, 1985). The medicine of systematic correspondence has evolved over the past two millennia and has integral links to Confucianism and the official culture of the Chinese state. Yin/yang dualisms in physiological, humoral (e.g., hot/cold, wet/dry, etc.) and other arenas parallel the dyadic relationships upon which Confucianism is premised. "Five phase" (*wu xing*) relationships between bodily organs and a variety of other substances parallel the emphasis on the complex network of relationships intrinsic to the bureaucratic structures of the Chinese state. These dyadic and quinary relationships are replicated at the levels of the individual body/self, the local level society/environment, and the state/cosmos. The goal for the maintenance of order/health and the avoidance of disorder/affliction is proper behavior/health care and balance. This discourse of proper behavior and the maintenance of balance reflects, among other things, an emphasis on prevention. Once out of balance, the medicine of systematic correspondence body is in a state of depletion and in need of some "bolstering" to renew its balance.

In addition to medicine of systematic correspondence references, informants sometimes also use germ theory references in their etiological definitions (i.e., in terms of concepts such as bacteria, viruses, and hygiene), but neither are these concepts used in keeping with their strict WM sense, nor are they viewed as primary agents in the cause of the afflictions just described (e.g., as would be particularly notable from a WM perspective in the cases of colds, influenza, dysentery, parasites, or stomach problems).

Many of the afflictions just outlined are treated by Lijiang basin village practitioners primarily with injections of antibiotics, intravenous infusions of glucose water, vitamins, and/or antibiotics, or tablets of WM pharmaceuticals. Some of the afflictions are treated primarily with Chinese medicine pharmaceuticals (which are administered through injections or IV infusions as well as in ways more conventional to Chinese medicine). And some of the afflictions are treated with a combination of both WM and TCM pharmaceuticals and techniques.

In all PRC official medical practices, an analytical distinction is made between the theory associated with a given practice, and the materia medica—or pharmaceuticals—associated with a practice. Techniques and substances associated with one practice (e.g., WM) can thus be "detached" and incorporated into another practice (e.g., TCM) without there being a perception that the integrity of a given practice has been violated. This logic also informs the local practice of integrated medicine in the Lijiang basin.

Additionally, when basin village practitioners and lay villagers are asked why they frequently utilize injections and IV infusions (of antibiotics or other substances), they explain that it is important to "bolster one's health." Both injections (usually of antibiotics—or, in other instances, of vaccines) and IV infusions (usually of glucose water combined with supplements such as vitamin C or iron or Chinese medicine infusions) are used to bolster the health of depleted bodies. Thus, with respect to the most common afflictions in the basin, WM treatment techniques and pharmaceuticals are being appropriated into an explanatory model primarily informed by a medicine of systematic correspondence discourse.

In addition to the much more prevalent medicine of systematic correspondence afflictions just addressed, village practitioners and lay villagers alike also refer to a number of afflictions that are much less common but that are highly stigmatized and consequently imbued with great symbolic importance. These afflictions include epilepsy, "madness," "leprosy," and *gu* witchcraft. The explanatory models used to talk about these afflictions reflect the discourses of two therapeutic practices. The first is the longstanding Chinese therapeutic practice of demonic medicine (Unschuld, 1985), in which are rooted concepts of "dirt," "pollution," and the need for the exorcism of malevolent and usually invisible agents. The other influence is WM, from which discourses of germ theory, hygiene, infectiousness, and genetic inheritability have been appropriated. Informants use these same discourses to talk about individuals with non-normative bodies as well.

Epilepsy, for example, is seen as both directly genetically transmitted and infectious. WM notions of germ theory and infectiousness, however, are appropriated into basin Naxi explanatory models of infectiousness that are rooted in both humoral concepts of "bad air" and demonic medicine concepts of malevolent agents. "Madness" is seen as potentially (although not necessarily) both hereditary and infectious. "Leprosy" (Hansen's disease) is regarded, again by basin village practitioners and patients alike, as both genetically transmitted and extremely infectious. "Leprosy" has a long historical record as one of the most stigmatized afflictions in China, and continues to be extremely stigmatized. *Gu* "poisoning" a form of witchcraft that was prevalent in the rural basin prior to 1949, was banned as feudal superstition during the Cultural Revolution, and has undergone a gradual resurgence in the villages of the basin since the beginning of the post-Mao period. *Gu* is an affliction that is first described in the demonological literature of the late Zhou dynasty (770–221 BCE) (Unschuld, 1985). *Gu* transmitability is thus talked about in a discourse of "infectiousness." It is also talked about in a discourse of genetic inheritability since *gu* cultivation was seen as passed through family lines (usually from mother-in-law to daughter-in-law—Diamond, 1988). Finally, as is the case in clinical and popular contexts elsewhere throughout the PRC, basin villagers with bodies that fall outside cultural definitions of "normality" are highly stigmatized. Should they choose to marry and have children, these individuals are generally criticized by their fellow villagers (including practitioners) and sometimes their own immediate families. The stigma draws upon fears of the inheritability of their disabilities.

While this analysis has been based primarily upon research in the village contexts of the Lijiang basin where the practice of integrated medicine is the prevailing epistemology informing therapeutic practices, it is reasonable to assume that this analysis is also reflective of therapeutic practices throughout much of the rural PRC; it is reasonable to make this assumption given the powerful influence of state policies (including cooperative health care and integrated medicine policies) during the Maoist period, and given that no new health epistemologies have been introduced to replace integrated medicine—notwithstanding the dissolution of cooperative health care as a brigade-based system. Additionally, it is important to note that in the town context of Dayanzhen and in the urban context of Kunming (the provincial capital), there is also a hegemony of Chinese medical epistemologies—most specifically TCM—that prevails in popular culture contexts, despite the institutional division of labor between WM and TCM that exists in urban contexts; this analysis is probably also generalizable to much of the urban PRC.

Implications for the Spread of HIV/AIDS and Hepatitis B in the PRC

The implications for these findings with respect to the spread of infectious diseases such as HIV/AIDS and hepatitis B are significant. As is the case everywhere in the world, popular culture and medical culture understandings of affliction and of treatment have profound effects on medical and lay responses to epidemics, and this is certainly the case with the emerging HIV/AIDS epidemic in the PRC. The frequent use of needles for injections of antibiotics that are a routine form of treatment in the basin (and elsewhere in the PRC) for many of the common afflictions described above can be a significant risk factor in the spread of both HIV/AIDS and hepatitis B. In the villages of the Lijiang basin as of 1990 (as was likely the case throughout much of the rural PRC at that time), needles were re-used multiple times for injections, with a ceremonious swish of a cleaning solution employed by village doctors between injections. Disposable needles have gradually become common in urban contexts over the past decade and a half, but the sterility of these needles has been compromised by the quality control issues that plague the contemporary PRC (see, e.g., Kristof, 1993).

SEXUALITY AND REPRODUCTION

It is taken for granted throughout the Lijiang basin, in keeping with the rest of the PRC, that the normative practice is to get married and then produce one or more offspring (depending on whether one is an urban or rural resident). In contrast to the dramatically pro-natalist policies of the Maoist period (Potter & Potter, 1990), since the early 1980s in Dayanzhen (as in other urban PRC contexts) the norm that has been strongly enforced by the government is the one-child policy—notwithstanding the fact that the majority of the population in Dayanzhen are classified as "minority nationalities" (a class of individuals who are theoretically exempted from the one-child policy). In the village contexts of the basin, as in other rural PRC contexts (where 80% of the population resides), there has never been a one-child policy. In basin villages that consider themselves more "cultured" and "advanced," most village families have two children, with the second child usually spaced two to three years after the first child (fines—sometimes considerable—are levied against those who exceed the two-child maximum or have their second child earlier than stipulated by the birth planning policy). In villages that are farther from the reach of the state, more than two children per family is the norm, and birth spacing is ignored.

In some contrast to other parts of the (Han Chinese) rural PRC—most notably southeastern China—not having male offspring to perpetuate the name of the lineage and to provide insurance of care in one's old age (as daughters generally marry out) is not considered to be completely devastating to basin families. Both in basin villages and especially in Dayanzhen, young men are frequently sought to marry into a son-less family and have at least one of their children take the surname of their wife's patrilineage. Nonetheless, the longstanding neo-Confucian ethos of "placing greater emphasis on males than females" does considerably influence the lives of both girls and boys as they grow up in the basin.

Young married women who do not become pregnant are stigmatized (in notable contrast to their husbands) as "without children." Of particular note in both town and village contexts of the Lijiang basin, the "stopping of menses" (i.e., among pre-menopausal women) is a much-cited causal factor cited by medical practitioners in their diagnoses of (female) infertility. How the basin compares statistically with other parts of the PRC is not clear, but basin practitioners attributed the frequency of amenorrhea to the ethos of hard physical labor and to the limited nutrition of most basin Naxi women.

One of the distinctive characteristics of the Naxi of the Lijiang basin—that is, in contrast to most other parts of the PRC—is the division of labor along gender lines. Even before the post-Mao feminization of agricultural labor in the PRC (Bossen, 2002), village basin Naxi women carried out most of the field labor, and both village and town basin women tend to do virtually all of the physical labor for their families. A Naxi woman's prestige is based on her ability to labor for her household, as well as to save and manage money for the household.

It is this ethos of ceaseless work that points to a major faultline between Naxi constructions of gender and the normative constructions of deficient female bodies reflected in pre-1949 neo-Confucian informed Chinese medical notions (Furth, 1986, 1987, 1999) as well as in the public health policies of the socialist Chinese state (including both WM and TCM paradigms). On one hand,

Naxi women work in their fields and in their homes everyday regardless of the rain or cold, blatantly ignoring the proscriptions of public health policy. On the other hand, there is nearly universal observance of postpartum "confinement" and special dietary practices among Naxi women in both town and village contexts of the basin (see below). The form and the spirit of Han-derived postpartum practices are observed, but the model of female physiological deficiency upon which this practice is based does not apparently carry over into any of the other practices proscribed by Chinese medical and public health policy.

The model of gender in Naxi society is one in which women are associated with production and men are associated with consumption. Epidemiologically, this dynamic is played out in the basin in the sense that the major chronic afflictions of women are those associated with activities of production. Naxi women, both rural and urban, suffer in highly disproportionate numbers (compared to Naxi men) from rheumatoid arthritis, chronic headaches and other pain, and a condition which they and local practitioners refer to as *shengjing shuairuo*, or neurasthenia (Kleinman, 1986). Naxi men, on the other hand, are plagued by diseases associated with consumption: liver and gallstone or kidneystone ailments from excessive consumption of grain alcohol, heart disease and strokes from excessive consumption of pork fat and meat in general (of which men get choice portions), and lung cancer and chronic bronchitis from excessive smoking. There is also a popular pun according to which the term for "vascular heart disease" is recast as "(government) official heart disease." This relatively new epidemic is ascribed to the consumption of too much rich food and too much grain alcohol among government officials at state-sponsored banquets (the overwhelming majority of whom are Naxi men).

This gendered division of labor also informs birth planning practices in the basin. As is also the case throughout the PRC, in the Lijiang basin, women's bodies are the primary objects of state birth planning policies and the objects of the Women's Federation representatives who monitor them; consequently women in both urban and rural contexts bear the primary responsibility for contraception. Intrauterine devices (IUDs) are the most common birth control technique utilized (approximately 40% of basin women), but many users and practitioners complain of often serious side effects (e.g., infections, bleeding) from these devices. Birth control pills and Depo-Provera type injections are popular methods in the basin (approximately 30% of basin women); condoms are not. Tubal ligation is also a strategy employed by some basin village women (approximately 20%) after their second child, if they have a hospital birth for this child (hospital obstetricians regularly recommend to their clients that they undergo this procedure during their hospital stays following their second birth). This latter strategy is not a popular one, however, as it is believed to weaken a woman's health, and vasectomies are rarely considered as a viable strategy for men for the same reason (see Potter & Potter, 1990). Abortion (which is free) is a common strategy of last resort for contraception.

HEALTH THROUGH THE LIFE CYCLE

In keeping with a medicine of systematic correspondence approach, basin residents and practitioners refer to their own body types as either "fire bodies" (intrinsically "hot"), or "water bodies," (intrinsically "cool"). Additionally, people who are plump are considered to be healthier than people who are thin. Babies' bodies up until their first birthday are considered to be particularly weak, as are women's bodies during the first month after having given birth.

Afflictions are seen as seasonal not only in terms of the annual cycle, but also in terms of the life cycle and gender. Village practitioners and lay informants alike usually present afflictions as "children's diseases," "elderly person's diseases," or "women's diseases." These categories of persons (children, the elderly, and women) are considered to be particularly vulnerable to affliction. In the neo-Confucianism-influenced medicine of systematic correspondence, female bodies are constructed as normatively out-of-balance and depleted (not unlike Victorian era women's bodies). Children in general are considered to be vulnerable, since they "eat too diversely," are always playing in water (therefore exposing themselves to cold), and do not have a regulated lifestyle. The elderly are presumed to be in a vulnerable state because of their advancing age. These divisions along lines of age and gender would seem to imply, by default, that the teen-aged to middle-aged male represents the normative body, at least in terms of lower inherent vulnerability to affliction. As noted earlier, children's diseases include colds, coughs, fevers, stomach aches, diarrhea, dysentery, and parasites. Elderly people's diseases usually include colds, tracheitis and chronic coughs, and rheumatoid arthritis. Older women in particular tend to be afflicted by rheumatoid arthritis, and post-menarche women of all ages seem to be afflicted with gynecological diseases, headaches, and neurasthenia.

Most babies in the basin are born between October and March, and health practitioners estimate that 80% of births are between November and February. This is because most young people get married during the Spring Festival (the lunar new year), and get pregnant during the first few months of marriage. Virtually all basin women—town and village alike—observe the longstanding Han Chinese tradition of a postpartum period lasting 30 to 40 days (and sometimes as long as 100 days) during which they rest under very proscribed conditions, and are supposed to consume specific "rich" foods.

REFERENCES

Bossen, L. (2002). *Chinese women and rural development: Sixty years of change in Lu village, Yunnan.* Lanham: Rowman and Littlefield.
Chao, E. K. (1995). Depictions of difference: History, gender, ritual and state discourse among the Naxi of Southwest China. Doctoral dissertation, University of Michigan.
Chao, E. K. (1996). Hegemony, agency, and re-presenting the past: The invention of dongba culture among the Naxi of Southwest China. In M. J. Brown (Ed.), *Negotiating Ethnicities in China and Taiwan* (pp. 208–239). Berkeley: Center for Chinese Studies and Institute of East Asian Studies.
Chao, E. K. (1999). The Maoist shaman and the madman: Ritual bricolage, failed ritual, failed ritual theory. *Cultural Anthropology, 14*(4), 505–534.
Croizier, R. C. (1968). *Traditional medicine in modern China: Science, nationalism, and the tensions of culture change.* Cambridge, MA: Harvard University Press.
Croizier, R. C. (1976). The ideology of medical revivalism in modern China. In C. Leslie (Ed.), *Asian Medical Systems: A comparative study* (pp. 341–355). Berkeley: University of California Press.
Diamond, N. (1988, January). The Miao and poison: Interactions on China's southwest frontier, *Ethnology XXVII, 1*, 1–25.
Farquhar, J. (1994). *Knowing practice: The clinical encounter of Chinese medicine.* Boulder, CO: Westview Press.
Farquhar, J. (1998). Re-writing traditional medicine in Post-Maoist China. In D. Bates (Ed.), *Knowledge and the Scholarly Medical Traditions* (pp. 251–276). Cambridge: Cambridge University Press.
Furth, C. (1986). Blood, body and gender: Medical images of the female condition in China, 1600–1850. *Chinese Science, 7,* 43–66.
Furth, C. (1987). Concepts of pregnancy, childbirth, and infancy in Ch'ing Dynasty China. *Journal of Asian Studies, 46*(1), 7–35.
Furth, C. (1999). *A flourishing Yin: Gender in China's medical history, 960–1665.* Berkeley: University of California Press.
Goullart, P. (1955). *Forgotten kingdom.* London: John Murray.

Hansen, M. H. (1999). *Lessons in being Chinese: Minority education and ethnic identity in Southwest China.* Seattle: University of Washington Press.
Harrell, S. (1995). Introduction: Civilizing projects and the reaction to them. In S. Harrell (Ed.), *Cultural Encounters on China's Ethnic Frontiers* (pp. 3–36). Seattle: University of Washington Press.
He, Z. (1994). Dui Mosuo Muxi Jiating de zai renshi (Rethinking the Mosuo matrilineal family). In Li Xiaojiang, Zhu Hong, & Dong Xiu Yi (Eds.), *Xingbie Yu Zhongguo (Gender and China).* Beijing: Shenghuo, Du Shu, Xin Zhi (Life, Schooling, New Knowledge Press).
He, Z. (1999). *Shengcun he Wenhua de Xuanze: Mosuo Muxizhi Jiqi Xianzai Bianqian (The choice of survival and culture: Mosuo matrilineal system and contemporary change).* Kunming, PRC: Yunnan Educational Publishing House.
Hsu, E. (1992). Transmission of knowledge, texts, and treatment in Chinese medicine and qigong: Three Settings in Kunming City, P.R.C. Doctoral dissertation, Cambridge University.
Kleinman, A. (1980). *Patients and healers in the context of culture: An exploration of the borderland between anthropology, medicine, and psychiatry.* Berkeley: University of California Press.
Kleinman, A. (1986). *Social origins of distress and disease: Depression, neurasthenia, and pain in modern China.* New Haven: Yale University Press.
Kristof, N. D. (1993). China is trying to stifle scandal over reused hypodermic needles. *New York Times,* May 5, 1993.
Mathieu, C. (1996). Lost kingdoms and forgotten tribes: Myths, mysteries and mother-right in the history of the Naxi nationality and the Mosuo people of Southwest China. Doctoral dissertation, Murdoch University.
Mathieu, C. (1998). The Moso Daba religious specialists. In M. Oppitz & E. Hsu (Eds.), *Naxi and Moso Ethnography* (pp. 209–234). Zurich: Volkerkundemuseum Zurich.
Mathieu, C. (2000). Myths of matriarchy, the Mosuo and the kingdom of women. In C. Brewer & A.-M. Metcalf (Eds.), Researching the fragments: Histories of women in the Asian context. Manila: New Day.
McKhann, C. F. (1992). Fleshing out the bones: Kinship and cosmology in Naqxi religion, Volume One. Doctoral dissertation, University of Chicago.
McKhann, C. F. (1995). The Naxi and the nationalities question. In Stevan Harrell (Ed.), *Cultural Encounters on China's Ethnic Frontiers* (pp. 39–62). Seattle: University of Washington Press.
Mueggler, E. (1991). Money, the mountain, and the state: Power in a Naxi village. *Modern China, 17*(2), 188–226.
Mueggler, E. (2001). *The age of wild ghosts: Memory, violence, and place in Southwest China.* Berkeley: University of California Press.
Potter, S. H., & J. M. Potter. (1990). *China's peasants: The anthropology of a revolution.* Cambridge: Cambridge University Press.
Rees, H. (2000). *Echoes of history: Naxi music in modern China.* Oxford: Oxford University Press.
The Revolutionary Health Committee of Hunan Province (RHCHP). (1977). *A barefoot doctor's manual.* Seattle: Cloudburst Press.
Rock, J. F. (1947). *The Ancient Na-Khi kingdom of Southwest China (Volumes 1 & 2).* Cambridge: Harvard University Press.
Rock, J. F. (1963). *The land and culture of the Na-khi tribe of the China–Tibet Borderland.* Wiesbaden: Franz Steiner Verlag.

Scheid, V. (1995). Transformation of the non-substantial: Developing traditional medicine in contemporary China. Paper presented at the Lu Gwei-djen Memorial Workshop on Innovation in Chinese Medicine. Needham Research Institute, Cambridge, U.K., March 1995.

Shih, C.-K. (1993). The Yongning Moso: Sexual Union, household organization, gender and ethnicity in a matrilineal duolocal society in Southwest China. Doctoral dissertation, Stanford University.

Shih, C.-K. (1998). Mortuary rituals and symbols among the Moso. In M. Oppitz & E. Hsu (Eds.), *Naxi and Moso Ethnography* (pp. 103–125). Zurich: Volkerkundemuseum Zurich.

Shih, C.-K. (2001). Genesis of marriage among the Moso. *Journal of Asian Studies, 60*, 381–412.

Sivin, N. (1987). *Traditional medicine in contemporary China*. Ann Arbor: Center for Chinese Studies, University of Michigan.

Unschuld, P. U. (1985). *Medicine in China: A history of ideas*. Berkeley: University of California Press.

Walsh, E. R. (2001a). The Mosuo—Beyond the myths of matriarchy: Gender transformation and economic development. Doctoral dissertation, Temple University.

Walsh, E. R. (2001b). Living with the myth of matriarchy: The Mosuo and tourism. In T. Chee-Beng, S. C. H. Cheung, & T. Hui (Eds.). *Anthropology, Tourism, and Chinese Society*. Bankok: White Lotus Press, 93–124.

Weng, N. (1993). The Mother House: The symbolism and practice of gender among the Naze in Southwest China. Doctoral dissertation, University of Rochester.

White, S. D. (1993). Medical discourses, Naxi Identities, and the state: Transformations in socialist China. Doctoral dissertation, University of California, Berkeley.

White, S. D. (1997). Fame and sacrifice: The gendered construction of Naxi identities. *Modern China, 23*(3), 298–327.

White, S. D. (1998a). State discourses, minority policies, and the politics of identity in the Lijiang Naxi People's autonomous county. *Nationalism and Ethnic Politics, 4*(1&2), 9–27.

White, S. D. (1998b). From "barefoot doctor" to "village doctor" in Tiger Springs Village: A case study of rural health care transformations in Socialist China. *Human Organization, 57*(4), 480–490.

White, S. D. (1999). Deciphering "integrated western and chinese medicine" in the rural Lijiang Basin: State policy and local practice(s) in Socialist China. *Social Science and Medicine, 49*(10), 1333–1347.

White, S. D. (2001). Medicines and modernities in Socialist China: Medical pluralism, Naxi identities, and the state in the Lijiang Basin. In L. H. Connor & G. Samuel (Eds.), *Healing powers: Traditional medicine, shamanism, and science in contemporary Asia* (pp. 171–194). Westport, CT: Bergin & Garvey.

Yan, R. (1982). A living fossil of the family—A study of the Naxi nationality in the Lugu Lake Region. *Social Sciences in China, 3*(4), 60–83.

Malagasy

Janice Harper

ALTERNATIVE NAMES

The people of Madagascar are known collectively as the Malagasy. This chapter is based on fieldwork conducted among the Malagasy living in the southeastern forests of Madagascar who are ethnically identified as both Tanala and Betsileo. Although treated as comparative and mutually exclusive ethnic groups, such social groupings are misleading. Tanala ("People of the Forests") is a performative identity; those who live in the forests and engage in the forest economy are "Tanala," whereas Betsileo is an administrative division of the pre-colonial Merina autocracy. Hence, for many living in the southeastern forests of Madagascar, Betsileo represents one's ancestry, while Tanala represents one's contemporary social identity. Moreover, ethnic identification is problematic because those so labeled do not necessarily think of themselves in terms of ethnicity, as much as they do in terms of ancestry or lineage. For further discussions on ethnicity in Madagascar in general, see Astuti (1995), Bloch (1989), Kottak (1971a), and Larson (1996, 2000). For further discussion of the Betsileo, see Dubois (1938) and Kottak (1971b, 1980), and on the Tanala see Beaujard (1983). For further discussion on lineage and ethnicity among the Tanala and Betsileo of the Ranomafana region, see Hanson (1997) and Harper (2002).

LOCATION AND LINGUISTIC AFFILIATION

Madagascar is the fourth largest island of the world, lying like a teardrop in the Indian Ocean, just off the coast of southeast Africa, approximately 400 km from

Mozambique. This geographical isolation from the main continent has contributed to its unique ecological biodiversity, as the island's flora and fauna have evolved independently from the rest of Africa. Despite its ecological diversity, the country is notable for its relative linguistic and cultural uniformity. The official language is Malagasy, originating from a Malayo-Polynesian language, and most, but not all, dialects are mutually understandable, despite significant grammatical differences. Many educated Malagasy also speak French, although this language is rarely spoken or understood in rural villages.

OVERVIEW OF THE CULTURE

Madagascar was first settled by Indonesian and Bantu seafarers in approximately AD 600. Although this human habitation is frequently noted as the chief cause of environmental degradation, less noted is the rich human diversity that has evolved from this settlement. There are now over 15 million Malagasy living on the island, half of whom are under the age of 18, and most of whom live in extreme poverty.

This poverty is related, in part, to the island's geography. A long mountainous terrain extends vertically down the center of the island, making agriculture and transportation particularly difficult in many parts of the island. Thus, agricultural lands, transportation of goods to and from major markets, and access to urban social services, are limited for many of the residents.

Pryor (1990) estimates that due to this topography, along with soil degradation, only 4% of the island is under cultivation. Nonetheless, agriculture comprises up to a third of the country's GDP, employs over three quarters of its labor force, and is the major source of its exports. The principal crops include rice, coffee, vanilla, sugar, spices, cotton, and tobacco. Up to half of the land is irrigated, although swidden agriculture (known as "*tavy*") is practiced extensively, particularly in the forests of the southeast, where *tavy* rice production is regarded by many as the principal cause of deforestation.

In addition to crop production, livestock and seafood contribute to important sources of domestic and international trade. The industrial sector includes food processing, clothing, textiles, soap-making, mining, petroleum refining, paper production, and tanning. More recently, efforts have been made to expand the island's service economy, most notably in ecotourism, but this objective is constrained by poor infrastructure, limited public transport and, in more remote areas, few hotels.

Despite its rich resource base, GNP per capita is approximately $250.00 annually; many have incomes far below this figure, although non-formal economic strategies are practiced widely to augment incomes. This poverty is related not only to the geographical constraints limiting agricultural production and transport, but also to a history of domination and exploitation that have contributed to present social divisions and marginalization.

Madagascar was colonized by the French from 1896 to 1960. Prior to colonization, however, a precolonial highland autocracy was established in the late 18th century, known as the Merina Empire. During the reign of the Merina, land was appropriated by the state, forced-labor of men was instituted, taxes imposed, and debtors or their families sold into domestic slavery. These objectives were facilitated by bestowing gifts, titles, and lands on indigenous leaders or others who would promote allegiance to the empire, a policy that furthered state aims to forge ethnic identities. Although the forging of a Merina ethnic identity was effective in the highlands (see Larson, 2000), the empire was never to gain authority over more than a third of the island. Nonetheless, a complex social hierarchy emerged (in varying respects) throughout the island, in which people were categorized not only in terms of ethnic affiliations, but in terms of caste as well, as they became identified as *andriana* (noble), *hova* (free), and *andevo* (slave). Moreover, within these categories there existed a range of social status. These social identities are no longer formally recognized, but they remain as significant markers of identity and status, contributing to contemporary social relations which Campbell (1985) suggests are best understood as impermeable categories of caste (see Bloch [1989] for further discussion on social rank).

Thus, ancestry and lineage are prominent in Malagasy culture as a means of establishing one's history and place in society. Kinship is generally bilateral, with one's ancestry traced through both the mother and the father. In most rural areas, where indigenous cultural practices are strongest, patrilocal residence patterns are the rule, though individual exceptions are common. Women have rights to land, including constitutional provisions (based on indigenous practice) of one-third inheritance. Women are active in decision-making, and men take an active part in child rearing (some areas, such as the southern Antandroy, are more patriarchal than others).

Throughout the 19th century, Lutheran and Catholic missionaries were active in the industrialization of the island, as well as the education of children, becoming exceptionally successful at converting Malagasy to Christianity. Thus, Protestantism, Catholicism, and indigenous belief systems continue to coexist and have fostered pluralistic belief systems that are particularly notable in the indigenous medical system.

By the end of the 19th century, the Merina Empire was replaced by French colonial rule, which replicated in many respects the strategies of the precolonial autocracy, including continued forced labor, taxation, land appropriation, and indirect rule through indigenous power structures. In addition, the French promoted a shift from subsistence agriculture to cash crop production, most notably in coffee, vanilla, and pineapple, with uneven success temporally and geographically. Although French occupation was noted for its intrusive control over the land and people of Madagascar, some practices had favorable results. For example, in an effort to augment the labor pool, maternal and neonatal hospitals were introduced in the early 20th century, which provided biomedical health services to women and infants. In order to limit the spread of infectious disease, manufactured pharmaceuticals were provided, in many cases, gratuitously. These and other health services contributed to improved health status and declining mortality rates, while compulsory education in the French language enabled many people to compete for administrative positions.

The oppressive nature of the colonial government, which was most characterized by its forceful rule over agricultural production and cultural practices, culminated in a peaceful transition to independence in 1960, followed by a period known as The First Republic, led by Philippe Tsiranana. Covell (1987) indicates that this period was effectively rule by a government of minority elites selected and supported by the colonial government, which continued to exert its influence throughout the decade. By 1975, however, the Second Republic was instituted when Didier Ratsiraka took power through a military coup, and promoted nationalist ideals including educating children in the public schools exclusively in the Malagasy language, banning foreign ownership of property, and drastically limiting foreign trade. Ratsiraka's rule was based on a Marxist-Lenninist economic philosophy, advocating public ownership of the primary means of production, and conversion of expropriated lands to agricultural co-operatives. At the same time, he borrowed heavily from the world market to finance poorly conceived national industries that failed to succeed, including soy and flour mills, despite the fact that the Malagasy do not produce soy or wheat. Consequently, notwithstanding his proclaimed resistance to foreign influence, Madagascar became the first socialist country to accept the economic liberalization criteria of the IMF. Structural adjustment policies included the continual devaluation of the Malagasy franc (including by nearly half in 1987), drastic cuts in the social sector such as health care and education, and privatization of industry, banking, and other financial institutions. These policies were aimed at expanding foreign economic investment in the country, but caused rapid inflation, severely limited access to biomedical health care, and contributed to the decline of educational standards in the rural areas (see Hewitt, 1992 on structural adjustment in Madagascar).

Ratsiraka's reign over the people collapsed following the massacre of peaceful demonstrators outside his palace in 1991. In 1993 physician Albert Zafy was democratically elected as president, but by 1996 he was impeached for abuse of power and Ratsiraka elected to return to office, this time as a "humanist ecologist," calling for international investment in the country's ecological resources. The island's species diversity has led it to be regarded as one of the world's "biodiversity hotspots," drawing considerable international aid and attention to its endangered flora and fauna, most notably the many species of lemur that exist no place else on earth.

In 2001, Ratsiraka lost a bid for re-election when the self-made millionaire mayor of Antananarivo, Marc Ravalomanana, narrowly defeated him. When Ratsiraka refused to concede defeat and declared martial law, mass protests ensued, leading to an economic paralysis of the country. A recount of the electoral results was ordered, and Ravalomanana was sworn in as president in May of 2002. The African Union refused to accept the presidency as legitimate and suspended Madagascar from the African Union, leading to a new election in December of 2002. When Ravalomanana clearly won a majority of the votes in the second election, Madagascar was reinstated into the African Union. Ravalomanana has made reducing poverty and inequalities in the distribution of wealth priorities of his presidency, and has continued to endorse the goals of environmental conservation. Toward these efforts, he has called for increasing attention to ecological tourism as a strategy toward economic self-sufficiency among the Malagasy.

THE CONTEXT OF HEALTH: ENVIRONMENTAL, ECONOMIC, SOCIAL, AND POLITICAL FACTORS

While the health of the island's lemurs and other endangered species receive frequent attention in the media and scientific literature, UNICEF (2002) reports that more than half of the population does not have access to safe water (including 15% in urban areas and 69% in rural areas), 40% of infants are moderately to severely underweight, and the infant mortality rate is 95 per 1,000 births, or nearly 10% (with far higher rates in rural areas). For every 100 children born who do live, more than 15 will die before their fifth birthday. For those who survive, life is likely to be short, for life expectancy at birth is a mere 54 years (New Africa 2001).

Illnesses include malaria, which is particularly prevalent in the rural areas, and has been increasing over the last two decades. This increase followed a 20-year decline in malaria, when an eradication campaign begun in 1948 proved to be very effective. Unfortunately, the effectiveness of the campaign led to decreased resistance to the disease. Hence, its return has been particularly devastating, as many Malagasy who cannot withstand the disease die. The United States Library of Congress (2001) reports that in 1985, 6,500 Malagasy died of malaria, increasing to 11,000 by 1987. The decreased resistance to the disease is further exacerbated by increasing malnutrition and parasitical disease.

Parasitical diseases include schistosomiasis (associated with the lack of safe drinking water and inadequate sewage), and chronic infestations of a variety of worms. Kightlinger (1993) found that 97% of residents of the Ranomafana National Park region in the southeastern region of the island were infested with multiple parasites. Other illnesses of concern in Madagascar include tuberculosis, leprosy, cholera, brucellosis, yellow fever, and bubonic plague. Of even greater concern was the prevalence of respiratory disorders and what appeared to be hepatitis. Respiratory disorders prevailed among children and women, and were probably related to both the rapid transmission of infectious disease in the closed quarters of local housing (with five to ten people sharing one to two bedroom houses no bigger than 10 by 20 ft), and the common practice of women cooking over wood burning fires inside their homes with only a tiny window and door for ventilation. In the village where I resided, several adults, both men and women, died following short illnesses that were locally diagnosed as *fefy*, *tazovony*, or *albumen*. Though having subtle differential diagnostic criteria, each of these illnesses included dark urine, yellowed skin, and swollen abdomens, suggesting that they were probably associated with hepatitis.

Malagasy are the world's greatest rice lovers, consuming more rice per capita than any other population in the world. *Tavy* rice is particularly nutritious, providing substantial protein and vitamins. The typical diet in the rice-growing regions includes two to three meals of rice per day, usually with some form of sauce (*laoka*), typically made of boiled greens. On occasion, rice is consumed with beans and, rarely, beef, pork, or chicken. Protein is also obtained from small fish caught in the irrigated fields, grubs and insects from the forest, and crayfish and eels caught in the rivers. Endangered species, such as tenrec, are also eaten, though consumption of these animals is not common, and more a matter of necessity than choice. Wild boar are also hunted and consumed occasionally. Manioc, cassava, boiled green bananas, and pineapple are additional staples of the forest diet. Beverages include coffee, rice tea, and *toaka gasy*, a locally distilled rum. Tobacco is chewed, primarily by women, as a means of curbing the appetite.

Malnutrition, however, is chronic, and exacerbates other health concerns such as malaria and respiratory problems. During my 14-month residency in a village adjacent to the Ranomafana National Park (1995–96), I witnessed the deaths of 18 men, women, and children (from a population of approximately 180). Of these, several were of small children who were extremely underweight for their ages (see Harper, 2002 for a detailed account of the death of a malnourished child). Villagers reported that in the ambiguously defined past, greater cooperation and exchange enabled less fortunate residents to remain well fed. As economic security declined, there was less willingness to provide food and cash to others, including close relatives. As a result, it was not uncommon to find weakened infants fed only breast-milk and watery rice while their healthier age mates received more substantial diets. Adults, as well, suffered from malnutrition, undoubtedly exacerbated by parasitical diseases, affecting both men and women.

MEDICAL PRACTITIONERS

A wide spectrum of medical practitioners can be found in Madagascar, but access to western biomedical practitioners, medicines, facilities, and technologies remains extremely limited to most Malagasy, particularly in the rural areas. Indigenous medical practitioners include *ombiasa* (shamans), who may generalize or specialize (in single illnesses, types of illness, or certain diagnostic or divination techniques), birthing specialists, herbalists, bonesetters, and veterinarians.

Ombiasa may be either men or women, though men are more common. The *ombiasa* may be trained from childhood, often learning the craft from his/her parent, or may assume the profession in adulthood. In the case of the latter, it is not uncommon that one's failure to have become an *ombiasa* as intended by the ancestors leads to illness. Upon diagnosis, the *marary* (sick person) accepts the mantle of *ombiasa*, learns the craft through spirit possession ceremonies and the assistance of an elder *ombiasa*, and becomes healed. More common, however, is the apprenticeship of the children or grandchildren of *ombiasa*. From childhood on, the apprentice is taught local medicines, diagnostic techniques, and ceremonial customs.

Diagnostic techniques may be limited to evaluation of physical symptoms, such as skin eruptions, and the prescription and preparation of local plant medicines. When such treatment fails, or the symptom constellation does not conform to a known illness category, or to an illness category regarded as originating in the environment, then other diagnostic techniques are employed to determine the source and treatment of the illness. Such techniques may include divination by mirror, water, seed, or sand. Linton (1933) indicated that divination by water was rare among the Tanala; my own fieldwork indicated that it was very common, perhaps because it was more easily learned by younger *ombiasa* than the more elaborate form of divination by seed, *sikidy* (see Linton, 1933 for a detailed account of *sikidy*). Less common is divination by sand, which involves pouring sand on a wooden tray and making marks in it with a stick.

Ombiasa also consult the ancestors through spirit possession ceremonies. They may either invite the ancestor to possess them, and thus, serve as a conduit through which the ancestor communicates to the living, or they may direct the possession of others. It is common for non-*ombiasa* to become possessed of spirits during such ceremonies, in which large numbers of participants may become possessed. (For ethnographic detail and analysis on spirit possession, see Sharp, 1993, 1994.)

More commonly, individuals with no particular professional standing may specialize in the treatment of a single disease or type of illness, such as an elder woman noted for being *mahay fanafody zazakely* (knowledgeable about children's medicines). Knowledge of a particular indigenous medicine may also be passed from generation to generation, such as a famed recipe for a measles vaccine made from local plants and chicken parts that I witnessed dispensed to children annually in the Ranomafana region. In many cases, men are the guardians of such knowledge, while women bring with them a more pluralistic medical strategy. This may be because the great variety of ecological niches in Madagascar have created differing native pharmacopoeia. In patrilocal settings, men are able to pass on to their sons the knowledge of their local indigenous medicines, while women, coming from other ecological zones, may bring with them broader knowledge of treating common health concerns. Thus, women's knowledge may be more generalized and innovative, while men's knowledge may reflect greater depth of the local pharmacopoeia available to them.

CLASSIFICATION OF ILLNESS, THEORIES OF ILLNESS, AND TREATMENT OF ILLNESS

While there is a wide range of understandings about illness classification and causation, illnesses tend to be classified as illnesses of God (*Zanahary*), illnesses of the environment, illnesses sent by the ancestors, those caused by witchcraft, and ghost or spirit illnesses. These categories are not mutually exclusive. For example, an ancestor can be displeased, and so send an illness of the environment, such as malaria (*tazo*—fever). It is understood that some illnesses are caused through infection, others through contagion, and still others through interactions with the environment. Nonetheless, if the illness is believed caused by acts of the ancestors, ghosts (*biby*), spirits (of which there are a variety), or witchcraft (*mosavy*), treatment of the symptoms will only lead to the recurrence of the disorder, or the onset of another health problem.

Malaria is prevalent and understood to be brought by mosquitoes. As an illness of the environment, treatment through pharmaceutical medication is preferable, but if not available, teas made from bitter plants are administered. *Hazomafaika* is the tea of choice in this case, but if it is not available, other bitter substances will be substituted. The concept of bitter is clearly associated with fevers, and in some cases over-consumption of bitter substances (such as unsweetened coffee) may be believed to *cause* the fever. When a child falls ill with malaria and dies, it is generally believed to be the cause of an illness of the environment or of God; if an adult falls seriously ill, it is more likely that other origins of the disease will be considered, such as witchcraft. If several family members fall ill or meet with misfortune, ancestral displeasure or witchcraft may be suspected, in which case, an *ombiasa* will determine the appropriate treatment.

Tazo vony (literally, "yellow fever"), *albumin*, and *fetsy* are diagnostic categories for illnesses associated with yellow palms, yellow eyes, dark urine, and swollen abdomens. Widespread disagreement exists among both *ombiasa* and lay people as to the causation and differential diagnoses of these three illness categories that probably represent hepatitis or other liver dysfunctions. Nonetheless, it was agreed that they were usually illnesses of God, and could not be cured by either indigenous medicines or by *fanafody vazaha* (medicine of foreigners).

Aretina biby ("ghost sickness") is often diagnosed when an illness is sudden, severe, and its symptom constellation does not correspond to any recognized illness category. Diagnosis is made by an *ombiasa*, and treatment, if successful, will require appeasing the offended ghost. *Biby* are considered so powerful, however, that treatment is often unsuccessful and the illness leads to death.

A classic example of an illness of the environment is *hantana* or scabies. *Hantana* is believed to be caused by poor hygiene and is very contagious; the prevalence of *hantana* in the village where I resided was so widespread that it was not even mentioned as a health concern, except when serious infections ensued. The treatment of choice is lindane, which is usually too expensive for most people. When unavailable, a number of plant medicines are used to make herbal infusions that are applied to the skin.

Mental illness is recognized, and the mentally ill are cared for and respected. In most cases I observed, mental illness was attributed to violation of a custom (*fomba*) on the part of the impaired person or his/her mother during pregnancy or while nursing. Ingestion of too much alcohol is believed to cause temporary madness, while smoking marijuana is believed to cause a person to become violent. It is also recognized that addiction to alcohol can occur, in which case the addicted ought never to drink again, corresponding in many ways to generalized views of alcoholism in the United States.

SEXUALITY AND REPRODUCTION

Sexuality is accepted as a natural experience for both men and women, although the consequences of sex, such as pregnancy at a young age or vulnerability to sexually transmitted diseases, is taken seriously. Single mothers are not stigmatized, although recognition of their economic vulnerability discourages women from being single mothers if they can avoid it. Adultery of both men and women is common, though probably no more so than in the United States, and responses range from passive acceptance to divorce. Women are more likely to be stigmatized for adultery, and may lose their children due to it, while men may be required by local elders (the *fokonolona*—village-level council) to make restitution to their wives (such as providing her with cash or livestock). Homosexuality is practiced with discretion.

The ideal family size has varied throughout Madagascar's history. Prior to colonization, marriage was often in the mid to late-twenties, and family size relatively small (see Harper, 2002). Colonial policies aimed at increasing the labor-pool included taxing unmarried people and waiving forced labor for fathers of seven or more children. In addition, maternal hospitals were introduced to improve women's reproductive fitness. All but one of the elder women with whom I lived had given birth in the colonial hospitals; their daughters all gave birth on dirt floors in the home, as access to safe and hygienic biomedical care was very limited (the local hospital lacked both running water and electricity).

Infertility is a grave concern to both men and women, in large part due to the shortage of labor and social support that one would suffer if he or she had no children. An elaborate system of fosterage has developed in response to infertility, and it is not uncommon for Malagasy to adopt the children of their neighbors or kin.

Sexually transmitted diseases are common and have been increasing, and include gonorrhea, syphilis, trichomoniasis, and candidiasis (U.S. Library of Congress, 2001). While the World Health Organization reported that HIV/AIDS cases were so rare in 1993 that it ranked the country as having a 0.0 case rate, by 1997 this rate of HIV had increased to 0.12%, an estimate that USAID reports is probably due to underreporting. Moreover, the rate is expected to accelerate, due to the high rate of other STDs, which include a 35% rate among sex workers and 15% among pregnant women. USAID researchers have found that up to 82% of all women have had at least one sexually transmitted disease, and that in rural areas, 21% of pregnant women had active syphilis. In my own research in the southeastern forest region near the Ranomafana National Park, I found that women were commonly not informed of their husband's STD infections, due to a belief that it was a "man's disease," that women without symptoms were not infected, and that men taking medication could not infect women.

HEALTH THROUGH THE LIFE CYCLE

Pregnancy and Birth

Birth control is commonly desired by both men and women, although birth spacing is regarded as more critical than curbing the number of children. Nonetheless, ineffective family planning campaigns which have provided only short-term supplies of birth control pills and/or no follow-up care, have led to resistance among Malagasy to participate in family planning campaigns. Concerns that have been expressed include the many side-effects associated with the pills, side-effects which people indicated were particularly aggravating to undernourished and chronically ill women. The use of contraceptive implants is of even greater concern, as follow up care is difficult to access, if it is provided at all.

Abortion is practiced, often with indigenous medicines, but it is highly stigmatized, particularly among those who have converted to Catholicism. Women have been known to accept birth-control pills from family planning agencies, which are then saved and dispensed discretely to women in need of abortions, who take several pills to induce spontaneous abortions.

Women often withhold public disclosure of their pregnancy, letting their swelling bellies announce the coming child. They continue to work up until they give birth if necessary, particularly if farming obligations find the woman hoping that she not go into labor until the harvest is in. When they are no longer able to work, their female friends and kin will assist them. When in labor, a woman is not expected to cry in pain. To do so would indicate weakness. Births usually take place in the home, on grass mats laid on dirt floors, with a fire burning in the home. Elder women assist, generally someone specialized in midwifery skills. When the child is born, it is washed and dressed, its head covered in a knit cap to protect it from entry of malevolent spirits (believed to enter through the fontanel). After it is dressed, it is shown to the crowd gathered outside the door, and quickly brought back inside. The birth of a girl is as joyously celebrated as the birth of a boy.

Infancy

The new mother will stay with the child in the confines of the home for up to 2 months, venturing outside only to relieve herself. This period of confinement has been growing shorter and shorter, as women's agricultural work must be resumed and the heat of the home may become unbearable (with little or no ventilation or light). I heard several stories of new mothers becoming ill from ghost sickness, requiring them to terminate their confinement early, in order to appease the ghosts.

Women try, if possible, to give birth in their natal home. When this occurs, there will be a celebration to mark the departure of the child and mother from the village, before their return to the woman's marital residence.

Children are nursed into their second year. They are toilet trained at a young age, often less than a year when diapers are not available. The urine of a baby is considered benign and laughter is the usual response to a child urinating in the home or on an adult. Although women are the primary caretakers of children, brothers, fathers, and grandfathers are very active caregivers, and do not hesitate to hold or watch young infants or children when the mother is unavailable.

Children who are seriously ill or handicapped are loved and cared for, although I did witness several instances of serious neglect of very ill children, with considerable disagreement as to why they were neglected (see Harper, 2002).

Measles are a grave concern for young infants, as is possession by ghosts. *Ombiasa* prepare special amulets for protection. Chronic disorders, such as scabies and respiratory problems, are so ubiquitous that they do not cause much alarm. Indigenous medicines may or may not be used. When pharmaceutical medicines are available, they are used.

Childhood

Children are well loved and cared for in Malagasy societies. Among the forest farmers of the southeastern region, childhood was a time of great joy, imaginative play, and hard work. Children learn at an early age to care for younger siblings, to help their parents with household chores, to labor in the fields, and to go to school. Learning the responsibilities and skills associated with rural living begins early. Children are taught to use large knives in many cases before they can even walk, and once walking, are quickly taught to master a multitude of household tasks. Gender distinctions are clear. Girls are taught to carry items on their heads and backs, building up their back muscles by carrying blocks of wood or small dolls, with the weight gradually increasing until they are able to carry an infant sibling by the age of 5 or 6. Girls also have small "tea parties," making small bowls and cups from leaves fastened together with bits of twig, and filled with various seeds, flowers, and leaves. And importantly, girls learn to help their mothers sort beans, and shake and pound rice, before the age of 3.

Boys' play includes various sporting activities such as fashioning a game of soccer from a grapefruit, or hunting for insects from the forest. Boys chop wood, harvest rice, and learn carpentry skills. They also assist in the care of their younger siblings and cooking if there are no older girls in the home.

Both boys and girls contribute to the household income when necessary by assisting in the rice fields (girls plant, weed, and harvest swidden rice, boys harvest both swidden and irrigated rice, and older boys may help prepare the rice fields). One very common way for children to earn extra income for the household or meals for themselves is by using their slingshots to scare away birds eating the rice fields. Boys use the typical y-shaped slingshot, while girls use a long macraméd type of slingshot.

Both boys and girls go to school starting at about the age of 6. Education is highly valued, but not always completed beyond the fourth or fifth grades for a number of reasons. The schools in the rural areas are virtually unfunded. What little funding they have goes to such costs as teacher salaries and building repair. Consequently, there are virtually no schoolbooks, and parents must purchase pencils, papers, and any other supplies. Teachers have decades' old texts to work with, chalkboards without chalk or erasers, and limited educations of their own. In addition, the agricultural needs of a farming community contribute to frequent absences. Schools close for harvest seasons, in many cases children's labor is needed on the fields for planting season, schools must be closed three days for funerals (during my fieldwork, this meant that the school was closed 18 times in a period of 14 months), and when teachers must hike to the provincial center for their monthly paychecks, schools may have to close. (Another factor contributing to the closing of the school in the village where I worked was the simultaneous birth of both teachers' babies, leaving the school closed for nearly 2 months while the teachers convalesced.)

The period of childhood generally lasts until the early teens, when an adolescents' growing independence and sexuality take hold.

Adolescence

Boys are circumcised from the ages of about 4 to 14. Circumcision is based more on availability of the circumciser than any culturally ascribed time period, although this may have been otherwise in the past, or may differ in other regions. I witnessed one circumcision ceremony, in which all uncircumcised boys over the age of about 4 were publicly circumcised in an unannounced ceremony. Following circumcision, they were given *toaka gasy* to drink and set down on the ground. Every boy thus circumcised developed serious infections, requiring the use of antibiotic creams and, in some cases, oral antibiotics (requiring the father and the infected boy to hike an hour and a half over the hilly terrain to the nearest hospital to get a prescription for antibiotics requiring a substantial portion of the household income).

Adolescence is a period of increasing responsibility and preoccupation with the opposite sex. Girls go to the market every week, in hopes of meeting a boy from another village. Boys also attend markets, and in many cases, begin drinking. Alcohol use and abuse can become a serious problem at this stage. Drinking, and sexual promiscuity, is a concern among a number of parents who regard these habits as new and potentially destructive to both the family and the culture, although it is unclear how new such habits are.

Many girls are mothers by their mid- to late-teens, and if they are unmarried, they remain in their natal household and receive considerable support from their families. Nonetheless, economic hardships are common, and adults recognize early pregnancy as increasing a daughter's economic vulnerability.

Adulthood

Domestic abuse does occur, but a woman is not expected to endure it. Should her husband abuse her, the community will support her and the *fokonolona* (village-level council) may intervene. Nonetheless, divorce may well have significant economic repercussions, leading a woman to endure an unhappy or abusive marriage despite its unacceptability. Divorce means that a woman is expected to return to her natal village. Although she will theoretically have rights to one third of her parents' land, the reality is that land shortages have led most farming families to intensifying production on lands they hold. Thus, a woman may well return home to find that "her" land is being farmed by her brothers, themselves likely to have limited land, and she cannot just reclaim it. Many women protect themselves from just such an event, by continuing to farm their land after marriage, returning frequently to their natal villages and enlisting the help of family members in agricultural production in exchange for giving them a share of the crops. In this way, a woman is able to augment the income of her own household income, and maintain her rights to her land in the event of divorce or the death of her husband.

For a woman to divorce, she must also accept the loss of custodial rights to her children in many cases. Although most groups in Madagascar recognize bilateral kinship, acknowledging one's ancestry through both the mother and the father, because children are important sources of labor, their custodial care remains with the father upon divorce, regardless of cause. A woman may take a nursing infant, but she is expected to return him or her to the custody of the father upon weaning. Visitation rights are generous and accorded both parents. In cases of child abandonment by a father (frequent with fathers, but rare in the case of mothers), a mother will retain custodial rights.

While access to medical care does not prioritize one gender over another, both men and women face limited health care options. Thus, chronic illnesses often become naturalized and suffered without complaint or treatment. Women are more likely to suffer respiratory disorders, which may well be related to cooking over wood fires in homes with little ventilation. Women also suffer serious back problems, related to the planting and weeding of rice that requires them to be bent over for several hours a day. In terms of nutrition, men may well eat more, but their rights to food are not considered a priority over women's. Women will be the first to wake in the morning, however, to prepare the morning meal and fetch water from the river, and the last to end work after washing the dinner dishes in the river.

The Aged

Age is regarded with considerable respect, and is considered a time when one's work in the fields is expected to give way to domestic responsibilities. Older people are important to the household economy by caring for children, cooking, cleaning the home and compound, and providing important medical care, such as finding and preparing indigenous medicines. More recently, elder people have often had to continue working in the fields for wages, if their children have died or left the village, or otherwise failed to care for them.

Elders are also important to the community as members of the *fokonolona* or otherwise contributing to important community decisions. Their views and wishes are regarded with respect and to defy the wishes of an elder is highly stigmatized. More recently, economic stratification has increasingly replaced the stratification that comes with age, as elders complain of having little voice in community matters that more prosperous but younger adults rule on.

Although it is not uncommon to hear of people living to be 115, 120 years (age often being a guess), as people age their health problems increase significantly. Unfortunately, with their increasing health needs, they also face decreasing access to health care. The older they become the more difficult it becomes to cross rugged terrain to reach public health facilities. Thus, elders living in rural areas must rely on traveling health professionals, if there are any (which are rare), indigenous medicine, or the pharmaceutical medicines younger adults bring to them.

While one's eyesight fades, in villages with only candles to light the night, there are no eyeglasses. Hearing aids are available only to those with the money, and who live or have access to larger cities. As arthritis sets in, it becomes more difficult to squat to urinate or defecate, or to cook over an open fire or get up from the floor where one eats and sleeps. As teeth are lost, there are no dentures to replace them. Thus, aging may bring greater respect, but it is also likely to bring greater pain and discomfort in carrying out the days' activities. It is also likely to bring economic hardship, particularly with the death of a spouse (elder women are not expected to leave the community upon the death of a spouse, but are more likely to remain in the household of a son).

Dying and Death

When a person is dying, the extended family is notified and the family prepares for their arrival. They will sit with the *marary* (sick person), caring for them, and visiting with kin. Women provide greater care, but if the dying is a man, male relatives will assist with personal needs. Upon death, the community will gather outside the home, and women will wail loudly, a wailing that continues for several hours.

A child's death is regarded as a great tragedy. In many cases, however, the likelihood of a child dying is so great, that it is met with dignity and stoicism, but little outward grief (aside from the wailing that announces the death). The older a child is, or the younger an adult, the more profound the grief.

Upon death, a body will be cleaned by the same-sex kin of the deceased, and then wrapped in clean *lamba* (cloths). If a married woman, her body will be wrapped in grass mats and carried by men to her natal village, where her kin will carry out the remaining preparations for burial. Men will go to the forest and chop down a tree, which will then be stripped of its bark and hollowed out. While this is being done, the *maty* (deceased; the term also means "dead") is wrapped in grass mats and taken to the *trano-be* (literally, "big house" used for ceremonial and communal purposes). Women will attend the corpse throughout its rest in the *trano-be*, fanning flies from its face and guarding its soul. Both men and women, girls and boys, will visit to pay their respects. All non-essential work duties cease for three days.

(Cows are sometimes sacrificed at this point, if affordable, but economic hardships, at least in the village where I conducted my research, have made such sacrifices rare.)

When the coffin is prepared, the *maty* is wrapped in clean cloths again (the body fluids having soiled the corpse), and placed in the coffin. On the third day (or longer if the family has not yet arrived from other villages, although the decay of the corpse is rapid and its internment encouraged), a funeral is held. The village gathers outside the *trano-be*, a *kabary* (public speech) is recited in honor of the dead, and the offerings made from the various households are publicly listed. The coffin is then brought out of the *trano-be*, and placed over burning embers, in which vines have been smoking. The vines are then wrapped around the coffin and tightly tied, the *kabary* continues, and the coffin is then carried into the forests, along with a burning ember (to provide warmth to the *maty* in its afterlife) where it will be interred in caves (based on lineage). As the coffin is taken away, the crowd follows, while the embers are stamped out, placed onto a grass mat, and all the soiled mats and *lamba* added. They are then wrapped up and tightly bound with the vines, and taken away to a sacred spot. Failure to follow this custom may lead to death from ghost sickness (see Harper, 2002 for an account of such a death).

As the coffin leaves the village boundaries, it is carried high overhead where it passes over the crouching figures of the children (if any) of the deceased. In this way, it blesses their lives before departing.

Grief is expressed with loud displays of sobbing and feinting among the children and women at this point. In some cases, the tears immediately give way to festive partying, in other cases, a deep sadness permeates the village which then prepares to feed the mourners and visiting families.

REFERENCES

Astuti, R. (1995). *People of the sea: Identity and descent among the Vezo of Madagascar*. Cambridge: Cambridge University Press.

Beaujard, P. (1983). *Princes et paysans: Les Tanala de l'Ikongo: Un espace social du sud-est de Madagascar*. Paris: Editions L'Harmattan.

Bloch, M. (1989). *Ritual, history and power: Selected papers in anthropology*. London and Atlantic Highlands: The Athlone Press.

Campbell, G. R. (1985). *The role of the London Missionary Society in the rise of the Merina Empire 1810–1861 (A contribution to the economic history of Madagascar)*. Swansea: University College.

Covell, M. (1987). *Madagascar: Politics, economics and society*. London and New York: Frances Pinter.

Dubois, S. J. (1938). *Monographie des Betsileo (Madagascar)*. Paris: Institut d'Ethnologie.

Hanson, P. (1997). *The politics of need interpretation in Madagascar's Ranomafana National Park*. Doctoral Dissertation, University of Pennsylvania.

Harper, J. (2002). *Endangered species: Health, illness and death among Madagascar's people of the forest*. Durham: Carolina Academic Press.

Hewitt, A. (1992). Madagascar. In *Structural adjustment and the African farmer*. Alex Duncan & John Howell (Eds.). London: Heinemann

Kightlinger, L. K. (1993). *Mechanisms of Ascaris lumbricoides overdispersion in human communities in the Malagasy rainforest*. Doctoral Dissertation, University of North Carolina at Chapel Hill, North Carolina.

Kottak, C. (1971a). Cultural adaptation, kinship, and descent in Madagascar. *Southwestern Journal of Anthropology, 27*(2), 129–147.

Kottak, C. (1971b). Social groups and kinship calculation among the southern Betsileo. *American Anthropologists, 73*, 178–193.

Kottak, C. (1980). The past in the present. Ann Arbor: The University of Michigan Press.

Larson, P. M. (1996). Desperately seeking the "Merina" (Central Madagascar): Reading ethnonyms and their semantic fields in African identity histories, *Journal of Southern African Studies, 22*(4), 541–560.

Larson, P. M. (2000). *History and memory in the age of enslavement: Becoming Merina in highland Madagascar, 1770–1882*. Portsmouth, NH: Heinemann Social History of Africa Series.

Linton, R. (1933). The Tanala: A hill tribe of Madagascar. Chicago: Field Musuem of History.

New Africa (2001). *Madagascar Health and Population*, available online at: www.newafrica.com/profiles/healthpopulation

Pryor, F. L. (1990). *The political economy of poverty, equity, and growth: Malawi and Madagascar*. Washington: Oxford University Press (for the World Bank).

Sharp, L. (1993). *The possessed and the dispossessed: Spirits, identity, and power in a Madagascar migrant town*. Berkeley: University of California Press.

Sharp, L. (1994). Exorcists, psychiatrists, and the problems of possession in northwest Madagascar. *Journal of Social Science and Medicine, 38*(4), 525–542.

UNICEF (2002). Madagascan statistics, available online at www.unicef.org/statistics/country. Last updated Feb. 1, 2002.

United States Library of Congress. (2001). Madagascar: Public Health. *Country studies and area handbooks*, available online at http://lcweb2.loc.gov/frd/cs

Malays

Ronald Provencher

ALTERNATIVE NAMES

There are no alternative names for Malays, but many Malays have sub-ethnic and regional identities based primarily on descent, dialect, customary ritual, and typical foods. Such identities are prefaced in conversation by the word *orang* (person), for example: *orang bugis*, *jawa*, *johor*, *kedah*, *kelantan*, *minang*, *patani*, or *selangor*. *Melayu* means "Malay" in the Malay language, the paramount trade language of Southeast Asia for centuries. It was the name of a seventh century AD trading state on the Jambi river in Sumatra, near the Straits of Melaka. The royal family of the port city of Melaka, which claimed descent from the Jambi monarchy, was among the first to convert to Islam, an effective foil against conversion of common people to the Christianity of European colonial regimes. From that time, converts to Islam who could speak Malay, whatever their natal ethnic identity, were said to have become Malays (*masuk Melayu*).

Who are the Malays? In Malaysia, according to federal law and everyday usage, the term "Malay" applies to anyone who habitually speaks Malay language and is a Muslim. Malay has been the primary language of commerce and the lingua franca of the Malayan Archipelago (modern Indonesia, Malaysia, southern Thailand, and Brunei) for many centuries because of the dominance of Malay speakers in trade.

LOCATION AND LINGUISTIC AFFILIATION

There are large populations of Malays in Malaysia, Singapore, the independent Malay nation of Brunei in northern Borneo, and in the southernmost provinces of the nation of Thailand (Pattani, etc). Other significant populations of Malay speakers live in many areas of Indonesia (especially in Sumatra, north coastal cities of Java, and Indonesian Borneo). Malay is the official national language of Malaysia, Indonesia, and Brunei; and one of the official languages of Singapore.

Malay language belongs to the Western Malayo-Polynesian sub-group of the Austronesian family of languages, which includes closely related languages spoken in Madagascar, Vietnam, the Malayan peninsula, island Southeast Asia, and western Micronesia (Bellwood, 1985, pp. 107–108). Most scholars believe that present-day Indonesia (particularly Sumatra) was the original homeland of speakers of Malay language.

Malay language has several registers (levels of sophistication of vocabulary and grammar) and a number of regional dialects. Regional dialects of the everyday register of Malay commonly coincide with sub-ethnic identities, and other languages and cultures closely related to Malay have been strongly influenced by it. For example, the everyday language of the Minangkabau of western Sumatra (in Indonesia) and of the Minangkabau in the state of Negeri Sembilan in Malaysia are dialects of Malay.[1]

OVERVIEW OF THE CULTURE

Modern Malay culture, like other complex cultures of the present, has absorbed much from neighboring civilizations, which in turn borrowed from Malay culture. As a particular civilization, it is at least as old as existing European cultures. Malay traders were probably involved in the maritime trade between East Asia and South Asia by the first century AD. Four centuries later they were described as long distance traders in Persian court literature. Malays may have sailed as far as the East African coast by that time. They traded luxury goods and ordinary commodities. Through a period of twenty centuries, as they interacted with other Southeast Asian, South Asian, East Asian, Middle Eastern, African, and European peoples (especially Portuguese, Dutch, and British) involved in long distance trade, they borrowed and adapted aspects of these other cultures. The result is a very complex culture and society that has long included urban as well as rural elements.

Probably, the beginnings of this civilization were based on the cultivation of irrigated taro (*keladi*) and rice (*padi*) combined with river and coastal fishing and trade, economic activities still pursued by many. Chiefdoms developed at junctures of large rivers with the ocean, and strong chiefdoms conquered weaker, leading to the development of small states. As Malay kingdoms came into contact, through trade and warfare, with states of greater scale they absorbed knowledge of "world religions" such as Hinduism, Buddhism, and later Islam. Traditional Malay medical knowledge bears the imprints of all three of these world religions (especially in representations of spirits, curing rituals, word charms, healing prayers, sacred numbers, and therapeutic objects). Traditional medical knowledge also retains older, aboriginal elements such as the concepts of soul substance (*semangat*), and of intrusion of a foreign soul (*badi*), or loss of one's own soul (*jiwa*) as causes of illness and death.

At the beginning of the 16th century AD, first the Portuguese and then the Dutch and English colonial regimes pressed European ideas about health upon the Malays. Those ideas, of course, were not clearly superior to those of traditional Malay medicine of that time. European medicine was still based largely on the humoral concepts of Greco-Islamic medicine (itself previously introduced to Malays by Arab traders) and certainly not those of modern cosmopolitan medicine, which was barely in its infancy. Cosmopolitan medicine came into being gradually, not becoming nearly what it is today until the beginning of the 20th century AD. It became available to colonial subjects in the Malayan Peninsula and the Indonesian Archipelago in small steps, about the same time that it became available to Europeans.

Colonial regimes promoted the continuation of Malay as the language of commerce and public communication. Many different kinds of people, with different languages and variations in medical beliefs, had become Malays by the beginning of the 20th century. Descendants of Indonesian immigrants to the Malayan Peninsula, such as the closely related Minangkabau as well as the more distinctive Bugis and Javanese, eventually accepted being members of that broader ethnic category used by colonial officials—"Malay". They took advantage of economic prerogatives, such as privileged access to land reserved for Malays. As members of the Malay communities they had joined, they contributed stories of health, illness, and curing appreciated by their Malay neighbors, adding to and reinforcing present-day Malay traditional medicine, which coexists with cosmopolitan medicine.

One result of the intensity and duration of relations with many other civilizations over a period of many centuries was the development of a traditional medical pharmacology, comparable to traditional Chinese pharmacology, too extensive and complex to begin to describe in an article of this length (Gimlette, 1971, p. 207).

THE CONTEXT OF HEALTH: CULTURAL, ENVIRONMENTAL, ECONOMIC, SOCIAL, AND POLITICAL FACTORS

The natural environment is not well reflected in traditional Malay medicine. Traditional Malay folklore does not glorify human settlement in the rainforest or jungle.[2] Very few houses in rural communities are isolated. They are usually within voice distance of neighboring houses, clustered in a hamlet, itself one of several hamlets connected by roads or broad trails to a central mosque (*masjid*) and school (*madrasah*). And/or the houses are within sight and hearing of their neighbors and clustered along highways, riverbanks or seacoast. Rural Malays prefer to be in their houses by nightfall. And given a choice, many prefer to live near or in market towns and cities. A young man should travel (*merantau*), usually to urban places, in order to gain experience before beginning a family. And if possible, every Malay should travel to Mecca at least once in their lifetime. Malays have been deeply involved in trade and urban places for centuries.

Basic concepts of health and illness in Malay culture have many origins. Some are clearly aboriginal and ancient. The concept of *semangat* ("soul substance"), for example, is clearly related to *mana*, *mano*, *manu*, etc. in many other Malayo-Polynesian languages of Southeast Asia and the islands of the Pacific Ocean. In these cultures, absence or loss of soul substance and intrusion of soul substance of another being are believed to be major causes of illness, especially mental illness.

Humoral and sensual concepts relevant to health have native Malay terms (*angin*, "air" or "wind"; *panas*, "heat"; *sejuk*, "cool"; *kotor*, "dirty"; *bersih*, "clean"; *manis*, "sweet"; *asam*, "sour"; *pahit*, "bitter"). These conceptual foci have been reinforced through early and continuous contact with other societies participating in the ancient long-distance sea trade involving China, South Asia, the Middle East, and the West. Some premodern aspects of medical knowledge systems, especially humoral concepts, from these other culture areas have merely reinforced native concepts.

Major themes in Malay culture are strongly reflected in traditional medicine. These themes include profound enjoyment of the details of: (1) food and drink; (2) registers (vocabulary and grammatical levels) and dialects of Malay and related languages; (3) systems and levels of courtesy in their own and other social systems; and (4) one's own social rank compared to others. One's rank vis-a-vis others depends upon such factors as relative age, ancestry, wealth, and authority in the community. This emphasis on expressing differences in rank has become more complex as Malays have become part of the modern industrial world. Basic consciousness of rank is embedded in sibling kinship terminology, in which gender is not noted in the term for younger siblings, but is noted in the terms for older siblings: *adik* for younger brothers and sisters; but *abang* for older brothers, and *kakak* for older sisters. Age is respected through use of a gender-specific term. Respect is good, but being the eldest of one's siblings often involves serious psychological trauma. Generally, a Malay baby is born into a world of adoring adults, and is spoiled by adults until entering school, where the child first meets serious discipline from adults. Male children are more spoiled than females, who by the time they enter school are already involved in responsibilities for the whole household. Also, the firstborn baby (*si long*) is fussed over and spoiled in ways that subsequent children are not. Nonetheless, the firstborn is, at a tender age, given responsibility for the younger siblings who have replaced him or her in the immediate affections of adults in the family. These family lessons concerning rank and its problems mark the lives of most Malays, because kinship terms are used in the everyday register of Malay language as seniority terms to address and refer to non-relatives in the community and workplace. Symbols of social rank are commonly important parts of curing rituals, where they are combined with symbols of "inappropriate" or "contaminated" food and social relationships.

Food and drink are connected to social relationships in all societies, of course. But the connection in Malay society is especially strong. Malays have an intense interest in food and drink. Their cuisine is enriched with fresh spices and many kinds of tasty vegetables and fruits. Rice is served not by the cup but by the plate. Malays routinely eat Hindu vegetarian and Muslim Indian foods, and they occupy most of the tables in Chinese Muslim restaurants. At the same time, an overwhelming majority of Chinese Malaysians are not Muslim, their food and drink are not *halal* (ritually pure), and they are viewed by many Malays as economic and political competitors. Malays love Western "junk" food, and eat at franchise restaurants that serve *halal* (ritually pure) versions of burgers and fried chicken. But more important are the traditional feasts (*kenduri*), which mark the major social events in everyday life: birth, coming of age, marriage, death, economic and social advancement, and the many Islamic holidays.

The greatest feasting occurs at the end of the Fasting Month (*Ramadan*) which is the annual celebration of God's greatest gift, the Holy Koran. Most individuals visit the homes of their parents, other relatives, neighbors, and friends. There, they feast on arrays of the great delicacies of Malay cuisine. They, too, in their own homes, must host relatives, friends, and neighbors. Hours are spent preparing and consuming food and drink. The monthlong fast that precedes this ceremonial weeklong feast is not absolute, of course. It is a fast (*kuasa*) that begins with the first daily prayer (just at first light in the morning) and ends just after sundown each day of the month of Ramadan. Evenings, into the early morning hours, are spent drinking iced sweet drinks, eating especially tasty foods, and performing and listening to the message and beauty of Koranic verses. Very few people lose and many gain weight during the fasting month.

Social rank, ritual, manners, and food are closely linked every day of the year. Traditional houses have formal "front regions" where guests are received and where rank behavior is celebrated. When a guest enters the front region (a verandah or front room), something should be offered, even if only water. The higher the rank of the visitor relative to that of the householder, the more elaborate the offering. Verbal greetings, gestures, and postures signal degrees of difference in social rank. The more formal the occasion or circumstance the more likely appropriate gestures and postures will be displayed by junior persons. The highest-ranking persons in social situations are not really required to perform the gestures and postures of rank courtesy, which is the work of persons of lower rank. Close friends and relatives come to the back door, the kitchen, a place of incredible informality.

Formal occasions are accompanied by formal serving of drink and food, by servers of low rank. Such feasts, although occasions of immense pleasure, are fraught with danger, because as a celebration of rank differences as well as a celebration of food, it is a "natural" setting for crimes of sorcery and affection magic through food (*santau*). Usually, the actual purpose is to make people incredibly fond of you, not to harm them.

Food and drink can be used to control individual behavior. A woman can force a man to be sexually attracted to her by adding word charms and her vaginal fluids to the rice that she cooks for him (*nasi uap*, "perfumed rice"), and a merchant can influence a customer to buy from him, even though his goods are bad and costly, by slipping a chemically potent and be-spelled concoction (*santau*) into his meal or drink. Illness or even death may result, but the intention is simply control over the customer.[3] It is the responsibility of the host of a ritual feast (*kenduri*) to be certain that the cooks are clean (*bersih*), which includes the notion of good intentions toward guests. Guests routinely ask who the cooks are.

The Islamic requirement of religiously pure (*halal*) food and drink is a common and deep concern, which is strengthened by the presence of a large population of non-Islamic Chinese that is economically dominant and politically competitive in Malaysia, and simply dominant in Singapore.

There are other concerns about food and drink. Children's illnesses, for example, are often attributed to their excessive appetite for "cooling" (*sejuk*) foods, especially soft fruits, and slightly sweet and weakly acidic iced drinks. Western friends are routinely warned not to eat durian, a "heating" (*panas*) fruit, while consuming alcoholic beverages, which are not only "heating," but also forbidden (*haram*) by Islamic law.

MEDICAL PRACTITIONERS

There are four commonly recognized categories of traditional medical practitioners: *bidan*, *bomoh*, *dukun*, and *pawang*. A *bidan* is comparable to a "midwife" in cosmopolitan medicine, but she is more. She helps pregnant women prepare for giving birth, attends and directs the birth, and deals with postpartum problems of the mother and child. She treats children's illnesses and female health problems. Any *bidan* knows a great deal about diet and the humoral qualities of foods. Most urban and many rural Malay women now give birth in clinics or hospitals; but even urban women consult with traditional *bidan* in such matters as diet, bathing, and positioning the fetus for easier birth, or preventing pregnancy after intercourse. In southern Thailand, Malaysia and Indonesia, traditional *bidan* are often incorporated into national health programs through additional training, licensing, and involvement in community clinics.

The *dukun*, who may be male or female, is a "general practitioner" in terms of knowledge; but may be recognized as a "specialist" (*tukang*) who is skillful in particular treatments—for example: *tukang urut* (massage), *tukang sunat* (circumcision), *tukang tulong* or *tukang patah* (setting broken bones). Also, a commonly used term for a *bidan* ("midwife") is *dukun beranak* ("birthing healer"). In some communities, successful students of a *bomoh* are made *dukun* in a formal ceremony led by the *bomoh* who instructed them. And the ritual is attended by members of the community. "*Dukun*" is a general term for a traditional healer of any sort.

*Bomoh*s are more knowledgeable and powerful than *dukun*s, especially in treating illnesses caused by social, psychological, and spiritual problems. They are general practitioners with a specialization in therapies similar to psychiatry in cosmopolitan medicine. Most *bomoh*s utilize a powerful theatrical therapeutic routine, *main puteri* ("playing the princess"), which varies depending on details of the patient's personal history. But they treat ordinary health problems as well. Teachers of "traditional personal self-defense" (*bersilat*) and master puppeteers (*dalang*) are also considered to be *bomoh*s because they utilize the same classical Malay knowledge system as medical *bomoh*s.

A *pawang* ("wizard") is more powerful and knowledgeable than even a *bomoh*. He can heal all kinds of illnesses, and he can conduct public rituals that ensure an excellent rice harvest or protect the whole community from a plague. Most *pawangs* and *bomohs* began their formal careers as *dukun*s.

All of these kinds of traditional healers tend to treat patients rather than illnesses, sometimes attempting to help clients gain the affection of other persons, or even win the public lottery. Successful healers and needy patients are not necessarily place bound and many travel great distances, even crossing international borders between Thailand, Malaysia, and Indonesia.

Cosmopolitan medicine did not begin to successfully challenge traditional *bomoh* medicine until after the successful independence movements following World War II, when nationalized health facilities and services and university training of natives in cosmopolitan medicine expanded rapidly. Since then, from time to time, government publicity campaigns have been directed against traditional medicine. *Bomoh* medicine survives, even prospers, but many practitioners have become more selective in the health problems they accept for resolution; perhaps, in response to the rising proportion of patients whose ordinary illnesses are resolved expeditiously and inexpensively through cosmopolitan medicine at government clinics and hospitals. However, psychological disorders of almost any sort are still likely to be dealt with by a traditional healer, and it is in the treatment of these disorders that traditional medicine continues to be an especially important health resource for Malays, and even for some members of other ethnic groups.

CLASSIFICATION OF ILLNESS, THEORIES OF ILLNESS, AND TREATMENT OF ILLNESS

Many aspects of traditional Malay medicine are known to ordinary persons. Malay adolescents and adults usually know how to stop the flow of blood from wounds and how to clean and dress minor wounds so that the risk of infection is reduced. Those who live along the seashore know how to treat the stings of jellyfish by applying a dressing made from crushed fish or crustaceans that are naturally immune to jellyfish stings. And ordinary Malays know about some plants and parts of plants that cause illness. Some ordinary individuals know how to set and splint simple fractures, although a bone-setter (*tukang patah, tukang tulong*) is more skillful. Also, the bone setter would have a method for speeding the mending of the bone, for example, by teaching the patient to relate the feeling of the fracture to a rhythmic sound produced by the patient.

As noted above, illnesses can result from a diet that is not balanced according to commonly known humoral (non-thermal but sensual) concepts of "heating" (*panas*) and "cooling" (*sejuk*) foods; but the most basic foods in traditional diets, rice, and fresh fish, are "neutral." This relates to the common knowledge prescription of fish soup with rice as part of conservative therapy for illnesses caused both by excessive "cooling" or by excessive "heating" elements in the diet. Virtually everyone knows that foods that are "heating" cause a feeling of warmth and foods that are "cooling" cause a sensation of coolness, but that the effect of particular foods and substances on individual patients may vary. Generally, "cooling" foods include juicy, sour (*masam*), or "almost bitter or unripe" (*kelat*) fruits and vegetables. Really bitter (*pahit*) foods, as well as "filling" foods (carbohydrates, fats, and animal proteins) and salty foods are "heating" (*panas*). Spicy foods, such as chili are "pungent" (*pedas*) but not necessarily *panas*. Also, the terms "hot" (*panas*) and "cold" (*sejuk*) are regularly used to characterize social relationships. This classification of kinds of food, and the emphasis on balancing the perceived effects of the different kinds of food, usually results in a balanced diet in terms of cosmopolitan medicine (Laderman, 1987).

Food and drink are also perceived as media that can be manipulated in order to control attitudes and actions of other persons. Word charms and potions put into refreshments or meals can affect the emotions and thinking of the person who consumes them. A *dukun* or *bomoh* may help a client to become more attractive to potential spouses, business associates, or even customers by "treating" the client or by providing the client with a prescription to "treat" other persons—a potential spouse, business associate or customer. When these ingredients are appropriately effective, it is an instance of "affection magic";[4]

but if they cause serious illness or death, it is a case of "poisoning" (*racun*), or "slow poisoning" (*santau*). There are traditional poisons for killing one's rivals and tormentors. But mostly, as in some cases of "perfumed rice" (*nasi uap*), illnesses or deaths are seen as instances of an "over-dose" and/or a result of the victim's resistance to magical persuasion.

Malays enjoy delicious foods and drinks, but they are deeply concerned with purity. This is related to Islamic rules regarding diet in which alcoholic beverages and pork are forbidden as well as the flesh of other mammals and fowl that have not been ritually slaughtered. Additionally, animals that are seen as "powerfully unnatural" (e.g., a fish with legs) or known to cause health problems if eaten are classified as *bisa*. *Bisa* also refers to deadly poisons applied to the blades of knives, spears, and swords; and to speech that causes others great pain or death (Gimlette, 1971).

There are some variations in the degree to which individuals apply *halal* (ritual purity) rules to their own diets; but all are deeply aware of the rules. Non-*halal* food is often referred to as "dirty" (*kotor*), just as *halal* food is said to be "clean" (*bersih*). And the distinction between dirty and clean food is also extended to include food or drink to which word spells or magical substances have or have not been added. Also, food and drink are necessary parts of the celebration of religious, familial, and organizational feasts, in which the complex formal rules of salutation and comportment regarding differences in social rank are prominent. Food, drink, and speech—clean, dirty, or poisonous—lubricate the celebration of rank differences. All three are strongly related to beliefs about health.

In Malay language, *angin* has many meanings (wind, air, intestinal gas, rumor, attitude, desire, or temperament), depending upon the conversational context. As a very common descriptive medical symptom it is usually associated with intestinal gas, or with swelling, or strong pain. It can also refer to patients' attitudes, desires, and temperament. All of these symptoms and problems are common among Malay patients. Some scholars have suggested that *angin* is a concept borrowed from Medieval European, Middle Eastern, South Asian or East Asian, ideas about the so-called four (or five) basic elements of material reality: earth (wood & metal), air, fire, and water.

Two behavioral syndromes, *amuk* (amuck) and *latah* (similar to LaTourette's syndrome), are relatively common in Malay populations. Both are probably related to the pressure of complying with complex rules of traditional social manners, which vary according to levels of formality in different behavioral settings and according to the ranks of different participants. The lower one's rank in a given setting the more one must know and perform proper terms of address, forms of gesture and posture, and registers of language. Symptoms of these two disorders differ markedly, but both are probably expressions of psychological depression and of individual resistance to the pressures of the traditional Malay social order.

Amuk is viewed as a temporary state of physically aggressive insanity. In a suicidal attack, the *amuk* person attempts to maim or kill virtually everyone present. Traditionally, if the *amuk* survived the defensive attacks of others and returned to normal behavior, he was allowed to go free on the basis of having been temporarily insane. This, of course, is no longer true. Cases of murderous *amuk* have been rare for almost a century; but instances of unseemly and abusive verbal and physical aggression by males and females toward members of their communities, rural and urban, occur and are often referred to as instances of *amuk*.

Latah continues to be fairly common, especially among low ranking women. Unlike *amuk*, *latah* does not involve physical aggression. Rather, the affected person seems to satirize traditional manners and to mimic the words and gestures of others with whom they are interacting. Many Malays are less offended than entertained by this stereotypic behavior, which can be initiated by startling the *latah* person, forcing the *latah* person to "perform" until she or he is exhausted. Most traditional healers do not consider *latah* or *amuk* to be curable health problems.

There are apparent instances of *latah* in which an individual who has never had *latah* behavior begins to yell obscenities and becomes aggressive, as in *amuk*. These instances are often diagnosed as a case of *tuju*, a victim of a magical curse (*jampi*) by a rival or person who feels jealous hatred (*dengki*) toward the victim, and who has directed an evil spirit to enter the body of the victim. The victim may be treated almost immediately by a knowledgeable person, who subjects the victim to excruciating but not damaging pain, while asking, "who are you." The question is directed to the evil spirit who has entered the victim's body. It is the spirit who feels the pain and cries out its name. Repeating the spirit's name forces it to answer all questions truthfully; so the *dukun* or *bomoh* asks it "who sent you." The truthful answer forces the intruding spirit out of the victim's body, and the victim recovers.

Hantu dadah (the evil spirit of drugs) has become a scourge of Malay youth and young adults in the last four decades. Traditional formulae for love magic and murder contain opium, and *ganja* (marijuana) is a common ingredient of traditional health elixirs consumed by Malays. Until 1952, in Malaysia, opium smokers had to register with the government, but were not otherwise prosecuted. Drug addiction was not seen as a serious public health problem among Malays until the late 1960s, when large numbers of adolescent Malay individuals began to migrate to urban centers to seek employment and adventure. Movies, television, and magazines sponsored the idea of personal freedom and individualism. Possession of addictive drugs became a capital offense, and many young Malays were hanged. *Ganja*, smoked (combined with tobacco, and/or filtered through a water-pipe device), became the usual drug of choice among most young Malay addicts, although there is some use of heroin and even cocaine, especially among the more affluent.

Traditional treatment usually involves cleansing the patient's body of the drug's "spirit" (*hantu dadah*). The first meeting of patient and *bomoh* is like that for most kinds of illnesses. A close family member or friend of the patient contacts the *bomoh*, describing the patient and symptoms, to set a date for the first visit. Then, patient, family members, and friends meet with the *bomoh*, who examines the patient to determine what the health problem is and whether the *bomoh* can cure it. The patient or a family member may have been instructed to purchase three, four or seven limes (*limau nipis*) for the latter purpose. Physical examination includes inspection of the eyes, mouth and ears, and palpitation of limbs and body. The patient's pulse is usually taken in a special way, with the first three fingers of the right hand of the *bomoh* on the radialis blood vessel of the left wrist of the patient. The "reading" of the first finger relates to health conditions from the feet up to the navel, the second from the navel to the shoulders, and the third to the neck and head. Other examination procedures may include inspection of the curve of the spine, elasticity of forehead skin, examination of nails and skin of the hands, neck pulse, thumping the back of the rib cage, listening to stomach noises, clotting time, and appearance of a drop of blood (humors). After the examination and some discussion with patient and family members, the *bomoh* may cut a thin slice from each lime, in turn, which fall into a bowl of water that has absorbed powerful words spoken by the *bomoh*. Some slices float peel-side up and others float peel-side down. The sequence and proportion of each of these two possibilities signals whether or not the *bomoh* will take the case. The details vary, of course (Provencher, 1984; Werner, 1986).

If the *bomoh* decides to take the case, she or he usually prescribes a routine of prayer and bathing, using *bomoh*-blessed water and herbs as additives to the bath water, and the patient is given instructions concerning diet, and scheduling of work and other activities. The patient or a relative is delegated to purchase ingredients (*ramuan*) to be used for treatment at home and for the next session. In subsequent sessions, the *bomoh* will probably: (1) administer herbal medicines that purge the contents of stomach and bowels; (2) provide other blessed ingredients to be added to bath water; (3) outline a course in religious devotion; and (4), if the patient completes the course of treatment successfully, celebrate success with a small ritual feast (*kenduri nasi guru*) with patient and family providing a special tray to feed any good spirits involved and a gift/payment to the *bomoh*.

The beginning and ending of treatment, the divining of whether or not the *bomoh* can cure the victim's illness and the payment/celebration in successful cases are standard. Details of diagnosis/treatment vary according to the perceived cause of the illness, and according to the routines of particular *bomoh*s. There are many different Malay "psychiatric" therapies, sometimes glossed in the literature as "playing the princess" (*main puteri*), in which the *bomoh* begins to pray, goes into a trance, speaks in another voice, collapses, revives, and magically extracts "dirty things" (*kotoran*) from the body and soul of the patient. The "dirty" extracted things vary: oddly shaped artifacts and thorns, parts of insects, hair, dirt, and rusty needles. And the means of extraction vary too, including: grasping odd shaped objects and insect parts out of thin air, collecting hundreds of red hairs by rubbing a ball of semi-dry dough over the body of the patient, collecting rusty needles and dirt by rubbing the body of the patient with an unbroken raw chicken egg, and collecting dirt and objects by rubbing the body of the patient with a previously unbroken betel nut pod. In all of these, the symbolism is that of "unnatural," "intrusive," or "dirty" things and food. From a Malay perspective, these broken dirty things symbolize broken and improper social relationships. *Bomoh*, patient's relatives and friends, and even the patient may join in trying to explain the "evidence." The explanation is the most important part of the cure. It works to the extent that the *bomoh* and her/his helpers know quite a bit about the patient and the patient's relationships with others. Successful practitioners usually have helpers, often knowledgeable practitioners, who gather information about the patient, the patient's family, work mates, other acquaintances, and friends. And the most successful of these *bomohs* write their cases in notebooks, which they read and reread for comparison with new cases.

SEXUALITY AND REPRODUCTION

Knowledge of sexual activity is gained early and naturally in the context of familiar and relaxed back region behavior of the Malay house, where dress, speech, and behavior are extremely relaxed and casual. Older children, especially girls, take care of their younger siblings. If a woman has no daughter, a son often takes the role of mother's helper, especially in the care of babies and infants. Malay children know the anatomy of gender and where babies come from. Most Malay couples share active and imaginative sex lives, and joke about the formality of what they imagine to be the standard position for sexual intercourse in English culture. Nonetheless adolescent girls are secluded and their virginity is protected, as much as possible, until they are married. Pregnancy before marriage is a matter of scandal. And, under Islamic law, a private meeting of a man and woman who are not married to each other constitutes the crime of *khalwat* ("close proximity"), which is punished under the jurisdiction of the Islamic courts. Adolescent boys and men, however, are expected to be sexually promiscuous before, and even after marriage.

Marriages are supposed to be arranged, and sometimes they are. But beginning in the 20th century, when Malays and most of the rest of the world began to experience rapidly increasing secularization through mass media, individuals more frequently exercised personal choice regarding marriage. For example, interviews in the 1960s of elderly Malay women who had migrated to Kuala Lumpur from Indonesia in the first three decades of the 20th century revealed that many had married more than once and that most of their marriages had been with men of their own choice. Nonetheless, even in the 1990s, the ritual details of most Malay marriages were traditional and they projected images of arranged marriage.

Some *bidan*s perform very early abortions for adolescent girls and unmarried women and for married women who do not want more children. Several common traditional methods do not utilize surgical instruments and are applied as soon as possible, usually just after the first missed menstrual period: (1) deep massage of the abdomen by a *bidan*; (2) bidan-made pills containing a mixture of drugs and other ingredients that are *panas* "heating"; and (3) the juice of plumbago root, consumed as a drink.[5] Often, these methods and others are applied in a single case.

HEALTH THROUGH THE LIFE CYCLE

Pregnancy and Birth

Conception is easier if both male and female are "cool" (*sejuk*), so both prospective parents should watch their diets if they want to have children. Traditionally, conception is said to be possible only during the first three days, the three middle days, and the last three days of each month ... a sort of "Malay roulette." At first the embryo is just a blood clot, a combination of the parents' seeds. Later, in the third month, it is said that the fetus's "elder sibling," the placenta, comes to join the fetus in its journey toward birth. As the pregnancy progresses into the seventh month, there is a formal ritual inspection and "rocking the belly" (*lenggang perut*), by the *bidan*, to verify or correct the position of the baby in the womb, so that its legs are up and its head is down, ready for an easy birth. If it is not in the proper position, the mid-wife, through massage and manipulation, can maneuver it into the proper position. It is said that a good *bidan* is more expert in this procedure than a doctor or midwife trained in cosmopolitan medicine. A male *bomoh* is not called to pregnancy cases unless there are serious complications, and his task is that of spiritual assistance. The *bidan* does the work of massaging the mother's abdomen to relieve pain and of applying gentle pressure appropriately to facilitate a difficult birth (Laderman, 1987). Traditionally, the umbilical cord is supposed to be cut with a bamboo knife, but nowadays steel scissors are used. The umbilical cord and placenta are set aside, to be buried with a viable coconut near the house where it will grow into a tree, from which the "elder sibling" can easily participate in the birth of future "younger siblings" (*adik*). Throughout the Malay world, since the 1960s, the frequency of hospital births has increased, but many, perhaps most Malay women still go to a *bidan* during pregnancy even if they intend to enter hospital for the actual delivery. A traditional 40-day postpartum period in which the new mother follows a restricted diet is still common in rural areas, but many (perhaps most) urban women follow diets compatible with cosmopolitan medicine.

There is no particular preference for boys or girls. Daughters perform more service to the family. Sons provide more opportunities for ritual display. Women often prefer the first baby to be a girl, so that the mother has a good helper early in the marriage (Provencher, 1971).

Infancy

A woman's first baby, female or male, receives a great deal of attention from adult relatives, neighbors, and friends of the family. Subsequent babies receive less attention, depending on the spacing of births, and they are tended by older siblings. Roles of boys and girls begin to diverge at five or six years of age. Psychological problems in later life are more common among the eldest and the youngest of a sibling set, perhaps because as infants and children their experiences in the ranked relations between siblings were less balanced, lacking deep experience either of being junior or of being senior in a society where rank is the most basic aspect of social relationships (Provencher, 1999).

Female babies are brought symbolically into the fold of Islam at a much earlier age and with less physical trauma and social fanfare than males. Forty days after a Malay baby girl is born, adult female neighbors and relatives gather at the mother's home to celebrate the end of the postpartum period of the new mother and to perform a ritual clitorodectomy (actually a clitoridotomy) of the newborn baby girl (Laderman, 1987). In this Malay ritual, unlike the equivalent ritual in some other Islamic societies, the clitoris is not removed or even damaged. It is barely scratched. The physiologically equivalent traditional entrance to Islam for Malay boys, circumcision (*sunat*), comes at a later age and involves more physical and psychological trauma.[6]

Childhood

By the mid 1960s most of the serious epidemic diseases of childhood had been controlled or mitigated by public health systems of the post-colonial governments of Southeast Asia, and treatments for simple ailments with commercial patent medicines were easily available and inexpensive. Nonetheless, intestinal ailments remain common among Malay children. After they can walk, they move easily from one neighbor's or relative's house or apartment to others', eating and drinking as they go. Especially in instances of prolonged illnesses that have been treated but not cured by cosmopolitan medicine or patent medicines, traditional healers are consulted.

Infancy ends with the beginning of schooling, which can be a traumatic psychological experience for some Malay children, especially those who are spoiled (*manja*). In Malay folklore and popular culture, the spoiled child (*si manja*) is one who does not have serious responsibilities. In some instances, this is a female with many older sisters who do most or all of the household chores; or it may be a male who is the eldest in a sibling set (*si long*), but whose sisters are almost his age, so that he would not have done the household chores. Sometimes such a favored child, usually a boy, has great difficulty adjusting to the new disciplines of schooling, even to the point of having tantrums. A *bomoh* or *bidan* may be consulted if tantrums continue.

Adolescence

Adolescence is not a clearly marked age in traditional Malay culture. For example, most males are circumcized after religious instruction, sometime between 9 and 13 years of age, but the range is 6 to 20 years of age. Traditionally, the circumcision ritual and feast (*kenduri bersunat*) is a public event, sometime hosted by a single household with one or several boys and sometimes by several households in a neighborhood. In the latter case, an arbor or open-wall tent may be erected in an empty space near the street to ease access for on-lookers and guests. This also reduces costs and labor for each household, and the ritual of the feast and the actual operation of circumcision of each boy can be viewed by all who attend. After prayer, the circumcision expert (*tukang sunat* or *mudim*) pulls the foreskin over the end of the glans of the penis, clamps the foreskin with a traditional implement, and cuts off the excess foreskin with a bamboo knife. Then he applies a medicinal compound (*obat tasak*) and bandages (*tali kundang*). The boys, now fully Islamic males, lie down on cots or bedding dressed only in their sarongs, and are on public view the first day. Night-lights are lit in their houses until the wounds heal, several days later. Health problems attending circumcision are rare, although newspapers occasionally print stories about a slip of the knife by an incompetent *tukang sunat*.

There is no similar coming of age celebration for girls. The onset of menarche does not define the beginning or ending of adolescence for Malay females. However, a part of marriage ritual involves an exhibition of the bride's ability to read and recite the Koran; and some female *bomoh*s have said that this is more equivalent to male circumcision than the cutting of the clitoris, because it is evidence of being fully Muslim. School systems in southern Thailand, Malaysia, and Indonesia provide another, institutional means, by which Malays discuss the stages (*darjah*) between childhood and adulthood.

The virginity of girls until their first marriage is closely protected through modest dress and careful chaperoning. And the possibilities of being attractive to boys through fashionable clothes are still dampened by the common requirement of school uniforms for public as well as private schools in Southeast Asia. Nonetheless, adolescents are regarded to be naturally *gatal* ("itchy" or easily sexually aroused), flirtations occur, and girls become pregnant. The pregnancy may lead to marriage, an early abortion, or birth, and informal adoption of the child by an adult relative. Adoption is not encouraged in Islamic law, but as a traditional practice among relatives in Malay society it is fairly common (McKinley, 1975).

Young unmarried men may be seduced or be driven sexually mad by a woman through her feeding him "perfumed rice" (*nasi uap*), a common diagnosis of young men caught engaged in lewd sexual behavior in public. Adolescent girls, too, are thought to be driven crazy by love magic. The cure may involve ritual discovery of the "real" perpetrator and/or sending the victim to live with a relative in a distant place.

Prostitution is common in towns and cities, and venereal diseases and AIDS are serious health problems. But these are dealt with almost entirely within the context of the cosmopolitan health services.

Adulthood

Marriage marks the beginning of adulthood, both for women and men. But full status as adults is attained only after one's first child is born. Rural women have more children than urban women. This is a choice based on the economic value and costs of children and on access to inexpensive birth control. A husband, to the extent that he is interested in girlfriends who might become secondary wives, is often viewed by his first wife as a health risk (venereal diseases and AIDS). According to law, a man must obtain formal consent of his present wife or wives in order to marry subsequent wives (only four wives are allowed). Most Malay women do not want co-wives because it reduces their own household income (by law, each wife is entitled to a separate household and reasonable household expenses) and they also understand the health risk. Sometime, however, as the first wife of a wealthy man ages she may welcome a husband's younger second wife, especially a friendly kinswoman without health problems who would also be an ally. Malay men who were born poor in rural places and have nonetheless become very successful in the modern urban world often suffer seriously from hypertension, obesity, and early death. This may be a matter of richer diet and less exercise than their ancestors.

The Aged

The aged are revered in formal courtesy and are commonly cared for by their children in their last years. This care of elderly parents is often given as a reason for having children. Men are viewed as more likely than women to become senile, and women who live beyond childbearing years usually live long lives.

Dying and Death

Death of a young or middle age person is often suspected to have been the result of sorcery or poison. Death of the elderly is viewed as natural and even a blessing. Funerals and burials follow standard Islamic practices. Notice is immediately given to close relatives and friends who will come to the wake, and the appropriate Islamic religious functionary is contacted to arrange for burial in an Islamic cemetery.

NOTES

1. The basic difference between Malay and Minangkabau cultures is in rules of inheritance. Malay kinship organization is bilateral and Minangkabau kinship organization is matrilineal. This difference may be related to different regional strategies for effectively

combining in a single community two major and different economic activities: wet rice cultivation and commerce.
2. There is anthropomorphic folklore about the trickster mouse deer (*kijang*) that resembles that of the trickster rabbit of Africa, but this may even support the other evidence of traditional Malay disinterest in the rainforest.
3. One of the fairly common ingredients in such concoctions is lead arsenate, which even in small accumulative doses can eventually cause death.
4. "Love magic" does not clearly suggest the range of kinds of attraction that bomohs claim to achieve for their clients: *kasih sayang* = familial love, as in a mother's love for her children; *suka* = preference, as for a particular restaurant; *kasihan* = pity, as in pity for an unfortunate person; *cinta* = romantic love; or *syahwat* = sexual lust.
5. See, especially, Gimlette, 1971.
6. There is mention in the literature of ritual haircutting of boys at birth and also at forty or forty-four days after birth. I witnessed several of the latter during my dissertation research in 1964–65, but none during later ethnographic fieldwork among Malays. In classical Malay literature, relevant to pre-16th century AD, heroes of Melaka wore their hair at shoulder length. A century ago, a boy's hair was shorn, except for a top lock, shortly after birth, and was often not cut until just before circumcision or marriage. And in modern Malaysian comic books a certain child-like mischievous spirit is pictured naked and with a top knot [Provencher, 1999; Skeat, 1900].

References

Andaya, B. W., & Andaya L. Y. (1982). *A history of Malaysia*. London: Macmillan Publishers.
Bellwood, P. (1985). *Prehistory of the Indo-Malaysian Archipelago*. North Ryde, N.S.W.: Academic Press Australia.
Endicott, K. M. (1970). *An analysis of Malay magic*. Oxford University Press.
Gimlette, J. D. (1971). *Malay poisons and charm cures*. Kuala Lumpur: Oxford University Press.
Golomb, L. (1985). *An anthropology of curing in multiethnic Thailand*. Illinois studies in anthropology no. 15. Urbana & Chicago: University of Illinois Press.
Hamidy, U. U. (1986). *Dukun Melayu Rantau Kuantan Riau*. Pekan Baru, Riau Islands: Bagian Proyek Penelitian dan Pengkajian Kebudayaan Melayu (Melayulogi). Departemen Pendidikan dan Kebudayaan.
Hartog, J. (1972a). Sibling rank of Malay psychiatric patients and juvenile delinquents. *The Southeast Asian Journal of Tropical Medicine and Public Health*, 3.1, 124–137.
Hartog, J. (1972b). The intervention system for mental and social deviants in Malaysia. *Social Science & Medicine, 6*, 211–220.
Heggenhougen, H. K. (1980). Bomohs, doctors and sinsehs-medical pluralism in Malaysia. *Social Science & Medicine, 14B*, 235–244.
Heggenhougen, H. K., & Navaratnam, V. (1979). *A general overview on the practices relating to the traditional treatment of drug dependents in Malaysia*. Penang: National Drug Dependence Research Centre, Universiti Sains Malaysia.
Karim, W.-J., (1984). Malay midwives and witches. *Social Science & Medicine, 18*.2, 159–166.
Kasimin, A. (1995). *Santau sebagai satu cabang ilmu sihir*. Kuala Lumpur: Percetakan Watan Sdn. Bhd.
Laderman, C. (1987). *Wives & midwives: Childbirth and nutrition in rural Malaysia*. Berkeley: University of California Press. (first published 1983).
Laderman, C. (1992a). Malay medicine, Malay person. In M. Nichter (Ed.), *Anthropological approaches to the study of ethnomedicine* (pp. 191–205). Amsterdam: OPA.
Laderman, C. (1992b). A welcoming soil: Islamic humoralism on the Malay Peninsula. In C. Leslie & A. Young (Eds.), *Paths to Asian medical knowledge* (pp. 272–288). Berkeley: University of California Press.
McKinley, R. H. (1975). *A knife cutting water: Child transfers and siblingship among urban Malays*. Doctoral dissertation in anthropology. The University of Michigan.
Provencher, R. (1971). *Two Malay worlds: Interaction in urban and rural settings*. Research monograph no. 4. Berkeley: University of California.
Provencher, R. (1979). Orality as a pattern of symbolism in Malay psychiatry. In A. L. Becker & A. A. Yengoyan (Eds.), *The imagination of reality: Essays in Southeast Asian coherence systems*.
Provencher, R. (1984). "Mother Needles": Lessons on inter-ethnic psychiatry in Malaysian society. *Social Science & Medicine, 18*.2, 139–146.
Provencher, R. (1999). Order in the Malay house: Malay kin categories, ethnic ranks, and the unconventional behavior of toyols. In L. W. Aragon & S. D. Russell (Eds.), *Structuralism's transformations: Order and revision in Indonesian and Malaysian societies*. Tempe, AZ: Program for Southeast Asian studies monograph series.
Provencher, R. (2001). Malaysia's mad magazines: Images of females and males in Malay culture. In J. A. Lent (Ed.), *Illustrating Asia*. Honolulu: University of Hawaii Press.
Pauka, K. (1998). Theater and martial arts in West Sumatra: Randai and silek of the Minangkabau. *Monographs in international studies. Southeast Asia series*, No. 103.
Resner, G., & Hartog, J. (1970). Concepts and terminology of mental disorder among Malays. *Journal of Cross-cultural Psychology, 1.4*, 369–382.
Skeat, W. W. (1900). Malay magic: Being an introduction to the folklore and popular religion of the Malay peninsula. London: Macmillan and Co. Ltd.
Suparlan, P. (1991). *The Javanese dukun*. Jakarta: Peka Publications. First published in Masyarakat Indonesia, Vol. 5, no. 2, 1978, pp. 195–216.
Werner, R. (1986). The practices and philosopies of the traditional Malay healer. Studia Ethnologica Bernensia, No. 3. Berne: The University of Berne, Institute of Ethnology.

Maori

Mason Durie

ALTERNATIVE NAMES

New Zealand Maori, Tangata Whenua o Aotearoa.

LOCATION AND LINGUISTIC AFFILIATION

Maori are the indigenous people of New Zealand. Of Polynesian descent, the first voyagers arrived in New Zealand around 1000AD in a series of planned migrations from Eastern Polynesia (probably Tahiti), by way of Rarotonga. Although there are minor tribal dialectal differences, there is a single Maori language that has similarities with other Polynesian languages especially Hawaiian, Tahitian, and Rarotongan.

OVERVIEW OF THE CULTURE

Maori society was essentially tribal. There were forty or fifty major tribes (*iwi*) and many more smaller tribal groups (*hapu*) who occupied specific territories and had distinctive customs and dialects, though subscribing to a single common language. While tribes were largely autonomous, complicated social networks were formed to gain political and strategic advantages and so leading to a dynamic set of interacting alliances.

Within all tribal narratives emphasis was placed on a close affiliation with the surrounding landscape and distinctive landmarks or waterways often reinforced tribal identity. This symbiotic relationship with the natural environment was reinforced in the creation story. *Rangi* (the sky father) and *Papa* (the earth mother) were forced apart by their children (the forests, oceans, winds, plants, fish) who subsequently challenged parental authority. By personifying the laws of nature, human motivation was attributed to the elements and by the same token, the human condition was linked to the often harsh natural environment.

Perhaps for that reason tribes were constantly on guard for the unexpected and placed great store on maintaining mana (authority) over their own territories and people. Attempts by other tribes to diminish mana, either through plunder or trading insults, led to war and, not infrequently, cannibalism. The ultimate act of asserting mana was to devour the heart of a foe. But there were also more subtle forms of retaliation including the imposition of a makutu or spell on an individual or group.

Loss of life through warfare was a common hazard, greatly exacerbated by the 19th century musket. Changes to the definition of Maori make it difficult to draw comparisons over time, but there is strong evidence of a substantial and sustained increase in the Maori population since 1900 when, at 45,000, extinction had been widely predicted. For the past three census takes it has been possible to determine the number who are descended from a Maori as well as the number who elect to identify as Maori. Both are valid measures though identity is regarded as the more meaningful measure. In the 2001 census 604,110 people indicated they were descended from a Maori, and 87%, 526,281, actually identified as Maori (Statistics New Zealand, 1998).

While accounting for some 14% in 2001, by 2051 the Maori ethnic population will almost double in size to close to a million, or 22% of the total New Zealand population. By 2051 33% of all children in the country will be Maori, and Maori in the working age group, 15 to 64 years, will increase by 85% (Statistics New Zealand, 1998).

Yet though the younger age groups will continue to grow, an equally significant change will be an increase in the number of older Maori. By 2051 the proportion of Maori elderly would have risen from the current 4% to 15%. Though still youthful, the population will have a larger cohort of over 60 year olds. At ages 65 and over, the growth is projected to be in excess of 300%.

Like many New Zealanders, Maori are mobile. Following World War II urbanization resulted in major migrations from country areas to towns and cities and by 1976, more than 8% of Maori were living in urban settings, a quarter in the greater Auckland area. Emigration overseas has also become a significant trend, some 30,000 Maori now being recorded as residents in Australia. More recently still, there has been a shift in internal migratory patterns away from urban areas where unemployment is high and back to tribal areas such as Northland, from where grandparents had moved some thirty or forty years earlier.

Over the past two decades, in addition to demographic change there has been a dramatic revitalization of Maori language and culture with a renewed sense of commitment to indigenous values and knowledge. It has been accompanied by a demand for increased autonomy and a parallel rejection of policies of assimilation and dependency. Maori providers of health and education services have been part of the trend and their emergence has resulted in pressure for theoretical and methodological frameworks that can incorporate Maori perspectives as well as scientific practice.

THE CONTEXT OF HEALTH: ENVIRONMENT, ECONOMIC, SOCIAL, AND POLITICAL FACTORS

Epidemiological Trends

Prior to European colonization of New Zealand, the Maori population was thought to have reached a steady state, around 150,000 to 200,000. The major life threatening diseases were associated with pneumonia, physical injury, child birth, and gastro-intestinal disorders. In the 19th century when settlers from Europe began to arrive, there was a dramatic epidemiological shift. Infectious diseases, such as typhoid fever, tuberculosis, measles, and influenza took a heavy toll. A lack of natural immunity combined with the impacts of new lifestyles and diets led to high mortality rates (in 1854 there were over 4,000 deaths during an epidemic of measles) and depopulation. The situation was aggravated by a high number of deaths attributable to muskets—an innovation that altered the nature and scale of tribal encounters and battles with the British imperial troops.

As depopulation was arrested and population increases began to occur disease patterns changed again. Although still a major problem until the middle of the 20th century, tuberculosis as a major cause of illness and death was eventually to give way to heart disease, cancers (especially cancer of the stomach, lung, cervix), non-infectious respiratory disorders, and metabolic disorders especially diabetes and obesity. By 1976 it was also apparent for the first time that mental health problems were a significant cause of morbidity, reflected in high hospital admission rates, a high prevalence of conduct disorders, alcohol and drug misuse, and quite recently, high rates of youth suicide.

Although Maori life expectancy has improved substantially over the past century and is now 68 years for men and 73 years for women, Maori and non-Maori disparities in health are evident in almost every disease category and in admission rates to hospital. Similarly, the increased burden of disease experienced by Maori, as measured by disability adjusted life years (DALYS) is approximately 75% greater than the age-standardized DALY rate for other New Zealanders (excluding Pacific peoples) (Ministry of Health, 2001). At the same time, if Maori standards of health are benchmarked against the health standards of earlier generations of Maori and the health standards of other comparable first peoples, it is clear that major improvements have occurred so that the Maori population is not only more numerous than at any time in history but also enjoys greater longevity and a higher overall standard of health.

Socio-Economic Circumstances

Collectively Maori are over-represented in lower socio-economic groupings, but between Maori individuals there is also considerable variation that is not immediately obvious when comparisons between Maori and non-Maori are made. At an aggregated level the gap between Maori and non-Maori is wide but there is also an emerging gap between Maori who are employed and well qualified and those who are unemployed with poor prospects of employment. Health status and housing standards are likely to be reflected in that differential. On the other hand, it is unusual for middle class Maori to live entirely apart from wider family networks. Each Maori family group is inevitably represented across the social strata (Henare, 1994). In this regard it is often difficult to separate socio-economic conditions from cultural and historical factors, even though some individuals may be relatively well-off.

Educational achievement is probably the most significant determinant of socio-economic advancement and although achievement levels lag behind non-Maori, there are signs that Maori are making gains. The establishment of alternatives such as Kōhanga Reo (Maori language early childhood education centers) have provided an incentive but within the mainstream higher Maori participation rates in early childhood education have also been evident, growing by over 30% between 1991 and 1993. However, despite the fact that over 40% of all Maori children under 5 years of age are enrolled in early childhood services, the growth has failed to keep pace with the increase in enrollments of non-Maori so that the disparity is not reducing and may even be increasing (Ministry of Education, 1998).

Educational underachievement is directly linked to work force participation and Maori levels of unemployment bear out the relationship. As well, and perhaps of greater significance, Maori unemployment has mirrored macro-economic swings and has been exacerbated by the move toward free market policies that occurred in New Zealand after 1984. Up until 1987 Maori were more likely than other New Zealanders to participate in the labor force. However, by 1988 a sudden rise in unemployment had occurred, and continued for a further 4 years reaching 27% in 1992. Non-Maori unemployment had also increased but the rates were relatively low (4% rising to 7%) so that the disparity was high, and remains so (Maori unemployment is now close to 14%, non-Maori around 5%). Of particular concern is the relatively high youth unemployment rate; in 1996 the overall unemployment rate was nearly 8% for all New Zealanders but 20% for 15–19 year olds and 30% for Maori aged between 15 and 19 years (Childrens Agenda, 1999).

Because of their dependence on employment, income levels for Maori are significantly lower than for non-Maori. But there are other reasons that contribute to Maori economic disadvantage including long term unemployment, disability and sole parent households (New Zealand Government, 1994). Maori are over-represented in all categories of beneficiaries, have uptake rates of more than three times the non-Maori rate for the domestic purposes benefit and are twice as likely to receive an unemployment benefit. In 1996 one in every two Maori women aged fifteen and over received a government benefit compared with one in five non-Maori women, and 45% of Maori women lived in households where the annual income before tax was $30,000 or less (Te Puni Kokiri, 1999).

Political Position

In 1840 the British Crown signed the Treaty of Waitangi with Maori tribes. The Treaty guaranteed continuing property rights and a degree of tribal autonomy. However, it was largely ignored and by 1975 Maori discontent with the appropriation of land, forests and fisheries, together with concern about the erosion of Maori language and culture, led to the establishment of the Waitangi Tribunal. The Tribunal has the authority to investigate grievances against the Crown for breaches of the principles of the Treaty and is able to recommend remedial action to the Government. Over 800 claims have been lodged. A continuing emphasis on the Treaty of Waitangi has created irritation for many non-Maori New Zealanders but, in the absence of a written constitution, it remains a pivotal focus for Maori especially in defining their position in modern New Zealand and their relationship with the government. The Treaty is included in health legislation and imposes an obligation of government agencies to actively address Maori health issues.

Maori have been active participants in the indigenous rights movement and are increasingly seeking opportunities for greater autonomy, not necessarily as an independent nation-state but in decision-making over Maori resources, the delivery of services to Maori, and full participation in the affairs of the nation. Progress has been considerable. In the current parliamentary system, out of 120 members of parliament, 17 are Maori and seven seats are reserved for members who represent Maori electorates. Maori are also represented in the judiciary, within the state services, in all professions, and in the diplomatic corp.

Medical Practitioners

Traditional healing was practiced by *tohunga* who were skilled in the use of *rongoa* (treatments derived from plant products) and the recitation of *karakia* (chants used to encourage healing through spiritual pathways). The Tohunga Suppression Act 1907 prohibited traditional healing methods but after the repeal of the legislation in 1964 there has been a resurgence of traditional methods and a number of healers have established large practices. There is a greater spirit of co-operation between medical and customary approaches to healing with mutual agreement that all treatments have limitations and a combined approach may heal mind and spirit as well as body. Traditional healing as part of the modern health care system is now recognised by government and some contracts have been offered for traditional healing services (Ministry of Health, 1999a).

The earliest Maori medical practitioner was Maui Pomare. He graduated from the American Medical Missionary College in Chicago in 1899 and was followed in 1904 by Te Rangi Hiroa, the first New Zealand trained Maori doctor (Durie, 1998). There are now 198 Maori medical practitioners, a little over two percent of the total registered medical practitioners (Health Workforce Advisory Committee, 2002). Important to the recruitment of Maori doctors have been affirmative action programs, first introduced at the University of Otago Medical School in 1900, and continued at the University of Auckland. Some 10 to 15 positions are reserved for Maori students each year. To be eligible for the scheme, students are required to demonstrate Maori descent, some involvement with Maori communities, and support from their own tribe.

In 1999 Maori medical practitioners formed an association, Te ORA, that provides a professional focus for Maori doctors and arranges scientific and cultural enrichment programmes. It also acts as a conduit for information relating to Maori health including opportunities for practice within Maori communities. Maori medical practitioners occupy a range of positions in hospital and private practice. Generally their clinical skills are indistinguishable from other practitioners but in addition they tend to be competent in Maori language and custom and are able to modify practice to accommodate Maori health perspectives. Their active involvement in Maori communities in a variety of roles inevitably confers expectations that will provide leadership beyond the health sector and not infrequently places demands on them that are difficult to meet.

CLASSIFICATION OF ILLNESS, THEORIES OF ILLNESS, AND TREATMENT OF ILLNESS

Customary Maori disease classification systems recognized *mate tangata* (i.e., diseases caused by human error or misadventure) and *mate atua* (diseases that could not be accounted for in rational terms). Underlying both types of illnesses was a system of knowledge based on *tapu* and *noa*. There are many interpretations of *tapu*. Now, most emphasize a sacred quality and are linked in some way to gods or divinities. Anthropologists and missionaries for example conceptualized *tapu* as a product of religious observations, highly spiritual and somewhat apart from everyday life. They were inclined to overlook the more practical goal of survival and environmental adaptation.

A more utilitarian view of the purpose of *tapu* was proposed by Dr. Te Rangi Hiroa. He drew a connection between the use of tapu and the prevention of accidents or calamities, implying that a dangerous activity or location would be declared *tapu* in order to prevent misfortune. In his opinion, the conferment of *tapu* was linked to healthy practices. *Tapu* was a type of public health regulation, basically concerned with the avoidance of risk, protection of the environment and its resources, and the promotion of good health. *Noa* was a term used to denote safety; harm was less likely to come to anyone who entered a *noa* location, ate food rendered *noa* by cooking, or touched a *noa* object. In Maori society the concepts of *tapu* and *noa* remain integral to Maori world views and color contemporary Maori attitudes to illness and injury.

Of equal importance has been the emergence of Maori health perspectives. The best known, te whare tapa wha, compares health to a four-sided house in which there is balance between spiritual, physical, intellectual/emotional, and family domains (Durie, 1985). This particular model has been incorporated into health planning as well as the delivery of health services and the construction of monitoring tools, such as outcome measures. Maori health workers adopt a more holistic approach to treatment and healing though as a result are often in conflict with the narrower goals of clinical agencies and funding arrangements that fail to recognize spiritual dimensions, or the links between health status and wider socioeconomic environments.

SEXUALITY AND REPRODUCTION

Sexuality in early Maori society was not treated with the same level of prudishness that Victorian settlers displayed although marriage and conception were more closely scrutinized. Arranged marriages were not uncommon, and certain women (puhi) were carefully cosseted until political alliances could be negotiated. Sexual unions between members of different tribes required permission from elders although were by no means infrequent and were sometimes actively encouraged for strategic purposes. Because birth rights were jealousy guarded and lines of descent prized for their connections to illustrious ancestors, parenting was far from arbitrary and coupling required tacit community approval. From genealogical records it is possible to conclude that childbirth occurred at relatively young ages, and that polygyny was practiced, though within narrow circles and usually to maintain a secure system for child care or to uphold family honor. The failure of a young wife for example to bear children might lead her husband to marry her sister, with the sanction of her own family. Sometimes a second wife or husband was procured for political reasons and occasionally prisoners of war were forced to enter into marriages both as a sign of victory but also as a token of reconciliation. The new spouse, though a former enemy, was usually afforded the full rights of marriage.

In modern times sexual activity among young Maori is common by fourteen years of age and contraception is used less frequently than for non-Maori. Sexual risk taking stands in sharp contrast to other elements of lifestyle and development and may be exacerbated by alcohol and drug use (Ministry of Health, 1997b). Termination of pregnancy is still regarded as offensive since it breaks a line of descent, though it is by no means infrequent. There is also some circumstantial evidence that homosexuality was practiced in pre-European times without attracting discrimination, and homosexuals were known as takatapui (Te Awekotuku, 1991).

HEALTH THROUGH THE LIFE CYCLE

Pregnancy and Childbirth

Fertility rates for Maori have declined sharply since 1964 and rates between Maori and non-Maori women are now similar (2.4 and 1.9 children respectively). However, Maori mothers are considerably younger. The average age for the first birth is 19 years, compared to 29 years for non-Maori. The potential health risks associated with younger mothers is compounded by the increasing number of sole parent households. In 1996 about 43% of Maori women with dependent children were sole parents (Te Puni Kokiri, 1999). Maori children under 5 years are more likely to live in one-parent households than non-Maori; in 1996 there were as many as 23% (Davey, 1998). Although that does not necessarily mean the children are more at risk than they would have been if there were two parents, children in two parent families are more likely to have a mother who is in the labor force. As a consequence, income levels for sole parent families are low. As it is, over half of all Maori children under 5 years live in households with incomes in the bottom two quintiles.

Interestingly, hospitalizations for complications of pregnancy, childbirth, and the puerperium are only slightly higher among Maori than non-Maori (1.2) and are lower for perinatal conditions (0.8) (Ministry of Health, 1999b).

Infancy

Although the Maori infant mortality rate declined more rapidly over the past half century than the rate for non-Maori, it remained significantly higher at 11.6 per 1000 compared with 5.3 per 1000 in 1996. Sudden infant death syndrome (SIDS) accounted for the major difference but there are signs that the mortality rate is falling for both Maori and non-Maori, due mainly to a reduction in SIDS since 1996 (Ministry of Health, 1999b). During an active campaign to reduce SIDS a number of risk factors were highlighted including smoking, lack of breast-feeding, the prone position, and sleeping with the baby. Prematurity still remains a significant problem for Maori infants. It appears to be associated with younger mothers, smoking, poorer antenatal care, and economic hardship.

Childhood

In Maori children under the age of fifteen years there is an over-representation in disability statistics; 16% as against 11%. And under the age of 5 years, Maori children are more than twice as likely to be hospitalized, most often because of respiratory diseases. Hearing impairment, though less prevalent than a decade ago is still significantly more common among Maori pre-schoolers, about twice as many fail hearing tests; and hearing loss is usually detected later at about 4 years compared to 21 months for a European child (Ministry of Health, 1998a). Most telling, in 1994 the mortality rate for Maori under the age of 15 years was 125 per 100,000 compared with 68 per 100,000 for non-Maori. Rheumatic fever, an uncommon disease in most developed countries continues to affect Maori children at rates that approximate those of industrialized countries a hundred years ago. Current rates for Maori are about 50–70 per 100,000 per year for Auckland children aged 5 to 14 years (compared with 2 per 100,000 for European) (Ministry of Health, 1997b).

There is also some evidence that the rates of physical abuse toward Maori children are high and that the capacity of family to provide adequate care is taxed. The large number of Maori children who are admitted to hospital because of accidents, including burns, for example is one indicator while the fact that fewer than a half of all Maori children under the age of 5 years are accessing early childhood care and education or that a third of Maori school leavers will have no qualification are others.

Adolescence

While mortality rates and hospital admissions tend to be lower for youth and young Maori adults, the health risk behaviors are high. They include smoking, alcohol and drug misuse, motor vehicle accidents, suicide, and attempted suicide. Between Maori and non-Maori the risks are similar but the rates show disproportionately high consequences for Maori, especially Maori males. The major cause of death remains motor vehicle accidents, often associated with alcohol use. The 1997 mortality rate for all New Zealanders was 14.1 per 100,000 population (Ministry of Health, 1998b); Maori rates, previously higher, have now converged with non-Maori. But the Maori and non-Maori rates for suicide have moved in different directions. Suicide was rare among Maori but over the past decade the rates for both groups have increased. An analysis of the suicide rates for the period 1957–91 showed an overall lower Maori rate; it was not until 1987 that Maori youth suicides had reached similar levels to non-Maori (Skegg, Cox & Broughton, 1995). By 1990 Maori male rates had risen to 10 though non-Maori rates had risen even more to 35. But by 1993 the differences in youth suicide between Maori and non-Maori males had virtually disappeared, both rates being around 33 per 100,000 (Skegg, 1997). Maori rates have escalated even further so that in 1997 the Maori male suicide rate (26.8 per 100,000) was 28% higher than non-Maori while the Maori female rate (8.6 per 100,000) was almost 60% higher than for non-Maori females (Ministry of Health, 2001). Factors thought to be relevant to the increase include deculturation, family adversity, social disadvantage, and a significant mental health problem in adolescence (Durie, 2001).

There have been few estimates of mental disorders within the community but a cohort study among 18 year olds concluded that the prevalence of mental disorders among Maori youth was exceptionally high. The mental health state of 115 Maori 18 year olds was assessed. Higher risks of disorder than non-Maori on all measures of disorder were shown. Overall 55% of Maori met criteria for at least one disorder in comparison to 41% of non-Maori. Maori males emerged as the group with the highest rate of disorder attributable to the elevated rates of conduct disorder and substance abuse disorders (Horwood & Fergusson, 1998).

Although striking, neither hospital admission rates nor community estimates of mental disorder give a comprehensive picture of either the prevalence of poor mental health or mental health status generally. Other indicators should also be taken into account. Young Maori are, for example, disproportionately represented in prisons, forensic services, child health camps, supervisory care, women's refuges, alcohol and drug services, and injury services (Dyall, 1997).

Adulthood

Deaths from stroke have been relatively stable for most New Zealanders but rates for Maori have fluctuated in an upward direction since 1988 and are now about a third as high as for non-Maori. Diabetes occurs more frequently amongst Maori, with a death rate of 47.4 per 100,000, compared to 10.3 per 100,000 for the total population (Ministry of Health, 1998b). For Maori and other New Zealanders the main causes of death are heart disease and cancer. While Maori mortality rates for both conditions have improved, the disparities are still significant. Mortality rates from ischemic heart disease are 257 per 100,000 for Maori and 150 per 100,000 for all New Zealanders (including Maori) while the gap between Maori and non-Maori death rates due to cancer has actually widened since 1988 to 348 per 100,000 compared to 250 per 100,000 (Te Puni Kokiri, 1998).

Since the mid-1970s mental health problems have emerged as a major health concern. Maori first admission rates to psychiatric services had surpassed non-Maori rates for all age groups by 1974. Not only are the rates of admission different but Maori patients also have different needs, receive different diagnoses, enter hospital through different pathways, and have higher rates of readmission. Between 1984 and 1993 the rates for first admissions for Maori men and women had been steady, 120 per 100,000 for women and 180 per 100,000 for Maori men. But readmission rates increased, greatly for men (64%) and significantly for women (28%). While Maori women and non-Maori women had similar rates of first admission over the decade, Maori male rates have been consistently higher by about a quarter than non-Maori rates.

Drug and alcohol abuse and psychosis accounted for 32% of all Maori first admissions. Admissions for schizophrenia had increased as well and by 1993 were two or three times higher than non-Maori rates and rates for Pacific peoples. Schizophrenia, affective disorders, and other psychotic disorders made up 40% of Maori first admissions and 78% of readmissions (Te Puni Kokiri, 1996).

The Aged

Despite several generations of Western influence, Maori society generally retains a positive view toward aging and older people (kaumatua), affording them status and at the same time expecting them to fulfill certain defined roles on behalf of the whanau (family) and hapu (tribe and community). In order to meet those obligations, however, kaumatua must contend with a range of issues that impact on their health and material well-being. In other words, the cultural role cannot be isolated from the conditions in which older Maori live.

Although the great majority of older Maori are not in dire circumstances, there is nonetheless a relatively high rate of disadvantage, poverty, and material hardship levels being around three or four times those of non-Maori. This has major implications because the proportion of older Maori is going to increase quite rapidly over the next two or three decades.

Age related disability is more likely among older Maori than non-Maori, affecting one in three (Ministry of Health, 1994). Major causes of death for this age group include coronary artery disease, cancer, and respiratory diseases, while the common reasons for hospitalization include respiratory disease, cancer, hypertensive disease, coronary heart disease, cataracts, stroke, and diabetes. In a survey of 400 older Maori men and women, self-assessed health status suggested a generally positive attitude toward aging, despite high levels of disability and a myriad of health problems (Te Puni Kokiri, 1997). The finding may reflect the positive roles older Maori play in their communities as carriers of culture and representatives of family and tribe. In the survey, higher standards of health were significantly associated with active participation in tribal affairs and strong cultural affiliations.

An important consideration for Maori society will be how to maintain positive roles for older people in the face of increased numbers and higher levels of deculturation. If kaumatua are valued for their cultural leadership, an urban cohort may struggle to meet obligations if their material circumstances do not permit full participation, or their health prevents active involvement, or they have not been inducted into the culture and are therefore not able to provide leadership. For many now in the 40–60 years age group, traditional kaumatua roles may never be seriously entertained. That in turn will have implications for the ways in which Maori elderly might participate in society, both Maori society and the wider New Zealand society.

Dying and Death

Maori views on death and dying tend to be philosophical. Death is an essential part of the life cycle and even though grief is unconstrained there is a sense of inevitability. Hospice care for the terminally ill is an increasingly acceptable alternative, but families prefer to nurse relatives at home, especially when it is obvious that death is imminent. If death has occurred away from family oversight there is sometimes a sense of shame and a fear of reproach for not providing adequate care.

Following death the body is usually taken back to a tribal cultural centre (a marae) for a three or four day period of mourning (the tangihanga). It is important that the body is intact. Unless the coroner insists on autopsy, permission for post-mortem is likely to be withheld; if it does occur the return of all body parts is imperative. For similar reasons, the harvesting of body parts remains highly controversial, possibly a reflection of earlier times when an enemy could inflict a final insult by desecrating a slain corpse. However, because Maori are over-represented as potential recipients of donor organs, recent attempts to encourage a more permissive attitude have led to a reconsideration of that position, guidelines have been developed that will offer some safeguards against cultural offence (Te Puni Kokiri, 1999).

During the tangihanga, the deceased is addressed as if still alive and close family members, as well as others, spend their time sitting and sleeping around the coffin. Grief is openly expressed as elders offer farewell laments, encouraging the hovering spirit of the deceased to join other family members who have died. A sense of union with the wider world of the departed is established, diminishing somewhat the impact of loss and reinforcing the boundary between the living and the dead. Most families have traditional burial sites (urupa) and avoid cremation. But urbanization has often made it difficult to travel back to rural areas and in many cases contact with tribal relatives has been lost, so burial in a local cemetery occurs.

As a final tribute, children born around the time of death are often named for that person. Continuity has been restored to the family life cycle.

REFERENCES

Children's Agenda. (1999). *Child policy briefing paper*. Auckland: New Zealand Children's Advocacy Trust.

Davey, J. (1998). *Tracking social change in New Zealand from birth to death iv*. Wellington: Institute of Policy Studies, Victoria University.

Durie, M. (2001). *Mauri Ora the dynamics of Maori health*. Auckland: Oxford University Press.

Durie, M. (1998). *Whaiora Maori Health development* (2nd ed.). Auckland: Oxford University Press.

Durie, M. (1985). 'A Maori perspective of health'. *Journal of Social Sciences and Medicine, 20*(5), 483–486.

Dyall, L. (1997). 'Maori'. In Pete M. Ellis & Sunny C. D. Collings (Eds.), *Mental Health in New Zealand From a Public Health Perspective, Public Health Report Number 3*. Wellington: Ministry of Health.

Health Workforce Advisory Committee. (2002). *The New Zealand Health Workforce A Stocktake of Issues and Capacity 2001*. Wellington: Author.

Henare, D. (1994). 'Social policy outcomes since the Hui Taumata'. In *Kia Pūmau Tonu, Proceedings of the Hui Whakapūmau Maori Development Conference*. Palmerston North: Maori Studies, Massey University.

Horwood, L. J., & Fergusson, D. M. (1998). *Psychiatric disorder and treatment seeking in a birth cohort of young adults*. Wellington: Author.

Ministry of Health. (1994). *Four in ten: A profile of New Zealanders with a disability or long-term illness*. Wellington: Author.

Ministry of Health. (1997a). *Primary prevention of rheumatic fever*. Wellington: Author.

Ministry of Health. (1997b). *Rangatahi sexual wellbeing and reproductive health*. Wellington: Author.

Ministry of Health. (1998a). *Our children's health key findings on the health of New Zealand children*. Wellington: Author.

Ministry of Health. (1998b). *Progress on health outcome targets 1998*. Wellington: Author.

Ministry of Health. (1999a). *Standards for traditional Maori healing*. Wellington: Author.

Ministry of Health. (1999b). *Our health our future Hauora Pakari, Koiora Roa the health of New Zealanders*. Wellington: Author.

Ministry of Health. (2001). *Priorities for Maori and Pacific health evidence from epidemiology*. Wellington: Author.

Ministry of Health. (2001b). *Suicide trends in New Zealand 1978–98*. Wellington: Author.

New Zealand Government. (1994). *Report submitted by the New Zealand Government to the United Nations world summit for social development*. Wellington.

Skegg, K. (1997). Suicide and Parasuicide. In Ministry of Health (Ed.), *Mental health in New Zealand From a public health perpsective*. Wellington: Ministry of Health.

Skegg, K., Cox, B., & Broughton, J. (1995). Suicide among New Zealand Maori: is history repeating itself? *Acta Psychiatrica Scandinavia, 106*, 1–7.

Statistics New Zealand. (1998). *New Zealand now Maori*. Wellington: Department of Statistics.

Te Awekotuku, N. (1991). *Mana Wahine: selected writings on Maori womens art, culture and politics*. Auckland: New Womens Press.

Te Puni Kokiri. (1996). *Ngā Ia o te Oranga Hinengaro Maori—trends in Maori mental health 1984–1996*. Wellington: Ministry of Maori Development.

Te Puni Kokiri. (1997). *Oranga Kaumatua: the health and wellbeing of older Maori people*. Wellington: Ministry of Maori Development.

Te Puni Kokiri. (1998). *Progress toward closing social and economic gaps between Maori and Non-Maori*. Wellington: Ministry of Maori Development.

Te Puni Kokiri. (1999). *Hauora o te Tinana me ona Tikanga: A guide for the removal, retention, return and disposal of Maori body parts, organ donation and post-mortem*. Wellington: Ministry of Maori Development.

Te Puni Kokiri, Ministry of Womens Affairs. (1999). *Maori Women in Focus Titiro Hāngai Ka Mārama*. Wellington: Ministries of Maori Development and Womens Affairs.

Matsigenka

Carolina Izquierdo and Glenn H. Shepard Jr.

ALTERNATIVE NAMES

Machiguenga, Matsiguenka, Kogapakori, Kugapakori.

LOCATION AND LINGUISTIC AFFILIATION

The Matsigenka are people of the *montaña*, the rugged rainforests of the upper Amazon fringing the eastern slope of the Andes. They currently number about twelve thousand people inhabiting the Urubamba, upper Madre de Dios, and Manu River basins in south-east Peru. Matsigenka belongs to the pre-Andine group of Arawakan languages that also includes Ashaninka (Campa) and Yine (Piro).

OVERVIEW OF THE CULTURE

Traditionally, the Matsigenka have lived in small, scattered, highly autonomous settlements organized around the household and the residence group showing a strong preference for matrilocal residence. Kinship follows a Dravidian pattern prescribing bilateral cross-cousin marriage (Johnson & Johnson, 1975). Polygamy was common in the past but has become less frequent. The Matsigenka continue to make important decisions within the household and residence group. At certain historical moments, political and economic integration under tyrannical leaders called *kurakas* has emerged, always in response to outside forces (Camino, 1977; Renard-Casevitz, 1991).

The Matsigenka subsist on a combination of fishing, hunting, forest foraging, and long-fallow swidden agriculture. They grow the staples of sweet manioc, plantains, and bananas alongside diverse other crops, medicinal plants, and fruit trees that mature as gardens are abandoned to forest regeneration. Women spin native cotton and weave tunic-like garments on backstrap looms. Women spend tremendous time and effort in preparing manioc beer (*ovuiroki*), the centerpiece of Matsigenka social life, consumed in great quantities at cathartic drinking parties.

The principal rites of passage—birth, adolescence, death, and mourning—are private, family matters accompanied by quiet, symbolic acts: dietary and behavioral restrictions, a degree of social isolation, and the use of special medicinal plants. Codes for good and bad behavior, though not expressed in legal or religious institutions, are reflected in folklore, interpretations of illness, and many aspects of daily life. Traditional medicine addresses many kinds of misfortune and serves as an arena for expressing and resolving social strife.

Missionary activity throughout the 20th century provoked major changes. Catholic missions were established at strategic points along the main river courses, serving as hubs of commerce and points of departure for colonization and development projects. Beginning in the 1950s, Protestant missionaries of the Summer Institute of Linguistics (SIL) began evangelical work in the hinterlands, contacting and settling dispersed Matsigenka households into large, permanent communities. Although SIL's main goal was evangelical, their work also included health care, linguistic, and ethnographic study (Snell, 1964; Snell, 1998; Snell & Davis, 1976), community organization, and bilingual education. Communal land rights and democratic representation were established formally in 1974 by the populist "Law of Native Communities." Currently there exist some thirty-five legally recognized Matsigenka communities with populations ranging from a few dozen to over three hundred inhabitants.

THE CONTEXT OF HEALTH: ENVIRONMENTAL, ECONOMIC, SOCIAL, AND POLITICAL FACTORS

The health status of the Matsigenka depends upon tightly interwoven cultural, environmental, epidemiological, economic, and historical factors. Their dependence on scattered forest resources provides a centrifugal force that is opposed by the centripetal forces of agriculture, social life, and more recently, education, health care, and economic opportunities available in permanent communities. Each family and community strikes a unique balance, affecting health, nutrition, and livelihood in complex ways. Protein in the Matsigenka diet comes mostly from high quality, animal sources, though it accounts for only 10% of caloric intake. Over three quarters of dietary calories come from the staple starches of manioc and bananas. Research among the Matsigenka has judged child health to be compromised by protein deficiency (Cueva, 1990), although deficiencies were not as marked as in other Amazonian populations. Game depletion around sedentary communities exacerbates existing deficiencies, prompting migrations or alternative economic strategies.

The health and destiny of the Matsigenka have been shaped by powerful exogenous forces. During the "rubber fever" (1895–1917), native people throughout the Amazon were enslaved, brutalized, and exposed to devastating epidemics. Von Hassel (1904) notes that the Matsigenka were highly sought after as laborers in Manu's infamous rubber camps, and estimates that more than 60% of those conscripted died. Slave raiding, debt peonage, violence, and tribal dislocation continued throughout much of the 20th century (Alvarez-Lobo, 1996; Lyon, 1984; Shepard & Chicchon, 2001; Townsley, 1987). Many native groups, including many Matsigenka, survived these grim years only by isolating themselves from outsiders, some through the present (Shepard, 1999a, p. 47). Yet isolation has severe consequences: interethnic warfare; fragmentation of intermarrying units; inability to obtain trade items; reduction or abandonment of agriculture; and high rates of mortality when Western contact is re-established.

The efficacy of antibiotics proved to be crucial in the missionaries' success in settling the Matsigenka in permanent communities. Today, contacted Matsigenka depend on Western medicines for their survival. Medicinal plants are effective for treating some common ailments, but epidemic diseases, especially the feared and deadly grippe (*merentsi*), have overwhelmed the capacities of traditional medicine. Moreover, some missionaries have been intent on stamping out traditional medicine, especially shamanism.

Gastrointestinal and respiratory conditions are the most common causes of disease and death among the Matsigenka and other Amazonian indigenous populations (see Figure 1). Respiratory illnesses cause extremely high rates of mortality in the initial years of Western contact (note Panoan cases in Figure 1). Even in acculturated

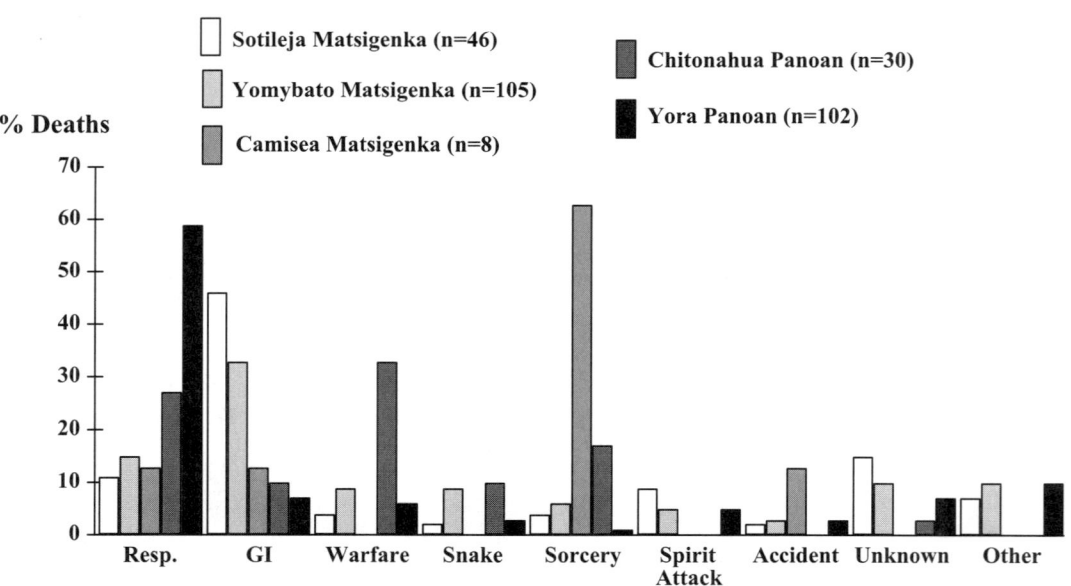

Figure 1. Attributed causes of mortality, all ages.[1]

communities, new cold and flu strains arrive continually, often developing into virulent pneumonia. In communities without adequate latrines or potable water, fecal contamination of soil and water leads to a high incidence of roundworm, amoebas, and other parasites, especially when seasonal flooding exacerbates existing sanitation problems. Snakebite and other accidents (e.g., falling from a tree) are not infrequent and sometimes fatal among children and adults, and interethnic warfare was a significant cause of death in some populations until recently. Tuberculosis has become prevalent in some areas, while malaria, sexually transmitted diseases, and rabies break out sporadically in the Urubamba region, but are not yet documented in Manu. Eye and ear infections, dental caries, and a variety of skin conditions (mycosis, leishmaniasis, scabies, boils, bot fly larvae, chigoes, infected wounds) are common but do not contribute to mortality.

Studies with other Amazonian groups suggest that health status declines with initial Western contact (Jelliffe, 1966; Neel, 1974). Nonetheless, improvements in sanitation and health care in acculturated communities may be effective in eventually reversing this trend. A biomedical study at Kamisea (Izquierdo, 2001, 2002) following up on a similar study 20 years prior (Strongin, 1982) suggests that objective health indicators have improved significantly. Yet ironically, the people of Kamisea report a subjective experience of increasing illness, social strife, and distress for the same period. This discrepancy illustrates the complex and sometimes contradictory health implications of rapid social change. The proliferation of sorcery accusations in Westernized communities testifies to the coincidence between exotic illnesses and new kinds of social and economic stress (Izquierdo 2001, p. 250; Shepard 2002c, p. 9). Western media images also exert a perverse influence on traditional concepts of health and beauty (Yu & Shepard, 1998). Most Peruvians and some Matsigenka welcome the economic development that promises to accompany the exploitation of the Kamisea gas fields in the Matsigenka heartland. Others, however, are worried about negative social, environmental, and health consequences (Rivera, 1991). Echoing traditional cosmological and etiological notions, some Matsigenka fear that the perforation of wells might unleash demons and illness from the bowels of the earth (Shepard,1999b). The plight of indigenous peoples in Ecuador, Colombia, and Nigeria at the hand of transnational petrochemical companies appears to justify these fears (see Kimerling, 1991).

MEDICAL PRACTITIONERS

Shaman-healers among the Matsigenka are known as *seripigari*, "tobacco-intoxicated ones" (Baer, 1992). Seripigari use tobacco and psychoactive plants to enter into a trance, communicate with spirits, and treat spirit-based illnesses (Shepard, 1998). They are usually men, but women are sometimes mentioned. The seripigari undergoes an apprenticeship with an experienced shaman and gains special powers by fostering a relationship with benevolent spirits known as *Saangariite*, "Invisible/Pure Ones." The seripigari sing esoteric songs to invoke spirits or call wayward souls, and use their magical breath to locate and suck sorcery objects (spines, stones, herbs, etc.) from the patient's body. More than healers, the seripigari are ambassadors to the cosmos who negotiate with the forces of nature to ensure the availability of game, acquire new crop varieties, and control the elements. The seripigari are heroes in Matsigenka mythology, however in actual practice, they are beset by moral ambiguity. The same powers required to cure illness can also be used to cause it. The Matsigenka distinguish between healing shamans and illness-causing sorcerers (*matsikanari*), however any shaman may be a suspect for sorcery accusations. This inherent moral ambiguity has been fueled by Christian missionaries' denunciations of shamans as "Devil worshipers." Furthermore, shamans were ineffective in halting the devastating epidemics. The seripigari have been apparently driven into hiding or extinction in many communities. Recent ethnographies note that the Matsigenka deny the existence of seripigari in their communities (see Bennett, 1991, p. 380; Shepard, 1999a, p. 82; Izquierdo, 2001, p. 233). Yet after years of such denials, Shepard (1999a) discovered that seripigari were in fact active (albeit discreetly) in several communities in the Manu and Madre de Dios Regions.

A new breed of practitioners known as "steam bath healers" (*itsimpokantavagetira*) has emerged in acculturated communities of the Urubamba, apparently filling the void left by the absent (or secretive) seripigari. Like seripigari, steam bath healers are able to remove intrusive pathogenic objects from patients suffering sorcery and other spirit illnesses. Unlike the seripigari, steam bath healers do not use tobacco or psychoactive plants and do not enter trance. Instead, they subject the patient to a steam bath by placing scalding hot river stones in a pot of water with special herbs layered in the bottom. The patient stands over the pot covered in a cotton tunic as the healer fans the hot steam. When the water cools, the healer removes the contents of the pot and discovers a variety of objects (nails, sorcery herbs, plastic containers, etc.) that have apparently emerged from the patient's body. The healer converses in detail with the patient about the history of the illness, focusing on social conflicts and possible sorcery suspects. Multiple treatments are usually required, and the healers charge a moderate to considerable fee. The technique of steam-bath healing is open to men and women, is learned from a practitioner for a large fee, and was introduced to the Matsigenka by the neighboring Ashaninka.

Another medical treatment alternative is the government health post. Each full-scale health post (Posta de Salud) is manned by one doctor and two nurses who remain (in theory) for twelve-month stints as part of a required rural service program, making trips to outlying communities for check-ups and vaccinations. Health posts are generally well stocked with pain relievers and antibiotics, though they lack laboratory facilities. There are three health posts in the Urubamba region, serving approximately 25 communities, and two in the upper Madre de Dios serving about 10 communities. Some communities have small village health posts (Posta Sanitaria) manned by regional Peruvian or resident Matsigenka *sanitarios*, health care workers with basic training who receive a small government salary and minimal medical supplies. Native health promoters (*promotor de salud*) work on a volunteer basis, sometimes combining biomedical with traditional healing methods. Health posts are largely underutilized because of the urban-educated professionals' typically paternalistic views, begrudging attitude, and poor communication with community members. Few complete the full twelve-month stint, leaving lesser-trained regional nurses, *sanitarios*, or *promotores* in charge for extended periods of time.

CLASSIFICATION OF ILLNESS, THEORIES OF ILLNESS, AND TREATMENT OF ILLNESS

Classification of Illness

The Matsigenka distinguish and classify illness according to a number of criteria: symptoms, body part/organ system affected, acuteness, duration, and population at risk (see Table 1). Other salient criteria include external (visible) versus internal (invisible) pathology, native versus foreign origin, and natural versus supernatural causes. Some illness terms correspond with Western categories such as leishmaniasis (*tsirivaito*), colds and flu (*merentsi*), malaria (*mogekari*), and measles/pox diseases (*saarontsi*). Other syndromes rely on culturally particular understandings of illness and etiology that are not translatable into biomedical terms. A simple sorting exercise resulted in the following general classification scheme: gastrointestinal, gynecological, respiratory, heart and chest conditions, fevers, body pains, ear/eye/dental, animal bites/stings, animal/plant spirit revenge, spirit attack and sorcery, and miscellaneous signs and symptoms (Shepard, 1999a). A great deal of overlap across categories was observed as illustrated in Figure 2. Multidimensional scaling of pile sorts carried out with 43 common illness terms (Izquierdo, 2001) revealed seven illness groups. The low stress (0.135) for the 43 items suggests that people distinguish illness along two major dimensions. We interpret the dimensions as acute and chronic and serious and non-serious. However, these categories often are associated with each other. For example, in some cases, duration and etiology are closely related: and ongoing or chronic illness is generally associated with spiritual etiology (sorcery, spirit attack).

Theories of Illness

The Matsigenka concept of well-being is summarized in the verb *shinetagantsi*, which means to be happy, productive, and well-fed as well as free of illness. Concepts antithetical to well-being include illness (*mantsigarentsi*), suffering (*tsipereagantsi*), thinness or weight loss (*matsatagantsi*), sorrow or worry (*kenkisureagantsi*), anger (*kisatsi*), and soul loss (*gasuretagantsi*). Health and well-being and, conversely, illness and malaise, embrace physical, emotional, and spiritual states as well as harmony (or lack thereof) in productive, social, and environmental interactions. In the Matsigenka cosmos-as-ecosystem, illness, misfortune, and death are often interpreted through the ecological metaphor of predation: just as humans hunt for sustenance, so also demons, illnesses, and dangerous animal spirits look on human beings as game animals (Shepard, 2002c).

Matsigenka theories of illness demonstrate complex notions of etiology and efficacy that challenge Western

Classification of Illness, Theories of Illness, and Treatment of Illness

Table 1. Matsigenka illness classification

Illness category/gloss	Matsigenka name	Etiology	Severity
Gastrointestinal			
"Intoxication," vomiting w/ fever (Yellow fever?)	*Pigarontsi, kepigari*	falls from sky, sorcery	Highest
Diarrhea w/ blood	*Shiarontsi*	(as *tseritsi*), from foreigners	High
Diarrhea	*Tseritsi*	falls from sky, "worms", mixing foods, dirty or spoiled food	Medium
Vomiting	*Kamarankagantsi*	(as *tseritsi*)	Medium
Indigestion	*Sametsi*	(as *tseritsi*)	Low
Stomachache	*Katsimotiatagantsi*	(as *tseritsi*)	Low
Gynecological			
Retained placenta	*Terira okontetake iranonta*	taboo activities (arrow resin use)	Highest
Difficult childbirth, post-partum bleeding	*Okatsitake oananekite, voaasetagantsi*	no reason, taboo activities (cutting bamboo)	High
Miscarriage (fetus)	*Omechotaira ogairi*	sorcery	High
Miscarriage, induced abortion (early pregnancy)	*Oseriakotake oananekite*	animal spirit, plant-induced	High
Menstruation (painful, excessive bleeding)	*voaasetagantsi katsiri / kogapage*	no reason, sorcery, animal spirits	Medium
Menstruation (normal)	*Voaasetagantsi, seriagantsi*	moon	Low
Respiratory			
Bloody cough, tuberculosis	*Voreagantsi iraatsi*	foreigners	Highest
Cold, flu	*Merentsi*	foreigners	Highest
Throat abscess (tonsillitis)	*Vompotsanotagantsi*	falls from sky, mixing foods, trauma	High
Cough	*Voreagantsi*	falls from sky, mixing foods, foreigners	High
Sore throat	*Katsitsanotagantsi*	falls from sky, mixing foods, trauma	Medium/High
Nasal congestion, runny nose	*Shirinkasetagantsi*	no reason	Low
Nosebleed	*Voaatagantsi iraatsi girimashiku*	mixing foods	Low
Heart /Chest			
Heart failure, cardiac arrest	*Vitantagantsi negiku*	lethal illness	Highest
Chest pain ("needle pain")	*Kentarontsi, kitsogirontsi*	animal spirits, demons, ghosts	Highest
Heart pain, epigastric pain	*Katsinegitagantsi*	bee spirit, demons, ghosts, sadness, falls from sky, mixing foods	High
Heart palpitations ("fear in the heart")	*Tsaronegintagantsi*	fright, love filtres, demons, ghosts	High
Fevers			
Shivering, chills (malaria)	*Mogekari, shigekari*	(as *kovaagantsi*), foreigners	Highest
Fever w/ chills	*Anatiri, janatiri*	(as *kovaagantsi*), cold weather	High
Fever	*Kovaagantsi*	falls from sky, demons, ghosts	Medium

Table 1. *Continued*

Illness category/gloss	Matsigenka name	Etiology	Severity
Body pain			
Bone pain, rheumatism	*Shinkogiitagantsi*	demons, ghosts, sorcery	High
Body pain, (back, knee pain)	*Katsipagetagantsi, (katsitishitagantsi, Katsigeretotagantsi)*	trauma, old age, demons, ghosts	Medium
Headache, migraine	*Katsigitotagantsi*	falls from sky, demons, ghosts, w/ fever	Medium
Rib pain	*Katsimeretagantsi*	trauma, bee spirit	Medium
Waist, kidney pain	*Katsitsakitagantsi, tsatsakirontsi*	trauma, demons, ghosts, falls from sky, w/ urinary pain	Medium
Ear/Eye/Dental			
Earache (w/ pus, abscess)	*Sakempitagantsi, sompokempitagantsi*	hygiene, "worms", foreigners	Medium/High
Earache (simple)	*Katsikempitagantsi*	hygiene, "worms", falls from sky	Medium
Toothache	*Okatsitake itsi (jitsi)*	"worms", food	Medium
Eye pain, infection	*Katsiaari*	hygiene, trauma, falls from sky, foreigners	Medium
Eye opacity (cataracts? trachoma?)	*Tsororoaari*	old age, animal spirit, trauma	Medium
Skin			
Wound-like			
Abscess, boil	*Sompotsi*	insect bite, botfly larvae, wound	High
Cut, wound, (arrow wound)	*Teretsi*	wound, hygiene	High/Medium
Sore, leishmaniasis	*Tsirivaito, katsinori*	insect bite, insect spirit	Medium
Pustule, infected wound	*Shomporentsi*	falls from sky, insect bite, wound, hygiene,	Medium Medium
Burn, blister	*Saatagantsi, meregagantsi*	heat, friction	Medium/Low
Scab	*(Teretsi)*	Wound	Medium/Low
Inflammatory			
Inflammation, swelling	*Nonarontsi*	no reason, w/ infection, fever	High/Medium
Hives, urtication	*Kepigisetagantsi*	plants (urticating), animal spirits, mixing foods,	Medium
Peeling, shedding of skin	*Patsaatagantsi*	falls from sky, wounds, hygiene,	Medium
Rash	*Piikitagantsi, pirikitagantsi*	hygiene, urticating plants	Medium
Itching, (mycosis)	*Kaenitagantsi, tsomiri*	"worms", water, foreigners,	Low/Medium
Scabies	*Patsetsi*	foreigners	Low
"White spots", (mycosis)	*Kutatagantsi*	no reason, worms	Low
Oral (esp. in Children)			
Candida, sores (thrush)	*Kotsetsi*	no reason, animal spirits, w/fever,	Medium
Fever blisters	*Patsaavagantetagantsi, teregantetagantsi*	(as *kotsetsi*)	Medium
"White tongue", (thrush)	*Kutanenetagantsi*	(as *kotsetsi*)	Medium
Measles, Pox			
Measles, smallpox	*Saarontsi, "sarampion"*	demons, foreigners,	Highest
Chicken pox	*Morokisetagantsi*	animal spirit (cowbird)	Medium
Animal bite/sting			
Snakebite	*Maranki yatsikanti*	bad dream, demon	Highest
Jaguar bite	*Matsontsori yatsikanti*	accident, sorcery	High

Classification of Illness, Theories of Illness, and Treatment of Illness

Table 1. *Continued*

Illness category/gloss	Matsigenka name	Etiology	Severity
Peccary bite	*Shintori yatsikanti*	—	Medium
Spider bite	*Jetyo yoganti*	—	Medium
Stingray sting	*Inaro ikentanti*	accident, bad dream	Medium
Paraponera ant sting	*mushi yoganti*	—	Low
Caterpillar sting	*Soromai iporonganti*	—	Low
Wasp sting	*Sani yoganti*	—	Low
Animal, plant spirit revenge	*Pugasetagantsi*	causes gastrointestinal, fever, skin conditions (esp. children)	Medium
Mixing foods improperly, animal spirit revenge (food allergies?)	*Tamampegagantsi*	causes gastrointestinal, fever, skin conditions	Low (except in ill, old)
Sorcery, demon attack			
Sorcery (general)	*Matsitagantsi*	due to envy, jealousy, personal conflict; causes severe or chronic illness	Highest
Footprint sorcery	*Ampaseri, gagitetagantsi*	causes swelling, numbness in legs, legs turn black, "burned"	Highest
Sudden death, demon attack	*Komutagagantsi*	causes sudden death	Highest
Rape by demon (deer)	*Itsitanti kamagarine (maniro)*	causes high fever, body pains, severe illness, wasting death	Highest
Demon, ghost (seeing)	*Tsavitetagantsi*	causes high fever, body pains, severe illness, quick death	Highest
Demon, ghost (hearing)	*Amumpava*	causes high fever, chills	High
Demon, ghost (dreaming)	*Kisanitagantsi*	causes headache, fever	Medium
Epileptic fit	*Kamakamatagantsi*	eyes roll back, foam at mouth, fainting; spirit attack, sorcery	Medium
Miscellaneous signs/symptoms			
Muscle cramps	*Tsoritsi*	w/ severe GI illness, trauma, falls from sky	Highest (w/ GI) Low (alone)
Pallor, anemia, jaundice ("yellowness")	*Kitetagantsi*	w/ severe GI, heart illness	High
Fatigue, difficulty in breathing	*Shigopirentsi*	w/ severe respiratory, GI, heart illness	High
Liver pain	*Katsiriraapanatagantsi*	no reason, w/ severe illness	High
Spleen pain	*Taratagantsi*	trauma, w/ severe fever (malaria)	High
Urinary pain, STD's	*katsitsinitagantsi, tsomiri*	no reason, foreigners, "worms"	Medium

dichotomies such as mind/body, individual/society, culture/nature, and natural/supernatural. Pneumatic (see Wilbert, 1986), germ-like, personalistic, and biomedical etiology models are found side by side, as are allopathic, homeopathic ("doctrine of signatures"), and spiritual models of efficacy (Shepard, 1999a). Illness episodes for the Matsigenka are not bounded by time or defined by symptoms. A series of apparently unrelated ailments are, for the Matsigenka, a single, ongoing illness linked by a common cause, often of supernatural origin. Accidental trauma is a frequent problem, but the most severe accidents (snakebite, falling from a high tree) are blamed on

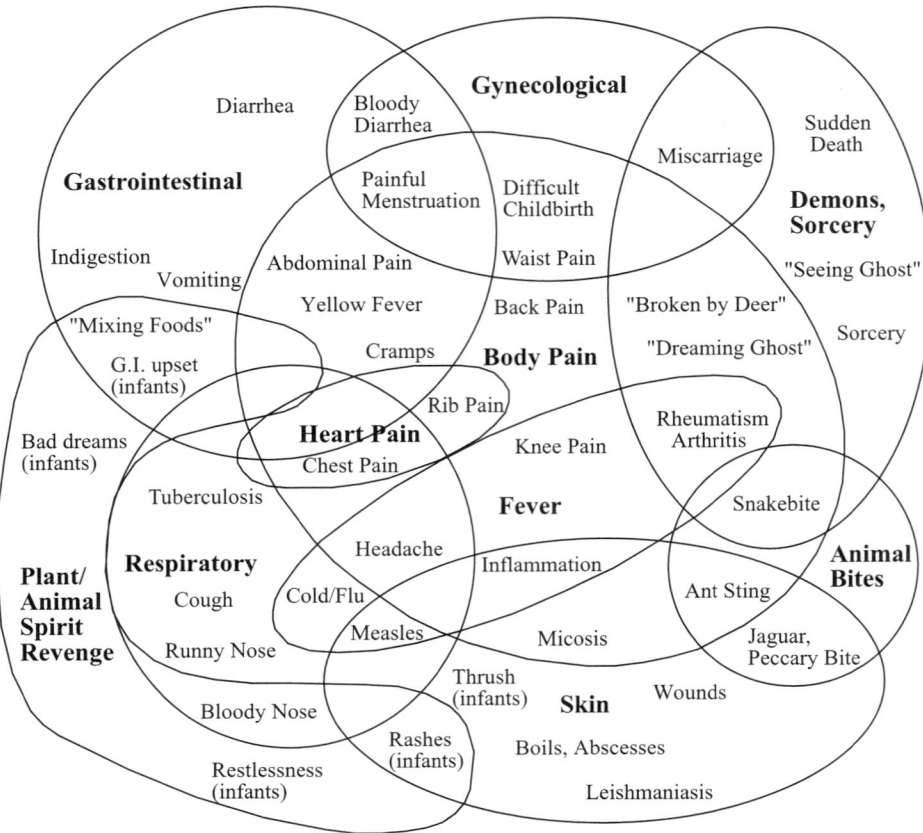

Figure 2. Matsigenka illness classification.

supernatural causes. Even common, self-limiting illnesses may include cosmological or spiritual elements: for example, one type of bee (*yamposhto*) is believed to cause diarrhea by feasting on uneaten food, thereby stealing a piece of the person's soul.

Many illnesses are conceived of as foul vapors that rise from the bowels of the earth where demons reside. They are released into the atmosphere by landslides or earthquakes, causing "yellow vapor in the sky" (*okiterienkatake*), rainbows, and other unusual sky coloration. With rainfall or fog, illness "falls as mist from the sky" (*oparienkatake*), adhering to the skin and entering the body through the nose, mouth, and eyes. This etiology pattern is associated with epidemic and seasonal illnesses (notably *pigarontsi*, apparently yellow fever), and implies a naturalistic, germ-like theory of contagion that is nonetheless linked with cosmological and spiritual notions.

The Matsigenka attribute most acute gastrointestinal conditions to the presence of "worms" (*tsomiri*) in the gut. Though aware of helminthic parasites (also *tsomiri*), the Matsigenka are not necessarily referring to these when they invoke "worm" etiology. Tsomiri is a broader concept, resembling Western germs or microbes, and is also used to explain toothache, earache, conjunctivitis, and various skin infections. Threadlike or like maggots, and so small as to be invisible, tsomiri are found in overripe fruits, spoiled food, and garbage. With poor hygiene, tsomiri enter the skin, ears, eyes, mouth, or gut. In the Urubamba, sexually transmitted diseases are likewise referred to as *tsomiri*.

Certain transitory gastrointestinal, skin, and other conditions (notably food allergies) are attributed to the improper mixing of foods (*tamampegagantsi*). Tapir should not be eaten with armored catfish (*etari*) to avoid scabies, characterized by dry, scaly skin like *etari*. Eating carnivorous fish such as piranha with other meats causes bloody diarrhea. These beliefs are closely associated with the concept of revenge (*pugasetagantsi*) by plant and animal spirits. Strong-smelling, carnivorous, and other salient fish and game species are said to have vengeful spirits that attack young children if either parent eats their meat. (The notion implies the existence of a shared "social body" that transcends individual body boundaries.) Vengeful spirits frighten or steal the soul (*yagasuretakeri*) of children, causing a wide range of illnesses including colic, restlessness, rashes, nosebleed, and sudden death. Children are painted with red dye on the face and crown of the head, the portal of the soul, to dispel harmful spirits. Plants with noxious, toxic, or other symbolically dangerous properties can also take revenge. It is impossible to separate spiritual from empirical aspects of these beliefs, since the spirit or soul (*suretsi*) of an organism encompasses supernatural as well as nutritional, toxic, or medicinal properties.

Many conditions, especially those associated with emotional disturbances, affect the heart. Death occurs when illness "grips the heart" (*avitantake negiku*). Heart palpitations, (*tsaronegitagantsi*, "fear in the heart") can be caused by bad dreams, sadness, gossip, unrequited love, sorcery, or a love potion gone awry. *Katsinegitagantsi*, "heart pain," combines in a single category a number of symptoms that in English would be described separately as "heartburn," "heartache," heart palpitations, anxiety, chest pain, pneumonia, and heart attack (Shepard, 2002a).

The Matsigenka are especially fearful of apparitions of demons (*kamagarini*) and ghosts (*kamatsiri*), believed to cause among the most severe illnesses. Dreaming (*kisanitagantsi*) about the ghost of a close family member causes fever, headache, and body pain, not necessarily life-threatening as long as treatment is undertaken. Dreaming in general for the Matsigenka involves a vulnerable state where the soul wanders and may encounter dangerous spirits or omens of illness and misfortune. Seeing a ghost while awake, *tsavitetagantsi* ("to see something that later vanishes") is the most serious illness recognized by the Matsigenka, leading to almost certain death. The Matsigenka believe that the deer (*maniro*) is a demonic seducer who appears as an alluring member of the opposite sex to a person walking alone in the forest or in a garden. The seducer offers food and entices the victim to have intercourse. Those who fall victim to the temptation suffer severe illness (fever, chills, body aches) or death, and those who resist must tell no one of the

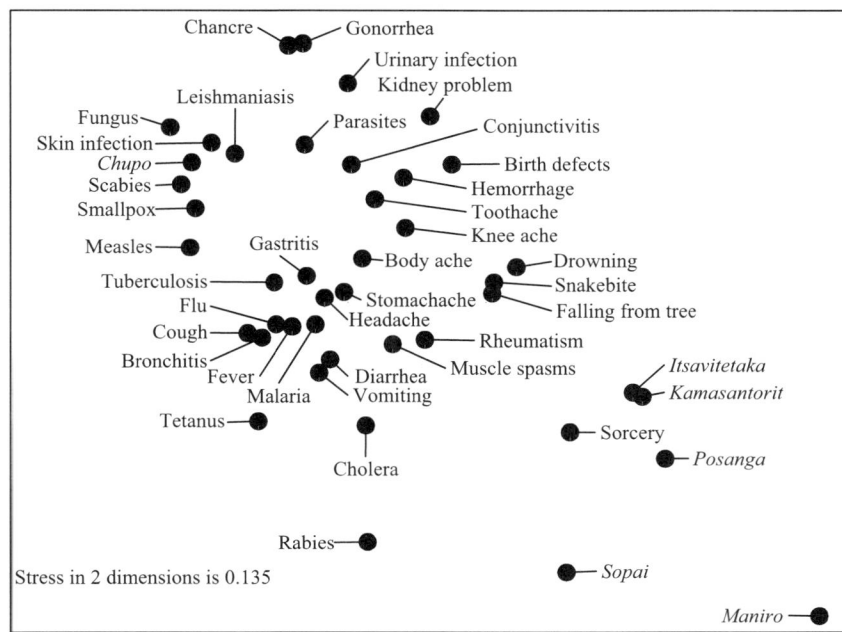

Figure 3. Multidimensional scaling of pile sorts of 43 Matsigenka illnesses (N=30).

encounter for three days and take numerous precautionary measures and remedies. Paralysis or sudden death in old people is attributed to assault by the deer demon: *otingarantira maniro*, "broken by the deer." People who are ill or at vulnerable junctions in the life cycle do not eat deer meat, and some people avoid it altogether.

Whereas predatory demons, animal spirits, and ghosts were blamed for inexplicable illnesses in the past, sorcery is frequently invoked in Westernized communities. Footprint sorcery (*gagitagantsi*) is carried out by collecting dried bits of mud from the victim's footprint and boiling them with special herbs, causing the victims leg to swell, become tingling or numb (*ampasetagantsi*), and turn black as if burned (*shinkogitagantsi*). The Matsigenka believe that love magic (*posanga*) is a kind of sorcery brought by outsiders, causing physical–emotional symptoms in its victims (dizziness, heart palpitations, sadness, insanity), as well as infidelity and social strife in the community. Sorcery accusations are often aimed at those same groups of outsiders (Ashaninka, Piro, Shipibo) who served as intermediaries for Western economic interests throughout the 20th century. However, in increasing numbers other fellow Matsigenka are being accused and identified as sorcerers. Ethnically diverse communities and mission towns are thought to be especially dangerous, and Matsigenka who have lived among them are prime sorcery suspects. Thus sorcery theories reflect Matsigenka constructions of interethnic relations and notions of culture change (Izquierdo & Johnson, 2003).

Treatment of Illness

Treatment decisions often relate as much to social dynamics as to the choice of the appropriate remedy, and may involve self or family care, consultation with kin, informal networks, local healers, and medical personnel. The Matsigenka incorporate biomedical and local knowledge in assessing health care options. Some 300 species of medicinal plants—mostly primary forest shrubs and herbs but including some cultivated plants and weeds—were identified in an ethnobotanical survey of seven communities (Shepard, 1999a), and the average informant knows 80 or more species (Izquierdo, 2001). Medicinal plant knowledge is widely shared, though a degree of specialization is found along gender lines: women are more knowledgeable about plants for child care and fertility control, while men specialize in hunting medicines and treatments for wounds and snakebite.

Medicinal plants are attributed sensory properties (taste, odor, color, etc.) that are directly related to notions of etiology and efficacy (Shepard, 2002b). Plant extracts are administered as herbal infusions, eye drops, inhalants, poultices, or warm baths. Bitter medicines are administered externally for skin conditions or internally for gastrointestinal conditions in order to "embitter," hurt, and expel the "worm" pathogens, a clearly allopathic model of efficacy. Fragrant plants are used to repel the foul odor of demons, ghosts, and vengeful animal spirits, or to undo the "intoxicating fragrance" of love potions. Painful eyedrops are used to improve a man's aim while hunting and to dispel ghosts, headache, sadness, or social strife.

Medicinal plants appear to be somewhat effective in addressing many common health problems, notably gastrointestinal and skin conditions, broken bones, snakebite, and other hazards of the forest. Hunting medicines, some of which may have tonic or stimulating properties, reflect the uncertainty and importance of game animals in the diet. Core Matsigenka values throughout the life cycle are reflected in plant medicines used to treat excessive crying in babies, promote weight gain, instill desired personality traits in children (industriousness, honesty), give young men good eyesight for hunting, improve garden productivity, make women attentive and careful at spinning and weaving, and reign in dangerous emotional excesses such as anger, grief, and passion.

The health post is an important treatment alternative, especially for illnesses deemed to be of mundane or foreign origin. Visits to the health post show apparent group differences across gender and age (Figures 4 and 5). Boys under five are taken to the health post twice as often as girls while among teenagers, the number of visits by girls increases, likely for gynecological inquiries. Among the middle-aged and elderly, females are much more frequent visitors. Several older men said they had never been to the health post because it was "a place for dying." In all age groups, respiratory and digestive problems were most frequently reported. Nonetheless, it is difficult to capture actual disease incidence because record keeping is not reliable, and because of the complexities of local diagnostic categories and treatment options.

Patients consult a traditional healer when specific symptoms or ongoing illness signals sorcery or spirit attack. Still, those who seek healers may continue using multiple other approaches (medicinal plants, dietary restrictions) to cover all possible vulnerabilities. In narrative accounts of such illnesses, the Matsigenka revealed

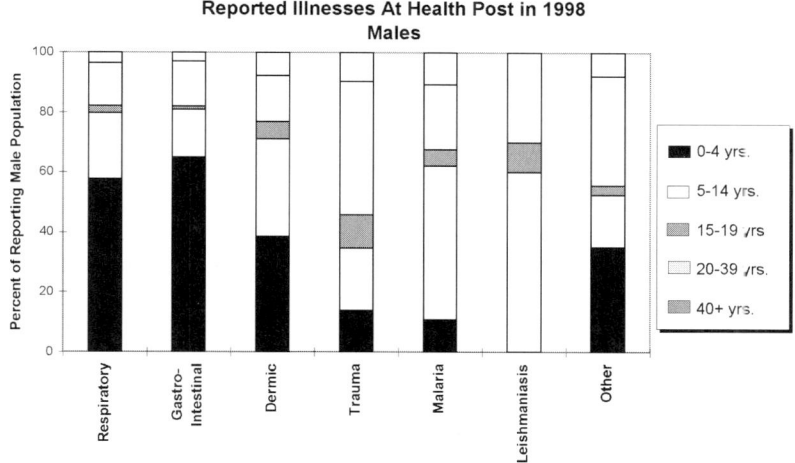

Figure 4. Illnesses treated at Kamisea health post in 1998: Males (percentages).

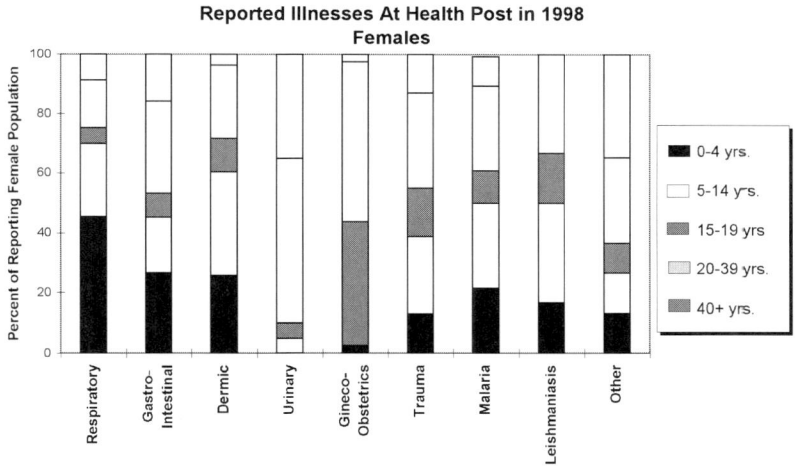

Figure 5. Illnesses treated at Kamisea health post in 1998: Females (percentages).

a great deal about bodily, emotional, and social distress in communities undergoing cultural and economic change (Izquierdo, 2001).

Despite Strongin's (1982) prediction, biomedicine has not replaced traditional medicine. More than just a therapeutic alternative, traditional healing is an expression of indigenous religion, values, and world-view. Not only do local therapies provide cultural and moral meaning to illness, they also function very effectively in alleviating symptoms.

SEXUALITY AND REPRODUCTION

Although typically sex-positive in their orientation, Amazonian societies reveal an undercurrent of ambiguity regarding the dangers posed by improper sexual relations (Gregor, 1985). Incest for the Matsigenka is among the most shameful offenses, and is thought to result in punishment for the perpetrators as well as the offspring, who show physical deformities and moral perversions. Several Matsigenka myths relate mother–son, father–daughter, and brother–sister incest, which all result in exile or death (Johnson, 2002). Other tales tell of the tragicomic outcomes of sexual relations between humans and animals. Demons including the deer (*maniro*) are hypersexual, have gigantic genitalia, and engage in perversions such as anal sex, thereby causing severe illness or death in their victims.

Hunting lore includes a series of sexual and behavioral taboos that establish an ethic of proper conduct between husband and wife, and depict a balance between reproductive and productive responsibilities of the family. Matsigenka men avoid sexual relations the night before hunting. Contact with menstrual blood, said to smell like carrion or raw meat, takes away a man's hunting skill, as does eating raw meat or improperly cooked food. Taboo violations accumulate in the hunter's body/soul, making him smell of carrion like the vulture and thus lose his aim. A man restores his aim by using hunting medicines taken as emetics, purgatives, or as painful drops in the eyes. Notable are the *ivenkiki*, cultivated varieties of sedge (*Cyperus* sp.).

Sedge varieties cultivated by women are used to control fertility. Depending on the variety and dosage, sedges are said to temporarily or permanently lower fertility, induce abortions, facilitate difficult childbirth, or reduce postpartum bleeding. Cultivated sedges have been found to host fungal parasites that produce ergot alkaloids (Plowman et al., 1990), whose physiological activity is consistent with many Matsigenka uses of sedge (fertility, headache, tonic, wounds).

Decisions about fertility are generally left to women. Women who want to terminate a pregnancy may prepare an herbal abortifacient or ask an experienced older women to do so. There is no stigma attached to abortion practices, though improper use is very dangerous. Western birth control is now available in some communities, though use is not yet widespread. Though the reasons are not clear, a significant change in average family size occurred in the late 1960s, from 6.7 in 1965 to 4.7 in 1972 (Strongin, 1982), and remaining at 5.1 in 1998 (Izquierdo, 2001). Sexually transmitted diseases were first brought to the Urubamba region during petrochemical exploration by Shell Oil in the early 1980s.

HEALTH THROUGH THE LIFE CYCLE

Pregnancy and Birth

Pregnancy and childbirth are a special source of concern. Expectant parents avoid a wide range of vengeful fish and game species, notably carnivorous fish. Expectant fathers are prohibited from manufacturing arrows: cutting bamboo is thought to cause uterine hemorrhage, and sticky arrow-fletching resin (*taviri*) causes retained placenta. Parents avoid specific foods and behaviors believed to result in unwanted physical as well as personality traits (anger, laziness) in the developing fetus.

Traditionally, births were attended by elder female relatives but in many communities today, midwives are chosen based on experience and given basic training by the medical personnel. Women give birth in a squatting position, gripping a vertical support such as a house post. Birthing in the health post—prone and surrounded by lights, instruments, and strangers—is viewed as unattractive and frightening.

The immediate concern upon birth is the passing of the placenta and the cutting of the umbilical cord. Medicinal plants are used to aid in the passing of the placenta (*iranonta*) to avoid dangerous complications. Traditionally, the umbilical cord is cut with a grass-like bamboo with razor edges (*tiposhi*). Metal implements are now used, though improper hygiene has been known to cause tetanus. The cut cord is treated with arrow-fletching resin (*taviri*) and other "umbilical cord plants" (*shirimogutopini*), or Western antiseptics when available. The mother and baby are washed in warm water and herbs. Both the placenta and later, the umbilical cord are buried to avoid contact with carrion animals or evil spirits.

Infancy

The time between birth and the falling of the umbilical cord is especially dangerous. Both parents remain indoors and follow strict dietary and behavioral restrictions. Through the first few months of life, parents maintain certain restrictions to avoid mishaps and illness in their infants. Strong-smelling and carnivorous fish and game are avoided because of their vengeful spirits, as are urticating or caustic plants, thought to cause rashes in newborns. To dispel vengeful spirits, the mother bathes the baby daily (once or more times) in warm herbal mixtures including sedges as well as fragrant, viscous, and colorful plants. In addition to their role in hygiene, these practices reflect the general anxiety of parents about newborns in a setting where infant mortality is considerable.

Quickly, the couple must return to the productive routines of daily life to provide a good example and prevent the child from growing up lazy or inactive. Babies are carried at all times by their mothers in a woven cotton sling (*tsagompirentsi*), nursing on demand. Babies and young children share bedding in close contact with both parents. Mothers masticate boiled manioc and feed it mouth-to-mouth (*aviakeri*) to infants as young as 3 months to facilitate weight gain. Sweet or tart herbs are sometimes mixed with this mash to stimulate appetite. Broth, and later meat of bland, non-vengeful fish and game, is introduced soon thereafter. Babies are breast-fed an average of 2 years, and may be weaned using chili pepper if they insist on nursing beyond that time.

Childhood

Diarrhea, especially when combined with vomiting, quickly dehydrates infants and young children, and is a frequent cause of death in this age category. Increasingly, health workers are teaching oral rehydration therapy. The extended abdomen visible in most children and many adults is evidence for a moderate, practically normative parasite load. Extreme parasite loads are known to cause anemia and even life-threatening intestinal blockage, especially in children. *Ficus* latex is a common and effective folk remedy for intestinal worms, though some Matsigenka fear the powerful purgative properties of an improper dose. (Matsigenka men's use of purgative and emetic plants may also help reduce parasite loads.) Pharmaceutical antihelminthic treatments are distributed with increasing frequency, however without sanitation and potable water, reinfection is rapid.

Mothers are the main care providers for small infants, though fathers come to take on an important role in education and discipline in later childhood. Socialization of appropriate emotions as well as competence, autonomy, and sociability is a strong focus even in infancy. Children are cherished and protected by an ever-present network of family members. However, once children have gained a strong footing, they are given great freedom. Children learn from a young age to participate in subsistence activities. Young girls help their mothers in cooking, cleaning, childcare, and gardening, while boys roam a bit more freely, playing at hunting, fishing, building shelters, and making fires.

Children may be given a series of herbal treatments to ensure fast growth, protection from illness, and positive behavioral traits (honesty, industriousness, emotional control) in later life. Children are also taught to show respect toward particular animals and plants, because mocking, handling, or bothering them may provoke spirit revenge. Children are taught to control their expressions of pain and other negative emotions, leading to a marked stoicism among Matsigenka adults.

Adolescence

By about age 15, children should be proficient in subsistence responsibilities and know right from wrong. Adolescents who act inappropriately are publicly humiliated and punished. Stealing is considered a serious offense. Adolescents are also punished for bullying, laziness, and overly precocious sexual behavior. Girls are generally permitted to engage in an open sexual relationship at a somewhat younger age than boys, who must first prove their abilities in hunting and gardening. Teenage boys may be given a series of purgatives and eye-stinging remedies to develop their hunting skills. They are not allowed to touch or eat their first several kills, and are taught to avoid bragging about or being selfish with meat, lest they lose their stamina and aim. Teenage girls may learn herbal remedies for learning to spin cotton, gaining weight, having strong and durable teeth, removing pubic hair (considered unaesthetic), or regulating fertility.

Transition into womanhood begins with first menstruation, marked by an extended ritual of seclusion (*oantarotira*, "she becomes an adult") during which the girl lives in a small hut, eats special foods, avoids the sun and the gaze of males, receives special instructions from close female kin, and spins and weaves cotton to produce a tunic for her future husband. She emerges from seclusion after completing the tunic, shaving her head to mark the transition. Her emergence is generally greeted by a manioc beer party held in her honor, during which suitors may express their intentions to her and her parents.

Adulthood and Aging

Adult life is characterized by the extreme self sufficiency of the Matsigenka household and the division of labor between husband and wife. Complementary gender roles are reflected in medical botany: whereas men use emetic and painful plant remedies to improve their hunting skills, women use fragrant herbs to protect children from vengeful game animal spirits. People with deficiencies (e.g., cleft palate, atrophied limb) are not ostracized, and usually engage in a fairly normal economic and social life, sometimes compensating for their deficiency in one area with special skill in another.

Couples remain autonomous in most subsistence activities through middle age and well into old age, and only a very small number of school teachers and health workers receive government pensions. As a man ages and loses his stamina, he tends to leave the more strenuous hunting and garden-felling to a son-in-law, but continues active garden work, fishing, and hunting from blinds. Women, too, remain active and productive through old age. Given the many hazards of life in remote rainforest villages with minimal or non-existent Western medical care, it is surprising that as many Matsigenka make it to old age as they do. The term for old age is *gatavaigetagantsi*, which translates literally as "to be done, to be satisfied after a long life," implying a sense of completion and fulfillment that contrasts with Western notions of old age as decay and decrepitude. The elderly among the Matsigenka are highly respected for their knowledge of medicinal plants, mythology, history, and folklore, and for their entertaining yarns about a past that was more difficult, more treacherous, and more mysterious than life today.

Dying and Death

Life and death for the Matsigenka are not mutually exclusive states. Instead, death belongs at the far end of a continuum of more and more permanent degrees of separation of the soul (*suretsi*) from the body (*ivatsa*). The soul leaves the body through the crown of the head (*vankagantsi*), nostrils, and eyes during illness, sadness, sleep, trance, and unconsciousness (the verb *kamagantsi* means both "to fall unconscious" and "to die"). The soul is a life force that infuses and activates the bodies of humans, animals, and animate beings with growth, appetite, and purposeful action. At variance with Western metaphors, skin and bones, rather than flesh and blood, are considered to be the inert, soulless aspects of the body. Flesh (*ivatsa*), blood, fat, and muscle are physical manifestations of the soul's presence in a healthy body. A person who is extremely emaciated due to illness, sadness, or old age is as good as dead: the soul has already left the body, and all that remains are the lifeless skin and bones. In the past, frail, elderly people who could no longer care for themselves simply walked off alone in the forest, since their souls had departed and they were already as good as dead. The dead were exposed in fetal position in the buttress roots of a large tree or in the floodplain down river from the settlement, though burial has been adopted since missionary contact. Shamans are believed to be immortal, awaking from death to join the benevolent spirits in safeguarding humanity.

Matsigenka mourning practices invert commonsense Western understandings of grief, since it is the dead who grieve for the living (Shepard, 2002a). The living sit in stoic silence while the dead roam in the vicinity of the village for several days, gathering up lost and discarded belongings and seeking loved ones to take as companions to the Land of the Dead. Grief is potentially lethal because the dead are still emotionally attached to the living and capable of stealing their souls to keep as company. Pensiveness, inactivity, and loss of appetite in grief are early signs of soul loss; seeing a ghost in dreams or apparitions is a very dangerous sign, requiring special treatment. The short mourning period (about 3 days) aims at protecting the living from the dangerous attentions of the ghost, speeding it on to the Land of the Dead as quickly as possible. No one wanders far to avoid encountering the ghost, which is feared, shunned, and dehumanized. Family members of the deceased shave their heads, paint their face with red or black dye, burn chili peppers, and may wear the tunic or garments of others, hoping to disguise themselves from the ghost. Often, family members temporarily or permanently abandon the dwelling or residence group where the person died. Crying and overt expressions of grief are strongly repressed. After 3 days, the family slowly returns to normal life, taking warm, medicinal herbal baths and eye-stinging drops to dispel sadness and visions of the ghost.

NOTE

1. Data from three Matsigenka populations (Sotileja, Yomybato, & Kamisea), organized from least to most acculturated, and two neighboring Panoan-speaking populations, the Yora, contacted forcefully in 1985 and the Chitonahua, contacted forcefully in 1995. Data are taken from extensive genealogical interviews (Shepard 1999a: 105–107), except Kamisea which represents deaths occurring during fieldwork conducted from 1996-1999 (Izquierdo, 2001).

REFERENCES

Alvarez-Lobo, R. (1996). *Sepahua: Motivos para crear una misión católica en el Bajo Urubamba*. Lima: Misioneros Dominicos/ENOTRIA S.A. Colección Antisuyo Vol. 1.

Baer, G. (1992). The one intoxicated by tobacco: Matsigenka shamanism. In Matteson-Langdon, J. & G. Baer (Eds.), *Portals of power: Shamanism in South America* (pp. 79–100). Albuquerque: University of New Mexico Press.

Bennett, B. Y. (1991). Illness and order: Cultural transformation among the Machiguenga and Huachipaeri. Doctoral Dissertation, Cornell University.

Camino, A. (1977). Trueque, correrías e intercambio entre los Quechas Andinos y los Piros y Machiguenga de la montaña Peruana. *Amazonia Peruana, 1*(2), 123–140.

References

Cueva, N. (1990). Un acercamiento a la situación de salud en la provincia de Manu—Departamento de Madre de Dios (Manu: Un gran reto en la selva). Manu: AMETRA/Centro Bartolomé de las Casas.

Gregor, T. (1985). *Anxious pleasures: The sexual lives of an Amazonian people*. Chicago: University of Chicago Press.

Izquierdo, C. (1995). The illness experience: Health care choices among the Mapuche living in Santiago. M.A. Thesis, Department of Anthropology, University of California at Los Angeles.

Izquierdo, C. (2001). Betwixt and between: Seeking cure and meaning among the Matsigenka of the Peruvian Amazon. Doctoral Dissertation, Department of Anthropology, Los Angeles: University of California.

Izquierdo, C. (In review). When "health" is not enough: Subjective, societal, and biomedical indicators of well-being among the Matsigenka of the Peruvian Amazon. *Social Science and Medicine*.

Izquierdo, C., & Johnson, A. (In review). Desire, envy and punishment: Matsigenka emotion schema in illness narratives and folk stories. *Culture, Medicine and Psychiatry*.

Jelliffe, D. B. (1966). Assessment of the nutritional status of the community. Geneva: World Health Organization.

Johnson, A. W. (1983). Machiguenga gardens. In R. Hames & W. Vickers (Eds.), *Adaptive responses of Native Amazonians* (pp. 29–63). New York: Academic Press.

Johnson, A. W. (2003). *Families of the forest: The Matsigenka Indians of the Peruvian Amazon*. University of California Press: Stanford.

Johnson, A. W., & Behrens, C. A. (1982). Nutritional criteria in Machiguenga food production decisions. *Human Ecology, 10*, 167–189.

Johnson, O. R. (1978). Interpersonal relations and domestic authority among the Machiguenga of the Peruvian Amazon. Doctoral Dissertation, Department of Anthropology, Columbia University.

Johnson, O. R. (1980). The social context of intimacy and avoidance: A Videotape Study of Machiguenga Meals. *Ethnology, 19*, 353–366.

Johnson, O. R., & Johnson, A.W. (1975). Male-female relations and the organization of work in a Machiguenga community. *American Ethnologist, 2*(4), 634–638.

Kimerling, J. (1991). *Amazon crude*. New York: Natural Resources Defense Council.

Lyon, P. J. (1984). Change in Wachipaeri marriage patterns. In K. M. Kensinger (Ed.), Illinois Studies in Anthropology No. 14. *Marriage Practices in Lowland South America* (pp. 52–263). Urbana and Chicago: University of Illinois Press.

Neel, J. V. (1974). Control of disease among Amerindians in cultural transition. *Bulletin of the Pan American Health Organization, 8*(3), 205–211.

Plowman, T. C., Leuchtmann, A., Blaney, C., & Clay, K. (1990). Significance of the fungus *Balansia cyperi* infecting medicinal species of *Cyperus* (Cyperaceae) from Amazonia. *Economic Botany, 44*(4), 452–462.

Renard-Casevitz, F. M. (1976). Notes sure la pharmacopée des Matsiguenga. *INSERM (Les Colloques de l'Institut Nacional de la Santé et de la Recherche Médicinal)* 63, 129–144.

Renard-Casevitz, F. M. (1991). *Le banquet masqué: Une mythologie de l'etranger chez les Indiens*. Paris: Lierre & Coudrier Distribution, Eadiff.

Rivera, L. (1991). *Territorio Indígena: El area de influencia del proyecto gas de Camisea*. Lima: Centro para el Desarrollo de Indígena Amazónica (CEDIA). Documento de Trabajo-Libreta de Campo.

Shepard, G. H. Jr. (1998). Psychoactive plants and ethnopsychiatric medicines of the Matsigenka. *Journal of Psychoactive Drugs, 30*(4), 321–332.

Shepard, G. H. Jr. (1999a). Pharmacognosy and the senses in two Amazonian societies. Doctoral Dissertation, Department of Anthropology, University of California at Berkeley.

Shepard, G. H. Jr. (1999b). Shamanism and diversity: A Matsigenka perspective. In D. A. Posey (Ed.) (pp. 93–95). *UNEP Global Biodiversity Assessment* Supplement 1. *Cultural and Spiritual Values of Biodiversity*. London: United Nations Environmental Programme and Intermediate Technology Publications.

Shepard, G. H. Jr. (2002a). Three days for weeping: Dreams, emotions and death in the Peruvian Amazon. *Medical Anthropology Quarterly, 16*(2), 200–229.

Shepard, G. H. Jr. (2002b). Nature's Madison Avenue: Sensory cues as mnemonic devices in the transmission of medicinal plant knowledge among the Matsigenka and Yora of Peru. In J. R. Stepp, F. S. Wyndham, & R. K. Zarger (Eds.), *Ethnobiology and Biocultural Diversity: Proceedings of the 7th International Congress of Ethnobiology* (pp. 326–335). Athens, GA: University of Georgia Press.

Shepard, G. H. Jr. (2002c). Primates in Matsigenka subsistence and worldview. In A. Fuentes & L. Wolfe (Eds.), *Primates Face to Face: The Conservation Implications of Human and Nonhuman Primate Interconnections* (pp. 101–136). Cambridge, U.K.: Cambridge University Press.

Shepard, G. H. Jr., & Chicchon, A. (2001). Resource use and ecology of the Matsigenka of the eastern slopes of the Cordillera Vilcabamba. In L. E. Alonso et al. (Eds.), RAP Working Papers No. 12. *Biological and Social Assessments of the Cordillera de Vilcabamba, Peru* (pp. 164–174). Washington, DC.: Conservation International.

Snell, B. E. (1998). *Pequeño diccionario Machiguenga-Castellano*. Lima: Instituto Lingüístico de Verano/Summer Institute of Linguistics. Documento de Trabajo Vol. No. 32.

Snell, B. E., & Davis, H. (1976). *Kenkitsatagantsi Matsiguenka*. Pucallpa: Instituto Lingüístico de Verano/Summer Institute of Linguistics. Comunidades y Culturas Peruanas.

Snell, W. W. (1964). Kinship Relations in Machiguenga. M.A. The Hartford Seminary Foundation.

Soto, J. C. (1982). Ecología de la salud in comunidades nativas de la Amazonia Peruana. *Amazonia Feruana*, 3,13–26.

Strongin, J. (1982). Machiguenga, medicine, and missionaries: The introduction of Western health aids among a native population of Southeastern Peru. Doctoral Dissertation, Department of Anthropology, Columbia University.

Townsley, G. (1987). The outside overwhelms: Yaminahua dual organization and its decline. In: H. O. Skar & F. Salomon (Eds.), Vol. 38 (pp. 355–376). *Natives and Neighbors in South America: Anthropological Essays*. Gotzborg: Goteborgs Etnografiska Museum. *Etnologiska Studier*.

von Hassel, J. M. (1904). Los varaderos del Purús, Yurúa y Manu. *Boletín de la Sociedad Geográfica de Lima, 15*, 241–246.

Wilbert, W. (1986). Warao Herbal Medicine: A Pneumatic Theory of Illness and Healing. Doctoral Dissertation, Department of Anthropology, University of California at Los Angeles.

Yu, D. W., & Shepard, G. H. (1998). Is beauty in the eye of the beholder? *Nature, 396*, 321–322.

Maya of Highland Mexico

Elois Ann Berlin, Brent Berlin, and John R. Stepp

ALTERNATIVE NAMES

None.

LOCATION AND LINGUISTIC AFFILIATION

The Highland Maya of Mexico are located in the central highland region of Chiapas, the southernmost state in Mexico. The two largest groups in the Highlands include the Tzeltal and the Tzotzil (see linguistics). These groups are closely related both culturally and linguistically. Beginning in the 1970s some Highland Maya began to emigrate into the lower elevation Selva Lacandon region of Chiapas, to other parts of Mexico, and to the United States. This migration has been largely due to economic reasons and land shortages but is also often related to religious conflict between the traditionalist catholic religious hierarchies, newly converted evangelical Protestants, and modern Catholics. Other Maya groups in Chiapas are the Chol in the northern mountains of Chiapas; Tojolabal in eastern Chiapas; the Zoque in western Chiapas and the Lacandon in the Selva Lacandon rainforest on the eastern Guatemalan border. In addition, there are approximately 25 other Maya socio-linguistic groups living in Guatemala, Belize, and the Yucatan Peninsula of Mexico.

The Maya language family is comprised of approximately 25 distinct languages distributed throughout southern Mexico, Guatemala, and Belize. Five of these languages represent the principal linguistic groups of the family presently found in Chiapas—Tzeltal, Tzotzil, Tojolabal, Chol, and Lacandón. Three Maya languages originally spoken in Chiapas are extinct or close to extinction—Coxoh, Chicomuceltec, Mocho, and Motozintlec (see Figure 1, after INEGI, 2001). Speakers of several other Maya languages Kanjobal ≈ 6,000 speakers, Mam ≈ 5,500 speakers, Chuj 1,500 speakers, Jacaltec and Cakchiquel ≈ 500 each, Maya ≈ 100, Chontal and Kekchi <100 each) several of these represent recent migrants from Guatemala.

Of the native Chiapas Maya groups, Lacandón (with perhaps fewer than 900 speakers) is linguistically most distinct, sharing close affinities with the Yucatecan branch of the family (Yucatec, Mopán, and Itzá). Chol (≈ 141,000 speakers), is the next most distinct, followed by Tojolabal (roughly 38,000 speakers). Tzeltal (≈ 279,000 speakers) and Tzotzil (≈ 292,000 speakers) are the most closely related of the group, having separated as little as 800–1200 years ago. These two languages are spoken throughout the Central Highlands region (Mexican national census data, INEGI 2001).

Since the early 1950s, speakers of each of the four larger Maya groups have migrated in large numbers and are today found dispersed throughout the southeastern portion of the state.

OVERVIEW OF THE CULTURE

Chiapas has been the site of some of the most intensive anthropological research in the world and there is abundant literature describing most aspects of Highland Maya

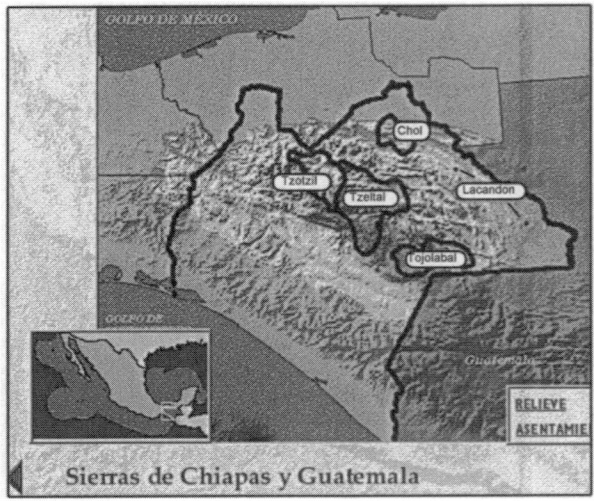

Figure 1. Map of Maya linguistic groups in Highland Mexico.

culture. With colonization the Spanish encouraged and cultivated ethnic distinctions between indigenous populations leading in large part to the diversity of dialects and municipal affiliations found today. Most Highland Maya continue to be subsistence farmers; although many have become involved in wage labor that has become available in the last two decades. Over the last several years, environmental degradation and political conflict have intensified in Chiapas, leading to political unrest and violence although there are currently negotiations taking place as part of a peace process in the region.

The Highland region was sparsely populated prior to Spanish conquest in the early 1500s. Even during the classic Maya period the Highland remained a relative backwater and there are few archaeological sites of significance in the area. Population grew at a rate of between 1–2% after contact up until the 20th century. Recent times have seen a much higher growth rate. Between 1970 and 1990 the indigenous populations of Chiapas grew at an astounding annual growth rate of 4.5%, however this rate has subsided with the rate from 1990–2000 at 2.0%. The average number of children born per woman is 3.4. The current population of Highland Maya over the age of 5 is approximately 571,000 (279,000 Tzeltal Maya and 292,000 Tzotzil Maya). Infant mortality is extremely high (9.5%) among the Tzeltal and Tzotzil Maya. This is double the rate for the country of Mexico as a whole.

THE CONTEXT OF HEALTH: ENVIRONMENTAL, ECONOMIC, SOCIAL, AND POLITICAL FACTORS

Environment

Chiapas is one of the richest areas of biodiversity in the world with more than 9,000 species of vascular plants and more than 1,150 species of vertebrates. The Highlands of Chiapas range in elevation from 700 m to 2,900 m above sea level and as such display a wide range of habitats from broad leafed evergreen tropical forest to temperate pine oak forest.

Because of imminent extinction threats to both flora and fauna, environmental nongovernmental organizations have given the area highest priority for conservation. Recently, many relatively heavily forested regions in the proper Highlands have undergone extreme disturbance due to new settlements and clearing for traditional swidden agricultural fields. Clear-cutting for timber and charcoal has occurred widely, in spite of government regulations prohibiting such practices. The region is best viewed as a patchy mosaic of a wide variety of different vegetational succession stages. These habitats are widely utilized for procurement of medicinal plants (Stepp & Moerman, 2001).

The primary influences of the environment on health relate to geomorphology, climate, including seasonal variation, and conditions of sanitation. The broken terrain of rugged mountain slopes and valleys provides for some variation in health risks. One would predict a higher rate of heat preferring bacterial diseases in the lower climes and more cold tolerant viral conditions in the higher altitudes. The primary leveling factor in the disease risk patterns across elevational boundaries is the quantity and quality of water supplies. Although some programs have been undertaken for construction of water storage and distribution facilities, water probably still represents the most serious environmental factor in current disease patterns. Inadequate water supplies inhibit adequate hygienic practices. Contaminated water resources promote disease transmission. Sanitary waste disposal is the second greatest environmental risk factor. Sewer systems are almost non-existent in the Maya communities. Indoor plumbing for water and sewage is available normally only in some houses in the municipal centers. Latrines are sometimes but not always used in the more remote hamlets and some houses of the municipal centers.

As more and more families rely on some cash-based economic activities, and as road construction makes transport of construction supplies more feasible, house construction types tend to provide greater protection from the elements of nature. Cement block walls replace wattle and daub, or bound-board walls that are permeable or open to drafts. Corrugated tin, cardboard, or asbestos, and in some cases, cement and/or ceramic tiles are replacing thatch roofing that can harbor disease-transmitting fauna. Ventilation is better in the newer constructions that provide windows and flues for light, air circulation, and smoke release. Gas stoves are slowly replacing wood cooking fires and further reducing irritating smoke inhalation, which can be a risk factor for respiratory problems. Cement floors provide a barrier between soil based organisms and human hosts. Figure 2 compares the highland Maya, the State of Chiapas, and Mexico national death rates from gastrointestinal and respiratory diseases.

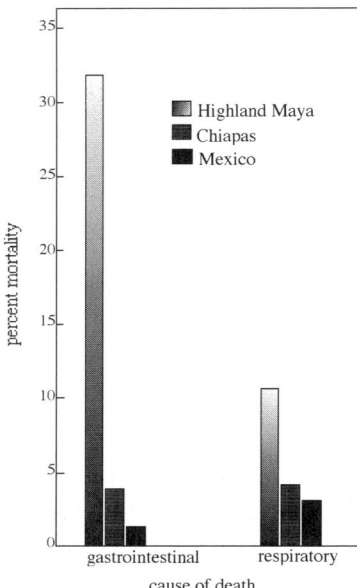

Figure 2. Gastrointestinal and Respiratory Disease Mortality Rates Reported for Maya Communities, Chiapas, and Mexico.

Diet

The Maya peoples of Highland Chiapas have a long tradition of swidden agriculture based primarily on corn, beans, squash, and chili peppers. Small flocks of chickens provide a generous quantity of eggs. Turkeys, pigs, and cattle provide occasional meat. Some sheep are raised but these play a very minor role in indigenous diet as their primary value is the wool they produce for warm clothing and a few tourist items. Availability and consumption of these cultivated and domesticated species is complemented by a rich variety of non-domesticated or "lightly managed" species. In more recent times, the non-domesticates have predominantly been of plant origin as population pressures have virtually eliminated the large game animals (such as deer, agouti, peccary etc.) and restricted the number and distribution of smaller ones (e.g., field rodents, squirrels, rabbits, gophers, and lizards) in the region. The small edible creatures are often opportunistically captured in traps, though a few people may still "go hunting." In a few areas, streams provide fish, snails, and frogs on occasion. A few wild birds are sometimes eaten. Some people still keep nests of native bees for their honey. The rainy season provides an abundance of wild mushrooms and there is growing cultivated mushroom production, primarily for market sales.

The period between last harvests, when agricultural stores are lowest is also the period of greatest consumption of non-cultivars. Subsistence agriculture is so marginal in the area that it is not terribly uncommon in bad years for some families to run short on beans or corn and be forced either to eat some of their seed stock or find a means of supplementing supplies through market purchase or exchange.

As in all such contexts of marginal resources, it is the children who are most affected. Chronic childhood nutritional stress is evident in the generally short stature of the population. Those families who are most well-off economically follow the universal pattern of short stature and overweight for first generation adults. Continued economic and dietary stability results in a second generation of (sometimes dramatically) increased stature.

There is also potentially a hidden danger in the change of lifestyles that frequently accompanies economic success. These typically include a shift from traditional food, especially wild species to foods of higher "prestige" such as beverages and snack foods with high sugar, salt, and/or fat content that are low in nutrients. Fortunately the classic shift from breast to bottle, with all the attendant health consequences, has not become a major problem for the highland Maya.

When peoples whose dietary problem has normally been getting enough to eat achieve food sufficiency and the potential for excess, the concept of restriction or balance is not operative. The usual pattern elsewhere is a trend toward obesity and an increased prevalence of diabetes. Such dietary consumption and nutritional health changes may be beginning to occur in the Highland Maya. People frequently report *asukcar* (< Spanish *azucar* "sugar") as a diagnosis. That there is no Maya term for diabetes (although they do recognize such signs as ants being drawn to urine) suggests that it is a relatively new problem. Based on the number of informal reports of the disease, a systematic study and potential early intervention program would be appropriate.

The weedy species (listed in Appendix 1) that contribute to traditional Maya diet provide high quality nutritional supplements, especially when agricultural resources are lowest. They appear to be the primary sources for vitamin A and thiamine and to add an important amount of calcium to the diet. Vitamin A deficiency is a major problem in many third world populations. It has not been reported as a problem for this group, although the association of vitamin A deficiency with increased susceptibility to diarrhea could potentially play a role in the high rates of diarrheal disease in this population. An early symptom of vitamin A deficiency is poor night vision. There appears to be no term in Maya for such a condition.

Iodine deficiency, represented by goiter and possibly dwarfism, is often reported in populations inhabiting mountainous regions, including Maya populations of Highland Guatemala. For reasons that are not as yet clear, this does not seem to represent a major health problem in the Highlands of Chiapas. Discovering the sources of this positive deviance in health pattern could possibly provide a model for other highland peoples.

Recent research (Luber, 1999) suggests that protein calorie nutrition is identified by a native term, *cha'lam tsots* "second layer of hair" and studies have shown that this is a significant health problem in the highland region (Berlin & Berlin, 1996). In a diet based on corn and beans, it is frequently difficult for small children to consume sufficient amounts to meet total nutritional needs. Simple addition of a fat source is usually sufficient to compensate for this. For both children and adults food quantity is currently more of a problem than the nutritional quality of the foods themselves. This has the potential to change for the worse as life-styles change and especially if diet shifts toward higher prestige, nutrient poor foods.

Disease[1]

As seen in Figure 3, gastrointestinal diseases are the greatest cause of both morbidity and mortality. The high frequency of gastroenteric conditions in both the ethnoepidemiological and epidemiological data is a reflection of the virtual universal parasitism and frequent diarrheal episodes typical of populations living in the socio-economic conditions characteristic of these communities.

Respiratory infections are the second most frequent cause of illness. Respiratory problems are surely related to the altitude (on average 2,000 m), to cold weather and traditional house construction which affords little protection from cold wind, as well as the smoke of cooking fires. Respiratory problems represent the fourth most frequent reported cause of death. This would suggest that many of these illness are more minor colds and coughs with low mortality.

Fever or, to use a bio-medical gloss "fever of unknown origin", is mentioned as the third most frequent

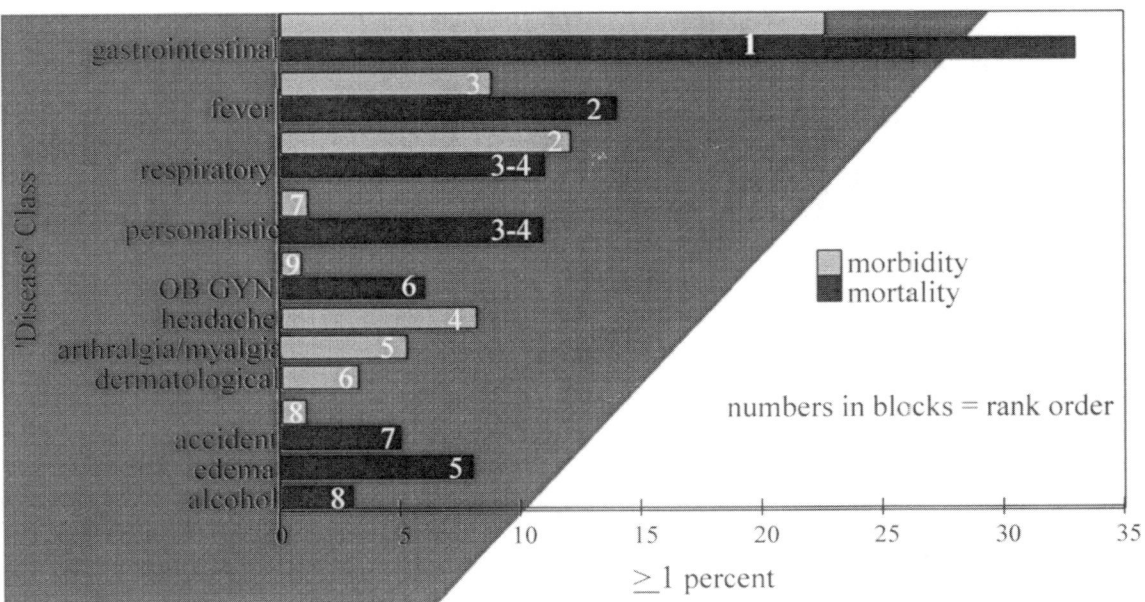

Figure 3. Rates for Morbidity and Mortality Reports in Maya Communities.

illness but the second ranked cause of death. Headaches and general aches and pains (glossed here as "arthralgias and myalgias") are the fourth and fifth most frequently reported health problems. People relatively seldom mention dermatological inflammations and infections (ranked sixth) as a cause of sickness. The lack of access to water and generally poor hygienic conditions promote high rates of dermatological problems, as evidenced by the variety of named dermatological conditions, probably making them so common as to not merit mention.

Personalistic health problems (ranked seventh in morbidity etiology) cross-cut many categories of disease. Despite the rather low frequency of report of so-called "personalistic" conditions in the morbidity files (ranking seventh as a cause of morbidity), conditions with an attributed personalistic etiology tie with respiratory disease as third and fourth causes of death. This supports the pattern, one that holds across many cultures, that most health conditions are considered at least initially to be naturalistic in nature and those that are more life threatening or more resistant to treatment may become re-diagnosed as personalistic. Further analysis of the conditions reported in association with personalistic deaths shows that more than half of the personalistic deaths were preceded by gastrointestinal symptoms and about 25% were associated with edema. This pattern of reported high fatality from gastrointestinal disease and edema of personalistic etiology form a significant cluster that merits further evaluation. Culturally, it reaffirms the attribution of personalistic etiology to serious illness. It also demonstrates a culturally defined link that also makes sense from a scientific physiological perspective.

The data for sex adjusted death rates due to accident or violence (which is the eighth most frequent cause of morbidity) show about 85% of males and 39% of women were attributed to homicide. A large proportion of both statistics were likely due to violence initiated by males and probably while drinking.

Eye problems in most municipalities consist of standard conjunctivitis and infections. However, in a few areas, particularly the municipality of Oxchuc, trachoma is a serious health problem that can lead to blindness if untreated. A diagnosis of mental illness may encompass advanced cysticercosis affecting the brain and this may account for the relatively high ranking of mental illness (Castille, 1996).

General "women's problems" are not often reported as a cause of morbidity. Death associated primarily with obstetric problems occurs at much higher frequency than is reported for morbidity. This is an artifact of the absence of reports of uncomplicated live births in a population with a very high birth rate. It also quite possibly reflects the fact that mortality is high when complications do occur and women might be more remiss to mention more minor problems of this nature. Death precipitated by accidents or violence is reported at about equal frequency with deaths related to obstetric/gynecological problems. The precipitating events for both either occur at very low frequency but carry high fatality or are relatively unremarkable when death is not the outcome.

The general distribution of reported health problems by age group, shows that more cases are reported in the lower age groups, with decreasing illness events with age. However, when adjusted for age, the infants and pre-school aged children have a lower report of illness than most other groups, in relation to the total number represented in the sample (Figure 4).

MEDICAL PRACTITIONERS

It is more accurate to define areas of medical specialization in Maya medicine and note that aspects of one or more of these specialties may be incorporated in any given healer's armamentarium of skills. These specialties include diviner/healer, one who prays to the mountains, pulse-reader, mid-wife, bone-setter, massager, and (in Yucatan) herbalist (*tsak xiu* lit. "herb grabber").

Diviner/healers incorporate the more traditional shamanic healing practices. Prayers to the mountains might best be considered as preventive medicine specialists in that they pray to the hills and in caves for the protection of the population and for the prevention of periodic and epidemic diseases (such as the classic childhood diseases). Pulse-readers are able to diagnose and monitor the progress of disease based on the detection of

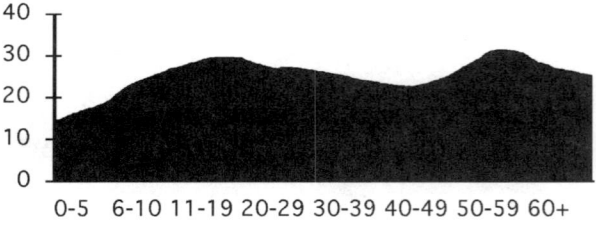

Figure 4. Proportion of population reporting illness by age group.

the type and quality of the pulse at various points on the body. The term mid-wife is somewhat misleading as the knowledge and skills of these practitioners generally is more inclusive of women's health problems and, like other medical specialists, may include various aspects of the other healing specialties. It is interesting to note that there is no specialized role with a distinct term for "herbalist" among the Maya of Highland Chiapas. Individuals with special knowledge of plants are sometime referred to with the descriptive phrase *ja' mach'a ya sna ta poxil wamale* "one who knows medicinal plants." In this region, knowledge of medicinal plants is broadly, if not equally, shared. Many people can identify a plant as medicinal and name the condition it treats, even if they have never used it and/or do not know exactly how it is prepared. Interest in medicinal plants and traditional healing has stimulated an emerging role for herbalists among the Tzeltal and Tzotzil and a widely accepted name for this role may eventually be coined.

Classification of Illness, Theories of Illness, and Treatment of Illness

General Principles

The Maya medical system is holistic in that there is a close relationship between the health of an individual and his/her relationship with other people, various environmental factors and forces and members of the spiritual realm as well as elements of contagion. Both problems of health and their treatments can be divided into two distinct systems (following Foster, 1976);

1. The naturalistic system encompasses a majority of health conditions and is the diagnosis of first resort even for many conditions that are later classified as personalistic.
2. The personalistic system comprises a set of conditions that are intentionally or conditionally specific to the individual and/or incorporate extranatural origins.[2]

Naturalistic Health Problems

Diagnosis of naturalistic health problems is founded on complex ethnomedical understanding of anatomy, physiology, and the symptomatology of specific health conditions. Because these conditions are symptom-based, people readily group conditions with similar symptoms together although these sickness clusters do not have standard Maya names, they can be glossed as follows: gastrointestinal conditions (alimentary paths), respiratory conditions (airways), dermatological conditions (skin), fevers, headaches, arthralgias and myalgias (general aches and pains), urinary problems (urinary paths), women's conditions, malaise (weakness and wasting), mental conditions, edemas, breaks/sprains, ophthalmic problems (eyes), buccal-dental problems (mouth), emotional conditions (emotion), bites/stings (Figure 5).

These classes, being symptom-based, relate roughly to physiological organ systems. Notably lacking for comparative purposes with biomedical medicine is the cardiovascular system. The close relationship between the heart and lungs (blood circulation and respiration) in Maya ethnophysiology and the general imprecision of cardiovascular symptoms probably accounts for this.

Any sickness that begins with a naturalistic diagnosis may eventually require reanalysis and referral to a personalistic healer. This reevaluation occurs normally when a condition is exceptional in that it does not respond to normal treatment (often with herbal remedies), or alternatively, if symptoms become very severe or life-threatening.

Causes of illness and disease in the naturalistic system range from the mundane life events to more existential reasons. Consumption of spoiled, contaminated, or inappropriate (e.g., causing humoral imbalance) foods and beverages, and prolonged hunger are particular sources of alimentary tract problems. Hard work can

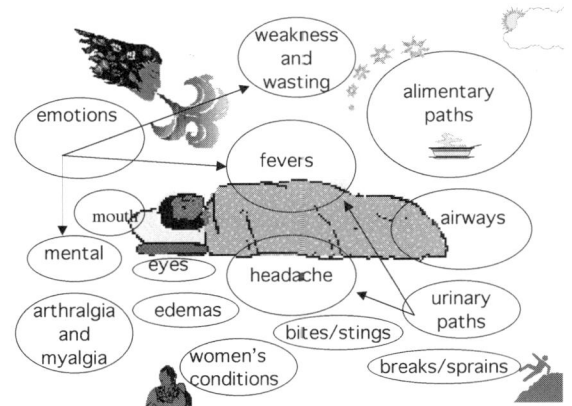

Figure 5. General Classes of Naturalistic Health Problems for Highland Maya.

cause health problems. In addition to breaks and sprains, falls can cause organs to dislodge and malfunction. Sorrow, embarrassment, and anger can all trigger physical illness as can environmental conditions such as rain, cold, heat, and contact with (being struck by or stepping on the landing site of) St. Elmo's fire.[3] Airs and wind are a prominent source of illness, especially among the Maya of Yucatan. These may be literal breezes or metaphorical winds that harbor substances that enter the body and cause disease. Airs and winds can also cause humoral imbalances in the body, as can many other substances and phenomena.

Treatment of naturalistic health problems is founded on a complex ethnopharmacoepia of primarily plant-based natural products. Appendix 2 lists the plant species most commonly used by the Highland Maya of Mexico for each major set of health conditions, as well as some of their known pharmacological properties and the medicinal virtues attributed to them in Maya medicine. These plants are readily available, usually within a 1 km radius of the household, and grow along trailsides and fallow fields (Stepp, 2001). Dietary and behavioral restrictions and recommendations accompany herbal treatments. Many of these are based on humoral balance, however, other factors such as digestibility may also be considered. The most frequent behavioral recommendation is rest, either in bed or by avoiding strenuous effort. Infrequently a patient may be prohibited from going outside and being exposed to sunlight.

The majority of Maya formulary are administered orally as teas, infusions, and concoctions, however other therapeutic measures include plasters, massage, blood letting/acupuncture,[4] bone-setting, and heat application.

Personalistic Health Problems

Personalistic health problems are directed specifically at the individual and originate from numerous sources that can be divided into the following classes: human given, God-given,[5] fright, problems of the soul/spirit, destiny. Human given illness include those that are intentionally caused (witchcraft), and those that are unintentional (e.g. evil/strong eye). Problems of the soul or spirit include soul loss and effects to one's animal companion soul. Deities (such as the Spiritual Ancestors, Sacred Father Sun, Sacred Mother Moon, Earth Lord, Thunderbolt, Mother Wind) control numerous aspects of life and can punish evil-doers as well as become irritated or angry and cause illness. Destiny (primarily determined by date and time of birth) is more significant in Guatemala as a health determinant than in Mexico. For serious believers, a considerable amount of time may be spent in maintaining good relations with the various deities. This is, in many respects, preventive medicine.

Diagnosis of personalistic conditions is made on the basis of divination and pulsing. Such diagnosis requires intervention of a specialist healer who first of all diagnoses the true cause (ultimate cause) of illness. Appropriate interventions can then be undertaken on behalf of the patient. These interventions include prayers and incantations on behalf of the patient, including the sacrifice (such as a black chicken in substitution for the patient's soul). Calling the soul's return, sweeping disease from the body with herbs, blowing liquids over the patient to dispel agents of disease and the offering of incense, candles, and liquor to the spiritual ancestor or deity in supplication, atonement, or appeasement. Natural medicines may also be administered at this time, but the primary intervention is non-secular.

Until very recently the concepts and methods for scientifically evaluating personalistic healing were lacking. As yet, few studies have been undertaken to elucidate the psycho-neuro-physiological processes involved in effecting healing through such practices. Recent developments in biomolecular medicine verify that personalistic healing can effect cures, precisely in those cases that are most resistant biomedical treatment, just as our traditional practitioner colleagues have previously informed us.

SEXUALITY AND REPRODUCTION

Pregnancy is said to be the result of the combination of the sexual fluids (strength) of both male and female. Sex of the child is said by some to be determined by the father because a woman produces only female substances but a man may produce male and female substances. Others say that the child's sex is determined by whether the substance of the man or the woman is strongest at the time of conception. Certain plants that can be used to ensure the birth of a male or female. All of these notions seem to conflict with the practice of the mid-wife to count the veins and arteries in the umbilical cord to determine the number of children that a woman will have and, by some reports, the darker vessels signify a male child and the lighter indicate a female child.

HEALTH THROUGH THE LIFE CYCLE

Pregnancy and Birth

Most children are born at home with the aid of a mid-wife. The birth attendant may administer plant and/or animal derived natural medicines as birth aids. Where the traditional steam bath is still in use, the birth may take place inside the sweathouse and the mother and infant may remain there for a while afterward. In other areas the woman is placed over a steaming pot of water and herbs. After childbirth, a woman is given warm foods and beverages. The mid-wife may return to administer baths and massages to the new mother over a period of days post-partum.

Dietary and behavioral proscriptions and prescriptions are intended to ensure a healthy baby and a safe delivery. Many of these have strong humoral elements. Others simply ensure that the mother not jeopardize the pregnancy through risky behavior.

Infancy

A new child of either sex is generally welcome, although there may be a slight preference for boys. Prayers and rituals are begun shortly after birth to make certain that the ancestral spirits and deities protect the child and a variety of preventive measures are undertaken to keep the baby's soul intact. During the first few weeks, the mother and child avoid social contacts and stay close to home to prevent illness, especially of the infant. The infant's soul is considered to be weakly attached and various precautions are taken to prevent the baby's soul from wandering, becoming lost, or returning to the cosmic pool. The infant's wrists and ankles may be tied with string or small bracelets to tether the soul. An infant may wear amulets, such as amber bracelets to prevent illness, especially from evil eye. When mother and child begin to go out, the mother must place branches or other items to across all paths crossing where her pathway to guide the baby's soul safely home.

Within a few months a baby is often be in the care of older siblings who learn from a very early age to carry a child in a shawl on their backs. Birth defects and undesirable physical and mental characteristics may result when a pregnant woman makes fun of a person or animal who has the particular characteristic. Sometimes just looking at an animal, especially a dead animal, will cause developmental problems for the child to be born. An envious woman may disrupt a pregnancy and take the fetus into her own body. A fetus resembling a frog, or other creature (probably abortion in early stages of development) may be sent to a woman through witchcraft.

Childcare practices are reflected in the morbidity patterns for infants and toddlers. Inability of very small children to communicate symptoms dictates that health problems must be interpreted on the basis of overt signs. Figure 6 depicts the cumulative frequency of sickness reports. Fever, which may accompany any number of conditions, is the primary diagnosis in small children. The early months and years breast-fed children are relatively protected by maternal antibodies and from dietary contaminants. The number of reports of all conditions, but especially fever, peak during the third year range when children are weaned and consuming regular foods, achieve greater freedom to wander around and come into contact with ambient pathogens. Since toddlers still have limited language skills, fever remains a frequent diagnosis. By five years of age, children begin to approach the adult pattern. Mothers report fewer illnesses overall and gastrointestinal and respiratory become the most significant health problems.

Childhood

Childhood is a training period for adulthood. Children are included in all aspects of life and, as their understanding and abilities grow, they gradually assume a greater role in daily household activities as well as special occasion rituals. Pre-pubescent girls are initiated into the art of tortilla making by wrapping their hands with hot tortillas to toughen them as they begin their training in this fundamental activity. Pre-pubescent boys begin accompanying their fathers in their agricultural duties. As they assume the roles of adulthood, their health risks are those of adults and 'hese are reflected in their health patterns. Forces of change for these patterns derive from the increased numbers of children attending school for longer periods, as well as increased involvement in cash-based

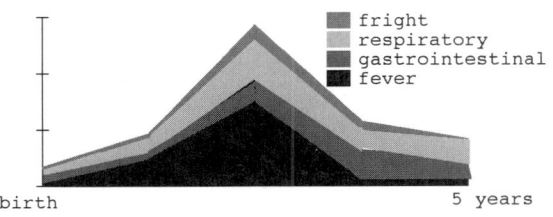

Figure 6. Relative Rates of Sickness Reports for Children 5 Years and Younger.

economic activities and a switch to pre-processed, machine-made tortillas.

Adolescence

Perhaps because adolescents are inclined to high risk behavior, there is an overall higher report of illness during this period (Figure 4 above).

Adulthood

While the causes of illness are similar for men and women, the causes of death vary considerably. Although mid-wives seem to have a high success rate for live births, maternal mortality is high when complications occur. Many women still know a number of herbal formulae for birth control reportedly that may effect temporary or permanent infertility, including *Diascorea floribunda*, the plant source of the primary modern contraceptive prescribed for hormonal suppression of female fertility. Biomedical health care providers and well-meaning family planning programs have discouraged the use of herbal controls. However, the Maya are joining the alternative medicine revolution and returning to their traditional medical formulations.

Death precipitated by accidents or violence is reported at about equal frequency with deaths related to obstetric/gynecological problems. Males, on the other hand, have a disproportionate mortality due to these causes. Over 85% of men's deaths due to accident or violence were attributed to homicide (39% for women). The old patterns of violence toward women are changing very slowly. It is possible that a large portion of those were due to violence initiated by (inebriated) males, which plays a tragic role in morbidity and mortality among the Highland Maya.

The Aged

As can be seen in demographic data from the Highlands of Chiapas, the elderly comprise a small percentage of the population. The extended household appears to be common enough to accommodate those few who reach advanced age. It is recognized that the older generations are sources of cultural knowledge that is not being learned by many youths. When occasions arise in which this knowledge is attributed special value, the aged gain in prestige. However, many of the young aspire to the knowledge and skills of the national and international community and their appreciation of traditional knowledge tends to be lessened. Those who have reached advance age appear to suffer no greater frequency of health problems than their other adult peers.

Dying and Death

Any loss of consciousness, such as fainting is viewed as a near-death experience. When illness strikes, the first recourse is to attempt self treatment, either with traditional remedies or, when cash is available, with medicines purchased at a local pharmacy, store or occasionally venders in the local markets. Serious illness or one resistant to treatment is likely to require intervention by a personalistic healer. Consultation at a clinic or hospital usually comes late in the course of sickness, if at all, primarily because of the high cost and lack of respect with which Indians are frequently treated by biomedical personnel. Persons who suffer a life-threatening illness are cared for and kept warm by family members. Precipitous death of the young and relatively healthy invites a causal attribution of personalistic intervention by deities or ill meaning humans.

The soul of the recently deceased is thought by some to linger for a while after death and to pose some risk to the living, especially family members and those whose souls are easily lost, such as children. A pregnant woman may be at risk of having the soul of the deceased enter her unborn fetus, which seems to be undesirable.

CHANGING HEALTH PATTERNS

Analysis of changes over time from "traditional" to "contemporary" involves an essentially arbitrary decision concerning a starting point. Most of the texts on the Ancient Maya that would have given us detailed information on their medical system and general health were destroyed by the Catholic priests accompanying the conquering Spanish armies. Early writings describe many aspects of culture, including medical beliefs and practices and even some diseases. Since these early texts were written after Spanish contact and under the guidance of priests who themselves administered to the sick, we cannot be sure that there is no old world medical influence.

One thing that is clear from all the studies that have been undertaken to date. Many of the plants that are currently in use in Maya medicine are native to the region and have most likely been employed for centuries, if not millennia.

APPENDIX A

Species Contributing Nutritional Supplements to Traditional Maya Diet

Acacia spp.
Amaranthus hypbridu
Amaranthus spinosus
Bidens pilosa, B. bicolor
Brassica campestris
Bromelia plumieri
Byrsonima crassifolia
Chenopodium ambrosioides
Cirsium horridulum
Diastacea micanthra.
Erythrina chiapasana
Eugenia acapulcensis
Eugeniatenejapensis
Galinsoga caracasana
Guazuma ulmifolia
Lantana camara
Lantana hirta
Linum nelsoni
Lycopersicon esculentum
Mangifera indica
Morus celtidifolia
Optuntia guatemalensis
Parathesis chiapensis.
Physalis gracilis
Physalis gracilis
Piper auriatum
Piper umbellatum
Portulaca oleracea
Psidium guajava
Psidium guineense
Rubus spp.
Saurauria scabrida.
Solanum nigrum
Solanum douglasii
Solanum nodiflorum
Solanum nudum
Sonchus oleraceus
Tagetes filifolia
Vitus bourgaeana

APPENDIX B

The Tzeltal-Tzotzil "Cuadro Básico" of Medicinal Plant Species for use in Treatment of the Twelve Major Classes of Health Conditions

Major illness category (and sub-class)	Primary species (in order of importance by sub-class)	Demonstrated pharmacological properties and Maya medicinal virtues
Gastrointestinal conditions		
general diarrhea	*Verbena litoralis* (Verbenaceae)	strong spasmolytic and antimicrobial activity, antifertality, pergative "*bitter*," "*warm*"
	V. carolina (Verbenaceae)	
	Ageratina ligustrina,	antibiotic, spasmolytic, anticancer (both spp.)
	A. pringlei (Asteraceae)	
	Baccharis trinervis,	antiseptic, digestive
	B. serraefolia (Asteraceae)	
	Lantana camara (Verbenaceae)	antibiotic, antispasmodic, carminative, diaphoretic, digestive, expectorant, hemostatis, nervine, pectoral, sedative, stomachic, tonic, vulnerary
	Cissampelos pareira	diuretic, emmenagogue, expectorant, stimulant
bloody diarrhea	*Calliandra grandiflora,*	astringent
	C. houstoniana (Fabaceae)	
	Psidium guineense (Myrtaceae)	strong spasmolytic, antimicrobial, and hemostatic, activity, "*cold*"
	Byrsonima crassifolia (Malpighiaceae)	alexiteric, astringent,
	Sonchus oleraceus (Asteraceae)	diuretic
	Acacia angustissima (Fabaceae)	antibacterial, antifungal
	Crataegus pubescens (Rosaceae)	
	Nicotiana tabacum (Solanaceae)	diuretic, anodyne, anorexiac, anthelmintic, anticonvulsive, antidote (various insect, arachnid venom), cathartic, CNS stimulant, cyanogenic, diaphoretic, emetic, entheogen, expectorant, hemostatic, intoxicant, laxative, narcotic, parasiticide, pediculicide, purgative, sedative, stimulant

Major illness category (and sub-class)	Primary species (in order of importance by sub-class)	Demonstrated pharmacological properties and Maya medicinal virtues
abdominal pain	*Baccharis vaccinioides* (Asteraceae)	
	Stevia ovata (Asteraceae)	
	Tagetes lucida (Asteraceae)	anesthetic, entheogen, strong spasmolytic
epigastric pain	*Chenopodium ambrosioides* (Chenepodiaceae)	abortifacient, amoebicide, analgesic, anodyne, anthelmintic, antiseptic, diaphoretic, diuretic, emmenagogue, febrifuge, fungicide, lactogogue, narcotic, nervine, parasiticide, stimulant, stomachic, tonic, vermifuge *"warm"*
intestinal worms	*Helianthemum glomeratum* (Cistaceae)	strong anthelminthic, antimicrobial, anti-candida, *"caustic"*
Respiratory problems		
cough, "colds"	*Litsea neesiana, L. glaucesens* (Lauraceae)	antimicrobial and spasmolytic activity, *"warm"*
	Salvia lavanduloides (Lamiaceae)	alopecia, *"warm"*
"croup" ("whooping cough")	*Hibiscus unicellus* (Malvaceae)	emollient
	Malvaviscus arboreus (Malvaceae)	
	Myrica cerifera	anthelminthic, astringent, deobstruent, emetic, febrifuge, laxative
	Liquidambar styraciflua	stimulant, tonic
Skin Infections		
"skin eruptions often with pus"	*Clematis dioica* (Ranunculaceae)	carminative, diuretic, expectorant, vesicant "rubefacient and blistering agent" [Morton 1981: 215], *"caustic," "cold"*
boils, open sores	*Lycopersicon esculentum* (Solanaceae)	antiseptic, aperient, digestive, pectoral
	Anoda cristata (Malvaceae)	
	Solanum nigrescens (Solanaceae)	"antifungal alkaloid tomatine" [ibid., 795], *"cold"*
	Erigeron karwinianus (Asteraceae)	
	Monnina xalpensis	
	Kearnemalvastrum lacteum (Malvaceae)	
	Rumex crispus	analgesic, astringent, blood cleanser, blood purifier, cathartic, diuretic, emetic, emollient, hemostatic, laxative, parasiticide, refrigerant, preproductive aid, stimulant, strengthener, suppurative, tonic
	Rhus schiedeana	
	Ranunculus petiolaris (Ranuculaceae)	antiseptic
Wounds		
due to accidents, violence	*Solanum lanceofolium* (Solanaceae)	antimicrobial, "leaves and fruit of *P. icosandra* contain unnamed alkaloid; sap is acrid and irritates the skin" [ibid. 200], *"warm"*
	Phytolacca icosandra (Phyt.)	
	Croton drago (Euphorbiaceae)	
	Pinaropappus spathulatus (Asteraceae)	antibiotic, astringent, hemostatic
	Asclepias currisavica (Asclepidaceae)	astringent, emetic, hemostatic, pectoral, purgative, sudorific, venereal, vermifuge, vulnerary
Breaks and Sprains	*Brugmansia candida* (Solanaceae)	antimicrobial activity, antiinflammatory, *"warm"*
	Ricinus communis	bactericide, cathartic, contraceptive, cyanogenetic, emollient, fatal poison, lactagogue, laxative, purgative, tonic, vermifuge
Fever	*Sambucus mexicana* (Caprifoliaceae)	Anodyne, expectorant, febrifuge, hydragogue, sudorific, *"cold"*
	Satureja brownei (Lamiaceae)	
	Bryophyllum bipinnata	
Headache	*Tagetes nelsonii*	
Teeth and Mouth Infections	*Rhus terebinthifolia* (Anacar.)	antimicrobial, *"cold"*
	Rubus coriifolius (Rosaceae)	(*"warm"*)
	Nicotiana tabacum (Solanaceae)	(see above)
Eye Infections	*Solanum americanum* (Solanaceae)	"in Panama' dropped in inflamed eyes" [ibid. 799] *"cold"*

Major illness category (and sub-class)	Primary species (in order of importance by sub-class)	Demonstrated pharmacological properties and Maya medicinal virtues
Edema	*Gaultheria odorata* Willd.	antibiotic, anticancer
	Buddleia americana L.	analgesic, anodyne, diuretic, hypnotic
	Buddleia crotonoides A. Gray	
Obgyn	*Salvia cinnabarina* M. & G.	
	Salvia polystachya Ortega	
	Ranunculus petiolaris H.B.K. *ex DC*	antiseptic
Aches and Pains	*Senecio salignus* DC.	anodyne
	Ricinus communis L.	(see above)
	Salvia cinnabarina M. & G.	
	Salvia rubiginosa Benth	

Sources: Morton 1981, Johnson 1998

NOTES

1. Data from two ethnoepidemiological surveys, conducted by E. A. Berlin in 1987–88 and J. R. Stepp in 1999–2000, form the basis of this report. These patterns generally fit the epidemiological reports from health clinics of the region.
2. The original proposition of Foster and Anderson limited personalistic illness to those directed specifically at an individual, however, it is clear that there are other non-natural or beyond natural etiologies that would not fit within a naturalistic classification and are therefore included here in the personalistic category.
3. St. Elmo's fire is the term used to refer to the atmospheric phenomena that produces ball-lightening that seems to shower on earth.
4. This aspect of Maya curing has been poorly studied except among the Maya of Yucatan, where García, Sierra, and Balán (1996) have made an interesting comparison between Maya and Chinese Medical systems.
5. In at least one municipality of the Highlands of Chiapas, this classification has come to be associated with naturalistic conditions, primarily through the influence of a local priest who has emphasized illness and disease as coming from God but more in the nature of things. This in the sense of God's will be done as opposed to particular punishment or penalty for transgression.

REFERENCES

Berlin, E. A., & Berlin, B. (1998). Biodiversity of Highland Maya diet. Annual meetings of the American Anthropological Association. Philadelphia, December 1998.

Berlin, E. A., & Berlin, B. In press. Dietary biodiversity of the Highland Tzeltal Maya of Chiapas, Mexico. In P. Townsend (Ed.), *Biodiversity and health*. Ames, Iowa: University of Iowa Press.

Berlin, E. A., & Berlin, B. (1997). Non cultivated plant foods Of The Highland Maya: Preliminary explorations. Paper presented at the II International Congress of Ethnobotany, Mérida, Yucatan.

Berlin, E. A., & Berlin, B., (1996). *Medical ethnobiology of the Highland Maya of Chiapas, Mexico: The gastrointestinal diseases* (pp. 557). Princeton University Press, Princeton: NJ 557.

Berlin, B, & Berlin, E. A. (1994): Anthropological issues in medical anthropology. In: (Ed.), Prance G. *Ethnobotany and the search for new drugs* (pp. 240–258). Ciba Foundation Symposium No. 185, New York: John Wiley.

Castille, D. (1996). Psychological affliction and mental illness among Maya Indians of Highland Chiapas, Mexico. Doctoral dissertation. Berkeley: University of California.

Fabrega, H. & Silver, D. B. (1973). Illness and shamanistic curing in Zinacantan. Palo Alto: Stanford University.

Foster, G. M. (1976). Disease etiologies in Nonwestern Medical Systems. *American Anthropologist*, 78(4), 773–782.

García, H., Sierra, A., & Balán, G. (1996). Medicina Maya tradicional: Confrontación con el Sistema Conceptual Chino. Mexico: Educación, Cultura y Ecología, A. C. (EDUCE) Johnson, T. (1998). *CRC Ethnobotany Desk Reference*. Boca Raton, FL, CRC Press.

Luber, G. E. (1999). An explanatory model for the Maya Ethnomedical Sundrome Cha'lam tsots. *Goergia Journal of Ecological Anthropology*, 3, 14–23.

Morton, J. F. (1981). *Atlas of medicinal plants of Middle America: Bahamas to Yucatan*. (Two Volumes), Springfield, Illinois: C. C. Thomas.

Stepp, J. R., & Moerman, D. E. (2001). The importance of weeds in Ethnopharmacology. *Journal of Ethnopharmacology*, 75(1): 25–31.

Stepp, J. R., (2001). Highland Maya medical Ethnobotany in ecological perspective. Doctoral dissertation, Athens: University of Georgia.

(UNICEF), United Nations Children Fund (1991). The state of the World's Children. London: Oxford University Press.

Mongolia

Daniel J. Hruschka and Brandon A. Kohrt

ALTERNATIVE NAMES

Mongolian, Mongol.

LOCATION AND LINGUISTIC AFFILIATION

The Mongolian People's Republic (MPR) is a landlocked country in northern Central Asia flanked by Russian Siberia to the north and surrounded by China to the east, south, and west. Many authors have referred to the region as Outer Mongolia in contrast to Inner or Southern Mongolia, which is now an autonomous region within the People's Republic of China. The majority (> 90%) of the MPR's 2,374,000 inhabitants (NSO, 2000) are ethnically Mongolian, speakers of an Altaic language distantly related to Turkic. Although Mongolian speakers also reside in Russia (308,000; SCRFS, 1994) and China (4,800,000; Tianlu & Rongqing, 1996), this article will concentrate on Mongolians living within the MPR.

OVERVIEW OF THE CULTURE

Three cultural systems—Central Asian mobile pastoralism, Tibetan Buddhist lamaism, and a Soviet model of development—have profoundly shaped Mongolian society. Although it is difficult to trace the ethnic term "Mongol" much farther back than Genghis Khan's 13th century confederacy of nomads (Morgan, 1996), today's Mongolian pastoralists maintain cultural patterns that are characteristic of many traditional Central Asian societies (Humphrey & Sneath, 1999). Most economic activity revolves around livestock grazing. The steppe's sparse vegetation constrains population density and requires frequent moves between pasture sites. Horses, which are the most prized of domesticated animals, have historically served to connect the territory's sparse human population. To accommodate annual mobility in a rugged climate, traditional Mongolian dwellings (the circular felt-lined *ger*) combine structural stability, heat insulation, and a flexible framework that can be dismantled and rebuilt in hours. The annual cycle revolves around herd animals, and involves long stretches of relaxed activity punctuated by intense periods of work during the birthing, shearing, branding, and slaughter seasons.

The household and labor organization of Mongolian pastoralists is surprisingly consistent across regional space and historical time (Cheney, 1968; Humphrey & Sneath, 1999; Shombodon, 1996). Nuclear households, consisting of parents, children, and elder family members, serve as the primary economic unit (Neupert, 1995). These households own herds, while higher social bodies administer the rights to grazing land. Settlements rarely grow larger than several *ger* households that converge for the sharing of person and animal power (*khot ail*) or circles of *gers* that are set up for defensive purposes (*khuree*; Bold, 1996; Vreeland, 1962). Marriage, generally monogamous, is deeply tied with the establishment of a new *ger* household and further reinforces a complex gendered division of labor (Cheney, 1968; Neupert, 1996; Shombodon, 1996). Senior males hold the highest authority (Cheney, 1968). Numerous Mongolian women, however, have possessed political or military authority, and socialist policies in the 20th century have further encouraged the economic and political participation of women in Mongolian society (Cheney, 1968; Skapa, 1995).

Tibetan Buddhism firmly infiltrated Mongolian culture in the 17th century when religious missionaries established a network of monasteries throughout the current MPR's territory. Monasteries served both as religious centers and as sites of education, and with 1/3 of all male Mongolians educated as lamas in these monasteries, the philosophy, art, ceremony, and medicine of Tibetan Buddhism soon became embedded in Mongolian life (Cheney, 1968; Heissig, 1980). Although the new religious regime persecuted non-Buddhist knowledge specialists, their indigenous forms of knowledge and practice eventually merged with Tibetan Buddhist forms to generate a uniquely Mongolian syncretism (Heissig, 1980).

During the three hundred years that Tibetan Buddhism reigned in Mongolia, the Manchu Qing dynasty administered the region. When the Qing dynasty fell in 1911, Mongolia quickly declared independence, and after a series of failed governments and foreign occupations, the Mongolian People's Republic was formed in 1924. Under the guidance of its Soviet neighbor, the MPR embraced a centrally planned socialist model for society. Buddhism and other traditional practices were suppressed, with a series of purges killing tens of thousands of Tibetan lamas and other political dissenters in the late 1930s (Batbayar, 1999). Key goals of the socialist regime were universal education and health coverage and by the 1970s, all Mongolians had access to free health care and all children were provided with at least 8 years of full-time education (Smith & Lannert, 1995). In the late 1980s, Mongolia, with many of its former socialist allies, transitioned from a socialist to a free-market system (Bawden, 1989; Bruun & Odgaard, 1996).

As heirs of mobile pastoralists, Tibetan Buddhist, and socialist traditions, Mongolians today engage in a wide range of lifestyles. Less than one third of the population remains in rural areas, while another third lives in urban settlements, and another third resides in the MPR's capital city, Ulaanbaatar (NSO, 2000). A universal mandatory education system with post-secondary training opportunities has generated a highly literate (> 80% literate) and specialized society. At the end of the socialist period, for example, only 33% of the population was involved in agricultural production, with 17% in manufacturing and mining, 7% in specialist areas such as medicine, law, and teaching, and 33% in the service industry (NSO, 2000). Although, Mongolia is classified as a low-income country, with per capita gross national income hovering from $350 to 450 (World Bank, 2000), the country ranks surprisingly well on a number of worldwide indicators. Women have achieved a relatively high level of equality (Skapa, 1995), universal education in the second half of the 20th century has raised literacy rates to higher than 95% (NSO, 2001), and a comprehensive health infrastructure extends to most parts of the country. Mongolians stay tuned to modern forms of media, with 80% getting information from the radio, 75% from newspapers, and 68% from televisions (Avirmed & Marckwardt, 1999). In a post-socialist atmosphere of religious freedom, two thirds of the population is Buddhist and one third claims no religious affiliation (Avirmed & Marckvardt, 1999). Despite the eclectic set of cultural resources now available in the MPR, Mongolians continue to maintain a relatively coherent set of traditions and attitudes that places them in contrast to their international neighbors (China and Russia) and to other cultural systems.

THE CONTEXT OF HEALTH: ENVIRONMENTAL, ECONOMIC, SOCIAL, AND POLITICAL FACTORS

At the border of dense Siberian forests and the vast Central Asian plains, the MPR is the most sparsely populated state on the planet (1.4 persons per km^2). Life expectancy at birth is 64 years for men and 68 for women (WHO, 1999). The leading causes of mortality are cardiovascular disease (31%), cancer (20%), and respiratory infections (13%) (WHO, 1999). The infectious diseases with highest incidence in the 1990s have been viral hepatitis (22–39 per 10,000), tuberculosis (11–14 per 10,000), brucellosis (4–6 per 10,000), gonorrhea (9–14 per 10,000), and syphilis (3–9 per 10,000) (NSO, 2000). As of 1998, there were only 2 confirmed cases of HIV/AIDS (WHO, 1999). However, the dramatic increase of HIV/AIDS in the Russian Federation and China coupled with an increased incidence of other STIs in Mongolia suggests that the country may soon experience its own increase in HIV/AIDS incidence (Purewdawa, 1997). Women suffer from increased rates of depression and indigenous forms of mental illness (*yadargaa*) (Kohrt et al., 2003a), but men seem to have been hit hardest (in both mortality and morbidity) by the recent shift from socialism to a market economy (see section on sociocultural change).

Mongolia's culture, society, and environment have deeply influenced the country's health profile. This section will discuss several of the most important influences on health: environment, mobile pastoralism, diet, and infrastructure. Sociohistorical changes dramatically transformed the health profile of Mongolians in the last century, and these changes will be discussed in the final section of this article.

Environment

Mongolia has an extreme continental climate, with low precipitation, long, cold winters (average January temperature ranging from $-35°C$ in the north to $-15°C$ in south) and short, hot summers (average June temperatures ranging from 15–30°C). Temperatures are generally below freezing for 7 months out of the year (Academy of Sciences, 1990). The country's low population density and cold climate have limited potential disease vectors (humans or otherwise), and Mongolians have had only rare encounters with a number of infectious diseases. Cholera, for example, was first reported in Mongolia in 1996 (WHO, 1999). Similarly, Mongolia experienced smallpox epidemics much later than did its southern neighbors (Fisher, 1988; Serruys, 1980). At the same time, the extremely cold, dry environment provides a fertile bed for respiratory infections, which contribute to nearly half of all newborn, infant, and child deaths (Neupert, 1995; WHO, 1999). In contrast, diarrheal diseases contribute to fewer than 15% of infant and child deaths (WHO, 1999).

Mobile Pastoralism

Mongolia's unique environmental regime has further influenced health by constraining the population's subsistence activities. Several aspects of the climatic regime—low precipitation, poor water supplies, and short summers—render crop agriculture a risky venture, while salty soils and an abundance of good feed grasses make most of the country ideal for sustained livestock production. The general reluctance of Mongolians to "cut" or disturb the earth, to fence the earth into agricultural plots, or to eat "grass" likely also contributes to the reliance on pastoral production (Williams, 1996). Although fewer than half of Mongolians now engage in herding (NSO, 2000; Shombodon, 1996), the lifestyle continues to affect the health of a sizeable portion of the population (Foggin et al., 1997).

Mongolians who maintain a rural lifestyle breed and raise different proportions of the traditional five domesticated animals (*tavan hoshuu mal,* horses, cattle and yak, sheep, goats, and camels). One of the greatest health concerns for herding populations is the threat of losing herd animals, either to poaching, to natural predators, to severe weather, or to infectious disease. Perhaps the most serious threat to Mongolian herds is extremes in winter weather, or *dzud*. In response to the extreme climate, herd animals are adapted to foraging in snow, surviving long periods of famine, and raising young in the severe weather conditions. Mongolian horses, for example, are adapted to the climate, with a short stocky build. They can survive solely on grass and can dig through the snow in winter to find grass. Despite these adaptations, severe *dzuds* can devastate herds with two consecutive *dzuds* in 1999 and 2000 killing over 4,000,000 heads of livestock (more than 10% of all livestock). Although little is known about the human health impact of losing substantial portions of a herd in Mongolia, a study of child nutritional status after the 1999–2000 *dzud,* suggests that herd loss has a deleterious effect on child growth (CDC, 2002).

In addition to affecting well-being indirectly through herd loss, several zoonotic diseases are capable of infecting human populations. Most notably, tuberculosis and brucellosis, which comprise 20% of all reported infections in the MPR (NSO, 2000), can be contracted from herd animals via contact with animals or ingestion of animal products (Foggin et al., 1997).

The traditional Mongolian *ger* also structures health risks. In 1990, 60% of Mongolians still lived in traditional *gers*, including 20% of the urban population (Mongolian Ministry of Health, 1992). Although a *ger*'s outer felt layer and circular design provide an insulated environment even in the coldest months, the close quarters, hot and arid air, and smoke from the central fire may explain the increased deaths from respiratory illness among infants living in *gers* (versus apartments) (Neupert, 1995).

Diet

Early Chinese chroniclers confirm that the Mongolian diet has for centuries consisted of meat and dairy products, with rice, grain, and wine either traded or captured from agricultural areas in China (Cheney, 1968; Jagchid & Hyer, 1979). Fat is accorded special value, with sheep bred for their fatty rumps and honored guests given the first portion of solid white fat from a cooked sheep's tail.

Mongolians prepare numerous milk derivatives (*tsagaan idee*)—cream from boiled milk (*orom*), yogurt (*tarag*), fermented milk (*airag*), raw milk (*suu*), varieties of dried milk curd (*aaruul*), butter (*tos*), cream (*tsotsgii*), dried cheese (*eetsgii*), and distilled sour milk (*nermel*). Salted milk tea is a mainstay. Mongolians consume a limited range of vegetables (potatoes, carrots, turnips, cabbage, onions, and garlic), a restriction that is due partly to availability, and partly to an ethic that "Grass (which can mean any plant food) is for cattle and meat is the food for man." Mongolians consider rice and grain flour as staples and most dishes involve some use of rice, noodles, or flour dumplings.

The diet follows an annual cycle that rural residents adhere to by necessity and urban residents by habit: meat is the staple for most of the year, while dairy products predominate in the summer months (Foggin et al., 1997). Mongolians traditionally slaughter animals in large numbers in December to prepare dried meat (*borts*) for the coming winter (Foggin et al., 1997; Jagchid & Hyer, 1979; Seidenberg, 1991). Dairy replaces meat in the summer due to ample supplies of milk products and the high spoilage rate for meat. Although there is geographical and individual variation in this cycle, it is recognizable throughout Mongolia (Foggin et al., 1997). *Tsagaan sar* (the Mongolian new year) and *Naadam* (a national holiday in July) are the exceptions to the rule. During *Tsagaan sar* frozen milk products are thawed for consumption. Conversely, in the summer heat of Naadam, large amounts of meat are consumed.

The country's cold climate and high altitude (average elevation = 1,580 m) exact significant metabolic requirements on organisms. A high-energy diet, as well as relatively high rates of excess fat and central fat deposition, may serve as buffers against these environmental challenges (Beall & Goldstein, 1992). However, this consumption combined with relatively high salt intake may also put Mongolians at risk for later cardiovascular disease. Although cardiovascular disease currently accounts for over 30% of deaths (WHO, 1999), it is difficult to place responsibility for these high rates on the diet, since these rates have increased dramatically in the last decade with few equally dramatic changes in eating habits.

Infrastructure

The central challenge for all health systems in Mongolia has been to equitably distribute medical knowledge and health resources to a thin mobile population ranging over huge areas. One grass roots solution to this problem has been a distribution of medical knowledge whereby households maintain moderate levels of know-how concerning local medicines and illnesses, with more experienced traditional practitioners dispersed but common enough to be accessible in times of need (Hruschka, 1998).

Both Tibetian and socialist medicine also employed some form of dispersed care (Academy of Sciences, 1990; Roerich, 1933). Today rural and urban populations differ most significantly in reproductive health, with maternal mortality twice as high in rural as in urban areas (WHO, 1999), and infant and child mortality also elevated in rural areas in the 1990s (Tserendulam, 1999).

Alcohol and Tobacco Use

Many Mongolian and foreign observers have noted an increase in alcohol consumption during the social transformations of the 1990s. Unemployment, poverty, lowered living standards, and a visibly widening gap between rich and poor have all been posited as reasons for the trend. The availability of alcohol has also changed considerably, with lax state regulations facilitating an influx of cheap foreign alcohol, an increase in domestic production, and a rise in the number of public drinking establishments. In addition to recent social pressures and increases in availability, Mongolians' cultural attitudes toward alcohol have likely also played a role in the increase in alcohol use. Although Russians are often blamed for the introduction of heavy alcohol consumption, Mongolians had developed a culture around drinking alcohol well before the introduction of Russian practices. For centuries, Mongolians have produced *airag*, or fermented mare's milk (2–5% alcohol) (Montell, 1937), and a brandy distilled from sour milk (Underdown, 1977).

Alcohol has long been a central part of gatherings, with an emphasis on compelling others to drink as much as possible that was established prior to extensive Russian influence (Montell, 1937). However, the speed with which such liquor spoiled (2 to 3 days in summer) and natural limits on the production of milk (Cheney, 1968; Montell, 1937) likely inhibited chronic alcoholism. Mongolians writing from previous centuries noted the drink's destructive side. Indeed a number of laws and proverbs discouraged excessive alcohol consumption (Bawden, 1976; Underdown, 1977). What likely changed with the socialist regime was the introduction of highly alcoholic mass-produced distilled spirits. Controlled during communism by ideological indoctrination, strict cell discipline, and limitation of sales, alcohol became a rare and treasured item and common means of transaction in the underground economy (Bulag, 1998). Although no longer controlled, alcohol still retains a precious image that may also drive its current consumption.

A public health concern which will likely come to the forefront in coming years is the high prevalence of smoking among Mongolians—both youth and adults. In 2001, 68% of adult males and 26% of adult females smoked regularly (Kirby et al., 2001). A more recent study of children (9–17 years old) in 17 East Asia and the Pacific countries and territories indicates that Mongolian children have the highest prevalence of smoking (> 55%), with 50% of Mongolian children listing that they had friends who wanted to quit, but could not (UNICEF, 2001).

MEDICAL PRACTITIONERS

The majority of medical practitioners in Mongolia belong to one of three major traditions—modern Russian medicine, Tibetan Buddhist medicine, and pre-Buddhist traditional medicine.

Pre-Buddhist Traditional Health Specialists

Healers, trained neither at government medical schools nor from Tibetan Buddhist sutras, have dominated much of Mongolia's medical history and continue to participate in healthcare today. These traditional healers specialize in distinct, but overlapping, arenas of knowledge rather than comprising a homogenous set of practices and theories (Hruschka, 1998; Humphrey & Onon, 1996). *Bariach (barishi)* generally perform massage and musculoskeletal manipulation to treat maladies including concussions, broken bones, headaches, and bone or organ displacements (Hruschka, 1998). *Boo* (male) and *udagan (idughan, yadgan)* (female) employ ecstatic trance to master spirits with power over maladies and misfortunes (Humphrey & Onon, 1996). *Ekh barigch (bariyachi)* hold knowledge of female birth-giving and are comparable with traditional birth attendants. *Otoch (otoshi)* have had different roles in different Mongolian contexts. Among the Dagur Mongols of Manchuria, *otoshi* were charged with female fertility and child development (Humphrey & Onon, 1996), while in other contexts the term has signaled a general medical specialist (Kriegel, 1997; Jagchid & Hyer, 1979) or someone skilled with medicinal plants (Lessing, 1982). Each of these specialists is believed to possess the power of spirits and forces related to his or her field of knowledge, much as blacksmiths (*darkhan*) in the same areas are believed to have access to spirits controlling metal-smithing (Hruschka, 1998; Humphrey & Onon, 1996).

Mongolians generally feel that traditional practitioners whether they are *boo, bariach,* or *eck barigch* do not choose to become healers, but rather are chosen by deceased ancestors (*ug*) who once practiced the specialty. An individual is usually "called by the ancestors" in the form of an untreatable illness, and the only way to cure the illness is to ritually take the ancestral talent *(udam)* and begin healing (Hruschka, 1998; Humphrey & Onon, 1996). These healers were persecuted under both Tibetan Buddhist and socialist regimes, and only since 1990 have they begun to experience a resurgence in public practice (Hruschka, 1998).

Buddhist Medicine

With the destruction of monasteries and the mass purging of monks in the 1930s and 1940s, the socialist regime effectively erased Tibetan Buddhist medicine as an institution (Bawden, 1989). Before that time, however, prominent Tibetan Buddhist monasteries in Mongolia often maintained a college of medicine (*mampa datsan*) where future physicians (*maaramba, emch*, or *otoch*) trained in the diagnosis and treatment of illness (Jagchid & Hyer, 1979). The Tibetan Buddhist medical canon in Mongolia was a vast collection of sutras, theoretical commentaries, and case studies compiled and revised over centuries of practice. To become skilled in this field of knowledge, a student was obliged to study at least 20 years in the *datsan* before becoming a physician (Cheney, 1968). In recent decades, Tibetan medicine has returned as a legal form of healing.

Russian Modernist Traditions

Although Russian-trained doctors and paramedics are the most recent addition to the Mongolian medical landscape, they quickly came to dominate healing under the 20th century socialist regime. Within two decades of the revolution in 1924, the MPR had established a school of modern Russian medicine to train nurses and paramedics, and by 1961, a central medical school was training physicians to staff the growing network of clinics throughout the country. During the first years after the revolution when doctors and clinics were still in short supply, Tibetan-trained *emch* and other healing specialists were allowed to continue practicing. This soon changed, however, when in 1938 the government extended a universal ban on non-modern forms of medicine. Coincident with a set of purges, this ban ensured that most medical care took place within Soviet-styled clinics and hospitals staffed by paramedic feldshers (*baga emch*) and doctors *(ikh emch)* (Academy of Science, 1990; Bawden, 1989).

CLASSIFICATION OF ILLNESS, THEORIES OF ILLNESS, AND TREATMENT OF ILLNESS

The theories and treatment of illness in Mongolia derive primarily from three medical traditions—pre-Buddhist traditional, Tibetan Buddhist, and Russian modern. Mongolians do not generally envision these predominant modes of health care as mutually exclusive, coherent wholes. Rather, individuals synthesize the basic tenets and classificatory systems of each tradition into frameworks that fit their past and current illness experience. Although it is impossible to completely disentangle the "original" theories and treatments from these traditions, this section will attempt to identify aspects particular to each tradition.

Pre-Buddhist Traditions

The pre-Buddhist, or shamanic, medical tradition deals with a broad range of misfortune ranging from inconveniences as benign as losing things to maladies as deadly as the plague (Bawden, 1960; Humphrey & Onon, 1996). Misfortune in its many forms (truck accidents, depression, miscarriages) is explained by recourse to a field of powers personified as malignant humans, animals, or ancestors (e.g., *ongon, lus savdag, elriye, chidkun, tuidker, ada, eliye* [Bawden, 1960; Heissig, 1980; Humphrey & Onon, 1996]). These forces can be set in motion by external actions, such as a curse, or by an individual's own actions, such as the breach of a taboo.

Treatments under this system usually invoke the powers of specific spirits. *Tsagaan ovgon* (the white old man), for example, was known among the Buriats to protect humans from poxes and various feverish illnesses (Heissig, 1980). The task of a healer is to prevent misfortune by appealing to specific spirits or to correct (*zasa-*) a misfortune or malady by severing the oppression (*dara-*) of specific forces. Different spirits require different ritual activity such as ecstatic trance induction (Humphrey, 1996), manipulation of a patient's bones and muscles (Hruschka, 1998), or more material forms of treatment (Cleaves, 1954). The workhorses of this ritual repertoire are effigies that absorb an individual's illness or misfortune (Heissig, 1986; Humphrey & Onon, 1996). Another force that has been especially consistent across space and time has been the purifying fire, which cleanses all forms of polluted individuals and items (Bulag, 1998). Although the practices and theories associated with this collection of forces maintain some theoretical consistency, the orientation is foremost pragmatic, incorporating those practices and concepts that work and eschewing those that do not (Heissig, 1980; Humphrey & Onon, 1996).

Tibetan Medicine

Tibetan medicine in Mongolia, although predominantly influenced by traditional Tibetan/Ayurvedic thought, has some unique facets related to Mongolia's prior theories of illness. The three humors of Tibetan medicine—*khii* (wind/air), *badgan* (phlegm), and *shar* (bile)—constitute the categorical backbone of the medical theory (Kohrt et al., 2003a; Kriegel, 1997; Matignon, 1895). Diseases are classified according to the interruption of flow or imbalance of these three humors (Cheney, 1968). In the Mongolian context, these humors map onto concepts that may have pre-dated the influx of Tibetan Buddhism. *Badgan* (derived from Tibetan *bad-kan*) and *shar* (the Mongolian word for yellow and jaundice) are opposed according to a cold-hot (*seruun-khaluun*) set of dichotomies (moon-sun, slow-fast, female-male, water-fire, and *arga-bilig* which is often described as yin-yang) (Kriegel, 1997). *Khii*, resembling both the Tibetan humor "wind" and the Chinese concept of *chi*, dominates the humoral triad, and is responsible for neurological and local idioms of distress, such as *yadargaa* (Kohrt et al., 2003).

Initial diagnosis involves taking a history, reading a person's pulse, and examining the patient's urine. A combination of pulse attributes reveals type of illness, location of pathology, personal temperament, and childhood history of disease. Urine, illuminates humoral constitution when assessed by smell, taste, temperature, color, and foaming potential (Cheney, 1968; Kriegel, 1997; Matignon, 1895).

Treatment depends upon the type of illness. Plant and mineral compounds (*tan*) are often taken in hot water or tea. Since the humors are strongly affected by climate, behavior, and state of mind, patients are often told to change their lifestyle for improved health (Cheney, 1968). Visits to the countryside, spending less time outside in cold weather, cessation of drinking or smoking, and mental relaxation are often encouraged. Illnesses related to karmic offenses in a previous life require the reading of sutras and *tarin* (short sutras). Treatments could include acupuncture (*zuu*), bleeding (*khanuur*), moxibustion (*toonuur*), and cupping (*bumba*) (Kriegel, 1997; Matignon, 1895). Dietary practices have both preventative and curative impacts. Foods possess different humoral and cold/hot properties and are consumed to adjust a person's constitutional imbalance (Kriegel, 1997). Seasonal dietary shifts accommodate climactic changes, for example "hot" foods, such as mutton, warm the constitution in winter, while "cold" foods, such as fermented mare's milk, cool the body's constitution and eliminate toxins taken from "hot" foods during the summer (Bulag, 1998).

Russian Medicine

The biomedical approach employed in Mongolia derives from the Soviet model. Physicians currently diagnose and treat according to the International Classification of Disease-10 (ICD-10). Mongolian biomedicine also emphasizes lifestyle and integrates therapies such as cupping, acupuncture, massage, and various forms of naturopathy. Injections (glucose, vitamin complexes, calcium chloride, and saline) are a popular form of treatment for myriad health problems (Kohrt et al., 2002).

SEXUALITY AND REPRODUCTION

Sexuality

Pre-socialist Mongolia is stereotyped as having promiscuous sexual norms, and consequently high rates of venereal disease (Bawden 1989; Cheney, 1968; Rupen, 1964; Jagchid & Hyer 1979). During socialism sexual activity outside of marriage was discouraged, and sex is not normally an acceptable topic in public discourse. Most education about sex and sexuality occurs through peers and via the media. Sexual or gender forms different from a standard heterosexual dichotomy are little understood and treated with either confusion or disdain. Consequently little information is available on the subject. Post-socialist changes, such as increased freedom of movement, less strict partner tracing, new attitudes about sexual behavior, and increases in sex workers, have been implicated in a marked rise in sexually transmitted diseases (syphilis, gonorrhea, and trichomonas) since 1990 (Purevdawa et al., 1997; WHO, 1999).

The Fertility Boom

For the last half century, Mongolia has witnessed a dramatic demographic transition. With decreased mortality and increased fertility achieved during the socialist period, the country witnessed a growth spurt that more than tripled the population in 60 years (Neupert, 1995; Randall 1993). Consequently, Mongolians are relatively young, with 32% of the population under the age of 15 years, and only 4% over 65 in 1999 (NSO, 2000). In 1970, however, Mongolia began to see fertility rates decline (Neupert, 1995; Randall, 1993), and it is likely that the population structure will include a much larger proportion of older individuals in the coming decades. Although the fertility boom has been linked with pronatalist policies of the MPR, Randall argues that improved medical treatment, especially for sexually transmitted diseases, provides a more consistent explanation for the sudden burst in fertility (Randall, 1993).

The strict pronatalist policy under socialism included strict control of abortions and contraceptive technologies. Since 1989, however, both abortion and contraception have become legal forms of fertility control. Since that time, nearly 2/3 of women with an unwanted pregnancy take steps to stop pregnancy, with one of every six pregnancies ending with an abortion. The termination of pregnancy is much more common among urban and educated women, with only 1/3 of rural women and uneducated women with unwanted pregnancies terminating the pregnancy (Chagnaadorj, 1999; Panday, 2002). The trend in abortions has been paralleled by changes in contraception. By 1995–1998, 60% of women aged 15–49 (compared to less than 15% in the 1980s) were using some form of contraception, with IUD the most common form (32%), followed by periodic abstinence (13%) (Mongolian Ministry of Health, 1993; Luvsantseren, 1999).

Ideal Household

The ideal Mongolian household, which is captured in socialist art and contemporary movies, is a heterosexual couple living in either a city apartment or traditional *ger*. It includes several children and potentially the couple's parents. The ideal number of children has changed over the years, with younger women wanting on average fewer than 3 children, compared with older women (> 35) who state a preference of around 4. These ideals are surprisingly uniform across region, education, and sex (Dashtseren & Marckhvardt, 1999). Despite the ideal of a couple participating in the production of a household, about 10–20% of households are headed by single women (NSO, 2000), and out of wedlock births seem to be accommodated well (Randall, 1993). Cohabitation is common with 48% of registered couples living 2 or more years together before marriage (NSO, 1999). Such cohabitation often occurs within the *ger* or apartment of other family members.

Reproduction and Risk

Due to the socialist emphasis on reproductive health, 98% of women were receiving prenatal care and 87% of one-year-olds full immunizations by the mid-1980s (Griffin, 1995). However, although maternal mortality in Mongolia has decreased considerably in the last century, pregnant women, especially in rural areas, have a significant risk of dying with rates hovering around 120–240 per 100,000 live births over the last two decades (Randall, 1993).

HEALTH THROUGH THE LIFE CYCLE

Pregnancy and Birth

The course of pregnancy for Mongolian women has changed dramatically in the last century. Prior to the establishment of maternal clinics in the mid-20th century, births occurred at home, usually with the aid of a traditional birth attendant (*ekh barigch, kuisutu eej*, or *udghan*) (Jagchid & Hyer, 1979). With the advent of universal socialist health coverage and countryside birth clinics, the incidence of home births delivered by midwives declined. In the 1990s, only 0.1% of births were attended by traditional mid-wives (Lhagvasuren et al., 1999). In addition to providing free antenatal and postnatal care, the state also allowed at least a month of leave both before and after birth (Goldstein & Beall, 1994).

Infancy

Infancy (*balchir nas*) in Mongolia, as in many other regions of the world, is an especially risky period of the life span. Even with the expansion of a socialist medical system that made child health one of its primary concerns, infant mortality rates are still high, with 50–60 of every 1000 live births not surviving past the first year of life (NSO, 2000). At this sensitive period of development, efforts are made to protect children from all kinds of disturbances, including states of fright or "soul loss." Among the Buriats, for example, black ash is smudged on a child's forehead before leaving a *ger* at night in order to protect the child from being frightened by wandering spirits. This protects the child from "loss of soul," a disorder accompanied by listlessness or recurrent illness, and if untreated, potentially death (Humphrey, 1996).

Breast-feeding is nearly universal, with over 96% of children born in the 1990s fed with breast milk. Mongolian mothers continue breast-feeding their infants far longer than world standards, with a median age of cessation at 25 months. One fifth of children receive breastmilk up to three years. The median duration (8 months) of postpartum amenorrhea is consequently quite long. This pattern of sustained breast-feeding does not differ by education or by residence (rural vs urban) (Dashtseren, 1999). Supplementation generally included animal milks and boiled mutton soup (*bantan*) (Randall, 1993).

Rickets is widely reported in Mongolian children. Due to concerns about cold weather, infants are rarely taken outside in winter, and even then only when heavily swaddled. The consequent lack of exposure to sunlight during winter combined with depleted maternal supplies of calcium may contribute to this problem (Randall, 1993). Respiratory illnesses are by far the largest cause of death in both infants and children (see Environment section).

Childhood

Childhood generally begins when a child reaches 3 "Asian" years of age, which overestimates years from birth by an average of 18 months (Beall & Goldstein, 1992). In Ulaanbaatar, children (*okhin* refers to girls and *khuu* to boys) are generally cared for by their mother (57%) and/or grandmother (31%); fathers are caregivers in 26% of homes (Kohrt et al., 2003b). A big step in rural enculturation is reached when children (at about age 5) begin to ride a horse alone with the parent galloping alongside, holding the reins (Jagchid & Hyer, 1979). The post-socialism privatization of herds has led some families to keep children, especially boys, at home to help with herding, instead of attending school (Humphrey & Sneath, 1996; World Bank, 2000). Parents do not generally condone physical force as a means of enculturation (Kohrt et al., 2003b). Survival for children is much better than it was prior to socialism, when frequent smallpox and influenza epidemics could lead to the death of half of all children before the age of 4 (Cheney, 1968).

Adolescence

Today Mongolians enter school at around the age of 8, a time when children are beginning to have a sense of the world around them, and when they would traditionally begin to ride horseback independently and to assist in herding large animals (horses, oxes and camels) (Cheney, 1968). Little data exists on sexual maturation among Mongolians, but a recent study in the rural West recorded a median age of menarche at 14 years of age (Beall & Goldstein, 1992). There is no clear rite of passage related to sexual maturation, but once children begin looking and acting more like adults, they take on a set of gendered terms different then those assigned to children (*hovguun* for boys and *busgui* for girls).

Knowledge of sexual behavior most often comes from friends, with surprising homogeneity of attitudes given this horizontal mode of transmission. A number of these attitudes engender behaviors that increase the chance of sexually transmitted infection (STI) (Hruschka, 2001), which could partly explain the high rates of STIs among youth aged 15–25 (Purevdawa et al., 1997).

Adulthood

A number of criteria are associated with "someone who has come of age" (*Nasan hursend hun*). In the past, these criteria were related to the establishment of a household—finding a spouse, having children, and developing a herd. Today, they can also be associated with getting a job. Women normally marry around 20–24, and marriage is nearly universal for women, with only 1% of women at age 49 unmarried (Tsendjav & Marckvardt, 1999). Men marry a few years later than women (NSO, 2001). Most women report the end of menses around 45–48 years old (Tsendjav & Marckvardt, 1999).

The Aged

Ancient Chinese chronicles noted with condescension that the nomads to the north "honor the strong and despise the old" (Jagchid & Hyer, 1979). Upon closer inspection, however, Mongolians maintain entrenched cultural patterns devoted to the aged. Those individuals who live a long life (*nastai khun*) are treated with great respect and often referred to with honorifics (*guai, ahai, avgai*). This deference to the aged is perhaps most apparent during the Mongolian new year, *Tsagaan sar*, when younger members of families and communities visit their elders and present them with gifts. During the socialist regime, the state provided generous pensions and state services to the elderly. Since 1989, however, pensions have not kept up with inflation, and the elderly must depend on family members for support (Briller, 2000).

Dying and Death

Death is treated delicately in the Mongolian language. Although there is a word for the death of animals or unhonored individuals (*ukhekh*), when discussing the death of a close individual, Mongolians use euphemisms—*ongorokh* or *burkhan bolokh* (to become a spirit) or *nasu bolokh* (to come of age) (Humphrey & Onon, 1996; Jagchid & Hyer, 1979). Great concern is taken to ensure that the essence of the dead person departs appropriately. A lama may be called in to chant sutras while the spirit of the dying person leaves. Lama or other ritual specialists decide whether the body should be cremated, interred, or abandoned in the steppe (Kler, 1938; Jagchid & Hyer, 1979).

CHANGING HEALTH PATTERNS

Tibetan Buddhist Medicine

Little epidemiological data exists for Mongolians before regular health censuses were begun in the 20th century. It is difficult to judge, therefore, where the Tibetan Buddhist medical system succeeded or failed in terms of public health. Bawden has suggested that the services of Buddhist lamas were beyond the reach of most people (Bawden, 1989). Considering, however, that the population of Outer Mongolia was about 500,000 people (NSO, 2000) at the turn of the century and assuming conservatively that one lama at each monastery was medically trained (Roerich, 1933), there would have been about 14 Tibetan trained doctors per 10,000 inhabitants. Contrary to Bawden's judgment, this ratio compares favorably with physician-to-population ratios during much of the later socialist period (Academy of Science, 1990). Indeed, several scholars have hypothesized that the success of Tibetan Buddhism in Mongolia was due in part to the ability of Tibetan doctors to treat or prevent illnesses, including smallpox (Fisher, 1988).

Despite the hypothesized health benefits of Tibetan medicine, the accounts of foreign travelers, researchers, and medical missions in the 19th and early 20th century often noted the poor health of the Mongolian people (Cheney, 1968; Rupen, 1964). The most common theme in these accounts was the high prevalence of venereal and skin diseases, especially syphilis, among the general population (Bawden, 1989; Cheney, 1968). Reports indicate that prior to the revolution, smallpox and influenza periodically rose to epidemic proportions (Cheney, 1968; Rupen, 1964). Official statistics further suggest that at the eve of the revolution the mortality rate totaled 25–30 person per 1,000 year, and half of newborn babies died before the age of 1 (Academy of Sciences, 1990). These statistics, however, must be considered as the estimates of a regime that had a vested interest in painting a bleak portrait of pre-socialist times.

Socialism

Given official statistics, the development of socialism beginning in the 1920s seems to have improved the health of Mongolians in profound ways. The socialist goal was to spread the benefits of modern medical technology to all citizens of Mongolia free of charge—a plan that mirrored the larger socialist platform of equitable distribution of economic, social, and cultural fruits of modern development (Academy of Sciences, 1990; Farkas, 1993). Based on a rational Soviet model of strong central planning, the system grew rapidly after World War II to form a comprehensive hierarchy of health care. Ideally, medical outposts staffed by a nurse practitioner catered to residents within a 35 km radius, county clinics with 1–3 doctors (general practitioners and maternal/child health specialists) provided primary and maternal care, and more specialized hospitals in cities treated referrals from more diffuse clinics. To staff these new posts, the system made extensive use of *baga emch* (equivalent to Russian feldshers) as paramedical personnel for scattered rural populations, complemented by more specialized practitioners located in central settlements (Academy of Sciences, 1990).

The expansion of socialist medicine is partly responsible for the dramatic changes in health profiles observed during the 60 years of socialist rule. Immunization programs and a comprehensive system of maternal child health played a role in the reportedly 8-fold decrease in infant mortality from the early 20th century to 1990 (Academy of Sciences, 1990). Overall, mortality rates dropped 3-fold, and average life expectancy at birth rose steadily from 38 in the early 20th Century to 65 years by 1991, in large part due to a reduction in child mortality (Academy of Sciences, 1990). The introduction of penicillin in the postwar era combined with the establishment of venereal disease clinics in the late 1940s dramatically reduced rates of syphilis and other infections diseases (Bawden, 1989; Randall, 1993; Rupen, 1964).

Post-Socialism

By 1990, in step with other post-socialist countries, Mongolia had begun a series of social and political transitions toward multi-party governance and a decentralized economy. The initial years of the transition were difficult for Mongolians. With the subsequent loss of Soviet subsidies (30% of its GDP prior to 1990), the state was not able to continue funding its massive social support system (Smith & Lannert, 1995). Inflation skyrocketed from near zero during the socialist period to 325% in 1992 (NSO, 2000). Unemployment rose as collectives and public corporations were privatized or simply ceased operations (Griffin, 1995). GDP per capita dropped 1/5 from 1990 to 1998 (UNDP, 2000). At the end of the 1990s, however, many of the same indicators now suggest that Mongolians have weathered the transitional storm (NSO, 2000). On the other hand, several disturbing trends have continued their course. With reductions in state expenditures in Mongolia, education and health infrastructure continues to deteriorate (Humphrey & Sneath, 1999; World Bank 2001), while a clearly visible gap continues to grow between the rich and the poor (Griffin, 1995).

Reductions in state funding have had serious consequences for the medical delivery system (Medvedeva, 1996).

Between 1989 and 1992, most of the MPR's maternity-waiting homes were closed, while the referral system that had relied on jeep ambulances and planes for critical emergencies went into disuse (Randall, 1993). There is evidence that increased dependence on women's work coupled with decreased state control is further leading pastoralist women to have home births and return to work quickly (Mongolian Ministry of Health, 1999). Rural feldsher's posts, which cost about 3 to 4 times as much per bed as any other health unit were discontinued in 1990 (Center for Health Statistics, 1993).

Disruption of health services in the first years of the transition has been implicated with the increase in maternal mortality from 120–175 per 100,000 live births (1985–1990) to a peak of 240 in 1993. With a stabilization of the health system, this rate has subsequently declined to around 150 in 1997 and 1998 (WHO, 1999). Nonetheless, the difference between maternal mortality in Ulaanbaatar and the rest of the country has increased in the last few years (WHO, 1999), suggesting increasing disparity in access to perinatal care. An opposing trend is that of infant and child mortality, with estimates of infant mortality actually decreasing (NSO, 2000; Tserendulam, 1999).

In contrast to infant and child mortality, adult mortality rose dramatically after the end of socialism with a 25% increase for all age groups between 15–44 years (1990–1993) (WHO, 1999). This increase in mortality disproportionately effected young males (15–34). Indeed the mortality of young men continued to increase throughout the 1990s, while that for young women experienced a slight decline after an initial increase (WHO, 1999). These gender differences in mortality rate changes are reflected in life expectancy trends, with male life expectancy at birth having stayed around 64 since the 1980s, and female life expectancy increasing from 64 in the 1980s to 68 years from birth in 1997 (WHO 1999). The incidence of several infectious diseases has also increased in the last decade. Perhaps related to the collapse of veterinary services, brucellosis and tuberculosis incidences have both doubled (NSO, 2000; WHO, 1999), while transitions in sexual behavior are likely implicated in the recent increases in sexually transmitted diseases (Purevdawa et al., 1997).

The stress of the socio-economic transition has also had a marked impact on the mental health of many Mongolians. For example, Mongolians who self-reported more changes in the past decade suffered from more depressive symptoms and higher rates of one indigenous idiom of distress—*yadargaa* (Kohrt et al., 2003). In tandem with social changes in the early 1980s, the prevalence of psychiatric disorders more than tripled, (Center for Health Statistics, 1993), while rates of suicide have increased dramatically with a rate in 1997 (17 per 100,000 per year) reaching nearly double the global average (WHO, 1999). Not all health transitions began abruptly in 1990s, however. In the early 1980s, for example, the mortality profile began a profound transformation, as the proportion of deaths due to respiratory illness decreased steadily from 50% to less than 15%, while deaths due to cardiovascular diseases increased steadily from 6% in 1980 to more than 30% of total mortality. In the same period, the proportion of the deaths due to cancers increased from one in 20 to one in five of all deaths (WHO, 1999).

Post-Socialist Medical Pluralism

Although the socialist regime suppressed the open practice of non-modern or religious forms of healing, Tibetan-trained physicians as well as non-Buddhist healing specialists continued to practice behind closed doors. Many practitioners received modern Russian medical training and transformed their practice into suitably modern forms, such as one *bariach* who trained as a nurse and was known for practicing "strengthening massage" (Hruschka, 1998; Kriegel, 1997). With the change of political climate since 1989, Mongolians have gained a renewed interest in "traditional" forms of healing (Hruschka, 1998; Humphrey & Sneath, 1999; Kriegel, 1997; Medvedeva, 1996). The Buddhist medical traditions have returned most strongly, with the renovation of traditional Buddhist monasteries and the re-establishment of several Tibetan medical colleges (Kriegel, 1997). Although many Mongolians show an aversion to pre-Buddhist practitioners, much as they did prior to socialism (Jagchid & Hyer, 1979), the growing number of pre-Buddhist traditional specialists suggests that Mongolians are beginning to view these traditions as viable forms of healing. Within this pluralistic environment, it is also becoming more difficult to discern to which traditions practitioners belong. Traditional specialists may take pulses similar to the Tibetan methods, while Tibetan-trained healers send individuals to sites an academic would likely classify as "shamanic" (Humphrey, 1993). Similarly, private clinics now prescribe combinations of traditional bleeding and vitamin or saline injections, while shamans may supplement trance therapy with prescriptions for factory-produced pain relievers.

REFERENCES

Aberle, D. F. (1961). "Arctic Hysteria" and Latah in Mongolia. In Y. Cohen (Ed.), *Social Structure and Personality: A Casebook.* New York: Holt, Rinehart and Winston.

Academy of Sciences of the Mongolian People's Republic. (1991). *Information Mongolia.* Oxford: Pergamon Press.

Avirmed, A., & Marckwardt, A. M. (1999). Household and respondent characteristics. In G. Altankhuyag & A. M. Marckwardt (Eds.), *Mongolia Reproductive Health Survey 1998.* Ulaanbaatar: National State Office of Mongolia.

Batbayar, B.-E. (1999). *Twentieth Century Mongolia* (D. Suhjargalmaa, S. Burenbayar, H. Hulan, & N. Tuya, Trans.). Cambridge, U.K.: The White Horse Press.

Bawden, C. R. (1960). The supernatural element in sickness and death according to Mongol Tradition. *Asia Major, 8,* 215–257.

Bawden, C. R. (1976). On the evils of strong drink: A Mongol tract from the early 20th century. In W. Heissig (Ed.), *Tractaca Altaica* (pp. 59–79). Wiesbaden: Otto Harrassowitz.

Bawden, C. R. (1977). A note on a Mongolian burial ritual. *Studia Orientalia, 47,* 25–35.

Bawden, C. R. (1989). *The Modern History of Mongolia* (2nd revised ed.). London: Kegan Paul International.

Bawden, C. R. (1994). *Confronting the supernatural.* Wiesbaden: Harrassowitz.

Beall, C., & Goldstein (1992). High prevalence of excess fat and central fat patterning among Mongolian pastoral nomads. *American Journal of Human Biology,* 4, 6, 1992: 747–756.

Bold, B.-O. (1996). Socio-economic segmentation—Khot-ail in nomadic livestock keeping of Mongolia. *Nomadic Peoples, 39,* 69–86.

Briller, S. H. (2000). *Whom can I count on today? Contextualizing the balance of family and government old age support for rural pensioners in Mongolia.* Unpublished Doctoral dissertation, Case Western University, Cleveland.

Bruun, O., & Odgaard, O. (Eds.). (1996). *Mongolia in transition: New patterns, new challenges.* Richmond, Surrey: Curzon Press.

Bulag, U. (1998). *Nationalism and Hybridity in Mongolia.* Oxford: Clarendon Press.

Center for Health Statistics and Information. (1993). *Health Statistics of Mongolia.* Ulaanbaatar, Mongolia: Ministry of Health.

Centers for Disease Control and Prevention. (2002). Nutritional Assessment of Children after severe Weather—Mongolia, June 2001. *Morbidity and Mortality Weekly Review, 51*(01), 5–7.

Chabros, K. (1992). An East Mongolian ritual for children. In G. Bethenfalvy (Ed.), *Altaic Religious Beliefs and Practices* (pp. 59–63). Budapest: Research group for Altaic Studies.

Chagnaadorj, B. (1999). Unwanted pregnancy and abortion. In G. Altankhuyag & A. M. Marckwardt (Eds.), *Mongolia Reproductive Health Survey 1998.* Ulaanbaatar: Mongolian National Statistical Office.

Cheney, G. A. (1968). *The Pre-revolutionary culture of Mongolia.* Bloomington, Indiana: The Mongolia Society.

Cleaves, F. W. (1954). A medical practice of the mongols in the thirteenth century. *Harvard Journal of Asiatic Studies, 17,* 428–444.

Dashtseren, A., & Marckwardt, A. M. (1999). Fertility Preferences. In G. Altankhuyag & A. M. Marckwardt (Eds.), *Mongolia Reproductive Health Survey 1998.* Ulaanbaatar: National Statistical Office of Mongolia.

Dashtseren, A. (1999). Breastfeeding. In G. Altankhuyag & A. M. Marckwardt (Eds.), *Mongolia Reproductive Health Survey 1998.* Ulaanbaatar: Mongolian National Statistical Office.

Farkas, O. (1993). *"Post-Modern" health services in Mongolia: On the integration of traditional medicine to the national health care system.* Unpublished Doctoral dissertation.

Field, M. G. (1976). The Modern Medical System: The Soviet Variant. In C. Leslie (Ed.), *Asian medical systems: A comparative approach* (pp. 82–101). Berkeley: University of California Press.

Fisher, C. T. (1988). Smallpox, Salesmen, and Sectarians. *Ming Studies* (Spring), 1–23.

Foggin, P., Farkas, O., Shiirev-Adiya, S., & Chinbat, B. (1997). Health status and risk factors of seminomadic pastoralists in Mongolia: A geographical approach. *Social Science and Medicine, 44*(11), 1623–1647.

Goldstein, M., & Beall C. (1994). *The Changing World of Mongolia's Nomads.* Berkeley: University of California Press.

Griffin, K. (1995). Preface. In K. Griffin (Ed.), *Poverty and the Transition to a Market Economy in Mongolia* (pp. viii–xii). New York: St. Martin's Press.

Hiessig, W. (1953). A Mongolian source to the lamaist suppression of shamanism in the 17th century. *Anthropos, 48,* 1–29, 493–536.

Heissig, W. (1980). *The Religions of Mongolia* (G. Samuel, Trans.). Berkeley, CA: University of California Press.

Heissig, W. (1986). Banishing illnesses into effigies in Mongolia. *Asian Folklore Studies, 45*(1), 33–44.

Hruschka, D. J. (1998, Winter). Baria Healers among the Buriats in Eastern Mongolia. *Mongolian Studies, 21.*

Hruschka, D. J. (2001). *The Stigma of Syphilis in Mongolia and its Public Health Implications.* Paper presented at the Central Eurasian Studies Conference, Bloomington, Indiana, March 2001.

Hruschka, D. J., Kohrt, B. A., Kohrt, H. E., & Tsagaankhuu, G. (2001). Yadargaa: A culturally and biologically bound illness in Mongolia. *American Journal of Human Biology, 13*(1), 37.

Hu, J., & Stuart, K. (1992). Illness among the Minhe Tu, Qinghai province: Prevention and etiology. *Mongolian Studies, 15,* 111–135.

Humphrey, C. (1993). Avgai Khad: Theft and social trust in post-communist Mongolia. *Anthropology Today, 9*(6), 13–16.

Humphrey, C., & Onon, U. (1996). *Shamans and elders: Experience, knowledge, and power among the Daur Mongols.* Oxford: Clarendon Press.

Humphrey, C., & Sneath, D. (1996). *Culture and environment in inner Asia.* London: Paul and Company.

Humphrey, C., & Sneath, D. (1999). *The End of Nomadism?* Durham, NC: Duke University Press.

Jagchid, S., & Hyer, P. (1979). *Mongolia's Culture and Society.* Boulder, Colorado: Westview Press.

Jagchid, S. (1988). Shamanism among the Dakhur Mongols. In S. Jagchid (Ed.), *Essays in Mongolian Studies*. Provo, Utah: Brigham Young University.

Kirby, E., Oyuntugs, S., Uranchimeg, D. (2001). *Smoking in Mongolia: Prevalence, knowledge and attitudes: Results from an urban survey, 2001*. Australia: Adventist Development and Relief Agency (ADRA) Australia.

Kler, J. (1938). Sickness, death, and burial among the Mongols of the Ordos Desert. *Primitive Man, 9*, 27–31.

Knaus, W. A., & Petroff, N. A. (1982). *Inside Russian medicine: An American doctor's first-hand report*. Boston: Beacon Press.

Kohrt, B. A., Kohrt, H. E., & Tsagaankhuu, G. (2001). Vulnerability and stress in Mongolia: the adaptive significance of *yadargaa*. *American Journal of Physical Anthropology, 32 Supplement*, 91–92.

Kohrt, B. A., Hruschka, D. J., Kohrt, H. E., Panabianco, N. L., Tsagaankhuu, G. (2003a). Distribution of distress in post-socialist Mongolia: a cultural epidemiology of *yadargaa*. *Social Science and Medicine*. In press.

Kohrt, H. E., Kohrt, B. A., Waldman, I., Saltzman, K., & Carrion, V. (2003b). An ecological-transactional model of significant risk factors for child psychopathology in outer Mongolia. *Submitted for publication*.

Korsunkiyev, T. K. (2001). Ancient Oirat books about oriental medicine. In Y. Bregel (Ed.), *Papers on inner Asia* (Vol. 36, pp. 33). Bloomington, Indiana: Indiana University, Research Insitute for Inner Asian Studies.

Kriegel, A. (1997). *Tradition et modernity dans la medicine en mongolie*. Unpublished Maitrise, Universite de Paris X, Nanterre.

Lessing, F. (1982). *Mongolian-English Dictionary*. Bloomington: The Mongolian Society.

Lhagvasuren, M., Tumurtolgoi, N., & Marckwardt, A. M. (1999). Reproductive and child health. In G. Altankhuyag & A. M. Marckwardt (Eds.), *Mongolia Reproductive Health Survey 1998*. Ulaanbaatar: National Statistical Office of Mongolia.

Luvsantseren, Z. (1999). Family Planning. In G. Altankhuyag & A. M. Marckwardt (Eds.), *Mongolia Reproductive health survey 1998*. Ulaanbaatar: National Statistical Office of Mongolia.

Matignon, J. J. (1895). La medicine des Mongols. *Archives cliniques de Bordeaux, 4*(11), 101–110.

Medvedeva, T. (1996). Medical Services and health issues in rural areas of inner asia. In C. Humphrey & D. Sneath (Eds.), *Culture and Environment in Inner Asia*. Cambridge: White Horse Press.

Mongolian Ministry of Health (1992). *Health sector review*. Ulaanbaatar, Mongolia: Ministry of Health.

Mongolian Ministry of Health (1995). *Contraceptive knowledge attitude and practice survey*. Ulaanbaatar: Mongolian Ministry of Health.

Montell, G. (1937). Distilling in Mongolia. *Ethnos, 2*(5), 321–332.

Morgan, D. (1996). *The Mongols*. London: Blackwell Publishers.

Narsu, & Stuart, K. (1988). Insects used in Mongolian medicine. *Journal of the Anglo-Mongolian Society, 11*(1), 7–13.

Neupert, R. F. (1995). Early-age mortality, socio-economic development and the health system. *Health Transition Review*, 35–57.

Neupert, R. F. (1996). Population and the Pastoral Economy in Mongolia. *Asia-Pacific Population Journal, 11*(4), 27–46.

NSO (1997). *Demographic Survey*. Ulaanbaatar: National Statistical Office of Mongolia.

NSO (1999). *Reproductive Health Survey*. Ulaanbaatar: National Statistical Office.

NSO (2000). *Mongolian Statistical Yearbook 1999*. Ulaanbaatar: National Statistical Office of Mongolia.

NSO (2001). *The 2000 Population and Housing Census*. Ulaanbaatar: National Statistical Office of Mongolia.

Panday, R. N. (2002). The use of induced abortion as a contraceptive: the case of Mongolia. *Journal of Biosocial Science, 34*, 91–108.

Purevdawa, E., Moon, T. D., Baigalmaa, C., Davaajav, K., Smith, M. L., & Vermund, S. H. (1997). Rise in sexually transmitted diseases during democratization and economic crisis in Mongolia. *International Journal of STD & AIDS, 8*, 398–401.

Randall, S. (1993). Issues in the Demography of Mongolian nomadic pastoralism. *Nomadic Peoples, 33*, 209–229.

Robinson, B., & Solongo, A. (2000). The gender dimension of economic transition in Mongolia. In F. Nixson, B. Walters, B. Suvd, & P. Luvsandorj (Eds.), *The Mongolian Economy: A manual of applied economics for a country in transition*. Massachusetts: Edward Elgar Publishing, Inc.

Roerich, G. N. (1933). La Medicine au Thibet et les lamas Gueriseurs. *L'Ethnographie, 27*, 97–98.

Rupen, R. (1964). *Mongols of the twentieth century* (Vol. 37). Bloomington, Ind.: Indiana University Press.

SCRFS. (1994). *Statistics of Russia*. Moscow: State committee of the russian federation on statistics.

Seidenberg, S. (1991). The horsemen of Mongolia. In P. Carmichael (Ed.), *Nomads* (pp. 96–125). London: Collins and Brown.

Serruys, H. (1980). Smallpox in Mongolia during the Ming and Ch'ing dynasties. *Zentralasiatische Studien, 14*(1), 41–63.

Shombodon, D. (1996). The division of labor and working conditions of herdsmen in Mongolia. In C. Humphrey & D. Sneath (Eds.), *Culture and Environment in Inner Asia*. Cambridge: White Horse Press.

Skapa, B. (1995). Mongolian women and poverty during the transition. In K. Griffin (Ed.), *Poverty and the transition to a market economy in Mongolia* (pp. 90–103). New York: St. Martin's Press.

Smith, S., & Lannert, J. (1995). Human capital: The health and well-being of the population. In K. Griffin (Ed.), *Poverty and the Transition to a Markey Economy in Mongolia* (pp. 77–89). New York: St. Martin's Press.

Strickland, S. S. (1993). Human nutrition in Mongolia: Maternal mortality and rickets. *Nomadic Peoples, 33*, 231–239.

Tianlu, Z., & Rongqing, H. (1996). *China's population in transition and development*. Beijing: Gaodeng Jiaoyu Chubanshe Shatan Houjie.

Tsendjav, B., & Marckwardt, A. M. (1999). Other proximate determinants of fertility. In G. Altankhuyag & A. M. Marckwardt (Eds.), *Mongolian Reproductive health survey 1998*. Ulaanbaatar: Mongolian Statistical Office.

Tserendulam, T. (1999). Infant and Child Mortality. In G. Altankhuyag & A. M. Marckwardt (Eds.), *Mongolia Reproductive Health Survey 1998*. Ulaanbaatar: National Statistical Office of Mongolia.

Underdown, M. (1977). The Mongols and wine (attitude to alcohol as revealed in Mongolian literature). *Journal of the Oriental Society of Australia, 12*, 11–113.

UNICEF. (2001). *Speaking out! voices of children and adolescents in East Asia and the Pacific*. Bangkok.

United Nations Development Programme. (2000). *Human Development Report 2000*. New York.

Vreeland, H. H. I. (1962). *Mongol community and kinship structure*. New Haven: HRAF Press.
World Bank (2001). *Interim Poverty Reduction Strategy*. Report. Ulaanbaatar: World Bank.
World Health Organization (WHO) & Ministry of Health. (1993). *Mongolia Health Sector Review*. Ulaanbaatar.
WHO (1999). *Health Sector Review*. Ulaanbaatar: World Health Organization.
Williams, D. M. (1996). The barbed walls of China: A contemporary grassland drama. *The Journal of Asian Studies, 55*(3), 665–691.
World Bank (1992). *Mongolia: Toward a Market Economy: A World Bank Country Study*. Washington, DC.
World Bank (2000). *World Development Indicators 2000*.

The Nahua[1]

Brad R. Huber

ALTERNATIVE NAMES

Aztec, Nahuat, Nahuatl, Mexicano, Mexijcatl.[2]

LOCATION AND LINGUISTIC AFFILIATION

The Nahua are the largest Native American group in contemporary Mexico. Approximately 1 to 1.5 million people speak Nahuatl[3] or one of its dialects. Nahuatl is the southernmost member of the Uto-Aztecan language family. As Dow and Van Kemper (1995, p.182) observe, most Nahua currently live around the periphery of what was once the core of the Aztec Empire in the modern states of Puebla, Veracruz, San Luis Potosí, Hidalgo, and Guerrero. They can also be found living in smaller numbers in the Federal District, the states of Mexico, Tlaxcala, Morelos, Michoacán, Durango, Nayarit, Jalisco, Tabasco, and Oaxaca, and the country of El Salvador (Dakin, 1995).

OVERVIEW OF THE CULTURE

The Nahua generally live in rural areas where they cultivate subsistence and cash crops such as maize, beans, chili peppers, tomatoes, squash, maguey, sugarcane, and coffee. Maize is the most important caloric component of their diet. It is a good source of complex carbohydrates, but is low in niacin, calcium, riboflavin, and protein, especially the amino acids lysine and tryptophan. These deficiencies are offset by the custom of combining maize with beans at the same meal, and boiling maize in water to which ground mineral lime has been added. The Nahua also raise chickens, turkeys, pigs, and to a much lesser extent, sheep, goats, cattle, horses, and mules. They generally consume animal proteins very sparingly, usually eating meat on festive occasions only. Many Nahua are also known for their production of crafts such as woven goods and pottery. In addition, young men and women may periodically work as migrant farm workers in Mexico and the United States or as masons and domestic servants in nearby towns and cities.

Nahua houses are often one- or two-room, rectangular structures with bamboo, wood, adobe, concrete block, or stone walls, thatch, tar-paper, concrete, or tiled roofs, and packed-earth or cement floors. Personal hygiene and household sanitation vary from community to community. In some communities, people bathe regularly, wear clean clothes, and sweep their homes frequently. The opposite is true in other communities. Regardless, the majority of rural residents do not have indoor plumbing or use a latrine. Furthermore, pigs, sheep, and fowl are kept in a shelter close to or adjoining the home. Fleas are common in houses and lice infest some school-age children and their parents. Because many people are poor, they cook using a floor level hearth, sit on low stools, and sleep on a *petate* (woven mat) directly on the floor or on wood planks supported by concrete blocks. These living conditions permit the spread of infectious diseases of the skin, and of the respiratory and gastrointestinal systems.

Households are composed of nuclear or patrilocal extended families whose adult members may own arable land or have access to *ejidos* (government land grants). In general, the Nahua trace descent bilaterally, avoid marrying a blood relative or *compadre* (ritual coparent), and marry within the community or municipality. C*ompadrazgo* ties are often very important socially and economically, and may be established on the occasion of a birth, marriage, severe illness, or death as well as for a first communion, confirmation, and the acquisition of a saint's image, car, or house. Fiestas that accompany these life-cycle events have banquets in which abundant amounts of food rich in carbohydrates, proteins, fats, vitamins, and minerals are served. They are also occasions for the ritual sharing of cigarettes and alcohol. The extent to which adults are addicted to nicotine and ethanol is not known, but fiestas are occasions during which some men and a few women become very inebriated.

Nahua social organization is generally based upon men agreeing to perform for little or no financial remuneration civil, political, and religious works (*cargos*) that benefit the community. Cargo holders and other community authorities participate in the distribution of *ejido* lands, the settlement of minor disputes, and the sponsorship and celebration of saints' feast days. Community social organization is variable with respect to the presence and influence of a council of elders, state-level political officials, teachers, biomedical personnel, Catholic priests, and Protestant missionaries.

As is the case with other aspects of contemporary Nahua culture, religious beliefs and practices are a complex blend of Native American and Spanish elements. Pre-hispanic religious beliefs and practices can be found in relatively unacculturated communities where native religious practitioners celebrate rituals coinciding with the winter-solstice, planting, and harvesting, and preside over disease-prevention and curing rituals, ceremonies petitioning rain, and divinations. The influence of Spanish Catholicism is evident in more acculturated Nahua communities where Spanish-speaking priests have encouraged community members to observe the Catholic liturgical calendar for nearly 500 years.

THE CONTEXT OF HEALTH: ENVIRONMENTAL, ECONOMIC, SOCIAL, AND POLITICAL FACTORS

Nahua settlements can be nucleated or dispersed, and may consist of Nahua-speakers only, or of speakers of Nahua, Spanish, and other Native American languages such as Otomí, Tepehua, or Totonac. When Spanish-speaking *mestizos* (people of mixed Spanish and Indian ancestry) are present, they tend to occupy town and village centers. They may denigrate Nahua culture, and try to dominate local and regional political, economic, and religious activities. Locally powerful *mestizos* form ties with Spanish-speaking Mexicans living in other towns and cities of the region and state. Together they implement state policies that benefit them, and which often exploit the Nahua and integrate, assimilate, and subordinate them to the nation's social, economic and political processes.

As a result, land tenure is precarious and the soils the Nahua cultivate have deteriorated. The Nahua have experienced illegal encroachments on their lands, land shortages and conflicts, and low prices for cash crops and the artisan products they produce. Many Nahua are illiterate and impoverished, and have restricted access to biomedical health services.

MEDICAL PRACTITIONERS

Nahua medical specialists are usually consulted only when the herbal remedies and special diet recommended by members of an individual's family (e.g., wife, mother, grandmother) are ineffective in treating an illness. The most common Nahua medical practitioners are the shaman (*tepahtihqui*,[4] *tepahtiani, tlamátiquetl, pajchijquetl, curandero/a*), midwife (*tetejquetl, partero/a*), bonesetter (*texixitojquetl, huesero/a*), herbalist (*xiutepatique, hierbero/a*), and massager (*tlatitilanki, sobador*). Less common are specialists such as curer of fallen fontanel, spirit exorcist, prayer leader, snake bite healer, sucking healer, injectionist, and spiritualist. The majority of medical practitioners heal on a part-time basis, devoting most of their energy to farming, housework, child care, wage labor, and crafts work. A small minority of medical practitioners practice two medical specialties, and a few undertake three or more. Nahua medical practitioners range in age from approximately 15 years old to more than 90, and are often recruited from the most impoverished households.

There is a sexual division of medical labor in many communities. Nahua midwives are almost always female; only 5% are men. Shamans are also generally women, but shamans who combine healing with the performance of public, communal rituals are male. Male shamans-priests officiate at rites concerning animal and crop fertility, the control of the weather, the installation of public officials, and events associated with the Catholic church calendar. Roughly an equal proportion of men and women become massagers and herbalists. Approximately 75% of Nahua bonesetters are men.

Each of the above five specialties have practitioners who claim to have received a divine call to the healing role. However, it is generally the case that the more often a practitioner is required to deal with the supernatural, the more likely it is that the healer claims a sacred calling. Shamans and many midwives claim to have been recruited after being attacked by supernatural beings in the guise of lightning bolts, large snakes, or a *mal aire*, taking hallucinogenic drugs, or more commonly, after experiencing premonitory dreams. Supernatural beings are encountered during these experiences and they tell future healers of their special destiny to heal as well as provide instruction in the medicines they should use. These premonitions are often followed by severe and recurrent illnesses which are thought to be punishment for their not immediately heeding their calling. They only begin to heal reluctantly, usually after another healer reveals to them that they must either begin treating clients or die. This may be followed by an optional apprenticeship to an experienced healer that varies in length from several months to several years, the novices' successful healing of his or her first few patients, and a corresponding increase in clientele and prestige.

Some midwives, and most bonesetters, massagers, and herbalists do not claim a supernatural calling. Instead, they are recruited to the role after apprenticing themselves to an established healer or because an emergency arose (e.g., someone broke a bone, went into labor), and they decided to help out because an experienced practitioner was not available. Regardless of the mode of recruitment, Nahua healers tend to have had opportunities to observe and talk to relatives who are also healers.

The shaman is the practitioner who is generally accorded the most prestige. A shaman is respected for his or her knowledge of medicinal substances and for the considerable skill and courage it takes to negotiate with powerful, supernatural beings. Requests for a shaman's services are always accompanied with gifts of food or drink, and compensated with gifts of food and cash that are generous by local standards. A shaman is also respectfully greeted when met on the street. The status of a shaman, however, is ambiguous. Sometimes, spouses, other family members, neighbors, and residents privately criticize shamans as people who are lazy, interested only in financial gain, and sexually promiscuous. Shamans are also feared because they are thought to have the ability to supernaturally harm as well as heal people. On rare occasions, shamans are killed by a group of angry residents who blame them for the misfortunes they are experiencing.

Midwives receive some of the same criticisms that shamans do but are not thought to intentionally harm people. The better known midwives are respected, and compensated well by local standards. For prenatal visits, attending a birth, and providing postpartum services, they receive the equivalent of 3 to 47 US$, approximately one to fifteen times the daily wage paid to laborers in rural regions. On average midwives make two to three house calls per week. The bonesetter's skills are appreciated but their caseload is generally light (several cases per month or per year). The bonesetter is essentially a farmer who has acquired a medical skill that is valuable at times.

The five main types of Nahua healers can be placed on a scale according to the kind of conceptual model they tend to use to explain and treat illness. Shamans are found near the sacred end of the scale, and bonesetters, herbalists, and massagers are near the secular end. Midwives fall in the middle since they employ both supernatural and naturalistic models for understanding and dealing with pregnancy, childbirth, and the postpartum period.

CLASSIFICATION, THEORIES, AND TREATMENT OF ILLNESS

A formal ethnoscientific study of the way(s) the Nahua classify and conceptualize illnesses has not been undertaken. Nevertheless, Alan Sandstrom (1978, 1989, personal communication) has identified several principles the Huasteca Nahua use to explain and understand disease and health.[5] First, is the notion that balance and harmony in social, psychological, and spiritual matters are necessary for health. Closely related to this idea is a second principle: Extreme and antisocial behaviors that disrupt the social fabric or threaten an orderly universe (e.g., drunkenness, spouse abuse, too much aggression or sex, gossiping, lying, cheating) are believed to attract *ejecamej* (*malos aires*, bad winds) which are disease-causing spirits. A number of the *ejecatl* spirits have compound names that include the word *tlasoli* (filth, refuse, trash) and are thought to originate in tangled underbrush and filthy water. Moreover, the *ejecamej* that cause illness (*cocoliztli*, *enfermedad*, *pestilencia*) can also cause infertility, afflict domestic animals, cause crop diseases, droughts, and floods.

Nahua shamans deal with the above kinds of supernaturally-caused maladies. For example, they treat *mal aire* (a type of spirit intrusion), object intrusion due to sorcery, *mal ojo* (evil eye, a type of witchcraft), and *ahmo tonalcahua* (*perdida de alma*, soul loss). They also treat illnesses with naturalistic causes, such as those due to incorrectly mixing activities or foods that are considered hot or cold. For example, fever (*atonahuistli*, *hueytotonqui*, *calentura*), one of the most frequent illnesses (or symptoms of illnesses) treated by Nahua healers, is often attributed to being exposed to a sharp change in temperature or remaining too long in wet clothes. As can be seen, illnesses with naturalistic causes are also governed by the principle of balance and harmony.

Collectively, the illness symptoms shamans treat include: lack of energy, sadness, fever, vomiting, diarrhea, headache, coughing, soreness, aches, dizziness, swelling, bleeding, rash, alcohol abuse, cramps, depressed fontanel, and lack of appetite. From a biomedical perspective, Nahua shamans treat maladies that include gastrointestinal, respiratory, and cardio-vascular diseases, skin lesions and sub-cutaneous infections, and female reproductive problems.

Shamans diagnose by verbally, visually, and physically examining patients, and through divination. Divination techniques include entering trance-like states after becoming intoxicated with rum, marijuana, and hallucinogenic substances, egg, corn kernel, water vapor, crystal, and dream divination, and pulsing the wrist, neck, temples, waist, and chest. Illnesses with supernatural causes are treated primarily by using rituals (e.g., ritual cleansings, soul callings), during which the shaman may claim to make face-to-face contact with supernatural beings (e.g., rain dwarfs, lightening-bolt spirits), the lost souls of their patients, and spirits of deceased people. In the Huasteca, shamans use cut-paper figures to portray the *yolotl* (heart-soul) of various spirits. Shamans control spirits for the benefit of clients by making offerings to the spirits' heart-souls during curing and disease-prevention rituals. The length of these rituals vary from a few hours for the simplest ones to several days for the most complex. In addition to ritual, shamans prescribe rest, baths, poultices, and teas and infusions made from a wide variety of plant, animal, and mineral substances (Argueta Villamar et al., 1994). They are holistic healers who treat body, mind, and spirit.

Uncomplicated fractures, dislocated joints, musculoskeletal pain and dysfunction, and sprains, cuts, and bruises are generally thought to have a naturalistic cause such as an unfortunate slip or an accidental fall. Nahua bonesetters treat these kinds of maladies by using massage, joint manipulation, herbs, splints, and casts. To arrive at a diagnosis, bonesetters palpate for the point of fracture or dislocation, and in the case of a bilateral anatomic structure, such as a shoulder, may use the healthy shoulder as a normal control against which the painful shoulder can be compared. Bonesetters treat fractures by setting them. Upon being set, uncomplicated fractures, such as those to an arm, may be immobilized with a cast made of reed splints, cloth, and pine resin. Dislocated or painful joints are treated by inducing motion in them. Inducing motion in a joint involves a thrust or a pull which pressures the joint slightly past its normal range of motion in an effort to set a dislocated joint, decrease pain, and increase range of motion. This kind of joint manipulation may be followed by joint massage during which preparations made from medicinal herbs, rum, rubbing alcohol, cooking oil, and iodine are applied to the painful area.

Less research has be undertaken with Nahua massagers and herbalists. Massagers appear to treat the less severe conditions that are also treated by the midwife, bonesetter, and shaman. They treat stomachache, headache, sore throat, and fallen fontanel in addition to massaging pregnant women, women with fallen uterus, and people with bruises and sprains. Some evidently use massage to treat culture-bound syndromes such as soul loss, evil eye, and *mal aire*. Herbalists specialize in making herbal remedies and using them to treat a variety of symptoms including diarrhea, vomiting, fever, tooth ache, malnutrition, coughing, intestinal parasites, dysentery, and some culture-bound syndromes such as soul loss and bewitchment.

SEXUALITY AND REPRODUCTION

The contemporary Nahua disapprove of both premarital and extramarital sexual relationships. Sexually promiscuous men and women are likened to dogs that mate indiscriminately. Premarital sex is disapproved of much more strongly for girls than for boys, and parents and brothers are vigilant of their daughters and sisters. Nevertheless, premarital sexual relationships are not uncommon, and a couple may decide to live together or make plans to marry as the result of a girl becoming pregnant.

Some men regard women as sexually voracious and suspect their wives will be unfaithful if given a chance. Women may suspect the same about their husbands and complain about promiscuous husbands to municipal authorities. Some men seek out prostitutes and find lovers when working outside of their community as migrant laborers and a few women find a lover in their home community (Taggart, 1997).

The Nahua think of planting as analogous to sexual intercourse. Men make holes with a dibble stick in the feminine earth into which they place seed. The Nahuat word to plant (*tatoca*) connotes sexual intercourse. They are aware that pregnancy can result from sexual intercourse. For some Nahua, conception is regarded as seating an infant in the womb (Taggart, 1997, pp. 135–136). In communities with government-sponsored clinics, some women are provided with contraceptives including birth control pills, IUDs, implants, and condoms.

HEALTH THROUGH THE LIFE CYCLE

Pregnancy and Birth

A small percentage of women give birth without the assistance of midwives. The majority, however, request that midwives attend them during childbirth and ask them to provide prenatal and postnatal care. Midwives massage the abdomen and legs of pregnant women during their second and third trimesters in order to make them more comfortable, to determine the position of the fetus and, when necessary, to change it. The latter procedure is known as external version. Some may also use a sash to perform a more rigorous massage known as a *manteada*. A *manteada* consists of placing a sash underneath a pregnant woman's waist while she is lying down. Straddling the client, the midwife pulls firmly on each end of the sash and rocks the woman back and forth. Some clients ask midwives to enter a *temazcal* (sweatbath) with them in order to heat their bodies, and make the bones, ligaments, and muscles of their pelvis more flexible. *Limpias* (ritual cleansings) may also be performed.

Pregnancy is thought to be an illness of sorts by many of the Nahua. Pregnant women may experience lack of energy, headaches, nausea, and chills. Occasionally, husbands of pregnant women have similar symptoms. However, there is no report of Nahua fathers customarily going to bed at the birth of children and simulating the symptoms of labor and childbirth (i.e., the couvade).

Nahua midwives generally attend births in their clients' homes. One or more additional adult women (e.g., the mother-in-law, mother) and, in some cases, the clients' husbands may assist at birth. Herbal teas are administered to speed delivery. Delivery generally takes place with the pregnant woman kneeling or squatting on a *petate* or blanket on the floor. Midwives position themselves behind, beside, or in front of their clients, and apply pressure to their abdomen with their hands. After delivery, the newborn's eyes, nose, and mouth are cleaned, the umbilical cord is tied and cut with a small knife, scissors, or a sharp piece of cane, and the infant is wrapped in a blanket. Following the delivery of the placenta, midwives clean and dress the mother, bind her waist with a sash (*ilpicat*), and place a ball of cloth (*fiador*) underneath it so that additional abdominal pressure is applied. In cases of problematic births, midwives administer medicinal teas made from the dried tail of an opossum, perform a *limpia*, or seek the assistance of other midwives or biomedical practitioners. If the placenta is slow in coming, midwives stimulate their patient's gag reflex by placing the end of a woman's braided hair in her mouth.

Midwives provide post-partum care during two or three visits to the home of the mother and newborn. They bathe the infant, massage and rebind the woman's abdomen with a sash, and wash soiled blankets and clothing. Midwives may also accompany their patients to the sweat bath and use bunches of herbs, whose species varies from community to community, to fan and warm a woman's body. Or, they may perform ritual cleansings instead. In addition to using herbs in the *temazcal*, midwives prepare herbal teas to stop heavy post-partum bleeding and to stimulate milk production as well as to facilitate labor, ease labor pains, and to treat sterility. In some communities, there is a 40-day period of sexual abstention and rest after childbirth known as the *cuarentena*. In practice, rest and post-partum sex taboos vary in length from 40 days to six months or longer (Taggart, 1997, p. 234). In the Sierra Norte de Puebla and the Sierra de Zongolica (Veracruz), recently delivered women eat special foods to help keep them warm and regain their strength.

The Nahua of the Sierra Zongolica thank midwives for attending births at the end of the *cuarentena* during a hand- and house-cleansing ceremony. In the Huasteca, Nahua midwives make offerings to the earth spirit after a woman has given birth. Offerings are made because there is a concern about offending this spirit with the afterbirth, blood, and amniotic fluid of a recently delivered woman. Nahua midwives from this region also play an important role in village-wide religious ceremonies such as the *Tlacatelilis*, a winter solstice ceremony meaning to cause to be born, and the *Xochitlalia*, a crop fertility ritual (Sandstrom, 1991; Sandstrom & Sandstrom, 1986). Xochitlalia can be literally translated as to place flowers, but its metaphorical meaning is probably to seat something delicate or precious, thus paralleling the Nahua view of conception.[6]

There are a number of food and behavioral taboos related to pregnancy. Pregnant woman should not travel about at night, especially during a lunar eclipse, since the latter can lead to an infant being born with a cleft palate. Nor should pregnant women view dancers at a fiesta because their children may be born with a face similar to the masks that dancers wear. Working hard while pregnant is to be avoided; it can lead to a miscarriage. Weaving or embroidering while pregnant is also dangerous because it can result in a child being born with its umbilical cord wrapped around its neck. Collectively, these taboos, if observed, would encourage pregnant woman to remain close to home at night, work in moderation, and avoid fiestas.

Some pregnant women have food cravings that husbands are encouraged to satisfy. These cravings are attributed to the fetus. If they are not satisfied, a miscarriage is thought to result. Giving birth to twins is attributed to a pregnant women eating twin fruits, such as double bananas or plums. The color and number of bumps on the umbilical cord can be counted to divine the number of boys and girls a woman will bear in the future.

The sex of a child can be divined by feeling for the position of the fetus while massaging a pregnant woman. It is thought that a couple's adult children will always live close to them if their placentas are buried beneath the floor of the house. The sheer number of beliefs surrounding childbirth, and this is by no means an exhaustive list of them, show the Nahua's concern with having many healthy children.

Infancy

Babies are carried with a shawl on their mothers' backs and breast-fed on demand for the first six months of their lives. In some parts of the Sierra Norte de Puebla, a mother applies a bitter herb (*chichicxihuit*) to her nipples to discourage a child from breast-feeding. Weaning foods (e.g., *atole*, tortillas, beans) are gradually introduced until the child is fully weaned, which occurs about the sixth month of a woman's next pregnancy (Taggart, 1997, pp. 234–235). By age 2 or 3, children eat the full repertoire of local foods, and begin to talk and walk. There is a widespread belief that a woman's children, especially the youngest one, may become *tzipititoc* when she becomes pregnant again. A child who is tzipititoc is said to ache, lack an appetite, and cry constantly. These symptoms are attributed to the jealousy a young child feels toward the expected baby.

Some Nahua mothers bottle-feed infants and serve cow's milk to their children. However, many Nahua children are probably lactase-deficient to some degree, and experience diarrhea, abdominal pain, flatulence, and bloating after consuming cow's milk (Cifuentes & Limón, 1985).

Nahua healers indicate that there are a number of illnesses that infants are likely to suffer. Those characterized by respiratory problems include *opatzmigui* (*oguío*, bronchitis), *mitetaxis* (*tos ferina*, whooping cough), and *pulmonía* (pneumonia). Others are characterized by fever (e.g., *aferecía*, epilepsy), and diarrhea, vomiting, and other kinds of gastro-intestinal distress including *ahuetzi* (*caída de mollera*, fallen fontanel), *caída de cuajo* (fallen or dislocated cuajo, an organ believed to be inside the stomach), *cólicos* (colic), *ocuiloua* or *chincual* (worms), *mal de ojo* (evil eye), and soul loss.

In a number of cases, these illnesses are attributed to something that happened to the mother or something the mother did while she was pregnant, for example inappropriate conduct, becoming angry frequently, eating excessive amounts of lard or sweets. In other cases, such as in Tlaxcala and the Sierra Norte de Puebla, a child's illness or death is attributed to nocturnal animals or witches who suck the blood of newborns. Babies and small children of all regions are thought to be especially at risk for *mal de ojo*, a culture-bound syndrome brought on by the strong or envious gaze of admirers. The affliction is caused involuntarily.

In communities that government medical teams visit or have government-sponsored clinics, infants and children receive vaccinations against diseases such as measles, tetanus, and diphtheria, and are given vitamin A. Nevertheless, the infant mortality rate is still relatively high during infancy and many Nahua parents intentionally wait to name their infants until they are reasonable sure of their survival, for example, at 6 to 12 months of age.

Childhood

Childhood begins around age 4 and lasts until the child reaches 13 or 14 years of age. In some villages, anthropometric data suggest that 50 to 70% of preschool and school-age children are underweight and malnourished. Intestinal parasites are also common in children. Biomedical practitioners treat children (and adolescents and young adults) for acute respiratory conditions (e.g., pneumonia, bronchitis, tuberculosis), diarrhea, and wounds sustained in accidents that occur at home, at school, at work, and while traveling by taxi, truck, or bus. Males are treated for wounds at a higher rate than are females.

Nahua healers indicate there are a number of culture-bound illnesses that children suffer, including *apisuilo* (*tiricia*, jaundice), which is characterized by yellowish skin and eyes and attributed to a variety of causes including being reprimanded continually or being a victim of envy. *Netatil* (*quemadas*, burns) is another culture-bound syndrome that children as well as adult men are thought to suffer. It is attributed to close contact with a woman who recently gave birth, and characterized by a rash or yellowish skin. Urinary infections (*mal de orín*) and eye irritations (*ixtemtoleulixli*, *mal de los ojos*) are also common among children.

Adolescence and Adulthood

Adolescent boys and girls are generally given no parental instruction in or preparation for sexual activity and the sexual aspects of adulthood. Girls begin to menstruate around age 15, and menstrual blood is thought to be particularly dirty. At this age or even before, a girl's mobility, both within and outside of her natal village is severely restricted in some communities. Nevertheless, some adolescents have their first children when they are in their early and middle teens. Most marry and begin to have children in their late teens and early 20s.

Men find medium-weight women to be more attractive than thin ones because the former are considered healthier, stronger, and better able to stand up to demanding household chores. Both men and women tend to gain weight as they mature, but men tend to be thinner due to the more strenuous physical activity required by the kind of work they do. There is no mention of healers treating people in Nahua communities for obesity but doctors do treat a small percentage of adult Nahua patients for diabetes.

In general, both men and women desire children; couples with 6–8 children are not unusual. In some communities it is believed that God punishes women who remain childless. Childless women may be likened to men and may be pitied or severely criticized. Women are often blamed for infertility, a condition attributed to a woman bathing with cold water or burying a doll while she was a young girl. Infertility is grounds for divorce in some communities.

Women are thought to suffer vaginal hemorrhages because they lift heavy loads when pregnant or due to placenta previa. *Caída de matriz* (fallen uterus) is attributed to excessive work. It is thought to be a "loosening" and "displacement" of the uterus from its normal position in the abdomen. *Recaída de parturienta* ("post-partum relapse") is an illness experienced by young women, especially first-time mothers, after having given birth. It is characterized by fever, headache, diarrhea, and joint and stomach pain, and is attributed to working or being frightened during the *cuarentena*.

Some of the illnesses that adults experience are attributed to their working very hard, often under difficult conditions. *Envaramiento* (stiff tendons) is an illness thought to result from bathing an excessively hot body too soon after working hard for a long period of time. It is characterized by muscular pain throughout the body. *Estiramiento de cuerdas* (stiff muscles) is characterized by intense muscular pain due to straining oneself while working. *Garrotillo* ("croup") is an illness that adults suffer after they have worked outside for a long time under a strong sun. Its symptoms include an intense headache, fever, a "dry" cough, sore throat, and leg and arm pain. *Cabeza abierta* ("open head") is attributed to carrying heavy sacks with a tump line or receiving a heavy blow to the head.

The Aged

People are considered older adults when they are in their late 50s and 60s. Most men and women remain active until their deaths, and it is not unusual to see 70 or 80-year-old men farming or elderly women weaving, cooking, and washing clothes. Men and woman expect their children to care for them when they become elderly. The very old (and the very young) are thought to be relatively weak, and especially prone to illness, including *mal aire* and soul loss. Sometimes, couples are faced with the difficult decision of allocating scarce financial resources to treat a sick elderly parent or grandparent, or to purchase food and clothing for their children.

Dying and Death

When a child or adult dies, a doctor may be summoned to determine the cause(s) of death, especially if the municipality has a government-sponsored clinic. The deceased usually lies in state for a day, and is washed and dressed by a widow. A married woman does not handle the deceased. If she handles a corpse, she might bring the spirit of the deceased to her home where it could attack her husband and children, and cause *mal aire*. The deceased is generally buried the day following death, after a wake has been held during the night. If a municipality has a cemetery, burials within it are regulated by a councilman, topiles (local police), or relatives of the deceased.

Beliefs concerning the afterlife are variable and depend upon the extent to which a community has been influenced by Catholicism over the past 500 years and by 20th century Protestant missionaries. In many acculturated communities, the fate of the soul is thought to depend upon an individual's conduct during life, including his or her meeting of religious obligations. In the Sierra Norte de Puebla, for example, souls of people who conducted themselves well go to heaven. Souls of infants who die before being baptized go to limbo and cannot see God; sinners become the slaves of the Devil.

In less acculturated communities, beliefs surrounding death and the fate of an individual's soul show more continuity with those of the prehispanic period. In the Huasteca, a person's *yolotl* soul generally travels to *Mictlan*, an underworld place of the dead. However, souls of those who die from water-related causes (e.g., drowning) go to a watery paradise (*Apan*). Disease-causing wind spirits are thought to be the spirits of people who die prematurely.

CHANGING HEALTH PATTERNS

There are a number of changes related to health and illness in Nahua communities. Visitations to biomedical personnel, acceptance of the germ theory, and the use of pharmaceuticals are increasing. In some communities, shamans prescribe medications that their clients are instructed to purchase in a pharmacy. Healers are also now much more likely to charge a fixed fee for their services than they were in the past. In the past, healers accepted as compensation whatever their patients could afford to give them. In addition, there have been changes in beliefs and practices. The use of the temazcal has declined dramatically, so much so that some people are no longer repairing or building them. Many people question the belief that burying the placenta inside the home will assure them that their children will always remain closeby. They observe that adult children are migrating to and living in distant cities and towns regardless of how carefully they dispose of the placenta.

In general, doctors and nurses are reluctant to work with shamans. Biomedical practitioners as well as schoolteachers, priests, missionaries, and village authorities often belittle or seek to eliminate shamans and shamanistic practices and beliefs. This is leading to a reduction in the number of Nahua shamans and a narrowing of the scope of their healing role. However, Mexico's National Indian Institute (*Instituto Nacional Indigenista*, or INI) has encouraged Native American groups throughout the country to form organizations for traditional medical practitioners. These organizations are designed to reinforce the knowledge of traditional healers, legalize their practice, and provide people with an alternative to the formal health sector. There are at least nine such organizations to which Nahua shamans and other types of healers belong: four in the state of Puebla, three in the Huasteca region of the states of San Luis Potosí and Veracruz, and one each in the states of Guerrero and Michoacán. In the case of the organization whose center is in Cuetzalan, Puebla,

INI doctors work with Nahua and Totonac healers in a hospital where both biomedical and traditional medicine are practiced. It remains to be seen how much impact INI and these organizations will have on the maintenance and promotion of shamanic healing practices.

Nahua childbirth beliefs and practices have been changing dramatically over the past 25 years. Mexico's Ministry of Health (*Secretaría de Salubridad y Asistencia*, or SSA) and Social Security Institute (*Instituto Mexicano de Seguro Social*, or IMSS) have established a large number of health clinics and hospitals in rural and urban areas. It is increasingly common for pregnant Nahua women to seek out the services of these institutions. Some Nahua women prefer doctors and nurses because they administer injections to control pain during childbirth and because they charge relatively little for their services. At the same time many Nahua midwives have received biomedical training from the SSA and IMSS. As a result, the use of a sharp piece of cane to cut the umbilical cord, cauterization of the cord end, and the kneeling birth position are declining in popularity while the use of pincers to cut the umbilical cord, a sterile tie, gauze, and tape to wrap the cord, and alcohol to sterilize hands and equipment are increasing. INI has promoted the continued use of the *temazcal* and teas made from local medicinal plants for use before, during, and after childbirth. Nevertheless, Nahua midwives are gradually being incorporated into Mexico's biomedical health care system. Many Nahua midwives are essentially auxiliary health workers whose primary role is to funnel pregnant women to SSA and IMSS clinics and take orders and training from doctors and nurses.

Notes

1. The author would like to thank Alan R. Sandstrom (Anthropology, Indiana University-Purdue University Fort Wayne) and James M. Taggart (Anthropology, Franklin & Marshall College) for critically commenting on this article. The author, of course, is solely responsible for any errors of fact or interpretation.
2. The way in which Nahua terms were rendered in the original sources has been preserved. There is no standard orthographic system used by all Nahua scholars.
3. Many of the Nahua, especially school-aged children, young adults, and men speak Spanish as a second language.
4. The Nahua, Spanish, and English terms used for different types of healers and illnesses are terms of reference rather than strict translations.
5. To varying degrees, these principles are applicable to Nahua living in other regions and to many other Mesoamerican groups as well.
6. The author thanks James M. Taggart for pointing out that "*tlalia*" means "to seat or place", and that a parallel might be drawn between conception and the Xochitlalia ritual.

References

Álvarez Heydenreich, L. (1987). *La enfermedad y la cosmovisión en Hueyapan, Morelos. (Colección INI, No. 74)*. México: Instituto Nacional Indigenista.

Álvarez Heydenreich, L. (1992). Tipos de curanderos en Hueyapan, Morelos. In R. Campos Navarro (comp.), *La antropología médica en México, Volume 2* (pp. 127–138). México, DF: Universidad Autónoma Metropolitana.

Argueta Villamar, A., Cano Asseleih, L. M., & Rodarte, M. E. (1994). *Atlas de las plantas de la medicina tradicional mexicana. (3 Volumes)*. México, DF: Instituto Nacional Indigenista.

Bonfil Batalla, G. (1968). Los que trabajan con el tiempo: Notas etnográficas sobre los graniceros de la Sierra Nevada, México. *Anales de Antropología, 5*, 99–128.

Cifuentes, E., Flores, J. J., & Limón, N. E. (1985). Deficiencia de lactasa intestinal en un pueblo Nahua: Alternativas para los programas de intervención nutricional en la región. *La Revista de Investigación Clínica (Mexico), 37*, 311–315.

Dakin, K. (2001). Nahuatl. In D. Carrasco (Ed.), *The Oxford encyclopedia of Mesoamerican cultures: The civilizations of Mexico and Central America, Volume 2* (pp. 363–365). New York: Oxford University Press.

Dow, J. W., & Van Kemper, R. (1995). Nahua peoples. In J. W. Dow & R. Van Kemper (Eds.), *Encyclopedia of world cultures, Volume VIII: Middle America and the Caribbean* (pp. 182–183). Boston: G. K. Hall.

Emes Boronda, M., Ochurte Espinoza, C., Castañeda Silva, G., & Peralta González, B. et al. (1994). *Flora medicinal indígena de México: Treinta y cinco monografías del Atlas de las plantas de la medicina tradicional mexicana. (3 Volumes)*. México, DF: Instituto Nacional Indigenista.

Huber, B. R. (1990). The recruitment of Nahua curers: Role conflict and gender. *Ethnology, 29*, 159–176.

Huber, B. R., & Anderson. R. (1996). Bonesetters and curers in a Mexican community: Conceptual models, status and gender. *Medical Anthropology, 17*, 23–38.

Huber, B. R., & Sandstrom, A. R. (2001). The Recruitment, training, and practice of midwives from the United States-Mexico border to the Gulf of Tehuantepec, In B. R. Huber & A. R. Sandstrom (Eds.), *Mesoamerican healers* (pp. 139–178). Austin, Texas: University of Texas Press.

Huber, B. R., Sandstrom, A. R., & Toribio Martínez, A. (forthcoming). Transformations in the recruitment, training, and practice of midwives in a Nahuat-speaking community of Mexico, In M. Good Maust & M. Güémez Pineda (Eds.), *Mexican midwives: Change, continuity and controversies*. Austin, Texas: University of Texas Press.

Instituto Nacional Indigenista. (2001). *Información básica sobre los pueblos indígenas de México; National profile of the indigenous peoples of Mexico; Pueblos Nahuas de la Huasteca*. México, D. F. (March 19, 2001); http://www.ini.gob.mx.

Lewis, Oscar. (1963). *Life in a Mexican village: Tepoztlán restudied.* Urbana: University of Illinois Press.
Madsen, C. (1968). A Study of Change in Mexican Folk Medicine. In M. S. Edmonson, C. Madsen, & J. F. Collier (Eds.), *Contemporary Latin American culture. Middle American Research Institute, Publication 25* (pp. 92–137). New Orleans: Tulane University.
Madsen, W. (1955). Shamanism in Mexico. *Southwestern Journal of Anthropology, 11,* 48–57.
Mellado Campos, V., Sánchez Reyes, A., Femia, P., Navarro Magdaleno, A., Erosa Solana, E., Bonilla Contreras, D. M., & Domínguez Hernández, M. del S. (1994). *La medicina tradicional de los pueblos indígenas de México (3 Volumes).* México: DF: Instituto Nacional Indigenista.
Mellado Campos, V., Zolla, C., & Castañeda. X. (with Antonio Tascón Mendoza). (1989). *La atención al embarazo y el parto en el medio rural mexicano.* México, D.F.: Centro Interaméricano de Estudios de Seguridad Social.
Montoya Briones, J. de J. (1964). *Atla: Etnografía de un pueblo Nahuatl.* México: D.F.: Instituto Nacional de Antropología e Historia.
Nutini, H. G., & Forbes de Nutini, J. (1987). Nahualismo, control de los elementos y hechicería en Tlaxcala rural. In S. Glantz (Ed.), *La heterodoxia recuperada en torno a Angel Palerm* (pp. 321–346). México, DF: Fondo de Cultura Económica.
Nutini, H. G., & Isaac, B. L. (1974). *Los pueblos de habla Nahuatl de la región de Tlaxcala y Puebla.* México, DF: Instituto Nacional Indigenista.
Ramírez Celestino, C. (1991). *Plantas de la región Nahuatl del centro de Guerrero.* Tlalpan, México: Centro de Investigaciones y Estudios Superiores en Antropología Social.
Redfield, M. P. (1928). Nace un niño en Tepoztlán: A child is born in Tepoztlán. *Mexican Folkways, 4,* pp. 102–108.
Redfield, R. (1930). *Tepoztlán, A Mexican village: A study of folk life.* Chicago: University of Chicago Press.
Robinson, D. F. (1961). Textos de medicina Nahuat. *American Indígena, 21,* pp. 345–353.
Sandstrom, A. R. (1978). *The image of disease: Medical practices of Nahua Indians of the Huasteca. Monographs in anthropology. no. 3.* Columbia: Dept. of Anthropology, University of Missouri-Columbia.
Sandstrom, A. R. (1983). Paper dolls and symbolic sequence: An analysis of a modern Aztec curing ritual. *Folklore Americano, 36,* pp. 109–126.
Sandstrom, A. R. (1989). The face of the devil: Concepts of disease and pollution among Nahua Indians of the southern Huasteca. In G. Stresser-Péan & D. Michelet (Eds.), *Enquêtes sur l'Amérique moyenne: Mélanges offerts à Guy Stresser-Péan* (pp. 357–372). México: Instituto Nacional de Antropología e Historia, Consejo Nacional para la Cultura y las Artes, & Centre d'études mexicaines et centraméricaines.
Sandstrom, A. R. (1991). *Corn is our blood: Culture and ethnic identity in a contemporary Aztec Indian village.* Norman: University of Oklahoma Press.
Sandstrom, A. R. (1995). Nahua of the Huasteca. In J. W. Dow & R. Van Kemper (Eds.), *Encyclopedia of world cultures, volume VIII: Middle America and the Caribbean* (pp. 184–187). Boston: G. K. Hall & Co.
Sandstrom, A. R. (2000). Contemporary cultures of the Gulf Coast. In J. D. Monaghan (Ed.), *Supplement to the Handbook of Middle American Indians. Volume 6, Ethnology* (pp. 83–119). Austin: University of Texas Press.
Sandstrom, A. R., & Sandstrom, P. E. (1986). *Traditional papermaking and paper cult figures of Mexico.* Norman: University of Oklahoma Press.
Signorini, I. (1989). *Los tres ejes de la vida: Almas, cuerpo, enfermedad entre los nahuas de la Sierra de Puebla.* Xalapa, Veracruz, Mexico: Universidad Veracruzana.
Taggart, J. M. (1995). Nahuat of the Sierra de Puebla. In J. W. Dow & R. Van Kemper (Eds.), *Encyclopedia of world cultures, volume VIII: Middle America and the Caribbean* (pp. 190–193). Boston: G. K. Hall & Co.
Taggart, J. M. (1997). *The Bear and his sons: Masculinity in Spanish and Mexican folktales.* Austin: University of Texas Press.
Taggart, J. M. (2001). Nahua. In D. Carrasco (Ed.), *The Oxford encyclopedia of Mesoamerican cultures: The civilizations of Mexico and Central America, Volume 2* (pp. 359–363). New York: Oxford University Press.
Unidad Regional de Acayucan. (1983). *Ciclo de vida de los nahuas. Cuadernos de Trabajo, Acayucan, No. 22.* México: Dirección General de Culturas Populares.
Vexler, M. J. (1981). Chachahuantla, A blouse-making village in Mexico: A study of the socio-economic roles of women. Doctoral dissertation, University of California, Los Angeles, 1981.

Navajo

Joanne McCloskey

ALTERNATIVE NAMES

Diné, means "The People" in Navajo, and is often preferred throughout the Navajo Nation.

LOCATION AND LINGUISTIC AFFILIATION

Located in the Four Corners area of the Southwestern United States, the Navajo Nation occupies 26,649 m^2 on the Colorado Plateau in portions of Arizona, New Mexico, and Utah. The language spoken is Navajo, an Athapaskan language.

OVERVIEW OF THE CULTURE

Relative to the Pueblo Indians, the Navajo were latecomers, arriving in the Southwest sometime before or around 1500 AD (Brugge, 1983). Originally hunter–gatherers, they adopted agriculture from the Pueblos and sheep and horses from the Spanish. The Navajos thrived as pastoralists and agriculturists, moving their sheep with the changing seasons between winter camp and summer camp where they raised corn, squash, beans, and melons.

After their release from imprisonment at Bosque Redondo in Ft Sumner, New Mexico in 1868, the Navajo continued their reliance on livestock and crops until the 1930s. The initiation of livestock reduction by the commissioner of Indian Affairs, John Collier, in 1933 sounded the death knell for the land-based economy. Unable to support themselves with the drastically reduced herds, Navajos had no alternative but to seek work in the labor force. At first Navajo men worked, often at off-reservation jobs on the railroad, ranches, in construction and mining, and as migrant field hands. Women stayed at home to maintain the fields and reduced herds, while weaving rugs to sell at trading posts. With the increase in reservation schools, hospitals, agencies, and stores after mid-century came expanded opportunities for Navajo women to enter the labor force.

The foundation of Navajo social organization rests on the matrilineal kinship system. The basic principle of k'é, meaning relationships comprised of kindness, love, cooperation, thoughtfulness, friendliness, and peacefulness (Morgan, F., 2002; Witherspoon, 1983), guides interactions among family members in the extended matrilineal kinship network and among clan members. Besides the primary affiliation with his mother's clan, a Navajo also identifies with his father's clan as his "born for" clan and with the clans of his maternal and paternal grandfathers. Egalitarian relationships exist between Navajo men and women, a reflection of the premise that all entities and aspects of life consist of male and female components.

Fundamental to Navajo philosophy and religion is the concept of hózhǫ́. Not easily translated from Navajo, hózhǫ́ involves everything that a Navajo thinks is good or favorable in the world (Morgan, F., 2002; Wyman, 1983). Hózhǫ́ means goodness everywhere and beauty in the sense of harmony and balance (Morgan, F., 2002). The fundamental purpose of Navajo traditional ceremonies is to restore a state of hózhǫ́ to the individual through songs, prayers, and ritual activities. Navajo ceremonies focus on healing the patient, "the one sung over," who may have a physical illness, a mental ailment, or distress brought about by improper contact with the natural or spiritual world.

Formerly, traditional leaders, naat'áanii, were selected on the basis of their wisdom, traditional knowledge, community citizenship, membership in a respected family, and the ability to successfully live in harmony and to conduct at least one ceremony (Office of Navajo Government Development, 1998). Replacing the traditional political system, the election of a tribal council began in 1923 as a way to authorize oil leases to extract oil discovered on Navajo lands. The Navajo Nation consists of 110 grassroots organizations known as chapters, which elect representatives to the tribal council.

THE CONTEXT OF HEALTH: ENVIRONMENTAL, ECONOMIC, SOCIAL, AND POLITICAL FACTORS

Three distinct climatic environments contribute to the striking beauty of the Navajo landscape: humid, steppe, and desert. Mountains rising between 7,500 and 10,416 ft characterize the high elevations of the relatively small humid area, including the Chuska and Carrizo mountains, the Fort Defiance Plateau, the highest ridges of Black Mesa, and Navajo mountain. At the intermediate elevations between 6,000 and 8,000 ft, the pinon and juniper-covered mesas and high plains dominate the steppe area, which makes up about one third of the total land area. Over half of the Navajo Nation lies between 4,500 and 7,000 ft, where sagebrush and grasses grow in a desert climate (Linford, 2000; Young, 1961).

The abundant energy resources on Navajo land—oil, natural gas, coal, and uranium—constitute a double-edged sword for the Diné. Although the extraction of energy resources provide an important source of tribal revenue and employment (Choudhary, 2001), major profits benefit large corporations, which operate under leases negotiated with the Navajo Nation. "In sum, Navajos have no control over their nonrenewable energy resources; profits from extracting and processing them flow elsewhere" (Aberle, 1983a, p. 651).

The 2000 U.S. census counted 173,987 Navajos residing within the Navajo Reservation (Choudhary, 2003), making the Navajos the largest reservation-based tribe in the United States. The Navajo population is young, with a median age of 24.0 years. Despite significant recent gains in educational attainment, unemployment and poverty plague the Diné. The 2000 census reported that 54.22% of the Navajo Nation population 25 years or older had a high school degree yet 42.90% of persons live below the poverty level (Choudhary, 2003). The unemployment rate for the Navajo Nation in the year 2000 was 43.4% (Choudhary, 2001). The annual per capita income for 1999 was $6,217 compared with $28,542 for the entire United States (Choudhary, 2001).

In 1990 most Navajos, 82%, spoke their native language at home. Also in 1990, 51.6% of housing units lacked complete plumbing, and 77.5% of occupied housing units were without a telephone. The average life expectancy at birth is 72.7 years compared with 75.8 years for the entire U.S. population (Navajo Area Indian Health Service, 2002).

MEDICAL PRACTITIONERS

Navajo medical practitioners include diagnosticians, healers, and herbalists. Three types of traditional diagnosticians are stargazers, listeners, and hand tremblers, of which hand tremblers are by far the most common. After praying to Gila Monster, one of the Holy People, a hand trembler goes into a mild trance (Milne & Howard, 2000). Hand tremblers may be either male or female, with a predominance of females among practitioners who function only as diagnosticians (Adair, Deuschle, & Barnett, 1988). In addition, road men in the Native American Church also act as diagnosticians. Although not all road men diagnose, many use charcoal gazing, the second most common method of diagnosis (Milne & Howard, 2000).

Three primary types of Navajo healers are medicine men, or singers, who conduct traditional ceremonies; road men who lead Native American Church meetings; and pastors of fundamentalist Christian churches who perform faith healing. In addition, western biomedical practitioners may be used alone or in combination with religious forms of healing. Similar to the many years of education required to prepare medical doctors, medicine men undergo many years of apprenticeship to learn a single ceremony. Depending upon the apprentice's other activities and the ability to memorize hundreds of songs, prayers, and rituals, the learning period may last as long as 15 years (Morgan, W., 1977).

Herbalists, who use their vast knowledge of local flora for medicinal as well as ceremonial purposes, are usually elders. They may use their herbal remedies only for members of their extended families or their cures may be in demand for a wider group. As an example, an herbalist from the Crownpoint area on the eastern portion of the reservation uses her knowledge of plants to prevent births, to facilitate an easy childbirth, and to treat sores, bruises, the common cold, and broken bones.

Some Navajos seek cures from healers belonging to other tribes. For example, on the western portion of the reservation, Navajos may travel to Hopi Indian villages for treatment to extract objects from the patient's body that cause illness and problems (Levy, 1983; Morgan, F., 2002).

Classification of Illness, Theories of Illness, and Treatment of Illness

Classification of Illness

Etiologic agents believed to cause disease include animals, insects, natural phenomena, mistakes made during healing ceremonies, contact with ghosts of deceased humans, and contact with living or dead foreigners, particularly enemies (Wyman & Kluckhohn, 1938). Among the animals and insects that cause illness, the most frequently identified are the bear, coyote, porcupine, snake, eagle, moth, ant, long-horned grasshopper, and camel cricket. The most commonly named natural phenomena responsible for illness are lightning and whirlwinds (Levy, 1983). Failure to observe the taboo regarding abstinence from sex during or four days after a ceremony can cause illness. Illness can result from contact with the dead or from the location where a person died. Because pregnant women and their husbands are particularly vulnerable, they cannot attend funerals.

These etiologic factors may lead to a variety of symptoms and illnesses. There is no direct correlation between a causative agent and a disease process. "A single cause may produce any of several symptoms and conversely, a single symptom may be caused by any of several etiologic agents." Likewise, "No Navajo disease is known by the symptoms it produces, or by the part of the body it is thought to affect" (Levy, 1983, p. 132). For instance, bear sickness and porcupine sickness have been caused by contact with these animals. The time frame for symptoms to appear after exposure may loosely be sooner or later. For example, a diagnostician may find that an adult's health problem originated during gestation because either parent came in contact with lightning or a dead person.

Theories of Illness

The ultimate source of illness is a lack of harmony between an individual and the natural or supernatural world. To restore health, evil must be driven out and replaced with good (Adair, Deuschle, & Barnett, 1988; Reichard, 1974). Rather than treating the physical symptoms of an illness, a ceremony focuses on exorcising the intrusive factors (Wyman, 1983). Purification rites, such as cleansing, sweating, taking an emetic, and brushing, remove the evil from the patient's body (Reichard, 1974). Also, the unraveling portion of a ceremony symbolizes a patient's release from evil and danger (Wyman, 1983). The removal of evil from the patient and the depiction of deities in a sandpainting attract the deities, the Holy People, to help the patient.

If a singer recites prayers, sings songs, and performs rituals according to the dictates of the Holy People, "the Holy People are compelled to participate and restore the one-sung-over to hózhǫ́" (Frisbie, 1987, p. 6). Each ceremony recounts the story of the exploits of one or more of the many Holy People who acquire the specific ceremony from the supernaturals and then make it available to the earth people (Spencer, 1957). These powerful stories guide and empower the patient as he or she identifies with the deities in the stories.

Treatment of Illness

Three types of Navajo healing, traditional ceremonies, Native American Church, and Christian faith healing, together with western biomedicine make up the contemporary Navajo health care system (Csordas, 2000). Traditional Navajo healing employs a didactic approach through rituals, songs based on stories of obstacles overcome by the Holy People, and prayer that immerses the patient in "the contextualization of life experience within a cosmological and physical 'home,' the Navajo land and people" (Csordas, 2002, p. 167). Crucial to the successful performance of ceremonies is the perfect recitation of prayers. Themes found in prayers, songs, and sandpaintings include the directions, colors, jewels, and simple numbers like four and its multiples (Reichard, 1944). Besides reciting prayers and songs and performing or directing the ceremonial rituals, medicine men talk to patients, emphasizing the positive (Carrese & Rhodes, 2000). As one man told how a medicine man had helped him, "They don't just sing. They also talk to you" (McCloskey, 1998b, p. 22). All participants in the hogan pray with corn pollen by taking a pinch, then putting some in their mouths, some on top of their head, and trailing the remainder in an upward gesture. Knowledge of the Navajo language is pivotal to understand and appreciate the progression of events during an all-night ceremony. One participant stated that he was "too absorbed in the beauty of what's going on even to think about sleep" (Csordas, 2002, p. 168).

Navajo ceremonies may be divided into two major groups, chantways, in which the singing is accompanied by a rattle, and all others, which are rites. Three kinds of chantways are Holyway, Evilway, and Lifeway ceremonies. While Holyway ceremonies are concerned with the attraction of good, Evilway ceremonies must first expel evil influences before invoking the good. Lifeway ceremonies treat injuries suffered from accidents (Wyman, 1983). Although the Blessingway and Enemyway are grouped together as rites, these two ceremonies are distinct from each other. In the Blessingway rite is found the backbone of Navajo philosophy, and from the Navajo perspective it is identical with Navajo culture (Farella, 1984). The purpose of a Blessingway rite is to promote the positive blessings of life, while an Enemyway rite exorcises alien ghosts (Wyman, 1983).

The Navajo pantheon of supernatural beings, the Holy People, Diyin Dine'é, are the protagonists of stories on which ceremonies are based, and they figure prominently in the songs and prayers of all ceremonies. Changing Woman, unlike other Holy People who have good and bad qualities and must be persuaded to help restore the patient to a state of good health, epitomizes goodness exclusively. Blessingway recounts the story of Changing Woman's discovery by First Man, her puberty ceremony, and the adventures of her twin sons, Born-For-Water and Monster Slayer, who were fathered by the Sun. During the Blessingway ceremony, the patient holds the mountain soil bundle, a medicine bundle containing stones and soil from each of the four sacred Navajo mountains.

The second major Navajo healing tradition is the Native American Church (NAC), also known as the peyote religion because of the sacramental use of peyote, a cactus that contains mescaline. The NAC employs the therapeutic principle of confession and seeks to build patients' self-esteem (Csordas, 2002). NAC road men work with patients in the same manner as do medicine men, talking to them in a positive, supportive way.

The NAC combines elements of Christianity with aspects of Plains Indian religions. Not until the 1930s did Navajos begin embracing the NAC, when Shiprock area Navajos were exposed to NAC meetings held by Ute Indians. The new religion appealed to the Navajo after the trauma of livestock reduction. The early Navajo NAC faced opposition from the Tribal Council, which made the NAC illegal in 1940, a statute that was not overturned until 1967 (Aberle, 1983b). However, "By the 1970s, the Native American Church was seen by most Navajo people as simply another chantway, 'azee'jí or Medicine Way" (Wyman, 1983, p. 536). By the 1980s, estimates of the number of Navajos participating in the NAC were as high as 40–50% of the population (Quintero, 1995). Although some conservative traditionalists do not participate, many traditional Navajos attend both ceremonies and NAC meetings.

A peyote meeting may be held in a specially constructed teepee or in a hogan also used for ceremonies. Aided by the power of peyote, participants communicate with God, the supreme being said to be the same God worshipped everywhere. A meeting may be held to cure a sick patient or to solve an economic, social, or legal problem, but also meetings are held to assure success in school, to marry a couple, or to celebrate birthdays and holidays.

Beginning in the late evening and lasting until morning, an NAC meeting consists of four segments: the opening, the midnight water ceremony, the morning water ceremony, and the peyote breakfast. Besides the road man in charge of the meeting, other officiants are the fireman, the drummerman, and the cedarman. Events throughout the night include prayers, songs, ritual smoking, and ingestion of peyote in the form of a tea, whole, dried, or powdered. The early portion of the meeting focuses on sin, suffering, and shortcomings, which later shifts in the early morning hours to an emphasis on hopes for the future and the right way to live (Aberle, 1983b).

The pairing of male and female pervades all aspects of Navajo life and thought.

In brief, the male aspect of men, women, mountains, storms, or just about anything else is considered to be "powerful" and in need of restraint. As a corollary, the female aspect of everything in the universe is characterized as restraint. A balance may be achieved by the proper coupling of these two essences. (Quintero, 1995, p. 79)

An NAC meeting, likewise, incorporates male and female elements. The early part of the ceremony, when the focus is on the patient's problems, is the male part of the ceremony, while the latter female portion shifts emphasis to renewal and healing. Consistent with this symbolic division, the male fireman conducts the midnight water ceremony whereas a woman brings in the water for the morning water ceremony (Quintero, 1995).

Christian faith healing by Navajo practitioners is a significant presence on the Navajo Nation. At fundamentalist churches and in revival meetings held in tents during the summer months, Navajo pastors lead services that incorporate faith healing. The congregation becomes the family that helps to effect a cure through spontaneous prayers to God. Besides prayer, healing techniques include the laying on of hands, anointing with oil, reading from the Bible, and counseling (Lewton & Bydone, 2000). Fundamentalist healers demand that Navajo members give up sinful traditional and NAC practices, even requiring that participants destroy their jish, or medicine bundles (Csordas, 2002).

Common elements in all three types of Navajo healing and practitioners' approaches are the goal of helping the patient to achieve a culturally shared way of understanding, the lack of division between physical and mental problems, and the emphasis on the family. Navajo practitioners work to achieve success with patients by talking to them so they understand in the sense of incorporating a Navajo specific view of life experiences. Instead of sharp distinctions between physical and mental afflictions, Navajo illness may arise from a single source that may be expressed either physically or mentally. The participation and support of family members are extremely important in ceremonies and NAC meetings.

Before 1955, when the United States Public Health Service (USPHS) took over the responsibility for American Indian health care throughout the United States, the Bureau of Indian Affairs faced barriers to their efforts to provide western biomedical care to Navajos due to limited facilities, inadequate funding, and early resistance from Navajo patients. After the transition to care by the Indian Health Service, the most significant result has been the reduction of morbidity and mortality rates (Trennert, 1998). Replacing the threat of infectious diseases, contemporary health care challenges include cancer, accidents, diabetes, and alcoholism (Davies, 2001). Most medicine men recognize the value of western medicine and distinguish between diseases best treated by medical doctors such as gall bladder disease, appendicitis, and tuberculosis and those best treated by Navajo ceremonies such as lizard sickness (Adair, Deuschle, & Barnett, 1988). Western medical practitioners today recognize the need to take the time with Navajo patients to communicate effectively, particularly when discussing negative information (Carrese & Rhodes, 2000). Increasingly, medical personnel incorporate culturally appropriate elements into western medical practices. For example, both Navajo and Anglo personnel in a mental health clinic integrate the principle of k'é in treatment to relate more effectively to patients (Willging, 2002).

Theories of Efficacy

From a western perspective, three main theories of efficacy attempt to explain the success of Navajo healing. These theories interpret Navajo healing as psychotherapy, as social acts, or as cultural performances. As a form of native psychotherapy, Navajo healing relies on a powerful practitioner. A medicine man is an imposing figure who "becomes the epitome of success, spiritual harmony, and culturally appropriate behavior" (Topper, 1987, p. 221). He is respected as someone who, like a grandparent, can resolve conflict. Psychotherapeutic elements of healing such as confession, therapeutic alliance, transference, suggestion, and the manipulation of symbols unite to effect a cure (Topper, 1987).

The social dimension of Navajo healing offers a second explanation of efficacy. The presence and support of extended family, clan members, and friends act as powerful affirmation of belonging in a support network, identifying the patient as a significant link in a web of relationships. When an individual suffers a crisis that indicates a disruption of harmony, the family rallies to make arrangements for a ceremony. Enlisting aid from family and friends acknowledges the patient's importance to the group. The patient becomes the center of supportive attention (Spickard, 1991, p. 201). The Navajo educator, Ruth Roessel (1981), also emphasizes the social support integral to a ceremony in her discussion of the Kinaaldá, the Navajo girls' puberty ceremony. The planning, gathering of items to be used such as the grass brush and cornmeal, cooking meals for guests, tending the fire while the corncake cooks, and participating in the all-night ceremony require the cooperation of many helping hands.

The third explanation for the efficacy of Navajo ceremonies is that they are cultural performances, a reenaction of cultural values and beliefs. The role of sandpaintings in ceremonies dramatically portrays their performative nature. A ceremony tells the story of a specific adventure of the Holy People, which is retold in the songs and prayers as well as in the sandpaintings. When a patient sits on a sandpainting, he or she identifies with the Holy People who have been attracted by their likenesses in colored sands. "It is too little to say that for the person, this identification must be a very significant event" (Gill, 1987, p. 55). The healing power of the supernaturals absorbs the sickness and restores the health of the patient seated on the sandpainting (Griffin-Pierce, 1992).

SEXUALITY AND REPRODUCTION

Accompanying the dramatic changes in Navajo lifestyle during the 20th century are changes in sexual attitudes and practices as well as ideas about reproduction. The former admonition to girls to stay away from boys and men was relatively easy to enforce when children and adolescents herded sheep in a rural environment. After an arranged marriage, women began giving birth to children until they reached menopause. A pronatal ideology favored large families that were considered a gift from the Holy People (McCloskey, 1998a). Since the 1960s, when 90% of Navajo children were enrolled in school (Bailey & Bailey, 1986), boys and girls have been in daily contact. During the middle decades of the 20th century, arranged marriages disappeared and women accepted the educational messages about the benefits of family planning, limiting their family size to three or four children.

The limitation of family size signaled an abrupt change from former fertility patterns when women had large families, often ranging between 5 and 11 children with a mean of seven children (McCloskey, 1998a). Contemporary birth rates continue to decline, from 55.4 births per 1,000 in 1965 (Broudy & May, 1983), to 34.4 per 1,000 in 1988, and 21.7 per 1,000 in 1997 (Navajo Area Indian Health Service, 2002).

Respect for Changing Woman, the beloved deity who celebrated her ability to bear children with the first Kinaaldá, and veneration of the earth's capacity to sustain life and growth promote a pronatal ideology. At the same time, women bestow matrilineal clan membership at birth. In a society that reveres motherhood, infertility is a burden, but frequently women who cannot bear children will raise a sister's child or children, who, in any case, refer to their aunt as shimá, or mother.

HEALTH THROUGH THE LIFE CYCLE

Pregnancy and Birth

In the past Navajo men and women observed an array of pregnancy taboos to ensure an easy birth and a healthy baby (Bailey, 1950). The most commonly mentioned contemporary taboo is the proscription against tying knots, which can contribute to a difficult birth or having the umbilical cord wrapped around the baby's neck. A variant of this belief is that weaving a rug or crocheting during pregnancy can prolong labor (Milligan, 1984). Another current taboo prohibits parents from coming into contact with the dead, even at funerals.

Although Navajo women have accepted birth control as an effective way to plan their families, they remain strongly opposed to abortion. With few exceptions, women, regardless of age and marital status, bring a child to term. If a woman does not feel that she can raise a child, a preferred option is to give the child to a sister or another close female relative. Matrilineal clan membership for all children contributes to the assertion that "No such thing as a bastard or illegitimate child exists in traditional Navajo society" (Roessel, 1981, p. 38).

In the past, Navajo women had a Blessingway ceremony toward the end of pregnancy. In effect, prenatal care at the Indian Health Service has replaced ceremonies during pregnancy (Hartle-Schutte, 1988). However, if a woman has problems during her pregnancy, she will arrange to have a ceremony performed by a medicine man. Traditional births took place at home where Navajo women kneeled on a bed of sand while holding on to a woven belt looped over a roof beam. The initial resistance to hospital births (Bailey, 1948) gave way to acceptance as a safer and more comfortable way to give birth. By the 1980s, 99% of women gave birth in the hospital (Boyce et al., 1986).

Infancy

When an infant's umbilical cord dries and falls off, parents save it for burial in a nearby location related to traditional economic pursuits and appropriate to the sex of the child. A boy's umbilical cord is buried near a corral to promote his successful work with livestock while a girl's umbilical cord is buried close to a loom to assure her future success as a weaver.

A baby's first laugh marks him or her as a social being and is celebrated with a gathering of family and friends. The person who made the baby laugh provides the sheep to be butchered. The baby, sitting on the mother's lap, "hands out" small gifts to guests, including salt. Distributing these gifts assures that a child will grow up to be a generous adult.

Many infants spend much of their first year of life in the protective and comforting environment of a cradleboard, which in Navajo means "baby diaper." In his study of Navajo infancy, Chisholm (1983) found that infants are quieter and less irritable at birth than white babies.

Parents use cradleboards as a soothing, quieting device until their behavior indicated that they no longer wanted to be there. He observed that mother–infant interaction was slower and less intense than that of white mothers, a pattern of behavior adaptive to the relatively dense and crowded environment of extended family residence groups.

Contemporary mothers' patterns of breast-feeding differ markedly from those of grandmothers who used breast-milk as the sole source of nutrition for a young infant. The majority of Navajo mothers now use a "combination feeding" pattern in which they feed the infant both breast-milk and formula. Mothers breast-feed their infants as a way to promote the baby's growth and to protect it against illness while enhancing a sense of security and closeness between mother and child. More traditional beliefs expressed were that breast-feeding makes a child well behaved and that mother's milk transmits the mother's values and attitudes (Wright, Bauer, Clark, Morgan, & Begishe, 1993).

Childhood

The traditional attitude toward children equated them with wealth. Not only were children important sources of labor in the land-based economy, but also they were looked to as companions and sources of help in old age. Although children now spend their childhood years in school, they nevertheless are viewed as blessings and as evidence of the sacred process of increase (Farella, 1984). Despite the trend toward nuclear family households, children often benefit from the influence of grandparents, who may care for children for extended periods during childhood while parents complete their education or work in an urban area. Approaches to discipline vary, but the ideal continues to be talking to children, calmly explaining the difference between right and wrong.

A major health concern among Navajo children is the increasing tendency toward obesity. This secular trend is a culmination of changes in diet and activity level, with traditional foods and high levels of physical activity replaced with foods high in fat and calories and a sedentary lifestyle. The trend toward obesity constitutes a risk factor for future health problems. Findings from a nutritional survey carried out in 1989 (Sugarman, White, & Gilbert, 1990) contrasts markedly with earlier studies. Earlier studies in 1955 and 1968 found some cases of malnutrition among Navajo children but few cases of obesity. In the 1989 study, among Navajo children between 5 and 17 years of age, twice as many children exceeded the 95th percentile of weight-for-age as among other U.S. children. Among all age groups, mean weight had increased 28.8% among boys and 18.7% among girls.

Disabled Navajo children are likely to be accepted and integrated into the family without great disruption of the family's lifestyle and alteration of parents' roles (Joe, 1982). Respecting their individuality, parents saw children's limitations as an indication of personal distinctiveness rather than as a deviation from normal. Regardless of chronological age, children with disabilities were "perceived as 'becoming persons' in the process of defining their own identities" (Conners & Donnellan, 1998, p. 175).

A study of Navajo middle school students conducted in 2000 found that the percentage of students who had ever used cigarettes, alcohol, and marijuana sharply increased between the sixth and eighth grade years. Among sixth graders, 43.9% had tried cigarette smoking compared with 73.8% of eighth graders. Of Navajo area sixth graders, 34.2% had had a drink of alcohol compared with 63.7% of eighth graders. Of sixth graders, 28.4% had ever used marijuana compared with 58.1% of eighth graders. (Navajo Area Indian Health Service, 2000, pp. 13, 18, 20).

Adolescence

For Navajo girls menarche is a notable event, celebrated by family, kin, clan relatives, and friends with a Kinaaldá, the Navajo puberty ceremony. Acknowledging her ability to bear children and ushering in her status as a woman, a girl's Kinaaldá is a 4-day event. Major events are running to the east at dawn; grinding corn for the alkaan, or corncake, to be distributed to guests; and participating in the all-night ceremony conducted by a medicine man. In the words of Mary Shepardson, "One cannot overestimate the importance of this rite in creating a positive self-image in a young girl" (1995, p. 164). The Kinaaldá reenacts the first Kinaaldá held for Changing Woman, the beloved Navajo deity and model for Navajo womanhood (Frisbie, 1993). The ceremony continues to thrive today, sometimes being arranged by mothers who themselves did not have one when they were students away from home at boarding school (Frisbie, 1993; McCloskey, 1998a). Compared to the Kinaaldá held for Navajo girls, the performance of a ceremony for boys at puberty is rare.

Harry Walters described a ceremony held for boys when their voices change, consisting of a sweat bath that may be followed by a Blessingway ceremony. Some activities, such as running to the east, are similar in the boys' and girls' ceremonies (Schwarz, 1997).

Overweight and obesity continue to be prevalent among Navajo adolescents, and risk factors for coronary heart disease and diabetes mellitus appear in this age group. In the Navajo Health and Nutrition Survey, among 160 adolescents (between 12 and 19 years) about 35% of boys and 40% of girls were overweight as defined by having a body mass index (BMI) exceeding the 85th percentile for their sex and age. They also found that these adolescents had low median HDL cholesterol levels, high median triglyceride levels, and a high prevalence of impaired glucose tolerance (Freedman, Serdula, Percy, Ballew, & White, 1997). Eight percent of the adolescents had either impaired glucose tolerance or diabetes. In another study of 234 Navajo high school students, 24% were overweight as defined by having a BMI exceeding the 95th percentile for their sex and age. However, only one (0.4%) had diabetes mellitus, and eight (3%) had impaired glucose tolerance or impaired fasting glucose (Kim, McHugh, Kwok, & Smith, 1999).

In a study of Navajo alcohol abuse, most men and women reported that they first started drinking during their adolescent years. However, only Navajo men who met the American Psychological Association criteria for alcohol dependence began drinking regularly when they were adolescents, while women and alcohol nondependent men did not begin drinking regularly until they were in their 20s (Kunitz & Levy, 2000).

Adulthood

The transition from a physically demanding lifestyle while caring for livestock and crops and a diet based on corn to a sedentary lifestyle and high fat diet has affected the health of the Navajo people. Findings from the 1991–92 Navajo Health and Nutrition Survey of a representative sample found a high prevalence of coronary heart disease risk factors, particularly overweight, hypertension, and diabetes mellitus (Mendlein et al., 1997). Among persons aged 20 years and older, there was an age-standardized prevalence of diabetes mellitus of 22.9% (Will et al., 1997). A high proportion of Navajo women who experience gestational diabetes mellitus later develop impaired glucose tolerance or non-insulin dependent diabetes (Steinhart, Sugarman, & Connell, 1997). The age-standardized prevalence of hypertension was 19% (Percy et al., 1997). Among Navajo adults, major illnesses leading to Indian Health Service hospitalizations are respiratory infections, digestive system disease (gallbladder), and genitourinary system diseases (NAIHS, 2002). Dietary analysis suggests that "much of the increase in chronic diseases among the Navajo is due to nutritional factors" (Byers & Hubbard, 1997, p. 2075S). Common foods include fry bread and tortillas, home-fried potatoes, mutton, bacon and sausage, soft drinks, and coffee and tea, which provided 41% of the energy and 15 to 46% of the macronutrients in the diet. Dairy products, fruits, and vegetables were each consumed less than once per day per person (Ballew et al., 1997).

A random sample of the adult Navajo population indicates that alcohol dependence constitutes a major health and social problem. Among men, 70.4% have a lifetime history of alcohol dependence, and 29.6% of women have such a history. The consequences of alcohol misuse include domestic violence and other family problems, health problems, alcohol-related deaths, and legal and economic problems (Kunitz & Levy, 2000).

Traditional Navajo women, like newly immigrated Latina, experienced fewer negative associations and physical complaints about natural menopause (Mingo, Herman, & Jasperse, 2000). Overshadowing menopause as a life event, the achievement of the status of grandmother, a venerated role in matrilineal society, assumes greater importance.

The Aged

In Navajo society elders enjoy high status, respected for their knowledge of ceremonies, traditional foods and clothing, livestock, clan relationships, and many customs such as the shoe game, the first laugh celebration, and burying the baby's umbilical cord. Navajo elders actively participate in ceremonies, in family activities and decisions, and often continue to make an economic contribution with their traditional activities such as weaving or by working in programs such as Foster Grandparents.

Chronic diseases increasingly affect Navajo elders. The rapid rise of the incidence of diabetes and its serious complications constitute a serious health problem for Navajo elders. Among Navajos 65 years of age and older, 41.3% had diabetes (Will et al., 1997). Among those 60 years or older, 44% of men and 28% of women had hypertension (Percy et al., 1997). Both diabetes and hypertension constitute risk factors for coronary heart disease (Mendlein et al., 1997).

A retrospective study of 28 Navajo uranium miners with a mean age of 66 years found that these men were adversely affected by chronic illness. Despite historically low rates of lung cancer for Navajos, former uranium miners have abnormally high rates of lung cancer. Other health problems include eye disease, hearing loss, and psychological trauma (Servilla, 1997). "A lower health-related quality of life was related to more years worked as an underground miner" (Servilla, 1997, p. 193).

For many Navajo elders, the expectation that children will care for them in old age may not be fulfilled. Kunitz and Levy (1991) documented the transition from family to institutional support on the Navajo Reservation. Although many elders live with family members, social service programs contribute by serving meals at the local Senior Citizens Centers and by visits from Indian Health Service Community Health Representatives. Because the Navajo Nation has limited on-reservation nursing care facilities, the majority of elders in need of nursing home care must go off-reservation. Elders at an on-reservation nursing home appeared content with culturally sensitive care in contrast with elder Navajos in off-reservation homes who reported loneliness, depression, and isolation (Mercer, 1996).

Dying and Death

According to the Navajo Area Indian Health Service (2002), the five leading causes of death in 1998 were accidents, heart disease, cancer, influenza/pneumonia, and diabetes. Of these, only accidents disproportionately affect young people. Death rates for accidents involving motor vehicle, diabetes, and pneumonia are many times higher than rates for the United States as a whole (Navajo Area Indian Health Service, 2002).

The traditional fear of death dictated that a person close to death be moved to a shelter apart from the family hogan (Mitchell, 2001). If a person died while in the family hogan, family members abandoned the home and moved elsewhere. After a death, family members did not eat or wash for four days. If these patterns of showing respect for the dead are not adhered to, the dead can become angry and cause ghost sickness (Mitchell, 2001).

In accordance with traditional burial practices, family members buried the dead as soon as possible and without any public ceremony. Those who prepared and buried the deceased participated in a purification ritual and a four-day mourning period before resuming normal activities. With many Navajos more recently converting to Christianity, funerals and burials in local cemeteries have become commonplace (Levy, 1978; Shepardson, 1978).

REFERENCES

Aberle, D. F. (1983a). Navajo Economic Development. In Alfonso Ortiz (Ed.), *Southwest* (pp. 641–658). Vol. 10, *Handbook of North American Indians*. Washington: Smithsonian Institution.

Aberle, D. F. (1983b). Peyote Religion among the Navajo. In Alfonso Ortiz (Ed.), *Southwest* (pp. 558–569). Vol. 10, *Handbook of North American Indians*. Washington: Smithsonian Institution.

Adair, J., Deuschle, K. W., & Barnett, C. R. (1988). *The people's health: Anthropology and medicine in a Navajo community*. Revised and Expanded Edition. Albuquerque: University of New Mexico Press.

Bailey, F. L. (1948). Suggested techniques for inducing Navaho women to accept hospitalization during childbirth and for implementing health education. *American Journal of Public Health, 38*, 1418–1423.

Bailey, F. L. (1950). *Some sex beliefs and practices in a Navaho community with comparative material from other Navaho areas*. Papers of the Peabody Museum of American Archaeology and Ethnology, Harvard University, Vol. XL, No. 2. Cambridge, Massachusetts: Peabody Museum.

Bailey, G., & Bailey, R. G. (1986). *A history of the Navajos: The reservation years*. Santa Fe, New Mexico: School of American Research Press.

Ballew, C., White, L. L., Strauss, K. F., Benson, L. J., Mendlein, J. M., & Mokdad, A. H. (1997). Intake of nutrients and food sources of nutrients among the Navajo: Findings from the Navajo health and nutrition survey. *The Journal of Nutrition, 127*, 2085S–2093S.

Boyce, W. T., Schaefer, C., Harrison, H. R., Haffner, W. J. H., Lewis, M., & Wright, A. L. (1986). Social and cultural factors in pregnancy complications among Navajo women. *American Journal of Epidemiology, 124*, 242–253.

Broudy, D. W., & May, P. A. (1983). Demographic and epidemiologic transition among the Navajo Indians. *Social Biology, 30*, 1–16.

Brugge, D. M. (1983). Navajo prehistory and history to 1850. In Alfonso Ortiz (Ed.), *Southwest* (pp. 489–501). Vol. 10, *Handbook of North American Indians*. Washington: Smithsonian Institution.

Byers, T., & Hubbard, J. (1997). The Navajo health and nutrition survey: Research that can make a difference. *The Journal of Nutrition, 127*, 2075S–2077S.

Carrese, J. A., & Rhodes, L. A. (2000). Bridging cultural differences in medical practice: The case of discussing negative information with Navajo patients. *Journal of General Internal Medicine, 15*, 92–96.

Chisholm, J. S. (1983). *Navajo infancy: An ethological study of child development*. NY: Aldine Publishing Company.

Choudhary, T. (2001). *2000–2001 Comprehensive Economic Development Strategy*. Support Services Department, Division of Economic Development, The Navajo Nation. Window Rock, AZ: Navajo Nation.

Choudhary, T. (2003). *Data from Census 2000*. Support Sciences Department, The Navajo Nation. Window Rock, AZ: Navajo Nation.

Conners, J. L., & Donnellan, A. M. (1998). Walk in beauty: Western perspectives on disability and Navajo family/cultural resilience. In H. I. McCubbin, E. A. Thompson, A. I. Thompson, & J. E. Fromer (Eds.), *Resiliency in Native American and immigrant families* (pp. 159–182). Thousand Oaks: Sage Publications.

Csordas, T. J. (2002). *Body/Meaning/Healing*. NY: Palgrave Macmillan.

Csordas, T. J. (2000). The Navajo healing project. *Medical Anthropology Quarterly, 14*, 463–475.

Davies, Wade. (2001). *Healing ways: Navajo health care in the twentieth century*. Albuquerque: University of New Mexico Press.

Farella, J. R. (1984). *The main stalk: A synthesis of Navajo philosophy*. Tucson, AZ: The University of Arizona Press.

Freedman, D. S., Serdula, M. K., Percy, C. A., Ballew, C., & White, L. (1997). Obesity, levels of lipids and glucose, and smoking among Navajo adolescents. *The Journal of Nutrition, 127*, 2120S–2127S.

Frisbie, C. J. (1987). *Navajo medicine bundles or Jish:Acquisition, transmission, and disposition in the past and present*. Albuquerque: University of New Mexico Press.

Frisbie, C. J. (1993). *Kinaaldá: A study of the Navaho girl's puberty ceremony*. Salt Lake City:University of Utah Press.

Gill, S. (1987). *Native American religious action: A performance approach to religion*. Columbia, South Carolina:University of South Carolina Press.

Griffin-Pierce, T. (1992). *Earth is my mother, sky is my father: Space, time, and astronomy in Navajo sandpainting*. Albuquerque:University of New Mexico Press.

Hartle-Schutte, M. (1988). *Contemporary usage of the Blessingway ceremony for Navajo births* (Master of Arts thesis, University of Arizona, 1988). Ann Arbor, Michigan: University Microfilms International.

Joe, J. (1982). Cultural influences on Navajo mothers with disabled children. *American Indian Quarterly, 6*, 170–190.

Kim, C., McHugh, C., Kwok, Y., & Smith, A. (1999). Type 2 Diabetes mellitus in Navajo adolescents. *The Western Journal of Medicine, 170*, 210–213.

Kunitz, S. J., & Levy, J. E. (2000). *Drinking, conduct disorder, and social change: Navajo experiences*. NY: Oxford University Press.

Kunitz, S. J., & Levy, J. E. (1991). *Navajo aging: The transition from family to institutional support*. Tucson, AZ: The University of Arizona Press.

Levy, J. E. (1978). Changing burial practices of the western Navajo: A consideration of the relationship between attitudes and behavior. *American Indian Quarterly, 4*, 397–405.

Levy, J. E. (1983). Traditional Navajo health beliefs and practices. In S. J. Kunitz (Ed.), *Disease change and the role of medicine: The Navajo experience* (pp. 118–145). Berkeley, CA: University of California Press.

Lewton, E. L., & Bydone, V. (2000). Identity and healing in three Navajo religious traditions: Sa'ah Naagháí Bik'eh Hózh. *Medical Anthropology Quarterly, 14*, 476–497.

Linford, L. D. (2000). *Navajo places: History, legend, landscape*. Salt Lake City: University of Utah Press.

McCloskey, J. (1998a). Three generations of Navajo women: Negotiating life course strategies in the eastern Navajo agency. *American Indian Culture and Research Journal, 22*, 103–129.

McCloskey, J. E. (1998b). Traditional components of treatment for Navajo alcohol abusers: Ethnographic case studies. Report for the Diné Center for Substance Abuse Treatment and the Navajo Nation. Center for Alcoholism, Substance Abuse, and Addictions, University of New Mexico.

Mendlein, J. M., Freedman, D. S., Peter, D. G., Allen, B., Percy, C. A., Ballew, C. et al. (1997). Risk factors for coronary heart disease among Navajo Indians: Findings from the Navajo health and nutrition survey. *The Journal of Nutrition, 127*, 2099S–2105S.

Mercer, S. O. (1996). Navajo elderly people in a reservation nursing home: Admission predictors and culture care practices. *Social Work, 41*, 181–189.

Milligan, B. C. (1984). Nursing care and beliefs of expectant Navajo women (Part 1). *American Indian Quarterly, 8*, 83–102.

Milne, D., & Howard, W. (2000). Rethinking the role of diagnosis in Navajo religious healing. *Medical Anthropology Quarterly, 14*, 543–570.

Mingo, C., Herman, C. J., & Jasperse, M. (2000). Women's stories: Ethnic variations in women's attitudes and experiences of menopause, hysterectomy, and Hormone Replacement Therapy. *Journal of Women's Health & Gender-Based Medicine, 9*, S27–S38.

Mitchell, R. (2001). *Tall woman: The life story of Rose Mitchell, A Navajo woman, c. 1874–1977*. C. J. Frisbie (Ed.). Albuquerque: University of New Mexico Press.

Morgan, F. (2002). Personal Communication. December 28, 2002.

Morgan, W. (1977). Navajo treatment of sickness: Diagnosticians. In David Landy (Ed.), *Culture, disease, and healing: Studies in medical anthropology* (pp. 163–169). NY: Macmillan Publishing Co.

Navajo Area Indian Health Service. (2000). *2000 Navajo Middle School Youth Risk Behavior Survey*. Shiprock, New Mexico: Navajo Area Indian Health Service.

Navajo Area Indian Health Service. (2002). *Health Profile*. Office of Program Planning and Evaluation. Window Rock: AZ.

Office of Navajo Government Development. (1998). *Navajo Nation Government* (4th ed.). Window Rock, Arizona: Navajo Nation.

Percy, C., Freedman, D. S., Gilbert, T. J., White, L., Ballew, C., & Mokdad, A. (1997). Prevalence of hypertension among Navajo Indians: Findings from the Navajo health and nutrition survey. *The Journal of Nutrition, 127*, 2114S–2119S.

Quintero, G. A. (1995). Gender, discord, and illness: Navajo philosophy and healing in the Native American Church. *Journal of Anthropological Research, 51*, 69–89.

Reichard, G. A. (1944). *Prayer: The compulsive word*. American Ethnological Society Monograph, 7. Seattle: University of Washington Press.

Reichard, G. A. (1974). *Navaho religion: A study of symbolism*. NY: Bollingen Foundation Inc., 1950. Reprint, NY: Bollingen Series XVIII, Princeton University Press.

Roessel, R. (1981). *Women in Navajo society*. Rough Rock, AZ: Navajo Resource Center, Rough Rock Demonstration School.

Schwarz, M. T. (1997). *Molded in the image of Changing Woman: Navajo views on the human body and personhood*. Tucson, AZ: The University of Arizona Press.

Servilla, M. J. (1997). *Describing the health-related quality of life of former Navajo uranium miners with spirometry abnormalities*. Masters of Science thesis, University of New Mexico.

Shepardson, M. (1978). Changes in Navajo mortuary practices and beliefs. *American Indian Quarterly, 4*, 383–395.

Shepardson, M. (1995). The Gender Status of Navajo Women. In L. F. Klein & L. A. Ackerman (Eds.), *Women and Power in Native*

North America (pp. 159–176). Norman, OK: University of Oklahoma Press.
Spencer, Katherine. (1957). *Mythology and values: An analysis of Navaho Chantway myths*. Memoirs of the American Folklore Society, 48. Philadelphia: American Folklore Society.
Spickard, J. V. (1991). Experiencing religious rituals: A Schutzian analysis of Navajo ceremonies. *Sociological Analysis, 52*, 191–204.
Steinhart, J. R., Sugarman, J. R., & Connell, F. A. (1997). Gestational diabetes is a herald of NIDDM in Navajo Women. *Diabetes Care, 20*, 943–947.
Sugarman, J. R., White, L. L., & Gilbert, T. J. (1990). Evidence for a secular change in obesity, height, and weight among Navajo Indian schoolchildren. *American Journal of Clinical Nutrition, 52*, 960–966.
Topper, M. D. (1987). The traditional Navajo medicine man: Therapist, counselor, and community leader. *The Journal of Psychoanalytic Anthropology: A Quarterly Journal of Culture and Personality, 10*, 217–249.
Trennert, Robert A. (1998). *White Man's medicine: Government doctors and the Navajo, 1863–1955*. Albuquerque: University of New Mexico Press.
Will, J. C., Strauss, K. F., Mendlein, J. M., Ballew, C., White, L. L., & Peter, D. G. (1997). Diabetes mellitus among Navajo Indians: Findings from the Navajo health and nutrition survey. *The Journal of Nutrition, 127*, 2106S–2113S.
Willging, C. E. (2002). Clanship and K'é: The relatedness of clinicians and patients in a Navajo counseling center. *Transcultural Psychiatry, 39*, 5–32.
Witherspoon, G. (1983). Navajo social organization. In Alfonso Ortiz (Ed.), *Southwest* (pp. 524–535). Vol. 10, *Handbook of North American Indians*. Washington: Smithsonian Institution.
Wright, A. L., Bauer, M., Clark C., Morgan, F., & Begishe, K. (1993). Cultural interpretations and intracultural variability in Navajo beliefs about breastfeeding. *American Ethnologist, 20*, 781–796.
Wyman, L. C. (1983). Navajo ceremonial system. In Alfonso Ortiz (Ed.), *Southwest* (pp. 536–577). Vol. 10, *Handbook of North American Indians*. Washington: Smithsonian Institution.
Wyman, L. C., & Clyde Kluckhohn. (1938). *Navajo classification of their song ceremonials*. Memoirs, 50. Menasha, Wisconsin: American Anthropological Association.
Young, R. W. (1961). *The Navajo yearbook: 1951–1961, A Decade of Progress*. Window Rock, Arizona: Navajo Agency, Bureau of Indian Affairs.

Nepal

Gregory G. Maskarinec

LOCATION AND LINGUISTIC AFFILIATION

The Himalayan constitutional monarchy of Nepal is a small landlocked country located in South Asia between India and the Tibetan Autonomous Region of China. Nepal is roughly shaped as a rectangle that extends approximately 500 miles (800 km) from southeast to northwest with an average width ranging from 90 to 150 miles (140 to 240 km) from north to south. Nepal is home to seven of the world's highest mountains, including the highest, Everest, known in Nepal as Sagarmatha. In 2000 Nepal's population was estimated to be 24 million. Nepal's significant cultural and linguistic diversity is indicated by the 120 distinct languages that have been recorded in the country, of which at least 60 are still used as the primary language of a local community. Most inhabitants understand the national language, Nepali, an Indo-Aryan language derived from Prakrit resembling Bengali and Gujarati but with an increasing number of loan words via Hindi of Persian or Arabic origin. Other Indo-Aryan languages of Nepal include Maithili, Rajbansi, and Bhojpuri. Major Tibeto-Burman languages in Nepal include Newari, Tamang, Gurung, Magar, Sherpa, Thakali, Tharu, and various Kiranti languages.

OVERVIEW OF THE CULTURE

The modern state of Nepal emerged in the late 18th century as the king of one local hill principality, Gorkha, began a successful campaign of conquest, conquering the sophisticated Newari city-states of the Kathmandu Valley in 1769. Military interventions by the British established the present borders of the kingdom but preserved Nepal's nominal independence, an autonomy that was strengthened by Nepal's strong support of the British during the Indian sepoy mutiny of 1857. Democratic government was introduced in 1991 after years of absolute monarchy, but the contemporary political system is extremely factionalized, with frequent changes of government. Since 1996, an increasingly violent Maoist insurgency has spread throughout the country.

Never having been colonized and rigorously outlawing all missionary proselytizing, whether by Christians or Muslims, Nepal has preserved vibrantly its diverse cultural traditions, which includes many varieties of shamans, oracles, and spirit mediums. In keeping with Nepal's self-proclaimed identity as "the world's only Hindu state," more than nine tenths of the population is classified as Hindu, a designation that officially includes Buddhists as well as the many local variations of ritual practice, so long as they observe in some way the Hindu caste system. Outside of this system, there are also very small groups of Muslims and an even smaller scattering of Christians, mostly repatriates from elsewhere in South Asia.

As a result of its geographical inaccessibility and a repressive political system sustained by inequitable land distribution and exploitive methods of agricultural taxation, Nepal is one of the least developed countries of the world, with an extremely low standard of living. Nepal's per capita annual income remains less than U.S.$200, with more than half of the population living below the absolute poverty level. Average life expectancy is around 55 for both males and females. No more than half of the population is literate. (All statistics regarding Nepal are only reasonable guesses; no accurate demographic data exists.)

The climate of Nepal ranges from subtropical monsoon conditions in the Tarai (the Gangetic plains in the south of the country) to alpine conditions in the Great Himalayas. Throughout the country, subsistence agriculture and animal husbandry are the most common economic activities, employing most of the work force. Productivity is low, leading to chronic food shortages and periodic famines in parts of the country. Rice is the leading staple, but is a luxury in more remote parts of the country, where corn (maize), wheat, or millet are the staple crops. Most houses are small, two storied buildings made of mud-bricks, without electricity or running water. Wood and dried dung are the standard fuels for cooking. The Nepalese economy, such as it is, is sustained by foreign aid and tourism; trade is dominated by India.

Despite a relatively high death rate along with migration to India, Nepal's population continues to grow rapidly, with 40% of the population younger than 15 years of age. Except for the urban concentration of the capital, Kathmandu, and a scattering of provincial towns, nearly all the population remains in small rural villages. Complicating its economy as well as its foreign policy, Nepal provides refuge to nearly 100,000 recent refugees from Bhutan, as well as to more settled communities of Tibetan refugees, who began arriving after the Dalai Lama fled Tibet in 1959.

Family structure in Nepal remains very traditional; most marriages are arranged and are caste-endogamous but frequently village-exogamous. Girls tend to be married by age 16, boys slightly older. In some groups, however, such as the Newars of the Kathmandu Valley, both men and women tend to be married at later ages. Newari girls are ritually married at a young age to a *bel* fruit (*Aegle marmelos*), a ceremony that de-emphasizes both the urgency and the solemnity of later marriages. Most Brahman girls, in contrast, are married by age 10, since Brahmans obtain religious merit by giving away very young daughters—according to tradition, the preferred age for a Brahman girl to marry is six, although modern families no longer follow this custom. In groups with Sanskritic influenced cultures, girls are expected to be virgins at marriage, but this stricture is ignored in most Tibeto-Burman communities. The status of Hindu wives may be inferred from a rite within the marriage ceremony during which the bride must wash her husband's feet (considered the most impure part of the human body) and sip the water. In traditional families, a wife always greets her husband by touching his feet.

The estimated average number of births per woman for the country is six. Most families are patrilocal and patrilineal. Depending on the ethnic group, both polygyny and polyandry are found in Nepal; variations on dowry or bride-price are also found in different communities. Divorce remains uncommon, as does widow remarriage, although some groups, including Kiranti populations and low caste groups, practice marriage by levirate.

The Context of Health: Environmental, Economic, Social, and Political Factors

Given the widespread poverty, poor nutrition, overpopulation, unfair land tenure systems, exploitive wage labor, political corruption, poor sanitation, and lack of clean drinking water, the overall health situation in Nepal is extremely poor. Overall infant mortality is at least 100 per 1,000. There are widely varying estimates of this and all other health-related statistics, all unreliable and all concealing enormous disparities between the relatively rich and the desperately poor. Personal observation over the past 25 years suggests that in some of the poorest villages of the western hills, the mortality of children under five is as high as 400 per 1,000.

While it is officially reported that there are 150 hospitals in Nepal (two of them ayurvedic, one homeopathic), only a few of these have sufficient staff, equipment, or stocks of medicines to provide a reasonable standard of care. Outside urban areas, the 1,000 health posts established throughout the country are poorly managed, corrupt, and sensibly avoided by much of the population until all other options have failed.

The prevalence of tuberculosis in Nepal is estimated at 15 per 1,000, with some 50,000 new cases a year and 20,000 annual deaths. Nepal has the second highest incidence of iodine deficiency disorders in the world with cretinism rates as high as 10% in some remote areas, where goiter rates can reach 80%. After substantial declines in the 1960s, malaria is resurgent and increasingly resistant to treatment. There are some 30,000 registered cases of leprosy in the country. There are periodic outbreaks of typhoid. The most recent cholera outbreak was in 1995. Plague may have occurred as recently as August 1994, when there was an outbreak in India; the most recent well documented outbreak in Nepal was in 1967/1968. Other severe diseases that are common include encephalitis, meningitis, hepatitis, and gastrointestinal disorders. It has been estimated that all of the adult population suffers from intestinal parasites. Partial blindness as a result of vitamin A deficiency affects at least half a million people. Half of females of childbearing age are anemic.

Medical Practitioners

His Majesty's Government of Nepal (HMG) officially acknowledges four systems of medicine: allopathy, Ayurveda, homeopathy, and Unami as being practiced in Nepal, although the latter two are not widespread. Ayurvedic practices tend to be localized in Kathmandu or the areas adjacent to India. Locally, Ayurveda is considered most efficacious in the treatment of hepatitis. HMG reports that there is one biomedically trained physician for approximately 19,000 persons and one nurse per 5,000 persons, but these numbers, bad as they are, are severely misleading unless it is concurrently noted that nearly all physicians and nurses are to be found in urban areas, leaving the rest of the country either to poorly trained health workers or to traditional practitioners. Nepal has one 40-bed acute care hospital for the mentally ill, and 13 trained psychiatrists, all in Kathmandu. Even in urban areas, health care delivery is systematically biased in favor of the wealthy and the middle class, while the poor are denied services both because of the cost and the structure of the traditional society. Although Nepal is a signatory to the 1978 Alma Ata Declaration and despite the rhetorical commitment of successive governments toward committing resources for better health, improved health care remains a low priority in Nepal, as it would necessarily require a more equitable social system.

Outside the Kathmandu Valley, drug therapy is almost inevitably mismanaged. Nevertheless, injections and capsules have acquired a nearly magical status, so that villagers who do consult a "modern" medical practitioner, even if it is only the lowest paid health post peon who may lack even a high school education let alone any medical training, expect their treatment to include these. Unrealistic expectations of biomedical efficacy coupled with poor patient education contributes to many villagers' disillusionment with the government's promotion of Western medicine. Self medication is common, as "medical halls" (drug stores) that lack trained pharmacists supply pharmaceuticals on demand. In rural areas, these medications are often expired stock from India.

Traditional practitioners, a diverse group including oracles, shamans, spirit mediums, Buddhist lamas, Tantrics, astrologers, wandering ascetics, and herbal healers, all remain more familiar, less frightening and less intrusive than is Western-style medicine. Traditional practitioners are less condescending and far easier to understand than are allopathic physicians, and their success rates appear no worse than those of their competitors. The variety of "traditional healers" is considerable. Defining the category of "shaman" fairly strictly as ritual intercessors whose divination, exorcism, and healing practices incorporate command over well-defined groups of spirits, who use elaborate paraphernalia, and who often enter trance states to conduct ritual journeys, one finds Gurung *poju* and *hlewri*, Tamang *lambu* and *bonpo*, Yolmo *bonpo*, Tharu *gurau*, Rai *bijuju*, *padem*, and *lambu*, Limbu *yeb*, Sunuwar *puimbo* and *ngiami*, Chepang *pande*, Sherpa *lhapa* and *mindung*, Magar *ramma*, and the low caste *jhangri* throughout the country. While each of these terms can be translated as "shaman," each has distinct characteristics, including different ritual practices, different costumes and paraphernalia, and different oral texts. Still, some generalizations are possible. Most shamans are male, although highly motivated women occasionally overcome the biases of patrilineality and gender discrimination to become shamans. Many future shamans undergo possession crises as adolescents that prompt their interest in the profession. For most groups, preparation of new shamans concentrates on their learning oral texts. Most important is to memorize the secret mantras that control the spirits that shamans must manipulate in their rituals, since, unlike other circumstances that involve involuntary or spontaneous spirit possession, shamans must control the spirits that they summon to their ceremonies, rather than be controlled by them. For the shamans of western Nepal, the variety of spirits that they use include local gods and goddesses, spirits of animals and inanimate forces, souls of human suicides (particularly vengeance spirits—men or women who were unable to find justice in their lifetime and committed suicide to become such a spirit), souls of other dead humans, especially those who died untimely deaths, and the souls of other shamans. Some of these groups are protective and therapeutic, others are malicious, oppressive, and threatening. Nearly all demand blood sacrifices, a key element of every major shamanic ritual. Communicating with these spirits allows shamans to diagnose problems, treat afflictions, and restore order and balance to the lives of their clients and their communities.

Distinct from shamans are the oracular spirit mediums found throughout the greater Karnali drainage area of western Nepal. Each oracle is associated with one particular spirit located at one particular shrine to that spirit. The oracle is regarded as the passive mouthpiece of the spirit, in contrast to the control over a wide variety of spirits expected of shamans. Only upon the death of one oracle does a new one emerge to take the position, as distinct from shamans, who study with a teacher to learn the profession. Oracles most often treat poorly differentiated misfortune, often for entire families. Such cases may also be taken to less well-established spirit mediums, who sometimes emerge without a well defined social position, often women who become spontaneously possessed and who temporarily exhibit healing powers, whose popularity rapidly waxes and then equally rapidly wanes.

Many villagers use herbal remedies for a wide variety of somatic complaints, and many may also informally consult other villagers about such treatments, although there are not well defined "herbalists" as such. Many herbal treatments are used in conjunction with mantras, knowledge of which is also widely diffused. Nepal has a significant source of medicinal herbs from its higher altitudes, which are exported to India for use in ayurvedic preparations.

Astrologers and various fortune tellers are also frequently consulted for a wide range of problems, including medical conditions. They offer both diagnosis and advice, but do not perform healing roles.

In the Tibetan refugee community, and to a small extent among the Tibeto-Burman groups within Nepal, classic Tibetan medicine is practiced, which is based to a considerable extent on the use of herbs and natural substances within a theory of Buddhist religious teachings, and is considered most effective for chronic conditions.

CLASSIFICATION OF ILLNESS, THEORIES OF ILLNESS, AND TREATMENT OF ILLNESS

Nepalis tend to be extremely fatalistic, observing that in this Age of Darkness (Kali Yuga), inevitable deterioration of the world, at all levels, personal, social, and cosmic, is all that can be expected. All concepts of life, death, and suffering are permeated by the theory of Karma, with its cycles of rebirth that connect morality with retribution and reward. One's caste and gender, for example, are generally recognized as reaping what has been sown in one's previous lives. Karma, then, is ultimately responsible for all illness and affliction. Nevertheless, Nepalis distinguish proximate causes to their illnesses, and most accept a classification of their problems into either *rog* (illness) or *dokh* (spiritual affliction/misfortune). Illness has many natural causes, including imbalances of hot and cold foods, violation of the laws of purity and pollution, or unpredictable accidents loosely connected to one's Karma.

One of the most elaborate local etiologies is found among the shamans of western Nepal. Shamans diagnose their patients' afflictions as commonly involving some combination of seven possible causes: (1) curses, and spells, particularly those of witches and other shamans; (2) misfortunate astrological configurations, foremost of which are dangerous planetary configurations called star obstructions; (3) the intrusion of alien substances into the body, whether by acts of sorcery, by violation of sacred space, or by violating rules of ritual purity, such as coming in contact with a woman's menstrual blood; (4) weakened life forces, including soul loss, lost wits, dullness, and several recognized forms of madness; (5) social disorder, especially disputes within families and communities; (6) fevers of autonomous origin; (7) the activities of spirits, ghosts, or demons, provoked by acts of neglect, pollution, or disrespect. When spirits are involved, the most common types are patrilineal family gods, recently deceased relatives, or the numerous minor local spirits who inhabit particular hilltops, trees, waterfalls, springs, and rivers. At other times, major local gods, minor deities of non-human origin, quasi-spirits thought to have a one-dimensional degree of corporeality, or even shaman tutelary spirits may be diagnosed as causing affliction (Maskarinec, 1995).

Much of the treatment for these conditions is through logotherapy, using ritual language, with each intervention involving a distinctive ritual. Most interventions can be described as either propitiatory or symbolic. Propitiatory rituals include sacrifices and offerings, while symbolic ones include rituals of binding and burying, sucking or blowing, raising the patient toward the heavens, and diverse acts of divination.

SEXUALITY AND REPRODUCTION

Large families are seen as insurance in old age, and as a guarantee of sufficient agricultural labor for the future. Many families wish to have as many children as possible, in hopes that some, at least, will survive to adulthood. Sons are expected to maintain the family's lineage, inherit property, and provide support to parents when they are old. For Hindu families, a son is necessary to complete the proper funeral rites for his parents, without which they are condemned to become restless spirits. "Ideal" family size (first reported by Worth and Narayan, 1969) remains in the range of four or five children, which is slightly higher than the number of children who survive to adulthood in most rural families. Daughters, who leave the family when they marry, tend to be seen as liabilities. However, there is no evidence that female infants are less likely to survive than are males, since Nepalis regard all children as divine gifts, but girls are certainly less likely to be sent to school, and are given more household chores, at an earlier age.

Knowledge of family planning is widespread, having been promoted by the government for decades with extensive media campaigns. Family planning itself is slowly spreading, primarily among the educated urban population but with some impact at the village level. The 1996 Nepal Family Health Survey (Pradhan, Ajit et al., 1997) found that sterilization accounted for more than half of current contraceptive use. Abortion is considered sinful; a woman who has an abortion will be reborn, it is widely believed, as a dog. There have been recent efforts to decriminalize abortion, but as of March 2002 these have so far been unsuccessful. Nepal's abortion law only permits abortions to be performed on the ground of "benevolence," but does not indicate which abortions are covered by this term, and few, if any, pregnancies have been terminated on this ground. A three-year study conducted in the early 1980s revealed that two thirds of all female inmates in Nepal were imprisoned on charges related to abortion and infanticide. (Population Policy Data Bank; Department for Economic and Social Affairs, United Nations Secretariat); it is estimated that illegal abortions cause the death of hundreds of women every year. Miscarriages are regarded as shameful; women try to conceal them.

Inability to have children, to have miscarriages or stillbirths, or for children to die in infancy are all blamed on a wife's Karma. All are extremely damaging to her status, and may lead to the husband marrying a second wife. Appeal to spirits held responsible or who may intervene against witchcraft are the most common attempted remedies. One oracle, that of Bijuli Masta in Jajarkot District, is considered particularly effective, with many families that have long been childless returning to dedicate their firstborn sons in the temple.

Health through the Life Cycle

Pregnancy and Birth

There are no traditional prenatal ceremonies, nor do most pregnant women receive medical care. Pregnancy is socially recognized by the fifth or sixth month, when it is held that the life-breath has entered the embryo. From this point, the woman, regarded as being two individuals, may not participate in religious ceremonies.

Traditional midwives assist in many deliveries in Nepal, but it remains common for a woman to deliver alone or with only a close relative or female neighbor assisting. There have been government efforts dating back to the 1920s to provide midwife training, though resources are limited. Most deliveries take place in the husband's home. The umbilical cord is usually cut on the same day that birth takes place, but it may be tied and the cutting postponed to avoid astrologically determined complications, or to allow a scheduled ritual or wedding within the immediate family to take place, as these must otherwise be postponed for 11 days. The placenta of a male child is buried under the kitchen floor near the hearth, that of a female child outside. A section of the umbilical cord may be dried to be used to treat colic. The newborn is washed and rubbed with mustard oil, regarded as essential to "warm" the infant. Women remain in seclusion after giving birth, with both mother and child treated as untouchable until the eleventh day. Female relatives do massage and oil both baby and mother, but they must bathe to purify themselves afterward. Breast-feeding begins as soon as milk is available. No other foods are ordinarily given to infants.

Medical risks associated with birth include, predictably, low birth weights, premature deliveries, neonatal tetanus, sepsis, and other infections. Multiple births and birth defects are regarded as inauspicious, and may lead to the infants being neglected or, in rare cases, abandoned.

Infancy

As soon as possible after a birth, an astrologer is consulted, to calculate the child's future and to choose an auspicious name, which is based on the time of birth. The most serious of astrological disturbances is that of *mul*, which occurs when certain planets of both the child and either of its parents occupy the *mul nakshatra*, one of 27 subdivisions of the lunar elliptic. *Mul* is a crisis that lasts for a finite, calculable length, sometimes a lifetime, sometimes for just a few moments. The parent who shares the configuration is fated to die quickly if he/she sees the child within *mul*'s duration, so that a *mul* birth may have to be placed outside the family to be raised as an orphan. Only after an astrologer has been consulted does the baby's father see the child.

On the sixth night after birth Bhabi, Goddess of Fortune, comes to write a child's fate on its forehead. Parents traditionally leave a light burning all night in the room with the baby so that she makes no mistakes, sometimes supplying a pen and inkpot.

Among many groups a first rice feeding ceremony is held six months after birth, though this is not to be confused with weaning. Breast-feeding often continues for two or three years. This is regarded as a natural contraceptive measure. Male children tend to be breast-fed longer than are female children, sometimes for five or six years if no sibling is born. Special consideration is given to the mother's diet when she is breast-feeding, with foods regarded as having "cooling" properties most often avoided.

Childhood

Nepal's last major smallpox epidemic was in 1963–64. Official eradication of smallpox by 1974 is concrete evidence of what a well-focused public health campaign might do for the six vaccine preventable diseases of childhood—diphtheria, whooping cough, tetanus, poliomyelitis, tuberculosis, and measles—all of which contribute significantly to childhood morbidity and mortality.

For high caste, "twice-born," males, the most important childhood ceremony is to receive the sacred thread. Brahman boys usually receive this at age seven or nine, while Chetris may wait until their teens. Once invested with a thread, a boy has many new ritual responsibilities and privileges, such as now being allowed to eat in the company of adults. There is no parallel ceremony for girls.

Children are, in general, treated with considerable affection and are rarely disciplined. They are gradually integrated into the family's daily chores. Older children assume responsibility for caring for their younger siblings.

Adolescence

As most Nepalis are married by their mid-teens, there is no real "adolescence" in Nepal, just a quick transition from childhood to adulthood.

Adulthood

First menstruation marks the beginning of adulthood for women. No special ceremonies are performed, but in some communities the woman is secluded for 15 days, during which time she may not be seen by males. Menstruation is considered ritually unclean, and there are many rules that a menstruating woman must follow. She is treated as an untouchable, even for other women, and should maintain physical isolation for the first four days of her period; she must not cook nor fetch water; she may not perform any religious rite.

The Aged

In all cultures of Nepal, to reach an old age is a guarantee of respect. A special ceremony is held to honor a person who reaches the age of 84. Elderly persons continue to contribute to household labor and childcare, to the extent of their abilities, and take major roles in family decision-making. Care for the elderly is regarded as a responsibility of the extended family.

The major medical problems of the elderly in Nepal are not significantly different than those found throughout the world.

Dying and Death

All Nepalis prefer to die at home, though some indigent elderly are cared for in hospices maintained by charitable organizations, primarily in the Kathmandu Valley.

A coin is placed under the tongue of a dying person; if not spat out, death is considered immanent. A drop of water purified by having had gold immersed in it is poured through the dying person's lips. After death, the body must be handled properly, touched only by persons of equal or higher caste, or the deceased may later trouble his family. Immediately following the last breath, a body is wrapped in a white or saffron shroud and tied to two bamboo poles to be carried to the cremation or burial site. Cremation on a riverbank is the preferred method for funerals, although both earth and "sky" burials are practiced in some communities, the latter found in some Tibeto-Burman communities who expose the body on an elevated platform to be consumed by vultures, graphically "paying back" to nature the debts incurred in a lifetime of dependence on the world.

For Hindu funerals, the eldest male circumambulates the pyre and is responsible for lighting it. Mourners remain in attendance until the skull splits, releasing the soul. Ashes are scattered in the river. If a family can afford it, a piece of bone is preserved, taken to Benaras in India, and cast into the sacred Ganges River there.

Those who suffered from leprosy may not be cremated, as the disease is traditionally regarded as a curse that denies human rebirth. Smallpox victims were also buried, a practice explained by their having been possessed by the smallpox goddess, Sitala. Children under the age of five are ordinarily buried, with little ceremony, as are those too poor to afford cremation. Muslims and Christians in Nepal also bury their dead, in graveyards.

Mourning by the immediate family is marked by fasting, including abstinence from salt, oil, and meat for 13 days. More distant relatives observe one day of fasting. Males have their head shaved as a sign of mourning, while women dress in white, without jewelry, and leave their hair uncombed. The death of progenitors is commemorated by annual *sraddha* rituals, in which they are symbolically fed balls of rice offered by all male descendants.

In Nepal, the Hindu custom of "sati," the immolation of widows with the husband's body, was never common except among the royal families. It was firmly abolished by Prime Minister Chandra Shumsher Rana on June 28, 1920 AD, at which time it became illegal and punishable as culpable homicide. The "sati" gate of Pashupati temple, the holiest site in Nepal, through which widows had been taken to their husband's pyres, was bricked shut at that time.

CHANGING HEALTH PATTERNS

Screening for HIV was begun in Nepal in 1986; the first case of HIV/AIDS was identified in June 1988. By September 1998 a total of 1,132 HIV positive cases and 108 deaths from AIDS had been recorded. Given the increasing number of young Nepalis who seek work elsewhere, as well as the widespread trafficking of Nepali girls into the sex trade of India, AIDS is expected to have major future consequences on the economics of health in Nepal.

It is too soon to predict whether the sudden flourishing of new medical colleges throughout Nepal, beginning in 1994, will have any major impact on the country, since standards of education are problematic and most students are expected to come from India.

It has been recently reported that Nepal hopes to become a center for inexpensive kidney transplants (Rising Nepal Daily Newspaper, 24 December 2001), projecting costs at less than half of those in India.

In the last 20 years Nepal has seen more rapid social change than at any time in its history, with large segments of the population increasingly marginalized economically and politically. Consequently, the overall prognosis for the state of health in the country is poor.

References

Ali, Almas. (1991). *Status of health in Nepal*. Kathmandu: Resource Center for Primary Health Care, Nepal and South–South Solidarity, India.

Bista, Dor Bahadur. (1972). *People of Nepal* (2nd ed.). Kathmandu: Ratna Pustak Bhandar.

Dixit, Hemang. (1999). *The quest for health: The health services of Nepal* (2nd ed.). Kathmandu: Educational Enterprise.

Maskarinec, Gregory G. (1995). *The rulings of the night. An ethnography of Nepalese shaman oral texts*. Madison: University of Wisconsin Press.

Pradhan, Ajit et. al. (1997). *Nepal family health survey 1996*. Kathmandu: New Era.

Sigdel, Shailendra. (1998). *Primary health care provision in Nepal*. Kathmandu: GTZ/Primary Health Care Project.

Tausig, Mark & Sree Subedi. (1997, August). The modern mental health system in Nepal: Organizational persistence in the absence of legitimating myths. *Social Science & Medicine*, *45*(3), 441–447.

Worth, Robert M. & Narayan N. K. (1969). *Nepal health survey 1965–1966*. Honolulu: University of Hawaii Press.

The Northwest Coast

Peter H. Stephenson and Steven Acheson

Alternative Names

The Northwest Coast is the standard name for the culture area. The main constituent cultures are from north to south, Eyak, Tlingit, Haida, Tsimshian, Haisla, Haihais, Heiltsuk (formerly Bella Bella), Nuxalk (formerly Bella Coola), Oowekeeno, Kwakwaka'wakw (formerly Kwakiutl), Coast Salish, Nuu-chah-nulth (formerly Nootka), Makah, Quileute/Chemakum, Chinookans, Takelma, Alsean, Siuslaw, Coos, and Athapaskans.

Location and Linguistic Affiliation

First coined by European explorers to the region in the late 18th century, "Northwest Coast" is now firmly entrenched in the anthropological literature to describe a culture area extending some 2600 km from Prince William Sound in the Gulf of Alaska, along the coast of British Columbia, to the California–Oregon border. At the time of contact with Europeans, an estimated 200,000 people lived in the region (Boyd, 1990) making it one of the most densely settled in the Americas north of Mexico (Suttles, 1990). Northwest Coast peoples do not constitute a single biological population or even a series of discrete populations, but are grouped together on the basis of shared cultural practices and geography.

This chapter focuses on the peoples north of California including southeast Alaska, Washington, and Oregon in the United States, and British Columbia in Canada. Most of the peoples of northern California were destroyed in violent encounters with gold miners and by infectious diseases by the late 1800s. Although the people of this region have long shared similar cultural patterns which stem from environmental factors, trade, and a system of raiding which involved captive workers, they belong to a dozen different language groupings. Typical of extremely mountainous terrain, cut by major rivers, the linguistic differences in the region are profound, and occur at the level of language families. Therefore the classification, conceptualization, and treatment of illness vary quite distinctively throughout the region. The Northwest Coast exhibited the greatest linguistic diversity within aboriginal North America, next to California. Over 40 languages, representing 12 language families, were once spoken on the Northwest Coast. Distribution of these language families, and particularly some discontinuities, indicate a complex settlement history within the region. Athapaskan speakers include the *Eyak*, the northern most Northwest Coast group, and two branches on the lower Columbia and southwest Oregon. The other 11 linguistic families ranging from north to south include: *Tlingit, Haida, Tsimshian, Wakashan*, which includes four major branches—*Haisla, Heiltsuk, Kwakwaka'wakw and Nuu-chah-nulth; Salishan*, represented by the *Coast Salish, Nuxalk, Tsamosan*, and *Tillamook*; *Chimakuan*; and five possibly related families belonging to the Penutian Phylum—*Chinookan, Takelman, Alsean, Siuslaw*, and *Coos* (after Thompson & Kinkade, 1990).

OVERVIEW OF THE CULTURE

Culture and Environment

There are at least 40 different culture groups in the Northwest Coast; we describe here the common and the variable features. The Northwest Coast is often described as a "classic" culture area in North America due to its distinctive and rich artistic traditions of monumental wood carving ("totem poles") and masked ritual performances of dance and song (See Drucker, 1965; Suttles, 1990). The importance of early anthropological fieldworkers and ethnographic writers (including Boas, Benedict, and Sapir) has also contributed to the visibility of the region and its peoples (Hawthorn, 1965). The Northwest Coast also stands apart as a distinct cultural region based on a number of widely recognized, shared cultural traits. Northwest Coast peoples were truly maritime, possessing a sophisticated fishing and sea mammal hunting technology, with river-run Pacific salmon of particular importance to many. Coupled with this was an efficient preservation technology for the long-term storage of foodstuffs. These technologies sustained dense population aggregates where people resided in semipermanent or permanent villages, with a system of rigid social ranking and stratification, and the creation of an elaborate art and architectural tradition. Social classes were recognized based on a combination of birth and wealth with chiefs and immediate kin forming the nobility, their followers or commoners, and slaves, who were acquired as property either by capture or purchase. Attention to social standing varied greatly, however, with class distinctions being more pronounced and rigidly maintained among the more northerly coastal groups. The importance of kinship shows a similar north–south trend with kinship ties being most rigidly defined among northern groups. As well, the kinship organization of people in the north is strongly matrilineal, and subsequently shifting to patrilineal descent as one moves southward.

The coastline of the Northwest Coast region is characterized by a chain of extremely rugged mountains which rise in places to over a thousand meters directly out of the sea. One need only precede a few kilometers inland to encounter even more precipitous mountains, glaciers, and dormant volcanoes rising 3,000 m and more. These mountain ranges are partially submerged in places, creating the many large island archipelagos that skirt the coast of Alaska and British Columbia. The mountains capture warm pacific air masses transiting from the Hawaiian Islands, and thus receive a heavy annual rainfall along the coast, with major snowfalls in the mountains during the winter. The region is also warmed by the Japanese current and so coastal climates are temperate and very moist, making large parts of the region a true rainforest—the only major one in a temperate region. Summers throughout the region tend to be very dry however, so the Pacific northwest is also the only major rainforest in the world comprised of evergreen trees—pines, cypress, hemlock, fir, and spruce. The deciduous trees are mainly found along rivers and streams and include wide variety of willows. Many fungi are found throughout the region (Suttles, 1990b). A wide variety of plants were collected by all of the peoples of the region and used as medicines and teas.

The major rivers of the region (the Stikine, Nass, Skeena, Bella Coola, Fraser, Columbia, and Klamath) and their systems of tributaries, as well as even very minor streams, are home to many varieties of fish: especially migrant salmon species, eulachon, lampreys, and cutthroat trout. There are still massive sturgeons in some of the larger rivers, although their numbers are seriously depleted, and their size reduced. A wide variety of fish are also available just offshore (especially many types of cod, halibut, rockfish, herring, and hake) and many varieties of shellfish can be collected (clams, oysters, crabs, octopus). Sea mammals—especially whales and seals—were traditionally hunted; and still are in some locations. Although a large proportion of the coastal traditional diet came from the sea, land animals and plants were and are also important foods; these include most large mammals, birds, and bird eggs. Plant foods included a wide variety of berries, a great many root vegetables (over 25), and seaweed. Seaweed covered in recently deposited herring roe is a widespread nutritious delicacy throughout the region (Hopkinson et al., 1995). Because the migratory salmon species required intensive processing and storage (gutting, drying, and smoking) during very brief annual periods, captives (especially women and children) were often taken in raids and used as labor (Donald, 1997).

Ritual practice and religion throughout the Pacific northwest was complex, and involved masked dancing, feasting, title giving, and property distributions (the "potlatch"). The mythology of the people throughout this region is elaborate and characterized by a belief in distant deities associated with the sun and sky and much more important proximal supernatural beings who co-inhabit the world of forest and beach with mortals. These supernatural beings may cause all manner of problems—including illness—as well as intercede to assist people in their daily lives. Many of these spirits act as guardians of the living and can guide one through life; they are often associated with animals and may appear in animal form. Other spirits are more monstrous and take the form of ogres, dwarfs, and giant man-eating cave dwelling birds. On the very highest peaks dwell "thunderbirds," said to keep lightning in the shape of large reptiles (Drucker, 1965).

To successfully deal with spirits, a person must be ritually "clean" by a combination of fasting, dietary proscriptions, bathing, and the use of purges or emetics. Maintaining a ritually cleansed state was a way to stay healthy, to receive good luck, and to avoid harm. If in the course of maintaining ritual purity a man encountered a supernatural being, he might receive the gift of a special power. In some groups these powers were associated with success in hunting or warfare, but in many they were associated with the power to heal (Drucker, 1965).

Historical contact wrought many changes to the indigenous peoples of the region, ranging from dislocation and the restructuring of the traditional social structure to outright extinction. Symptomatic of these events have also been changes in the ethnonymy for a number of surviving groups who have chosen to reidentify themselves by more traditional names from those assigned within the contact period. Among these are the *Heiltsuk*, who were formerly known as the Bella Bella, Kwakiutl have now become *Kwakwaka'wakw*, Nootka are now *Nuu-chah-nulth*, and the Bella Coola are now known as the Nuxalk.

Social Organization

Aside from their linguistic diversity, the ideas of family, ancestry, descent, and kinship differed greatly, though there are similarities across cultural boundaries. Small, localized groups of people formed the principle unit of production and consumption organized around a core of kin defined according to the local rules of kinship. Ownership of resources typically resided with the kin group, whose members associated themselves with specific localities and village sites. Traditional village life included permanent winter villages, which often grew in population size through the winter feasting and ceremonial season to include hundreds of people. In the summer, families often dispersed to annual settlement camps to collect shellfish, fish and hunt.

Among the *Eyak*, *Tlingit*, *Haida*, *Tsimshian*, and neighboring *Kwakwaka'wakw*-speaking *Haisla* on the northern coast, autonomous matrilineal households or lineages formed the basic units of a village. Typically "towns" were comprised of one or more matrilineal lineages. Each lineage was represented by a hereditary chief who acted as trustee of the lineage properties while house chiefs managed the affairs of their own individual houses. With the possible exception of the *Tsimshian*, these northern groups divided their societies into moieties, as a means of organizing relationships between individuals, families, and lineage groups. According to Halpin and Seguin (1990), traditional *Tsimshian* society at the village level was a moiety, like that found among the *Eyak*, *Tlingit*, and *Haida*, and not the four-fold structure commonly reported. The basic social unit was the corporate matrilineage called a "house" whose members, together with affines, children belonging to other lineages, and slaves, occupied one or more dwellings.

The *Eyak* represent the northernmost Northwest Coast group, and are linguistically related to the Athapaskan family and more remotely to the *Tlingit*. Within the historical period, the neighboring *Tlingit* were to exert considerable cultural influence on this group

Overview of the Culture

through trade, intermarriage, purchase of lands, or conquest (de Laguna, 1990). To the south, the *Heiltsuk, Nuxalk, Kwakwaka'wakw*, and *Nuu-chah-nulth* had corporate kin groups, described as "ancestral families," which functioned as crest-holding units based on extended bilateral descent. The neighboring *Haisla* are a northern Wakashan language isolate related to the *Kwakwaka'wakw*, who occupy the upper reaches of Douglas Channel and Gardiner Canal on the inner coast. The *Haisla* resembled the *Tsimshian* in technology and social organization, but maintained a set of secret societies typical of other Northern Wakashans (Hamori-Torok, 1990). The neighboring *Heiltsuk*, and related *Haihais* and *Oowekeeno*, occupy the shores of Queen Charlotte Sound from Price Island on Milbanke Sound to the southern shore of Rivers Inlet and inside channels and inlets. Though neither rigidly exogamous nor matrilineal, both residence and descent among these groups mirrored that of exogamous matrilineal northern groups (Hilton, 1990).

On the adjacent mainland coast are the *Nuxalk*, Salishan speakers who formally occupied a number of permanent villages alongside and at the mouths of major salmon rich rivers and creeks in the Bella Coola valley, North and South Bentinck Arms, Dean Channel, and Kwatna Inlet (Kennedy & Bouchard, 1990). *Nuxalk* society consisted of descent groups who held in common a set of ancestral names and prerogatives based on an origin myth. Though a number of descent groups might share a village site, the household remained the primary social and economic unit. Descent was traced ambilineally while residence was patrilocal which tended to reinforce bonds with the father's side (McIlwraith, 1948).

The *Kwakwaka'wakw* of the Wakashan language family occupy northern Vancouver Island, the adjacent mainland and intervening islands of Queen Charlotte and Johnston Straits. They once consisted of some 30 autonomous groups each consisting of several corporate kin groups or "*numaym*," who were the owners of resource sites, myths, and crests. The southern branch of this language family includes the *Nuu-chah-nulth*, the west coast people of Vancouver Island, and the *Makah* on the Olympic Peninsula within Washington State. Local kin groups linked by ambilineal descent held defined territories. Local kin groups did on occasion unite to form "tribes" and at one point the Northern *Nuu-chah-nulth* formed a number of local confederacies consisting of a series of distinct "tribes" whose ranked chiefs shared a common summer village. Political authority, however,

remained with the local group in the absence of any formal political office for these larger aggregations (Drucker, 1951, 1983).

The salmon rich waters between Vancouver Island and the mainland stretching from Johnstone Strait to the Strait of Juan de Fuca and adjacent Puget Sound are the home of the *Coast Salish*. The *Coast Salish* were similarly bound together by bilateral kinship ties (Duff, 1964). Clusters of villages were so closely identified with each other by virtue of locality, dialect, culture, and intermarriage as to become distinct units and bear a common name (Kennedy & Bouchard, 1990; Suttles, 1990). Like all Northwest Coast groups, the *Coast Salish* recognized three "classes" of people, but the distinction was neither as rigid nor as pronounced as that found among the more northerly groups (Suttles, 1990).

In addition to an enclave of *Coast Salish* (*Tillamook* and *Tsamosan*) and *Chimakean* speakers, a mosaic of smaller language groups, collectively assigned to the Penutian Language Phylum, occupy the coasts of Washington and Oregon. The basic social and political unit amongst these groups, as well as a neighboring group of *Athapaskan* speakers, was the autonomous winter village group, consisting of one or more residence groups of paternally related kin. Residing on the coast and major rivers of southwestern Oregon, this Pacific branch of the *Athapaskan* language family represents the southerly most Northwest Coast group.

Contact, Contagions, and Change

The introduction of various high-mortality, density-dependent diseases, previously unknown to pre-contact populations, contributed in no small measure to the erosion and restructuring of traditional lifeways. Their severity varied from group to group according to their specific contact histories. Some, like the *Haida* and *Kwakwaka'wakw* experienced abrupt, catastrophic population loss. For others the process of depopulation was more gradual. In some cases the outcome was outright extinction of local groups, while in other instances new creative cultural practices emerged to cope with the changes of the period such as nativistic movements like the Prophet Dance. Smallpox proved by far to be the most devastating disease, and in the wake of this contagion, which struck on at least three occasions between the late 1700s and the late 1800s, were a host of other serious illnesses. Measles, influenza, whooping cough, tuberculosis, and scarlet fever reached epidemic proportions among an increasingly weakened and vulnerable native community throughout the 1800s. In the case of the *Haida*, the *Nuu-chah-nulth*, and many others, a lethal combination of disease, warfare, dislocation, and stress worked to destroy many family groups with weakened or remnant groups joining more powerful neighbors to create new political structures (Acheson, 1998; Arima and Dewhirst, 1990). For small-scale societies everywhere, catastrophic population losses required, if they were to survive, new social, economic and political alignments.

The concept of "composite bands" and "tribes" as a product of the merging of various like bands due to the "initial shock, depopulation, relocation, and distribution of the early contact period" is certainly not new. Among the *Nuu-chah-nulth*, for example, declining populations and extended warfare throughout the late 1700s and early 1800s worked toward creating or maintaining tribes and confederacies (Drucker, 1951).

The move away from matrilineal kinship toward more patrilineal oriented descent observed among the *Haida*, for example, can in part be attributed to severe population fluctuations. The need to restructure social groupings as a result of population decline, coupled with the growth of a market economy, increasing economic cooperation between father and son, and the growing advantages of politically important rank in a father, conspired to break down the matrilineal descent group.

The implications of these trends are equally significant to traditional native experiences with illnesses and medicinal practices, the role of shamans, and religious practices. Shamans occupied an ambiguous position within the community in their role as both caregiver and a potentially malevolent force that possessed unique supernatural curing, divining, or witchcraft capabilities. They also had an extensive practical knowledge of pharmacologically active plants and herbs. The introduction of new, destructive illnesses not readily understood, placed shamans in a particularly precarious position. The discrediting of their role and diminishing value placed on traditional healing practices, and even the outright loss of such cultural knowledge, created a vacuum that gave rise to the messianic movements of the 1800s and enabled the missionary movement to make significant inroads by the late 1800s. The emergence of "revitalization" or "nativistic" movements, such as the Prophet Dance and Shaker religion among a number of coastal groups, underscores this relationship. But it also demonstrates resourcefulness on the part of the people confronting a new political and economic reality, who sought to relieve the afflictions of mind and body through a creative blend of traditional healing and Christian practices.

THE CONTEXT OF HEALTH: ENVIRONMENTAL, ECONOMIC, SOCIAL, AND POLITICAL FACTORS

Diseases of the Pre-Contact and Early Contact Periods

At least four infectious diseases—viral pneumonia, non-venereal syphilis, tuberculosis, and trachoma—were known to pre-contact peoples of the Americas. Most acute infections rarely involve bone, however, making the detection of disease among pre-contact populations extremely difficult to detect. The kinds of high-mortality density-dependent infectious diseases common to the Old World are thought to have been absent prior to cultural contact (Boyd, 1985).

Chronic bone infections likely due to tuberculosis or treponematosis, and malignant bone tumors (both primary and secondary) were present among coastal populations (Cybulski, 1990). Of these three disease groups, only treponemal infections, likely a form of endemic non-venereal syphilis, appear in the prehistoric record for the Strait of Georgia and Prince Rupert Harbour (Cybulski, 1994; Skinner, McLaren, & Carlson, 1988). A form of conjunctivitis or trachoma often resulting in blindness and leprosy is documented within the initial contact period, suggesting a pre-contact origin. While known to pre-contact New World populations, the presence of tuberculosis on the coast is so far confined to the historical period, first appearing among the *Nuu-chah-nulth* at Nootka Sound in 1793 (Boyd, 1990).

Cribra orbitalia (porotic hyperostosis), cited as a possible indicator of iron-deficiency anemia, has been detected in coast-wide prehistoric and historical populations at frequencies of 13–14%. Causes of iron-deficiency anemia are varied, but the low frequency of cribra orbitalia is seen to indicate a population's successful adaptation to the pathogen load of its local environment, including all fungi, viruses, bacteria, and parasites. Nutritional health among coastal populations generally appears to have been stable from 3500 BC to the post-contact period (Cybulski, 1994). Mortality rates, on the other hand, underwent an unprecedented increase with the introduction of a suite of highly infectious diseases following contact.

Described as "the most terrible single calamity" to strike the Aboriginal community (Duff, 1964), smallpox was a reoccurring pandemic resulting in the progressive and catastrophic decline in population from the outset of culture contact. Following in the wake of the major smallpox epidemics of the late 1700s, 1837–38, and 1862–63, were a host of other diseases, including measles, influenza, whooping cough, tuberculosis, and scarlet fever, that reached epidemic proportions among an unsuspecting and highly susceptible native community (Table 1). Lingering diseases such as influenza, scrofula, and syphilis, observed among later native populations, can also be inferred for the early contact period.

By far the deadliest of all the "virgin soil" epidemics to strike the Northwest Coast was smallpox, which appeared in cycles as a result of contact with outside carriers and a sufficiently large enough population of non-immunes for the disease to take hold and spread. The disease afflicted people of all ages, and not just successive generations of children, which had the effect of fragmenting families and dissolving kin groups. Caring for the ill was thus made much more difficult and mortality rates soared.

Historical accounts provide indisputable evidence of smallpox on the coast by the 1770s, appearing among the *Haida*, *Ditidaht* (*Nuu-chah-nulth*), the *Coast Salish*, and the *Chinookan*. It struck the *Tsimshian* in 1795 and the *Coast Salish* again a few years later. Though the severity of these early outbreaks can only be inferred in the absence of detailed population records for this period, it is reasonable to expect a catastrophic population decline equal to (if not greater than) the population losses of the late 1800s. Trade, warfare, and even attempts to escape an outbreak spread the virus rapidly. For the period 1835 to 1890 the number of lives lost on the coast is estimated to be in the realm of 62%, and possibly as great as 90% (Boyd, 1985, 1990, 1994) (Tables 1 and 2).

Within the first quarter of the 19th century another epidemic, variously identified as smallpox, measles, or simply "the mortality", was reported among the *Cowichan* on southern Vancouver Island (British Colonist, 1862). On the North coast, smallpox reappeared among the *Tsimshian* at Fort Simpson in the early autumn of 1836, spreading north to the Nass by December and by the spring of 1837 had reached the *Haisla*, *Haihais*, *Heiltsuk*, and *Nuxalk* in the south (Boyd, 1990; Tolmie, 1963). Hudson's Bay Company officials maintained that the epidemic claimed a third of the population on the North coast, which corresponds closely with recent estimates based on the detailed work by Boyd (1985, 1990) (Table 2), while the *Haida*, *Nuu-chah-nulth*, and *Kwakwaka'wakw* were seemingly spared.

The measles epidemic of 1848, and a subsequent outbreak of smallpox in 1862, spread throughout most of the coast with the possible exception of the *Nuu-chah-nulth*. The *Ditidaht*, a subgroup of the *Nuu-chah-nulth*, however, were struck by smallpox in 1853 and again in 1874 along with the Central *Nuu-chah-nulth* around Barkley Sound (Boyd, 1990; Drucker, 1951). This, and the increased presence of tuberculosis and respiratory diseases, pushed the population to a low of 1605 in 1939 (Duff, 1964). Their immediate neighbors to the north, the *Kwakwaka'wakw*, suffered a second outbreak of measles in 1868 (Boyd, 1985). The 1862–63 smallpox epidemics, while only marginally affecting the Central *Coast Salish*, due in part to the widespread use of smallpox vaccine around missions and white populated areas, devastated the Northern groups. From a preepidemic population of 6,693 around 1840, the *Haida* numbered just 741 in 1881, reaching a low of 588 in 1915 (Canada Census, 1881; Duff, 1964). For the same period, the *Tsimshian* experienced a decline approaching 35% due to the combined impact of the 1848 and 1868 outbreaks of measles and intervening outbreak of smallpox in 1862–63. The *Tlingit* had fallen to 4501 in the same period, representing less than a third of their estimated pre-contact population. The *Haihais* were reduced to just one village site by 1870, having vacated all their villages for a former seasonal camp at Klemtu (Hilton, 1990). The same pattern was repeated among the *Heiltsuk*, who united at the one settlement at Bella Bella by the 1870s (Olson, 1955). Of the more than 45 known *Nuxalk* villages inhabited in 1793, they numbered half that by 1889 and had declined to three when McIlwraith (1948) conducted his fieldwork in 1922–24. At the end of the 1800s there was a comparable decline in the number of *Kwakwaka'wakw* villages and kin groups with only 19 *Kwakwaka'wakw* of some 30 "tribes" surviving in just a handful of communities (Codere, 1990).

The scale of destruction in terms of population loss is only part of a complex picture concerning the impact

Table 1. Coastal Population History

	Pre-contact projection (Boyd 1990)	1780 (Mooney 1928)	1829 (Green 1915)	1835 (Tolmie n.d.)	1836–41 HBC Census (Douglas n.d.)	1858 (Duncan n.d.)	1881 Canada Census	1889	1890 (Canada 1891)	1895 (Canada 1896)	1896 (Canada 1897)	1906 (Mooney 1928)	1930
TLINGIT	14,820	10,000			9,880								4,462
TSIMSHIAN	14,500	5,500											
Nishga					1,615 (1,625)			726	2,075	1,333	782	814	
Coast Tsimshian			5,500		2,815 (2,827[b]; 2,495[c])	2,500		1,869			1,364	1,383	
Gitksan					409	2,352		1,462	1,119	1,087	1,071	1,130	
Southern Tsimshian				1,225	1,429	2,500		284					
HAIDA	14,500	8,000	3,000									599	
Massett					3,291 (3,285[c])		315	445	438	364	354		
West Coast					1,086								
Skidegate					2,316		426	285	292	229	244		
HAISLA	1,200				409		233	364	392	393	377		
HEILTSUK	4,500	2,700		1,628[a]	1,834 (2,871[b]; 1,628[c]; 1,250)		662	481	415	578	533	852	
NUXALK	3,000	1,400		1,940	509		491	378	204	340	340	288	
KWAKWAKA'WAKW	19,000	4,500		8,500	32,460[a] (40,998[d])		2,252	1,900	1,797	2,853	1,639	1,257	
NUU-CHAH-NULTH	8,000	6,000		7,500	7,093[a]		3,593 (3,613[c])	3,093	3,084	2,834	2,750	2,159	
COAST SALISH													
Northern	4,000	3,100						1,012	868	813	770	567	
Central	14,000	17,400		5,025	8,216			5,899	5,364	5,309	4,640	3,658	
Southern/Southwest	24,399	9,300											
CHEMAKUAN	780	900											
PENUTIAN SPEAKERS	11,990	8,600											
ATHAPASKAN (Lower Columbia; Oregon)	8,640	7,500											
Bands not visited									8,522				

Some discrepancies in the figures exist due to differences in identification and ethnolinguistic grouping as well as calculation errors in some of the original tables. Entries prior to the mid-1830 are estimates only.
[a]Douglas, 1856; [b]Kane, 1859; [c]Martin, 1848; [d]Curtis, 1915; [e]Dominion of Canada 1882:164.

Table 2. Projected North Coast Population Losses from Smallpox, 1836–37

Group	Pre-epidemic population	Post-epidemic population	Total loss	%
Tlingit	9,980	7,255	2,725	27
Nishga	2,423	1,615	808	33

of these diseases on coastal populations. These losses not only posed a threat to the stability and continuity of social institutions through the loss of cultural knowledge, but the occurrence of these diseases also posed a direct challenge to traditional religious beliefs, placing even greater duress on the community. Demoralization itself undoubtedly worked to compound the clinical impact of the epidemics. At the same time we see evidence of a creative accommodation by the native community to new circumstances that drew on traditional medicinal practices and beliefs.

MEDICAL PRACTITIONERS

The practice of 'medicine' embodied an array of ideas and concepts along with highly practical and effective remedies and treatment. Alongside the skillful practical treatment of injuries such as fractures and wounds, there is a long history on the Northwest Coast of shamanic practices, which invoke the spiritual both as the cause and in the treatment of the suffering.

Shamans

Shamans act as a medium for the supernatural, possessing unique curing, divining, or witchcraft capabilities acquired through birth, visionary experiences, contact with some form of supernatural force and, in some cases, through vocational training. Among most Northwest Coast groups a person who received healing powers had to apprentice with an established shaman in order to learn to control his newfound powers. The period of apprenticeship varied, but normally ended with a major healing performance. Illnesses cured by shamans were (and are) of essentially three types: soul loss, spirit disease, and spirit or object intrusion. Shamanic practice is now much reduced, but elements of it can be found in many groups, and some much respected shamanic practitioners can be found throughout the region.

Traditional Ethno-Biological Knowledge: Herbalists

Shamans also possessed an extensive knowledge of pharmacologically active plants and herbs. Herbal healing is seen, however, as somewhat distinct though not entirely separate from the magical or supernatural healing practices by shamans. This distinction between "ritualists" and "herbalists" is near universal in the region. Certain plants were considered "personal property" while other medicines, such as the cascara bark laxative, were universally known (Bouchard and Turner, 1976). The Green or Indian hellebore of the Lily Family (*Veratrum viride*), for example, had widespread use in North America for curing wounds and alleviating toothaches (Vogel, 1970). An extremely poisonous plant due to a combination of toxic alkaloids, hellebore was widely known to coastal peoples for its medicinal properties "as a blistering agent, local anaesthetic and decongestant, and internally as a physic" (Turner, 1978). The effectiveness of many herbal medicines, however, was dependent on a level of secrecy about the type of plant used and its application, and in practice most treatments combined herbal use with shamanism.

As privately held secret knowledge, plant remedies retain potency and also secure some respect and power for those who maintain the knowledge. Much of this practical knowledge was lost over generations confronting severe depopulation from waves of epidemic diseases like smallpox, tuberculosis, influenza, and pneumonia. Confidence in shamans to be able to deal with these terrible epidemics has also meant a loss of knowledge in dealing with psychological anguish.

Today, many people with plant knowledge confront a difficult ethical dilemma; if they choose to share their knowledge the plants upon which they depend may become desirable commercial commodities and be overharvested (as happened with Pacific Yew in the treatment for breast cancer). Losing control over traditional knowledge therefore may not benefit local people and may harm their environment. Yet local indigenous people often wish to assist a wider humanity with their knowledge. As well, the efficacy of teas and decoctions may vary greatly between locally prepared and ritually administered herbs and synthesized commercial preparations.

CLASSIFICATION OF ILLNESS, THEORIES OF ILLNESS, AND TREATMENT OF ILLNESS

The treatment of externally caused injuries, such as fractures, dislocations, wounds, skin irritations, and the like was highly practical and effective. The use of splints to set fractures, for example, was universal and included the steps of straightening and use of pain-allaying medicine. Evidence of the cauterization of teeth to alleviate severe periodontal disease and trephination in accordance with the surgical principle of fracture decompression, is cited for pre-contact populations. Drawing on the affected part effectively treated snakebites and the cleaning of wounds. Such a practice extended to the treatment of internal ailments attributed to supernatural agencies or spirit intrusion.

Generally, disease was the result of either a human or supernatural agent, though natural causes could be considered. Acts such as sorcery, taboo violation, disease-object intrusion, spirit intrusion, and soul loss called for culturally prescribed treatment by a shaman. Confession often forms part of the treatment, serving as a powerful catharsis in helping to alleviate sickness due to broken taboos.

Soul loss is often associated with what could be glossed as depression, possibly including a degenerative disease as well. Although soul loss does not kill its victims immediately, it can be lethal if the soul is not retrieved in a reasonable period of time. Only the most experienced shamans treated (and continue to treat) soul loss. In traditional ceremonies the shaman, or his spirit helper, retrieves the lost soul. Often the shaman brings the soul back to the person in a bed of eagle down; he may also have to struggle against many hostile spirits to return the soul to the person. In our experience soul loss is often associated with people whose close friends and relatives have recently died in a sudden and unexpected manner—indeed, the departed may wish the individual to come and live with them in the realm of the dead. Spirit diseases are generally associated with ritual impurity and the afflicted individual may either become possessed by the spirit itself, contract a fatal disease, or become contaminated by the spirit. Although some shamans have claimed to be able to treat spirit illnesses, most appear to have regarded them as terminal. Spirit and object intrusion is associated with belief in witchcraft: malevolent spirits or shamans hired by others or acting on their own can send illness producing objects (bone slivers, etc.) into a person's body. The shaman, in a dramatic performance, removes these objects. This kind of classic shamanic performance transforms a person from the subject of hatred by unknown spirits or people into the object of community concern and assistance. It is in essence restorative whereby sick individuals become the focus of family and community concern.

Aside from, though often combined with these practices, is the widespread medicinal use of plants and plant products. Commonly utilized plant medicines throughout the region include various rejuvenating teas made from aquatic plants that grow in streams; treatments for relief of pain obtained from tree bark (willow, as the source of aspirin is known) or especially Devil's Club root (Turner, 1982). Diuretics created from stinging nettle teas are widely used. Bark preparations taken from a wide variety of trees are also used for respiratory, digestive, gynecological, and dermatological ailments throughout the area. Tree barks have also been used to treat fevers, diabetes, kidney problems, sore eyes, and hemorrhaging, and also as general tonics. In most cases, infusions or decoctions of barks are used. The medicines are drunk or applied externally as a wash and are particularly well described for Coast Salish people on Vancouver Island (Turner, 1990).

As well, at least 20 species in *Ranunculaceae,* the buttercup family, are reported as having been used medicinally by 19 different groups of native peoples in British Columbia and adjacent areas. These species are known to contain the skin-irritating, blister-causing compound, protoanemonin, in their fresh state and it is probably the active principle involved in many of these medicinal applications. Most groups utilized the plants as external poultices for boils, cuts, abrasions, and other skin sores. Other disorders treated with anunculaceous species include: muscular aches, colds, and other respiratory ailments (Turner, 1984). The slime from slugs, which numbs skin surfaces and serves to protect the animal from predation, also appears to have been widely used as a topical anesthetic.

Many plants or plant products used on the Northwest Coast have made their way into Western medicine, and include a variety of laxative, diuretic, emetic, and febrifuge drugs (Vogel, 1970). Examples include sarsaparilla (*Aralia nudicaulis*) used by the *Nuxalk* and the *Kwakwaka'wakw* for stomach pains.

The use of Oregon grape root (*Berberis aquifolium*) as a bitter tonic among the *Kwakwaka'wakw* was adopted by Europeans. Cascara bark was similarly valued by Europeans as a tonic and laxative long recognized by coastal peoples (Boas, 1932; Turner, 1975; Vogel, 1970).

An array of medicinal treatments was added to an already formidable native pharmacopeia to cope with introduced diseases, including smallpox, whooping cough, tuberculosis, and syphilis. Attempts to arrest these diseases, however, met with little success. One of the cruelest ironies for a community exposed to new contagions was the universal use of sweat baths as a panacea for most diseases. Certain plant medicines were felt to be most effective when taken in the sweathouse. The use of steam baths to fight febrile infections such as smallpox, however, only exacerbated the illness and aided in its spread. Complications arising from this practice, such as pneumonia and pleurisy, greatly increased the probability of death.

HEALTH THROUGH THE LIFE CYCLE

Pregnancy and Birth

Birth traditionally necessitated the seclusion of both mother and father and enforced dietary restrictions throughout the Northwest Coast. Infants have also long been regarded as particularly vulnerable to illness through the machinations of malign spirits, including ghosts, which may cause sickness. Twins were often thought to be special people but there is little evidence of their receiving special treatment except among the *Nuxalk* where Kennedy & Bouchard (1990, p. 331) report that during the 1970s it was believed that twins resulted from the spirit of Salmon entering the body during pregnancy. In later life *Nuxalk* twins also had to be careful not to offend the spirit of the Salmon, but were said to be able to induce salmon runs. Likewise the birth of people with defects, while felt to be unfortunate, does not normally lead to poor treatment. In British Columbia death from congenital anomalies among infants is actually slightly lower for Native populations than the rest of the province (Foster et al., 1995, p. 69). Infanticide and abortion was originally reported by McIlwraith (1948, pp. 702–712) among the *Nuxalk*, and is suggested for the *Tillamook* as well (Seaburg & Miller, 1990, p. 564). These practices appear to have been associated with the disgrace of illegitimate births in the past. The practices, however, are not widely reported throughout the region. The extent of breastfeeding is unknown for the region although Aboriginal midwifery and its revival after colonial suppression is well described in British Columbia among coastal peoples (Benoit & Carroll, 1995).

Infant care ideally takes place within extended families, and children often spend a great deal of time with grandparents, and "aunties" and "uncles" of their own clan. Sudden infant death syndrome is disastrously high in many Northwest Coast groups and well documented in British Columbia but is often associated with older birthmothers among Native people, rather than younger ones as is more common in the general population (Foster et al., 1995, p. 70). Adolescent pregnancy is currently relatively high as are the numbers of low birth weight babies (Foster et al., 1995, p. 58).

The umbilicus was traditionally treated with great ceremony, as well as early childhood belongings—especially the cradle and infant clothing. The belongings of children who die are often ritually burned, often on an isolated strand of beach. When newborns die in hospital, ritual destruction may include sonograms, clothing, toys and the like and can be very helpful to grieving mothers. Ritual purification, through bathing in private ceremonies for parents ended their seclusion, and is still practiced in modified form in some groups today.

Adolescence

Seclusion of girls at puberty accompanied by dietary restrictions is traditional and premarital sex has long been frowned upon throughout the region. In many areas where potlatch ceremonies and feasting continues, a major public ceremony for the daughters of high status chiefs may take place announcing their changed status as a mature woman, ready for marriage. Boys throughout the region did not have a traditional feast to announce their arrival at manhood but often went on quests to obtain a guardian spirit. This practice made them vulnerable to potentially dangerous supernatural powers. Some features of this quest still survive throughout the region and include ritual bathing in the sea during the early morning. Visits to hot springs and arduous hikes or canoe trips are also relatively commonplace along with travel to cities, or distant relatives, and friends. Elevated causes of "external" death among young people throughout the region include motor vehicle road accidents, suicide, drowning, poisoning, homicide, fires, and falls. Reliable data on these exist for British Columbia, especially (see Cooper, 1995; Foster et al., 1995, p. 74).

The Aged

The aged are widely respected for their knowledge, and if they are especially wise and known for their rectitude they are designated as "elders" throughout the region. Because early mortality is commonplace, and the Native population bears a young profile due to high birth rates, there are not a large number of elderly people in the Native population of the region. However, serious degenerative illness is commonplace among those seniors who survive into old age, and it increasingly centers on Non Insulin Dependant Diabetes Melitus (NIDDM), and all its *sequalae*, including limb loss, organ damage, heart disease, and blindness (Heffernan, 1995). Rates of autoimmune diseases are also elevated for Native populations on the Northwest Coast, especially rheumatoid arthritis and systemic lupus erythmatosis.

CHANGING HEALTH PATTERNS

The most important factors associated with health and illness among Aboriginal people throughout the region is related to economic poverty and a history of political and religious subordination. Following in the wake of waves of epidemic diseases which obliterated cultural knowledge and greatly reduced population sizes, surviving children were sent to residential "schools" in the United States and Canada as part of a policy of assimilation in both countries. These were run by Christian religious denominations. At these institutions children were systematically stripped of their culture and language, inadequately fed non-traditional foods, and separated from cultural traditions during the winter ceremonial period. It has now been documented that abuse, both physical and sexual, was appallingly widespread. The impact of residential institutions on dietary change associated with rapidly rising rates of diabetes is well described for British Columbia's coastal populations (Hopkinson et al., 1995).

Impacts on health as a result of these trends have been far reaching. Death from digestive system diseases (associated with alcohol overconsumption) is excessively high for many groups, and gastrointestinal cancer rates are also especially high, as are neoplasms of the female reproductive system, and blood/lymph. Premature death from Ischemic heart disease, cerbrovascular strokes, and cardiomyopathies are also higher than for base populations throughout British Columbia. Mortality from infectious diseases, especially those affecting the respiratory system are uncommonly high and affect many age groups; these include: pneumonia, influenza, asthma, pulmonary fibrosis, and tuberculosis. Septicemia and viral hepatitis rates are also elevated, as are HIV infections, especially in urban areas such as Vancouver (Foster et al., 1995). No systematic comparative epidemiological or vital statistics data appear to have been published for American jurisdictions, where socialized medicine is not available, and statistical data are not comprehensively compiled.

Traditional belief emphasizes that death is associated with dangerous ghosts, whose powers include those which may make the living seriously ill. The great epidemics of the past, and current high mortality from many causes, have certainly provided a steady supply of ghostly spirits for over a century. Throughout the region the dead were traditionally disposed of in myriad ways: by suspension in trees, in mortuary boxes set atop memorial poles, in caves or in canoes suspended on scaffolds, or by cremation. After the creation of missions, cemeteries were widely used and stand in sad testimonial to the loss of young lives during the waves of epidemic diseases that struck many Native villages throughout the region. Many gravestones are those of children. Today, both burial and cremation are commonplace, along with a major memorial ceremony for high status and chiefly persons where the title is passed on and where the participants receive gifts of food and clothing.

The last half of the 20th century also witnessed a massive migration into cities throughout the region and consequently Anchorage, Vancouver, Victoria, Seattle, Prince Rupert, Bellingham, and Portland all have large populations of Native People. These populations come from the Northwest indigenous groups but are also migrants from other areas including the interior plateau, the Arctic, the Canadian prairies, American Middle West and the Southwest. The traumas—physical and psychological—associated with epidemic diseases, residential institutions, urbanization, and poverty are also deeply connected to a loss of traditional lands and water. The destruction of forests, salmon streams, pollution from mining and paper mills, and a great reduction in marine resources have all contributed to widespread despair, impoverishment, high rates of stress related illnesses, and deaths associated with substance abuse and addiction. Attempts to deal with these problems are now quite extensive (Stephenson & Elliott, 1995) and include the use of innovative and traditional healing techniques (Harris, 1995; Wade, 1995), diets (Hopkinson et al., 1995), and a shift in decision-making over health planning to local Native communities (Modeste et al., 1995; Read, 1995). However, ceding control over health care by paternalistic federal powers in both the United States and Canada, to local Native communities with conflicting interests, has proved frustrating, and difficult to achieve.

REFERENCES

Acheson, S. (1998). In the wake of the *ya'aats' xaatgaay* ['Iron People']: A study of changing settlement strategies among the Kunghit Haida. BAR International Series 711, Unpublished dissertation, University of Oxford.

Arima, E. & Dewhirst, J. (1990). Nootkans of Vancouver Island. In W. Suttles (Ed.), *Handbook of North American Indians: Northwest Coast* (Vol. 7, pp. 391–411). Washington, DC: Smithsonian Institution.

Benoit, C. & Carroll, D. (1995). Aboriginal midwifery in British Columbia: A narrative untold. In P. H. Stephenson, S. J. Elliott, L. T. Foster & J. Harris (Eds.), *A persistent spirit: Towards understanding aboriginal health in British Columbia* (pp. 223–248). Victoria: University of Victoria, Western Geographical Press.

Boas, F. (1932). Current beliefs of the Kwakwaka'wakw Indians, *Journal of American Folk-Lore, 14* (176), 177–260.

Bouchard, R. T., and Turner, N. J. (1976). Ethnobotany of the Squamish Indian People of British Columbia. Unpublished manuscript, British Columbia Indian Language Project, Victoria.

Boyd, R. (1985). The introduction of infectious diseases among the Indians of the Pacific Northwest, 1774–1874. Unpublished Doctoral dissertation, Pullman, Washington: University of Washington.

Boyd, R. (1990). Demographic history, 1774–1874. In W. Suttles (Ed.), *Handbook of North American Indians: Northwest Coast,* (Vol. 7, pp. 135–148). Washington, DC: Smithsonian Institution.

Boyd, R. (1994, Spring). Smallpox in the Pacific Northwest: The first epidemics. *BC Studies, 101,* 5–40.

British Colonist 12/9/1862, British Columbia Archives and Records Service, Victoria.

Canada. (n.d.) Federal Census Returns, 1881. Microfilm copy, British Columbia Archives and Records Service, Victoria.

Canada, Dominion of. (1882). Annual Report of the Department of Indian Affairs for the year ended 31st December, 1881. Sessional Papers No. 6. Ottawa: MacLean, Roger.

Canada, Dominion of. (1886). Annual Report of the Department of Indian Affairs for the year ended 31st December, 1885. Sessional Papers No. 4. Ottawa: MacLean, Roger.

Canada, Dominion of. (1891). Annual report of the Department of Indian Affairs for the year ended 31st December, 1890. Sessional Papers No. 18. Ottawa: Brown Chamberlin.

Canada, Dominion of. (1896). Annual Report of the Department of Indian Affairs for the year ended 30th June, 1895. Sessional Papers No. 14. Ottawa: S.E. Dawson.

Canada, Dominion of. (1897). Annual Report of the Department of Indian Affairs for the year ended 30th June, 1896. Sessional Papers No. 14. Ottawa: S.E. Dawson.

Canada, Dominion of. (1882). Annual Report of the Department of Indian Affairs for the year ended 31st December, 1881. Sessional Papers No. 6. Ottawa: MacLean, Roger & Co.

Codere, H. (1990). Kwakiutl: Traditional culture. In W. Suttles (Ed.), *Handbook of North American Indians: Northwest Coast* (Vol. 7, pp. 359–377). Washington, DC: Smithsonian Institution.

Cooper, M. (1995). Aboriginal suicide rates: Indicators of needy communities. In P. H. Stephenson, S. J. Elliott, L. T. Foster, & J. Harris (Eds.), *A persistent spirit: Towards understanding aboriginal health in British Columbia* (pp. 207–221). Victoria: University of Victoria, Western Geographical Press.

Curtis, E. S. (1915). *North American Indian* (Vol. 10). New York: Johnson Reprint Corporation [reprinted 1970].

Cybulski, J. (1990). Human Biology. In W. Suttles (Ed.), *Handbook of North American Indians: Northwest Coast* (Vol. 7, pp. 52–59). Washington, DC: Smithsonian Institution.

Cybulski, J. (1994). Culture change, demographic history, and health and disease on the Northwest Coast. In C. S. Larsen (Ed.), *In the wake of contact: Biological responses to conquest,* (pp. 75–85). New York: Wiley-Liss, Inc.

de Laguna, F. (1990). Tlingit. In W. Suttles (Ed.), *Handbook of North American Indians: Northwest Coast* (Vol. 7, pp. 203–228). Washington, DC: Smithsonian Institution.

Donald, L. (1997). *Aboriginal slavery on the Northwest Coast of North America.* Berkeley: University of California.

Douglas, J. (nd). Private papers, second series 1853. A/B/40/D75.4, British Columbia Archives and Records Service, Victoria.

Douglas, J. (nd). Indian Population Vancouver's Island 1856. Enclosure No. 1 in Governor Douglas' Dispatch No. 24 of the 20th October 1856, Colonial Office 305/7, Microfilm 393A, British Columbia Archives and Records Service, Victoria.

Drucker, P. (1951). The Northern and Central Nootkan tribes. *Bureau of American Ethnology Bulletin, 144.* Washington.

Drucker, P. (1965). *Cultures of the North Pacific Coast.* Scranton, Pennsylvania: Chandler.

Drucker, P. (1983). Ecology and political organization on the Northwest Coast of America. In E. Tooker (Ed.), *The development of political organization in Native North America,* Proceedings of the American Ethnological Society, 1979. New York: J. J. Augustin.

Duff, W. (1964). The impact of the White Man: The Indian history of British Columbia, *Anthropology in British Columbia,* Memoir No. 5, Victoria.

Foster, L., Macdonald, J., Tuk, T. A., Uh, S. H., & Talbot, D. (1995). Native Health In British Columbia: A vital statistics perspective. In P. H. Stephenson, S. J. Elliott, L. T. Foster, & J. Harris (Eds.), *A persistent spirit: Towards understanding aboriginal health in British Columbia* (pp. 43–94). Victoria: University of Victoria, Western Geographical Press.

Halpin, M. M. & Seguin, M. (1990). Tsimshian peoples: Southern Tsimshian, Coast Tsimshian, Nishga, and Gitksan. In W. Suttles

(Ed.), *Handbook of North American Indians: Northwest Coast*, (Vol. 7, pp. 267–284). Washington, DC: Smithsonian Institution.

Hamori-Torok, C. (1990). Haisla. In W. Suttles (Ed.), *Handbook of North American Indians: Northwest Coast* (Vol. 7, pp. 306–311). Washington, DC: Smithsonian Institution.

Harris, J. (1995). Rebuilding: Listen with the ears of our eyes. In P. H. Stephenson, S. J. Elliott, L. T. Foster & J. Harris (Eds.), *A persistent spirit: Towards understanding aboriginal health in British Columbia* (pp. 357–385). Victoria: University of Victoria, Western Geographical Press.

Hawthorn, H. B. (1965) Introduction. *Cultures Of The North Pacific Coast*. Scranton, Pennsylvania: Chandler.

Heffernan, M. C. (1995). Diabetes and aboriginal peoples: The Haida Gwaii diabetes project in a global perspective. In P. H. Stephenson, S. J. Elliott, L. T. Foster & J. Harris (Eds.), A persistent spirit: *Towards understanding aboriginal health in British Columbia* (pp. 261–269). Victoria: University of Victoria, Western Geographical Press.

Hilton, S. F. (1990). Haihais, Bella Bella, and Oowekeeno. In W. Suttles (Ed.), *Handbook of North American Indians: Northwest Coast*, (Vol. 7, pp. 312–322). Washington, DC: Smithsonian Institution.

Hopkinson, J., Stephenson, P. H., & Turner, N. J. (1995). Changing traditional diet and nutrition in aboriginal peoples of Coastal British Columbia. In P. H. Stephenson, S. J. Elliott, L. T. Foster & J. Harris (Eds.), *A persistent spirit: Towards understanding aboriginal health in British Columbia* (pp. 129–166). Victoria: University of Victoria, Western Geographical Press.

Kane, P. (1859). *Wanderings of an artist among the Indians of North American from Canada to Vancouver's Island and Oregon through the Hudson's Bay Company's territory and back again*. London: Longman, Brown, Green, Longmans, and Roberts.

Kennedy, D., & Bouchard, R. (1990). Bella Coola. In W. Suttles (Ed.), *Handbook of North American Indians: Northwest Coast* (Vol. 7, pp. 323–339). Washington, DC: Smithsonian Institution.

Martin, R. M. (1848). *The Hudson Bay Territories and Vancouver's Island*. London: T. Brettell.

McIlwraith, T. F. (1948). *The Bella Coola Indians*. 2 vols. Toronto: University of Toronto.

Modeste, D., Elliott., D., Gendron, C., Greenwell, B., Johnny, D., Payne, H. (1995). S'huli'u*tl* Q'uw'utsun/The spirit of Cowichan: A journey through the *Tse*wultun Health Centre/Huy Tseep Qu Nu Siiye'yu Kwun's 'I M' I Ewu'u Tuna *Tse*wultun. In P. H. Stephenson, S. J. Elliott, L. T. Foster & J. Harris (Eds.), *A Persistent Spirit: Towards understanding aboriginal health in British Columbia* (pp. 331–356). Victoria: University of Victoria, Western Geographical Press.

Read, S. (1995). Issues in health mangement promoting First Nations wellness in times of change. In P. H. Stephenson, S. J. Elliott, L. T. Foster & J. Harris (Eds.), *A persistent spirit: Towards understanding aboriginal health in British Columbia* (pp. 297–331). Victoria: University of Victoria, Western Geographical Press.

Seaburg, W. R. & Miller, J. (1990). Tillamook. In W. Suttles (Ed.), *Handbook of North American Indians: Northwest Coast* (Vol. 7, pp. 560–567). Washington, DC: Smithsonian Institution.

Skinner, M., McLaren, M., & Carlson, R. L. (1988). Therapeutic cauterization of periodontal abscesses in a prehistoric northwest coast woman. *Medical Anthropology Quarterly, 2*(3), 278–285.

Stephenson, P. H., & Elliott, S. (1995). Preface. In P. H. Stephenson, S. J. Elliott, L. T. Foster & J. Harris (Eds.), *A persistent spirit: Towards understanding aboriginal health in British Columbia* (pp. 43-94). Victoria: University of Victoria, Western Geographical Press.

Suttles, W. (1990). (Ed.) *Handbook of North American Indians: Northwest Coast,* Vol. 7. Washington, DC: Smithsonian Institution.

Suttles, W. (1990b). Environment. In W. Suttles, (Ed.) *Handbook of North American Indians: Northwest Coast* (Vol. 7, pp. 267–284). Washington, DC: Smithsonian Institution.

Thompson, L. C., & Kinkade, M. D. (1990). Languages. In W. Suttles (Ed.), *Handbook of North American Indians: Northwest Coast* (Vol. 7, pp. 30–51). Washington, DC: Smithsonian Institution.

Tolmie, W. F. (1963). *The journals of Willian Fraser Tolmie, physician and fur trader*. Vancouver: Mitchell Press.

Turner, N. J. (1975). Food plants of British Columbia Indians, Part 1/ Coastal Peoples. *British Columbia Provincial Museum Handbook*, No. 34, Victoria.

Turner, N. J. (1978). Food plants of British Columbia Indians, Part 2/ Interior Peoples. *British Columbia Provincial Museum Handbook*, No. 36, Victoria.

Turner, N. J. (1982). Traditional use of Devil's Club (*Oplopanax horridus*; Araliaceae) by native peoples in Western North America. *Journal of Ethnobiology, 2*(1), 17–38.

Turner N. J. (1984). Counter-irritant and other medicinal uses of plants in Ranunculaceae by native peoples in British Columbia and neighbouring areas. *Journal of Ethnopharmacology, 11*(2), 181-201.

Turner N. J., & Hebda, R. J. (1990). Contemporary use of bark for medicine by two Salishan native elders of southeast Vancouver Island, Canada. *Journal of Ethnopharmacology, 29*(1), 59–72.

Vogel, V. J. (1970). *American Indian medicine*. Norman: University of Oklahoma Press.

Wade, A. (1995). Resistance knowledges: Therapy with aboriginal persons who have experienced violence. In P. H. Stephenson, S. J. Elliott, L. T. Foster & J. Harris (Eds.), *A persistent spirit: Towards understanding aboriginal health in British Columbia* (pp. 167–206). Victoria: University of Victoria, Western Geographical Press.

Ojibwa

Linda C. Garro

ALTERNATIVE NAMES

"Anishinaabe; Ojibway, Ojibwe; Chippewa (U.S.); Mississauga or Southeastern Ojibwa (southern, central Ontario), Nipissing, Algonquin, Plains Ojibwa (sometimes known as Bungi); Northern Ojibwa; Saulteaux or Saulteurs (Manitoba); Ojicree or Oji-Cree, Southwestern Chippewa (based on Goddard, 1978, p. 583 cited in Brown, 1997)." Geographically, this entry primarily concerns present-day Canada, particularly Manitoba and Ontario. Anishinaabe (plural Anishinaabeg) is the preferred ethnonym in Manitoba and Ontario (Peers, 1994, p. xvii).

LOCATION AND LINGUISTIC AFFILIATION

Most of the estimated nearly 50,000 speakers of Ojibwa are found in the area ranging from southwestern Quebec through Ontario, Michigan, northern Wisconsin and Minnesota, southern and central Manitoba, and southern Saskatchewan (Mithun, 1999, p. 334). The eight Ojibwa dialects belong to the Algonkian language family (Rhodes & Todd, 1981). Ojibwa and Cree are linguistically, culturally (Rogers, 1969), and genetically related (Young, 1988, p. 12), with considerable intermarriage starting in the fur trade period.

OVERVIEW OF THE CULTURE

Records of Ojibwa contacts with Europeans date from the 1640s (Hallowell, 1992). Within a century after European contact, the Ojibwa began to migrate into southern Ontario and Michigan from a "homeland" located "somewhere within an area extending from the eastern shore of Georgian Bay (Lake Huron), west along the north shore of Lake Huron, and a short distance along the northeast shore of Lake Superior and onto the Upper Peninsula of Michigan (Rogers, 1978, p. 760)." It is unclear how unified Ojibwa culture was prior to the 1600s (Hallowell, 1992, p. 20).

A reasonable starting point is a glimpse at late 18th century life in the region northwest of Lake Superior—a forested region with numerous lakes, rivers, and swampy regions. During this historical period, the Ojibwa complemented their subsistence activities of hunting, fishing, and gathering with participation in the fur trade, which had been ongoing for over a century in this region (Peers, 1994, pp. 22–26). The effective unit of social and economic organization was the extended family, typically composed of several close relatives and their spouses, children, and other kin, which ranged in size from six to about a dozen people (see Hallowell, 1992, pp. 56–57 on bilateral cross-cousin marriage as the prototypical pattern and pp. 29–30 on polygyny). Generally there were patrilineal, exogamous clans (descent groups named after totemic animals)—see Rogers (1978, p. 763) and Brown (1997). The seasonal harvesting round emphasized fish and wild rice (Peers, 1994). In early spring, leaving their winter hunting grounds, several extended families would gather to make maple sugar. Families subsisted on small game, and caches of meat and wild rice from the previous summer. Men might leave on short trapping excursions lasting several days. After the sugar run, people from many camps gathered together at sturgeon-fishing sites for a month. Sturgeon was an important food source, both fresh and processed, to be stored for future use and as sturgeon oil (Holzkamm, Lytwyn, & Waisberg, 1988). As the largest gatherings, spring fishing villages ranged in size from several hundred to over a thousand. Trading exchanges of furs for goods occurred at local trading posts in close proximity to the fishing sites, or with visiting traders. Maple sugar and sturgeon might also be traded.

Activities associated with the Midewiwin, a ceremonial organization concerned with curing (Hallowell, 1992, p. 11) only took place during large gatherings. Hierarchically organized in degrees or grades of membership, initiates were instructed under the leadership of Mide priests (Steinbring, 1981, p. 251). As seeking a cure for illness was one means for entering the society, almost all were inducted into at least the lowest rank. Novices were taught to identify and prepare botanical medicines and to commune with other-than-human persons in order to enhance individual power to heal (Waldram, Herring, & Young, 1995, p. 112). Advancement in the Midewiwin's structured organization contrasts (see below) with the individualized "gifts" of "power" granted by other-than-human persons through private sensory encounters.

After spring fishing, the Ojibwa dispersed for the summer to engage in fishing, berry picking, and hunting. At the end of the summer, larger groups gathered to hunt migrating waterfowl and collect wild rice. Harvested wild rice was stored for winter use and traded. Autumn activities included visits to trading posts to obtain goods for winter activities. Individual extended families would move to winter hunting grounds and carry out a fall hunt for summer-fattened deer, bear, and moose. The amount of cached food and availability of small game determined how much effort was directed to trapping activities rather than fishing or hunting larger animals. As spring approached, several families might come together for a last intensive period of trapping before rejoining other families to make maple sugar (Peers, 1994).

As Peers (1994, p. 25) notes, "the Ojibwa moved within a world that was at once spiritual and physical.... The forest, streams, and lakes were alive with supernatural [other-than-human] 'persons' and powers, all of whom had to be treated with respect to prevent misfortune, accident, and illness from befalling the human persons who moved among them." Feasts and other offerings conveyed respect and gratitude to other-than-human persons (Hallowell, 1992). Everyday objects and clothing were designed to tap into beneficial powers and protect from harmful powers. Sent alone into the forest around the time of puberty, young men and women actively sought to enter into relationships with other-than-human persons, fasting and begging other-than-human persons to take pity on them and to help them throughout their lives. The "gifts" and "objects of power" bestowed by other-than-human persons, who occupy the "top rank in the power hierarchy of animate being" (Hallowell, 1960, p. 377), enabled human being" to do things that would otherwise not be possible. In entering into lifelong personal relationships with their attendant gifts, individuals incurred specific and general obligations founded on respectful behavior toward the other-than-human persons. Examples of gifts include: ability to change the weather, ability to call wild game, unusual success in warfare, knowledge of healing skills, and the ability to communicate with other-than-human persons to learn things that would otherwise be unknowable (Hallowell, 1955, p. 104, Peers, 1994, p. 25). At times some communications took place in public settings (Hallowell, 1942). Black (1977a) has proposed an "Ojibwa power belief system" underlying interpersonal relationships, including relationships with other-than-human persons.

Both learned skills and practices, as well as knowledge and abilities granted by other-than-human persons, were important in the maintenance of well-being and care of illness. Teachable skills and practices included bloodletting (Pettipas, 1977; Steinbring, 1981, p. 249), enemas (Pettipas, 1977), a range of herbal preparations, and sweat baths (Peers, 1994; Waldram et al., 1995).

From the woodlands northwest of Lake Superior, the Ojibwa subsequently fanned out in a westerly direction into areas rich in furs, large game, fish, and other resources (see Ray, 1974, ch. 2), coming into increasing conflict with the Dakota. By the last decades of the 18th century, small groups of Ojibwa began to move even further west into the prairies and parkland (Peers, 1994, p. 3).

Peers maintains that the fur trade "may not have been as central to their lives as it seems" (1994, p. 27) and notes (p. 43): "Where some trade goods reinforced and elaborated many aspects of Aboriginal cultures, others caused rapid cultural change and damage. Some trade goods did both." For example, alcohol and the "destructive nature of fur-trade drinking parties—even when these were conducted as redistribution events or ceremonies—hints at great stresses generated by the fur trade and at cultural changes resulting from participation in the fur trade" (see also Waldram et al., 1995, pp. 137–140).

Later sections of this entry concern the contemporary context of health based on field research carried out in the 1980s in a southern Manitoba community. Classified by Howard (1965) as plains Ojibwa, the community is located in the "transition parkland belt" which historically provided "most of the resources of both the forest and grasslands" (Ray, 1974, p. 28). It is an Anishinaabe community where members report the long-standing annual observance of the Thirst (Sun) Dance found historically in the plains region and closely associated with the plains Cree (see Pettipas 1994 for a description). Yet, as noted below (and complementing an analysis by Peers, 1994, pp. 152–153, 179, 210–211), there was also considerable commonality with work carried out to the east by Hallowell (e.g., 1955, 1976, 1992) and Black (e.g., 1977a, b). To help situate the discussion of the contemporary community, the next section begins with an abbreviated synopsis of some key social forces starting around the time that the fur trade began to wane.

THE CONTEXT OF HEALTH: ENVIRONMENTAL, ECONOMIC, SOCIAL, AND POLITICAL FACTORS

Missionaries, Treaties, and Newcomers

As the importance of the fur trade declined and the animal resource base became depleted, missionary activities by Roman Catholics, Anglicans, Methodists, and others intensified during the 1830s and 1840s (Brown, 1997; Hallowell, 1992; Peers, 1994). The response to missionization varied (see Peers, 1994, ch. 5). Early missionary schools did not attract large numbers of Ojibwa students. But for some, baptism was seen as protective against the covert use of evil powers by others and/or providing a connection to new other-than-human persons that could prove helpful in the changing times (Peers, 1994, pp. 162 and 169). Accepting Christianity did not necessarily entail renouncing what has been referred to as the "Ojibwa power belief system" (see Black, 1977a). Peers (1994, pp. 167–169) writes of the turmoil and uncertainties, both social and personal, occasioned by these missionary efforts, of missionary attacks on the Midewiwin and of struggles between missionaries and Midewiwin leaders to attract adherents. And although the Midewiwin began to wane, and continued to do so as time passed, it remained visibly active up through the 1930s, with the report of a ceremony in 1942 (Steinbring, 1981, pp. 252–253; see also Hallowell, 1936). In some communities, the Midewiwin may have continued without coming to the attention of outsiders after that time (see Black-Rogers, 1989).

The intensification of missionary activities heralded the increased intrusions, restrictions and encroachments of the coming years, which escalated following the formation of Canada in 1867, and subsequent treaty negotiations. Hallowell (1992, pp. 35–38) asserts that the signing of treaties and creation of bands and chiefs through the treaty process was more drastic than the earlier impact of the fur trade and Christianity in that the Ojibwa were asked to give up all legal claims to the great majority of lands they occupied, as well as become subject in the future to decisions of a national state. Efforts to negotiate continued access to land resources were ultimately disregarded (Peers, 1994, p. 204). Lands were opened up to settlement and cleared for agriculture by Euro-Canadian settlers in the southern regions and the First Nations groups became increasingly constrained by the boundaries of reserve lands. Off-reserve involvement in wage labor, often in agriculture, became more common. Hunting and fishing rights were restricted while newcomers engaged in the commercial exploitation of natural resources, such as logging and fishing operations. Destructive environmental impacts include: the expropriation and overexploitation of sturgeon fishing by commercial non-Indian fishermen which led to the near-extinction of the sturgeon; impacts on fishing, trapping, and wild rice caused by hydroelectric projects; the clear cutting of forested lands; and mercury poisoning from pulp and paper mills raising individual health concerns while also curtailing fishing and jobs as fishing guides.

The Indian Act of 1876 and later amendments gave sweeping powers to federal Indian administrators to "transform 'Indians' into 'Canadians'" (Pettipas, 1994, p. 17). While First Nations negotiators wanted schooling provisions included in the treaties to adapt to the future, the educational policy developed by the Canadian government became entwined with the efforts of missionaries to "christianize and civilize" (Reading, 1999) and the day-to-day management of the schools was placed under the control of churches. Children were often removed to schools distant from their homes and, among other hardships, often forbidden to speak their own language, while the ways of their parents were disparaged. Missionary aims were furthered when government representatives enforced repressive statutes aimed at eliminating indigenous religious ceremonies, including the Midewiwin (Pettipas, 1994, p. 157). Since the 1950s, fundamentalist Christianity has attracted a significant following in many communities (Steinbring, 1981, p. 253; Young, 1988, pp. 16–17), with messages censuring alcohol and Aboriginal healing traditions. However, healing traditions, like the Midewiwin and the sweat lodge, have been held once again in communities after a hiatus lasting many years (e.g., Pettipas, 1994, p. 7). Some hospital administrators accommodate healers and their practices within hospital settings (O'Neil, 1988; Waldram et al., 1995, p. 205; see for hospital-based case studies illustrating some of the complexities involved).

While missionaries dispensed medicines (like traders before them) and cared for the sick, government increasingly took a role in providing health services (see Young, 1988, pp. 101–102). A recurring debate throughout much of the history concerns responsibility for the provision of health care, with "status Indians" from groups who signed treaties contending that health services are a treaty right and Federal government officials claiming that health services are provided on a humanitarian and not on a legal basis. This controversy lessened with the introduction of universal health insurance in Canada in the 1960s, but First Nations and Aboriginal organizations refer to federal responsibility for health in proposing efforts to enhance, as well as in voicing concerns about reductions to, health programs and existing services for status Indians. Leaders of First Nations bands and political organizations have long asserted that the improvement in health for First Nations people requires radical changes in the historical relationship between Canada's first peoples and those who came later. For example, Phil Fontaine, formerly the Grand Chief of the Assembly of Manitoba Chiefs and later of the national Assembly of First Nations, stated: "The present state of ill health among aboriginal people is a reflection of their placement in the social and political structure of Canadian society" (Fontaine, 1991, p. 21).

Health in Historical Perspective

While no reliable information is available about the size of the pre-contact population, Bishop (1999) states that from "perhaps 10,000 persons at contact whose descendants are now called Ojibwa, the population grew to around 140,000 (registered) in 1996." The trend is not one of a smooth linear increase. The population figures at the time of contact declined through time and reached their lowest points at some point around the transition from the 19th to the 20th century (Ubelaker, 1988).

The pre-contact and the early fur-trade periods were characterized by a relatively healthy population (Young, 1988). Acute infectious diseases swept through the region at intervals starting in the 18th century. The smallpox epidemic of 1780–82 (Young, 1988, pp. 35–36) not only caused extremely high mortality rates, but also fear and despair contributing to suicides and other premature deaths, as well as the vulnerability of survivors because of a reduced labor force: "In a hunting-and-gathering society...the death of a single hunter or female worker could be a threat to the survival of an entire extended family" (Peers, 1994, p. 20; Young, 1988, p. 38). Indirect effects—reduced fertility and post-epidemic mortality linked to malnutrition and dehydration—may have contributed significantly to population decline (Thornton, 2000; Waldram et al., 1995, p. 59). Around 1825, as westward expansion progressed, epidemic diseases such as measles, influenza, and scarlet fever began to strike with more frequency. Disease impact was amplified by malnutrition and undernutrition associated with the depletion of local game and fur resources around the mid-1800s (Waldram et al., 1995).

By the early 1900s, relocation onto reserves or other semipermanent settlements with insufficient food, inadequate sanitation, and crowded housing ensured the continued prominence of infectious diseases, as did the concentration of children in residential schools. Infant mortality was extremely high; for every 1,000 live births there were over 200 deaths during the first year (Young, 1988, p. 124). Social conditions fostered the emergence of tuberculosis—a "genuine plague of enormous proportions": a review of residential schools over a 15-year period indicated that from 25–35% of all children had died, primarily of tuberculosis but also from other diseases like measles (Waldram et al., 1995, p. 156). Tuberculosis "continued unabated throughout the 1930s and 1940s as dwindling economic resources kept the population at mere subsistence level" (Young, 1988, p. 124).

Since World War II health has improved with national universal health insurance, improvements in accessibility to biomedical health services in rural and remote communities, economic assistance programs, and greater control of infectious disease. Still, morbidity and mortality rates for infectious diseases remain at a persistently higher level when compared to the rest of the Canadian population (Hallett, 2000; Waldram et al., 1995; Young, 1988) and are associated not with the quality of or access to health services, but with continuing socioeconomic marginalization.

According to Young (1994, p. 94): "Over the past several decades Native Americans have undergone the 'epidemiological transition' characterized by the decline, though not disappearance of infectious diseases and the increasing importance of the chronic, noncommunicable diseases, accidents, and acts of violence as causes of mortality and morbidity." For the chronic diseases, particularly striking has been the emergence of the so-called "diseases of modernization" or "diseases of westernization," such as type II diabetes, hypertension, obesity, cardiovascular disease, and cancer. Many chronic diseases are seen to be "the result of rapid changes in lifestyle, particularly in dietary habits and physical activity levels" (Young, 1994, p. 216). The prewar situation of low body weights has changed to one where obesity is common (Young, 1988). While type-II diabetes was rare before the 1950s, the current situation is described as "an epidemic in progress," one that is "still on the upswing, with a trend toward earlier age at onset" (Young et al., 2000, p. 561). Type-II diabetes is also known as maturity-onset diabetes as it typically develops later in life, but it is increasingly diagnosed in children and adolescents in Manitoba and northwestern Ontario (Young et al., 2000). In recent years, the discrepancy in life expectancy for Aboriginal peoples compared to others in Canada has steadily decreased but still remains, with recent reports indicating 8 years less for males and 6.7 less for females (Hallett, 2000, p. 34) and primarily reflects more deaths occurring relatively early in life (Young, 1994, p. 37).

With those under the age of 5 years suffering disproportionately from infectious disease (Young, 1988, 1994), the most important causes of morbidity and mortality for the younger age groups (upper limit of 45 years of age) are injuries sustained as a result of accidents and violence (Hallett, 2000; Young, 1994). In the years from 1972–81, over one third of the deaths in the Sioux Lookout region of northwestern Ontario were attributed to accidents and violence, mostly involving alcohol. Suicide rates were almost twice as high as for Canada as a whole. The rates for accidents and violence, with the exception of suicide, have declined since the 1980s (Hallett, 2000; Young, 1994).

Consider the case of Grassy Narrows, one of the more southerly Anishinaabe First Nations in northwestern Ontario (Shkilnyk, 1985). Looking at the 20-year period from 1959–78, Shkilnyk examines the detrimental impact of the government's decision to relocate the community in 1963 and the discovery in 1970 of mercury poisoning in the river that was so central to the life and livelihood of the community. Prior to the relocation, from 1959–63, 91% of all deaths were due to natural causes and 9% were attributed to violence, including suicide. From 1964–68, the frequency of violent deaths rose to 14%, and from 1969–73 to 49%. For the period from 1974–78, violent deaths associated with alcohol or drugs accounted for 75% of all deaths. Although presented as relatively unique (Shkilnyk, 1985), Grassy shares commonalities with many other Ojibwa and First Nations communities and that serves to illuminate the development and nature of the relationship between the first peoples and the wider society (see Garro, 1993 for a more extended discussion). Like Grassy Narrows, the traditional resource base of other communities has been eroded with the resulting dependence on government-sponsored welfare programs and other forms of economic assistance (see also Waldram, 1985). In addition, separating children from their families and removing them to off-reserve locations has increased the feelings of demoralization and hopelessness. Exposure to mercury poisoning and forced relocations, often because of flooding of land by hydroelectric dams, has occurred with other communities. Many First Nations communities, like Grassy Narrows, have in recent years worked toward achieving a greater degree of self-determination which involves greater involvement in such institutions as education, child welfare, health care, and the justice system. Actions directed at regaining control over a land resource base, through settling land claims or by other means, are another important part of this ongoing process (see Garro, 1993). The struggle for self-determination, for resource rights and lands rights, is also a struggle for better health (Culhane Speck, 1989, p. 90)

The Community Context

The remainder of this entry draws on field research carried out in the 1980s in an Anishinaabe First Nations community in southern Manitoba. It is a community where a substantial majority of adults and children, and essentially everyone over 40 years of age, speaks their own language, Anishinaabemowin; it is the language predominantly used in most social interactions in the community. Many individuals identify as Catholics.

There is insufficient land on the reserve or surrounding it to allow a subsistence or trapping-based means of livelihood, except for a few fishermen. Over time, much of the land surrounding the reserve has been acquired for cultivation, although not by reserve members. A few families farm some small tracts of reserve lands and some households have large gardens. Wild game and fish supplement the diets of most families, and some rely heavily on such products. Because of the lack of on-reserve employment, young people used to move back and forth between the reserve and work sites in agriculture or in the city. This still occurs to some extent. Some elders state that a major shift occurred in the late 50s when welfare assistance became available, making it possible for people to live full-time on the reserve and to purchase store-bought goods. Like many other First Nations communities, increasing numbers of registered members have relocated, permanently or semipermanently, to urban settings over time (Thornton, 2000).

The majority of fulltime on-reserve employment opportunities are through local First Nations government or related services—mainly positions in teaching, administration, and health care services. There is high unemployment and many depend on economic assistance. Until relatively recently, the highest level of schooling attainable without leaving the community was the eighth grade. Many older individuals attended residential boarding schools where progression through the grades was slow. Proportionally, few adults have completed many years of formal schooling. However some young adults in their twenties and thirties, many of whom work for the local government, spent considerable time in boarding schools off the reservation and then often proceeded to obtain a university degree or take college-level courses. On the reserve, much of the housing is overcrowded and of substandard construction; there is a lack of plumbing and water services with many homes relying on water delivered to outside barrels.

Physicians, other biomedical health professionals, schools, television, printed material, radio, and alternative forms of medicine, including diverse Aboriginal healing traditions, have all influenced local understandings. It was not uncommon for community members to contrast the health-sustaining characteristics of the foods and lifestyle of the Anishinaabe past and sickness-inducing propensities of contemporary ways of living from which the Anishinaabeg cannot escape.

MEDICAL PRACTITIONERS

Outside of urban centers, few communities have resident physicians. The health center was staffed by visiting physicians approximately three afternoons a week with other physicians and hospitals located in towns approximately an hour's drive distance. All expenses of biomedical care, including transportation costs and prescription drugs, are either covered by universal health insurance or the federal government. Public health nurses and community health workers (residents of the community who have participated in training courses and whose work is supervised by nurses) focus on public health, health education, and prevention. They also make home visits to provide ongoing care and guidance to new mothers and others deemed in need of help with a health problem.

In this community, the most respected healers see themselves as descendants, in spirit if not through direct familial connections, of other Anishinaabe healers reaching back into the community's past. Two general types of Anishinaabe healers are important. First, there are individuals, usually women, who "know how to make (herbal) medicine," the special knowledge of which is learned from another healer and generally passed down through the family. Some are known for their skill in making one specific preparation, for example, medicine to help teething infants, dissolve gallstones, or to control diabetes. Others treat a variety of conditions. Herbal remedies may be used by women for a variety of reproductive health functions, for example, to control menstrual cramping, and to prevent miscarriages or premature birth. For services outside the family, herbalists receive tobacco along with some form of payment, usually money. Second, there are curers who are regarded as "gifted" with special powers. Most often the "gifts" of these medicine persons are manifested by their ability to communicate with other-than-human persons, through dreams, visions, or other altered states of consciousness to diagnose, establish cause, and prescribe and/or carry out remedial actions. If truly gifted, these medicine persons are able to ascertain whether a condition is best treated by a physician or can only truly be resolved through the mediation of a medicine person with other-than-human persons.

Although youths no longer routinely go into isolation and fast for blessings, healing gifts are still granted through visions and dreams. Medicine persons may also guide an individual to interpret his or her own personal experiences as indicating the likelihood of special powers that may manifest themselves or can be called upon in the future. Most medicine persons are not recognized as gifted and consulted by others until they have reached middle age. While it is not as common for women to be recognized as medicine persons, when it does occur it is after menopause.

Medicine persons do not charge for their services but they are always given tobacco and usually a sum of money or goods. Although what is given is at the discretion of the individual seeking help, the value tends to reflect the complexity of the case and the affluence of the giver. Families may incur considerable travel expenses to consult with a respected medicine person in another community.

An important component of being helped by a medicine person is a belief in their "gift" and the culturally embedded rationale for Anishinaabe sickness. This is not true of treatment by an herbalist. While individuals in this community can generally be said to share knowledge of Anishinaabe theories about the causes of illness and misfortune, there is considerable variation in the extent to which individuals espouse these theories and rely on them in evaluating events and making treatment decisions. Indeed, some individuals, particularly those who went to boarding school, express strong skepticism about the existence of "Anishinaabe sickness" and the gifts attributed to medicine persons. (Sometimes a household member consulted a medicine person without informing other family members who were not favorably disposed toward Anishinaabe healers [Garro, 1998, p. 350].) In this community there is no perceived contradiction between accepting a medicine person's abilities and being a Christian. The Midewiwin is not present in this community.

Black (1977a, p. 149) notes that those who heal also potentially have the power to cause death, illness, or misfortune at a distance. Uncertainty about the potential to misuse power may enter into interactions with medicine persons.

CLASSIFICATION OF ILLNESS, THEORIES OF ILLNESS, AND TREATMENT OF ILLNESS

Illness Treatment Decisions

Patterns of care-seeking are diverse and complex (Garro, 1998). During the course of fieldwork, thirteen separate visits were made to 61 randomly selected families over a six to eight month period. The total number of illness case histories is 468. Of these, 189 (40%) were cared for at home without recourse to any other source of care. Among the remaining 279 cases, 225 (81%) were seen by a physician, 98 (35%) were seen by medicine persons who were consulted as mediators with other-than-human persons, 21 (8%) sought remedies from a herbalist, and 10 (4%) consulted another type of practitioner (non-Anishinaabe herbalist, chiropractor, acupuncturist, psychologist). The percentages add up to more than 100% because some cases involved multiple treatment alternatives outside the home. Of the 225 cases seen by a physician, 66 (29%) also consulted a medicine person.

In cases where both physicians and medicine persons were consulted, physicians tended to be seen first. All but 13 families reported consulting a medicine person in the past five years. Six families (10%) reported never seeing a medicine person.

Classification and Explanatory Frameworks for Illness

In Anishinaabemowin, there are a couple of generally applicable terms indicating the state of being sick that can be used to refer to any illness condition. These "sickness" terms can be modified in a way that connects them with two broad causal categories or explanatory frameworks—"white man's sickness" and "Anishinaabe sickness." Named illness conditions (e.g., diabetes, measles, arthritis) are not fixed exemplars of any one of the categories (see Garro, 1990). Uncertainty is common, an explanatory framework may be framed in terms of possibilities (or probabilities) open to debate, and an illness may be linked to multiple explanatory frameworks. One mother with a sick infant consulted both a physician and a medicine man, who diagnosed an Anishinaabe sickness. When asked which interpretation she thought was correct she replied: "I think they both were right. I believe them both."

Sickness. Sickness encompasses a diverse range of explanations, many of which are widely found throughout Canada and the United States. For example, overeating or eating the wrong thing, including "food poisoning" due to spoilage can lead to illness such as stomachaches or diarrhea. Colds, the flu, and tuberculosis could be talked about as something "caught" from someone else and transmitted by means of "a bug," "germs," or a "virus." Similarly, measles was talked about as a common childhood illness that could spread to anyone. An unchecked illness can develop into something more serious, such as a child's fever developing into convulsions. Another concept likely based in biomedical teachings is that of an illness being inherited or passed down through the family. Talk about the possibility that an illness could "run in the family" was generally understood, but heredity is not a common explanation (Garro, 1995, 1996).

Colds, fevers, and respiratory conditions, like bronchitis and sometimes tuberculosis, may be attributed to excessive cold, or to being overheated and catching a chill. A minor illness, like a cold, may leave one in a weakened state and thus more susceptible to the effects of cold and the danger of having the illness worsen into something more serious such as pneumonia or bronchitis. Rheumatism and arthritis were also attributed to getting wet or being exposed to the cold; fishermen were considered to be particularly vulnerable. Women are considered vulnerable to the effects of cold during their menstrual periods (but unlike other illnesses linked to the cold, remedies from physicians were not seen as providing any benefit).

There are also a number of things that, if in excess, have the potential to stress the body and tip the balance toward ill health. Smoking too much can lead to cancer as well as other problems involving breathing and the lungs; eating too many greasy foods can lead to gallstones; drinking too much precipitates a variety of problems, including damage to one's liver and birth defects, such as fetal alcohol syndrome; and being under too much emotional stress can cause sickness by itself as well as exacerbate most illness conditions. To provide a more complex example, high blood pressure is commonly seen to result from any number of sources of bodily imbalance (Garro, 1988). Although the most frequently cited cause for high blood pressure is that of being under too much emotional stress or having too many worries, a number of other catalysts included overexertion or working too hard, drinking too much alcohol, eating too much of particular foods (e.g., salt or greasy foods), and being overweight. These can act either singly or jointly. The same set of multicausal possibilities may be used to explain diabetes, though for this illness the most commonly given explanation centered on ingesting too much sugar through either foods or alcohol, or being overweight as a consequence of eating too much, particularly sweets.

White Man's Sickness. "White man's illnesses" are seen as occurring after Europeans came to North America. This phrase is commonly coupled with measles, chickenpox, tuberculosis, high blood pressure, diabetes, and cancer. To use the label of "white man's sickness" is to make a statement, at times with overt political overtones, about the social epidemiology of these diseases, embedding their presence in the community within the continuing disruption and destruction of the Anishinaabe way of life. Reference to "white man's sicknesses" affirms that these diseases were introduced by European settlers. Health problems linked to alcohol may also be referred to as "white man's sickness" since alcohol was not present before contact with Europeans. Discussions about diabetes, high blood pressure, and cancer often highlighted strongly articulated contrasts between the healthy and fortifying foods obtained through Anishinaabe subsistence activities in the past and the comparatively unhealthy reliance on the store-bought foods of the Anishinaabe present. For some, it was simply the inferiority of the present day foods that predisposed one to illness whereas "wild food" was inherently fortifying. There were several explicit comments targeting the large amount of junk food and sugar laden foods eaten by the Anishinaabeg in the present day. By far the most commonly cited source of bodily disturbance associated with "white man's sickness" was the contaminating and rather insidious omnipresence of "poisons" in comestibles. These "poisons" included chemicals and other substances sprayed on crops and injected into animals as well as those added during food processing and canning. Referring to diabetes, high blood pressure or cancer as "white man's sicknesses" shifts responsibility away from the individual (e.g., for eating too much sugar) to a societal etiology that is based in outside actions and is consistent with the local perception that contemporary Anishinaabeg are powerless to reverse such trends. Actions taken in response to illness, including "white man's sicknesses," relate to causal understandings (Garro, 1996, 2000a). Still, only a few individuals (Garro, 1995) consistently affirmed the primacy of "white man's sickness" in accounting for a given illness. It was more common for individuals to mix this explanation with other potential sources of bodily imbalance.

Anishinaabe Sickness. Hallowell (1960, p. 410) maintained that the "causes of illness are sought by the Ojibwa within their web of interpersonal relations, rather than apart from it." This assessment remains relevant to an understanding of Anishinaabe sickness. Three features recurred across the case histories and apply to the two explanatory frameworks mentioned below: (1) sickness originates in discrete and identifiable, or potentially knowable, actions of human beings, alive or dead; (2) it involves breaches of certain rules that govern social relationships within the Anishinaabe behavioral environment; and (3) effective redress can be achieved only through the guidance of other-than-human persons.

Most visits to a medicine person take place when one or both of two explanatory frameworks are suspected.

The first is "bad medicine"; its use contravenes the high cultural value on an individual's right to autonomy (Black, 1977a, p. 150). Although "bad medicine" often strikes the intended target, there is some unpredictability in its use; victims may accidentally come into contact with "bad medicine" and suffer unintended consequences. In addition to causing direct harm to others, "bad medicine" can be used, for example, to win a competition, to induce someone to fall in love with the user, or to influence a courtroom judge to dismiss charges. Other misfortunes include flat tires, losing a talent contest, a house fire, minor physical problems such as backaches and headaches, striking alterations in an individual's behavior, loss of a spouse's affection, seemingly accidental injuries, acute illnesses and prolonged illnesses, "twisted mouth," miscarriages, and sudden death. A series of misfortunes or even relatively minor illnesses and/or events may raise suspicions that "bad medicine" may be involved, suspicions that may not be acted upon, at least not immediately, but kept in mind and reevaluated in the future in light of other events (Garro, 1998). The ability of a medicine person to remove "bad medicine" from a person or otherwise counteract its effects depends on the relative powers of the sender and the medicine person. The ability to wield "bad medicine" is hidden and not advertised; it is understood that a skilled user's private evil intent may be masked by public amiability.

The second explanation for Anishinaabe sickness links illness and other problems to transgressions that lie outside the realm of everyday and observable interactions between normal living human beings. Participation in the Anishinaabe behavioral environment requires that one enters into respectful relationships with other animate beings, with obligations to behave appropriately toward sources of power. In addition, it is imperative that obligations to other-than-human persons be fulfilled. When these respectful relations and obligations are breached, illness and misfortune may result. The term *ondjine* is used to indicate an illness or misfortune has occurred "for a reason," with the reason attributable to something that someone did at some point in the past. It is typically under the guidance of a medicine person that the individual gains insight into the specific circumstances under which normative expectations were breached. Recalling the past incident is an integral part of treatment; through the medium of the medicine person the other-than-human persons convey what needs to be done to make amends for the past incident and may make other requests for the future. These behests may include such things as holding an annual feast, preparing offerings, or participating in an annual Sun Dance.

There are many different actions with the potential to result in *ondjine* (see Garro, 1990, 2000b, 2001, 2002). The most common explanation for *ondjine* involves improper relationships between human beings and animals, such as causing an animal to suffer or to otherwise interfere with its autonomy. Such interference is considered a violation of norms governing social relationships. Animals may be killed to sustain human life, but this must be done gratefully and without causing undue suffering. *Ondjine* is also an inevitable consequence of the use of "bad medicine." Any crippling illness or painful death may be seen as a confirmation that the afflicted individual or a close family member has used "bad medicine" (see Garro, 1990; Hallowell, 1942, p. 77).

HEALTH THROUGH THE LIFE CYCLE

Pregnancy and Birth

The local health center tries to ensure the health of the expectant mother and her unborn child through prenatal classes, home visits, and by "arranging prenatal care and other help during pregnancy. Canadian government actions and resources relating to maternal and child health have contributed to reduced infant mortality. Almost all births take place in hospitals about an hour's drive away.

For those who recognize Anishinaabe sickness, during pregnancy, an unborn child is considered vulnerable to *ondjine* and birth defects may be attributed to something the mother did when pregnant (or, rarely, the father). Interaction with animals is to be avoided, as is staring at or making fun of a mentally or physically handicapped person (which could lead to one's child being born with the same disability). If steps to acknowledge and redress a transgression are not taken, *ondjine* can continue to affect subsequent births. A miscarriage may also raise concerns about "bad medicine".

Life is respected and comments made about contraception and abortions censured such practices. Actions with the potential to interfere with beings who wish to be born can result in *ondjine*. This was the explanation converged upon by several women discussing another woman who had an abortion in her teens and was now unable to carry other pregnancies to term. A woman with a very large family was told by a medicine person that her chronic lower back pain was due to *ondjine* caused by the sterilization procedure carried out by the doctor after the birth of her last child. Another expressed similar concerns about taking birth control pills. (Yet, birth control pills were dispensed at the health center and one herbalist said that she was at times asked to make a medicine that could help bring on a menstrual period for a woman who was "late.")

Infancy

The health center encourages breast-feeding through home visits. This is seen as an important public health concern for the Canadian Aboriginal population because of the potential impact on elevated morbidity rates for infectious diseases (Martens & Young, 1997) and rates of breast-feeding for Aboriginal women in Manitoba are considerably lower than non-Aboriginal women (Hildes-Ripstein, 1998). Another activity to which health center staff devote considerable effort is a vaccination program for infants and children. A yearly, and popular, baby show is sponsored by the health center with the judging based on vaccination records, breast-feeding, and other health promoting activities.

When infants cry and cannot be comforted a number of possible reasons may be considered—teething, a temperature, or simply a child just "born" that way. One often-considered possibility for the fussiness of a "colicky" infant is that this signals the need for the child to be given an Anishinaabe name. An Anishinaabe name confers some protection on the infant and serves to establish relationships between the child and other-than-human persons. The power to bestow names is a gift given to some individuals in the community. A feast for those invited is given at the time that the name is bestowed and commemorative feasts are given in subsequent years.

Childhood

A central goal for Ojibwa individuals is "not to be controlled by one's 'environment' including other people" with "acts that are attempts to control others ... negatively evaluated" (Black, 1977, p. 145). Even in childhood, autonomy is respected and parents tend not to interfere with what children choose to do. Children are allowed considerable scope to discover the world for themselves and learn from mishaps. Providing food and shelter for children is the responsibility of the parents, but bedtimes, eating times and what is eaten are essentially under the control of the child. Parents have expressed concerns about health education programs that instruct parents about what children should eat or what children should do. Physical punishment of children is rare, and like spousal abuse, the instances recorded involved occasions when the perpetrator was seen to be under the influence of alcohol or "bad medicine". Adults help children learn about responsibilities to others and standards for respectful behavior through their actions but do not overtly control the actions of children, though there may be some verbal commands to desist when children are particularly noisy or troublesome. Many transgressions that later in life result in *ondjine* are remembered as occurring during childhood.

Many children still speak Anishinaabemowin as their first language and others receive instruction through the school system. Like many other First Nations schools, an original language program is an integral part of education efforts.

Adolescence

Suicide, alcohol, the use of drugs, sniffing (solvents, glue, and gasoline), break-ins, and vandalism among adolescents are significant concerns for community members. Living in a community with a high level of unemployment and low incomes dependent on social assistance, many youth do not see the future as having much to offer. Manitoba has the lowest rate of school attendance among Aboriginal youth of any province or territory in Canada. At the time of the 1996 census, only 44.1% of those aged 15–24 reported attending school either full or part-time, with a considerable gap in school attendance between Aboriginal and non-Aboriginal youth (Hallett, 2000). Many of the youths who do not attend school are also not employed (37.5% of Aboriginal youths) and Aboriginal youths in Manitoba are 3.5 times more likely to be unemployed as non-Aboriginal youths. Aboriginal youths are over-represented in correctional facilities and in urban gangs (Hallett, 2000). Teen births also occur at a rate three times higher for "status Indians" than for others in the province.

Adulthood

National figures show Aboriginal women having fewer children, steadily declining from an average of 5.7 births per woman in 1970 to 2.55 in 1995 (Hallett, 2000). In recent years, many men and women return to the classroom as adults to make up for missed educational opportunities or to acquire new skills.

The start of menstruation adds an additional expectation for women's behavior. Menstrual blood is considered to have the power to cause illness in males and there were several cases involving young boys who were inadvertently affected. Women are expected to respect this power through appropriate behavior. While the danger is greatest when a woman is menstruating, generally speaking, women should not step over a male or even over a piece of clothing that will later be used by a male, and women should take care that males do not use the same bath towels. There are other associated proscriptions. Hunting may be unsuccessful if women touch hunting equipment, such as guns, while they are menstruating. The objects used by a medicine person should also not come near or into contact with menstruating women. Women who are gifted only come to be known as medicine persons after menopause.

When adults in the community talk about the "old days," reference is being made, not to some time beyond individual memory, but to time within an individual's lifetime, before the introduction of social assistance programs like welfare in the 1950s. The time before is recalled as one when people worked much harder and had little material wealth. Young people assumed more responsibility and had chores to carry out every day. Gardens were more common, as was home canning. Game was more available as were plant foods that could be gathered in the nearby bush. Simply prepared fish, stews, and bannock (bread) were everyday mainstays. But, by those who remember, this harder life was also considered a healthier life. There were no sodas or other junk foods. Wild meats, fish, and gathered foods were seen as more fortifying. Diabetes and high blood pressure were unknown diseases.

The Aged

Elders are treated with respect and are often consulted for advice. For example, as part of the transfer of control process, the health center's director and staff decided to consult with a group of elders on a regular basis and especially before implementing any changes in their services or policy. Although being a respected elder is a matter of recognition rather than something achieved simply by age, a party given for elders invited everyone over the age of 55. At the same time, an increasing number of elders are living on their own, although if grown children are still living in the community they are often nearby and visit often. Increasingly, the long-term consequences of diabetes are evident—including blindness, renal failure, and amputations.

Dying and Death

Most individuals have wide kinship connections throughout the community and a funeral often led to the cessation of most activities because many were in attendance. Prior to the burial, wakes can last for three days and three nights. Offerings, of food or drink, are often left on top of graves. Graves must be treated respectfully or it is said that *ondjine* may result. A man said to steal whisky and tobacco from graves became an alcoholic and his death was the result of drinking. A man who stepped on a grave over the location where the head was buried was said to have become bald because of this.

Interpersonal relationships may continue after death. It is not uncommon for an individual to dream of someone who is dead and be informed about something new through this encounter. Such interactions can become problematic, however, when the dead person desires to be joined by another family member, usually a beloved child. The dead person's efforts to bring this about results in the living person's enduring despondency over the preceding death and typically the individual regularly dreams of the dead person. These indicators of the dead person's continued and quite real presence in the living person's daily life suggest the need for the intercession of a medicine person lest the dead person's desire be realized.

REFERENCES

Bishop, C. A. (1999). Ojibwa. In J. H. Marsh (Ed.), *The Canadian Encyclopedia 2000* (p. 1702).Toronto: McClelland & Stewart.

Black, M. B. (1977a). Ojibwa power belief system. In R. D. Fogelson & R. N. Adams (Eds.), *The anthropology of power: Ethnographic studies from Asia, Oceania, and the New World* (pp. 141–151). New York: Academic Press.

Black, M. B. (1977b). Ojibwa taxonomy and percept ambiguity. *Ethos, 5*, 90–118.

Black-Rogers, M. (1989). Dan Raincloud: "Keeping our Indian Way." In J. Clifton (Ed.), *Being and Becoming Indian: Biographical Sketches of North American Frontiers* (pp. 226–248). Dorsey Press: Chicago.

Brown, J. S. H. (1997). Ojibwa, cultural summary. *Human Relation Area Files.* Yale University.

Culhane Speck, D. (1989). The Indian health transfer policy: A step in the right direction or revenge of the hidden agenda. *Native Studies Review, 5,* 187–213.

Fontaine, Phil. (1991). Health for First Nations in the 1990s. *Manitoba Medicine, 61,* 21–22.

Garro, Linda C. (1988). Explaining high blood pressure: Variation in knowledge about illness. *American Ethnologist, 15,* 98–119.

Garro, Linda C. (1990). Continuity and change: The interpretation of illness in an Anishinaabe (Ojibway) community. *Culture, Medicine & Psychiatry, 14,* 417–454.

Garro, Linda C. (1993). Perspectives on the health and health care of Canada's first peoples. *Culture, Medicine & Psychiatry, 17,* 145–157.

Garro, Linda C. (1995). Individual or societal responsibility? Explanations of diabetes in an Anishinaabe (Ojibway) community. *Social Science and Medicine, 40,* 37–46.

Garro, Linda C. (1996). Intracultural variation in causal accounts of diabetes: A comparison of three Canadian Anishinaabe (Ojibway) communities. *Culture, Medicine and Psychiatry, 20,* 381–420.

Garro, Linda C. (1998). On the rationality of decision making studies: Part 1: Decision models of treatment choice. *Medical Anthropology Quarterly, 12,* 319–340.

Garro, Linda C. (2000a). Remembering what one knows and the construction of the past: A comparison of cultural consensus theory and cultural schema theory. *Ethos, 28,* 275–319.

Garro, Linda C. (2000b). Cultural meaning, explanations of illness, and the development of comparative frameworks. *Ethnology, 39,* 305–334.

Garro, Linda C. (2001). The remembered past in a culturally meaningful life: Remembering as cultural, social and cognitive process. In H. Mathews & C. Moore (Eds.), *The psychology of cultural experience* (pp. 105–147). Cambridge: Cambridge University Press.

Garro, Linda C. (2002). Hallowell's challenge: Explanations of illness and cross-cultural research. *Anthropological Theory, 2,* 77–97.

Goddard, Ives. (1978). Central Algonquian languages. In B. Trigger (Ed.), *Handbook of North American Indians* (Vol. 15, *Northeast* pp. 583–587). Washington, DC: Smithsonian Institution.

Hallett, Bruce. (2000). *Aboriginal people in Manitoba 2000.* Winnipeg, MB: Human Resources Development Canada & Manitoba Aboriginal Affairs Secretariat.

Hallowell, A. Irving. (1936). The passing of the *Midewiwin* in the Lake Winnipeg region. *American Anthropologist, 38,* 32–51.

Hallowell, A. Irving. (1942). *The role of conjuring in Saulteaux Society.* Philadelphia: University of Pennsylvania Press.

Hallowell, A. Irving. (1955). *Culture and experience.* Philadelphia: University of Pennsylvania Press.

Hallowell, A. Irving. (1960). Ojibwa ontology, behavior and world view In S. Diamond (Ed.), *Culture in history: Essays in honor of Paul Radin.* New York: Columbia University Press. (Reprinted in Hallowell, A. I. (1976). *Contributions to Anthropology* [pp. 357–390]. Chicago: University of Chicago Press.)

Hallowell, A. Irving. (1976). *Contributions to anthropology: Selected papers of A. Irving Hallowell.* Chicago: University of Chicago Press.

Hallowell, A. Irving. (1992). *The Ojibwa of Berens River, Manitoba: Ethnography into history.* J. S. H. Brown (Ed.) (and author of Preface and Afterword). Fort Worth, TX: Harcourt Brace.

Hildes-Ripstein, E. (1988). Infant care practices in First Nations Peoples of Manitoba: Are there any modifiable risk factors for SIDS. M.Sc. Thesis, University of Manitoba, 1988.

Holzkamm, T. E., Lytwyn, V. P., & Waisberg L. G. (1988). Rainy River Sturgeon: An Ojibway resource in the fur trade economy. *The Canadian Geographer, 32,* 194–205.

Howard, J. (1965). *The Plains Ojibwa or Bungi.* Vermilion: University of South Dakota, Anthropological Papers No. 1.

Martens, P. J., & Young T. K. (1997). Determinants of breastfeeding in four Canadian Ojibwa communities: A decision-making model. *American Journal of Human Biology, 9,* 579–593.

Mithun, M. (1999). *The languages of North America.* Cambridge: Cambridge University Press.

O'Neil, J. (1988). Referrals to traditional healers: The role of medical interpreters. In D. Young (Ed.), *Health care issues in the Canadian North* (pp. 29–38). Edmonton: Boreal Institute for Northern Studies, University of Alberta.

Peers, Laura. (1994). *The Ojibwa of Western Canada, 1780 to 1870.* Winnipeg: University of Manitoba Press.

Pettipas, Katherine. (1994). *Severing the ties that bind: Government repression of indigenous religious ceremonies on the Prairies.* Winnipeg: University of Manitoba Press.

Pettipas, Katherine Autumn. (1977). Ojibwa pharmacopoeia. *Manitoba Nature,* 12–19.

Ray, A. J. (1974). *Indians in the fur trade: Their role as hunters, trappers, and middlemen in the lands southwest of Hudson Bay, 1680–1870.* Toronto: University of Toronto Press.

Reading, J. (1999). An examination of residential schools and elder health. In *First Nations and Inuit Regional Health Survey.* Ottawa: First Nations and Inuit Regional Health Survey Steering Committee.

Rhodes, R. A. & Todd, E. M. (1981). Subarctic Algonkian languages. In June Helm (Ed.), *Handbook of North American Indians* (Vol. 6, *Subarctic* pp. 52–66). Washington, DC: Smithsonian Institution.

Rogers, E. S. (1969). Natural environment–social organization-witchcraft: Cree versus Ojibwa—A test case. In David Damas (Ed.), *Contributions to Anthropology: Ecological Essays* (pp. 24–39). Bulletin No. 230, Anthropological Series No. 86, Ottawa: National Museums of Canada.

Rogers, E. S. (1978). Southeastern Ojibwa. In B. Trigger (Ed.), *Handbook of North American Indians* (Vol. 15, *Northeast* pp. 760–771). Washington, DC: Smithsonian Institution.

Shkilnyk, A. M. (1985). *A poison stronger than love: The destruction of an Ojibwa community.* New Haven & London: Yale University Press.

Steinbring, J. H. (1981). Saulteaux of Lake Winnipeg. In June Helm (Ed.), *Handbook of North American Indians* (Vol. 6, *Subarctic* pp. 244–255). Washington, DC: Smithsonian Institution.

Thornton, Russell. (2000). Population history of Native North Americans. In M. R. Haines & R. H. Stecke (Eds.). *A Population History of North America* (pp. 9–50) Cambridge: Cambridge University Press.

Ubelaker, D. H. (1988). North American Indian population size, AD 1500 to 1985. *American Journal of Physical Anthropology, 77,* 289–294.

Waldram, J. B. (1985). Hydroelectric development and dietary delocalization in northern Manitoba, Canada. *Human Organization, 44,* 41–49.

Waldram, J. B., Herring, D. A., & Young T. K. (1995). *Aboriginal health in Canada: Historical, cultural, and epidemiological perspectives.* Toronto: University of Toronto Press.

Young, T. K. (1988). *Health care and cultural change: The Indian experience in the central Subarctic.* Toronto: University of Toronto Press.

Young, T. K. (1994). *The health of Native Americans.* New York & Oxford: Oxford University Press.

Young, T. K., Reading, J. Elias, B., & O'Neil, J. (2000). Type 2 diabetes mellitus in Canada's First Nations: Status of an epidemic in progress. *Canadian Medical Association Journal, 163,* 561–566.

Oklahoma Choctaw

Joseph Neil Henderson and Linda Carson Henderson

ALTERNATIVE NAMES

The official name for the tribe is the Choctaw Nation of Oklahoma. Other terms that are locally used include "Choctaws," and sometimes "Chocs." The linguistically correct name is "Chattah." This is occasionally used by native speakers in oral discourse and for some official occasions such as tribal ceremonies and, on occasion, at political speeches.

The two largest Federally recognized Choctaw tribes are the Oklahoma Choctaw and the Mississippi Choctaw who live on a Federal reservation north of Jackson, Mississippi. They are descendants of families that were not removed to Indian Territory. These Choctaw tribes are not politically affiliated with each other, although there are friendly relations between the two.

LOCATION AND LINGUISTIC AFFILIATION

At European contact time, the Choctaws were on lands mapped today as Mississippi, and had dispersion that also reached into present day northern Florida, Georgia, Alabama, Louisiana, and Arkansas. Today, the Oklahoma Choctaws are located in the southeastern quadrant of Oklahoma, with the administrative headquarters in Durant, Oklahoma. At Tushkahoma, Oklahoma, is the tribe's council house used into the early 1900s, but now occupying a largely symbolic tie to the past. The Choctaw Nation of Oklahoma is bounded on the west by the territory of the Chickasaw Nation, and to the north is the Creek Nation.

The current location is the product of the creation of Indian Territory in the early 19th century. The earliest mapping of Indian Territory, in preparation for "receiving" the southeastern natives, assigned the Choctaws the entire lower portion (east to west) of the land. Later, this was divided in half so that the western half of the former Choctaw land was made available for the natives from the plains. Then, later, the Choctaw land was cut in half again to assign the Chickasaws adjacent land (McKee, 1989).

Today, the former Indian Territory is the State of Oklahoma. As a result of its Indian Territory history, it has 37 Federally recognized tribes. However, the Oklahoma Choctaws, like the other Oklahoma tribes, have only tiny fractions of their former Indian Territory land space that is reservation status. The old tribal boundary lines still exist, but the State of Oklahoma is co-mapped onto them.

Linguistically, the Oklahoma Choctaws are Muskogean speakers (Swanton, 1946). There are numerous native speakers living. However, language extinction is a real risk. The tribe has language courses offered on the internet, and at one time required those working at the tribal headquarters to take Choctaw language courses during the work day, if they did not speak the language.

OVERVIEW OF THE CULTURE

Before contact time, Choctaws lived in the southeastern woodlands and were intensive horticulturists. Primary food items included corn, beans, squash, pumpkins, and meat items from deer, rabbit, turtle, birds, fish, and others. Their neighbors were the Chickasaw, Creek, Cherokee, and Seminoles. The Choctaws were encountered by the DeSoto Expedition and were reported to have extensive body tattoos, forehead flattening, and short hair styles. They also used bow and arrows, blow guns, and snares for trapping land animals and for fishing. Ceramics and baskets were also commonly used for storage, cooking, and carrying. Extensive foot trails facilitated travel and communication. The society was organized as matrilineages and politically structured as a chiefdom at contact time (Galloway, 1995; McKee, 1989; Swanton, 1946).

Today, the population of the Oklahoma Choctaw living in Oklahoma is about 40,000. The total enrollment of the tribe is over 200,000. The southeastern portion of Oklahoma is woodlands with rivers and rolling hills whose elevations reach about 2,500 feet. The highest elevations are in the east, and the lower, flatter lands are in the west. It is in the higher elevations with lower population density where tribal members note that the most traditional Choctaws live. "Traditional," or colloquially, "following the old ways," is indicated by speaking the native language, living on old allotment lands, being as separate from the local white society as possible, going to an "Indian Church," eating traditional foods, and engaging in Choctaw rituals. However, two communities, considered by many to be "traditional," are located inside the former Choctaw boundaries which are now designated as Chickasaw. In spite of the category "traditional" and its unstated opposite "progressive," the best approach to understand contemporary Choctaw culture is to assume great heterogeneity in all aspects of culture.

Religion is based on exposure to Protestant missionaries from the early 18th century. While many Choctaws go to the conventional Protestant churches in the Oklahoma towns of the area, many Choctaws go to "Indian Churches." These are located remotely, may or may not have an ordained pastor, and often are conducted in Choctaw language only. The Indian Church is actually a hub of community cultural dynamics. In addition to the church building which looks like a small Christian chapel with steeple, the surrounding few acres have family "camp houses" used for weekends or longer stays. The Indian Church provides an exclusive place away from the otherwise dominant white society. It is here that cultural identification is fully reinforced, communications and alliances maintained, and children enculturated in Choctaw ways.

It should also be noted that many non-Christian beliefs are present and actively practiced. These include beliefs about cosmology, ghosts, disease-object intrusion, love medicine, use of traditional "Indian doctors" which are called "*Chattah alikchi*" (Choctaw "doctors"), belief of medicinal (i.e., supernatural) power in eagle feathers, releasing "medicine" in sage and cedar through smudging, and use of numerous herbal remedies.

THE CONTEXT OF HEALTH: ENVIRONMENTAL, ECONOMIC, SOCIAL, AND POLITICAL FACTORS

Many of today's late middle aged and older aged Choctaws had been forced to go to Indian Schools operated by the Bureau of Indian Affairs or by denominational interests. In these schools, the tribal culture was devalued and punishment meted for speaking their native language. The chiefs, at the time of removal from their homelands, were rewarded for being "progressive" by the government of the United States. The Chief today is elected according to a tribal constitution modeled after the U.S. constitution (Faiman-Silva, 1997). However, there are some who still prefer to have a model of government that is based more on the pre-removal form of governance.

The Oklahoma Choctaw tribe is now also a multi-million dollar corporation. The Chief serves much like a CEO (Chief Executive Officer) with the responsibility of many diverse for-profit businesses. The tribe is proud of its economic successes and has received awards of commendation from the U.S. government for innovation in enterprises. The Chief is informed and counseled by an Assistant Chief, and a 12-member tribal council, all of whom are elected. It is fortunate that the success of the economic engine of the tribe has been, in part, applied to the health and social service provisions owned and operated by the tribe.

The economy is one based on participation in the area's commerce that includes, labor, farming, manufacturing, lumber, and other ordinary jobs. The tribe also engages in business development of its own. The tribe now operates several gaming facilities, several "travel plazas" which are convenience stores with fuel and gaming, "smoke shops" with discount tobacco products, and several other smaller operations including pre-fabricated housing manufacturing.

Since 1985, all Choctaw health care facilities have been managed by the tribe under the auspices of the Indian Self-Determination Act (Indian Health Service, 2000). There is a contract between the Oklahoma City Area Office of the Indian Health Service and the Health Service Authority of the Choctaw Nation. Federal funds are provided for tribal use each year and technical assistance is provided by Indian Health Service personnel. Physicians and other health care professionals may be either commissioned officers of the Public Health Service or direct employees of the Choctaw tribe. Currently, the majority of providers are direct employees of the Choctaw Nation, and many are members of the tribe.

The Choctaw Nation of Oklahoma is the first tribe in the United States to build its own hospital. The Choctaw Health Services Authority manages the hospital at Talihina, Oklahoma, and four health centers located in the towns of McAlester, Hugo, Poteau, and Broken Bow. The Broken Bow Health Center has two staff physicians to service 10,800 outpatient visits annually while support staff care for an additional 5,000. The Hugo Health Center and McAlester Health Center each have one physician and one physician's assistant. They manage 11,000 outpatient visits a year. Information about staff and outpatient visits at Poteau was not available at the time of this writing.

The hospital and the clinics annually log about 48,000 patient visits. The patient seeking services must be a member of any federally recognized tribe and, must reside in the Choctaw Nation Health Service Area. Tribal facilities cannot provide a full range of health care, and it may be required that the patient be treated elsewhere. Unfortunately, prejudice remains toward Choctaws. For some, the prospect of obtaining needed services in the White community is considered undesirable and may result in a demeaning experience. Nonetheless, non-tribal care is an extremely important element of the health care delivery system although care must be authorized through the Contract Health Service Program. Costs remaining after Medicare, Medicaid, VA, Worker's Compensation, Title XIX, and/or group health insurance is used, is paid by the tribe.

Diabetes mellitus II (adult onset) is a particular problem for most tribes, and the Oklahoma Choctaw are no exception. The American Diabetes Association states diabetics should receive their care from a physician-coordinated team. Such teams include, but are not limited to, physicians, nurses, dieticians, and mental health professionals with expertise in diabetes (American Diabetes Association, 1998). The Diabetes Clinic at the Choctaw Nation Hospital provides high-risk diabetic patients with a comprehensive multidisciplinary program. Staff includes physicians, nurses, a podiatrist, a nutritionist, certified diabetes educators, and mental health practitioners.

The Choctaw Nation also has numerous other health and social services. These include emergency medical service, drug and alcohol testing, substance abuse recovery program, mail order pharmacy, eyeglasses, dentures, and hearing aid clinics. Also, other services include child care assistance, day care, food distribution, Head Start programs, Indian Child Welfare programs, low income home energy assistance, nutrition services for elders (55+), vocational rehabilitation, and Women, Infant, and Children (WIC) programs.

Living conditions vary widely among the Choctaw, from one-bedroom homes in need of repair with a wood stove, cold water, and outhouse, to three-bedroom centrally heated homes with contemporary furnishings. Many elders reside in small "Indian homes," which have running hot and cold water and adequate plumbing. These "Indian homes" are federally subsidized housing provided under strict eligibility requirements. Living rooms sometimes doubled as sleeping quarters for extended family (Faiman-Silva, 1997). The Choctaw nation is committed, however, to providing support to elders in order to improve living conditions. Aid includes the availability of affordable rental housing, Indian Housing, the tribally funded Housing Improvement Program, and the Bureau of Indian Affairs Housing Improvement Program. Renovations may be done with tribal assistance. Heat assistance is also available.

MEDICAL PRACTITIONERS

Since about 1955 when the Indian Health Service (IHS) was established, most tribes have had allopathic physicians providing basic primary care at modest clinics and hospitals on tribal lands. In recent years, many tribes have elected to take over the cost of operating their own health care facilities independent of the Indian Health Service. Yet, the ability to do so is uneven across tribes and leaves some with excellent services and some with suboptimal care. The IHS itself estimates that it can meet only half of the needed primary care for tribes.

Other types of health care providers are numerous. These include licensed nurses, physician assistants, psychologists, social service staff, nutritionists, and podiatrists. Podiatric service is connected to the high rate of diabetes mellitus and consequent foot trauma. Nutritional counseling is also available through programs of the Women, Infant, Children early nutrition and child care program of the Federal government.

In addition to physicians and other biomedically trained staff, an important part of health provision is done by tribal Community Health Representatives (CHRs). These CHRs are paraprofessionals that are members of the tribe for whom they provide services. They receive training from an Indian Health Service program. Their functions are to conduct disease screening (e.g., hypertension, diabetes), deliver health education (e.g., diet, exercise, safety), conduct in-home visitation, broker information/resources, and provide transportation to health care services. The number of CHRs for the Oklahoma Choctaw is about thirty and these are mainly women. Their value as part of an ongoing health promotion program for family members of all ages is extremely high.

Each CHR travels many thousands of miles per year because of the predominately rural nature of the Choctaw Nation. The Indian Health Service provides a four-door, medium sized car for each CHR. This vehicle is used to go to clients' homes, no matter how remote, and to transport clients to the tribal clinics and hospital. CHRs also have required in-service education to maintain expertise on the topics germane to their paraprofessional role. On a regular basis, the CHRs will conduct health fairs which are done in communities large and small. The purpose of CHR tribal health fairs is to bring the health education, screening, and resource information sharing to a community-level population base.

CHRs are evenly distributed geographically across the Choctaw Nation based on population. The land base comprises ten and one-half counties of the State of Oklahoma. Each of the ten counties has a tribal "Field Office" which is the place for many activities such as senior meals, speakers, and other social functions. Also, the Field Offices serve as office space for the Women, Infant, and Children programs, CHRs, and workers in other programs. The CHRs generally live in the area that they serve. In towns that are larger that most of the rural towns, there may be two CHRs working. However, most of the counties have one CHR. Their case loads range from about 30 to 60 clients per month.

The "cultural distribution" of the CHR is not absolutely fixed, but has a trend toward having a CHR who is fluent in Choctaw in the regions where more people are most comfortable speaking Choctaw. In these areas are found those with the strongest cultural identification with the old ways and some elders who are monolingual Choctaw speakers. Since CHRs are educators and recipients of information related to their clients' personal health and family related matters, having CHRs whose cultural identification and language skills are most similar to the client base is seen as helpful.

Outside the establishment provider groups is the traditional healer. This person is the "*Chattah alikchi*," or "Choctaw doctor," in today's idiom. The *alikchi* occupies the role of the spiritual healer and probably, in some instances, is close to the definition of shaman. Originally, it did not mean "doctor" in the contemporary sense that a doctor is a physician who treats biomedical disease based on allopathic theory. *Alikchi* is a morpheme that means "supernaturally connected." It was applied to a person, male or female, who had specific supernatural connections via training as a child by the "little people." These people are considered to be inhabitants of the local woods and have supernatural capabilities. Many Choctaw people believe that these entities are still present. The "little people" were said to capture a child at play, take them to their home in the woods, and train them to be *alikchi*. Their skills would be applied only as an adult.

The notion of healer is conceptualized today in three separate categories. First, is the *Chattah alikchi* who is specifically a Choctaw person with supernatural healing capacity. There are vague references indicating that *Chattah alikchi* can use their power to do harm as well as good. Second, there is the category of *alikchi* meaning a Native American person who may be Choctaw but could be a member of another neighboring tribe such as the Creek or Chickasaw. Sometimes, however, the designation of *alikchi* is applied to someone more culturally removed from Choctaws, such as a Comanche person in southwestern Oklahoma. The distinctive features for inclusion in this category are; being Native American, and having an accepted capacity to heal based on tribal ethnomedical constructs. Third, *alikchi* is used in Choctaw language discourse to refer to allopathically trained, non-Native American physicians who are encountered in the tribal health services or surrounding community. In this usage, *alikchi* is synonymous with "doctor," as used by the non-Native American population.

In contemporary Oklahoma Choctaw life, *Chattah alikchi* occupy a somewhat vague position in terms of contemporary practice. From data collected about degrees of cultural identification by one of the authors (LCH), 14 of 30 elder interviewees acknowledged the existence of *Chattah alikchi*. However, only eight discussed using *Chattah alikchi* themselves. Moreover, information of the specific whereabouts of *Chattah alikchi* was not forthcoming. In summary, the fully practicing *Chattah alikchi* may be rare in practice, but common in cultural consciousness.

CLASSIFICATION OF ILLNESS, THEORIES OF ILLNESS, AND TREATMENT OF ILLNESS

The Oklahoma Choctaw have experienced generations of contemporary American education and life experience. Most of the Choctaw population are subscribers to the biomedical model and are participants in it via the health care services provided by the Indian Health Service or the tribal health care operations staffed by conventionally trained personnel. However, medical pluralism is present as indicated by the use of culturally derived healers, as reported herein.

The contemporary Oklahoma Choctaws have a cultural history imbued with multiple explanatory models for health and disease. Although many of the "old ways" are shielded from outsiders, there are reported instances of practices derived from disease–object intrusion theory. For example, a report of shoulder pain was treated with superficial skin incisions over the site of pain, direct application of the mouth to the incisions, sucking of the incision site, and showing the disease–object (a "spider-like" mass) to the patient. Also, a type of "soul loss" is known to the tribal members, although no current reports of an instance of this have been reported. An entity known as "*impashilup*" (literally, "eater [of the] spirit") would enter the body and consume the soul. Prevention was to abstain from destructive or morose thoughts. Cure is not currently known.

Medicine bags (*baht inkwish*, literally, "bag" "medicine") made of animal skin and containing plant and mineral material considered to have "medicine" were widely used in the past, and, many Choctaw people have their own medicine bags today. They are used to ward off problems and give "medicine" (i.e., protective power) to the holder.

Although historically the use of communal sweat lodges was not found within the Oklahoma Choctaw culture (sweat lodges for one person were common), the use of communal sweat lodges has increased within the Choctaw Nation, as well as in other parts of Indian Country, in the post-World War II era. The sweat lodges have been especially useful in the treatment of drug and alcohol abuse, incorporating traditional healing with accepted counseling modalities.

Diabetes mellitus afflicts one out of every five American Indians. Yet, according to the Diabetes Quality Improvement Project, in 1996, the Indian Health Service received $1,578 per capita to care for its population compared with $3,920 per capita expended for the U.S. civilian population (Acton et al., 2001). Moreover, the rate and degree of severity of the disease within the American Indian population is higher than that of the rate and severity within the general U.S. population. Additionally, American Indian diabetics are four times more likely than their White counterparts to experience an amputation as a consequence of diabetes and six times more likely to experience kidney failure. In Oklahoma, where the Choctaws are one of the largest tribes of 37 federally recognized tribes, 24% of Oklahoma Indians have diabetic retinopathy (American Diabetes Association, 2002).

Improving diabetic health status among Oklahoma Choctaw could well be possible. Recent insights about the cultural dynamics of Oklahoma Choctaw diabetic health behavior constitute a crucially important component for the biocultural appreciation of the disease (Henderson, 2002). Sixty subjects (30 diabetics and 30 health care providers) were interviewed. All subjects were classified as either identifying strongly with Choctaw culture or identifying strongly with White culture. Subjects that identified strongly with the "old Choctaw ways" were *less* likely to understand the biology of the disease and to see the disease and its sequelae as an expected part of one's life. However, Choctaws subjects who had a strong cultural identification with the larger White community were more likely to know the basic biology of the disease and to comply with medical treatments and advice. Moreover, the majority of health care provider subjects that treated these patients did not consider the Choctaw patient's cultural identification position relevant to their clinical work. However, the few health care providers who are themselves Choctaw and who identified strongly with Choctaw culture recognized the relevance of cultural differences to treatment practices within the Choctaw community. The result is that while diabetes is one disease, there are at least four variant models for diabetes operating in the community.

The four models, based on this sample, first include those Choctaws that strongly identified with the "old Choctaw ways." This explanatory model normalized the symptoms of the disease, made compliance an undesirable connection to the White establishment, and rendered disease sequelae such as amputation and renal insufficiency expected life course events for "Indian" diabetics. Second, those Choctaws that strongly identified with mainstream American White society had an explanatory model of diabetes that is similar to "Whites" and were more likely to comply with interventionists' recommendations. Third, the dominant explanatory model of the providers was one which recognized that there were Choctaws that follow the "old ways" and those that do not, but believe that these two positions are not relevant to their practice, treatment, or intervention efforts. Last, those providers who were Choctaw and identified strongly with the "old Choctaw ways," considered the cultural identification position of Choctaw diabetics very significant in the development of treatment approaches for them. Accounting for these multiple models of diabetes in intervention and prevention efforts could greatly improve diabetic health status.

SEXUALITY AND REPRODUCTION

Partner selection mainly follows the larger American society practices. However, there is extant an admonition to preserve Choctaw ancestry as fully as possible. This promotes a pressure to court and marry only members of the tribe. Secondarily, there is acceptance of marriage with a member of another tribe. Nonetheless, it is common for Choctaws to marry members of the majority population.

Children are desired, but do not constitute a type of kinship wealth. They are generally well cared-for largely in the context of a nuclear family setting. However, there is some greater sense of corporate child care across family kindred groups than in the local White society. This is most observable when families have maintained their former allotment lands and live in clustered home sites. Such home sites typically have houses grouped together with about 20–75 yards separating them. These clustered home sites have produced good-natured jokes about those families that refer to them living on the "rez." This is considered a humorous analog to tribes that do live on Federal reservation land, unlike the Oklahoma Choctaw.

HEALTH THROUGH THE LIFE CYCLE

The items below represent main trends in overt behavior. Cultural heterogeneity allows for many conceptual and behavioral variations that may be linked to the earlier Choctaw practices but that are privately held or selectively revealed.

Pregnancy and Birth

The exposure to American educational institutions and the biomedical model via the Indian Health Service has lead to a Choctaw model of pregnancy and birth that is largely based on contemporary American patterns. Today, virtually all births occur in hospitals whether it is the tribal hospital or in a non-Indian local hospital. However, before the last decade, prenatal care was not easily available due to many barriers, such as rural distances to care, transportation limitations, financial limitations, and cultural barriers. It was also not uncommon for women in labor who could not reach the tribal hospital to go to a non-Indian hospital parking lot, stay in the car throughout labor, and when delivery was imminent, present themselves at the emergency room. In so doing, they were assured of receiving obstetrical care, albeit on an emergency basis.

Elder women today tell of labor and delivery in the early Choctaw hospital built in 1916. Women of that era did use indigenous herbal teas to ease and speed labor. These were brought by family members to the hospital, kept in a jar under the bed, and used by the patient. Also, the indigenously preferred position of delivery was probably a squatting position based on reports of women who would get in a squatting position in the hospital bed. However, these reports conclude with the expectant mother being wheeled from the ward to a delivery room where the patient was place in the supine position.

Infancy

The exposure to American educational institutions and the biomedical model via the Indian Health Service has lead to a Choctaw model of infancy that is largely based on contemporary American patterns. However, there is sufficient concern for some mother–infant pairs regarding nutrition and parenting that a strong Women–Infant–Children program is available throughout the tribal area. Mothers will come to one of ten "Field Offices" operated by the Choctaw Nation. These Field Offices typically are brick buildings with about 1,000 square feet of space. They have a common room for meals and meetings as well as offices for project coordinators who administer nutrition projects, family resources (e.g., parenting, domestic abuse) assistance, and Community Health Representatives. The infants can be brought to these locations for basic anthropometric measurements to screen for developmental indicators. The mothers also receive health education information from the staff and by print materials. They can also obtain nutritional supplements for the infant.

Childhood

All children receive education in the local public schools. There are Head Start programs that are well attended and to which tribally owned buses transport the children from remote rural locations. However, there is a problem with proper childhood nutrition that many believe results in adult onset diabetes mellitus, and an increasing concern about a rising incidence of Type II diabetes in childhood. The Choctaw Nation is currently expanding its diabetes education base to include elementary school children because primary prevention strategies are of utmost importance to diabetes control.

Adolescence

Choctaw adolescents attend local public middle and high schools where they are exposed to a typical American curriculum regarding the sciences and health. There is a significant problem with smoking rates that is compounded by Oklahoma's dubious distinction as the state with the highest use of tobacco products. Moreover, there is some cultural impetus for tobacco use that relates to tobacco as a sacred substance. Many advertisements in American Indian newspapers advocate "Traditional Use, Not Abuse." However, adolescent tobacco use remains high.

There is also an increasing prevalence of obesity in this age group. Consequently, Type II diabetes is as much a health concern for adolescents as it is for adults. There are new programs for adolescents that emphasize diet and exercise. Also, cultural messages that relate to the early days of American Indian life in which there was no obesity or diabetes are becoming more common. The current epidemic of obesity and diabetes is blamed on the "white man," or "civilization."

Adulthood

Nutrition, exercise, and substance abuse are the main health issues of adulthood. The health care program of the tribe has aggressively responded to these in the last several years. As with other non-Indian programs responding to these matters, it has proven difficult to have a total success with these health problems.

Substance abuse is a problem of sufficient magnitude that special programs have been developed. A special inpatient facility has been operated by the tribe for years and serves members of all tribes. This includes alcohol abuse as well as other substances. One component of the substance abuse program integrates the use of sweat lodges with the more conventional treatment protocols. It is considered to offer spiritual, mental, and physical purification.

The Choctaw Nation hospital has made strides to respond to adult onset diabetes (Type II) by operating a Diabetes Wellness Center. This center is staffed by endocrinologists, diabetes educators, specially trained nurses, and podiatrists. The center takes a family model approach to intervention and conducts family education for the prevention and treatment of diabetes.

Aged

Elderhood has many facets to it. First, "elder" may be a social status conferred to adults considered helpful, socially powerful, and having qualities of leadership regardless of age. Second, "elder" may also be used to indicate a person who is old regardless of social status. Last, "elderlies" is sometimes used to refer to all the old people in a tribe. In general, anyone who is considered old is accorded respect. Those who are both "old" and "elders" (i.e., the social status) receive even more respect.

Like many elders, Choctaw elders experience multiple jeopardy. They are members of a minority group (by government definition), many are poverty-stricken, and are in poor health compared to their general population counterparts. The Indian Health Service designates those persons age 55 and older as elders. American Indian elders are living longer, due to improved public health, but they are "sicker longer" with chronic disease when compared to the general population. Indian Health Service data indicate that elders are more disabled and experience health declines at an earlier age than their White counterparts.

Assistance for Choctaws elders is also available by the provision of hearing aids, eyeglasses, dentures, canes, wheelchairs, and other disability-related equipment to those who cannot afford to purchase them. Through the U.S. Department of Agriculture, the Choctaw Nation has a food distribution program which serves over 5,000 people per month throughout the 10 counties. In addition to this program, the Nutrition and Supportive Services Programs provide meals, information and referral, transportation, and arts and crafts to Indian Senior Citizens 55 years of age or older. The program provides one meal a week at the local community centers.

Dying and Death

The influence of general American society has lead to a contemporary view of dying and death that is very similar to those of the Christian church. However, there are some Choctaw people whose multi-culturalism is seen relative to death and spiritual matters. For example, some people will not walk near the cemetery of their own "Indian Church" after the sun sets because *shilup* (i.e., ghosts) come out then. An Indian Church is a remotely located church that is partly Christian and, yet, retains some Choctaw-specific beliefs. Most Indian Churches have adjacent cemeteries. For burial, a back-hoe from the county is brought out to dig the grave. Also, in addition to Christian beliefs and rituals, some will burn white sage, tobacco, and cedar and "smudge" the body prior to closing the casket lid. Then, they will smudge themselves as a rite of purification.

CHANGING HEALTH PATTERNS

Increased longevity results in the manifestation of diseases and conditions that are age associated, and typically, are chronic, incurable, and debilitating. One of these is late life onset cognitive dysfunction due to organic brain disease (e.g., Alzheimer's disease and vascular dementia). However, a recent genetic study suggests that Choctaws may be less likely to develop Alzheimer's disease than Anglo-Europeans (Henderson et al., 2002). The Apolipoprotein E4 gene is associated with developing Alzheimer's disease. From a sample of 70 people with Certificate of Degree of Indian Blood confirmation of "full blood" status, the frequency of this allele is less than 50% of White controls. However, there is speculation that vascular dementia may be more prevalent due to antecedent risk factors that are high prevalence conditions in this and other tribes (e.g., smoking, diabetes, obesity, hypertension, and alcohol abuse) (Henderson, 1994).

Linguistically, there are no Choctaw morphemes coding for the disease known in English as "Alzheimer's Disease" (Henderson & Henderson, 2002). The closest terminological connection is *"imanokfila kanea"* (intellect lost). Other candidate morphemes, such as, *"tasembo"* (crazy) fail since they can apply to non-humans as well (e.g., dog, horse). Also, from a cultural construction of disease perspective, there are families in the Choctaw tribe in which the hallucinatory symptoms of dementia are interpreted as the person communicating with the supernatural world in preparation for death (Henderson & Henderson, 2002). As one person said of her grandmother who suffers from dementia, "She sees people we don't see." The granddaughter does *not* say that her grandmother "sees people that aren't there." The afflicted person simultaneously exhibits special capacity, reveals details about the vaguely understood "other side," and brings a bit of mystical awe to the family. This view departs greatly from the usual medical model of pathology and consequent symptomatology due to brain cell death. Additionally, this "supernormal" construct of dementia offers an expanded window into the extensive range of interpretive models applicable to perceptible states of health and disease in the context of culture.

REFERENCES

Acton, K. J., Shields, R., Rith-Najaran, S., Tolbert, B., Kelly, J., Moore, K. et al. (2001). Applying the diabetes quality improvement project indicators in the Indian Health Service primary care setting. *Diabetes Care, 24*, 22–26.

American Diabetes Association. (2002). Diabetes among Native Americans. March 12, 2002. Available online at, http://www.diabetes.org

Faiman-Silva, S. (1997). Choctaws at the crossroads. Lincoln: University of Nebraska Press.
Galloway, P. (1995). Choctaw genesis. Lincoln: University of Nebraska Press.
Henderson, J. N. (1994). The epidemiology of American Indian gerontology. Paper presented at the annual American Public Health Association meeting. Washington, DC.
Henderson, L. C. (2002). The cultural construction of diabetes mellitus among Oklahoma Choctaw elders and health care providers: Discordance between models. *Association for Anthropology and Gerontology, 23*, 4–6.
Henderson, J. N., & Henderson, L. C. (2002). Cultural construction of disease: A "supernormal" construct of dementia in an American Indian tribe. *Journal of Cross-Cultural Gerontology, 17*, 197–212.
Henderson, J. N., Crook, R., Crook, J., Hardy, J., Onstead, L., Carson-Henderson, L. et al. (2002). Apolipoprotein E4 and tau allele frequencies among Choctaw Indians. *Neuroscience Letters, 324*, 77–79.
Indian Health Service. (2000). Trends in Indian health: 1998–1999. Rockville: Indian Health Service.
McKee, J. O. (1989). The Choctaw. New York: Chelsea.
Swanton, J. (1946). The Indians of the Southeastern United States. Bureau of American Ethnology Bulletin 137. Washington, DC: U.S. Government Printing Office.

Roma of the United States and Europe

Anne Hartley Sutherland

ALTERNATIVE NAMES

The Roma name is one that has become common currency today, replacing the designation Gypsy which has negative connotations. Since many Roma use the term Gypsy with outsiders, and there are contexts in which Gypsy is the broader term, its use is still applicable in certain contexts and certainly appears in the literature as well as Internet search engines. In Europe and the British Isles, terms such as Romanies, Travelers or Tinkers are also used. Many different groups form the Roma population based on a common sense of belonging although they may have very different characteristics and use different names.

LOCATION AND LINGUISTIC AFFILIATION

Roma can be found in significant numbers in the United States, Europe, Russia, Middle East, North Africa, India, Pakistan, and Central Asia. Some have migrated to Australia, Hawaii, and Alaska, but as far as we know they have not reached Japan, China, or Southeast Asia. The Roma are generally believed to have originated in Northern India, but they dispersed about 1,000 years ago and have lived in other places since. Today they are located in cities and towns.

The Roma speak Romanes which is a Sanskrit based language, that belongs to the Indo-Aryan branch of Indo-European languages. Because the Roma have lived in so many places over a long period of time, Romanes contains approximately 60% loan words. The closest language to Romanes is Hindustani. All Roma speak a second or third language from the country in which they live or have traveled.

OVERVIEW OF THE CULTURE

There is no census data on the Roma in the United States, but estimates of the Roma population range between 100,000 and 300,000 members of various diverse groups (Vlach Roma, Boyash, Irish Travelers, Hungarian Roma, etc.) living in all parts of the United States. The population of Roma in Europe is estimated between 4 and 10 million, with the largest numbers concentrated in Central European and Balkan countries (as much as 5% of the population). Not all Roma are Vlach-speaking Roma, but the designation of Roma has been given a more general meaning by intellectuals among the Roma to apply to all groups who self-identify as Gypsy. There are a number of diverse Roma groups. Not all of them speak the same dialect or language or have the same cultural practices.

Different groups have taken up different occupations, including music, metal work, buying and selling horses or cars, fortune-telling (primarily women), and selling craft items. Middle-class Roma have entered the professions, but this is still a small group. Most Roma work for themselves or are occupied in menial jobs.

The Roma migrated into Eastern and Western Europe through Persia en route from India over a period of approximately 1,000 years. Theories about the origins of the Roma in India have been based on linguistic and cultural information, but recent studies in genetics indicate that the Roma indeed originated with a small group of founders splitting from a single ethnic population in the Indian subcontinent (Gresham et al., 2001). Since leaving India, Roma have always lived within another culture or country as a minority and pariah group. They have been the subjects of extreme discrimination and persecution throughout history, especially in Western and Central Europe where they were enslaved in the Middle Ages. Between 500,000 and 600,000 European Roma perished under the Nazis in World War II. In the 19th century they migrated to North and South America where they continue to be a nomadic or semi-nomadic group. In the United States where discrimination is less severe than in Central Europe and the Balkans, they still suffer negative stereotypes. Police, for example, typically view the Roma as a criminal class rather than a culture.

Roma trace descent through both parents but take on patrilineal names and have a patrilocal marriage preference. They live in a large, close, patrilocal descent group called a *vitsa*. Authority in the *vitsa* is based on age with both older women and men having the highest status. Men are powerfully situated in the system of juridical authority, and women in the complex system of religious, spiritual, and medical authority. The lowest status kinship category is the daughter-in-law or *bori*. She is expected to serve her in-laws, take on a primary role as provider and wage earner for the extended family, as well as provide children for the grandparents. Roma have no religious specialists other than older women, but they use religious specialists from local churches to conduct baptisms. In the United States their own religion is punctuated by certain rituals including the baptism of a 6-week-old child, marriage, the *pomana* (death ritual), *slava* (saint's day feast) and some American holidays such as Easter and Thanksgiving (Sutherland, 1986).

Roma in the United States generally live in urban areas, usually on main streets and in the poorer parts of towns. They are not as easily recognizable to the American population as they are in Europe where they stand out more. In America they often prefer to represent themselves as a member of another ethnic group than Roma since it abates the stereotyping and discrimination against them. However, the women often wear long colorful skirts and low cut sleeveless blouses with a scarf on their head if they are married. One of their survival mechanisms is to keep to themselves and avoid contact with non-Roma except in work related circumstances.

THE CONTEXT OF HEALTH: ENVIRONMENTAL, ECONOMIC, SOCIAL, AND POLITICAL FACTORS

The Roma are not a healthy population. Life expectancy is up to 10 years less than in the non-Roma population, and infant mortality is up to four times higher (Braham, 1993). High levels of poverty, overcrowding and unemployment, and low levels of education are contributing factors. Their medical condition is in fact quite serious. In one study of 58 Roma in the Boston area, Thomas (1985) found that 73% exhibited hypertension, 46% diabetes, 80% hypertriglyceridemia, 67% hypercholesterolemia, 39% occlusive vascular disease, and 20% chronic renal insufficiency. Their diet is extremely high in animal fat, and their lives are very sedentary. Thomas also found that 84% were obese and 86% smoked cigarettes. All of these factors combined with high cholesterol levels, hypertension, and perhaps a genetic predisposition, result in the life expectancy of a Roma in the United States to be between 48 and 55 years (Thomas, 1987). Since births are often not registered and age at death is difficult to determine, life expectancy statistics are not exact in accuracy. However, during my fieldwork, of the eight deaths that took place, the oldest was 50 and the average age was 40 (Sutherland, 1992b). A study in Sweden noted a higher incidence of asthma and chronic bronchitis among Roma than the control group. A U.K. study also documented high incidences of obesity, heart disease, and diabetes (Lehti & Mattson, 2001). Globally their nutrition and access to clean water is poor as is their access to adequate income, housing, and education. In Russia and Central Europe, as well as in the Balkans, their health condition is seriously deteriorating.

MEDICAL PRACTITIONERS

The Roma in the United States seek medical help both from American doctors and their own *drabnari*, literally meaning older women who have knowledge of medicines. They are very assertive in seeking medical care from the medical system. Many are on welfare and have medicaid cards and therefore have access to good medical care. They have an unusual ability, given that most are illiterate, to understand our complex medical system and to get attention from medical personnel (Sutherland, 1992a). On the other hand, their insufficient knowledge of biology and medicine as well as their rudimentary vocabulary in English puts them at risk. They often do not practice preventative medicine or understand the full implications of doctors' instructions. While they may demand specific famous doctors or treatment they have heard about, they ignore preventative and long-term treatment. They have been known to share pills with each other, request specific colored pills and prefer older, physically bigger doctors over younger, thinner ones (Sutherland, 1992a,b). When a family member is sick, large numbers of relatives congregate at the hospital, sometimes camp on hospital grounds, and create confusion among medical personnel. However, they respect authority and are eager to learn about the best treatments and are supported by a huge network of relatives. One study in the United States concluded that Roma receive better medical care than other urban minorities because they are so effective at utilizing medical services (Salloway, 1973). These generalizations are not applicable to Balkan and Eastern European Roma where the discrimination is so much more severe, and where many Roma are denied access to medical care.

Some older women are medical practitioners for diseases they consider under the purview of their own group. These women may supervise treatments for the patient in addition to those ordered by the doctor. Their treatments deal with diseases they classify as "Gypsy diseases" which include convulsions based on possession by the Devil (the cure is the Devil's own dung or *Asafetida*); infection by a spirit called *Mamioro* which means "little grandmother" (the cure is her vomit prepared into tablets). Older women acting as midwives also attend childbirth to help the mother have an easier birth (a *Selaginella* plant, called St. Mary's Hand, is placed in water to open the womb—also at the deathbed to make sure the spirit leaves the room) (Sutherland, 1992a).

CLASSIFICATION OF ILLNESS, THEORIES OF ILLNESS, AND TREATMENT OF ILLNESS

The Roma view health and illness in the larger context of social order. Good health and good fortune (many children, money) are consequences of individual diligence in following purity rules. Conversely illness, bad luck, and impurity are also closely associated with being ritually unclean (*marime*). To be ritually clean, the Roma must follow a large number of rules about washing, eating, separation of parts of the body, separation of male and female, and separation from the polluting influences of non-Roma who do not follow these rules (Sutherland, 1986). They make a distinction between illnesses caused by contact with non-Roma and their own "Gypsy illnesses." For the former they go to the American medical system, and for the later they consult their own *drabarni*. Gypsy illnesses include those caused by *Mamioro*, a spirit who visits places that are unclean and brings illness, *tosca*, a disease they translate as "nerves" is caused by the devil, as are convulsions which are a sign that the devil has entered the body of the individual. Other illnesses and bad luck are caused by *mule*, the ghosts or spirits of dead relatives. *Mule* are not necessarily harmful, but relatives of the dead who do not properly observe *pomani* (death rituals) could be made ill by them. Mental illness is rare, although virtually every Roma person I knew who was tested by a psychiatrist was diagnosed with borderline personality disorder, that is, in relation to the American view of a healthy personality (Sutherland, 1986). The cure for mental illness is for a person to follow the purity rules and other social customs such as getting married. All *drabarni* carry a bag of medicines that may include Asafetida, *johai* (literally meaning, ghost) which are baked pieces of the vomit of *Mamioro* (most likely is a slime mold called *Fuligo septica*) and other herbal medicines such as a fungus called "the spoon," a special herb called *drarnego*, garlic, black pepper, gold coins, and pieces of the aprons of ancestors who were *drabnari*. Johai is the most valuable medicine a *drabnari* carries in her medicine bag. It cures fear of ghosts, hemorrhages, influenza, pneumonia, cholera, and epileptic convulsions and is administered in addition to medicines prescribed by a doctor (Sutherland, 1992a).

SEXUALITY AND REPRODUCTION

Roma attitudes toward sex and sexuality include separation of the sexes before marriage (even close relatives), and rules about sex and sexuality based on notions of purity. Bodily substances emanating from the lower body are polluting, including sexual fluids, menstrual blood, blood from childbirth, urine, and fecal matter. These substances must be separated in touching and washing from the upper body and particularly the head which must be kept pure. Thus sex can make a person impure and liable to illness if practiced wrongly. People who engage in sexual intercourse may also abstain from all animal products on Fridays as a cleansing act. Impurity may also adversely affect fecundity which is generally welcome, children being highly valued. Roma may practice contraception or have their tubes tied when they reach the limit of family size they desire. Fecundity increases a woman's status; infertility lowers it. Oral sex is considered the most polluting act one could perform, but its prohibition also has the effect of making it the most titillating.

HEALTH THROUGH THE LIFE CYCLE

Pregnancy and Birth

Most Roma go to the hospital to give birth but neglect the important period of pre-natal care because they do not want to undergo an internal examination by an obstetrician. If a *romni* midwife delivers the baby she cleans the umbilical cord with ashes, and an amulet prepared by the *drabarni* is sewn into the baby's clothes for protection. A hospital birth will usually include a mother or older woman as an assistant rather than the father who stays outside the birth room for cleanliness reasons. An important reason that women have chosen hospital births over a *romni* midwife is to leave all polluting substances from the birth in the hospital and lessen the period of separation and isolation after birth. This period is now about 9 days instead of the traditional 6 weeks. The incidence of miscarriages and infant mortality is high. In a 1967 sample of 400 American Roma, infant mortality was 25% of all reported past pregnancies and births. Since hospital births have risen since then, infant mortality is presumed to be lower now (Sutherland, 1992b) although still higher than the control population.

A recent study (Nemenyi, 1999) of 80 Hungarian Roma women has shown that different Roma groups have different attitudes toward family planning. The Vlach (Roma) group, the most "traditional" in customs have large numbers of children by choice and practice family planning (both contraception and abortion). The Boyash practice no family planning, but the Romungro (musicians) and urban Budapest "Gypsies" have smaller families by choice with the highest level of consciousness of family planning of any Roma group. However, the Romungro and urban Roma start giving birth at a very young age. One of the most interesting conclusions of this study is that the Hungarian health care workers seemed to view all Roma women as "wild," assuming that their higher fertility was due to lack of family planning. Nemenyi also noted that the poor health conditions of the mothers and their shorter life expectancy in relation to the Hungarian population, contributed to a higher incidence of premature births in connection with early, frequent, and late deliveries and a higher amount of dystrophy, mental and physical disabilities.

Infancy

In the first 6 weeks, an infant is viewed as highly vulnerable. At birth the baby is tightly swaddled and handled only by the mother. If the mother nurses, she is told to avoid certain foods considered to produce colic (green vegetables and tomatoes). Many women choose to bottle-feed. In the first weeks of life the mother and baby are isolated from all other family members, and the windows and door to her room are kept shut to keep out a spirit of death that may come to harm the baby. Another danger to the baby is the evil eye which causes the baby to fuss and become ill. The giver of the evil eye is then asked to make a cross with spittle on the forehead of the baby to counter it. The dangers of infancy are ameliorated when the baby is baptized at about 6 weeks of age. If a baby dies, it is considered *prikaza* (a polluting misfortune) for the parents who must avoid the body of the baby (Sutherland, 1992a, 1996).

Childhood

A child is protected from illness when he or she is baptized and considered free of pollution from birth substances. Children are much loved and enjoy freedom from the restricting rules of cleanliness. They are not susceptible to becoming defiled and are not expected to understand feelings of shame. Children, for example, may eat food prepared by non-Roma, food that adolescents must reject lest they become defiled by it. The entire extended family cares for children. They are not left alone and are included in adult social activities. They are allowed to eat sugar and fat and may develop tooth decay early. Children are indulged and allowed to run free with their cousins and only told to stop being noisy and boisterous if they bother the adults (Sutherland, 1992a, 1996).

Adolescence

A major change occurs at adolescence for both boys and girls. Adolescents are introduced to the notion of personal shame. Their bodies and their behavior are suddenly judged in terms of their control of polluting substances (menstrual blood, semen) and shameful actions (sexual contact). Girls in particular have to keep clean and observe the washing, dressing, cooking, and eating codes of women. Women's clothes are washed separately from those of men and children, and they cannot cook food for others during menstruation unless they avoid touching menstrual blood. Some girls fast during menstruation (Sutherland, 1986). Adolescents are expected to behave much like adults and to marry as early as possible. Those who experiment with drugs or sex are considered to have a social or mental problem for which marriage is a cure. Alcohol and smoking are not discouraged.

Adulthood

Married women are expected to support their husband's family and their own children. Men's income can be spent on themselves or their children. Women must serve food to their in-laws and show respect to men by not passing in front of them (unless they have a baby in their arms) or allowing their skirts to touch men. They must show modesty by wearing more traditional Roma clothes, and, if married, a head scarf (diklo). Adult women and men soon develop health problems such as obesity, diabetes, high cholesterol, and hypertension leading to an untimely death in many cases. At menopause or when they cease sexual relations, men and women become pure again and the cleanliness rules are eased. To encourage good health, they eat pepper, salt, vinegar, garlic, and onions (Sutherland, 1992a).

The Aged

Old people are highly regarded for their knowledge and special status as pure because menstruation and sexual relations are assumed to have ceased. The aged are politically powerful and exert a great deal of influence over the younger members of the extended family group. They are consulted on everything including arranging the marriages of their offspring and all political decisions by the group. Usually a younger person will not agree to medical procedures without the approval of their oldest relatives who often arrange the medical treatment and act as a mediator with the doctors. In the hospital an older authority figure must be involved in all decisions, and the older women may also treat the patient. The aged are surrounded by their family and cared for at home until death. They may not travel much because of illnesses, but they are at the center of all social and political life. Showing proper respect for the old is the best way of ensuring that their spirit will not carry a grudge to the grave and plague the relatives left behind. Even before a person dies their good deeds begin to take on a mythological stance, and the bad deeds are never mentioned (Sutherland, 1986).

Dying and Death

The period of dying and death constitute a personal crisis as well as a social crisis. When Roma become seriously ill, the entire community is galvanized, and doctors, hospital staff, reporters, social workers, and the police are made aware of a great happening. The Roma flock into town, camp on hospital grounds, and gather in large numbers. Dying in old age when one has prepared for death is a crisis for the *vitsa*, but it is accepted as natural. The ideal death is a "great death" for which one has prepared and that can be shared with large numbers of relatives and friends. The death of a young person, on the other hand, is an enormous tragedy and is thought to have happened because of some moral uncleanness. The relatives are so grief-stricken and fearful, they may become wild and threatening, especially to doctors. Relatives scratch their faces until they draw blood, beat themselves on the chest and head, and wail or shout plaintively to the deceased.

A person who is dying is never left alone. Relatives keep a constant vigil with the dying person to make sure they do not die with a grudge or curse on their lips, to assure that the spirit is released out an open window at the moment of death so it can roam free and leave the relatives left behind, or to frighten death away if they see him approaching the room. The last words of the dying are very important. When the person dies, embalming is considered good to remove the blood from the body. Relatives sit with the body day and night usually in the funeral parlor.

Relatives who are in mourning abstain from washing, shaving, combing hair, or changing clothes for 3 days, at the end of which they can wash and the body is buried. A *pomana* (death feast) is held 9 days, 6 weeks, 6 months and a year after death. The *pomana* may be held in a hall or at the end of the year at the gravesite. The spirit of the dead person (*mulo*) is present at each *pomana* and is set a place to eat and drink. Relatives generally try to avoid the spirit except at the *pomana* when it is important to show the deceased that they are receiving the proper respect.

CHANGING HEALTH PATTERNS

Published literature on the health of the Roma is sparse, and there is a need for more study of their health needs and conditions (Hajioff & McKee, 2000). Recent genetic medical research has identified nine Mendelian disorders in Roma populations caused by mutations including muscular dystrophy, galactokinase deficiency, primarily congenital glaucoma, congenital myasthenia, congenital cataracts, facial dysmorphism neuropathy syndrome, and hereditary motor and sensory neuropathies (Kalaydjieva et al., 2001).

Today the Roma in Europe and the United States are increasingly transnational. In the last 25 years in Europe, and since the fall of communism, Roma have either joined the middle class or become increasingly marginalized at the bottom of the social scale where their health conditions have seriously deteriorated. This altering of the traditional social structures in a demographically young population, in addition to the loss of traditional jobs (scrap dealing and traveling sales) combined with the obstruction they meet in trying to find new ways of making a living, mean that all too often the sale of drugs or prostitution is a primary source of income. Until recently drug use by Roma had been unknown. The rise in drug use has been accompanied by the appearance of HIV infection (Sastipen Network).

Poor health, overcrowded living conditions, lack of employment, and little access to health care have created serious health problems. A number of alarming trends in health have been noted in the European Roma community. One study in Spain reported a nine times greater prevalence of antibodies to hepatitis A in Roma children than in the non-Roma population (Cilla, Perez-Trallero, Marimon, Erdozain, & Gutierrez, 1995). Another Spanish study showed that Roma children are at particularly high risk for lead poisoning (Redondo, 1995). In Romania the Roma have one of the highest fertility rates in the region, but infant mortality is also higher than in the population at large. Life expectancy is significantly lower as well. Romani children have a higher rate of vitamin deficiency, malnutrition, anemia, dystrophy, and rickets than the non-Romani in Romania. Romania has the most AIDS-infected children in Europe due to the use of unsterilized needles, and the Romani community has also been dramatically affected. The number of cases of TB and AIDS has been growing at an alarming rate. Over half of the children in one Romani neighborhood were HIV positive according to interviews with one group of doctors (Zoon, 2001, pp. 79–80). In Bulgaria, doctors working in Romani neighborhoods identified widespread cardiovascular conditions, and kidney, liver, gastric and intestinal diseases. In one neighborhood 40 of 70 children tested positive for TB. The last reported cases of polio in Bulgaria were among Romani children (90 cases) and half were disabled by the disease. Romani children seem to have significantly low immunization coverage (Zoon, 2001, p. 90). In Macedonia the Roma infant mortality rate is twice the national average, and infants have extremely high incidences of diarrhea (52% of the women answered "yes" when asked if their infants had diarrhea in the last month) and respiratory diseases (Zoon, 2001, p. 102).

REFERENCES

Braham, M. (1993). *The untouchables: A survey of the Roma People of central and eastern Europe*. Geneva: UNHCR.

Cilla, G., Perez-Trallero, E., Marimon, J. M., Erdozain, S., & Gutierrez, C. (1995). Prevalence of hepatitis A antibody among disadvantaged gypsy people in northern Spain. *Epidemiology and Infection*, 115, 157–161.

Gresham, D. et al. (2001). Origins and divergence of the Roma (Gypsies). *American Journal of Human Genetics*, 69, 1314–1331.

Hajioff, S., & McKee, M. (2000). The health of the Roma people: A review of the published literature. *Journal of Epidemiology and Community Health*, Nov 54(11), 864–869.

Lehti, A., & Mattson, B. (2001). Attitude to care and pattern of attendance among gypsy women—a general practice perspective. *Family Practice*, 18, 445–448.

Nemenyi, M. (1999). Gypsy mothers and the Hungarian health care system. *Patrin Web Journal*, Jan 6.

Redondo, M. J., Guisasola, F. J. (1995). An unknown risk group of lead poisoning: The gypsy children. *European Journal of Pediatrics, 154*, 197–200.
Salloway, J. (1973). Medical care utilization among urban Gypsies. *Urban Anthropology, 2*(1), 113–126.
Sastipen Network. The European Roma community facing drugs and HIV/AIDS phenomena. Available online at, http://www.asgg.org.sastipen/doc2.htm
Sutherland, A. (1986). *Gypsies, the hidden Americans.* Chicago: Waveland Press.
Sutherland, A. (1992a). Health and illness among the Rom of California. *Journal of the Gypsy Lore Society, 5, 2*(1), 19–59.
Sutherland, A. (1992b, September). Gypsies and health care. Cross-cultural medicine (Special Issue). *Western Journal of Medicine, 157*, 276–280.
Thomas, J. (1985). Gypsies and American medical care. *Annals of Internal Medicine, 102*, 842–845
Thomas, J. (1987 August, 15). Disease, lifestyle and consanguinity in 58 American Gypsies. *Lancet*, 377–379.
Zoon, I. (2001). *On the margins, Roma public services in Romania, Bulgaria and Macedonia.* New York: Open Society Institute.

Samoa

James R. Bindon

LOCATION AND LINGUISTIC AFFILIATION

The Samoan Archipelago comprises nine volcanic islands from 13° to 15° south and 168° to 173° west. Since 1900, the islands have been divided into eastern and western moieties. The five islands in the east; Ta'u, Ofu, Olosega (the Manu'a Islands), Tutuila, and Aunu'u form the Territory of American Samoa, while the four islands to the west; Upolu, Savai'i, Manono, and Apolima make up the Independent State of Samoa (ISS). The largest population concentrations are in the Pago Pago Bay Area on Tutuila in American Samoa, and Apia on Upolu in ISS. Prior to colonial influence, there were two polities in the archipelago, separating the Manu'a Islands from the six islands to the west. All of the Samoan Islands are high volcanic islands with peaks up to 6,000 ft on Savai'i and partially encircling reef structures with nearshore lagoons.

The Samoan language falls within the Polynesian Group of the Austronesian language family. There are no significant dialect variations of Samoan, but there are important distinctions in the phonology from region to region within the islands. Samoan has a relatively elaborate lexical division between common words used for everyday conversation (Shore, 1982) and respect or chiefly vocabulary used both for speaking to chiefs (*matai*) and between non-chiefly individuals to denote formality and politeness. Within the chiefly vocabulary, there are also alternate words used for and by chiefs (*ali'i*) and talking chiefs (*tulafale*). English is the most common second language and Samoans in American Samoa are highly bilingual in English.

OVERVIEW OF THE CULTURE

Population

Demographers agree that there were no reliable estimates of population in the Samoan Islands until the 1920s. Reports of missionaries and administrators during the 19th century vary widely in quality, and supply widely varying population counts. Commodore Wilkes based his 1839 estimate of the population on reports from local missionaries to come up with a total of about 57,000. This early figure is viewed as the least reliable of the 19th century. The first reasonably reliable census yielded a total of about 34,000 in 1853. The discrepancy between these two censuses led to speculation about severe depopulation after European contact. However, most demographers maintain that the population of the islands was relatively stable between about 34,000 and 39,000 throughout the 19th century. This stability was the result of the typical pre-transition patterns of high fertility coupled with periodic high mortality. The 20th century saw the impact of the demographic transition on the population of the islands. Entering the century with a population of about 40,000, continued high fertility coupled with reduced mortality due to improved medical care led to a 600% increase in population by the turn of the century. The estimated population of the archipelago for July 2001 was about 250,000, with 179,058 in ISS (CIA, 2001b) and 67,084 in American Samoa (CIA, 2001a). In addition, as many as 100,000 Samoans live abroad, mainly in the United States, New Zealand, and Australia.

History

Voyagers of the Lapita culture settled in the Fiji-Tonga-Samoa area between about 1500 and 1000 BC. Contacts between Samoa, Tonga, and Fiji were maintained throughout prehistory and, later, relations were maintained with other Polynesian groups including Tokelau, Wallis, and Futuna. Significant European contact and acculturation began in 1830 when John Williams of the London Missionary Society (LMS) arrived and established his church through fortuitous political circumstances.

There was conflict for colonial control between Germany, Great Britain, and the United States until 1900 when the four islands in the west became a German colony and the five islands in the east were claimed as the Territory of American Samoa by the United States. As a result of the League of Nations actions during World War I, New Zealand assumed administration of the German colony from 1914 to 1962. Independence from New Zealand as a constitutional monarchy came in 1962 with the founding of the Independent State of Western Samoa, later changed to the Independent State of Samoa (commonly Samoa) in 1997. The eastern islands remain a United States territory, administered by the Office of Insular Affairs, U.S. Department of the Interior. The territory elected its first Samoan governor in 1977.

Economy

The traditional economy of the archipelago was based on root and tree crop farming supplemented by fishing. The economic development of the two Samoas (ISS, American Samoa) differed dramatically throughout the 20th century as American Samoa received substantially more developmental aid from the U.S. government than Western Samoa did from New Zealand. Family remittances from overseas remain a substantial part of the economy throughout the islands. Agricultural products, featuring coconut cream, coconut oil, and copra, are responsible for 90% of the exports from ISS, and 65% of the employment is in agriculture with another 30% in service, and 5% in industry (CIA, 2001b). About 50% of adult males and 20% of adult females are in the wage economy in ISS. The economy of American Samoa is tightly intertwined with that of the United States, with tuna canned at the local cannery forming the primary export. Unemployment was 16% in 1993, with the government accounting for 33% of the jobs, tuna canneries 33%, and a variety of other industries making up the final third of jobs in American Samoa (CIA, 2001a).

Social Organization

The fundamental unit of social organization in Samoa is the *aiga* which can mean the household, the nuclear family, or the extended family. As an extended family, each *aiga* is headed by a title holder or *matai*. Traditionally each *aiga* had a specific compound with several related families living under its *matai*. Communities or villages (*nu'u*) consist of several *aiga* with long-term associations. *Nu'u* are governed by councils or *fono* in which all *matai* have a voice. The *matai* titles are divided into two general types, chiefs or *ali'i* and talking chiefs or *tulafale*, and ranked within each of these divisions from lowest to highest. The *nu'u* are independent political units but they group together into regional alliances for some purposes. This community level of organization is still important in the ISS and American Samoa, although residence has become increasingly based on the nuclear family rather than the *aiga*.

Religion

There is little remnant of traditional religion other than generalized beliefs in ancestral spirits (*aitu*). Today Samoans are overwhelmingly Christian. In ISS 99.7% of the population adheres to Christianity, about half of which belongs to the LMS derived Christian Congregationalist Church of Samoa (CCCS). The rest are Roman Catholics, Methodists, Latter-Day Saints, and Seventh-Day

Adventists (CIA, 2001b). In American Samoa the CCCS accounts for about 50% of the population, the Roman Catholic Church about another 20%, and other Protestant and non-Christian denominations about 30% (CIA, 2001a).

THE CONTEXT OF HEALTH

Health Indicators

Many 19th-century sojourners remarked on the general good health of the Samoan population (MacPherson & MacPherson, 1990). The well being of the Samoans has been attributed to a hospitable environment, an excellent vegetable- and marine-based diet, limited opportunity for infectious disease exposure from outsiders, and a high standard of personal hygiene, including frequent bathing. This good health notwithstanding, the early visitors to Samoa heard of an epidemic illness introduced by "sailing gods" long before the arrival of Europeans. This epidemic is said by one observer to resemble cholera. Other than this relatively isolated prehistoric episode, Samoans appear to have enjoyed a long, stable period of good health prior to European exposure in the 19th century. Kramer (1903/1995), because of his medical training and extended stay in Samoa, provides the best early description of medical conditions. He notes the absence of malaria and comments on the occasional presence of leprosy, the frequent occurrence of respiratory ailments (much of which he classifies as consumption or tuberculosis), and the widespread incidence of elephantiasis (filariasis), which he estimates as afflicting 5% to 10% of the population. Many other 19th-century visitors commented on the presence of elephantiasis which manifests as extreme swelling of the legs, scrotum, or breasts. While the high visibility of this disease (it was not uncommon to see Samoans with the signature swollen legs and ankles in the 1970s) accounts for some of the remarks, it is clear that this was a significant pre- or early-contact disease.

A seminal event separating the health history of ISS and American Samoa occurred in 1918. An epidemic outbreak of influenza (the Spanish flu pandemic) arrived in Apia aboard the New Zealand ship *Talune* in November 1918. The official death rate in ISS was recorded as 22% (20% among Samoans, 33% among part-Samoans, and 2% among Europeans) although the impact was much greater as many deaths occurred in the succeeding year and the toll on the native leadership was particularly high. Seventy miles away, in American Samoa, there was no evidence of the flu. Death records for 1918 and 1919 from American Samoa show no increase in the number of deaths, no alteration in the age structure of mortality, and no cases attributed to flu. The Samoans blamed the New Zealand administrator for failing to quarantine the islands as had been done in American Samoa where the *Talune* was not allowed to dock. This incident colored much of the later New Zealand administration of the western islands.

Today health differs in the two Samoas based in part on differences in the resources available to American Samoa from the United States. The health care system, water treatment, and general sanitation in American Samoa have been ahead of those of their neighbors to the west for more than 50 years. As a result, American Samoa progressed through the epidemiological transition before Samoa. Health indicators for ISS are on par with the worst found in Polynesia, although health conditions are substantially better than in many developing areas (see Table 1). On the basis of life expectancy and infant mortality American Samoa is among the healthiest of Polynesian groups. Throughout the Samoan archipelago, the most common cause of hospital admission is respiratory disease, frequently flu or pneumonia. Tuberculosis, leprosy, and viral hepatitis are also present in significant numbers.

Chronic and obesity-related diseases became more frequent causes of death in both Samoas throughout the

Table 1. Health Characteristics of the Independent State of Samoa and American Samoa

Characteristic	ISS	American Samoa
2001 estimated life expectancy at birth, years	69.5	75.3
1995 crude birth rate/1,000 pop.	4.6	4.1
2001 estimated Infant Mortality/1000 births	31.8	10.4
1995 Mortality	deaths/100,000	Rank
Circulatory diseases	42	#1
Neoplasms	23	#3
Respiratory diseases	18	#2
Perinatal conditions	13	—
Infections and parasitic diseases	13	—
Kidney diseases		#4
Accidents and injury		#5

1995 sources: American Samoa (WHO, 1999a), ISS (WHO, 1999b). 2001 sources: American Samoa (CIA, 2001a), ISS (CIA, 2001b).

20th century, with circulatory diseases being the number one cause of death by 1995. Modernizing influences of engagement in the world economy in the second half of the 20th century brought about significant changes in diet, especially in American Samoa where the shift from subsistence agriculture to wage labor and purchased foods was most dramatic (Bindon, 1988). Changing economic conditions also altered patterns of activity, as wage employment for both men and women generally involves less demanding activities than prior to contact, resulting in massive obesity and its sequelae (Bindon, 1995). There appears to be a genetic predisposition to diabetes among Samoans with a diabetes prevalence of 3–12% in ISS and 27% in American Samoa (Bindon & Baker, 1997). The high diabetes rate in American Samoa shows up in the mortality statistics primarily through the complications that result in increased deaths from kidney and cardiovascular diseases (see Table 1).

Health Infrastructure

Prior to contact there was a limited set of personal practices and healers (*fofō*) available for the relatively few native health problems. Most medical consultation occurred within the village context where different healers were responsible for specific conditions. By the 1860s, there were trained physicians (*foma'i*) in Apia, brought in by the LMS and German companies. The German Navy established a hospital in Apia in the 1890s, but there was no Western medical attention paid to the eastern islands until 1900 (Gray, 1960). As the U.S. Navy began administration of American Samoa, plans were made for a dispensary near Pago Pago which turned into the local hospital and served until it was replaced by the current L.B.J. Tropical Medical Center in Faga'alu in the 1970s. In ISS, the genesis of the public health system of outreach dates to the experience with the Spanish flu. The New Zealand administration established women's committees (*komiti tumama*) in the villages to promote health and hygiene. By the 1940s there was a well established system of primary health care available throughout ISS, and many village hospitals were built.

Medical Practitioners

Prior to contact individuals offered their own prayers for ailments or they consulted one of the healers or *fofō* near their village. Missionaries brought a limited medical knowledge starting in the early 19th century and Western physicians and nurses began to arrive later in the 19th century. Throughout the 20th century increasing numbers of Western practitioners took part in health care in the Samoan Islands. Physicians, nurses, midwives, dentists, pharmacists, and an array of technicians currently populate the healthcare workforce (see Table 2). The training of most of the biomedical practitioners takes place overseas in the Fiji schools of medicine and dentistry as well as in schools of medicine and public health in New Zealand, the United States and Australia. Most nurses are trained locally although some are sent overseas for additional training. *Fofō* continue to train through apprenticeships and practice in most villages.

Table 2. 1996 Healthcare Workforce in the Independent State of Samoa and American Samoa

	ISS		American Samoa	
	Number	Per 10,000	Number	Per 10,000
Doctors	57	3.44	16	2.90
Dentists	6	0.40	N/R	N/R
Pharmacists	6	0.40	N/R	N/R
Nurses	257	15.50	146	26.20
Midwives	60	3.60	N/R	N/R

N/R = Not Reported.
Sources: ISS (WHO, 1999b); American Samoa (WHO, 1999a).

Classification of Illness, Theories of Illness, and Treatment of Illness

Precontact paradigm

Prior to the arrival of Europeans, Samoans attributed illness to the displeasure of the gods or to the work of spirits (*aitu*). Stair, who visited Samoa in the 1840s, classified the gods into four categories, the third of which was the *aitu* who were responsible for much of the mischief and illness that befell Samoans (Stair, 1897). These beliefs resulted in treatments that focused on identifying the act that caused the displeasure or the *aitu* responsible for the malady and prescribing behaviors to assuage the god or *aitu*. There are many prayers that are aimed at keeping various gods happy and preventing illness, such as this from George Turner who was an LMS missionary in Samoa in the 1840s and 1850s:

> In [one] family [the god Tuiali'i] was prayed to for life and health before the evening meal; an offering of a blazing fire was essential to the success of the prayer, which ran as follows: "This is our fire to you, it burns bright; other fires are dim and going out; send these families to the lower regions, but give us life and health." (Turner, 1884, p. 75)

Other prayers and offerings were made once a specific god or *aitu* was identified as being responsible for an illness:

> In a case of sickness, a cup of kava ['*ava*, beverage made from the root of *Piper methysticum*] was made and poured on the ground outside the house as a drink-offering, and the god [Salevao] called by name to come and accept of it and heal the sick. (Turner, 1884, p. 51)

In some cases the families could not approach the god directly for intercession, and instead they had to commission a priest or work through other human agents, as noted by Turner:

> At one place in Savai'i Salevao had a temple in which a priest constantly resided. The sick were taken there and laid down with offerings of fine mats. The priest went out and stroked the diseased part, and recovery was supposed to follow. (Turner, 1884, p. 49)

and

> In another place [Taisumalie] was incarnate in an old man who acted as the doctor of the family.... His principal remedy was to rub the affected part with oil, and then shout out at the top of his voice five times the word Taisumalie, and five times also call him to come and heal. This being done, the patient was dismissed to wait recovery. On recovery the family had a feast over it, poured out on the ground a cup of kava to the god, thanked for healing and health, and prayed that he might continue to turn his *back* towards them for protection, and set his *face* against all the enemies of the family. (Turner, 1884, p. 59, emphasis in the original)

In these ways, Samoans sought to cope with the ailments they encountered prior to contact.

Stair is not alone among 19th-century visitors in referring disdainfully to the limited native medical system of Samoa, comparing it unfavorably to that of Tonga and commenting that, "their remedies were few, and for the most part unreliable" (Stair, 1897, p. 164). MacPherson and MacPherson (1990) argue that the impoverished indigenous medical system of Samoa became elaborated upon contact with Europeans due in part to the introduction of new illnesses and in part to borrowing of practices from the missionaries and traders who gained more prominence in Samoa during the 19th century. By the time Kramer visited the islands in the 1890s, much of the original health belief system about *aitu* had been obliterated by the missionaries and the Samoans had over 50 years of experience in dealing with diseases and treatments introduced by Europeans.

Contemporary paradigm

MacPherson and MacPherson (1990, p. 79) characterize the contemporary Samoan paradigm as "two sets of medical belief and practice [existing] alongside one another in ... an arrangement aptly described ... as a 'collage'." That is, the biomedical and indigenous systems are both used to satisfy the medical needs of the population. Part of this coexistence may have resulted from the public health training of individuals who were already recognized as *fofō* within their villages. As the western medical establishment increased in scope throughout the 20th century, Samoans came to rely on biomedically trained physicians for *ma'i palagi* (European illness) and on *fofō* for *ma'i samoa* (Samoan illness). In American Samoa, where the establishment of medical care by the Naval administration was seen as beneficial and the epidemic of 1918 was avoided, traditional medicine has been less integrated into the overall medical paradigm. As a result, the tension between the systems is greater as noted by Holmes and Holmes in their quotation of the following letter from a medical center member published in the *Samoa News* of November 17, 1988:

> The most disturbing preventable problem has been the use in children of local Samoan bush medicine. By this I mean the plant and herbal medicines given by *taulesea* [*taulasea*, polite term for healer] or *fofō*. In the past year, we saw at least six children die after being given "Samoan medicine" by mouth from a *fofō*. The picture was not a pretty one. The children initially had mild cases of the "flu." They were then given "Samoan medicine" and soon developed seizures, kidney failure and increased acid in the blood. Despite intensive care at the hospital, these children died within three days. ... Many of the medicines given by a *fofō* are probably safe for children, but some are poisons and will quickly kill a child. In the first half of 1988, more children died in American Samoa from being given "Samoan medicine" than died from any other cause. (In Holmes & Holmes, 1992, p. 132)

In Samoa, as elsewhere, idiosyncrasies are certain to abound within the lay view of the biomedical paradigm, such as the attribution of diabetes to eating too much sugar. This may derive in part from a misapprehension of the Samoan term for diabetes (*suka* or *ma'i suka*). With regard to obesity, one of the first claims an acculturated Samoan will make is that, "I eat too much taro," possibly because of the full feeling that taro gives. In fact, both obesity and diabetes would benefit from a diet with more taro and less of the high fat, low fiber foods that have become common during the 20th century. Both conditions would benefit even more from cultivating the taro that is eaten, thus increasing the physical activity of the individual.

Mental Health

Few cultures have had as much written about mental health as have the Samoans, beginning with Mead's characterization of a people with few neuroses and little maladjustment (Mead, 1928) and accelerating to a fever pitch with Freeman's depiction of aggressive and suicidal Samoans (Freeman, 1983). Several volumes have since dealt with this controversy (see Caton, 1990; Holmes, 1987; Orans, 1996). Neither extreme viewpoint is likely to have much merit and a better characterization of Samoan mental health would lie somewhere between the two. A psychiatrist working in American Samoa in the 1970s hypothesized that the underreporting of mental illness by biomedical standards in American Samoa was a result of the social system, which provides a means of "curing" emotional disorder by family group process and ritual, by making the disorder less disruptive, and by absolving the affected individual of personal guilt (Walters, 1977). Thus, much of what would be diagnosed and treated as deviant under the biomedical paradigm is informally handled within Samoan families.

SEXUALITY AND REPRODUCTION

Sexual attitudes and practices in Samoa have been shaped in part by the open walled houses and the general openness of the village setting. Premarital sex is strongly discouraged for women, an attitude that continues to be supported by the church in Samoa and enforced by a woman's brothers. Abortion was practiced in cases where the girl was afraid of her family or ashamed and pressure on the abdomen was the primary method used (Turner, 1884).

In ISS and American Samoa, relatively high fertility rates have been maintained from pre-contact times. A poster aimed at promoting birth planning in the Ob/Gyn clinic at the LBJ Tropical Medical Center in American Samoa in the 1970s said "You space your coconuts, why not your children." Fertility continues to be high. The WHO (1999a; 1999b) provides total fertility rates of 4.76 for ISS in 1991 and 4.5 for American Samoa in 1995. Harbison (1986) described a differential effect of education on women in ISS and American Samoa. In ISS, education did not show a depressing effect on fertility. Harbison speculated that because education was provided within the village context, it tended not to shift attitudes on fertility and ideal family size, and since there was little opportunity for employment, education did not tend to remove women from their families. In American Samoa, wage jobs were available, especially for educated females, and attitudes about the cost of additional children and ideal family size were changed as a result of women's experience in schools and the work place (Harbison, 1986).

HEALTH THROUGH THE LIFE CYCLE

Pregnancy and Birth

During pregnancy, Stair (1897) tells us a woman would permit her hair to grow long to mark her condition and to assure a healthy child. After 2 or 3 months, food would be brought by the husband's 'aiga. A few months later the husband's family would present a gift of pigs called *o le popo* (of the child). The number of pigs would vary according to the rank of the husband, up to fifty pigs for the wife of a very important chief. One final gift of food called *o le taro fanaunga* (the taro of birth) was brought to the mother and then all of the gifts were divided up according to political connections of the wife's family (Stair, 1897). Turner (1884) describes rituals surrounding the birth of a child as follows:

[The woman's] father was generally present ... and either he or her husband prayed to the household god, and promised to find any offering he might require, if he would only preserve mother and child in safety.... "O Moso, be propitious; let this my daughter be preserved alive! Be compassionate to us; save my daughter, and we will do anything you wish as our redemption price." (Turner, 1884, p. 78)

He says that if the child was a boy the umbilical cord was cut on a club to help him be brave in war; if it was a girl her cord was cut on the *tapa* (bark-cloth) making board so that she would be useful to the family. Infanticide was unknown in old Samoa according to Turner.

Today most Samoans attend pre-natal clinics and give birth in the local dispensary or hospital with family in attendance or nearby, or at home with the assistance of a midwife. Birth weights of American Samoan infants averaged 3,250 g in the 1980s, a very high value, perhaps reflecting the high rate of diabetes in the population (Bindon & Zansky, 1986). Birth weights were somewhat lower in Western Samoa, but they were still in the mid-range of birth weights on a worldwide basis. Samoans are less likely than most populations to be at risk for low birthweight (less than 2,500 g).

Infancy

After children were born they were lovingly cared for. For the first 3 days, attention was paid to shaping the head by laying the child on its back and surrounding the head with flat stones. The hand was used to press the forehead and the nose to produce the desired shape (Turner, 1884). The mother's milk was examined by a woman every day for up to a week to determine if it were ready for the infant. During this time the infant was fed on the juice of the coconut, after which breast-feeding and supplementation with pre-masticated vegetables were started (Stair, 1897).

Contemporary infants are primarily breast-fed, but this varies between American Samoa and ISS. In American Samoa, where employment opportunities are better, women tend to wean their infants early to return to work. In ISS, breast-feeding is likely to continue longer. In both areas, infants grow very rapidly during the first 6 months of life after which time the infants in ISS, especially from remote rural areas, tend to slow in rate of growth (Bindon & Zansky, 1986). Bindon and Cabrera-Mereb (1990) found infants in American Samoa to be quite healthy, with respiratory complaints being the most common illness. They also found that exclusively breast-fed infants tended to be healthier at 3 through 9 months than infants who were at least partially bottle-fed.

Childhood

Stair (1897) describes child rearing as alternating between overly severe punishment for minor infractions and over-indulgence. Holmes and Holmes (1992) also note severe punishment in Manu'a in 1954. Similar discipline continues to be the norm, and is the source of some difficulties with child welfare organizations outside the islands. By 3 or 4, the gender roles are being shaped among the children and girls begin to assume household roles. By 5 or 6 a young girl may be baby-sitting and caring for her next younger sibling. Boys have more freedom during early childhood, although they may have to feed the chickens, fetch water, and accompany their mother in gathering food on the reef. By 7 or 8, both boys and girls have assumed an active role in the gender-appropriate tasks of the household. An exception to this training may occur in a family where a mother has no young daughters and cannot adopt a young girl to help with the housework. In such a case, a son may be recruited by the mother to fill the role of a young girl in helping the mother manage the household. In such cases, when the boy continues to adopt the female gender role into adolescence and perhaps adulthood, the name *fa'afafine* is applied. A *fa'afafine* may marry and have children but continue to play a female role in the family and the village. Since the institution of formalized schooling, the patterns of training and household assistance by children have been dramatically altered. Pre-school child care is more likely to be done by the elderly while school-age siblings are in class, and everyone shares in the household chores once done by the children.

Adolescence

Male circumcision is the norm in Samoa. At about 9 or 10, two boys arrange to go to a specialist to perform the surgery. They bring him a gift of food and *tapa*. The operation involves a single longitudinal incision on the foreskin which peels back and after healing looks as if the foreskin has been removed (Holmes & Holmes, 1992). There is no other ceremony associated with this milestone. No such female operations are conducted.

A young man joins the *aumaga* (society of untitled men) with the sponsorship of his *matai* at around 14 or 15 or, today, after he graduates from school. The *aumaga* serves the village at the direction of the *fono*. Adolescent girls used to join the *aualuma* (unmarried women) and sleep in a separate house serving as a court to the *taupou* (village princess). Today the group comes together only on special ceremonial occasions, frequently for dancing and singing.

Adulthood

Turner (1884) says that adulthood was marked for men by tattooing. Males spend their adulthood seeking a *matai* title and their wives support this attempt as their status is also enhanced. Today upward mobility through educational and occupational advancement increase a candidate's chances of gaining a title. This upward mobility has not been without added health cost. Bindon (1997) reported on an analysis of lifestyle and blood pressure in American Samoa that indicated attempts to present a higher material culture status than can be afforded caused a stress response in Samoan men, but not women, increasing the men's blood pressures. Women who were working outside their home had lower blood pressures, but husbands of women working outside the home had higher blood pressures. Some of the strongest impacts of modernization among Samoans have been exerted as a result of trying to adopt a Western material life style with insufficient resources.

The Aged

Old age has traditionally been identified in Samoa as the best time of life. The elderly, over about age 50, are treated with respect, they have minimal demands on their time and energy, and one can rely on the support of one's children and relatives. Many of these attitudes persist today, although times are changing. Since the 1970s, homes for the aged have been opened in both ISS and American Samoa, a result of the emigration of younger Samoans and the occasional failure of family resources to provide for the elderly who remain in the islands (Holmes & Holmes, 1992). A number of elderly Samoans suffer the depredations of diabetes, blindness being one of the most visible symptoms.

Dying and Death

The dying were traditionally looked after at home where friends and family could easily visit to pay respect. If wounded in battle or taken sick away from home, every effort would be made to return home prior to death. Today, Samoans who are terminally ill prefer to spend their time at home with their family rather than staying in one of the hospital or dispensary facilities. Local nurses assist with this desire by stopping in on the ill at home.

At death, while the body is being prepared by female relatives, word is sent to relatives in other villages. Family begin to arrive bearing gifts of fine mats (*afuelo*, woven of pandanus fiber) and *tapa*. After the funeral the gifts will be given to the female relative in charge of body preparation and she may make a further division to reinforce political ties. If the deceased is an important chief, the display of wealth may be great (Holmes & Holmes, 1992). Today, in addition to the fine mats and barkcloth, cash and purchased goods may also form part of the funerary wealth. A *matai* from the deceased's family will call on family members to make contributions (*fa'alavelave*) to support the funeral and burial.

Burial takes place on the family homestead with the erection of a small mound of volcanic rocks in the form of a monument to the deceased. The interment of individuals on family land was an important part of pre-missionary religion with the belief that they joined the ranks of the *aitu* that oversaw the health and well-being of the family. Today the burial also takes place on family land but the mounds are likely to be constructed of concrete and incorporate Christian icons.

REFERENCES

Bindon, J. R. (1988). Taro or rice, plantation or market: Dietary choice in American Samoa. *Food and Foodways, 3*, 59–78.

Bindon, J. R. (1995). Polynesian responses to modernization: Overweight and obesity in the South Pacific. In I. de Garine & N. J. Pollock (Eds.), *Social aspects of obesity* (pp. 227–251). London: Gordon and Breach.

Bindon, J. R. (1997). Coming of age of human adaptation studies in Samoa. In S. J. Ulijaszek & R. A. Huss-Ashmore (Eds.), *Human adaptability: Past, present, and future* (pp. 126–156). Oxford: Oxford University Press.

Bindon, J. R., & Baker, P. T. (1997). Bergmann's rule and the thrifty genotype. *American Journal of Physical Anthropology, 104*, 201–210.

Bindon, J. R., & Cabrera-Mereb, C. C. (1990). The health status of infants in American Samoa during the first year of life. *American Journal of Human Biology, 2*, 511–519.

Bindon, J. R., & Zansky, S. M. (1986). Growth and morphology. In P. T. Baker, T. S. Baker, & J. M. Hanna (Eds.), *The changing Samoans* (pp. 222–253). New York: Oxford University Press.

Caton, H. (1990). *The Samoa reader: Anthropologists take stock*. Lanham, MD: University Press of America.

CIA. The World Factbook—American Samoa. (2001a). Washington, DC (January 1, 2001); Available online at, http://www.cia.gov/cia/publications/factbook/geos/aq.html.

CIA. The World Factbook—Samoa. (2001b). Washington, DC (January 1, 2001); Available online at, http://www.cia.gov/cia/publications/factbook/geos/ws.html.

Freeman, D. (1983). *Margaret Mead and Samoa: The making and unmaking of an anthropological myth*. Cambridge, MA: Harvard University Press.

Gray, J. A. C. (1960). *Amerika Samoa: A history of American Samoa and its United States Naval Administration*. Annapolis: United States Naval Institute.

Harbison, S. (1986). The demography of Samoan populations. In P. T. Baker, T. S. Baker, & J. M. Hanna (Eds.), *The changing Samoans* (pp. 63–92). New York: Oxford University Press.

Holmes, L. D. (1987). *Quest for the real Samoa.* South Hadley, MA: Bergin & Garvey.

Holmes, L. D., & Holmes, E. R. (1992). *Samoan village: Then and now* (2nd ed.). New York: Holt, Rinehart & Winston.

Kramer, A. (1995). *The Samoa Islands, Volume II.* (T. Verhaaren, Trans.). Honolulu: University of Hawai'i Press. (Original work published 1903.)

MacPherson, C., & MacPherson, L. (1990). *Samoan medical belief & practice.* Aukland: Aukland University Press.

Mead, M. (1928). *Coming of age in Samoa.* New York: William Morrow.

Orans, M. (1996). *Not even wrong: Margaret Mead, Derek Freeman, and the Samoans.* Novato, CA: Chandler & Sharp.

Shore, B. (1982). *Sala'ilua, a Samoan mystery.* New York: Columbia University Press.

Stair, J. B. (1897). *Old Samoa: Or flotsam and jetsam from the Pacific Ocean.* London: The Religious Tract Society.

Turner, G. (1884). *Samoa a hundred years ago and long before.* London: Macmillan.

Walters, W. (1977). Community psychiatry in Tutuila, American Samoa. *American Journal of Psychiatry, 134,* 917–919.

WHO Country Health Information Profile: American Samoa. (1999a) . Manila, Philippines. (July 1, 1999); Available online at, http://www.who.org.ph/chip/ctry.cfm?ctrycode=ams&body=ams.htm&flag=ams.gif&ctry=AMERICAN%20SAMOA

WHO Country Health Information Profile: Samoa. (1999b). Manila, Philippines. (July 1, 1999); Available online at, http://www.who.org.ph/chip/ctry.cfm?ctrycode=sma&body=sma.htm&flag=sma.gif&ctry=SAMOA

Saraguros

Ruthbeth Finerman

ALTERNATIVE NAMES

The word "Saraguro" translates from the Quichua language as "Land of Corn." The indigenous community self-identifies as Saraguros, *indígenas*, or more rarely, as *runas* ("people"). Non-indigenous residents may speak of Saraguros with the pejorative *indio* or *chinita*. Saraguros occasionally make derogatory reference to acculturated individuals as *leichos* or *gente acomodado*.

LOCATION AND LINGUISTIC AFFILIATION

Most Saraguros continue to reside in Loja Province, in Ecuador's southern Andes. A majority populates intermontane valleys surrounding the township of Saraguro (3.7°S, 79.3°W, elevation 640 m to 910 m). In addition to the town, the region comprises croplands, managed pastures, residences, and scattered *altiplano* woodlands and *paramo* grasslands. In recent generations, intense competition for land encouraged migration to exploit farm and pasture lands in the Yacuambi Valley on the eastern foothills of the Andes, and wage labor opportunities in major metropolitan centers. In the late 1990s, Ecuador's economy collapsed, spurring hundreds of Saraguros to migrate to Spain for work as farmhands.

Virtually all elderly and most adult Saraguros speak both Spanish and Quichua, a southern Ecuadorian highland dialect of the Quechuan language (Harrison, 1989). Younger generations are less likely to speak Quichua, and few youths are proficient in their native tongue. Nevertheless, Saraguros retain a strong identity as *indígenas* and possess close cultural and political affiliations with all other Quichua speakers.

OVERVIEW OF THE CULTURE

Demographics and History

Although census data are unreliable, Saraguros are estimated to number at least 20,000. Historically, the region remained independent until Incan domination in the late 15th century AD. At that time, Saraguro town was founded as a *tambo* or way station between Cuenca and Loja. While their origins remain uncertain, most Saraguros accept the view that their current population was founded through the intermarriage of forced migrants (*mitimaes*) and local Cañari or Palta (Macas, 1995).

European rule began by the 1530s, but lands remained in the possession of indigenous Saraguros. With construction of the PanAmerican Highway in the 1940s, non-Indians took control over most of the town center, but Saraguros maintain land claims in town and most of the surrounding countryside. Although contact with a dominant outside culture has influenced aspects of their culture, the indigenous community sustains a fairly strong sense of tradition and identity, and a basic distrust of external forces.

Production

Like most Andean peasants, Saraguros are agropastoralists (Gade, 1999; Wilson, 1999). Bullock-pulled plows are used to plant maize, potatoes, beans, wheat, and barley. Cattle herding yields major capital, and families also raise sheep, guinea pigs, swine, chickens, and rabbits. Wool from sheep is handspun and woven to fashion traditional Colonial-period garments. Crafts, including beadwork, embroidery, and weavings are also manufactured for personal use and for sale.

Kinship, Marriage, and Family

Descent here is reckoned bilaterally and egocentrically, although patronyms take precedence in official records. While monogamous, divorce and abandonment can occur. Most voice a preference for neolocality, yet small extended households predominate. There is no clan or lineage affiliation, but they express solidarity within *barrios*, politically linked neighborhoods that support communal lands and labor cooperatives. There is also a clear inclination toward *barrio* endogamy. Still, Saraguros possess a strong sense of household autonomy, making *barrio* alliances relatively unstable.

Social Organization

Strong differences in wealth and status exist among Saraguros, and between indigenous and non-indigenous residents (Belote, 1978). Status also differs somewhat by gender, as males enjoy a higher public profile. However, Saraguro women can garner recognition by sponsoring religious festivals and, as in other parts of Ecuador, they wield ample power over household resources, decision-making, and family comportment, particularly as they relate to health (Finerman, 1995; Wayland, 2001; Weismantel, 1988).

Religion

While evangelical movements have made some inroads in the region, Catholic doctrine dominates religious life. Yet, as in much of the Andes (Allen, 1997; Dover, Seibold, & McDowell, 1992; MacCormack, 1991), Saraguro beliefs and practices retain many elements of pre-Conquest cosmology, including animistic reverence for the sun, moon, wind, rivers, and rainbows. Such spirits are also linked to personalistic illnesses.

The Context of Health: Environmental, Economic, Social, and Political Factors

Health Profile

Reliable data are unavailable, but Ecuadorian health officials agree that Saraguro morbidity and mortality well exceed national rates. As in many Andean nations, rates of infant mortality for the richest and poorest provinces vary by as much as 200%. One survey found that more than 40% of Saraguros reported at least one health complaint during the prior two months (Kroeger & Barbira-Freedmann, 1992, pp. 251–252).

Newborns in Saraguro are slightly more vulnerable than average to post-partum infection. Infants and children experience frequent if not chronic parasitic infection, diarrhea and dehydration, and acute respiratory infection, although infectious disease rates began to fall after 1960 and most dramatically after about 1980, as a result of intensified immunization campaigns (UNICEF, 1995). Adults are subject to work-related injury, and health workers report that a sizeable proportion (up to 30% in some *barrios*) carry tuberculosis. Many adults also suffer complications from alcohol abuse. Malnutrition is less common, as most Saraguros sustain a balanced diet and active lifestyle. Yet, the PanAmerican Highway has eased access to high calorie but nutritionally inferior processed foods (e.g., sugar, white rice, noodles), alcohol, and tobacco, posing greater risk of obesity, hypertension and diabetes, particularly in pregnancy. The incidence of other so-called "diseases of development" such as cancer, heart disease, and sexually transmitted diseases has also grown.

The indigenous population is also vulnerable to a range of traditional or culture-specific illnesses (Argüello & Sanhuenza, 1996; McKee & Argüello, 1988), such as nerves (*nervios*), envy (*envidia*), fright (*susto*), evil airs (*mal aire*), evil eye (*mal ojo*), soul loss (*espanto*), and witchcraft (*brujeria*). With the exception of *nervios*, children and elders are the most common victims of such disorders.

Environmental Factors

The natural ecosystem presents a mixed bag of health risks and benefits. The high altitude and cold temperatures produce several deaths each year from hypothermia. The climatic phenomenon *El Niño* produces devastating rains that wipe out crops and wash out roads, cutting off access to medical care. Wildlife—mainly venomous snakes and spiders—poses some threat, particularly since most children and older adults walk barefoot. Domesticated animals expose their owners to zoonoses including tuberculosis, rabies, flea and tick infestations, and a host of parasitic diseases. Nevertheless, the rich ecosystem also supports a diverse food supply and a vast array of medicinal plant species that are routinely exploited for health promotion in Saraguro and throughout the Andes (Bastien, 1987; Naranjo & Escaleras, 1995; Wilson, 1999).

Household ecology also yields risks, although these have shifted over time. Most old homes have dirt floors, and kin tend to share beds in one room, facilitating contact with contaminants and the spread of communicable disease. Recent home renovations and new homes invariably boast plank flooring and separate sleeping quarters, yet these upgrades mainly reflect the crowding of large extended families into a single home. Prior to 1990, most Saraguros cooked on open hearths fueled by firewood, risking burns and respiratory irritation. However, depletion of firewood supplies has since forced a shift to use of cooking gas. This has reduced hearth fire hazards, but poses a danger should gas cylinders rupture and explode. During the same period, most homes gained access to water pipelines, but the unpurified water continues to spread disease. The community also lacks effective sewage and waste disposal, but many homes now feature outdoor latrines and showers, slightly reducing waste levels indoors. Still, some Saraguros continue to collect urine in the home (for use in dye vats), and most residents come into contact with solid waste and contaminated soil and water when cultivating fields.

Social Factors

Saraguros associate social and economic productivity with health, thus disability, inactivity, or unemployment is considered both unhealthy and antisocial. Pressure to remain productive has fostered migration, particularly of unemployed males, forcing many women and young children to take charge over households, crops, and herds. Worse still, the migrants frequently return from the lowlands and cities carrying a host of new diseases (e.g., malaria, cholera) that spread through the community.

Interpersonal harmony is also viewed as fundamental to health, and some illnesses are attributed to anger, envy, despair, or discord. Thus, social cohesion is characterized as essential to well-being, and gossip stands as a powerful social control mechanism. Women here are concerned to avoid gossip should their children appear ill or malnourished and they take special pride when told that their children look robust. These women retain the primary social role in managing family health, and skillful healers gain substantial power in both the household and community (Finerman, 1995). Females are socialized into the family healer role during childhood, when they begin to assist their own mothers in preparing remedies for younger siblings. As they age and establish their own families, women continue to consult their mothers and female friends for health care advice.

Political and Economic Factors

Throughout most of their history, Saraguros remained at the social, political, and economic periphery of the state. However, the PanAmerican Highway expanded contact, and the adoption of cattle husbandry increased dependence on the national economy. In the final decades of the 20th century, depleted oil reserves reduced the nation's wealth, sustained *El Niño* rains devastated roads and infrastructure, and pervasive political corruption and massive international debt all combined to lead Ecuador into financial ruin. Saraguros, previously immune from economic downturns, were gravely impacted as banks, schools, and hospitals closed, the national currency (the *Sucre*) was abandoned, prices for pharmaceuticals and essential goods soared, and commercial ventures failed. As the crisis intensified, health levels began to deteriorate. While reliable statistics are not available, government officials claim that mortality among infants, children, and elders has begun to rise, and domestic abuse reports have increased.

MEDICAL PRACTITIONERS

Indigenous Saraguro curing specialists include herbalists, midwives, and shamans. Herbalists provide medicinal plant preparations and advice on diet for a range of ailments. Midwives advise women on fertility, prenatal and postpartum care, and may assist in labor and delivery, employing massage, baths, herbal remedies, and dietary guidance. *Curanderos* specialize in intractable conditions, particularly illnesses suspected of being caused by supernatural agents. Nevertheless, Saraguros make minimal use of these practitioners, treating almost all illnesses with home remedies (*remedios casseros*). Not surprisingly, only a few of these specialists practice here, and those who do are more likely to be patronized by non-indigenous town residents.

In addition to traditional curers, the town supports a few private physicians and dentists, and a hospital staffed by medical and nursing interns and community health workers. Several private pharmacies also operate, typically selling products without prescription. But, as with traditional healers, Saraguros have proven unlikely to consult biomedical providers unless all home treatment efforts are exhausted.

Home-based health care combines preventive activities involving hygiene, dress, dietary laws, and personal comportment, plus an extensive repertoire of herbal remedies. Women view their healing role as a natural extension of their duties in bathing, clothing, feeding, and generally nurturing kin. They invariably argue that they are the first to spot the signs of illness (e.g., appetite or sleep loss, pain, lethargy) and they usually possess years of healing experience. Most also reason that other practitioners could never match a mother's concern for her own family's welfare.

CLASSIFICATION OF ILLNESS, THEORIES OF ILLNESS, AND TREATMENT OF ILLNESS

Etiology

Saraguro views of illness concern the actions of both naturalistic and personalistic agents. One of the most pervasive beliefs here and throughout Latin America is that of humoral opposition (Foster, 1998). Specifically, "heat" and "cold" appear as ubiquitous elements, found in individuals, animals, plants, foods, seasons, and emotional states. A balanced state assures health, while an excess or deficiency yields illness and symptoms of heat (e.g., fevers, blisters, infection) or signs of cold (e.g., chills, cough, diarrhea).

Naturalistic etiologies here also reflect a view of the outside world as inherently dangerous (Finerman, 1987). Illness is considered most likely to occur when eating or socializing away from home, and especially when travelling a great distance (although this belief holds more strongly for women than for men). The introduction of germ theory merely reinforced opinions of the outdoors as infested with toxins, contaminants, and microscopic *bichos* or "bugs." Saraguros invariably complain that cheese, milk, and similar foods prepared by other families are inferior or "dirty" (*sucio*); food prepared in the town is particularly tainted. Consequently, little of the food served at local communal festivals is consumed on the spot; most is taken home and re-cooked to eliminate impurities. When offered refreshment, guests routinely hear the phrase, *Quizás no le hace mal* ("Maybe this will do you no harm") and respond with *Dios se lo pague* ("God will repay you"), a refrain that both reassures and warns the host.

Intense natural emotions can also cause harm (Tousignant, 1988). For example, adult Saraguro women suffer a high prevalence of *nervios* ("nerves"), a physical collapse resembling depression, which they usually blame on *pura sufrimiento*: "pure misery." *Nervios* is also common for women overwhelmed by the strain of family illness. While disabling, the condition offers relief from the stress of family care and affords a degree of control over kin (Finerman, 1988). Such ailments also reveal how conditions of mind and body are linked in Saraguro culture. Mental illness is not generally recognized here; instead, both physiological and psychological distress find expression in physical form.

In addition to these naturalistic etiologies, Saraguros retain several pre-conquest animistic beliefs about dangerous supernatural forces including sacred lagoons, winds, rainbows, rivers, the earth, moon, and sun. While few expressly describe these agents in anthropomorphic terms, they manifest their concerns in folktales and codes of conduct. For instance, many recount legends or personal accounts where mountaintop lagoons rise from their banks, seeking to abduct women. They use medicinal plant bundles to "sweep away" malevolent winds, and avoid rivers where the rainbow is said to impregnate female bathers. During feasts, hosts may spill small offerings of alcohol as a gift to the earth mother (*pacha mama*), and many cease planting during a new moon to avoid crop failure. Saraguros also wear distinctive hats while outdoors, to show respect for the sun and thus avoid its "vengeance." Of note, younger generations increasingly ignore these traditions, especially the precaution of wearing hats. Few minors have even heard of once common illnesses like *bao de agua* ("water fright illness"). Such youths leave elders fretful about the health and well-being of their families and community.

Despite a decline in some animistic beliefs, nearly all Saraguros retain conviction in the power of other human beings to cause harm through envy, evil eye, or witchcraft. Mothers here usually swaddle infants and carry them on their backs hidden under shawls, to protect them from envious gazes. Young children wear red bandanas, jewelry, or clothing to ward off witches. Even small animals (especially kittens) may be safeguarded with a red collar. Nonhumans (e.g., snakes, dogs) are also said to cause magical fright or soul loss sickness, although episodes here are relatively rare.

Treatment

As noted, Saraguros usually rely on their mothers or wives to manage health, despite access to various professional curers. This "popular" or informal sector of family-based care predominates cross-culturally (Kleinman, 1980); in southern Ecuador Saraguros enjoy a degree of fame for their expertise in herbal medicine.

Indigenous women here possess curing knowledge that is at once highly complex and syncretic, blending traditional curing elements with biomedicine. The richness of their knowledge derives from a lifetime of experience in treating sick kin and from continual expansion of their curing arsenal. Saraguro women routinely consult their own mothers and female friends for advice, and they gain fresh curing insights from health specialists. Women usually accompany spouses and children to practitioners, and providers tend to direct treatment instructions to these women. In the process, they learn new concepts and therapies that help them upgrade their curing routines. By contrast, specialists such as *curanderos*, midwives, and physicians almost never share insights among one another; indeed, they usually express scorn for different medical systems.

Family care in this population incorporates both preventive and primary care. Illness prevention is accomplished mainly through attention to dress, hygiene, and diet. Infants and children are clothed in layers to bundle them against malevolent winds and humoral disorders. As noted, children may also be dressed in red to guard against witchcraft. Hygiene acts to cleanse away contaminants, eliminate humoral excess, and thwart sorcery. Thus, medicinal plants may be added to bath water to both clean skin and prevent evil airs, while hairs lost during combing are meticulously gathered into bundles and burned to keep them out of the hands of witches. Meals include a mix of nutrients and medicinal plants, designed to ensure nutritional and humoral equilibrium. Some foods are also utilized as remedies. For example, humorally "hot" foods such as chilis, raw sugar (*panela*), and beef can alleviate ailments like extreme cold (*recaida*), while "cold" foods such as milk, white sugar, and chicken help to reverse ailments such as extreme heat (*gangrena*). Several foods are also expressly prescribed or prohibited during various stages of the reproductive cycle. For instance, avocado consumption is taboo for menstruating girls, as it is considered toxic at such times.

A majority of treatments are administered as herbal teas (*aguita* or *infusión*), however many remedies also take the form of a vapor bath (*baño vapor*), spray (*sopla*), flotation (*flotación*), rub (*sobado*), poultice (*cataplasma*), or plaster (*escayola*). Saraguro women take pride in their mastery of medicinal plant lore. They can name dozens or even hundreds of plants in free lists, and describe plant growth habits, uses for different plant parts, processing techniques, and applications. Women cultivate many of the more vital domesticated medicinal plants in kitchen gardens (*huertas*), located next to or near the house. These gardens vary in size and complexity, depending on several factors. Often, the largest and most complex gardens are cultivated by women with many children, many years of curing experience, and the wealth and labor pool necessary to support garden maintenance. Such gardens can contain hundreds of varieties of medicinal plants. Even young brides and elderly women tend medicinal plant gardens, although these are likely to be smaller and less diverse. Women add to their reserves by gathering hundreds of medicinal plants that grow wild in the region's diverse ecosystems (Naranjo & Escaleras, 1995).

SEXUALITY AND REPRODUCTION

Sexual Activity

Despite strong affiliation with Catholicism, Saraguros demonstrate only weak adherence to religious mores regarding sex. In particular, premarital sex is only mildly discouraged, and births outside wedlock are both common and readily accepted. Indeed, such births are a positive indicator of a female's fecundity and can enhance her prospects for marriage, though aspirations decrease with subsequent pregnancies. Extramarital affairs incur slightly more social disapproval, yet most residents profess tolerance, particularly in cases of absent spouses (e.g., wives of migrant workers, or abandoned spouses). Homosexuality is criticized, but almost never discussed. Greater censure is reserved for incest, which is called immoral and unhealthy. Local folklore claims that incestuous couples gradually transform into dogs, manifesting new canine attributes after each transgression. For the most part, however, locals regard sex as both normal and healthy, but also a private matter. Women follow select rules of sexual decorum; in particular, they cover their mouths when smiling and laughing, as showing teeth is vulgar. However, men and women are equally fond of sexual jokes and innuendo.

Fertility

In the indigenous community, the social transition from childhood to adulthood is not marked by initiation into sexual activity or, in the case of females, by menarche. Instead, childbirth is the main marker of maturity. Females retain the title "girl" (*niña*) or "child" (*hija*) even after marriage, until their first birth. Males also lose the moniker *niño* or *hijo* only after they have children. Childbearing is significant in most peasant societies; offspring are critical to expand a family's labor pool and provide care for aging parents; households also procure support through fictive kin networks (Mitchell, 1991, pp. 47–51).

Fertility itself is a sign of good health and humoral equilibrium. A woman who readily conceives and bears children is robust and "warm"; infertility and miscarriage may arise from various problems, including a "cold" metabolism, witchcraft, and envy. Barren women are also considered prone to weakness (*debilidad*) and heart failure.

Saraguro understandings of conception shifted in recent generations. Accounts from the 1960s cite preformationist views. At that time, locals claimed that males were preformed in sperm and implanted in the womb, while females generated inside the mother. By the 1970s they had abandoned such notions, as more youths learned biology in school. After that point, conception was explained as the fusion of egg and sperm, while the womb was characterized as a "basket" that carries the fetus.

Fertility Control

Ideal family size also evolved over the last half of the 20th century. Historically, Saraguro couples hoped to have many children to expand production and care for their aging parents. Yet, by the 1950s competition for land intensified and youths deserted Saraguro in search of acreage and wages. More couples also found that larger families could not afford to educate all of their children. By the 1970s, family planning campaigns began to actively promote the benefits of smaller families. Today, families with just one son and one daughter are the ideal. Even so, Saraguros face multiple barriers to effective contraception.

Church representatives in Saraguro condemn birth control, and many men fear that contraception facilitates adultery by women. Such antagonisms make discussion of the topic both sensitive and infrequent. Yet, most in the indigenous community regard pregnancy prevention as a woman's privilege, and as her burden. A majority of women rely on herbal teas to prevent or terminate a pregnancy, though such preparations are unreliable. Occasionally, women seeking abortion consult *curanderos*, or travel to larger cities to purchase contraception. Few see local physicians or druggists, fearing gossip. A small number have taken drastic steps to induce miscarriage, pounding their belly, falling down hillsides, or ingesting irritants such as wood ashes. Still, few attempt to terminate unwanted pregnancies; almost all reconcile themselves to the condition as "God's will" (*lo que Dios quiere*). Nevertheless, they frequently bemoan their large families and extol the good fortune of women with few children.

HEALTH THROUGH THE LIFE CYCLE

Pregnancy and Birth

As noted, pregnancy usually occasions happiness or, at worst, forbearance. In any event, Saraguros describe the condition as fraught with danger. Thus, they strive to diagnose pregnancy early, to ensure a successful outcome (Finerman, 1988). Pregnancy signs cited most frequently include interrupted menses and/or morning sickness. Some report other indicators such as a darkening of the nipples and changes in appetite, sleep, or energy levels. A few claim to detect pregnancy through the presence of white solids in expectorate or urine.

Once recognized, Saraguros take protective measures. Usually, the expectant mother is encouraged to fulfill food cravings; these are credited to the fetus, which may kick or harm its mother if frustrated. Exceptions to such indulgence are extremely "hot" or "cold" foods, which might exert additive or antagonistic effects on the hot state of pregnancy and induce illness. Women may also reduce workloads and avoid activities that shake the body (e.g., work at sewing machines or weaving looms, as well as sexual activity), as these could prematurely "wake the child." Emotional disturbances, including domestic arguments, accidents, shock, and envy are also harmful. Pregnant women are also warned to avoid cemeteries (e.g., funerals, All Soul's Day mass) as they readily succumb to sickness from evil airs (*aire grande*) around graves.

Across Latin America there is great variation in the use of birth attendants (Bolton & Bolton, 1978; Cosminsky, 1976). In Saraguro, indigenous women almost always give birth at home, alone or assisted by their spouses. Primiparous (first time) mothers are somewhat more inclined to seek additional assistance from female relatives or midwives. Hospital deliveries are rare; most result from protracted labor or other complications.

During accouchement, Saraguro women may soak in an herbal bath or drink medicinal plant teas to aid labor; eating is permissible, but few do so. Women rest in any position that feels comfortable, but avoid sudden movement. Husbands provide massage to reduce pain. Women usually deliver in a vertical squat position supported by spouses or female attendants, if present. The umbilical cord is cut immediately upon delivery; some claim that this practice prevents the womb from "reclaiming" the infant. Women may then smoke tobacco or datura, to encourage placental rejection. Customarily, the placenta is buried at the entrance to a home, without prayer or special ceremony.

Childbirth is followed by a 40-day post-partum confinement (*la quarenta*). This term of lying-in protects the debilitated mother and newborn from evil airs and other illnesses. During this stage, both take a series of herbal baths on prescribed days. Mothers wrap their abdomen with poultices made from various fats and medicinal plants, to aid recovery, and nourish the womb. New mothers also adhere to a strict traditional diet. They continue to avoid foods with extreme humoral qualities, especially cold foods, claiming these sicken a mother as her metabolism cools after the heat of pregnancy. They also eschew foods like citrus that might "cut off" lactation.

Infancy

Indigenous parents profess no gender preference among offspring. Both sexes retain their parents' names, and both can inherit. Twin births seem to cause no concern, but are extremely rare in this population. It is possible that, in the past, such births precipitated higher maternal mortality, leaving a population that is genetically less prone to multiple pregnancy. Birth defects and deformities do incite shame and are interpreted as a punishment from God. However, the child faces minimal risk for active or passive infanticide, since religious doctrine effectively prohibits such recourse. Still, a few families have hidden such offspring at home, so that their disgrace is less public.

Saraguros rightly regard newborns as especially vulnerable to illness and death; fears are understandable given that, prior to the 1960s, infant mortality at times reached 40%. Waves of measles, chicken pox, and whooping cough decimated families before immunization campaigns intensified. Infants still face substantial risk of diarrhea and dehydration, respiratory infection, and communicable disease, and are considered highly prone to evil airs, magical fright, soul loss, envy sickness, and witchcraft.

Here and elsewhere in Latin America, infants and children who die before their First Communion are said to become angels or *angelitos* (Lillo, 1942). The dead child is adorned with a paper crown and wings, and seated on a decorated altar in their home. Their death is marked by an extended period of feasting and dancing; the tradition helps parents to better cope with the loss of a child.

While such deaths are celebrated, Saraguros take pains to forestall the loss of offspring. Newborns are kept in virtual isolation for the first 40 days of life. Saraguro mothers breast-feed infants on demand, rather than on a schedule. Mothers customarily breast-feed for one year, or until the mother becomes pregnant again.

Solids (mostly cereals) are introduced gradually, usually beginning around 6 months of age. Infants wear little clothing while indoors and can defecate freely, but they are kept clean, warm, and protected. As they grow they are free to explore the home, under the watchful eyes of their mother or older siblings. Infants enjoy attention from multiple caretakers, including parents, siblings, and extended kin. By the age of five, most children here begin to help carry and tend younger siblings.

Baptism represents a crucial religious rite and supernatural safeguard against illness, but the ritual cannot take place until some months after birth. In Ecuador, infants must complete a preliminary course of immunizations before they can be baptized. Local community health workers usually assist by making house calls to finish the vaccination series. Infants must also be strong enough to leave the safety of home. Therefore, most parents arrange baptism after the infant is at least 6 months old and celebrate the occasion with only a private family feast.

When taken outdoors, an infant is dressed in layers of clothing and red cord, swaddled, and carried on the caretaker's back under one or (more often) two shawls. Some mothers explain that swaddling helps the infant grow straight, while others claim the child is more comfortable and sleeps more readily when bound. However, the practice also curbs movement under shawls that might attract unwelcome attention, thereby protecting infants from witchcraft and envy.

Childhood

The transition from infancy to childhood in Saraguro is gradual, marked mainly by enhanced mobility, communication, and autonomy. Once offspring begin to walk, talk, and play independently, they are given more responsibilities and tasks. By five years of age children begin to help with younger siblings, keep watch over sheep and cattle, and fetch items for their mothers. Daughters assist mothers in meal preparation, sibling care, housework, and preparation of medicinal remedies. Parents rarely discipline offspring, although they may threaten spankings or a thrashing with nettles. More often, parents simply assign more chores, reasoning that work keeps children too busy to cause trouble.

In Saraguro, children must be fully vaccinated to register for school. Nonetheless, sickness rates increase for students, as contagious and parasitic diseases spread and children suffer more accidental injury. Parents also fear that offspring will fall ill from foods prepared at school or purchased in town. Over time, though, children come to be viewed as relatively safe from illness, including personalistic conditions like evil airs, envy, and magical fright. Risk seems to decline around the time children complete the Catholic rites of Confirmation and First Communion.

Adolescence

Puberty and adolescence receive almost no recognition in this population. Typically, daughters quietly inform their mothers of the onset of menarche, but they receive little instruction and face few restrictions other than minor dietary taboos (e.g., consumption of avocado) and perhaps admonishment to avoid pregnancy. Adolescent sons garner still less attention, but may gain greater freedom; young Saraguro males frequently stay out all night attending fiestas or drinking with male companions. Such behavior risks few penalties other than mild parental rebuke. Youths are said to face few health risks, other than unplanned pregnancy or injury precipitated by intoxication.

Adulthood

Most Saraguro adults describe their lifestyle as exceptionally healthy (*sano*), comprising vigorous labor, and a balanced diet. Nevertheless, a majority report chronic, albeit minor complaints like muscle aches, fatigue, and eye strain, plus a range of acute respiratory and intestinal disorders. Other common concerns include pregnancy, physical trauma (usually related to work or alcohol intoxication), alcoholism, animal bites, and tuberculosis. Adults claim to suffer fewer traditional ailments like magical fright and soul loss, yet *nervios* is pervasive among adults (especially women), and *mal aire* strikes most adults who take cattle to pastures in the windy and wet high elevation *paramos*.

The Aged

Aging, like other social transitions, is a subtle process in this population. This is largely because there is no designated age for "retirement" or "social security" benefits; Most Saraguros remain economically productive throughout their lives, and provide supplemental aid with family care and curing expertise. Also, Saraguros show few signs of aging; hair tends to remain black until extreme old age, and years of sun and wind exposure reduce skin elasticity early in life, making aging or decline difficult to detect. Often, the sole mark of maturity is a subtle change in mode of address. Over the years, Saraguro men gain the title *Taita* or *Taiti*, and women come to be called *Mama*. These Quichua terms, meaning father and mother, are titles of respect used by all Saraguros, not merely kin. It is unclear precisely when and why the title is added to one's Christian name, but it correlates with advanced age and a high level of social recognition.

While elders are treated with deference, aging here is accompanied by a decline in overall health. Muscle pain and respiratory and digestive disorders common among adults advance into persistent or even debilitating conditions. The aged also manifest high rates of arthritis, visual impairment, wasting, more cases of active tuberculosis, and traditional syndromes, especially *mal aire*, and *nervios*. Extended kin are expected to offer care, but elders sometimes complain that relatives are of little help; a few have no surviving kin to look after them. After Ecuador's financial crash in the 1990s, many elders were abandoned, as kin left the community in search of work. In such instances, elders have turned to fictive kin, friends, and neighbors for health care and other assistance.

Death

Saraguros exhibit both respect for and fear of the dead. With the exception of children honored as *angelitos*, the deceased seem to be regarded in vaguely malevolent terms. Survivors are loath to speak of the dead by their Christian name; more often, the deceased is referred to obliquely as *el cadver* or *el difunto*. Mourners view corpses as corrupted flesh that pollutes the home and threatens "corpse sickness" (*aire grande*), which can rob the living of their souls and even their lives. Consequently, indigenous funerary customs venerate the memory of the dead, but also operate to protect the health and well-being of the family and community.

Soon after death, the body is bathed, dressed for burial, and laid out at home. A rosary is placed in the hands, and candles encircle the bier. Mourners "accompany" the corpse on an all-night vigil and pray for their soul. As the night progresses, attendants may consume alcohol to toast the dead and pass the hours. The coffin is then carried to church for funeral mass, and then up a hill to the local cemetery. At graveside, mourners remove their hats (but hold these over their heads out of respect for the sun) and recite final prayers. Saraguros assert that open graves emit dangerous airs and the essence of ghosts. Thus, mourners (especially those who handled the corpse, held vigil, or prepared the grave) return home immediately after the ceremony and bathe to remove any residue of death. Most also pour cologne in their hands and cup it to their face, rubbing it in and inhaling deeply as a treatment for supernatural illness.

Soon after mass and burial, family and friends strip the home of all possessions; these are carried to a local river, where all goods (e.g., clothing, bedding, and even furniture) are meticulously washed to purge the residue of death. The walls and floor of the house are also scrubbed, and doors and windows are left open for several hours to expel any lingering malevolent airs or spirits.

Rites may be interrupted if cause of death is in doubt (Finerman, 1984). Until recently, physicians performed autopsies in the cemetery just prior to burial, in full view of the mourners. Since proper instruments and lab analysis were not available onsite, results were unreliable at best. Worse still, this act delayed burial for hours and subjected mourners to the sight of a loved one's dissection (for which kin were billed). Today, bodies are usually autopsied at the hospital, then released for vigil, mass, and interment.

CHANGING HEALTH PATTERNS

In many respects, it is a mistake to characterize any health system as "traditional," because medical systems undergo constant transformation (Finerman, 1989). For Saraguros, contemporary beliefs and practices form a syncretic fusion of pre-conquest doctrines (e.g., animistic views of the sun, rainbows, rivers, and winds), Colonial theories (e.g., humoral equilibrium), and contemporary biomedicine (e.g., pharmaceuticals). Indigenous women here do not distinguish between these; all are selectively added or rejected on the basis of perceived efficacy. No matter the source (i.e., kin, friends, health specialists, or public education campaigns), new concepts that better explain illness causation, and innovative procedures that more effectively diagnose or cure, are readily incorporated into family care. Thus, explanatory models like humoral opposition or germ theory, and diagnostic labels like "parasites" are welcomed, as these explain how sickness can arise from agents too small to see. Similarly, new medications like anti-parasitic purgatives can yield dramatic results. Nevertheless, innovations are rejected if they contradict accepted tenets or render poor results. Consequently, doctors have failed to dissuade mothers from treating kin and to rely instead on hospitals, since such facilities rate as inferior to the experience and compassion mothers can offer.

Saraguro is experiencing other transitions that shape health, for better or worse. As noted, the collapse of Ecuador's roads, infrastructure and economy have devastated its citizens. At present, most Saraguros find themselves in debt, lowering quality of life, and risking increased malnutrition, morbidity, and mortality. Hundreds have migrated, robbing the community of many of its youngest and most productive members, and leaving many children and elders abandoned.

Despite such setbacks, agencies have stepped up support for aid and development. Programs include livestock vaccination, nutrition supplements, child immunization, and community health (Barreto, Barrera, Unda, & Carri, 1996). Alcoholics Anonymous has established chapters in a few *barrios*, helping Saraguros recover their productivity. Day care centers have opened, freeing women to expand economic activity (though centers struggle with funding cuts). Some *barrios* participate in new cooperative economic ventures (e.g., weaving mills, trout farms, greenhouses for cash crops, and incipient ecotourism) selling wares in and beyond the community. Such programs help offset some current threats to health. More critically, they represent an investment in the future, offering the best prospect for the continued wellbeing of Saraguros and their culture.

REFERENCES

Allen, C. (1997). When pebbles move mountains: Iconicity and symbolism in Quechua ritual. In R. Howard-Malverde (Ed.), *Creating context in Andean cultures* (pp. 73–84). New York: Oxford University.

Argüello, S., & Sanhuenza, R. (Eds.) (1996). *La medicina tradicional Ecuatoriana*. Quito: Banco Central.

Barreto, R., Barrera, A., Unda, M., & Carrión, A. (1996). *Ciudades y Pueblos Saludables*. Quito: OPS/OMS (PAHO/WHO).

Bastien, J. (1987). *Healers of the Andes*. Salt Lake: University of Utah.

Belote, L. (1978). *Prejudice and pride: Indian-White relations in Saraguro, Ecuador* (Ph.D. Dissertation). Urbana: University of Illinois.

Bolton, R., & Bolton, C. (1978). Concepcion, embarazo y alumbramiento en una aldea Qolla. *Antropologia Andina, 1–2*, 58–74.

Cosminsky, S. (1976). Cross-cultural perspectives on midwifery. In X. Francis, S. Grollig, & H. Haley (Eds.), *Medical anthropology* (pp. 229–248). The Hague: Mouton.

Dover, R., Seibold, K., & McDowell, J. (Eds.). (1992). *Andean cosmologies through time*. Bloomington: Indiana University.

Finerman, R. (1982). Pregnancy and childbirth in Saraguro: Implications for health care change. *Medical Anthropology, 6*, 269–277.

Finerman, R. (1984). A matter of life and death: Health care change in an Andean community. *Social Science & Medicine, 18*, 329–334.

Finerman, R. (1987). Inside-out: Women's world view and family health in an Ecuadorian Indian community. *Social Science & Medicine, 25*, 1157–1162.

Finerman, R. (1988). The price of power: Gender roles and stress-induced depression in Andean Ecuador. In P. Whelehan (Ed.), *The anthropology of women in health* (pp. 153–169). Granby, MA: Bergin & Garvey.

Finerman, R. (1989). Tracing home-based health care change in an Andean Indian community. *Medical Anthropology Quarterly, 3*, 162–174.

Finerman, R. (1995). Parental incompetence and selective neglect: Blaming the victim in child survival. *Social Science & Medicine, 40*, 5–13.

Finerman, R. (1998). Saraguro: Medical choices, medical changes. In M. Ember, C. Ember, & D. Levinson (Eds.), *Portraits of culture: Ethnographic originals* (pp. 59–83). Upper Saddle River, NJ: Prentice-Hall.

Foster, G. (1998). How to stay well in Tzintzuntzan. Reprinted in M. Whiteford & S. Whiteford (Eds.), *Crossing currents: Continuity and change in Latin America* (pp. 290–304). Upper Saddle River, NJ: Prentice Hall.

Gade, D. (1999). *Nature and culture in the Andes*. Madison: University of Wisconsin.

Harrison, R. (1989). *Signs, songs and memory in the Andes: Translating Quechua language and culture*. Austin: University of Texas.

Kleinman, A. (1980). *Patients and healers in the context of culture*. Berkeley: University of California.

Kroeger, A., & Barbira-Freedmann, F. (1992). *La lucha por la salud en el Alto amazonas y en los Andes*. Quito: Abya-Yala.

Lillo, B. (1942). El angelito. In *Relatos Populares* (pp. 219–234). Santiago, Chile: Nascimento.

Macas, F. (1995). Los Saraguros. In J. Vinneza (Ed.), *Identidades Indias en el Ecuador contemporaneo* (pp. 339–369). Quito: Abya-Yala.

MacCormack, S. (1991). *Religion in the Andes*. Princeton: Princeton University.

McKee, L., & Argüello, S. (Eds.). (1988). *Nuevas investigaciones antropologicas Ecuatorianas*. Quito: Abya-Yala.

Mitchell, W. (1991). *Peasants on the edge: Crop, cult and crisis in the Andes*. Austin: University of Texas.

Naranjo, P., & Escaleras, R. (Eds.). (1995). *La medicina tradicional en el Ecuador*. Quito: Universidad Andina.

Pan American Health Organization. (2000). *Health statistics from the Americas*. Washington, DC: PAHO.

Tousignant, M. (1988). La teoría quechua de las emociones: Un ejemplo de la Provincia de Bolívar, Ecuador. In L. McKee & S. Argüello

(Eds.), *Nuevas investigaciones antropologicas Ecuatorianas* (pp. 189–198). Quito: Abya-Yala.
UNICEF. (1995). *The State of the World's Children 1995*. Oxford: Oxford University Press.
Wayland, C. (2001). Gendering local knowledge: Medicinal plant use and primary health care in the Amazon. *Medical Anthropology Quarterly, 15*, 171–188.

Weismantel, M. (1988). *Food, gender and poverty in the Ecuadorian Andes*. Philadelphia: University of Pennsylvania.
Wilson, D. (1999). *Indigenous South Americans of the past and present*. Boulder, CO: Westview.

Shipibo

Warren M. Hern

ALTERNATIVE NAMES

Shipibo, Conibo, Chama.

LOCATION AND LINGUISTIC AFFILIATION

The Shipibo are located in the upper Peruvian Amazon. Their communities are found principally along the banks of the Ucayali River and its tributaries from Atalaya to Requena or on lakes off the main course of the river. The coordinates of the area are approximately 6° to 9° South and 75° West. Upstream on the Ucayali from the town of Pucallpa, the Shipibo are known locally as the Conibo, although the language spoken and the culture shared by the Shipibo/Conibo are essentially the same. The entire group will be referred to here as "Shipibo."

Shipibo is one of the Panoan languages of the Amazon. Panoan speakers tend to be concentrated around the upper Ucayali drainage basin in eastern Peru and western Brazil.

The word "Shipibo" is not generally used by the Shipibo to describe themselves. They tend to use the word "jonibo" (person [pl.]) in reference to themselves and the word "nahuabo" (foreigner/stranger/not-person [pl.]) in reference to others who are not Shipibo. The term "Shipibo" has been used by members of other tribes to refer to the group under discussion because of their custom of capturing and maintaining as pets specimens of the pygmy marmoset (*Cebuella pygmaea*), the smallest known primate, which is indigenous to the eastern Peruvian Amazon. In Panoan languages, this animal is called a "shipi." The suffix "bo" indicates plural.

The term, "Chama," is considered by the Shipibo to be a highly derogatory term that is used by the Peruvian mestizo population to portray the Shipibo as a subhuman underclass. The term is found in some early anthropological literature.

OVERVIEW OF THE CULTURE

The Shipibo are thought to have occupied the upper Peruvian Amazon area for about 1,000 years. They have principally and traditionally inhabited small settlements on the banks of oxbow lakes ("cochas") and small tributaries of the Ucayali, although there have been both riverine and interfluvial groups. The latter groups, living away from large rivers, maintained a cultural ecology adapted more to the forest with wild game as protein sources. The predominant riverine groups, on the other hand, depend on fish as a primary but not exclusive protein source.

The Shipibo have long had a highly developed ceramic tradition, and their contemporary pottery is internationally known for its beauty and craftsmanship. Pots are used for cooking, eating, and ceremonial purposes, and sometimes for decoration.

The Shipibo women are also accomplished weavers, using a native cultivated cotton for making yarn and thread. They use a backstrap loom for large bolts of cloth that may be 10–15 m long and require the weaver to secure the distal end of the longitudinal threads to a tree. Small exquisitely decorative pieces such as bracelets and anklets are woven in the same manner with tiny looms.

The geometric and repeating patterns used by the Shipibo are distinctive and are painted or engraved on pottery, cloth, faces, oars and clubs, and anything that might retain the figures.

Until the last few decades, the Shipibo economy has largely been one of subsistence supported by fishing, hunting, gathering, cultivation of high carbohydrate plants such as yucca (at least two varieties of manioc, *Manihot esculenta* sp.), plantain, and a large purple sweet potato (Bergman, 1980). More recently, rice and corn have been sown and cultivated for the market as well as for consumption. Chiclayo (black-eyed pea) is sown and cultivated on the exposed river beaches in the dry season.

Traditional Shipibo family patterns tend to be matrilocal and matrilineal, although this appears to alternate from generation to generation (Abelove, 1978; Eakin, Lauriault, & Boonstra, 1980; Hern, 1992b). Sororal polygyny was widely practiced in the past but is now much less common. Levirate and sororate are practiced: the brother of a man who has died accepts his deceased brother's wife as a second wife. She is often the sister of her new husband's first wife.

Cross-cousin marriages were the preferred marital arrangements in the past, particularly in polygynous families.

A typical Shipibo village consists of a single matrilocal extended family containing five or six nuclear families and representing as many as four generations. A specific example is a study community of 48 persons in which the oldest couple, parents of several daughters, have a household that includes several grandchildren in addition to the original couple. The most prominent man in the community is the husband of three of the couple's daughters, each of whom has her own hearth and household where her children reside. As a woman's daughters come of age, the woman's household may be joined by a young man who is a suitor or husband to one or more of her daughters. A son of hers is likely to leave the community to find a wife elsewhere.

Larger Shipibo communities appear to be the result principally of missionary activity in 19th and early 20th century (Myers, 1990). Catholic then Protestant missionaries induced aggregation of various family groups into small communities that were more susceptible to proselytization for religious purposes.

Missionaries describe some encounters with the Shipibo as extremely dangerous (Samanez Y Ocampo, 1980). This was particularly true for more isolated groups such as the Pisquibo: Shipibo who lived along the banks of the Pisqui river, a tributary of the Ucayali. Shipibo are depicted in various accounts as ruthless in the treatment of their enemies, especially the Cacataibo, or Cashibo as they are called by the Shipibo. In the early 20th century, Europeans and Peruvian nationals seeking to exploit the rubber and timber resources contracted with the Shipibo to drive the Cacataibo out of the areas of economic interest. The Cacataibo were driven deep into the forest and the Shipibo are now afraid to penetrate into the territory of the allegedly ferocious Cacataibo. Also during the rubber boom and during the late 1930s, some Shipibo were taken as slaves from the Ucayali region to other parts of the Amazon.

The traditional leader (*curaca*) of a Shipibo community is the male head of a large extended family, with an informal but permanent status until death, old age, or disability requires him to accede to a younger man. Currently, the Shipibo choose a village chief on a rotating basis and elect officials to conform with the national governmental structure. These leaders generally consist of an chief "jefe de la comunidad," or *curaca* (traditional), a *teniente gobernador*, official representative to the district government, and *agente municipal*. These are not traditional Shipibo designations, but the community process by which leaders are currently chosen resonates with traditional Shipibo methods of dealing with issues that affect the community. Community assemblies are attended by both men and women, and while male leadership predominates, women express their opinions vigorously and often prevail.

Traditional Shipibo religious views and cosmology are animistic. Spirits reside in various living things, specifically certain trees and animals, and in the stars. Fresh water dolphins are of particular interest and are not killed because of their intelligence and capacity to inflict harm on humans. The best published description of Shipibo cosmology is *The Cosmic Zygote* by Peter Roe (Roe, 1982).

THE CONTEXT OF HEALTH: ENVIRONMENTAL, ECONOMIC, SOCIAL, AND POLITICAL FACTORS

The Shipibo have survived culturally and demographically through hundreds of years of European colonization and missionary activity at the same time that other tribes such as the Cocama have lost their identities and languages within the Shipibo culture area. The Shipibo have apparently absorbed other groups such as the Setebo. The population numbers of the Shipibo during precolonial times is not known, but there is evidence that they, like all indigenous Amazonians, experienced catastrophic population losses following European contact due to the introduction of exogenous diseases, armed conflict with European settlers, slavery, and intertribal warfare.

In the mid-1960s, there were approximately 100 Shipibo settlements from Atalaya to Requena, including those found on Ucayali tributaries and interior lakes, comprising a total population of approximately 15,000. In the mid-1980s, there were 125 identifiable Shipibo settlements. The present Shipibo population is estimated to be about 40,000–45,000 in 150 or more settlements, and many Shipibo have moved permanently to larger Mestizo towns such as Pucallpa, which was originally a central Shipibo settlement.

In a baseline health study of the Shipibo village of Paococha in 1969, Hern (1971, 1977) found a population growth rate of 4.9% per year in a carefully defined population of 538. The annual birth rate was 69.3 per thousand, and the death rate was 20.4 per thousand. The difference ($69.3/1000 - 20.4/1000$ or $0.0693 - 0.0204$) yields the excess of births over deaths ($48.9/1000$ population $= 0.0489$) or 4.89% per year. This extremely high population growth rate means that the population doubles approximately every 14.3 years.

In this group, the Total Fertility Rate, which is the sum of age-specific birth rates, was 9.935, which means that the average woman had an average of 10 live births during her reproductive years. This and other measures of fertility showed the Shipibo to have the highest fertility ever recorded in a human group.

It is immediately apparent that this rate of population growth is recent since it would only have taken 90 years for a village the size of Paococha to have supplied the entire contemporary (15,000) Shipibo population, whereas this was only one village. In another perspective, the population growth rate of the Amazon population over the past 10,000 years or so is estimated at approximately 0.1% up until the time of European contact, and Myers estimates that there may have been as many as 10 million indigenous residents of the entire Amazon at the time of European contact (Myers, 1988).

This means that, following a massive population crash during the 16th–19th and early 20th century time span, the Shipibo experienced a rapid population recovery with population growth rates that exceeded pre-contact rates. In fact, it appears that the most rapid population growth of the Shipibo population occurred in the years immediately following World War II.

At the same time, there has been an aggressive immigration of other Peruvians into the upper Peruvian Amazon. The town of Pucallpa, which was principally a Shipibo settlement in the mid- to late 19th century, contained a Peruvian Mestizo–Criollo population of about 3,500 in 1944. The Trans-Andean "highway" reached Pucallpa at about that time, and immigration from the Andes and the Peruvian coastal cities began, as did increased commercial activities in logging, petroleum exploration, fishing, cattle ranching, and agriculture. The Shipibo were increasingly exposed to sources of rapid cultural change, and they also found themselves competing increasingly with other groups and immigrant populations for the same resources.

Rapid cultural change in the region was enhanced by the establishment in the late 1940s of the Summer Institute of Linguistics, an evangelical Christian group dedicated to translating the Bible into native languages. Their base was built on the shore of Lake Yarinacocha and included a landing strip for the use of the missionary planes as well as establishing a fleet of float planes capable of landing on the waterways and lakes. The missionaries also provided excellent medical care, both preventive and therapeutic, to all indigenous groups with whom they had contact.

Another important influence on both cultural change and the health of the Shipibo was the establishment in 1960 of the Hospital Amazonico (Albert Schweitzer) by Dr. Theodor Binder, a German physician who was dedicated to helping the indigenous people of the Peruvian Amazon. The hospital, located on a high bank overlooking Yarinacocha and near the village of Puerto Callao, was several kilometers upstream from the Summer Institute of Linguistics. This area was separated from Pucallpa by approximately five kilometers of canopy rain forest and an overgrown cacao plantation.

Dr. Binder found the Shipibo in the 1950s to be suffering from a wide variety of infectious and parasitic diseases, with tuberculosis as a major epidemic which was killing many adults. He established the hospital to treat the Shipibo and other indigenous people for free. He also employed Shipibo men in the agriculture and animal husbandry projects that provided the hospital staff and patients with food.

Extensive and intensive contact between the Shipibo and Dr. Binder, along with the rest of his hospital staff and support operation, has been a major source of cultural change for the Shipibo as well as major contribution to their improved health during that time. Shipibo families came from outlying villages to reside at the campsite near the hospital while family members received prolonged treatment for diseases such as tuberculosis and leishmaniasis. More recently, they lived in modest housing constructed by the hospital administration. This contact resulted in exposure to health education but also to European customs and a Spanish-language environment. These families then have taken some of their adopted customs, material culture, and language influences back to the home villages.

Yet another source of cultural change at the village level was the introduction of Western-style schools. These were primarily elementary schools sponsored by missionary groups such as the Seventh-Day Adventists or bilingual schools established by the Summer Institute of Linguistics. Some educational materials included reading and arithmetic, but the curriculum had a heavy emphasis on religious indoctrination and marching around the village plaza or soccer field in a goose-step military fashion. Later, these schools were replaced in all villages by government-sponsored bilingual schools with education levels through high school in some villages. The goose-step, which seems antithetical to the languid cultural ethos of the Shipibo, continues to be the prescribed mode of marching.

The health of the Shipibo has waxed and waned during the past 50 years depending on the availability of vaccination programs, local epidemics, the sporadic availability of altruistic young doctors and nurses, and climatic conditions. A smallpox epidemic swept through the upper Amazon in 1964, killing thousands of indigenous people in hundreds of villages of various tribes. Some local vaccination efforts carried out by missionaries and individuals such as myself protected certain groups from decimation. Both before and after 1964, epidemics of various other contagious diseases such as chickenpox, measles, whooping cough, and polio affected people throughout the region. An epidemic of measles in the upper Pisqui River in the early 1970s killed dozens of Shipibo children before a young German physician, Andreas Kaper, went into the remotest Pisqui villages with vaccine.

Prior to large recent population increases throughout the upper Peruvian Amazon, the Shipibo lived in the presence of spectacularly abundant food sources (Bergman, 1980). A few hours' fishing resulted in more than enough for a large family. Hunting wild game on high ground during the seasonal flood season often resulted in kills of deer, wild boar, tapir, monkeys, large birds, large rodents, and land turtles that provided excellent sources of protein. Gathered and cultivated fruit and vegetables resulted in a highly varied diet rich in vitamins and fiber. General levels of nutrition were excellent.

As regional population levels have increased, the Ucayali fishery has been subjected to great pressure. The town of Pucallpa, which had a population of approximately 3,500 in 1944, was estimated in 2002 to exceed 300,000 in its metropolitan population. The Ucayali is no longer merely the richest freshwater fishery in the world supplying small amounts of fish protein to a few Shipibo settlements living in a subsistence economy. Commercial fishing enterprises use large refrigerator ships with huge drift nets to capture all fish and other organisms 200 km downstream from Pucallpa. The catch not consumed by Pucallpa's growing population is shipped by land to towns on the flanks and high plains of the Andes.

The consequences of the decline in the Ucayali fishery for the Shipibo as well as the complexity of the ecosystem have been dramatic. Whereas a small group of men could leave the village before dawn and return by noon with several canoes full of fish, enough to feed a large extended family in several households for days, a pair of brothers may fish all day now to return at dusk with only a basin full of small fish—enough to feed a nuclear family for one day. The crash in the fish population has been most dramatic during the "mijano"—the upstream fish migration during the dry season months of July and August. In the past, abundant fish caught during this time could be salted, sun-dried, and preserved for weeks or months.

The decline in the fishery has been accompanied by other important changes in the ecology of the Ucayali. Extensive deforestation has resulted in severe changes in the forest climate and river flows. Slash-and-burn agriculture no longer affects small, isolated plots. Large tracts are purposely burned for ranching or cultivation, and large tracts are burned unintentionally and uncontrollably when the forest is exceptionally dry. The desiccation of the forest has accompanied dramatic drops in dry-season rainfall and river levels as local daytime temperatures increase with deforestation. Heavy rains in the wet season are no longer held in the canopy rainforest and its groundcover. Rapid runoff, erosion, and flooding produce extensive and prolonged inundation downstream that destroys all cultivated crops, domestic animals kept for food, and some forests not adapted to flooding. Wild game is driven deep into the interfluvial forest areas. Fish are hard to find. Nutrition suffers from lack of protein, carbohydrates, and fresh vegetables and fruit that are sources of vitamins. Poisonous spiders, snakes, and scorpions join the household. It is hard to find firewood. Hygiene is next to impossible, and there is no place to bury the dead.

In the dry season, lower water levels mean stagnant, contaminated water in ponds near villages, a higher incidence of gastrointestinal diseases, especially in children, and widespread respiratory diseases due to severe air pollution. Dust storms from the exposed beaches are more common, and this combines with severe smoke density from burning season slash in large plots and burning sawdust at sawmills to produce air that is often unfit to breathe.

The exact impact of these large system changes on overall Shipibo health is difficult to determine with certainty, but it appears that nutritional levels have decreased as a result of intermittent deficiencies in both protein and vitamins. Increasing dependence on a cash economy results in the sale of certain high food value items and produce for money used to purchase lower food value items such as sugar, cooking oil, and polished rice. The physical development of children is now less dependent on traditional activities such as hunting game, paddling canoes, carrying game and produce, and wrestling. Exercise is more dependent on school sports and programmed activities. Soccer is intensely popular and provides excellent cardiovascular exercise but not strength conditioning.

A study of Shipibo infants and children by Hodge and Dufour (1991) showed a tendency in 1984 to lower than normal development and diminished growth rates, especially after the age of 9 months. The authors suggested that mild to moderate protein deficiency malnutrition might be responsible for these growth disturbances. No time-comparison studies of Shipibo child development have been published, but recent unpublished field studies suggest that child development is showing no improvement and may be diminishing.

A study comparing the relative parasite burdens and hemoglobin levels in two radically different Shipibo environments found that people living in the upper reaches of the Pisqui River have much higher burdens of hookworm (*Ancylostoma* sp.), among other things, and lower hemoglobin levels, than Shipibo living on the banks of the Ucayali River (Hern, 1995, unpublished manuscript). In the upper Pisqui, the soil is composed primarily of poorly drained dense, sticky clay, whereas the soil along the Ucayali is principally river sand that is well-drained and quite dry most of the time. Hookworm larvae that penetrate bare feet survive much better in the poorly drained clay soil environment, where it is also somewhat more difficult to dig latrines.

MEDICAL PRACTITIONERS

Several kinds of traditional medical practitioners are recognized among the Shipibo, although the highest and most authentic status is accorded the *muraia*, or "seer." The *muraia* is usually a man who, in addition to his normal activities of fishing, hunting, and tending his chacra, practices the traditional Shipibo healing arts. A principal component of the *muraia*'s method of combating illness is *oni*, otherwise known widely as *aya huasca*, or "dead man's vine." *Oni* is the Shipibo preparation of *banisteria caapi*, a hallucinogenic drug known and used in different forms throughout the Amazon. It is derived from a certain vine that is harvested, cut up into small pieces, and boiled for 8–10 h in order to decoct the alkaloid that is the active hallucinogen. The practitioner takes large gulps of this liquid while entering a healing session (*jonibensuate*) with one or more patients. The songs he sings under the influence of the *oni* help him go into the underworld to fight off the *yushin*, or evil spirit trying to kill both the patient and the *muraia*, and help him summon the *ani yushin*, or great spirit. The *ani yushin* tells the *muraia* which songs ("*huihua*") to sing to heal the patient and where to look for the right herbal remedies.

In addition to singing the healing songs ("*jonibensuatehuihua*"), the *muraia* talks to the patient, sounding very much like a Western psychotherapist, blows smoke over the patient's body, and rustles bundles of dried plants over the affected area. These sessions go on for hours.

Another medical practitioner may be a "curiosa," (Spanish term), or birth attendant who is not trained formally as a midwife. Another Spanish term for such a person is "parteira," or someone who attends "partos" (births). The Shipibo do not seem to have a word of their own for this activity or defined role except as someone who helps with "*bacque picotash*" (birth). This can be just about any adult woman with birth experience herself or who has attended births in the past. Each woman helps her daughters and granddaughters, in particular, and a woman with many daughters and granddaughters may develop an earned reputation as someone with great experience in attending births.

Other practitioners may be informally recognized as men who have special knowledge or experience at setting bones. But orthopedic manipulation and setting fractured bones happens in broad daylight surrounded by the curious as distinguished from the true healing session conducted by the *muraia*, which is sparsely attended and happens very late at night into the early morning hours.

To some extent, all Shipibo are medical practitioners because, at least in the old days, such knowledge of herbal and folk remedies was available to everyone (Arevalo, 1994; Foller, 1990). The exception is the *muraia*, who has both special powers that are earned and ascribed.

For his part, the *muraia* becomes an expert practitioner by fasting, taking the *oni* over long periods of time, learning the sacred songs and when to sing them, and obtaining and controlling the pure vision possible only under the influence of *oni* that is precisely prepared.

In recent decades, a new kind of modern practitioner has appeared, the *sanitario*, or medical corpsman. Some *sanitarios* are well-trained Shipibo men who have relatively good diagnostic and therapeutic skills. Some non-Shipibo *sanitarios* sent by the Peruvian government have variable skills, but none of the *sanitarios* have had much to work with in the way of medical supplies, vaccines, or instruments.

The *sanitarios* sometimes play critical roles in the prevention and management of epidemic disease as in the case of the cholera epidemics of the 1990s and in vaccination programs. But missteps occur as new technologies are applied. In one instance, a poorly trained Shipibo *sanitario* decided to perform a needle aspiration of an abscess that he perceived in the umbilicus of an elderly woman, who had a little occasional discomfort in the bulging spot. She subsequently developed diffuse abdominal tenderness, which was diagnosed as peritonitis by a Western medical practitioner. The umbilical bulge was a herniation of a loop of small bowel that had been punctured by the *sanitario*'s needle.

CLASSIFICATION OF ILLNESS, THEORIES OF ILLNESS, AND TREATMENT OF ILLNESS

The Shipibo are practical people who live in a complex, dynamic, and potentially hazardous environment in which injuries and death can and often do occur quickly. Traumatic injuries such as snakebite, puncture wounds, fractures, burns, lacerations, sting ray wounds, bites by carnivorous fish and alligators, insect stings, accidental gunshot, and allergic contact with known poisonous plants are obvious sources of suffering and death for the Shipibo. They have innumerable herbal remedies (*rao*) for these injuries, and some of them work amazingly well. The Shipibo are generally, it seems, incredibly resistant to a wide variety of pathogens that would fell the average person of European descent almost instantly.

The Shipibo have a certain working knowledge of what Europeans would call comparative anatomy and physiology since they dissect animals of all kinds in the process of food preparation. They know, for example, that there are embryos and fetuses in the uteri of various female animals that they kill such as tapirs, wild boars, and monkeys and that these features are not found in male animals.

On the other hand, internal illnesses and conditions without obvious origins such as fevers, paralysis, birth defects, and difficult deliveries are traditionally ascribed to conditions in the spirit world. In particular, the condition of *cupia* is widely attributed as a cause for various ailments, and the term for experiencing a variety of *cupia* is *cutipado*. The first term may be an original Shipibo word, but the latter is the grammatically Spanish past participle of the verb, *cutipar*, which may be a Spanish adaptation of a Shipibo concept. It is a widely used expression throughout the Peruvian Amazon. In any case, *cutipado* roughly corresponds to "bewitched" in English, and *cupia* is a kind of instant explanation for an otherwise inscrutable condition or illness experience. A child who is albino or who is exceptionally light-skinned, but normal in all other respects, may be said to be *cutipado*. This characterization is also a useful deflection of potentially inconvenient questions about why the individual looks so different from his or her siblings or cousins.

If, for example, a woman is pregnant, there are certain foods she may or may not eat and certain acts she may not commit at certain times such as sexual intercourse. If she violates these restrictions, she may experience *cupia* or said to be *cutipado*, (Alvarez, 1990) a part of which is the process of being visited by and having sexual intercourse with a river porpoise or dolphin during the night. This state becomes the obvious post-hoc explanation for a difficult delivery, a stillbirth, delivery of a deformed child, death of the woman during childbirth, hemorrhage following childbirth, lack of breast milk, etc.

If a man kills a cayman (*Caiman sclerops*, variety of alligator found on the Ucayali), dolphin, or certain other animals, or if he cuts down a tree such as a lupuna that is sacred, he also may be *cutipado*, which is the obvious explanation for his prolonged inexplicable illness in which he experiences severe abdominal pain, headache, malaise, convulsions, or jaundice and dies an agonizing death. It could also be that he simply offended someone, who arranged for him to be *cutipado*.

A young woman who is *cutipado* may experience a dissociative, hysterical episode during which she is obviously out of control, having seizures, screaming, shaking, clenching her teeth, sweating, moaning, and beyond any social interaction. These episodes, which are uncommon, may go on for long periods of time up to an hour. In such a case, she has clearly been *cutipado* by an enemy, perhaps a social or sexual rival, with the help of someone who practices *brujeria* (Spanish for witchcraft). The person who practices black magic (*magica negra*) casts a spell (*cupia*) on the person who then experiences the dissociative state, which does not in any way resemble what is known in Western medicine as an epileptic seizure. I have observed these behaviors (with similar explanations) in other indigenous societies in Latin America.

SEXUALITY AND REPRODUCTION

The Shipibo like sex and are sometimes quite open about it, particularly in joking relationships. They love obscene and intimate sexual humor, especially when there is a cover of darkness to obscure the speaker (even though everyone knows from the voice who is speaking). Whether it is under the influence of Christian missionaries or reflects traditional Shipibo custom, sexuality is not flaunted or openly recognized during the day and in the midst of communal activities. Young people court and disappear into the bush. Although there is no formal marriage ceremony, young couples who are recognized as "newlyweds" are indulged as they spend long hours under the mosquito net together, even during the day. The young man is then likely to be ribbed mercilessly by his age-mates about his consequent alleged weakness and incapacity for any useful work. Most sexual activity, however, seems to occur in a more furtive fashion as husbands and wives bathe together in the river at dusk and meet secretively in a remote section of the *chacra*, or garden, distant from the village. It is customary for whole families to sleep under one mosquito net. Sexual activity between spouses occurs during the night under the family mosquito net when the children are asleep.

Shipibo women have a wide variety of herbal remedies that are thought to control reproduction. The general category of these remedies is *to-otirao* (pregnancy = *tooti*, and medicine = *rao* or "rau"). A kind of *tootirao* is taken in order to become pregnant.

Remedies to prevent pregnancy are more common. The most commonly known *tootimarao* (*ma* = negative) is *tootimahuaste*. *Tootimahuaste* is a grass-like plant (probably a sedge) that grows on the shores of a lake (*huaste* = herb). It is pounded and the juice squeezed into a cup of hot water. This tea is taken the first three days of two successive menstrual periods. This is alleged to result in permanent sterility.

Tootirao works by making the *baquenanuti* (uterus) moist, lush, and receptive to the seed of the man. *Tootimarao* works in the opposite way: by making the inside of the *bacquenanuti* hard, dry, and unreceptive to the male's seed.

Failing to observe the prescribed pregnant woman's diet or behavior restrictions may result in *cupia* and a complicated pregnancy in which the woman dies, suffers a stillbirth, or gives birth to a deformed baby.

In spite of the widespread knowledge and use of herbal contraceptives (Hern, 1976), fertility has been exceptionally high among Shipibo women. A possible explanation for this paradox is that the prescribed dietary/behavior regimen that accompanies the use of herbal contraceptives includes sexual abstinence. In epidemiology, a lack of pregnancy under these circumstances of herbal contraceptive use would be called a "secondary non-causal association."

Women living in monogamous, as distinguished from polygynous relationships, find it hard to observe a period of sexual abstinence. A later study of the relationships between polygyny and fertility showed that polygynous women had longer birth intervals and fewer pregnancies. Further, the measured community fertility rate had a straight-line negative relationship to the level of polygyny practiced in a particular village. The less polygyny, the higher the community fertility rate, and vice versa (Hern, 1988, 1990, 1992a, 1992b).

In a traditional family, sororal polygyny (in which all co-wives are sisters) is the preferred and prevailing family structure. In this setting, women are able to observe postpartum sexual abstinence for longer periods of time than women who are in monogamous unions. Births are fewer with more time between them. This has a positive effect on the health of both mothers and children.

The change from polygyny to monogamy began with the first contact with Christian missionaries. Polygyny is still severely criticized by Christian missionaries, and in one well-known instance, a disapproving official Peruvian census taker would not count members of polygynous families. The extremely high fertility and high rates of population growth that have been found in Shipibo communities are at least partly due to cultural change with a disruption of patterns that dampened fertility.

Although the Shipibo treat children with a great deal of gentle affection, they do not express any desire for more than two or three children. The rare woman who is infertile or subfecund is regarded as unfortunate, but not tragically so. Such women and their partners readily adopt children from other households, and the children have two homes and families, almost always harmonious.

The Shipibo express vague concern and unease about population growth as they see it directly affecting their daily lives. There are a bewildering number of children needing school classrooms and teachers, more each year, and there are increasingly scarce resources. But they do not see how they can influence events, especially the growth and intrusion of large population groups from outside the village. *Estaremos aplastado por los nahuabo*—"We will be squashed (and wiped out) by the (non-Shipibo)."

HEALTH THROUGH THE LIFE CYCLE

Pregnancy and Birth

Young Shipibo women learn about pregnancy and birth from close observation of their mothers, sisters, aunts, and cousins. Traditionally, the young woman has her first menarche at the age of 12 or 13. She has her first sexual experience and perhaps a permanent partner at 13 or 14, and has her first baby by age 15.

One type of herbal contraceptive is called *navashuaste*, which is taken by young women in order to postpone pregnancy instead of causing permanent sterility, the effect sought by taking *tootimahuaste*. There is no evidence that the use of *navashuaste* for this purpose is successful.

When a woman is pregnant, she must observe certain dietary laws and taboos that restrict activity and foods. She must not be subjected to a frightening experience such as encountering a snake or other wild animal.

Spontaneous abortions occur, but they are uncommon and accepted as a matter of fact. The term for a spontaneous abortion is *bacquencahuana* (literally "baby falls out"). Twins are rare but do occur. Induced abortion is almost unknown.

Birth occurs in the woman's home, and she is usually attended by her mother and/or close female relatives in the same age range as the woman's mother. A young woman may be surrounded by all the women neighbors in the case of a difficult delivery. Freely offered folk advice from this gathering of interested spectators is accompanied during labor contractions by a frantic chorus of "Push! ... Push! ... Push!"

After delivery, the placenta is usually buried under the woman's house.

Infancy

Newborn infants are not given a name until one or more months of age. The Shipibo name (*janecon*, "true name") is given first, although a Spanish name with paternal and maternal surnames may be given for birth registry. The reason for the delay in naming is the recognition of a high probability of neonatal death.

Newborn infants are traditionally painted with *huito*, a black vegetable dye (*Genipa* sp.), with their faces painted in a typical geometric Shipibo design. The explanation for the face painting is that the design wards off the *yushin*, or evil spirit that causes illness and death. It also looks nice. There is no distinction between painting a thing or person with designs to protect it from evil spirits or just for fun.

Shipibo children are engulfed in love from their first moments and are highly indulged by both parents. Children are breast-fed for at least 6 months, and complete weaning may not occur for several years.

Infants are carried on the hip in a shawl tied around the mother's neck as she goes about her various chores. Very young infants sleep during the day in a covered hammock that is watched by a grandparent, sibling, or other relative who may be sitting nearby performing such tasks as weaving and potting.

In the past, a multiple birth resulted in one twin being killed by suffocation, as was the case of a severely deformed infant. This practice appears to have been abandoned, especially for twins.

All children receive close supervision and attention from a variety of adults, but particularly the child's mother and the mother's immediate relatives. This is true for both healthy and sick infants. The child of a woman who has deficient or no breast milk is nursed by a female relative who is lactating.

When infants are weaned, they are given increasing amounts of stewed ripe mashed plantains, a preparation that has the consistency of liquid oatmeal. Unfortunately, the mother may add raw river water that is heavily contaminated with human feces to this otherwise healthful preparation. As an obvious result, between 10–40% of children in some villages die of gastrointestinal ailments before they reach one year of age.

Childhood

Young children up to the age of 10–12 are given great freedom to play with their peers and to explore the area around the house and village. They frequently accompany their parents or other adults on excursions such as going to the garden to cultivate or gather food, or, in the case of boys, to hunt and fish with their fathers.

Shipibo children begin learning the complex family relationships including kinship terms and avoidance patterns at a very early age (Abelove, 1978). A Shipibo child's social success is highly correlated with its mother's pattern of interactions with others.

Child abandonment is almost unknown. It is inconceivable to the Shipibo. "Only mestizos abandon their children." Discipline is gentle and consists principally of quiet talking and persuasion. Once in a while, a parent will spank a young child. A parent may gently scold or, rarely, have an angry exchange with an ill-behaved older child, but it does not appear that the physical abuse of children occurs among the Shipibo.

Sick children are given round-the-clock nursing care, and those with chronic illnesses are given special support, such as having someone specifically assigned to help that impaired person.

Adolescence

Although traditional patterns are changing rapidly with the introduction of formal schools through the high school level, Shipibo adolescents still assume many adult responsibilities that include subsistence activities and mating. Informal trial marriages occur with an adolescent couple cohabiting in the girl's parents' house, and these are generally regarded as temporary arrangements. If the girl becomes pregnant, however, a more stable relationship is likely to emerge.

Adulthood

In the absence of formal Western schools, Shipibo girls entered permanent sexual relationships at the age of 13 or 14 and immediately began having children. They were considered adults from that point on. Currently, girls are encouraged to attend school at least through grade school and beyond if possible. Sexual encounters occur, but adolescent cohabitation in the traditional patterns occurs much less than it did before the introduction of schools.

Once a nuclear family is established, the young husband builds a house for his own family, typically in the family compound or immediately adjacent to the home of the woman's parents. A separate structure containing the hearth and eating area is usually a few meters from the house. In the case of polygynous marriages, each woman may have her own house and hearth.

People over 40 years of age are considered "old" since life expectancy is under 50 for women and not much past 50 for men.

The Aged

People who are very old—those who reach the age of 70 or 80—are treated with great respect and veneration. An elderly adult woman suffering from tuberculosis, for example, is given her own hut next to the main family house, and young children are assigned to help her in every way. Her food is brought to her and she is accompanied to the lake or river for bathing.

Tuberculosis is a major killer of mature and elderly adults among the Shipibo—those who have the most knowledge of the ancient traditions, songs, and secrets of the forest. An inordinate number of Shipibo women die from cervical cancer.

Although autopsy studies are not possible, it appears that some adults die from the effects of migration of certain ingested parasites to vital organs such as the brain and liver.

As death approaches a very sick person, family members, mostly women, gather around the person who is *isin* (sick) and offer whatever support they can. This is especially true for a very elderly person with many progeny and other relatives. When the person expires, the gathered kin begin wailing a melancholy falsetto falling scale in a minor key spanning about an octave. This pattern is repeated over and over, with the high notes sometimes sustained to express the intense grief. It is a penetrating and unmistakable sound that carries far, especially over large expanses of water, and it communicates the loss to all within hearing range.

The deceased is typically laid out on a table, if a child, or on a mat on the ground under the kitchen roof. Candles are sometimes placed at the head and feet, a custom that may be Catholic in origin. Friends come to pay their respects, sit quietly for awhile, perhaps wailing, then leave, offering a few words of consolation to the survivors.

Adults are placed in an old canoe, which is covered with sawn boards, and the body is buried in the village cemetery. Infants and children are placed either in a child's canoe or in half of an adult canoe. Everyone takes turns throwing dirt onto the casket as the grave is filled.

Funerary jars uncovered by erosion at an ancient settlement that surely predated European contact indicates that deceased individuals were buried in very large pots at that time.

The custom of close contact with those who are ill, including the sharing of drinking and eating vessels, contributed to high Shipibo mortality during a cholera epidemic on the Pisqui and Ucayali in 1990–91. Whole families died within hours of each other.

When an adult man dies, his house is burned down with all his possessions. His widow crops her hair and dresses in black for one year. The possessions of a woman who dies are burned, but not the house in which she lived.

At the end of one year, a funeral wake is held and marks the end of the mourning period.

The Shipibo have a proverb that helps them cope with constant death and loss among their families and friends: *Huinata jahuequi moa shinantima*—"We no longer think about the things that are so sad they make us cry."

REFERENCES

Alvarez, J. A. (1990). *Diccionario de Peruanismos*. Lima: Libreria Studium Ediciones, p. 164.

Arevalo, V. J. (1994). *Medicina Indigena: Las plantas medicinales y su beneficio en la salud Shipibo-Conibo*. Lima: Ediciones AIDESEP.

Abelove, J. M. (1978). *Pre-verbal learning of kinship behavior among Shipibo infants of Eastern Peru*. Doctoral dissertation, City University of New York. Ann Arbor, Michigan: University Microfilms International.

Bergman, R. W. (1980). *Amazon economics: The simplicity of Shipibo Indian wealth*. Doctoral dissertation, Syracuse University. Ann Arbor, Michigan: University Microfilms International.

Eakin, L., Lauriault, E., & Boonstra, H. (1980). *Bosquejo Etnografico de los Shipibo-Conibo Del Ucayali*. Lima: Ignacio Prado Pastor Editorial.

Foller, M. (1990). *Environmental changes and human health: A study of the Shipibo-Conibo in Eastern Peru*. Goteborg: University of Goteborg.

Hern, W. M. (1971). Community health, fertility trends, and ecocultural change in a Peruvian Amazon Indian village, 1964–1969. M.P.H. thesis, Department of Epidemiology, University of North Carolina School of Public Health.

Hern, W. M. (1976). Knowledge and use of herbal contraceptives in a Peruvian Amazon village. *Human Organization, 35,* 9–19.

Hern, W. M. (1977). High fertility in a Peruvian Amazon Indian village. *Human Ecology, 5*(4):355–368.

Hern, W. M. (1988). *Polygyny and fertility among the Shipibo: An epidemiologic test of an ethnographic hypothesis*. Doctoral dissertation, Department of Epidemiology, University of North Carolina. Ann Arbor, Michigan: University Microfilms International.

Hern, W. M. (1990). Individual fertility rate: A new individual fertility measure for small populations. *Social Biology, 37*:102–109.

Hern, W. M. (1992a). Polygyny and fertility among the Shipibo of the Peruvian Amazon. *Population Studies, 46*:53–64.

Hern, W. M. (1992b). Shipibo polygyny and patrilocality. *American Ethnologist, 19*(3):501–522.

Hodge, L. G., & Dufour, D. L. (1991). Cross-sectional growth of young Shipibo Indian children in eastern Peru. *American Journal of Physical Anthropology, 84*(1):35–41.

Myers, T. P. (1988). El efecto de las pestes sobre las poblaciones de la Amazonia alta. *Amazonia Peruana, 8*(15):61–81

Myers, T. P. (1990). *Sarayacu: Ethnohistorical and archeological investigations of a nineteenth-century Franciscan mission in the Peruvian Montana*. Lincoln: University of Nebraska.

Roe, P. G. (1982). *The cosmic zygote: Cosmology in the Amazon basin*. New Brunswick: Rutgers University Press.

Samanez Y Ocampo, J. B. (1980). *Exploracion de los rios Peruanos Apurimac, Eni, Tambo, Ucayali y Urubamba Hecho por Jose B. Samanez Y Ocampo En 1883 y 1884: Diario de la expedicion y anexos*. Lima: Consuelo Samanez Ocampo de Samanez e hijas.

Sotho

Nancy Romero-Daza and David Himmelgreen

ALTERNATIVE NAMES

Basotho, Basuto, Southern-Sotho.

LOCATION AND LINGUISTIC AFFILIATION

The Sotho inhabit the southernmost region of Africa and are found across the international lines of three countries: South Africa, Botswana, and Lesotho. The Sotho constitute about 1% of the general population in Botswana and 16% of the total population in South Africa (Ikuska, nd). In both these countries, the Sotho are scattered around the territory and are not geographically distinguishable from other ethnic groups. On the other hand, there is a considerable concentration of Sotho in Lesotho, where they comprise 99.7% of the national population (Central Intelligence Agency, nd). Because of the close correspondence of ethnic and political boundaries, our discussion of cultural characteristics and their relation to health will be drawn from populations living in Lesotho. The Sotho are native speakers of Sesotho, one of the Bantu languages. Sesotho is closely related to other major languages spoken in southern Africa including Setswana (spoken in Botswana and South Africa), Sepedi (spoken in northern South Africa), and Siloze (spoken in Western Zambia) (Ambrose, 1976).

OVERVIEW OF THE CULTURE

Basotho society is patrilineal and patrilocal and politically organized as a chiefdom. Descent is traced exclusively through the father's lineage, as is the allocation of property rights (Gay, 1982). Upon marriage, women move to the husband's village, where they and their children reside for the rest of the woman's married life, usually in extended-family compounds. The Basotho people trace their ethnic origins back to the Bakoena clan, a group of cattle-owning agriculturalists who inhabited the Witwatersrand area of the Transvaal in present day South Africa as early as the 5th century (Ambrose, 1976). In the 19th century, several Sotho-speaking groups that were fleeing the Zulu chief Shaka came together under the protection and leadership of Moshoshoe, who would become the founder and first king of Lesotho. During the Lifaqane wars between 1820 and 1830, Moshoshoe and his people were forced to retreat to the Thaba–Bosiu mountainous fortress in the Maluti mountains (Schwager, 1986). From there, they continued fending off various attackers such as the Amangwane in 1828, the Batlakoa in 1829, and the Amandebele in 1831. Starting in the mid-1830s and continuing through the mid-1860s conflict over land rights between the Basotho and Boers from the Cape Colony in South Africa resulted in the loss of the most fertile lands and the relegation of the Basotho to the most arid and infertile mountain regions. In March 1868, Basutoland became a British protectorate, and in 1966 it gained its independence from Great Britain.

Although an independent entity, the Kingdom of Lesotho remains economically dependent on its all-encompassing neighbor, South Africa. At present, only about 10% of the land is arable and agriculture contributes about 4% of the Gross National Product (GNP). There are some flourishing industries such as textiles, clothing, and mohair wool, and some natural resources such as water, which is exported to South Africa. Nevertheless, the country's main source of revenue is migrant labor to the South African mines (Romero-Daza & Himmelgreen, 1998). Participation of Basotho men in the South African mining industry dates back to the 1860s when the first diamond mines were opened in the Cape Colony (Gay et al., 1991; Murray, 1981). During the last decade there has been a very significant reduction in the number of Basotho men engaged in migrant labor in South Africa. For example, between 1997 and 1999 the number of Basotho miners was reduced to almost half due to the declining price of gold on the international market (Global Policy Network, 2001).

The present-day Basotho population in Lesotho is predominantly Christian. The main denominations include the Roman Catholic Church, the Lesotho Evangelical Church, the Anglican Church, the Methodist Church, and several independent churches (Ambrose, 1976). However, many traditional beliefs and practices, especially those relating to ancestor worship are still prevalent among the Basotho. The ancestors, those relatives who have died, are considered to be guardians of their living relatives, and to afford protection, good health, and prosperity. However, if they are not honored in the appropriate way, or if they are neglected, they are likely to cause misfortune and illness (Sheddick, 1953).

THE CONTEXT OF HEALTH: ENVIRONMENTAL, ECONOMIC, SOCIAL, AND POLITICAL FACTORS

By the year 2000, the total population of Lesotho was estimated at around 2,033,000 inhabitants, with a population growth of 1.9% between 1999 and 2000 (WHO, 2002). Life expectancy at birth was calculated at 42 years, with no noticeable differences by gender. As of May 2002, infant mortality was estimated at 72 per 1,000 live births, and mortality for children under five at 99 per 1,000 live-births (WHO, 2002). However, accurate figures are very difficult to obtain. In fact, UNICEF reports considerably higher rates for both these indicators, 92 per 1,000 and 133 per 1,000 respectively as of February 2002 (UNICEF, 2002).

Lesotho's mountainous terrain and its location outside the tropics provide a relatively salubrious environment. As a result, many of the tropical diseases such as malaria and schistosomiasis that are commonly found in other parts of Africa are not common in Lesotho (Gish, 1982; Ministry of Planning, 1991). Rather, the main problems that affect the health of the population are directly related to widespread poverty and the lack of basic sanitation. Among these are diarrheal disease, typhoid, and dermatological infections such as bacterial skin sepsis, fungal infections, and scabies, which affect the population at large and especially young children (Ministry of Planning, 1991). Only about 62% of the total population have access to safe water sources, while the rest, especially those in mountainous areas, use unprotected springs and wells which, if contaminated with fecal matter or other waste, could become a major source of infection for many Basotho (WHO, 2002).

The marked seasonal differences found in Lesotho are closely associated with the health of infants and children. For example, the hot humid summer months are the prime time for diarrheal disease. The effects of diarrhea are compounded by the rise of malnutrition that results from the low availability of foods during this time of the year, which corresponds with the pre-harvest period. During the dry and cold winter months, when food is plentiful, the incidence of diarrhea decreases, but respiratory infections rise significantly (Himmelgreen, 1994). At present, it is estimated that about 16% of Basotho children under 5 suffer from acute malnutrition (WHO, 2002). For adults there are also seasonal differences in nutritional status, especially in the highlands and for households that rely mostly on their own agricultural production. For example, Huss-Ashmore and Goodman (1988) found a seasonal weight fluctuation of 4 kg (7%) among women living the Mokhotlong District, in the Northeastern Maluti mountains. In a later study, Himmelgreen and Romero-Daza (1994) also found seasonal differences among women from Mokhotlong but these differences were not as marked as the earlier study. However, women who relied on subsistence agriculture lost more weight seasonally than did women who purchased most of their food. Today, as many developing countries, Lesotho is also faced with increasing rates of overweight and obesity, partly as a result of rapid modernization and the increased consumption of sugars and fat. The traditional diet of the Basotho tends to be highly monotonous and consists mainly of maize meal ("papa") and wild vegetables ("moroho"). Consumption of animal protein is generally limited to that provided by eggs, while the much more expensive sources such as red meat are consumed much less frequently (Himmelgreen, 1994).

An issue of central importance for the overall health of the Basotho is the rapid spread of Sexually Transmitted Infections (STIs), including HIV/AIDS. Between 1983 and 1993, STIs were the second most common outpatient condition seen in Lesotho clinics and hospitals (Family Health International, 2001). It is estimated that about 24,000 Basotho are currently living with HIV, reflecting an infection rate of 24% among adults, one of the highest in the world. Over half of these cases (54%) are found among women. So far, the epidemic has left an estimated 17,000 orphans among Basotho children (Family Health International, 2001). Central among the factors that have contributed to the rapid explosion of the AIDS pandemic is the system of labor migration of men into South Africa

(Romero-Daza, 1994b; Romero-Daza & Himmelgreen, 1998). The widespread poverty that fuels the prostitution industry in South Africa, coupled with lack of preventive and curative services for underlying STIs, results in increasing rates of infection for both women and men. Once infected, miners spread the virus to their wives and other sexual partners when they return to their home countries, and from them to their unborn children.

In Lesotho both traditional and biomedical systems of care co-exist. The country has been divided into 18 Health Service Areas (HASs), each of which includes a hospital, at least one clinic for outpatient care, health centers run by a nurse clinician with the support of village health workers and traditional birth attendants, and village health posts which are visited by medical teams on a monthly basis (Ministry of Planning, 1991). However, the provision of biomedical services to the Basotho is restricted by overall low budgets, limited infrastructure, and shortage of personnel, and is especially deficient in isolated rural areas. According to estimates by the World Health Organization, as of 1995 there were 5.4 physicians, 60.1 nurses, 47 midwives, and 0.5 dentists per 100,000 inhabitants (WHO, 2002). Traditional medicine provides a major source of health care for the Basotho, especially those living in places where biomedical facilities are inadequate. The practice of traditional medicine is regulated and overseen by the Universal Medicinemen and Herbalist Council established in 1978. This body is responsible for the registration and licensing of traditional medicine practitioners, and promotes the development of new skills among practitioners (WHO, 2001). It is not possible to get an exact number of Basotho traditional healers still practicing in Lesotho, since many of them are not officially registered with the Council. As of 1991, there were 9,579 registered healers and an estimated 23,000 unregistered practitioners (Gay & Hall, 1994).

MEDICAL PRACTITIONERS

Basotho traditional healers can be categorized depending on the healing methods they use and the type of training received. Two of the major categories include herbalists and Bible readers (Romero-Daza, 1994a). Herbalists specialize in the treatment of diseases through the use of herbs and roots, while Bible readers usually belong to a specific church and treat their patients almost exclusively with blessed water and prayers. Depending on the severity of the disorder, some Bible readers mix their water with ashes obtained from animal bones or with specific types of soil. Bible readers usually diagnose through the use of biblical passages, by the "laying of the hands," or by praying over and examining water collected by the patient from a river or brook. Herbalists usually diagnose by casting bones, dice, or shells, and interpreting the way in which such objects fall on the ground. Regardless of healing or diagnostic differences, the call to become a healer follows a similar pattern. Usually the person becomes very sick (physically and/or mentally), and is said to have "moea" (literally wind, spirit), which is an indisputable sign of his/her call to be a traditional healer. The disease period is characterized by "thoasa," recurrent dreams in which the person's ancestors instruct the individual on their wishes and guide him/her to look for a tutor to teach the apprentice about traditional medicine (Romero-Daza, 1994a). The role of the traditional healers often extends beyond the treatment of physical and mental disorders, and covers the solution of interpersonal difficulties that may affect the harmony among family or village members. This role fits well with the overall Sotho conceptualization of "health" as a balance among physical, psychological, social, and spiritual aspects of life.

CLASSIFICATION OF ILLNESS, THEORIES OF ILLNESS, AND TREATMENT OF ILLNESS

While the biomedical paradigm is readily accepted by most of the Basotho, traditional beliefs regarding the cause of illness and disease are still commonly found among this population. Prominent among these is the belief that diseases are caused by the breech of taboo. For example, a person who has sexual relations while still mourning the death of his/her spouse may be afflicted by "mashoa" or "mahae" which roughly correspond to genital warts. According to traditional beliefs, as the body of the deceased decomposes, so do the sexual organs of the surviving spouse (Hall & Malahleha, 1989). Another common cause of disease is the use of witchcraft. For example, a person can cause another to become sick by placing a special magical substance (usually water containing powerful medicines) on the path to be traveled by his/her enemy (Romero-Daza, 1994a). Finally, disease can be caused when ancestors have been neglected by their living relatives, or when spirits such as the "thokolosi" enter the body. One of the main results of witchcraft or of spirit possession is the manifestation of mental health problems that can only be cured with traditional therapies. However, mental problems are not always considered negative. Quite often, a person who presents what would be classified as mental health symptoms (e.g., hearing voices, having hallucinations) is believed to have been chosen by their ancestors to become a traditional healer. In such cases, the ancestors communicate their desires and specific instructions through visions and dreams. Once the person becomes a healer, he/she may continue to communicate with the ancestors, thus continuing to strengthen his/her curative powers.

Regardless of the cause of disease, the placation of the ancestors' wrath is one of the main methods of treatment for the great majority of problems. Traditional Basotho culture prescribes the slaughtering of animals and the performance of specific rituals in honor of the ancestors. Other curative methods include the widespread use of medicinal plants both by traditional healers and by lay people. An important healing practice that might become significant in the spread of HIV is that of "scarification," in which traditional healers make small incisions on the patient's skin, and use their mouth to extract the disease-causing object from the patient's body (Romero-Daza, 1994a, 1994b). Given the great number of patients seen by traditional healers, many of whom may be HIV positive, the use of contaminated blades may provide an effective medium for the transmission of the virus from person to person (Romero-Daza & Himmelgreen, 1998).

SEXUALITY AND REPRODUCTION

Fertility rates among the Basotho tend to be lower that among other African populations, and were estimated at 4.6 in 2000 (WHO, 2002). Possible reasons for such low rates are the constant absence of men engaged in migrant labor in South Africa, the practice of extended breastfeeding, and the existence of postpartum sex taboos. In addition, the high rate of sexually transmitted infections and of maternal malnutrition may also decrease pregnancy rates and increase spontaneous abortions (Ministry of Planning, 1991). The rate of contraceptive use has been estimated at 23% (UNICEF, 2002). Nevertheless, Basotho greatly value children of both sexes and strive to have large families. Children provide a source of support for elderly parents, a source of labor for domestic, agricultural, and pastoralist activities, and a great source of prestige. Having children affords women the status of full womanhood (this is represented linguistically by the honorifics *M'e* (mother) and *Ausi* (sister) used to address women who have children and those who are childless, respectively). Moreover, infertility was considered a just cause for divorce. Traditionally, remedies for infertility included the use of special herbs provided by healers who specialized in the treatment of the condition, and the practice of carrying specially made dolls on the women's back, as if they were children (Ministry of Planning, 1991). If the combination of these two methods was ineffective, the woman's husband was allowed to take a second wife. However, any children born of that union would be considered to belong to the first wife (Ministry of Planning, 1991). At present, infertility is still regarded as a serious problem that needs to be treated by either Western or traditional doctors. Even among traditional healers there is a growing understanding of the fact that men, and not only women, can suffer from fertility problems, and need to be equally evaluated.

HEALTH THROUGH THE LIFE CYCLE

Pregnancy and Birth

A Basotho woman's first pregnancy is considered of great importance since fertility is an indication of harmony between the living and their ancestors, and pregnancy signifies the continuation of the father's lineage, and the establishment of the couple as a new family (Gay, 1980). The pregnant woman is expected to follow a diet that is rich in nutritious grains, and to avoid the consumption of fish, which is believed to cause health problems to the unborn child. Traditionally, Basotho women return to their paternal home around the seventh month of pregnancy and remain there for up to three months after the birth of their child. This transition from their husband's to their own group is marked by special rituals and feasts in which the two families participate (Gay, 1980).

Traditional Basotho considered spontaneous abortions, major abnormalities in infants and fetuses, and the death of young children to be the result of witchcraft

(Ministry of Planning, 1991; Sechefo, nd). To protect against such problems, a pregnant woman was required to remain in her house, avoid going out at night, and avoid walking in places where people could have placed harmful medicines. Women in labor were supposed to cover their breast with a special skin to ward off evil spirits that could harm her or her baby (Ministry of Planning, 1991). Placentas were traditionally placed in earthen pots and buried in the early morning or at night. Only women who had helped the mother during her pregnancy or labor were allowed to perform the burial rituals (Sechefo, nd). While most of these practices have disappeared, some are still common in the most remote rural areas of the country.

Infancy

During the first two or three months after the birth of a baby, mother and infant are secluded in a house whose entrance is marked with reeds projecting from the thatch right above the door. These reeds serve to warn strangers as well as men who are not immediate relatives to stay away (Gay, 1980). It is believed that this protects the new baby from evil that can be caused by the "bad conduct" of adults (Sechefo, nd). The new mother must remain inside as much as possible, but if she needs to come out, she must cover herself with a special blanket or shawl. Both mother and child receive direct care from the mother's female relatives, thus exposing the baby from the beginning of its life to direct interaction with many "mothers." Special ceremonies to end childbirth pollution are conducted when the baby loses the remains of the umbilical cord. These include the shaving of the baby's head and the washing of clothing and bedding. At this point, the baby's father is allowed for the first time to enter the house where his wife and child are (Gay, 1980).

Among Basotho, there is a strict taboo against sexual relations during the post-partum period and especially during the breast-feeding period, which may extend for up to two years. Engaging in sexual activity during the lactation period is believed to result in the contamination of the breast milk by the semen, which causes serious health problems to the infant (Romero-Daza, 1994a). "Senyeha" as the disease is called, is characterized by rapid weight loss, severe diarrhea, loss of appetite, and failure to thrive, and may be fatal to the infant. Weaning informally marks the end of the period of infancy, and signals that the mother is ready to resume sexual activity (Gay, 1980).

Childhood

The beginning of the weaning period, when the child is around two years of age, coincides with the expansion of the number of caretakers a child has. At this time, the child begins to spend much more time under the care of other female relatives, including very young girls, and starts the active process of socialization into the prescribed gender roles. Young girls spend most of their time in the family compound and begin contributing to domestic chores such as the collection of dry dung to be used for fire, or the collection of wild vegetables. Young boys, on the other hand, spend most of their time taking care of the animals. In fact, with the absence of adult men for most of the year, even very young boys often spend extended periods of time away from their homes while taking care of the family's animals. This early division of labor translates onto marked differences in school attendance. Unlike girls in many other developing countries, Basotho girls have a higher school attendance rates and higher overall literacy rates than do boys (Ministry of Planning, 1991).

Although minor corporal punishment as a disciplinary measure is not uncommon among the Basotho, the culture rejects the use of excessive force that may result in injuries (Ministry of Planning, 1991). Among Basotho children, the most common health problems are waterborne diseases such as diarrhea and typhoid, nutritional deficiencies such as kwashiorkor and marasmus, acute respiratory infections, and skin problems such as scabies and fungal infections (Gish, 1982; Himmelgreen, 1994; Ministry of Planning, 1991; WHO, 2002). Rates of HIV infection among Basotho children are increasing rapidly. With prevalence rates among pregnant women reaching up to 24%, many HIV positive children are being born. Even those who are not infected are experiencing the indirect consequences of the epidemic, as they often lose one or both of their parents to the epidemic. More and more, orphaned children are being forced to take the responsibility for younger siblings if they do not have grandparents or other relatives who can support them when their parents die.

Adolescence

In Basotho society, the end of childhood is traditionally marked by participation in initiation schools. Although the practice is diminishing in frequency, it still constitutes a major rite of passage for many young Basotho. During initiation school, groups of adolescents spend a period of up to several months in isolated areas of the country. During this time, which is surrounded by secrecy, teenagers learn about their traditions, their culture, and about what it means to be Basotho. At the end of the initiation school, boys are circumcised, and special ceremonial feasts are conducted to mark their entrance into adulthood and their new status as full members of the group. Young girls are not required to undergo any genital operation.

Basotho discourage the practice of premarital sex and encourage marriage at an early age. Traditionally, there existed taboos to regulate the sexuality of adolescents. For example, young women were not allowed to eat eggs, since these were believed to increase the adolescent girls' sexual needs and to make them more fertile than necessary. Nevertheless, premarital sex is on the rise, as is the number of children born out of wedlock (Ministry of Planning, 1991). Teenagers are also exhibiting increasing numbers of Sexually Transmitted Infections, and of HIV/AIDS. As of 1993, teens accounted for 9.4% of all individuals seeking outpatient care for STIs in public hospitals and clinics (Family Health International, 1997). One factor that contributes to the spread of STIs and undermines the efficiency of prevention programs is the cultural tradition that prohibits the open discussion of sexual matters before marriage (Romero-Daza, 1994a, 1994b).

Adulthood

The most common health problems among Basotho adults include high blood pressure, alcoholism, respiratory tract diseases, cardiovascular disease, skin conditions, STI (especially gonorrhea, syphilis, and AIDS), and injuries (Gay & Hall, 1994; Schumacher et al., 1990). Among women, reproductive problems and STIs including gonorrhea, syphilis, trichomonal vaginalis, and candida moniliasis account for most outpatient visits to health facilities throughout the country (Gay & Hall, 1994). Women accounted for more than half of the HIV infection cases in the country as of 2000 (Family Health International, 2001). The rate of infection for pregnant women who seek prenatal care has been increasing considerably in both urban and rural areas of the country, and ranges from 12.3% in the isolated mountainous areas of Mokhotlong to 42.2 % in the capital city of Maseru (Family Health International, 2001).

Among men, the major causes of morbidity and mortality are associated with their occupation as miners. For example, mining accidents are a major cause of injury, disability, and death among Basotho males. Likewise, the high prevalence of tuberculosis may be directly related to the overcrowding and overall poor sanitary conditions of the living quarters commonly found in the mining compounds. The high rate of sexually transmitted infections is directly associated with sexual relations with prostitutes and the low use of condoms. In addition, alcoholism is a very serious problem among adult Basotho. In 1993 it was considered the second most serious disease affecting the population at large (Gay & Hall, 1994). Alcoholism has been associated with very high rates of automobile accidents and interpersonal violence. The easy availability of liquor, including the homemade traditional "joala" beer and the general social acceptance of drinking contribute to the problem.

Domestic violence also constitutes an important cause of morbidity and mortality among modern day Basotho. Actual rates are difficult to determine because many cases of domestic violence are never reported to the authorities. However, admission and treatment records from hospitals, clinics, and other health care facilities report high rates of violence-related injuries, especially among women of reproductive age (Gay, 1980).

The Aged

Among the Basotho, the elderly are afforded special status and respect as pillars of the society. Traditionally, children and grandchildren offer a source of material and social support for their elders. However, as in many places in Africa, with the explosion of the AIDS pandemic, the traditional roles are rapidly changing, and the elderly are being forced to care for their younger relatives. Specifically, as adults of reproductive age become ill, their parents become the main care providers, who are also burdened with the responsibility of finding financial resources for the treatment of the diseased. As the younger generation dies, the elderly find themselves being responsible for the upbringing of any of their young grandchildren left orphaned (Romero-Daza & Himmelgreen, 1998).

Dying and Death

Basotho consider the death of an older person as a normal part of the life cycle and as a transition to the world of the ancestors. However, the death of young people, and especially of children is often believed to be the result of either witchcraft or of the failure to perform the necessary rituals or to fulfill obligations to the ancestors (Ministry of Planning, 1991). In the past, the death of a young woman at childbirth was especially alarming since it represented an omen of misfortune for the whole family. In such a case, the surviving child was considered to be highly vulnerable and was afforded special protection by his/her relatives throughout his/her life (Ministry of Planning, 1991). Also worrisome were any sudden deaths or those caused by lightning, which were believed to be the result of witchcraft. Regardless of the cause of death, the body was buried immediately following death, usually during the night. Only adults were allowed to participate in funerals, and there was a strict taboo against mentioning the name of the dead person, especially during the mourning period. Special ceremonies were performed immediately following the funeral. These included the cutting of the hair for all close relatives of the dead person, and the wearing of the "thapo" or mourning veil (Schefo, nd). The official mourning period lasted about one to two months for relatives other than the spouse of the deceased. The end of mourning was marked by the slaughtering of an ox in honor of the ancestors. Failure to do so could result in physical or mental problems for the surviving family. At present, many changes have occurred, including the common practice of holding wakes. However, prohibitions and taboos regarding widows remain unchanged. A widow is supposed to spend about one week after the funeral lying or sitting in the room where her husband's body was kept. After the funeral she is supposed to wear only black clothes, is not allowed to wear any new clothes, is not allowed to go out of her house after dark, and is forbidden from having sexual relations. After one year, the woman is supposed to return to her own family' home where her brothers perform a special ceremony to officially mark the end of the mourning period. The widow is expected to come back to her late husband's home where she remains the head of the household. The traditional practice of levirate by which a widow would become the wife of her late husband's brother has almost completely disappeared, as more and more widows choose to live by themselves or even to remarry (Gay, 1980).

REFERENCES

Ambrose, D. (1976). *The guide to Lesotho*. Johannesburg, South Africa: Winchester Press.

Central Intelligence Agency (CIA) (n.d.) The World Factbook, Lesotho. Available online at, http://www.odci.gov/cia/publications/factbook/geos/lt.html

Family Health International. (2001). Lesotho and Swaziland, HIV/AIDS risk assessments at cross-border and migrant sites in Southern Africa. Available online at, http:// www.fhi.org /en/aids/impact/impactpdfs/lesothoandswaziland.pdf

Gay, J. S. (1980). Basotho women's options. A study of marital careers in rural Lesotho. Doctoral dissertation, University of Cambridge, Lucy Cavendish College, London.

Gay, J. S. (1982). *Women and development in Lesotho*. Maseru, Lesotho: USAID.

Gay, J. S., Gill, D., Green, T., Hall, D., Mhlanga, M., & Mohapi, M. (1991). Poverty in Lesotho, a mapping exercise, Phase II report. Maseru, Lesotho: European Community Aid Counterpart.

Gay, J. S., & Hall, D. (1994). Poverty in Lesotho, 1994. A mapping exercise. Maseru, Lesotho: Secheba Consultants Publishers.

Gish, O. (1982). Economic dependency, health services, and health: The case of Lesotho. *Journal of Health Politics, Policy, and Law, 64*, 762–779.

Global Policy Network. (2001). Highlights of current labor market conditions in Lesotho. Available online at, http://globalpolicynetwork.org

Hall, D., & Malaheleha, M. (1989). *Health and family planning services in Lesotho: The people's perspective*. Maseru, Lesotho: World Bank for Lesotho Government Ministry of Health.

Himmelgreen, D. (1994). *Coping in a highly seasonal environment: A household study of changing nutritional status, health, and diet among women and children from highland Lesotho*. Doctoral dissertation, State University of New York at Buffalo, Amherst, New York.

Himmelgreen, D., & Romero-Daza, N. (1994). Changes in body weight in Basotho women: Seasonal coping in households with different socioeconomic conditions. *American Journal of Human Biology, 6*(5), 599–612.

Huss-Ashmore, R., & Goodman, J. L. (1988). Seasonality of work, weight, and body composition for women in highland Lesotho. In R. Huss-Ashmore, J. J. Curry, & R. K. Hitchcock (Eds.), *Coping with seasonal constraints* (Vol. 5, pp. 29–44). Philadelphia, MASCA.

Ikuska website (nd). Sotho. Available online at, http://www.ikuska.com/Africa/Etnologia/Pueblos/pueblos_fram.htm

Ministry of Planning. (1991). The situation of women and children in Lesotho. Maseru, Lesotho: Ministry of Planning, Economic and Manpower Development.

Murray, C. (1981). *Families divide: The impact of migrant labor in Lesotho*. Cambridge: Cambridge University Press.

Romero-Daza, N. (1994a). *Migrant labor, multiple sexual partners, and STDs*. Doctoral dissertation, State University of New York at Buffalo, Amherst, N.Y.

Romero-Daza, N. (1994b). Multiple sexual partners, migrant labor, and sexually transmitted diseases, the makings for an epidemic. Knowledge and beliefs about AIDS among women in highland Lesotho. *Human Organization, 53*, 192–205.

Romero-Daza, N., & Himmelgreen D. (1998). More than money for your labor. Migration and the political economy of AIDS in Lesotho. In M. Singer (Ed.), *The political economy of AIDS*, Critical Approaches in Health Social Sciences Series. New York: Baywood Press: 185–204.

Schumacher, R., Kamphorst, E., Morojele, S., Raditalope, M., & Motleleng, N. (1990). *Sexually transmitted diseases, family planning and condoms.* Mokhotlong, Lesotho. Mokhotlong Health Service Area.

Schwager, D. (1986). *Lesotho.* Maseru, Lesotho: Schwager Publications.

Sechefo, J. (nd). *Customs and superstitions in Basutoland.* Unpublished manuscript.

Sheddick, V. G. J. (1953). *The Southern Sotho.* London: International African Institute.

United Nations Childrens Fund (UNICEF). (2002). UNICEF Statistics. Available online at, wysiwyg://35/http://www.unicef.org/statis/Country_1page101.html

World Health Organization. (2001). Legal status of traditional medicine and complementary/alternative medicine: A worldwide review. Author. Available online at, http://www.who.int/medicines/library/trm/who-edm-trm-2001-2/legalstatus.shtml

World Health Organization. (2002). Lesotho. Available online at, http://www.who.int/country/lso/en/

Sudanese

Rogaia Mustafa Abusharaf

ALTERNATIVE NAMES

None.

LOCATION AND LINGUISTIC AFFILIATION

The Sudan, Africa's largest country in area, is a territory of incredible historical and political importance. It covers an area of about 2.5 million km^2 or almost one tenth of the total area of Africa. The country is located in the northeastern part of Africa and extends from latitude 3° to 23° north and from latitude 22° to 39° east. The Sudan shares borders with nine countries: Egypt, Libya, Central African Republic, Chad, Zaire, Kenya, Uganda, Ethiopia, and Eritrea. The country is on the whole a mammoth plain, divided by mountain ranges such as Jebel Marra in western Sudan and Mount Kinyeti Imatong, bordering Uganda. The Blue and White Niles come together in Khartoum forming the River Nile, which flows northwards through Egypt to the Mediterranean Sea.

The Sudanese land and its location at the crossroads of Africa have influenced the course of its politics and history, and the Sudanese people have ardently developed multifaceted identities across it. The characteristics of the land have influenced the life and social organization of its inhabitants, defining in a dramatic way the socioeconomic and political organization of people across this vast territory. As a result of the geographical and historical heritage of the land, the Sudan straddles Africa and the Middle East, thus consolidating its place in history as a meeting point of Arab and sub-Saharan worlds.

The Sudan is the place of birth of more than five hundred cultural and linguistic groups, all with distinctive outlooks on life, culture, cosmology, faith traditions, and experiential knowledge. Arabic is the lingua franca in addition to English, which is the main language of instruction in southern Sudan. A northern Sudanese Christian minority, the Copts, also speaks Arabic. In addition, numerous linguistic groups and dialects are spoken throughout the country.

Diversity rather than conformity remains a distinguishing quality of cultural and religious Sudanese life. For example, Arabized Sudanese cultures predominate in regions where Islam is the major religion (70% of the entire population). The branch of Islam practiced by Sudanese Muslims is known as Sunni. This tradition is distinguished by the pervasiveness of religious orders or brotherhoods, each placing its unique demands on its followers. What they have in common with the Muslim population worldwide, however, are the Five Pillars, which are constitutive of universal Islam. These pillars include the profession of the faith, daily prayer, almsgiving, fasting in the holy month of Ramadan, and fulfilling the pilgrimage to Mecca. In communities where Islam is the major religion, it is noticeable that it has adapted itself to preexisting beliefs and local cultural understandings. northern Sudan is, to a great extent, arabized in cultural traits, identity, and political alignment. Historians of the Sudan attribute the twin processes of Arabization and Islamization to immigrants coming to the Sudan from the Arabian Peninsula across the Red Sea, Egypt, and the Maghreb. A complex mix of conquest, migration, religious conversion, and miscegenation define the identity of the northern Sudanese as Arab. According to Ali Mazrui two parallel processes of social transformation swept the country in the course of Arab expansionism, "one linguistic and cultural, by which the people of the land acquired Arabic as their language and certain Islamic cultural conceptions and became connected with the Arab tribal system; and the other racial, by which the incoming Arab stock was absorbed in varying degrees, so that today a modicum of Arab blood flows in their veins" (Mazrui, 1973, p. 47).

According to Mudathir Abdel Rahim: "(Arab immigrants') readiness to mix, coupled with the matriarchal system of the Nubians on the one hand, and the Arabian patriarchal organization of the family and the tribe on the other, had the effect not only of facilitating the assimilation of the immigrants and the spreading of their culture and religion, but also of giving them the reins of power and political leadership in the host society" (Abdel Rahim, 1973, p. 31).

Southern Sudan, conversely, is home to countless ethnic groups, ranging from the Niolitics such as the Nuer, the Dinka, and the Shilluk; the Nilo-Hamites such as the Latuka; and the Sudanic comprising the Jur, Moru, and Azande. The Christian people of the Sudan's south and the Nuba Mountains are predominantly Roman Catholic and Anglican. In addition to Islam and Christianity, indigenous beliefs and religions are followed with equal devoutness and fervor. Indigenous beliefs are specific to particular ethnic groups. According to the Sudan country study report "the beliefs and practices of indigenous religions in Sudan are not systematized, in that the people do not generally attempt to put together in coherent fashion, the doctrines that they hold and the rituals they practice."

OVERVIEW OF THE CULTURE

In spite of the cultural complexity of the Sudan, the people of the Sudan are frequently classified into binary social categories on the basis of geography (north versus south), ethnicity (Arabs versus Africans), and religion (Muslim versus Christian). Nevertheless, an impressive array of ethno-linguistic and religious groups inhabits this vast territory, and they do not necessarily fall into this opposition. They are *Hadandwa, Mahas, Nuer, Danagla, Dinka, Shiluk, Nuba, Rubatab. Rikabia, Shayqia, Murle, Kababish, Manasir, Azande, Jaleen, Bori, Shuli, Joar, Anwak, Latuka,* and *Beja,* among other groups.

The variety of Sudanese cultural life has been imaginatively portrayed in the ethnographic literature. From the southern part of the country, Evans Pritchard brought the lives of the Nuer and the Azande peoples to international attention. In the Nuba Mountains in the west, James Faris enriched our knowledge about the intersections of politics and aesthetics in this region through his ingenious analysis of the Nuba personal art. Here, we learn a great deal of how bodily adornment, is employed as an effective means for articulating important social relations and self-expression in Nuba society (Faris, 1972).

The extraordinary variation in Sudanese cultural life, notwithstanding, some commonalities exist in some fundamental aspects pertaining to social organization. On the whole, patriarchal authority, patrilineal descent, and patrilocal residence distinguish the Sudan. Territorial endogamy remains the preferred type of marriage unions. Invariably, all Sudanese peoples accord considerable deference and loyalties to extended families and collectivities.

THE CONTEXT OF HEALTH: ENVIRONMENTAL, ECONOMIC, SOCIAL, AND POLITICAL FACTORS

Health care delivery systems throughout the Sudan are acutely under pressure. The high occurrence of devastating diseases has increased considerably in recent years as a result of economic, social, and environmental factors. Moreover, the pervasive political and economic volatility in the country had weighed heavily on the health and security of individuals and communities. Life expectancy is 53 years; infant mortality rate is estimated at 7.8% while literacy rate is 43%. The civil war, which has been raging since 1955, has taken an enormous toll on the population, which struggles against the effects of chronic malnutrition and contagious diseases. Health care in the Southern States is severely lacking. Exceptionally susceptible are the 1.25 million children and adolescents that constitute part of the Sudan's internally displaced people. According to United Nations Humanitarian Operations in Sudan, food shortage affected the health of war victims considerably. The report stated "Food insecurity in the Sudan in 2000 has been a result of continued general insecurity; population displacements, drought and floods, as well as the high prevalence of disease due to poor health and sanitation. In some areas where rainfall patterns were ideal for cultivation, insecurity drove people from their homes and fields" (2000, p. 14). In addition to the deleterious impact of the civil war, during 1984/85 the Sudan experienced one of the most devastating environmental disasters. Severe drought and desertification and crop failure in western Sudan led to a mass exodus to the capital city of Khartoum as well as to other parts of the country. Today, massive numbers of farmers and their families find their way to Khartoum's camps for the internally displaced persons along with their southern counterparts.

To remedy these serious problems, some measures such as consolidation of efficient health coordination structure, rehabilitation of all health facilities, provision of drugs and equipment, and staff training, particularly in the area of child and maternal health were recommended (UN Report, 2000, p. 16).

MEDICAL PRACTITIONERS

Medical practitioners in the Sudan include physicians, physician assistants, midwives, nurses, traditional birth attendants, traditional healers, *zar sheiks* and *sheikhas* (the Sheikhas are the women who officiate in *zar* or spirit possession ceremonies), and diviners among others. Modern medicine in the Sudan was officially introduced by the Anglo-Egyptian regime (1898–1956), a process that was thoroughly chronicled by historian Heather Bell (1999). During the colonial period, different sectors of Sudan's health care were developed including the introduction of modern midwifery training. Reliance on biomedical versus traditional healers varies vastly by education, residence, and gender differences. For instance urban dwellers are more likely to go to hospitals as opposed to rural populations who rely heavily on traditional or folk cures. Also significant are gender differences, where women seek the advice of religious leaders or *zar sheikhas* more than their male counterparts who often dismiss the importance of folk healing.

CLASSIFICATION OF ILLNESS, THEORIES OF ILLNESS, AND TREATMENT OF ILLNESS

Illness, physical or psychological, is subject to scrupulous theorization. Consequently, ways of dealing with illness are not confined to biomedicine. Instead, reliance on traditional healing is especially obvious. In fact, throughout Sudanese provinces and villages, traditional healing is the single mode of curing that is readily available for the community. According to Sudanese gynecologist Osman Modawi, modern medicine is commonly thought of as the last resort (1982). The term "traditional medicine" encompasses a variety of methods of protecting and restoring health that existed before the appearance of modern medicine. On the whole, traditional medicine refers to acupuncture, traditional birth attendants, mental healers, and herbal medicine (WHO Fact Sheet, 1999).

In the Sudan, one of the most important theories of illness is derived from the overwhelming belief in the power of the evil spirits. The spirits are thought to inflict injury as well as undermine one's health through the notion of *Ammal*, or the foul act. *Ammal* is typically inflicted through witchcraft causing substantial impairment and damage. Treatment is at all times sought through the counsel of shamans or witch doctors who alone have the God-given gift of counteracting the callous act. The shaman, who listens carefully to the symptoms of illness, performs a series of steps to identify types of spirits and methods for the cure. The Azande for instance, believe that witchcraft can be inherited and that a person can be a witch, causing others impairment, without realizing his/her power. Because of this threat, effective resources of diagnosing witchcraft are very important. One method is through the use by oracles, of *benge*, a poison, which is fed to little chickens. The chicken's death or survival provides the oracle's answer. Azande also use *benge* as proof to substantiate one's guilt before a court of law (Evans Pritchard, 1976). Notwithstanding the changing cultural life of Azande people and their forced displacement as a result of war, their beliefs on illness and adversity continue to hold witchcraft as the most important threat to someone's health and well-being. In an interview I conducted with Thomas, a 43-year-old Azande, who resides in ElSalam camp for the internally displaced, he stressed that "Kuguor still run rampant. People cannot forget everything about their lives at home before they came to this place."

Diseases, which are attributed to evil spirits in many Sudanese communities, are classified as *zar* or spirit possession. In the words of British anthropologist Susan Kenyon: "Zar is both category of spirit and the cult associated with possession by those spirits. Such possession can cause problems or illness, usually a form of mental illness, marud nufsi as the Sudanese describe it, and these are referred to as zar" (1991, p. 185). *Zar* spirits are generally appeased and kept at bay through ritual ceremonials including animal sacrifice, drumming, and dancing. The types of healers who officiate in these ceremonies are called shaikhas, generally women who are well-versed in the world of the spirits and can act as interlocutors between the spirits and the patient whom they possess.

Another powerful branch of traditional healing is known as *El-tibb*, *El-nabawi*, or "prophetic medicine". In Khartoum today large numbers of clinics are scattered all over the capital city. In May 2002 I interviewed Hajj Hassan, a man whose knowledge in this field is superb. In the words of Hajj Hassan, human beings should realize the complicated link between mind and body. Prophetic medicine does not target physical ill health as an isolated phenomenon. Instead through Quranic verses, the patient receives considerable mental consolation and calmness. He articulates the underlying principle of the proliferation of Prophetic medicine as follows: It is well known that the belief of people in God plays a greater role in their recovery, regardless of the seriousness of the disease. So, the religion of Islam inspires the person who is ill to make his way through the verses of the Quran and the instructions of the Prophet. For example Prophet Mohammad described to his followers the necessity of taking honey everyday for various illnesses. He, peace be upon him, described for those suffering from high blood pressure (*elfisada*) to let blood out of their veins. He emphasized the benefits of herbs like ginger and peppercorns etc. In addition, there are a number of verses that when read regularly can bring serenity and peace of mind for patients. That is why people believe that the Quran can provide ways of dealing with their anxieties, not readily available in modern medicine.

The connection of mind and body in Sudanese theories of illness is made with crystal clarity in situations where patients seek the opinion of modern medicine and traditional healing simultaneously. This procedure is not at all infrequent or rare. The mother of a 37-year-old woman who suffered from clinical depression told me, that the psychiatrist who treated her daughter used a combination of modern procedures with elements from prophetic medicine with impressive results.

SEXUALITY AND REPRODUCTION

Sexuality

Attitudes toward sexuality and reproduction are positioned at the heart of significant cultural and religious beliefs amongst the Sudanese. Open discussion of matters pertaining to sexuality is extremely proscribed by these beliefs. To a great extent, this interdiction is intimately linked to how society views sexuality in the first place. Largely seen as an ominous threat that looms largely over one's purity and morality if left unchecked, social and physical regulation is aggressively pursued. This view is of special relevance to female sexuality. One of the most important vehicles for dulling women's sexuality is achieved through female circumcision.

The origins of this practice are very mystifying and obscure. In exploring its ideology, we find varied oral accounts describing mythological rather than documented origins. Although the majority attributed the diffusion of the practice to pharaonic Egypt, some analysts argued that the custom might have originated from the Red Sea People who introduced it to neighboring peoples. Writers such as Abdalla Eltayeb (1964), a noted Sudanese linguist, see female circumcision as a legacy of the Arabization of the Sudanese people. What is important to remember is that female circumcision is not a universal practice in the Sudan, which is not surprising because the peoples and cultures of the country are very heterogeneous. According to Sudan Demographic & Health Survey the prevalence of the practice is as follows: northern Sudan, 98.7%, eastern Sudan, 86.5%, western Sudan, 95.5%. Because of the important link between female circumcision and sexuality, it is necessary to elaborate its ideological underpinnings. Understanding ideology is a step toward understanding strategies for its eradication. Indeed, Habermas (1971) argues that the only knowledge that can orient action is knowledge that frees itself from mere human interests and is based on ideas. The reasons for female circumcision and the age at which it is performed differ across Sudan by regional, ethnic, and class differences. As far as the operations among northerners are concerned, they have been practiced for a variety of cosmetic, religious, medical, and social reasons. For instance, people justify their support of the practice by arguing that it preserves virginity, enhances femininity, and increases purity and cleanliness. There is also variation in its prevalence, in the exact types of the practice, and in the rituals associated with it. The practice includes the following types: clitoridectomy, excision, and infibulation.

In spite of the cultural importance of this practice, it has been met with strong opposition since British colonial rule (1898–1956). In fact the practice became illegal in 1946, when the British colonial administration passed a law making female circumcision a crime punishable by imprisonment and fines. From the time of independence in 1956 until the end of the 1970s, new plans were formulated to stop the practice. Attention now focused on those social and cultural aspects of the practice that accounted for its endurance in the Sudan. More recently, several organizations, including the Sudan National Committee for Traditional Practices Affecting the Health of Women and Children, The Red Crescent, and the Mutawinat Group have identified the practice as an act of violence, and have devised research strategies, and adopted a variety of approaches to eradicate female circumcision. The most notable of these plans, include community outreach through the use of mass media, audio-visual aids, and publications. Another approach aims to incorporate efforts to end the practice into existing programs concerned with public health, family planning, anti-violence programs, maternal and childcare, midwifery training, and nutrition.

Fertility

Fertility and fecundity figure prominently in the cultural constructions of gender identities and conceptions of womanhood and manhood amongst the Sudanese. Conversely, fear of infertility prompts individuals and communities to invent effective measures to counteract the prospects of childlessness. In the Nuba Mountains, Modawi (1982, p. 81) described the effort of one community of dealing with infertility: "In the Nuba Mountains there is a small hill which looks like the male genital organs. Infertile women visit this hill to perform certain rituals to treat their infertility. In some of the other rituals a toy boy with prominent genitalia and made of hina is thrown into water. It is also thought that eating the male genitals of animals corrects infertility. The scarcer the animal, such as crocodile, the more potent the cure."

In other parts of the country, important rituals are fervently pursued to guarantee that a woman who enters into a marriage union is able to contribute to the productiveness of the community as a whole. One of the most joyous occasions for the public recognition of fertility is the *jirtig* ceremony. It is believed that this rite prevents evil spirits from undercutting the woman's ability to bear children, thus acquiring a quasi-religious quality and significance. A bride who does not take part in the *jirtig*, Sudanese say, will become infertile, for her reproductive organs are filled with evil spirits. The *jirtig* is redolent with symbolism and cultural meaning, and the values it celebrates are deemed essential to the life of the married couple, whether at home or abroad. During this ceremony, the *jirtig* bed is placed in the middle of the front yard of the house where tens of people gather to witness the ceremony. The bed is covered with a red velvet bedspread, which symbolizes fertility. The bride is also dressed in red, while the groom wears the traditional robe-like outfit, with a red band tied around his forehead. The ritual concludes with a procession to the Nile for blessing and good fortune. These rituals however are overwhelmingly female-oriented. In fact in several cases, it is almost impossible for a man to acknowledge infertility, which is more often than not equated with lack of virility and masculinity.

Apart from cultural notions about fertility and reproduction, fertility levels have declined sharply in the Sudan from an average of six children per woman to five children (Sudan Demographic & Health Survey, 1991). There is a great variation in fertility levels in light of educational attainment, socioeconomic backgrounds, and residential patterns. The great majority of Sudanese women are well aware of the various methods of family planning and contraception use (Swar Eldahab, 1996). Significant hurdles that mitigate against their involvement in family planning, however, loom large. In Dar Elsalam, Khartoum, I interviewed a group of women about their views on birth control. There was a significant consensus about the impact of their own economic situation on their ability to limit family size "Contraceptives are very expensive. We came here from Dar Fur when we used to have our farming land and we lost it because of the drought. We attended the talk of a group of health workers about birth control. They told us about pills and condoms. These items are very expensive. We cannot afford birth control pills. So, even if we want to have a small family, we cannot because of our financial situation." This sentiment echoes those of thousands of women who are constrained by poverty and inability to secure better lives for themselves and for their offspring.

HEALTH THROUGH THE LIFE CYCLE

Pregnancy and Birth

Cultural understandings about conception and birth are shaped by the notion that giving birth is not purely a result of sexual intercourse between men and women. One of the most compelling explanations of this view was discussed in Francis Deng's ethnography the Dinka of the Sudan in which he argues: "(Conception) is the creation of God and the blessing of the ancestral spirits. The cooperation of the father, the mother, and the spirits in the venture is verbally conceptualized. The word ahieth means, 'to beget' and 'give birth'. Dinka view of conception does not distinguish between the role of man and woman. They copulate to 'beget' jointly and 'give birth' while God intercedes to 'create' and the ancestors assist in protecting the creation from malevolent powers of destruction"(1972, p. 30). In other parts of Sudan children are considered to be wealth in and of themselves. To secure the health of the pregnant mother and her unborn child, Sudanese prescribe a variety of foods and herbs believed to be beneficial for the blood and for breast-feeding after delivery. In the Sudan many practices associated with breast-feeding are considered very beneficial. According to Shazali "Most mothers especially in rural areas, believe that mother's milk is best for the infant and are psychologically prepared to breast-feed" (1982, p. 103). In some situations however, local beliefs interfere with feeding such as the notions that boys should not be fed at sunset, because it is considered to be a bad omen and a threat to their intelligence (Shazali, 1982).

Infancy

In the Sudan infants are most warmly welcomed through an exciting assortment of rituals and rites celebrating their arrival in the world. One of these celebratory occasions is the *Simaya*, or the naming ceremony. In northern Sudan, and elsewhere this ritual has attracted the attention of anthropologists for its zestfully symbolic content. The symbolism of this ceremony is brilliantly captured by Fadwa El-Guindi, who argued, "A function of such a ritual is to emphasize and impress beyond the shadow of a doubt that one-passé of the life-cycle has been left behind and to announce and stress that henceforth he or she is in another phase. Another aspect is to make sure that the individual has received communion with the deepest cultural ideals"(1996, p. 3). The *simaya* solemnizes the naming of the child as an important rite of passage and a source of blessings (*baraka*) and good omens. Generally, anywhere from one to two weeks after the birth of a child, a sacrificial lamb is slaughtered for the ritual naming of the newborn. It is celebrated with feasts, drumming, remembrance ceremonies, and concerts and marks the rite of passage from birth to initiation into the wider community. Equally emblematic, are occasions celebrated all over southern Sudan where names signify important aspects of social organization (Deng, 1972). Infants are cared for by parents, grandparents, family members, neighbors, and hired help, who also provide invaluable help with the daily household tasks. Most infants are breast-fed for at least two years, a factor that seems to mitigate some of the problems of malnutrition and childhood diseases.

Mothers and infants are often believed to be the most vulnerable people to health problems and to the attacks of evil spirits. To deal with their inbuilt vulnerability, mother and child are obliged to remain indoors for 40 days. A Sudanese woman told the unusual story of her being visited by a ghost-like creature three days after giving birth "I will never forget the incident that happened to me shortly after the birth of my first son, who is now 41 years old. I was lying down, sleeping in the middle of the night when I was awakened by an unusual noise, almost like very heavy breathing. I opened my eyes and saw a Caucasian woman dressed in a nurse's uniform sitting in a chair across from me. I was horrified. I started to call my cousin who was sleeping outside in the hosh: "Fathia, Fathia, come quickly." The woman started to call Fathia with a masculine, hoarse voice. Fathia came running and she turned on the light; the women disappeared. But Fathia said she had heard the strange voice calling her name. "From that moment on, we had to read the Quran and put amulets on my son's arm to protect him. My neighbors told me that this woman wanted to steal my infant and put another deformed one in his place but Allah protected both of us." To a large extent these complex beliefs have a significant role to play in attitudes toward the health and sickness of infants. Official reports however, demonstrated that lack of safe water, undernourishment, insufficient sanitation, and poor hygiene are considered the most significant causes of high infant mortality and morbidity in Sudan.

Childhood

Children are seen as a source of security and spiritual immortality (Deng, 1972). Childhood is also a phase in which intensive efforts at gender bending and conditioning gets underway. For example, in many parts of the Sudan Sudanese boys and girls are circumcised during this period. Rites of passages that signify the acquisition of new identities are also observed during this time. However, children, like the rest of the Sudanese population are experiencing insurmountable problems relating to health, political conflict, economic insecurity, and social unrest. This situation is particularly clear in a UNICEF report (2001) on the situation of children in Sudan. The report provides detailed analysis of the impact of the civil war, internal displacement, landmines, economic exploitation, and abductions, on children. Contagious diseases and illnesses that infringe upon children's health in the Sudan are wide-ranging, including malaria, diarrheal and acute respiratory infections. Lack of drugs and inadequate medical care combine to affect the health of Sudanese children dramatically.

Adolescence

In the Sudan, the onset of adolescence is managed through culturally proscribed means and strategies. In fact, the socioeconomic circumstances prompt many parents and caregivers to entrust adolescents with responsibilities and obligations that are far beyond their years. Hence, social maturity is not measured by one's age as such, but by their productivity and contribution to the family's economics.

Adulthood

Achieving an adult status in the Sudan is contingent upon fulfillment of crucial steps in the social ladder such as marriage, attainment of employment, and property ownership. Adulthood, however, does not presuppose total individuality or independence from extended family networks. It is very common for adult children—married or unmarried—to continue to reside with their parents.

The Aged

Compassion and empathy toward the elderly population is one of the most important values for Sudanese people irrespective of region, ethnicity, or religion. Children and grandchildren view their relationship to aging parents as one of obligation and debt. Hardly ever do aging family members get sent to nursing homes.

Aging does not only mean a longed-for release from physical work, but it also means greater authority especially for post-menopausal women. Age frees women considerably from some of the most arduous tasks and societal expectations that they had had to adhere to during their reproductive years. It follows that menopause is looked at as a phase in life that is increasingly empowering, and is rarely discussed as a medical problem. A woman in her fifties recalled that when her mother was experiencing menopause and had all the symptoms associated with the condition such as hot flashes, her physician never explained to her that she was experiencing menopause. She added that in the Sudan "women never talk about menopause." This observation is corroborated by the fact that literature on fertility in the Sudan has rarely touched on attitudes toward menopause.

Dying and Death

The miscellany of Sudanese cultural worldviews and ideologies are manifestly reflected in how they deal with the question of death and dying. In the Sudan beliefs about death and dying are to a great extent molded by the deeply entrenched faith in *El-qaddaa' wa El-Qaddar* (God's will and fate). Death presents an opportunity for communities to come together to express their support and solidarity for the deceased person's surviving kin. In most parts of the Sudan, mourning rituals extend to 40 days during which the family of the deceased continue to receive tens of mourners a day.

CHANGING HEALTH PATTERNS

AIDS rates have escalated in the past few years due to a number of factors ranging from use of unsterilized syringes, unprotected sex, poor health in general. There is no comprehensive governmental policy regarding the mitigation of the escalating rate of the disease.

REFERENCES

Abdel Rahim, M. (1973). "Arabism, Africanism, and self-identification in the Sudan." In D. Wai (Ed.), *The Southern Sudan: The problem of national integration* (pp. 29–47). London: Frank Cass.

Bell, H. (1999). *Frontiers of Medicine in the Anglo-Egyptian Sudan 1899–1940.* Oxford: Clarendon.

Deng, F. (1972). *The Dinka of the Sudan.* Prospect Heights, IL: Waveland Press.

Eltayeb, A. (1964). The changing customs of the Riverain Sudan. *Sudan Notes & Records, 45,* 12–27.

Evans-Pritchard, E. (1976). *Witchcraft, oracles, and magic among the Azande.* Oxford: Clarendon.

Faris, J. (1972). *Nuba personal art.* London: Duckworth.

Habermas, J. (1971). *Knowledge and human interest.* Boston: Beacon Press.

Inter Press Service. (1997). *Sudan health: UN issues call for emergency aid.* Feb. 18th. (Doc. No. 970218)

Kenyon, S. (1991). *Five women of Sennar: Culture and change in central Sudan.* Oxford: Clarendon.

El-Guindi, F. (1996). *El Soubu: Egyptian celebration of life. Film Study Guide.* Los Angeles: El Nil Research.

Mazrui, A. (1973). "The Black Arabs in comparative perspective: The political sociology of race mixture." In D. Wai (Ed.), *The Southern Sudan: The problem of national integration* (pp. 47–83). London: Frank Cass.

Modawi, O. (1982). Traditional practices in childbirth in Sudan. In Taha Baasher (Ed.) *Traditional practices affecting the health of women and children.* Alexandria: WHO Technical Publication No.2.

Shazali, H. (1982). Nutritional taboos and traditional practices in pregnancy and lactation including breast-feeding. In Taha Baasher (Ed.). *Traditional practices affecting the health of women and children.* Alexandria: WHO Technical Publication No.2.

Sudan Country Study. Available online at http:// LCWEB2.loc.gov/cgi-bin/query/frd/CStudy

Sudan Demographic & Health Survey. (1991). Maryland: Micro International Institute.

Swar Eldahab, A. (1996). "Contraceptive use and fertility of women in Urban Sudan." *The Ahfad Journal 13*(2), 12–31.

United Nations Children Fund (UNICEF). (2001). *The situation of children and women in the Sudan.* Khartoum: United Nations.

World Health Organization (WHO). (1999). Fact Sheet. Geneva.

Thai

Chris Lyttleton

ALTERNATIVE NAMES

Siamese, replaced in 1939.

LOCATION AND LINGUISTIC AFFILIATION

The Thai live within the Southeast Asian nation-state of the Kingdom of Thailand which consists of five regions characterized by distinct landforms: the northern ranges and plateaus, the central fertile plains, the southeast coastal seaboard, the arid northeast plateau, and the Southern humid plateau. The Thai live in all regions but predominately in the central plains. The Thai are one of many *Tai*-speaking peoples inhabiting mainland southeast Asia. Thai is the official national language (often called Central Thai) and is gradually replacing other *Tai* dialects in regional parts of Thailand.

OVERVIEW OF THE CULTURE

The *Tai* peoples moved down from Southern China roughly a thousand years ago to populate most of the central portion of the Indochinese peninsula. Over many centuries, a Thai culture, civilization, and identity has emerged as a product of interaction between the *Tai* and prior inhabitants such as the Mons and Khmers.

The population of Thailand was 62.31 million in 2001. The Thai (Siamese) are the majority population living within the Thai borders and number more than 30 million. Other *Tai* ethnic groups also live within the Kingdom of Thailand, the most prominent being the ethnic Lao (*Isan*) with just over 20 million, however due to the impact of national administrative and development policies, ethnic differences are slowly being dissolved and Thai culture is becoming dominant throughout the country. Although the following summary of social, cultural, and economic aspects of health is focused on the Thai, statistical data (cited from MOPH, 1999 unless noted otherwise) is gathered at the national level and therefore includes other minority populations who nowadays regard themselves as politically Thai. Most national surveys identify all *Tai*-speaking peoples as Thai if they live within national borders—it is thus common to see reports that cite more than 80% of Thailand's population as ethnically Thai.

The Siamese emerged as a dominant regional identity during the 13th and 14th centuries when Khmer authority was defeated and the independent *Tai* Kingdom of Sukhothai was established in and around the central plains. By the 16th century a characteristically Siamese culture became associated with the styles, tastes, and values of the new court based at Ayutthaya that slowly distinguished itself from neighboring *Tai* kingdoms in the north and east (Wyatt, 1982). In the 18th century the capital of Siam was moved to its present day site in Bangkok and by early 1800 had extended the rule of Siam to cover much of mainland Southeast Asia.

Thailand is the only nation in Southeast Asia that was not colonized as successive Siamese monarchs preserved their right to independent rule with great political acumen. The current borders were formally established in 1896 when the French took control over all Siamese/Lao territories east of the Mekong River. In 1932, King Rama VII accepted that the country would be best served by a democratic government and a constitutional monarchy. In 1939, Prime Minister Phibun changed the name of Siam to Thailand to emphasize Thai characteristics of national identity (rather than the economically dominant Chinese) and as a gesture of solidarity with other *Tai*-speaking peoples within national borders. Since then there have been frequent changes of governments (53 cabinets) and military coups (11), however, a new constitution was promulgated in 1997 that enshrines far more rigorous controls on the electoral process. It is widely considered the "first constitution of the people."

Since the 1960s, the Thai have rapidly transformed from being a largely subsistent and agrarian, primarily rice-growing, population into an export-oriented industrialized society. They are now a highly urbanized and modernized society that, at the same time, still utilizes social and cultural understandings from pre-modernist ideologies and belief systems. The combination creates a complex mix of behaviors that influence illness and healing practices. From the early 1980s until the late 1990s, Thailand had one of the fastest growing economies in the world fostering rapid social and economic development. By 1996, 97.5% of the 70,000 villages nationwide had electricity, by 1997, 98.27% of households had sanitary latrines and by 1998, 95.47% of households had access to clean drinking water. A major impact of this economic transition has been the expansion of the urban sector. Rapid industrialization has promoted large-scale internal migration—more than 40 million Thai live in rural villages but a huge percent spend extended periods working in cities or larger agricultural provinces with serious impacts on the cohesion of the family unit. Bangkok has grown from 1.5 million in 1960 to over 9 million in the 1990s (Osborne, 1995). A further consequence is a growing disparity in income. By 1990, Greater Bangkok had 1,404 slum settlements housing over 1.2 million people (Pasuk & Baker, 1996). From 1960–90 the proportionate contribution of agriculture to the GDP dropped from 40% to 12% even though it still employed roughly 60% of the populace (Medhi, 1992). While overall poverty levels have declined, the gap between rich and poor has increased and the vast majority of those below the poverty line are farmers. In 1997, the "bubble" economy crashed and high levels of unemployment sent many urban workers back to their rural homes, average incomes dropped, poverty incidence increased, and health expenditure declined (Tangcharoensathien et al., 2000). In 2000, poverty incidence ranged from 0.3% in Bangkok to 50% in the poorest provinces, with a national average of 14.2% (UNDP, 2003).

Almost all Thai are Theravada Buddhists.

THE CONTEXT OF HEALTH: ENVIRONMENTAL, ECONOMIC, SOCIAL, AND POLITICAL FACTORS

Thailand's population growth rate dropped from 3.2% in 1970 to 1.16% in 1997. It is projected to decrease to 0.53% by 2020. Between 1964 and 1996, life expectancy rose from 55.9 to 69.97 years for males and from 62 to 74.99 years for females. It is projected that Thailand's annual birth rate will be approximately 15–16 per 1,000 population for some years to come or approximately 900,000–1,000,000 births per year (Pramote, 1998). The crude death rate in 1998 was 5–6 per 1,000. It is expected to rise over the coming years as the population's age structure changes. The infant mortality rate (deaths per 1,000 live births) has dropped from 84.3 in 1964 to 26.1

in 1996; however, the rate in urban areas is consistently lower than in the rural sector. The rate of low-birth weight infants (less than 2,500 g) dropped from 9.3% in 1991 to 7.9% in 1996 but it rose again to reach 8.9% in 1998. Maternal mortality rates dropped significantly from 374.3 (per 100,000 live births) in 1962 to only 10.6 in 1997. Many vector-driven disease threats to health have been reduced. The widespread immunization program has caused vaccine preventable diseases to decline dramatically. For example, tetanus, measles, diptheria, and poliomyelitis have all decreased to low incidence. Leprosy has been eradicated and encephalitis nearly so. Malaria incidence and mortality have also dropped markedly (from 200,000 cases in 1991 to 100,000 cases in 1996) except in border areas with large refugee populations where incidence has recently increased primarily due to parasite drug resistance.

Despite such overall improvements, social changes affecting the everyday lifestyles of most Thai have lead to a changing profile of health and illness. Main causes of death nowadays are heart disease, accidents, cancer, and AIDS. Non-communicable diseases such as cardiopulmonary vascular disease and cancer have become the leading causes of morbidity and mortality over the past 30 years; their prevalence generally attributed to changes in lifestyle and dietary habits. Heart disease prevalence rates (per 100,000 population) increased from 56.5 in 1985 to 168 in 1997, while the cancer rate rose from 53.8 in 1987 to 60.4 in 1997. During the same period, lung cancer prevalence rates rose from 3.96 to 14.2 related to both smoking (currently 38.9% of males over 11 smoke) and air pollution. Alcohol and drug use (both licit and illicit) is also considered to be on the rise. Liver disease rates have increased from 4.3 to 12.3 between 1977 and 1997. Alcohol is also linked to a high prevalence of road accidents.

Accidental death is currently the second highest cause of death for the Thai. The vast majority occurs as a result of traffic accidents, although drowning accounts for death rates between 4.4 and 6.7 (per 100,000). Traffic accidents have declined slightly since the economic crash, causing 11,000–13,000 deaths in 1997–98. The main cause of road fatalities is head injuries incurred in motorcycle accidents, despite a 1992 law making crash helmets mandatory.

HIV/AIDS (and associated tuberculosis) is the most rapidly rising cause of ill health and mortality in Thailand. Since 1984 when the first case was detected, it is estimated that just over 1 million Thai have been infected with HIV (nearly 2% of the adult population) and by 2000 more than 200,000 had died from AIDS. In provinces hardest hit by HIV infection there is a projected life expectancy decrease of 10–13 years for young boys and 5 years for young girls (Van Griensven et al., 1998). In 2003, the UNDP listed the life expectancy of the Thai as 68.9 years—a drop of 1.4 years because of AIDS. Acute respiratory diseases have remained important health problems with little overall improvement. Dengue hemorrhagic fevers are persistent threats with periodic outbreaks and no signs of consistent decline. In short, the Thai now face not only the burden of a huge caseload of HIV infected people, and non-communicable chronic and degenerative diseases but also the emergence of new health threats for instance, drug abuse, occupational hazards, and environmental pollution.

The changing complexion of the health status of the Thai can be directly linked to the social and economic changes of the past 40 years. Thailand's rapid industrialization occurred primarily at the expense of the agricultural sector through the provision of an agricultural tax and encouragement of a mobile workforce. Thai farmers are still the world's largest rice exporters, but they face increasing problems from debt required to stay technologically competitive and from diseases related to a reliance on pesticides. Data collected throughout the 1990s shows that 16–21% of farmers has abnormally high cholinsterase levels (pesticide poisoning). Meanwhile in the cities, occupational diseases directly linked to the industrial sector have emerged—incidence of silicosis, byssinosis (cotton dust disease), and lead poisoning have increased dramatically. Many studies show that health levels in Bangkok are severely compromised by toxic air and water pollutants. For example, it is reported that in Bangkok 27.4% of children had an average lead content in their blood higher than 10 mcg/dl (the level above which it is regarded as dangerous to the brain and nervous system).

More generally, we can link increased vulnerability (both voluntary and involuntary) to 'lifestyle' diseases with changes in the social and cultural life of the Thai. Forty years ago Thailand was 80% rural; for the newer generations traditional village based life has evolved to one oriented to migration and urban wage labor. In addition to a changed family structures, social values are now more deeply embedded in materialism and consumer consciousness than before. In the absence of adequate controls (as government policies do not immediately replace traditional community sanctions), rapid modernization has lead to a wide array of health damaging behavior, ranging from workplace risks and stress to vulnerability directly linked to forms of material purchase (e.g., road accidents, drug abuse, and risk of sexually transmitted diseases through commercial sex).

As a case in point, since the economic crash, mental illness has received greater attention. Rates of psychoneurosis, hypertension, and stress disorders have been steadily on the rise over the past 15 years, as have rates of suicide (up from 6 per 100,000 in 1978 to 11.7 in 1998). Initially it was felt this was a product of competition spurred by the rapidly growing economy but recent increases are associated with unemployment, retrenchment, and financial difficulties as well as the emotional and physical burdens of widespread HIV/AIDS. Stress relief clinics have recently been opened in 125 public and private hospitals.

Changing consumption practices promote an evolution in nutritional status. The overall development of the economy has improved certain aspects of diet: protein energy malnutrition (from eating mainly rice), anemia, and vitamin A deficiency have diminished to almost negligible levels (although the economic crash has caused cuts in the school lunches program and some families are unable to adequately feed their children resulting in the recent increase in low birth weight levels). Iodine deficiency (goiter) is down to a national average of 2.6%. In contrast, the marked shift to packaged and pre-cooked food has caused new nutritional disorders such as high blood cholesterol. Nutritional surveys in 1986 and 1995 showed increased obesity in all Thai age groups—in the 40–49 age group obesity rose from 19.1% to 40.2%. Greater levels of everyday consumption also lead to inappropriate use of pharmaceuticals, particularly antibiotics and analgesics. The Thai have one of the highest per capita drug consumption levels in the world (pharmaceuticals have numerous outlets including village grocery stores). In 1998, 28.5% of national health expenditure was for the purchase of drugs.

Health Infrastructure

Health services infrastructures in Thailand include government, non-government organizations (NGOs), and private sector clinics and institutions. Over the past 30 years the government has allocated slightly less than 5% of its annual budget to the Ministry of Public Health (MOPH). Since the onset of the Five-Year Development Plans in the early 1960s, the highly bureaucratized network of State health facilities has been extended across the country. Gradually patterns of health-seeking behavior have shifted away from self-medication and traditional remedies towards predominant patronage of government services although until the economic crash there was a parallel trend towards private clinics and hospitals where finances permitted. The first three Development Plans (1960–76) concentrated on the expansion of health facilities and trained staff to all provinces. The 4th Plan recognized the ongoing difficulties in the rural areas and proposed more institutional resources, hospitals in all districts and health centers in every sub-district. These goals were implemented in the Fifth Five-Year Plan (1982–86), along with one key change. Bottom-up planning from the village and sub-district level was encouraged to attack poverty and ill health more effectively. Integrating health care with a wider level of rural development, the Thai Government vigorously promoted essential elements of Primary Health Care (PHC) in the villages: health education, nutrition, mother and child health (including family planning), safe water supply and sanitation, immunization, prevention and control of locally endemic diseases, and provision of essential drugs.

As a core aspect of PHC philosophy, local participation in health management has been encouraged through the recruitment of Village Health Workers and Communicators. By 1990, virtually all villages in Thailand had some members trained to assume the liaison role between public health initiatives channeled through the sub-district health centers and their village community. The 6th and 7th Plans (1987–96) aimed to improve public participation towards achieving key WHO benchmarks of reduced mortality and morbidity. Health insurance schemes were begun during this period and the poor issued with health cards that allowed free consultations and care at government facilities. In face of enduring economic difficulties, the Government has recently replaced this with a more efficient and equitable "30 Baht" (U.S. 75 cents) scheme that guarantees subsidized medical treatment for all below certain income levels.

Despite widespread provision of services, analysis of health shows that achievements have not always been forthcoming in all sectors. The lip service given to fundamental social development fuelled by PHC programs is sometimes no more than the manipulation of statistics. This shrouds the enduring factors that promote ill health in a wide array of sectors, as the poverty figures from rural Thailand and the emergence of new forms of morbidity attest. Severe shortages of trained medics in some rural areas were exacerbated by the economic crash, when budget cuts caused medical staff to seek work in the private sector. Against this backdrop, NGOs throughout Thailand have made attempts to generate more fundamental changes to social and political systems including health. The increased recognition of civil society organizations as a fundamental foundation of a secure social environment has lead to the proliferation of alternative health movements and organizations. Historically such groups have included religious and ethnic welfare organizations. Following the 5th Plan, NGOs received more active support from the Government and a recent survey listed 513 organizations whose activities are related to health promotion. In 2000, 126 organizations applied for MOPH support including advocacy groups, AIDS networks, alternative and traditional medicine groups, consumer's forums, village volunteers groups and so forth (Amara et al., 2001).

Medical Practitioners

The Thai practice pluralistic preventive and curative strategies engaging a range of health practitioners. The medical system in Thailand was oriented to traditional Thai health regimens until the late 18th century. Over the past 100 years, Western medical practice has become dominant in terms of government support and local patronage. Health practitioners trained in (cosmopolitan) biomedicine are therefore the most prominent in Thai society.

In 1998, there were 19,500 practicing doctors (a ratio of 1 : 3,136 people, with a disproportionate number in the urban sector). Between 1971 and 1995, the percent of these doctors working in the public sector declined from 93.3% to 76.3% reflecting a move into the more financially rewarding private hospital sector (in 1998 there were 957 public and 473 private hospitals, some of which have become famous worldwide for cosmetic surgery). Since the economic crash this trend has been reversed. In 1998, there were 56,366 professional nurses (1 : 1,073) with the vast majority (roughly 90%) working in the public sector.

A large number of government doctors (and sometimes nurses in rural areas) also run private clinics during out-of-work hours. Here they offer a range of services from general practice to specialist medical services. It is estimated that as high as 69% of practicing specialists have not been certified by the Thai Medical Council. The medical doctors (and nurses) are able to increase their income not simply from their diagnostic services but also from pharmaceutical sales.

Community health workers at sub-district health centers form a crucial link between state policy and village level communities—they initiate government health programs at the village level providing technical support and supervision. In 1999, there were 9,689 health centers covering every sub-district in Thailand, staffed by 39,438 health workers who have received a 2-year tertiary training course in public health. Staff are equipped to handle minor ailments, monitor maternal and child health, and are licensed to prescribe commonplace pharmaceuticals and contraceptives. They supervise Village Health Volunteers who assist with the dissemination of public health knowledge and the distribution of certain pharmaceuticals and health provisions such as painkillers and condoms. The volunteers also monitor the health status of the villagers and report back to the sub-district health centers. They administer certain village organizations, such as drug cooperatives, rice banks, and health card funds.

Pharmacists are another key link in the Western style bureaucratic management of health. Many Thai self-medicate and pharmacists often provide diagnostic services for particular ailments, selling most pharmaceuticals without a prescription. A common (although lessening) practice in rural areas is the sale of combination therapies. This entails the dispensing of a number of pharmaceutical drugs in one package. Occasionally rural pharmacists will compile and then sell these packages to village stores. At times they include dangerous combinations of stimulants and depressants (Lyttleton, 1996).

Other health practitioners conduct therapies outside of the biomedical framework. After years of marginalization, traditional forms of Thai medical practice are undergoing resurgence as biomedical models are seen as not being the answer to all ills. Although fast disappearing, some villages still have elderly members who practice traditional arts of massage and prescribe herbal medicine therapies. In place of informal training Thai traditional massage is increasingly taught in national institutes, which receive both government and international support. Thai massage is widely practiced throughout the country, albeit with varying degrees of expertise. Traditional herbal medicine is also a form of local knowledge that has received government and foreign support. Traditional Thai medical institutes both train doctors and study plant forms and their efficacy in treating illness. Since the rapid increase in AIDS-related illnesses it is common for

HIV/AIDS support groups to both cultivate and prescribe local herbal remedies. International agencies are presently including a number of Thai plants known for their retroviral activity in their drug trials. Village midwives form a third group of alternative practitioners receiving official recognition although after the rapid economic development most village women give birth in local hospitals.

A further group of health practitioners receive no legitimization from government circles, however they form an important component of multiple health seeking strategies. Firstly, there is a large group of healers who utilize spiritual assistance. This is more common in rural areas although some well known 'divine' healers have large urban followings. These healers are usually of two types, those that rely on Buddhist or Hindu modalities of sacred power, and those that communicate with the host of animist spirits that are believed to be ever-present in everyday surroundings. Healing entails divination of causality and prescription of remedial action. Secondly, there are those that illegally practice forms of Western medicine. It is common in more remote rural areas for "injection doctors" to perform injections for villagers. This involves diagnosis, prescription and injection of pharmaceuticals, saline, and/or vitamins. Backyard abortion doctors are also found in both rural and urban areas. A small number use traditional techniques of severe massage intended to snap the fetus's neck along with herbal effusions to induce miscarriage.

CLASSIFICATION OF ILLNESS, THEORIES OF ILLNESS, AND TREATMENT OF ILLNESS

Strongly influenced by the Ayurvedic system of medicine in India, traditional Thai medicine classifies forty-two bodily elements into four groups—fire, wind, water and earth. Disease comes from changes in the balance of elements of fire, wind, and water—earth is very stable and seldom causes disease—and correspondingly illness is classified as a disease of either bile, wind, or mucus (Mulholland, 1979). Historic Thai medical texts report wind illness as the most common affliction which, following the Indian humoral theories of medicine, denotes any illness affecting the quality or faculties of movement, including such diseases as leprosy or epilepsy. In practice, it is highly likely that over the centuries a synthesis has occurred between imported Indian medical systems and indigenous beliefs in causes of ill health that include soul loss as the cause of a wide array of ailments. Traditional Thai medical practice is best considered as a continuum from oral folk medicine practiced at the village level and more closely linked to animistic beliefs through to more official elite versions associated with text-based Indian theories of disease (Bamber, nd).

In traditional Thai medicine, diagnosis involves careful consideration of environmental, behavioral, and physiological characteristics. Treatment is determined after considering symptoms, place of birth, age of patient, and time of year and primarily uses medicines concocted from a wide range of plants and occasionally animal and mineral substances (1,500 drug remedies are listed in handbooks). Massage and use of spells can also be enlisted to effect a cure depending on the extent to which malevolent spirits are believed to be involved. Towards the end of the 19th century, the adoption of Western medicine was spearheaded by the Royal Family's interest in "modernizing" indigenous medical practices at a time when cholera epidemics were widespread. The association of Western medicine with scientific knowledge, modernity, and royal patronage led to its rapid domination of the traditional Thai medical practice. It has been described as elitist (in terms of class and status), capitalist, drug oriented, urban centered, and assertive of its dominance over other health practices by means of subordination or exclusion (Cohen, 1989). Biomedicine is almost invariably the first option in seeking medical assistance. A majority of Thai self-medicate as first choice for minor ailments using pharmaceuticals although a recent study shows that self-medication has dropped as socioeconomic status has improved (Komatra et al., 2000). If medical expertise is required, private clinics are usually preferred over government health services (just as commercial pharmaceuticals are preferred over government-issue) although this will depend on practical issues of distance and cost.

Even though biomedicine dominates as a paradigm for diagnosis and treatment, many Thai still employ pluralistic explanatory models and therapeutic strategies. Magico-religious beliefs remain complementary components of health-seeking behavior. They provide answers to "why" an illness has taken place. In a social order strongly predicated on a belief in *karma* as a causal ontology, current situations are frequently interpreted as emerging from prior events. While Western etiology is almost universally accepted as explaining morbidity, spirit doctors might also be consulted to determine whether malevolent spirits were involved in its onset. Buddhist, Hindu, and animist beliefs commonly intersect in the prescribed rites and remedies. Recalling a lost soul is a crucial element both in preventive and healing rites. Persistent diseases or those without a clear pathogen, such as forms of mental illness, are still commonly treated by a divine or "spirit" doctor who will either divine, incarnate, or do battle with, the inflicting forces. Although belief in spirits is declining in favor of pharmaceutical treatment for most diseases, a Buddhist and Hindu background confers legitimacy on traditional Thai medicine that ensures a widespread faith in Thai herbal remedies. These are still widely sold and consumed.

SEXUALITY AND REPRODUCTION

Sexuality for the Thai has often been described as a "fluid" arena where what goes on in private is of less concern than the maintenance of public appearances. This has particular relevance to same-sex practices. Similarly, debates concerning Thai sexuality often draw on the tolerance shown for cross-gender displays and whether the widespread presence of *khathoey* (transgenders/transsexuals) constitutes an ontological third sex/gender. Despite an arguable lack of fixity in sex/gender identities, normative models of gender and sexuality still strongly shape the sexual and reproductive lives of the Thai.

Gender ideologies, based on predominant bilateral kinship patterns and Buddhist beliefs in the innate karmic superiority of the male, structure ideas about appropriate sexuality. By and large, Thai men are socialized to believe they have a "naturally" greater sex drive that requires consistent relief. Many men therefore have multiple and regular sexual interactions both prior and external to marriage most commonly within the context of commercial sex. As adolescents, male same-sex behavior carries no stigma if the male maintains his role as penetrator. Thai women, on the other hand, are taught to be more conservative and express sexuality solely within the bounds of marriage. These social values are changing rapidly with modernization and young Thai women are more sexually active than ever before. Gender roles have never entirely subordinated women—the Thai have one of the highest levels of female participation in the workforce (over 45% of total workforce) and within the household they have active economic and decision-making capacities including family size. Remnants of the widespread belief that female sexuality is dangerous and polluting still find contemporary manifestation particularly in early HIV/AIDS campaigning that targeted female prostitutes as the cause of its spread.

A direct result of both the prevalence of commercial sex interactions and the liberalization of youth's sexual mores, sexual health is a pressing issue for the health of many Thai. By the late 1980s, sexual transmission had become the major vector of HIV spread and by the late 1990s accounted for 82.6% of AIDS cases. Increasingly women are becoming infected, mostly from their husbands. By the end of 1999 UNAIDS calculated that of the 740,000 adults living with HIV/AIDS, 305,000 or 41.2% are women. In 2000, it was estimated that roughly 50% of new infections in Thailand would be women infected by their husbands or sex partners, 25% through injection drug use, and 20% among sex workers and their clients (World Bank, 2000). Due to effective campaigning, patronage of prostitution has plummeted throughout Thailand. But in its place there has emerged a more complex and ambiguous arena of negotiated relationships in which casual sex takes place. What is of current concern is that condom use frequently does not accompany sexual interactions that take place outside of institutionalized commercial sex raising the threat of continued HIV spread, particularly in light of burgeoning permissiveness and drug use amongst many Thai youth.

The inability of women to insist on condom use, within or outside marriage, highlights ongoing forms of sexual vulnerability. Over the past 30 years Thai women had actively embraced contraception as a means of limiting the family size but this seldom included condom use (average family size has dropped from 5.6 members in 1960 to 4.4 in 1990 and is expected to reach 3.4 in 2015). Easy access to a wide range of contraceptive measures without social or religious stigma coupled with extensive health promotion advocating the benefits of smaller families is behind the high prevalence of contraceptive use. At present, Thailand's total fertility rate, which was as high as six 30–40 years ago, has declined to only 1.9 in 1996. By 1987, 70.5% contraceptive prevalence nationwide had been achieved rising to 72.2% by 1996. The oral contraceptive is the most popular, followed by female sterilization and injectable hormonal contraceptives. These three account for 85% of contraceptive use. Male methods of family planning are almost negligible. Vasectomy is cited by 2.0% of married women of reproductive age and condoms by 1.8% as their methods of contraception (Gray & Sureeporn, 1999).

Declining fertility is also linked to delayed age of marriage and increasing percentage of women who remain unmarried (reaching 14% in 1990 for women between 30–34 years of age). This trend is a product of higher educational levels for women, higher levels of employment in the highly modernized occupational sector, increased urban migration and gender-specific acceptability for men to marry spouses of lower socioeconomic status (Guest & Tan, 1994). For women of higher socioeconomic status being unmarried carries no particular stigma, whereas more pressure is placed on rural women to marry even if the spouse is not a focus of affection.

Between 8% and 12% of women seeking gynecological care in government hospitals are found to be infertile. Adoption is not common amongst the Thai due to beliefs that an orphan has little karmic merit. In the past, particularly in rural communities, divorce has been regarded as a relatively straightforward occurrence (officially, it has increased from 8.3% in 1987 to 20.8% in 1998 although many rural marriages are not registered). In some instances, inability to have children may be an instigating factor in marital separation but I am unaware of any survey of its prevalence.

HEALTH THROUGH THE LIFE CYCLE

Pregnancy and Birth

Biomedicine has taken complete control over the management of pregnancy and birth—subdistrict health clinics provide antenatal care and women give birth in district or municipal hospitals. Cesarean births account for over 50% of all deliveries partly due to the desire to choose an auspicious birth date. For many Thai, marriage is less important than pregnancy in marking female transition into responsible adulthood although these days pregnancy is increasingly postponed as the married couple works to gain some material foundations. Unmarried parenthood is not widespread and if marriage is not feasible, unwanted pregnancies commonly end in abortions (54% of abortions are performed on women under 25 years old). Following biomedical models, miscarriages are understood in physiological terms however rites restoring potential or actual soul-loss are often performed. Both male and female children are valued—males for the karmic merit they can bring the family (particularly through ordination into the Buddhist monkhood) and females for their practical household assistance.

Infancy

Traditional postpartum confinement rites wherein the mother would lie by a fire in her house for 5 to 9 days, have largely disappeared. Coupled with specific food restrictions geared to restoring humoral balance, this was believed to restore and strengthen her body and soul and complete a first-time mother's transition into mature womanhood (Hanks, 1963, p. 71). Such practices have been dispersed by changing beliefs as to the nature of recuperation, practical issues such as lack of fireplaces in modern houses, lack of time, and the recommendations of doctors who discourage it especially after Cesareans and/or tubal ligations.

Lying by the fire was based on beliefs that females are soft natured/souled and spiritually vulnerable after birth. Such beliefs also apply to the child, and ritual specialists still typically perform soul protection ceremonies shortly after the child is born.

Breast-feeding has declined in favor of bottle and formula feeding in sync with the large presence of many young mothers in the workforce. Public health campaigns have recently advocated breast-feeding as the most desirable option for new mothers. Infants and young children are typically cared for by extended family members and older siblings, as it is common that both parents work. National rates of young children (ages 3–5) enrolled in nurseries and pre-school childcare rose from 39.3% in 1992 to 79.5% in 1998. On the other hand, child abandonment has also increased. A survey of abandoned infants at 40 health facilities showed a rising rate from 90/100,000 to 120/100,000 in 1998.

Childhood

In the past young rural Thai boys and girls would begin to work the fields at around the age of 11 or 12. From some parts of Thailand, teenage girls were sent into the sex industry shortly after reaching puberty. These days such practices are illegal and prosecuted as child abuse. Six years of state-sponsored secondary school education has recently become available nationwide ensuring a later entry into the workforce. Gender roles still strongly influence models of socialization with male qualities believed to be best nurtured through avoidance of strict discipline whereas girls are required to carefully observe parental rules. In early childhood, it is not uncommon for the young boy's genitals to be an item of attention (joking and tickling), young girls on the other hand are generally taught to be modest and compliant from an early age. Pneumonia is the main cause of death in children under five, although its incidence has dropped from 5.2% in 1995 to 2.73% in 1997.

Adolescence

The major health problems of youth (10–19 yrs—20% of population) are linked to social behaviors including substance abuse, teenage pregnancy, sexually related health problems (STDs), and mental health issues such as violence and suicide. Maternal mortality is highest in the 15–19 age group often resulting from unsafe and illegal abortions. Thai adolescents face large risks from STDs, in particular HIV/AIDS—47% of STD patients are between the ages of 15 and 24—and in the first decade of HIV spread from 1984 to 1996 single females under 24 years of age were the highest risk group for infection, largely through commercial sex (Pimpawun et al., 2000). Nowadays, for many Thai "safe sex" is as much an issue of *who* one has sex with as it is of using a condom or not. As sexual relations between Thai adolescents take place more commonly, and condom use is not prevalent as these relations are seen as safe because they are not commercial sex. Genital operations, the insertion of small round objects (pearls) under the foreskin, are found amongst a small number of Thai males but it is unclear whether this contributes to the spread of disease. Substance abuse, in particular of methamphetamines (ATS), has also increased dramatically amongst Thai youth. Over 6 months in 1995/96 a nationwide survey conducted a urine analysis of 118,375 students from primary through university level. Of the 79,671 secondary students sampled, 1.1% tested positive to methamphetamine use and 3.8% of the 17,082 vocational students tested positive (Vichai et al., 2001).

Adulthood

As the majority of pathogen-based diseases decline, social factors also contribute to the morbidity and mortality of Thai adults. Many of these can be linked to rising levels of stress. It is not only the youth who are susceptible to substance abuse. One survey reports that more than 300,000 Thai labourers (20%) in Bangkok use methamphetamines as a response to stringent quota demands incurred by the 1997 economic crash. The overall number treated for ATS abuse has risen from a paltry 97 in 1990 to 12,518 in 1999 (UNDCP, 2001). Domestic abuse is widespread. In a recent study (Kritiya et al., 2001) conducted with over 2,000 women in Bangkok and a Northern Province, 44% reported experiencing sexual or other forms of violent abuse from their partner (22% within the 12 months). Male suicide rates are currently 2.8 times higher than that of females and rank 6th in terms of years-of-life lost to premature deaths (AIDS is the highest for both men and women). AIDS also contributes to the socio-economic burdens felt throughout Thailand at the turn of the millennium. Many thousands of HIV-infected Thai women are widows who, having nursed their husbands, are left without economic resources to take care of themselves and their children. Hundreds of support groups throughout the country have come together to provide moral and social support for those affected by HIV and to date their membership is largely women.

The Aged

In accordance with Buddhist ideology the elderly generally have high status within the household; in rural areas they have traditionally resided with the youngest daughter and her husband. Home care is typical for most manageable forms of illness. The most common diseases of the elderly are muscular, tendon and skeletal disorders, followed by peptic ulcers, hypertension, and heart disease. Primary causes of mortality are cardio-pulmonary disease, cancer, diabetes, liver, and kidney disease. A 1995 survey (Chantaphen, 1995) of both rural and urban elderly showed 82.8% of those between 50 and 59 still worked which dropped to 42% for those 60 and over. The most common diagnosed ailment was back pain and 29.1% had been hospitalized in the previous year. Forty-two percent exercised regularly and 57% meditated daily. Illiteracy is high amongst this group and self-medication, especially analgesics and vitamins, by far the most common form of health-seeking treatment.

Dying and Death

Where possible, it is considered preferable for dying to take place at home rather than in a medical institution. Terminal patients are often sent home from hospitals in their last stages of life. The dead are cremated and Buddhist rites take place at the local temple and the deceased's residence to ensure smooth passage to the next life and to protect kin from potential malevolence associated with the dead spirit. The remains of the bones will sometimes be stored in receptacles within the temple walls. Although less common these days, distinction between normal and abnormal death is made that corresponds to "natural" versus unexpected, violent or accidental forms of death that prevent the completion of a life cycle. Bodies of those who have suffered an abnormal death will be buried for several years—the remains are subsequently exhumed and cremated—to prevent similar misfortune befalling kin. These beliefs are no longer widespread. There is usually no change to the living arrangements of the remaining spouse with the existing kin.

REFERENCES

Amara Pongsapich et al. (2001). Current status of civil society organizations in health sector in Thailand, unpublished report, Social Research Institute, Chulalongkorn University, Bangkok.

Bamber, Scott. nd. *Lom*: A term in association with Illness in traditional Thai medicine, unpublished paper.

Chantaphen Chupraphawan et al. (1995). *Health survey of Thai over 50 years of age* (in Thai). Bangkok: Health Research Institute.

Cohen, Paul. (1989). The politics of primary health care in Thailand, with special reference to nongovernment organizations. In P. Cohen & J. Purcal (Eds.), *The political economy of primary health care in southeast asia* (pp. 159–176). Canberra: Australian Development Studies Network.

Gray, Alan., & Sureeporn, Punpuing. (1999). *Gender, sexuality and reproductive health in Thailand*. Institute for Population and Social Research, Mahidol University, Thailand.

Guest, Philip., & Tan, Jooean. (1994). *Transformation of marriage patterns in Thailand*, Institute for Population and Social Research, Mahidol University, Thailand.

Hanks, Jane. (1963). *Maternity and its Rituals in Bang Chan*. New York: Cornell University Press.

Komatra, Chuengsatiensup et al. (2000). Community drug use in Thailand, unpublished report, Faculty of Pharmacy, Chiang Mai University.

Kritiya, Archawanichakul et al. (2001). Domestic violence against women: The scale of the problem, its impact and solutions, (in Thai). Research Report, Institute for Population and Social Research, Mahidol University, Thailand.

Lyttleton, Chris. (1996). Health and development: Knowledge systems and local practice, *Health Transition Review, 6*, 25–48.

Medhi, Krongkaew. (1992). The reawakening of the Thai economy, paper presented at the Thailand Update Conference, Sydney 16 October.

Ministry of Public Health (MOPH). (1999). *Health Situation 1997–1998* (in Thai). Bangkok: Ministry of Public Health (English version at http://eng.moph.go.th).

Mulholland, Jean. (1979). Thai traditional medicine—The treatment of diseases caused by the Tridosa. *The Southeast Asian Review, 3*(2), 29–38.

Osborne, Milton. (1995). *Southeast Asia: An introductory history*. Chiang Mai: Silkworm Books.

Pasuk Pongpaichit., & Baker, Chris. (1996). *Thailand's boom*. Chiang Mai: SilkwormBooks.

Pimpawun Boonmongkon et al. (2000). *Thai adolescent sexuality and reproduction: Implications for developing adolescent's health programs in thailand*. Center For Health Policy Studies, Mahidol University, Thailand.

Pramote, Prasartkul. (1998). Population projection for development planning of Thai children. *Family Planning and Population, 2*(1), 1–4.

Tangcharoensathien, Viroj et al. (2000). Health impacts of rapid economic changes in Thailand. *Social Science and Medicine, 51*, 789–807.

UNDCP. (2001). 'Global Illicit Drug Trends 2001'. Available online at, htttp://www.undcp.org

UNDP (2003). Thailand Human Development Report 2003. Bangkok: UNDP.

Van Griensven, F. et al. (1998). *The use of mortality statistics as a proxy indicator for the impact of the AIDS epidemic on the Thai population*. Bangkok: Institute of Population Studies, Chulalongkorn University.

Vichai, Poshyachinda et al. (2001, October, 3–6). The future outlook of HIV infection among substance abuse population in Thailand. Paper presented at the 6th International Conference on AIDS in Asia and the Pacific, Melbourne.

World Bank. (2000). Thailand's response to AIDS: Building on the success, confronting the future, (unpublished report), *Thailand Social Monitor V* World Bank: Bangkok.

Wyatt, David. (1982). *Thailand: A short history*. New Haven: Yale University Press.

Tongans

Barbara Burns McGrath

ALTERNATIVE NAMES

Tongans originate from the Kingdom of Tonga, formerly known as "The Friendly Islands." Due to out-migration, there are more Tongans living overseas than within the country, primarily settling in New Zealand, Australia, and the United States. Outside of Tonga, Tongans are included under the more general category of Polynesian or Pacific Islander.

LOCATION AND LINGUISTIC AFFILIATION

The Kingdom of Tonga is made up of 170 small islands, approximately 40 of which are inhabited. These are spread over 700,000 sq km of the South Pacific about two third the way from Hawai'i to New Zealand. They are characterized as high islands formed by geologically recent volcanic activity, and low coral limestone atolls. The climate is sub-tropical with temperate weather and high humidity. Rainfall is moderate and regular so that most crops are planted and harvested throughout the year. Cyclones and hurricanes occur every few years, and can result in extensive damage to buildings and crops. Tongan is in the Proto-Polynesian branch of the Austronesian language family. Tongan and English are the main languages spoken. The literacy rate in Tonga is 98%.

OVERVIEW OF THE CULTURE

Tonga is a relatively homogenous society (most of its population of approximately 105,000 are Polynesian) with a highly stratified social structure headed by the monarch at the national level, and by chiefs and chiefly lineages at the local level (Marcus, 1978). The first ruler, Tu'i Tonga reigned during the late 9th century. He was the son of Tangaloa, the god of the sky, and Va'epopua, an earthly mother. Succeeding holders of the title of Tu'i Tonga descend from Tangaloa and form the top of the pyramid-shaped social organization, the base of which is made up of commoners. In 1643, the Dutch explorer Abel Tasman noted the long-established peace and stability of the islands, which he felt were due to the system of reciprocal relations that existed between the chiefs and commoners. In 1845, Tupou I united the islands that were ruled by a number of competing chiefdoms (Campbell, 2001; Lātūkefu, 1974).

Much of the early history of Tonga is recorded in oral tradition with genealogies providing a chronology of events. They continue to be used to evaluate rank, seniority, and status, and are called upon to mediate disputes involving land allocation and title succession (Herda, 1990). Poetry and dance are another means used to record both culture and history. These often link events to specific localities of the Tongan landscape emphasizing the connection between the people and their environment. This ecology-centered concept of culture and history (*tala-e-fonua*) is being explored by contemporary Tongan scholars to examine past events from a new perspective (Hau'ofa, 2000; Mahina, 1993).

Despite a long history of European contact, Tonga is the only island group in the Pacific to avoid colonization. It became a constitutional monarchy in 1875 and a British protectorate in 1900. Tonga acquired its independence in 1970 and became a member of the Commonwealth of Nations. The government reflects the hierarchical social structure with decisions being made primarily by the monarch (King Taufa'ahau Tupou IV), the nobility, and a few prominent representatives of the commoners. Calls for constitutional change and government reform began in the 1970s and 1980s and coincided with increased efforts at economic development in Tonga. The 1987 general election is considered a turning point in modern political history with the addition of new representatives who then spawned a pro-democracy movement (Campbell, 1994). This political movement is challenging traditional assumptions about power and rank, and subsequently having far reaching consequences in the country beyond the political system.

Missionary influence, beginning in 19th century, was very successful with the result that today Tonga is a strongly Christian nation. The major religions are Free Wesleyan Church of Tonga, Catholic, Free Church of Tonga, Anglican, and Seventh-Day Adventists. More recently, The Church of the Latter-Day Saints has been growing in popularity with 46,000 members, giving Tonga the distinction of having the highest percentage of Mormons of any country in the World.

The economy is based on the cultivation of tropical and semitropical crops including squash, coconuts, bananas, and vanilla beans. Agricultural exports make up two third of total exports. Demand for imported food (primarily from New Zealand) and manufactured goods and products that are unavailable locally have resulted in a sizeable trade deficit. Overseas aid and remittances from Tongans living abroad are critical aspects of the economy, and in turn have great influence on the affairs within the country.

The *'api* or household is the basic unit of the society. *Kainga* is another important unit, and describes extended family, or in some cases, those persons loyal to a particular chief (James, 1990). Mutual support with formalized obligations and privileges are realized within *kāinga* social relations. Considerations of rank and status affect daily life and are organized along a strict hierarchy. Although the father is the head of the household (*'ulomotua*), sisters have higher status than brothers, with the father's eldest sister (*mehekitanga*) having great power within a family (Rogers, 1977). Other factors such as birth order, age, and mother's personal rank determine an individual's rank in relation to another person. Both religion and kinship determine the rank of chiefs (Campbell, 2001). Prestige that was dependent on birthright and rank in the past is now also available to those who are able to achieve individual success such as with an advanced college degree, a job with high status, or personal wealth.

Adoption and fosterage among kin is very common and parents who give up their children (perhaps to a childless relative) are looked upon as generous. Migration has had a more recent effect on adoption practices in Tonga with children living overseas being sent back to Tonga to live with relatives for varying lengths of time (Gailey, 1992; James, 1991). It may be a temporary solution until the parents have established a stable home, or an arrangement for children to live with relatives for a few years as a way to ensure that they learn Tongan ways and proper behavior. In other cases, if the child is doing poorly or in trouble with the law in the new home, he or she is sent back to Tonga in hopes that the family, often the grandparents, can exert some control over the child.

Tongans are highly mobile and have a long history of travel, both sea-faring voyages and for more permanent settlement. Today there is a pattern of kin based migration resulting in a very transnational population of Tongans. These diaspora communities have an impact on affairs in Tonga because, in addition to sustaining the economy through their remittances, they are a conduit for new ideas and practices from the outside. Because of the Internet, communication among all the diaspora communities and Tonga is very free flowing. Most of the web sites have chat rooms where dialogue and gossip can occur.

In Tonga, as in most places around the world, globalization is viewed with mixed feelings. There is popular appreciation and pride for the fact that certain cultural values and traditions are key to Tongan identity (love, respect, sharing, etc.). This is accompanied by explicit efforts to retain these aspects of the culture, and reflected in slogans or tourist marketing schemes that proclaim Tonga as the most traditional island in Polynesia. At the same time, participation in a global economy and having access to some of the benefits (material and nonmaterial) of modernization are also desired (ease of travel, health care, etc.). Some changes are being welcomed, but there are also unintended consequences that are cause for concern (alcoholism, suicide, unemployment).

THE CONTEXT OF HEALTH: ENVIRONMENTAL, ECONOMIC, SOCIAL, AND POLITICAL FACTORS

The health care system of Tonga may be characterized as a medical pluralism with coexisting paradigms. It includes more than one medical ideology and more than one system of services so that an individual has a choice of receiving care from a traditional Tongan healer (*faito'o fakatonga*), or a physician or registered nurse who has been trained and who practices within the biomedical tradition. There is a range in attitudes toward biomedicine and traditional practices with some families relying on a single system. Most however, use both approaches when treating illness or seeking health. Health is not simply the absence of disease, but also includes having a good life, being lucky or fortunate, and being satisfied that you are fulfilling duties to God, your country, and family. Health includes all of these states, and illness strikes when any one is out of balance.

William Mariner is a popular source of information regarding healing practices of early Tonga (Mariner & Martin, 1981). In 1806 he was aboard an English ship that was attacked and burned. Mariner's life was spared and during his four years in Tonga he recorded his observations of healing practices, including the skill used in performing operations, such as removing an arrow point lodged in a man's lung, and in setting broken bones. Historical accounts, together with interviews conducted in the first part of this century (Beaglehole & Beaglehole, 1941; Bott, Salote, & Tavi, 1982; Collocott, 1923; Ferdon, 1987; Gifford, 1929) create a picture of healing among the chiefly class in early Tonga with the supernatural at the center of the medical paradigm. In order for a cure to be successful, the god or the spirit of a deceased ancestor who was causing the trouble had to be appeased. This was done with sacrifice and invocation. Tongans had a reputation with the other islanders in the Pacific as being also knowledgeable about herbs and internal remedies (Macpherson & Macpherson, 1990). In 1990, it was estimated that there were over 200 healers practicing on the main island of Tongatapu.

The more recent history of biomedicine is intimately tied to the history of Christianity in Tonga, as it is in much of the Pacific. Missionaries quickly learned that villagers who would not attend church for a sermon, traveled great distances for medicines, and the dispensary was soon part of the mission structure (Lātūkefu, 1974). Although it is unlikely that missionary medical treatments of the time were any more effective than what was being practiced, they did offer an alternative (Shineberg, 1978).

Today there are two hospitals in Tonga with outpatient departments, dozens of public health clinics, and a few private physicians who practice primarily in the capital,

Nuku'alofa. Health care that is provided at a government facility is free for all citizens and administered by the Ministry of Health. Patients are often referred to New Zealand for medical care that is unavailable in Tonga. The leading causes of death are cardiovascular, neoplasm, diabetes and related conditions, respiratory, and hypertensive conditions. Life expectancy is 68 years.

MEDICAL PRACTITIONERS

Health care providers in Tonga may be divided roughly into two types: traditional healers (*faito'o fakatonga*) who prescribe herbal cures, set bones, and use therapeutic massage, and biomedical practitioners (physicians and registered nurses) who are located in the hospitals and clinics.

Traditional Healers

Both men and women can be healers, though there appear to be more women. Entry into the profession varies, but most individuals are young adults when they begin to practice. Recruitment is informal with a tendency to choose a relative. The person who is selected may be one who showed an early interest or aptitude, or one the healer decides is appropriate to receive the skill. When it is time for a healer to retire, he or she will transfer the knowledge and skill to the successor. This ritual has a name, *fanofano*, but may be as simple a saying a few words, "I give you my *faito'o.*" Other healers learn of their ability by accident or are told about it in a dream. If they are able to effect cures in people previously labeled as incurable, they are described as having *mana*. Healers live ordinary lives, receiving minimal material benefits. They receive gifts from patients, but these are more tokens of gratitude than payment. These can be as small as a few cigarettes or as valuable as woven mats or bark cloth for a particularly impressive cure. During the treatment, *kava* or food is often shared. Because the healer is merely the *vaka* ("vessel") for the healing power of God, abuse of this divine gift will result in the healer losing the ability to cure. The sins of greed or pride put the healer at risk for possession by the very spirit being fought (Cowling, 1990).

Healers tend to be specialists, not general practitioners, and those who claim to have cure-alls have little credibility (Parsons, 1985). As in the past, infusions and poultices made from herbs, leaves and bark are prescribed. New ones are always being added to the pharmacoepia, and in 1992, a botanist collected information on 77 that are commonly known for their curative properties (Whistler, 1992). The art of massage is practiced throughout Polynesia for a variety of external and internal ailments. Other specialties include pediatrics, midwifery, setting bones, and treating spirit possession.

Biomedical Practitioners

Following the influenza epidemic of 1918 that took the lives of 8% of the population, Queen Sālote established a Department of Health with free medical care for all. Most practitioners studied medicine in Fiji, and later in Samoa (Wood-Ellem, 2001). Today there are approximately 70 physicians, most of whom trained in Fiji, New Zealand, or Australia, and 340 nurses (Young Leslie, personal communication, 2002). Nurses attend the nursing school in Tonga for their basic education and study abroad for advanced training.

The oldest and largest hospital is on the main island of Tongatapu with 200 beds, running at 50–60% occupancy. The average adult duration of stay is 10.1 days. The reasons for hospitalization match the recorded leading causes of mortality in Tonga: diseases of the circulatory system, neoplasm, intestinal infectious disease, respiratory disease, and digestive disease.

When a person is admitted to the hospital, it is assumed that a relative will be available for personal care such as bathing and feeding. The caretakers of the patients create an informal system within the institution. Physicians and nurses receive respect as individuals who have gifts for healing, however, many people view the hospital with a mixture of relief and fear. Relief that Tonga now has a modern facility to care for its citizens, and fear because it seems that everyone who is admitted dies there. The hospital is referred to as *fale mahaki*, house of sickness. The medical response to the fear of death is that statistics do not confirm this perception, but what does happen is that people wait so long before coming in that it is too late to save them, and the mortality rates are higher than they need be. For their part, health care workers express ambivalence toward traditional practice. One would be hard-pressed to find an individual who has not experienced the therapeutic effects of massage when skillfully administered, but traditional methods are also described as dangerous, and should not be used in conjunction with biomedical therapeutics. Their use may also hinder a person from seeking medical care. This problem is being addressed in a very interesting public education campaign currently underway that is attempting to assist traditional healers make a differential diagnosis between spirit possession and schizophrenia in order to avoid the harmful delay of medical treatment for major psychotic disorders (Puloka, 1999).

CLASSIFICATION OF ILLNESS, THEORIES OF ILLNESS, AND TREATMENT OF ILLNESS

The concepts of *mana* and *tapu* are central to an understanding of health and illness. *Mana* refers to the degree to which objects and persons are endowed with supernatural powers. Having *mana* offers some protection against misfortune and bad luck, and it is the force that allows healers to act as a vessel for the healing power of God. *Tapu* is also a complex concept that combines a number of distinct notions (Radcliffe-Brown, 1952). *Tapu* is sacred, and describes a connection with the gods. It also describes those things that are forbidden or prohibited (the English word "taboo" is derived from this Tongan term). It is in this sense that the early Tongan chiefs and objects touched by them were *tapu*. These two dimensions of social control also have cosmic consequences, so that breaking a *tapu* or misusing *mana* may result in misfortune through the agency of spirits (Levy, 1973; Shore, 1989).

Illness is talked about in a number of different ways. One system distinguishes Western illnesses from indigenous ones. Conditions thought to be introduced by Europeans, or at least best treated with Western therapies, are *mahaki faka Palangi* ("European sickness"). The category *mahaki fakatonga* includes everything else (Parsons, 1985). Another classification that is used describes those misfortunes caused by "natural" events as *mahaki pe* or *puke pe* ("just sick"), as opposed to problems caused by spirits. The latter are *mahaki* (or *puke*) *faka tevolo* (McGrath, 2000).

The word *tevolo* is a complex term, most often glossed as "devil," but more accurately refers to ancestral and nameless spirits (the older term for *tevolo* is *fa'ahikehe*, "other side" (Ka'ili, personal communication, 2002). Puke *faka tevolo* implies supernatural agency. The cause of an illness is an action or behavior by the person, or someone in the family, that is offensive or disrespectful to God, an ancestral spirit, a chief, or to a superior on earth who is in a higher social position. In addition, failure to fulfill one's responsibility to family or church is implicated in cases of misfortune. Breaking a *tapu* leaves one open to punishment until forgiveness is requested and granted. In some cases even though forgiveness is given, the punishment will be a lethal illness. This does not negate the importance of correcting the wrong. If forgiveness is not granted, the punishment will be passed on to descendants and they will suffer for many generations. Spirit-caused illness is most often a result of improper behavior, but spirits also play tricks on people or act mischievously (McGrath, 2003). The actions of *tevolo*, and the way people use them, demonstrate their ambiguity and instrumentality. This allows for humorous explanations of the intentions of *tevolo* and offers the potential for human interpretation and deception (Poltorak, 2002). Sociopolitical purposes may also be served by *tevolo*. Mageo (1998) notes that in the Pacific, discourse on possession has a countercolonial aspect to it, as well being oppositional vis-à-vis Christianity.

While spirit-caused illness is not as common an ailment as in the past, every community has a healer who knows what to do, a *faito'o faka tevolo*. The treatment must not only cure the symptoms (with methods such as massage or herbal poultices), but deal with the cause of the trouble by first identifying the offended spirit, and then by righting the wrong. Prayers to God are often added to the therapy. A special case of spirit possession is *āvanga*, which seems to occur most often with young unmarried women. The person may act in ways that are considered socially improper, exhibit altered mental status, unusual strength, and increased sensory awareness (Gordon, 1996). Treatment provided by the *faito'o faka'āuanga* includes medicines as well as a discussion with the family concerning the cause.

There are certain conditions that are clearly spirit-caused, and going to a doctor will do little good, just as there are a number of conditions that can only be treated in the hospital (e.g., automobile accidents, tuberculosis). However, even in these situations, if the primary therapy is not progressing, there is some room for flexibility by adding alternative treatments as adjuncts. There is also movement between systems (going to see a healer or a physician) with the most common type of conditions, those described as *puke pe*. This category is used with a wide variety of conditions that are self-limiting or seem to respond in a predictable manner to treatment.

There are a number of traditional interventions available when a person is *puke pe*. Massage is frequently used for a range of ailments, both internal and external. It can be done gently with the tips of the fingers, superficially with the whole hand, or deeply for underlying muscles. Bone fracture reduction is also done with massage. Some healers work with dry hands, other use water or coconut oil ("Tongan oil") as a lubricant. Tongan oil may also be used as the medium for herbal medicines. The plant parts used in healing are the leaves, leaf buds or young leaves, the inner bark or trees, and the root or rhizomes of ferns and ginger (Whistler, 1992). These are crushed and placed in a bottle of Tongan oil, or boiled in water, to drink, or to place on the skin as a poultice.

Disease categorization is of interest, but does not necessarily dictate treatment. There are a number of options available to a person who is sick, and illness is usually accompanied by much discussion with the result that diagnosis is most frequently made retrospectively— what treatment worked defines the condition. Decisions about where to go for care then are based primarily on pragmatic factors. Even though the ideologies of biomedicine and traditional healing are very different, it is often practical considerations that determine where care will be sought. Family and friends' influence are important, as are past experiences with illness, and accessibility to care providers.

The official (government) position supports modern biomedicine, and encourages the best students to enter medical and nursing schools. There are also strong proponents in Tonga of the value of traditional methods, especially if they are used in a complementary manner with biomedical approaches. There has been a call for rigorous study of herbs so they can be standardized (Bloomfield, 1986; Finau, 1981) Tongan medicine is a common topic in the local newspaper, with interest in its survival. A concern is that many of the healers are getting old and their children seem uninterested in following in their footsteps.

Tonga, like other Pacific Island nations, is facing serious health problems as a result of changing diet and activity levels (Englberger, Halavatau, Yasuda, & Yamazaki, 1999; Evans, Sinclair, Fusimalohi, & Liava'a, 2001; Sinoue, 2000). The effects of "modernization" or "Western lifestyle" with a high-calorie diet and low physical activity lead to obesity-related conditions such as diabetes and cardiovascular disease (Crews & MacKeen, 1982). This is true in Tonga, where diabetes is so common, a Diabetes Centre was established in 1993 to focus and intensify efforts to prevent and treat the disease (Vivili, 'Eseta, Finau, & Lutui, 1999) and among Tongans living overseas who (along with other Polynesians) suffer disproportionately high rates of type 2 diabetes (Bathgate, Donnell, Mitikulena et al., 1994; Collins et al., 1994; Simmons, 1997).

Research from New Zealand suggests knowledge about the need for dietary change is widespread, but the changed behavior is difficult to achieve. In Polynesian cultures, the value of food is in its ability to preserve traditions and help develop the community's unique identity. Its social value is perceived to be greater than the health value, with certain foods having higher social values than others (i.e., imported food, or meat with high fat content) (Small, 1997). Food continues to be an important way social relations are managed between kin and neighbors as evidenced in the everyday greeting, "ha'u tau kai" come let us eat (Young Leslie, 2004).

Large bodies are a reflection of *mana* in the case of chiefs, but in general reflect personal and familial rank and power. Cultural role expectations also contribute to the high rates of obesity in women (39%). Value is placed on a large and motionless body so that women should remain still, indeed to stay seated whenever possible. Their prototypic forms of labor, such as beating tapa, weaving pandanus mats, washing clothes, and caring for children, is done sitting down. Schoolgirls participate in active sports, but for many, this stops as soon as they leave school and must begin to behave like proper adults (Tupoulahi, 1990; Young Leslie, 2004). Typically, men's lives include more physical activity such as working in the gardens or fishing. As more families engage in wage labor in the city, adopt a more sedentary lifestyle, and shift their diets to more convenient and available "fast foods," it can be assumed that diseases related to obesity will increase. Stress is thought to be another consequence of changing lifestyle. A study in Samoa using biomarkers to measure stress suggests that those adolescents who experience social status incongruity as a result of culture change also experience greater stress (McDade, 2002). A similar pattern may be assumed to exist in Tonga with nontraditional, Western ways of living often being at odds with long-standing practices, thus posing challenges for individuals who wish to participate in both (Foliaki, 1999).

SEXUALITY AND REPRODUCTION

Gender relations in Tonga follow the pattern of much of Polynesia where women enjoy high status, yet men have greater power (Schoeffel, 1978; Tongamoa, 1988). Women are *fahu*, they have specific rights and certain privileges over their brother's children, maternal uncles and their children (Moengangongo, 1988). This status is usually exercised at social occasions, such as weddings or funerals. While sisters exercise a controlling power (*mana*) over their brothers, and are thus honored and served by them, as wives they are expected to serve their husbands and submit to their authority. Women's sexuality is viewed as under the control of her father and brothers, who keep a close eye on the young women of the family.

The topic of sexuality and reproductive health are considered to be very private, and it is *tapu* to talk about these things when brothers and sisters (biologic or classificatory) are together. When there is discussion about sex, it tends to be in the form of humor using metaphor and double entendre.

Gender liminal men in Tonga, or *faka leitī*, and their counterparts throughout Polynesia have a long tradition and are variously accepted into their communities. Besnier defines *faka leitī* by their characteristics: the demeanor exhibits stereotypically feminine qualities; many cross-dress and engage in sexual activity with "straight" men; they are associated with domestic spheres of activity; and work in activities associated with women such as mat weaving, tapa cloth beating, keeping house, or work in the hotel or entertainment industry (Besnier, 1996). A form of transsexualism, *faka leitī*, resists easy generalization and is undergoing change as a result of globalization and increasing transnationalism. The influence of Western gay culture is a topic of discussion within the community. There is very little known about these gender identity issues among women.

Chiefly marriages in the past were arranged for politically strategic purposes. Social relations continue to be important considerations when choosing a partner, particularly the qualities of the potential in-law family. Although not all couples include their families in their plans to marry, most decisions are made after all sides have given their approval. Once this is obtained, the parents and other important relatives meet to discuss the wedding arrangements, including the exchange of gifts. If the couple chooses to elope, the marriage can still be acknowledged by attending the church services the following Sunday wearing ritual fine mats. Most marriages occur when the couple are in their 20s, and they often live, at least temporarily, with parents of one side or the other. A young wife who lives with her husband's family faces challenges as she is of lower status than her affines, especially her husband's sisters.

Having children is very important and fertility is valued as an expression of femininity. Although infertility is cause for concern, adopting a child can moderate the social stigma. Fertility rates vary, with some rural families having six or more children; in the city two or three children is more common. Overall the fertility rate is gradually decreasing. The government and a few of the churches promote family planning, but because of feelings of modesty about sexuality, and personal religious beliefs, few couples use contraceptives. Children are viewed as future providers for the family, and blessings from God, so there is pressure to have a large family. This presents a conflict for the increasing number of families with both parents working outside the home.

HEALTH THROUGH THE LIFE CYCLE

Pregnancy and Birth

Pregnancy is considered a special and potentially dangerous time. A pregnant woman is taught by other women about the multiple prohibitions (*tapu*) she must observe to ensure the health of her baby. She is told that her behavior, what she eats, her relationships with others, and even how she moves her body all influence the physical and mental being of the baby inside (Morton, 1996). The mother breaking a *tapu* may be offered as the explanation for a miscarriage, stillbirth, a physical deformity, or mental or behavioral problems of the child.

Prenatal care is available at government public health clinics in the villages, and by midwives (*mā'uli*). The majority of births (over 80%) occur in the hospital attended by nurse-midwives and physicians (Morton, 2002). Some women deliver their babies at home by choice; others do so because they are unable to travel the distance to get to a hospital or because they miscalculate the dates and find themselves in labor on an outer island.

Infancy

Traditional practices, such as burying the placenta near the home, are followed to a varying degree throughout the islands. The Tongan term for placenta, land, and grave are the same, *fonua*, signifying the cyclical nature of life, and a connection between person and place (Mahina, 1999). After childbirth, the mother and infant return home for a period of rest and seclusion. Ideally, female relatives will appear and provide help for the household. The Christening, which occurs at three months, marks the end of the seclusion, acknowledges the child's membership within the *kāinga*, and is accompanied by gift exchanges. Naming the child is an honor often given to a person of rank or a relative of high status, such as the *mehekitanga*.

During the first few years of life, the baby receives almost constant attention from both men and women of the household, and sleeps next to the mother at night (Spillius [Bott], 1960). Breast-feeding is the norm, and lasts until the baby is about 8 months old. Babies are frequently massaged with scented or herbed coconut oil to prevent ailments or deformities, or to treat fussiness or fever. This massage is done by the mother or other female relatives. Massage of the skull soon after birth is done by a *faito'o* to ensure the fontanel closes properly. The first birthday is cause for great celebration with an elaborate feast attended by the *kāinga*, the minister, and neighbors in the village.

Childhood

Most of the following data on childhood and adolescence in Tonga are from Helen Morton (1996). Children, both boys and girls, are highly valued and bring meaning to life and to social institutions such as marriage. Early childhood is marked by love and affection from indulgent adults. It is the responsibility of the mother to teach her children to act properly, but as the child grows, all adults of the community, as well as older sibling and relatives, act in parental ways. As they get older, boys have more freedom to wander, whereas girls are expected to stay at home and do chores, be better behaved, and act more responsibly.

Adolescence

The whole span of childhood and adolescence is regarded as a time for molding behavior to become a fully adult person. The importance of kinship relations and the values of love (*'ofa*), respect (*faka'apa'apa*), and obedience (*talangofua*) are described by many Tongans as the key to their cultural identity and inform proper behavior. These exert a powerful influence on children and are reinforced not only in the home, but also in the church and at school. Transgressions are viewed as bringing shame on the whole family. Physical punishment is a common form of discipline with children explaining their parents do it out of love. This issue poses problems for migrant communities where the distinction between discipline and abuse is not necessary shared (McGrath, 2002).

There are no initiation rites to mark a girl's passage to womanhood. Male circumcision is done as a boy enters puberty, either at the health clinic, hospital, or performed by an elder male.

Dying and Death

Death ideally occurs in the home, with the person surrounded by close family members. Although there is no formal stage of terminal illness, there are different terms to describe the seriousness of an illness. If a sickness that initially is described as *puke pe* or *puke faka tevolo* is resistant to cure and progresses, the person is then *puke lahi*, or very sick. Typically, he or she will remain in bed, or on a mat on the floor. A piece of bark-cloth or fabric may be hung up to separate the space where the person lies from the rest of the house. Children are not allowed to play nearby or make loud noises. During this time when recovery is not expected, last advice, *tala tuku*, is given. Not all families have this tradition, but parents are likely to make this their last formal act with their children. *Tau'aki* describes a more serious state, usually marked by loss of consciousness, with death occurring within hours or at most, days. This is a special time with sanctions against improper behavior imposed by the family. Only specific relatives are allowed near the dying person. *Tapu* prohibitions must be respected and a sister will not be close to her brother, and vice versa. They will remain on the other side of the cloth divider. Personal care is performed by females, usually daughters and other female relatives, who are of the same or higher rank as the individual. Expressions of strong emotions are to be avoided around the time of *tau'aki*. It is thought that it will be more difficult for the person to leave if those around are very sad or having a hard time. Similarly, physical contact such as touching also makes it difficult for the spirit of the person to leave when its time has come to depart. After death, the body will be washed with water, the skin rubbed with coconut oil, powder, and perfume. The body is dressed and preparations are made for the funeral.

Funeral rituals are very elaborate, and a point of pride among Tongans. Pulotu, the Tongan afterworld, was the focus of funeral rituals before widespread conversion to Christianity, and although heaven has replaced this image in many respects, there continue to be frequent references to it in song and poetry, suggesting that the image of Pulotu may not be far from the minds of many present at funerals (Filihia, 2001). Aspects of status and rank which are less important in everyday life become central at a funeral (Kaeppler, 1993). These are also times to reaffirm Tongan traditions and pass them on to the next generation. Principles of rank are enacted with those attending the funeral placed according to their relationship to the deceased. Those higher are *fahu* and have certain rights and obligations, those lower are *liongi*, and are expected to take care of other tasks. The rituals begin with the *'āpō*, or wake, in the home. Everyone who has a connection to the person will come by to pay respects. While some relatives are inside the house keeping watch over the body and welcoming mourners, the relatives who are *liongi* are outside preparing and serving food and drinks. The main activity outside is the gift exchange. *Koloa*, literally wealth or valuables, are ritually presented. These may include very finely woven mats (some of which are named) and bark cloth (Filihia, 2001). These will be re-distributed during the following days. The burial in the cemetery occurs within three days.

Graves in Tonga are very distinctive with shaped mounds of white sand that are then decorated by women. These decorations, or art forms, respond quickly to cultural change and prevailing taste so that now in addition to more traditional crushed coral and woven flowers one finds beer bottles or plastic syringes carefully arranged on the grave (Teilat-Fisk, 1990).

After the burial, the head of the household invites the entire community to a feast and announces how long the mourning period is to be. The time of mourning is marked by special clothing, *ta'ovala putu*, a mat wrapped around the waist, and prohibitions against sporting games, dances, music, and beating tapa in the village. Ten days of village mourning was the norm in the past, now one or three days is more common. There have been other changes in funeral rituals: the body may remain in the hospital morgue to wait for overseas relatives to arrive; coffins (rather than wrapping the body in tapa) are being used; and while funerals are still important social occasions to strength kinship ties, often those of the nuclear family are often emphasized over ties with the extended *kāinga*.

CHANGING HEALTH PATTERNS

A caution must be raised concerning the nature of this description of the Tongan culture and health care. Generalizations about beliefs and practices have the potential to smooth out all difference and present a picture of an essential Tonga that is unchanging and without variation. This is not the intent, nor is it an accurate representation of Tonga or of Tongans. One of the tenets of medical anthropology is that the body is a site for social action, so that healing becomes a useful way to examine cultural transformations. In Tonga, as everywhere, all individuals do not experience the culture in the same manner. Knowledge and behavior around health and illness are dynamic and reflect the diversity that exists based on gender, class, education level, and so forth. Research and analysis of these issues, as well as the influence of political, economic, and social change on health is currently underway by a number of Tongan scholars whose works will make important contributions to the literature based on their deep understandings of the culture.

REFERENCES

Bathgate, M., Donnell, A. Mitikulena, A. et al. (1994). *The health of Pacific Islands people in New Zealand. Analysis and monitoring report 2*. Wellington: Public Health Commission.

Beaglehole, E., & Beaglehole, P. (1941). *Pangai, village in Tonga*. Wellington, NZ: The Polynesian Society.

Besnier, N. (1996). Polynesian gender liminality through time and space. In G. H. Herdt (Ed.), *Third sex, third gender: Beyond sexual dimorphism in culture and history* (1st pbk ed., pp. 285–328). New York: Zone Books, Distributed by MIT Press.

Bloomfield, S. F. (1986). *It is health we want*. Fiji: University of the South Pacific, Suva.

Bott, E., Salote, T., & Tavi. (1982). *Tongan society at the time of Captain Cook's visits: Discussions with Her Majesty Queen Sāalote Tupou*. Wellington: Polynesian Society.

Campbell, I. C. (1994). The doctrine of accountability and the unchanging locus of power in Tonga. *Journal of Pacific History,* 29(1), 81–94.

Campbell, I. C. (2001). *Island kingdom: Tonga ancient and modern* (2nd rev. ed.). Christchurch, NZ: Canterbury University Press.

Collins, V. R., Dowse, G. K., Toelupe, P. M., Imo, T. T., Aloaina, F. L., Spark, R. A., & Zimmet, P. Z. (1994). Increasing prevalence of NIDDM in the Pacific island population of Western Samoa over a 13-year period. *Diabetes Care,* 17(4), 288–296.

References

Collocott, E. E. V. (1923). Sickness, ghosts, and medicine in Tonga. *The Journal of the Polynesian Society, 22,* 136–142.

Cowling, W. A. (1990). *Eclectic elements in Tongan folk belief and healing practice.* Paper presented at the Tongan Culture and History, Canberra.

Crews, D. E., & MacKeen, P. C. (1982). Mortality related to cardiovascular disease and diabetes mellitus in a modernizing population. *Social Science Medicine, 16*(2), 175–181.

Englberger, L., Halavatau, V., Yasuda, Y., & Yamazaki, R. (1999). The Tonga health weight loss program, 1995–1997. *Pacific Health Dialog, 6*(2), 153–159.

Evans, M., Sinclair, R. C., Fusimalohi, C., & Liava'a, V. (2001). Globalization, diet, and health: An example from Tonga. *Bulletin of the World Health Organization, 79*(9), 856–862.

Ferdon, E. (1987). *Early Tonga.* Tucson: University of Arizona Press.

Filihia, M. (2001). Men are from Maama, women are from Pulotu: Femals status in Tongan society. *Journal of Polynesian Society, 110*(4), 377–390.

Finau, S. (1981). Traditional medicine in Pacific health services. *Pacific Perspective, 9*(2), 92–98.

Foliaki, S. (1999). Mental health among Tongan migrants. *Pacific Health Dialog, 6*(2), 288–294.

Gailey, C. (1992). A good man is hard to find: Overseas migration and the decentered family in the Tongan Islands. *Critique of Anthropology, 12*(1), 47–74.

Gifford, E. W. (1929). *Tongan society.* Honolulu: Bishop Museum.

Gordon, T. (1996). They loved her too much: Interpreting spirit possession in Tonga. In J. Mageo and A. Howard (Eds.), *Spirits in culture, history and mind.* New York: Routledge.

Hau'ofa, E. (2000). Epilogue: Pasts to emember. In R. Borofsky (Ed.), *Remembrance of Pacific Pasts: An invitation to remake history.* Honolulu: University of Hawai'i Press.

Herda, P. S. (1990). *Genealogy in the Tongan construction of the past.* Paper presented at the Tongan Culture and History, Canberra.

James, K. E. (1990). *Gender relations in Tonga: A paradigm shift.* Paper presented at the Tongan Culture and History, Canberra.

James, K. E. (1991). Migration and remittance: A Tongan village perspective. *Pacific Viewpoint, 32*(1), 1–23.

Kaeppler, A. (1993). Poetics and politics of Tongan laments and eulogies. *American Ethnologist, 20*(3), 474–501.

Latukefu, S. (1974). *Church and state in Tonga: The Wesleyan Methodist missionaries and political development, 1822–1875.* Honolulu: University of Hawai'i Press.

Levy, R. (1973). *Tahitians: Mind and experience in the Society Islands.* Chicago: University of Chicago Press.

Macpherson, C., & Macpherson, L. A. (1990). *Samoan medical belief and practice.* Auckland New York: Auckland University Press, Distributed outside New Zealand by Oxford University Press.

Mageo, J. M. (1998). *Theorizing self in Samoa: Emotions, genders, and sexualities.* Ann Arbor: University of Michigan Press.

Mahina, O. (1993). The poetics of Tongan traditional history, *tala-e-fonua*—An ecology-centered concept of culture and history. *Journal of Pacific History, 28*(1), 109–121.

Mahina, O. (1999). Food *me'akai* and body *sino* in traditional Tongan society: Their theoretical and practical implications for health policy. *Pacific Health Dialog, 6*(2), 276–287.

Marcus, G. E. (1978). Nobility and the chiefly tradition in the modern Kingdom of Tonga. *Journal of the Polynesian Society, 87*(2), 74–120.

Mariner, W., & Martin, J. (1981). *Tonga Islands: William Mariner's account: an account of the natives of the Tonga Islands in the South Pacific Ocean, with an original grammar and vocabulary of their language* (4th ed.). Neiafu, Vava'u, Tonga, Southwest Pacific (P.O. Box 83, Neiafu): Vava'u Press.

McDade, T. (2002). Status incongruity in Samoan youth: A biocultural analysis of culture change, stress, and immune function. *Medical Anthropology Quarterly, 16*(2), 123–150.

McGrath, B. B. (2000). Swimming from island to island: Healing practice in Tonga. *Medical Anthropology Quarterly, 13*(4), 483–505.

McGrath, B. B. (2002). Seattle *fa'a Samoa*. *The Contemporary Pacific, 14*(2), 307–340.

McGrath, B. B. (2003). A view from the other side: Place of spirits in the Tongan social field. *Culture, Medicine and Psychiatry, 27,* 29–48.

Moengangongo, M. (1988). Tonga. In T. Tongamoa (Ed.), *Pacific women: Roles and status of women in Pacific societies.* Suva: Institute of Pacific Studies of the University of the South Pacific.

Morton, H. (1996). *Becoming Tongan: An ethnography of childhood.* Honolulu: University of Hawai'i Press.

Morton, H. (2002). From *ma'uli* to motivator: Transformations in reproductive health care in Tonga. In M. J. Vicki Lukere (Ed.), *Birthing in the Pacific* (pp. 31–55). Honolulu: University of Hawai'i Press.

Parsons, C. D. F. (1985). *Healing practices in the South Pacific.* Laie, Hawaii Honolulu, Hawaii: Institute for Polynesian Studies, University of Hawaii Press distributor.

Poltorak, M. (2002). *Aspersions of agency: Ghosts, love and sickness in Tonga.* Unpublished dissertation, University College, London.

Puloka, M. H. A. (1999). Avanga: Tongan concepts of mental illness. *Pacific Health Dialog, 6*(2), 268–275.

Radcliffe-Brown, A. R. (1952). *Structure and function in primitive society, essays and addresses.* London: Cohen & West.

Rogers, G. (1977). "The father's sister is black": A consideration of female rank and power in Tonga. *Journal of the Polynesian Society, 86*(2), 157–182.

Schoeffel, P. (1978). Gender, status and power in Samoa. *Canberra Anthropology, 1*(2), 69–81.

Shineberg, D. (1978). "He can but die": Missionary medicine in pre-Christian Tonga. In N. Gunson (Ed.), *The changing Pacific: Essays in honour of H. E. Maude* (pp 285–296). Melbourne: Oxford University Press.

Shore, B. (1989). *Mana* and *tapu*. In R. B. Alan Howard (Ed.), *Developments in Polynesian ethnology.* Honolulu: University of Hawai'i Press.

Simmons, D. (1997). Diabetes and its complications among Pacific people in New Zealand. *Pacific Health Dialog, 4*(2), 75–79.

Sinoue, S. a. P. Z. (2000). *The Asia-Pacific perspective: Redefining obesity and its treatment.* Melbourne: International Diabetes Institute, Regional Office for the Western Pacific, World Health Organization.

Small, C. A. (1997). *Voyages: From Tongan villages to American suburbs.* Ithaca: Cornell University Press.

Spillius [Bott], E. (1960). *Report on brief study of mother–child relationships in Tonga.* Geneva: World Health Organization.

Teilat-Fisk, J. (1990). Tongan grave art. In F. A. H. a. L. Hanson (Ed.), *Art and identity in Oceania* (pp. 222–243). Honolulu: University of Hawai'i Press.

Tongamoa, T. (Ed.). (1988). *Pacific women: Role and status of women in Pacific societies*. Suva: Institute of Pacific Studies of the University of the South Pacific.

Tupoulahi, C. (1990). *The socio-cultural antecedents of obesity in Tonga*. Unpublished PhD Dissertation, Flinders University, Australia.

Vivili, P. F., 'Eseta, Finau, S., & Lutui, T. (1999). Foot complications among diabetics in Tonga. *Pacific Health Dialog, 6*(2), 205–207.

Whistler, W. A. (1992). *Tongan herbal medicine*. Honolulu: University of Hawai'i Press.

Wood-Ellem, E. (2001). *Queen Salote of Tonga: The story of an era, 1900–65*. Honolulu: University of Hawai'i Press.

Young Leslie, H. (2004). Pushing children up: maternal obligation, health and illness in the tongan Ethnoscape. In U. Lockwood (Ed.) *Globalization and Culture Change in the Pacific Islands*. Englewood Cliffs, NJ: Prentice Hall.

Trobriand

Jay Bouton Crain, Allan Clifford Darrah, and Linus Silipolakapulapola Digim'Rina

ALTERNATIVE NAMES

Bweyowa, Kiriwinia, Kilivila, Bowoya, Trobrianders, Trobriand Islanders.

LOCATION AND LINGUISTIC AFFILIATION

The Trobriand Islands are located in the Massim region of Melanesia 120 miles north of the eastern tip of New Guinea (approximately Latitude: 8.30S, Longitude: 151E). The Trobriand archipelago, consisting of some 22 flat and partially raised coral atolls, sits at the intersection of the Coral and Solomon Seas and is politically situated within the Milne Bay Province of the Republic of Papua New Guinea. The population is centered in the larger islands of Kiriwina (Bowoya), Vakuta, Kaile'una, and Kitava. Boyowa's population, exceeding 26,000, is located in some 60 villages (Lepani, 2001; see also Weiner, 1987). The language of the Trobriands is Kilivila (Kiriwinan). Lawton (1993) lists 11 dialects. Five of these are spoken on Boyowa proper, three on each of the remaining large islands, and the last three in culturally related Luscancy Island (Simsimla) and the Marshall Bennet group (Iwa, Gawa, and Egum Atoll). Milke (1965) suggests putting Kilivila in the Oceanic subgroup of Austronesian. Senft (1986) treats Kilivila as a language family within the Austronesian "Papua-Tip-Cluster" defined by Capell (1976) and identifies three languages: (1) Budibud (Nada) on the Laughlan Islands, (2) Muyuw (Murua) on the Woodlark and Marshall Bennet Islands, and (3) Kilivila on the Luscancy Islands and Trobriand Islands (Kiriwina, Vakuta, Kitava, Kaile'una, Kuaiwa, and Munuwata).

OVERVIEW OF THE CULTURE

Trobriand cultural institutions share the forms of cultural practice which are characteristic of many Melanesian societies: religious concepts focus on the processes of growth and decay; male sorcery involves powers of life and death; magic controls food production; rank is asserted pro forma, but in reality is contingent on elaborate exchanges during birth, mortuary, and harvesting rites and, to a lesser extent, in association with kula exchanges (Young, 1983). Men are defined by their success in gardening and exchanges of yams and kula shells while women are distinguished through mortuary exchanges of skirts and leaf bundles (Weiner, 1976). Trobrianders are also defined by a system of rank tied to the alimentary history of their direct matrilineal ancestors, transmitted through mother's milk (Montague, personal communication). Differences in rank are articulated through violations of food prohibitions that adversely affect magic stored in the stomach.

Trobriand social life has been described as being organized around sub-clans (*dala*) each of which is assigned to one of four clans (*kumila*). Malinowski (1929) and to some extent, those who have followed, treated Trobriand kinship as exclusively matrilineal, viewing fathers as merely affines. Recently Montague (2001) has argued that Trobriand kinship has bilateral characteristics and that Trobrianders define their relationships in terms of dietary history and exchanges. However, bilateral relations attain their meanings only within the generation or lifetime of the individuals concerned, but not beyond it; matrilineal descent (*dala* principles) takes over in between generations in order to regulate new sets of relations once again.

Bweyowa subsistence is based on agriculture and fishing. The main crops are yams, sweet potatoes, taro, bananas, and coconuts. A food surplus is produced in normal years; however periodic droughts, attributed to the magic of Tabalu chiefs, result in famine (Digim'Rina, 1998). Young's observations that Goodenough Islanders' preoccupation with food dominates their symbolic idioms and cannot be completely explained by ecological exigencies, also holds for Trobrianders (Young, 1986). Trobrianders have gone to inordinate lengths to make food the measure of all things, particularly health and well-being.

Evidence for the initial occupation of the Massim, in the form of pottery deposits, suggests widespread settlement by Austronesian speakers about 2,000 BP (before present) (Bickler, 1998). Recent evidence dates the occupation of Kiriwina to about 900 years ago (Burenhult, 2002). Inter-island group trade in the form of *kula* may date from around 500 BP (Egloff, 1978). In 1793 French explorer Bruny D'Entrecasteux named the group after his lieutenant, Denis de Trobriand and in the century that followed there were occasional visits by traders and whalers. By 1894 pearl traders and Wesleyan Missionaries were living permanently on Boyowa (Campbell, 1984). Australian colonial officials set up a government station at Losuia in 1904. One year later a hospital was built in response to reports that there had been a serious population decline due to the spread of venereal diseases (Black, 1957).[1] In 1936–37, a mission and primary school were set up by the Sacred Heart Catholic Mission. During the Pacific War, Australian and American forces were stationed at two airstrips constructed on Boyowa. Local government and various business ventures were established in the 1950s and a high school in the 1980s. National independence and inclusion into Milne Bay Province took place in the 1970s (Young, 1983).

Bronislaw Malinowski's pioneering and widely influential works on Trobriand society made the islander's lives accessible to readers around the world. Trobriand society has been the focus of considerable ethnographic field research and the descriptions and analyses derived from these studies have themselves been the subject of voluminous debate and re-analysis. Each generation of scholars have brought to the Trobriand materials the insights and biases of their own training.[2] The corpus of Trobriand literature is enormous and, in an appropriate analogy for a work in medical anthropology, it resembles nothing less than the thick chart of a chronic patient: a series of not clearly connected narratives written at different times by different clinicians, ostensibly about the same person. What follows is a synthesis of this huge literature that focuses on the visual aspects of health.[3] However we recognize that such a synthesis is just that, synthetic.

THE CONTEXT OF HEALTH: ENVIRONMENTAL, ECONOMIC, SOCIAL, AND POLITICAL FACTORS

The islands have undergone sustained population growth. Pösch and Pöschl estimate that between 1913 and 1985, the population of Boyowa grew exponentially (Pöschl & Pöschl, 1985). They also report that pregnancy in younger girls is increasing in frequency and that family size is expanding (1985; see also Lepani, 2001). Darrah and Crain (personal communication) found the population still expanding in 2001.

The scarcity of arable land has led to a reduction in the yam planting cycle from seven to three years. Pöschl and Pöschl report an increase in deficiency diseases and malnutrition even in years with a good harvest. They also found tuberculosis in malnourished children and underweight adults and suggest that enteric diseases were to be found in larger villages with poor sanitation and inadequate water supplies (Pöschl & Pöschl, 1985). Lepani (2001) reports that malaria, skin disease, and pneumonia are the leading causes of morbidity, while perinatal conditions, meningitis, malaria, and tuberculosis are leading causes of mortality. However, Montague (1985) notes that the risk assessment for the effects of malaria, made by people in Kaduwaga, tends to be lower than those of health officers. Annual sprayings of DDT to control mosquitoes are opposed locally and available malaria drugs are used for treatment rather than prevention. Montague

Diet and Health

Food consumption is central to Bweyowa thought about society and the body. Eating articulates social divisions of rank (Malinowski, 1929), kinship (Montague, 2001), age, states of being, and definitions of self. Pre-colonial Trobrianders contrasted themselves with neighboring cannibals. Tudava, the great culture hero, eliminated cannibalism substituting yams for humans and lessened the effect of garden magic to make room for individual achievement. This magic provides a lattice of inferences that yam and human life cycles mirror each other (Brindley, 1984; see also Darrah, 1972). Today Trobrianders raise and exchange yams, which are metaphorical humans.

Malinowski's Trobrianders assumed that people ate for pleasure rather than to sustain the body (1929, 1935). However, Malinowski also notes famine was thought to produce a variety of illnesses which could lead to death (1929). Montague reports that death ultimately results from consumption of food that has lost its nourishment due to adulteration by a sorcerer (1989). Humans require a steady intake of yams or taro (*kaula* or *kauna*) to build their bodies and confer the hardness essential to withstand sorcery (Montague, 1989). Trobrianders, particularly during mourning, stress the obligatory side of ingestion making it a moral duty to eat food provided by others. Annual gifts of yams, totaling more than half a man's production, result in people being food-dependent.

A powerful expression of the obligation to "eat for others" was the former reciprocal obligation of the *kopoi* relationship where fathers pre-masticated food for their infants, and were repaid after death when son's symbolically ate from their father's bodies and vomited. This exchange should be viewed in the context of food prohibitions. Sons are responsible for their father's well-being and go to great lengths to show that they did not cause their deaths. A son refrained from eating his father's totems so that the father would not get sick when the son used his food utensils (Seligman, 1910). Alternately, fathers also had to maintain their children's food taboos.

Coupled with the obligation to eat with appreciation and share prohibitions is a public reticence to eat at all. Restraint in eating, particularly *kaula*, is a civic virtue. The intent of harvest magic was to control people's appetites so that yams rotted in storehouses rather than stomachs. The famous Trobriand magic of prosperity [*vilamalia*] worked directly on people's appetites (Malinowski, 1935). Obesity, which is evidence of the failure of magic to restrain appetite, as well as a sign of selfishness, is not desirable. Nutrition surveys have generally found the Trobriand diet to be a healthy one (Hipsley & Clements, 1950: see also Lindeberg et al., 1994).

Wives maintain their husbands' food taboos to protect the magic stored in their husbands' stomachs (Montague, 1974). Foods, mentioned in magical formula and/or which share defining characteristics with the magic's intent, are avoided to preserve the effectiveness of spells (Munn, 1986). Violation of a food taboo would be to "eat one's ancestor" (Munn, 1986). Spells list the ancestors who held the magic in their own bellies. "Eating one's ancestors" is to ingest and then eliminate the ancestor laden spells; magic should be ingested and then brought back up.[4]

MEDICAL PRACTITIONERS

The literature makes no note of specialized traditional positions or offices for medical practitioners. However, there are individuals in each village who have knowledge of curative spells and magic-producing materials. Malinowski noted that magic to prevent dangers in childbirth, to cause abortion, treat genital discharge, swellings of the limbs, and toothache was controlled by women, but men also possess spells to control various aspects of reproduction (Malinowski, 1929).

Noting that Trobrianders lack a tradition of local midwives, Pöschl and Pöschl (1985) called for the training of female birth attendants. Lepani (2001) indicates that 126 women have been trained as Village Birth Attendants. There is also a health center in Losuia and more than a dozen local aid posts, with the actual count of functioning sites varying due to staffing problems.

CLASSIFICATION OF ILLNESS, THEORIES OF ILLNESS, AND TREATMENT OF ILLNESS

Classifications of Illness

Bweyowa notions of the body, the beginning and ending of life, the nature of disease, and the various stages of life involve a complex assemblage of assertions and practices that are not easily, nor perhaps properly, separable from other aspects of life. Linking these, and used here as a guide through a brief discussion of Bweyowa views of biology, are local notions of agency elaborated through overlapping classificatory schemes involving properties of color (white/red/black), hot/cold, hard/soft, shiny/dull, wrapped/unwrapped, and mobile/anchored.

Agency and Illness. As in other areas of Melanesia, the body is mediated through a complex web of social relationships; its condition is enhanced, sustained, or endangered through a carefully monitored series of social and economic exchanges (Knauft, 1999). Bweyowa attribute most illness to four kinds of agency: *bwaga'u* (sorcerers), *mulukwausi* (witches), *tauva'u* (malignant spirits), and *gaga* (offenses against customary exogamy rules). Secondary forms of agency include *kosi*, a spiritual essence of recently deceased sorcerers, and *tokwai*, tree spirits who work in conjunction with sorcerers.

Sorcery. Illness (*silami*) results from the introduction of magically treated objects into the body. Vectors include tobacco, food betel, smoke, and charmed stones (Malinowski, 1929). The object is wrapped to preserve the magic's force. Feeling heavy, the recipient retires to his/her home to heal over the family hearth. In the dead of night the sorcerer attempts to intensify the illness by depositing herbs in the protective fire. The third, projective, phase of sorcery employs metaphors of spearing, stabbing, or otherwise piercing the body with a magical implement.

Tauva'u. Epidemics are attributed to the agency of *tauva'u*, spirits with the power to assume the shape of men, as well as crabs, snakes, and lizards, all of whom emerged from underground, in the mythic past, with the ability to rejuvenate by sloughing their integuments. A *tauva'u* taught sorcery to a man who then killed his benefactor. *Tauva'u* come to the Trobriands from the south, at the change of seasons, a time when people become ill from eating the first, unhealthy, "black" yams of the new garden. The yams are made unwholesome by Kitavan magic, which banishes evil influences, via the trade winds blowing toward the Trobriands (Malinowski, 1935). The invisible *tauva'u* kill people in epidemic proportions by striking each with a club. Not only are *tauva'u* the source of sorcery but they also are responsible for teaching witches, their consorts, how to kill people (Malinowski, 1929).

Flying Witches. Witches remove internal organs, particularly those associated with ingestion and speech, and eat them causing sudden death. Most witches live on other islands, but Trobrianders identify some local women as witches and are quick to include them in food distributions when a relative is close to giving birth. The actions of witches, including their rituals of parturition, are the inverse of proper Trobriand women. Witches are greedy for food, and eat raw meat. Witches are strongly associated with fire, heat, and hyper-sexuality. At night their inner spirit, *sans* clothes and skin, takes the shape of a flying animal and moves through the sky emanating fire.

Prohibitions. Breaches of exogamy [*suvasova*] result in skin disease, swelling, and wasting. The body swells, the skin turns white, and breaks out into sores. A small snake-like creature appears and moves about in the body causing symptoms of swelling and wasting (Malinowski, 1929).

Theories of Illness

The natural state of a person is health. Theoretically people die from old age; however, usually one of the above agencies is blamed. Weiner has argued that magic comes to the forefront, not because of risk, as Malinowski suggested, but rather when the outcome of an exchange is imperative. In Weiner's view, magic is about controlling men through exchanges rather than control of nature; control over natural forces, such as the wind and rain, is proof that one has the power to dominate others (Weiner, 1987). Weiner's starting point is different from that of Weber, who places the source of meaning in the crises of illness. Weiner's perspective does not focus on misfortune or its remediation, and tends to ignore revenge, the other half of reciprocity. It is also a top-down view of society given that magic is such a limited resource, a prerogative of rank, while suffering is the great common denominator.

Both illness and senescence are associated with blackening and also with heat/fire. Malinowski observed that magic which affects health comes in paired sets of spells, called *vivisa* for therapeutic magic and *silami* for magic of affliction (Malinowski, 1916). A sorcerer without the perceived capacity to heal would be deemed incompetent. *Vivisa* refers to the defensive or healing portion of magical formulae. It means to untie the knot of offensive magic which has the patient tied up in illness. The -*visi* suffix is also associated with peeling with the hands and to cool by fanning. Transformations of the skin are closely connected to health and therapeutics. The process of aging, visualized as darkening skin, results from accretions of black magic; *vivisa* and beauty magic reverse this process.

The general term for illness is *katoula* (e.g., *to-katoula*, "sick person," *eweya katoula*, "he caught an illness," *I kapilakeigu kala katoula*, "he gave me his sickness") (Baldwin, 1937). Specific Bweyowa disease categories include *kaivatokula* (wasting disease), *diega, bwawa, pwawa* (elephantiasis of the leg, arm, testicles, respectively), *kweyagola* (disease of the bones; rheumatism; tertiary yaws), *popoma* (disease causing swollen belly), *silaipwasa* (disease of stricture of the bowels), *lelia* (swellings; plague), *tobudawa* (abscess), and *silami* (incurable internal abscess) (Baldwin, 1937).

Fires of Life. *Momova* is life, and *mova* is life/alive; humans are *tomomova* [*to*- prefix for man]. *Yomova* and *yomovi*, literally "to cause life," means to heal. *Tokatumova* means healer. Scoditti (1996) translates *momova* as vital force. *Mo-* or *mwa-* are prefixes which impart maleness or humanity while Lawton notes that -*va* is a gentle, intimate action done with fire. *Mova* can be glossed as human fire. On the other hand, *mata* is both death and a dormant fire; *kimati* is both "to kill someone" and "extinguish a fire." The life as flame metaphor is elaborated through analogies between fire making and intercourse. The traditional form of starting a fire was rubbing a small pointed stick, on another stick, fire plough fashion (Silas, 1926; see also Fortune, 1932). The rubbing stick was called *kaikwila*; *kai* means wooden and *kwila* is penis. On neighboring Dobu, the fire stick is called *kekusi*, and the act of rubbing is *kusi* or *usi*, terms for the penis. Sexual intercourse is sometimes spoken of as "…we copulate and it flames up" (Fortune, 1932).

Fire originated through human reproduction. Malinowski (1929) provides a myth about the origins of fire in which a woman gives birth to fire, followed by the sun and moon. She conceals fire in her vagina bringing it out only to secretly cook her food. Intercourse produces heat that, at birth, ignites into the flame of life. Rituals that attempt to regulate the problematics of reproduction draw upon the quotidian acts of cooking. Trobrianders metaphorize alimentation as reproduction. Elements from the culinary domain are used to build scenarios for the manipulation of analogous aspects of reproduction thereby reinforcing the assumption that life is a fire.

Treatment of Illness

Bloodletting is an important therapeutic practice in the Massim (Baldwin, 1937; see also Munn, 1986; Senft, 1986; Villeminot, 1967). Blood, produced by eating heavy *kaula*, is itself heavy, and its removal has the effect of lightening a person in terms of both color and weight. Gawan dancers are bled to lighten themselves. In the Trobriands, cuts or scratches are made with a sharp stone in order to remove blood or relieve pain, headaches, lassitude, chills, hematoma, and bruises (Villeminot, 1967). Bruises which produce a discoloration, likened to those suffered by pigs when clubbed to death, are lacerated to remove the blackness. Blood, once released from the body, is dangerous to matrilineal kin, but can be a gift that can establish exchange relationships. Gawan fathers put blood on their children to inaugurate exchanges of betel and other edibles with them (Munn, 1986).

SEXUALITY AND REPRODUCTION

Malinowski's (1929) discussion of the sexual freedom enjoyed by young Trobrianders has led to a distortion of Trobriand sexual ideologies and practices. The popular appellation of "Isles of Love," and the mind-set which is attracted by this term, ignores the extensive limits to sexual expression which were also documented by Malinowski. Prior to marriage, adolescents enjoy great latitude so long as their actions are private and within the confines of exogamy as well as peer pressures to restrict their choices to local candidates. It is commonly believed that sexually transmitted diseases are usually contracted from individuals from outside the local community (Lepani, 2001).

Marriage, which is bound by strong expectations of monogamy, does not legitimize sexual relations; on the contrary, marriage requires the couple to maintain a public fiction that intimacy is nonexistent. It is a grave insult for someone to suggest to a married person's face that he or she is sexually active with their spouse. The Trobriand ideology of asexual reproduction furthers this fiction; otherwise children would be public evidence of their parent's private actions. However, Malinowski (1929) also observed that, during the harvest season, there were occasions when the strict norms of marital monogamy were relaxed. This was also a time when young people engaged in organized public competitions and dances, which were expected to lead to liaisons. Bellamy reported a large increase in births occurred 9 months following the harvest season (Black, 1957).

Instead of sex, marriage legitimizes alimentary behaviors. The first, trial phase of marriage, is marked by the couple sharing food in public, an activity that lasts for only the first year of marriage. During this period the couple consume food prepared by others as the bride is taught by a woman from the husband's home, how to maintain his dietary restrictions. Phase two begins at the end of the first year when the wife is given her own hearth; after this time, the couple will separately eat her cooking. Divorce is signaled by either party throwing the hearth stones out of their house (Montague, 2001).

HEALTH THROUGH THE LIFE CYCLE

Pregnancy and Birth

It has been frequently reported, and contested, that Trobrianders believe that human conception is asexual, yet is also clear that intercourse plays a part in reproduction (Malinowski, 1929: see also Austen, 1934–35; Leach, 1966, 1968; Montague, 1971, 1973; Powell, 1968; Spiro, 1968, 1972, 1973; Weiner, 1976). Montague (personal communication) reports that the "hammering" of intercourse closes the cervix, staunching the menstrual flow and thereby facilitates pregnancy.[5] Fertility is also aided by attachment to a particular partner. Conception occurs when a *waiwaia* [spirit child] is deposited on a woman's head by a deceased relative (Austen, 1934–35). Shortly after the *waiwaia* is deposited, an embryo [*veguvegu*] is brought into existence [*ebubuli*] (Austen, 1934–35). Blood and water from the uterus move to the woman's head, collects the *veguvegu*, and descends again, causing dizziness, headaches, and vomiting. At about 3 months, the *veguvegu* turns into a rat-like object rolled up in a mat like membrane.

Malinowski (1929) reports that *waiwaia* are reincarnations of ancestral spirits, created when rejuvenation, under the influence of magic, reverses spirits back to their pre-embryonic state. In 1918, Billy Hancock, Malinowski's expatriate friend in the Trobriand Islands, advised Malinowski that Trobrianders did not believe in reincarnation, having said so only to agree with Malinowski (Stocking, 1977). Montague (1971), Austen (1945), Campbell (personal communication), and Digim'Rina (personal communication) all report beliefs in reincarnation.

If we examine *ebubuli* we find a variety of usages for its root *bubula*, the noun form. It is the initial state in carving, when order is imposed (Senft, 1986). It also means to shine, and to adorn. *Bububula* refers to the process of creation and manufacturing (Lawton, 1993). In his transcription of the great creation myth of Tudava, Malinowski records a passage describing Tudava's creation of fine gardens with the phrase *'valu i-bubuli*, the countryside was made bright' (Malinowski, 1935, Vol. 2). *Ebubuli* is a construct of Trobriand ideas of order and the initial stage of a transformation. The highly valued condition of brightness is linked to transformations of age, beauty, and health. One of the more compelling themes in Trobriand ritual is the use of magic to contravene the visual effects of aging, brought about by black magic, by restoring lightness and a beautiful bright sheen to the body's surface.

Special ceremonies are performed to make a pregnant woman beautiful and white. She is given a special bath, her father's sisters perform beauty magic over her, and her skin is covered with coconut oil. She is covered with a special white cloak and avoids the sun and thoughts about sex, for both would darken her skin. In the third trimester she resides with her own family who protect her with armed guards. At delivery she is placed over a fire to cause her blood to liquefy. Magic is performed to prevent sorcerers from darkening her skin, cooling her reproductive organs, and sabotaging her passage toward an uncomplicated delivery. Montague (1985) noted that a woman was ideally secluded after delivery for up to six months, so that her skin would match the white color of her child's.

In the *veguvegu*, or nascent embryo stage of existence, the symbolic focus is a house cricket called *vegu*. The salient qualities of the *vegu* are its long sensuous feelers, a very beautiful, luscious abdomen, and a sun sensitive skin that causes this nocturnal creature to wrap itself in a leaf (Malnic, 1998). Used to attract birds, *vegu* also means bait. The characteristics of the *vegu* are dramatized in the rituals that transform a *prima para* into a mother. The equally vulnerable parturient, who also attracts flying predators (i.e., witches) must cloak herself from the sun and wait patiently in her house.

Infancy

Children are generally welcome additions to the family, however, both Austen (1934–35) and Lepani (2001) indicate that abortions, by both herbal and mechanical means, are also an option. Lepani also reports that pregnancy often results in marriage, an outcome that is far more frequent these days than 30 years ago. Powell (1980) mentions the use of traditional forms of contraceptives, called *kaikariga* [*kai-* is a prefix for wood and *kariga* is death].

Bellamy (Black, 1957), Assistant Resident Magistrate and doctor in the Trobriands, from 1906 to 1915, reports that female infanticide may have been practiced; however, Malinowski rejected this possibility (1929). Montague (1985) reports that Kaduwagans deny practicing infanticide even though it is within a mother's rights up until she has fed her child. A newborn, prior to its first meal, is the property of the woman who did the work of growing it, just as yams are the property of the man who raised them. A genetrix can elect to feed a neonate, give the child to someone else to feed, or the infant can just disappear, without ever entering the kinship system. Once the neonate has been fed, it is human rather than property, and its kinship ties, which follow milk rather than blood, are to the person whose dietary history it shares (Montague, 2001). All newborns have the same kind of undifferentiated blood. It is only when a child drinks milk, which incorporates the nurse's unique dietary history, that it becomes related through these shared differentiations. Weiner notes that wet nurses are given axe blades [*beku*] as repayment for their milk and to reclaim the infant for its natal *dala*.[6]

From the third month of her pregnancy through the first month after parturition, a *prima para* must eat hard dry foods [*kaula*], which produce blood and milk. If she eats soft and wet foods, the baby will suffer from a nonspecific disease called *gwemata* (Pöschl & Pöschl, 1985). *Gwemata* means cold and damp (Baldwin, 1937). As late as the1970s, women were encouraged to refrain from conjugal relations until their children entered the toddler stage in order to protect the quality of their milk. During postpartum seclusion, a major concern is to protect the mother's milk by keeping it warm. Cold or reheated foods are also avoided to prevent damaging fluctuations to the mother's internal temperature. Women also cover their heads at night out of fear that they will lose body heat, thereby cooling their milk (Montague, 1985).

Mothers nurse their infants for periods reported to be in excess of one year or up until the child starts walking. Colostrum may or may not be given. A 1980 Milne Bay weight-by-age nutrition study raised concerns about the diet and feeding practices for children from weaning up until 24 months of age (Nutrition Monitoring Group, 1980). The report found that a starchy diet, low in protein and high-energy foods, combined with a low frequency of food consumption made it difficult for young children to ingest amounts adequate to their energy and nutrient needs. Montague (1985) suggests Trobrianders exaggerate their dependence on *kauna* and underreport their consumption of other foods. In any case, Trobrianders of her acquaintance were far more concerned about emergency health care than issues related to child nutrition. In the Trobriand view it is good for babies to be plump but from the toddler stage on the preference is for thinness coupled with sturdy muscle development and high energy.

Any ambiguity surrounding the father's part in conception does not spill over onto his role as nurturer. Fathers have intense physical and emotional contact with their children starting several months after birth (Weiner, 1976). Fathers feed and care for their children even before their abrupt weaning. After this juncture, the child sleeps with its father until age 10 or until it is adopted by a family member. Powell (1969) suggests that young Trobriand children do not lack for support or guidance when their parents are absent.

Childhood

Children enjoy considerable independence. Malinowski (1929) reports that parents are more likely to request favors of their children, sometimes accompanied by threats, than they are to give them commands. They rely on the child's sense of fairness and obligation, which is instilled at a very early age. By age four, the child moves into the village play group and increasingly avoids parental discipline and oversight in a process of emancipation which Malinowski (1929) judged to be gradual and pleasant. At a point roughly coinciding with menarche, young girls are given a short red skirt symbolic of their capacity to evoke desire from others.

Adolescence

As a child ages it is expected to explore its sexuality and may receive adult approval and encouragement in this quest. First sexual encounters reportedly occur between the ages of 11 and 16 years and are associated with the onset of puberty (Lepani, 2001). It is commonly believed that the enlargement of a girls breasts and her menarche result from her being sexual active. Young women have great freedom in choosing their partners but if they have too many partners they run the risk of being labeled "tasteless." Young people who do not go out at night may be categorized by peers and adults as worthless. As girls age, they look for longer, monogamous relationships in preparation for marriage. These steady attachments are thought to enhance the chances of pregnancy, which reportedly occurs at an increasingly earlier age, resulting in growing numbers of unwed mothers (Lepani, 2001). Lepani suggests that the freedom of adolescence, combined with beliefs that liaisons within the community are safe, and expanding contact with people outside the local group, put Trobriand youth at an increasing risk for sexually transmitted diseases. Adolescence is a period devoted to honing skills of seduction, which employ decorations and magic provided by the father's family. Weiner (1976) notes that adults persuade others with magic and exchanges while the young must rely on their beauty augmented by magic. Mothers and aunts (both paternal and maternal) take an interest in their daughters'/nieces' relationships, which they monitor through the gifts provided by the girl's suitors (Lepani, 2001). Fathers, like mother's brothers, are properly not concerned with their daughters'/nieces' courtship.

In line with the general association of red with adolescence, the letting of blood provided an idiom through which Trobriand girls could communicate about their relationships with a particular boy. A girl could publicly express her interest by wounding a boy at a special harvest competition. During the foreplay, which would ensue, the couple would bite each other's lips to mingle their blood. Sometimes a girl would cut her lover's breast or upper arms as an indication of her commitment to him (Malinowski, 1929). This "red" period of the life cycle ends with marriage when the girl receives a white skirt with the expectation that pregnancy will follow, her menstrual flow cease, and her skin will be bleached through the rituals surrounding parturition.

Adulthood

The emphases on white symbolism, in *prima para* rituals, is best viewed in the broader context of Trobriand efforts to control time with beliefs and practices focusing on the chromatics of ontology. Campbell (1984) reports that white is associated with newness, cleanness, purity, immaturity, innocence, but not with semen or milk. It is with experience and history that the skin ages and darkens. As people age their bodies turn from the white of infancy to the sexually active red of youth followed by progressively darkening hues as they gain experience, acquire magic, and are affected by the magic of others. White, black, and red have both positive and negative aspects (Tambiah, 1968, 1983). On the positive side a youth's "red" glossy exterior is highly desirable and allows him/her to influence others. The *prima para* is at the point of exiting the red stage of her development cycle and in entering the black stage of maturity.

Young people lack social maturity and the ability to make important exchanges. Mature adults, who have darkened as the result of magic, must use magic to compel others to succumb to their own wills (Campbell, 1984). At maturity, the jet-black hair of youth turns white, a reversal of the body's progression. White hair, like dark skin, signifies knowledge of life, and either draws fear and respect or ridicule depending on the individual's conduct. Hair, the site of conception, may be a harbinger of reincarnation when it whitens.

Transforming Liquids. Magic that reverses or speeds the effects of time, whether a complete rejuvenation or less potent beauty magic, follows a common pattern. The person is bathed, exfoliated with leaves to remove the accreted darkening layers of magic, and then anointed [*putuma* or *vaputuma*] with coconut oil, to give the skin a youthful sheen (Lawton, 1993; see also Senft, 1986). Tuma Island is not only the residence of the dead but also a place of perpetual youth because the spirits who reside there are able to return to a youthful red stage of existence by bathing in a special brackish spring called S*opiwina*. *Sopiwina* is water used to wash off unwanted smells such as meat or fish, or to remove dirt. *Sopi* is water. Malinowski (1929) tells us that *wina* is an old form of *wila* [cunnus]. *Sopiwina* may be evocative of the transformative power of semen. Bathing is also associated with pregnancy; bathing in the lagoon can lead to impregnation by the *waiwaia* floating on the water.

Weiner's account of the bathing process in Tuma differs significantly from Malinowski's: "...the wrinkled skin is sloughed off and [the *baloma's*]...life continues as before. When this occurs, however, a ...*waiwaia*...is created" (1987, p. 54). In Weiner's version, the *waiwaia* is a by-product of rejuvenation and is either an entirely new entity or something that has broken off from an ancestor spirit, perhaps the sloughed blackened skin.

Dying and Death

Visual aspects of a child's identity are usually regarded as acquired from its father and his sisters as a result of close contact and applications of beauty magic. Baldwin (1945) records a passage that suggests that the dead endow the living with aspects of their appearance; he was told that a dying woman would "endow with her beauty [*bubula*] the child that was coming."[7] Shining beauty, *bubula*, the color and brightness of the skin, is a gift from one generation to the next. When persons die their major affines, in a reversal of beauty magic, blacken their bodies, cut their hair, wear old dark clothes, and forfeit their names. Just as there is a close association between the whiteness of *prima para* and embryo, there is also a close and analogous association between the blackness of deceased and the chief mourner, and to a lesser extent the deceased's other close affines (Weiner, 1987). The principal mourner is confined indoors, thus avoiding the sun but is made black rather than white. The *prima para's* diet is restricted to dry *kaula* [yams and taro] while the principal mourner avoids *kaula* and consumes wet foods. Cooked over a fire, and shielded from the sun, a *prima para* is made white, dry, and shining, so that her embryo will be white, beautiful, and strong. Mourners are denied fire, their transition into the damp blackness of death and decomposition mirror the condition of the deceased.

Beauty Magic. Weiner indicates that the gifts of *wageva* beauty magic, performed by the father's sisters [*tabu*], for a man's children, are linked to the yams he gives them. Children without fathers experience shame because their mother's kin cannot perform this essential rite for them. The ritual, refered to as *talilisi*, which means to wipe away, requires the *tabu* to dress in mourning skirts when administering this magic to young people (Weiner, 1976). Like mourning services, the *talilisa* is linked to the yam prestations from a man to his sisters but mourning and *talilisa* are alike in other ways. *Talilisi* is a compound of *tali*, to bid farewell and *lisi*, to lower or knock down or push away. *Lisaladabu* is payment for the mourners having shaved their heads, darkened their bodies, and otherwise given up their beauty, acts which are repayments triggered by the deceased gifts of yams. *Dabu* is heaviness and *lisaladabu* is "pushing away heaviness." *Lisaladabu* removes the heaviness of grief and the need to act under the heavy strictures of taboo but some of this heaviness is transferred to the individual who did not perform well in the competitive exchanges. Kasaipwalova notes that those who have done well in the exchanges will feel good and light while those who have not will feel heavy (Malnic, 1998).

But *lisaladabu* is not just about skirts, it is also very much about bundles (*nununiga*), which are metaphorical breasts (*nunu*) (Weiner, 1987). The visual representation of the heaviness of grief is the black skins of the mourners and their dark attire. To be light at a time of grief would be tantamount to claiming responsibility for the death. Instead kinswomen set about the task of creating thousands of bundles, the distribution of which, if successful, will remove their own heaviness, and which will pay others for assuming heavy tasks. Appropriately the creation of bundles involves a transformation from black to white which mirrors the hoped for change from heavy to light. After a death occurs, the women of the *dala* are conspicuously busy creating pseudo breasts, symbols of youth and whiteness, by a process that peels away the dark surface to reveal a white interior. The meaning of bundles lies, in part, in the details of their manufacture. The removal of the dark integument to expose a white interior is the very transformation that takes place in Tuma when a *baloma* sheds its darkened, aged skin to be reborn white and young again. White breasts are associated with nursing while black breasts mark the end of nursing; women will paint their breasts black to deter a child from nursing (Schiefenhövel & Schiefenhövel, 1996). In making and distributing symbolic breasts, women of the *dala* manipulate the chromatics of their cyclical model of ontology, to nurture the transfer of the *baloma* from life in this world to life in Tuma.

Ignorance speaks when those in the know are silent. Malinowski wrote his future wife of his reluctance to say anything about *kula* due to the fact that any Trobriander might know a great deal more than he (Wayne, 1995). As members of the initial culture to "benefit" from first-hand, in-depth, descriptions of their institutions, Trobrianders have, for the most part, silently suffered more than their share of mis-conceptualizations at the hands of anthropologists casting them in the role of definitive, and thus static, "other." Malinowski's descriptions of the great complexity of their society helped dispel ethnocentric notions of primitive simplicity but in revealing his "discovery" of an underlying order, and thereby testifying to his own understanding, he also inaugurated the facile process of essentualization which masks the great diversity of their often closely held opinions. Montague, who achieved a singular vantage of intimacy, was told by Trobrianders, who are generally careful to avoid offence, that they go along with the "tourist's" view of their reality (personal communication). We therefore offer the above discussion of the esoteric, "withheld other," situated in the highly private context suffering, as invitation for those who know to speak.

NOTES

1. Montague has collected oral histories that indicate that there was a severe population decline during the early 19th century due to post-contact diseases (personal communication).
2. *The major ethnographers of the Trobriand and related Massim cultures include*: C. G. Seligman, Bronislaw Malinowski, Henry Powell, Annette Weiner, Shirley Campbell, Ann Chowning (Fergusson), Debbora Battaglia (Port Moresby and Sabarl), Fredrick Damon (Woodlark Island), Linus Digm'Rina (Trobriands and Fergusson), Reo Fortune (Fergusson and Dobu), Edwin Hutchins, Susanne Kuehling (Dobu), Jerry Leach, Maria Lepowsky (Sudest), Luciana Lussu, Martha Macintyre, Susan Montague, Nancy Munn (Gawa), Giancarlo Scoditti, Carl Thune (Normanby), Karin Grossman, Stuart Berde (Panaeiti), and Michael Young.

    Medically trained observers include: R. L. Bellamy, R. H. Black, R. Pöschl and U. Pöschl, Wolf Schiefenhövel, and S. Lindeburg.

    Other relevant views of Trobriand life have been written by: Tom'Tavala, John Kasaipwalova, Chief Naributal, Juta Malnic (photographer and author), Leo Austen (Magistrate), Rev. Ralph Lawton (missionary-linguist), Father Bernard Baldwin (missionary), Kenneth Costigan (architect), and Ellis Silas (artist).
3. This work employed the DEPTH Database, a compilation of 8,000 pages of digital texts on Massim cultures, which is a greatly expanded version of the HRAF Trobriand Collection (*www.csus.edu/anth/trobriand/depth*). The authors wish to acknowledge the assistance of Caroline Gardner, Andy Connelly, Erin Caddy, and Sebastian Barbosa of the DEPTH team.
4. For magic to live in the belly, it must continually be retrieved by sub-vocalization; thus spells recorded by anthropologists are no longer active. The usual expression for a man who has completely learned all of his father's magical knowledge would be "Your father (his name) died empty handed, you have completely taken all of his stomach's contents." This is a direct reference to the man's magical knowledge and its transmission, whilst adding praise for the competence of the heir.
5. Digim'Rina notes that even though the benefits of "hammering" have been frequently cited in the literature, the idea is foreign to him.
6. It should be noted that Weiner (1976) views *dala* as a kinship category equivalent to a matrilineage, with membership being determined by shared blood. Montague, on the other hand, says that people are grouped into a *dala* based on shared mental capabilities which are determined by an individual's dietary history.
7. People are said to resemble their fathers rather than their mothers or other matrilineal kin. It is a grave insult to say that blood relatives resemble each other. In this passage, beauty refers to the brother's daughter.

REFERENCES

Austen, Leo. (1934–35). Procreation among the Trobriand Islanders. *Oceania, 5*, 102–113.

Austen, Leo. (1945). Cultural changes in Kiriwina. *Oceania, 16*, 15–60.

Baldwin, B. (1937). *The vocabulary of Biga Boyowa*. Unpublished manuscript, Pacific Manuscript Bureau, Reel 63.

Baldwin, B. (1945). Usituma! Song of Heaven. *Oceania, 15*, 201–238.

Black, R. H. (1957). Dr. Bellamy of Papua. *Medical Journal of Australia, 2*, 189–197, 232–238, 279–284.

Bickler, Simon H. (1998). *Eating stone and dying: Archaeological survey on Woodlark Island, Milne Bay Province, Papua New Guinea* (Doctoral dissertation, University of Virginia, 1998).

Brindley, Marianne. (1984). The symbolic role of women in Trobriand gardening. *Miscellania anthropologica, 5*. Pretoria: University of South Africa.

Burenhult, G. (2000). *The archaeology of the Trobriand Islands, Milne Bay Province, Papua New Guinea*. BAR International Series 1080. Oxford: Archaeopress.

Campbell, Shirley. (1984). *The art of Kula* (Doctoral Dissertation, Australian National University, 1984).

Capell, A. (1976). General picture of Austronesian languages, New Guinea area. In S. J. Wurm (Ed.), *Austronesian languages—New Guinea area languages and language study* (Vol. 2, Pacific Linguistics, Series C, No. 39, pp. 5–52). Canberra: Australian National University.

Darrah, Allan C. (1972). Ancestors in Trobriand ritual. Manuscript, Northwestern University.

Digim'Rina, Linus. (1998). An updated effect of the dreadful drought: The Trobriand experience. *APFT Briefing Note*, Brussels.

Fortune, Reo F. (1932). *Sorcerers of Dobu*. London: Routledge & Kegan Paul.

Egloff, Brain. (1978). The Kula before Malinowski: A changing configuration. *Mankind, 11*, 429–435.

Hipsley, E., & Clements, F. (1950). *Report on the New Guinea Nutrition Survey Expedition 1947*. Canberra: Department of External Territories.

Knauft, Bruce. (1999). *From primitive to postcolonial in Melanesia and anthropology*. Ann Arbor: University of Michigan Press.

Lawton, R. (1993). *Topics in the description of Kiriwina* (Pacific Linguistics Series D-84), Malcom Ross & Janet Ezard (Eds.). Australian National University: Research School of Pacific Studies.

Leach, Edmund R. (1966). Virgin birth. *Proceedings of the Royal Anthropological Institute* (pp. 39–49).

Leach, Edmund R. (1968). Virgin birth: Correspondence. *Man, 3,* 651–656.

Lepani, Katherine. (2001). *Negotiating open space: The importance of cultural context in HIV/AIDS communication models—A qualitative study of gender, sexuality, and reproduction in the Trobriand Islands of Papua New Guinea.* Masters thesis, University of Queensland.

Lindeberg, S. et al. (1994). Cardiovascular risk factors in a Melanesian population apparently free from stroke and ischaemic heart disease: The Kitava study. *Journal of Internal Medicine, 236,* 331–340.

Malinowski, Bronislaw K. (1916). Baloma: Spirits of the dead in the Trobriand Islands. *Journal of the Royal Anthropological Institute, 46,* 353–430.

Malinowski, Bronislaw. (1929). *The sexual life of savages in North-Western Melanesia: An ethnographic account of courtship, marriage and family life among the natives of the Trobriand Islands, British New Guinea.* London: G. Routledge & Sons, Ltd.

Malinowski, Bronislaw. (1935). *Coral gardens and their magic: A study of the methods of tilling the soil and of agricultural rites in the Trobriand Islands.* (Vol. I); *The description of gardening* (Vol. II); *The Language of Magic and Gardening* London: George Allan & Unwin.

Malnic, Jutta. (1998). *Kula: Myth and magic in the Trobriand Islands.* Wahroonga, NSW: Cowrie Books.

Milke, W. (1965). Comparative notes on the Australian languages of New Guinea. *Lingua, 14,* 330–348.

Montague, Susan. (1971). Trobriand kinship and the virgin birth controversy. *Man, 6,* 353–368.

Montague, Susan. (1973). Copulation in Kaduwaga. *Man, 8,* 304–305.

Montague, Susan. (1974). *The Trobriand society* (Doctoral dissertation, University of Chicago, 1974).

Montague, Susan. (1985). Infant feeding and health care in Kaduwaga village. In Leslie Marshall (Ed.), *Infant care and feeding in South Pacific.* New York: Gordon & Breach.

Montague, Susan. (1989). To eat for the dead. In F. H. Damon & R. Wagner (Eds.), *Death rituals and life in the societies of the Kula ring* (pp. 23–45). DeKalb: Northern Illinois University Press.

Montague, Susan. (2001) The Trobriand kinship classification and Schenider's cultural relativism. In R. Feinberg & M. Ottenheimer (Eds.), *The cultural analysis of kinship: The legacy of David M. Schneider* (pp. 168–186). Champaign: University of Illinois Press.

Munn, Nancy. (1986). *The fame of Gawa: A symbolic study of value transformation in a Massim (Papua New Guinea) society.* Cambridge: Cambridge University Press.

Nutrition Monitoring Group. (1980). *Report of the Nutrition Monitoring Group.* Provincial Health Office, Division of Health, Alotau, Milne Bay Province, Papua New Guinea.

Pöschl, R., & Pöschl, U. (1985). Childbirth on Kiriwina, Trobriand Islands, Milne Bay Province, Papua New Guinea. *Papua New Guinea Medical Journal, 28,* 137–145.

Powell, Harry A. (1968). Correspondence: Virgin Birth. *Man* (N.S.), 651–653.

Powell, Harry A. (1969). Genealogy, residence and kinship in Kiriwina. *Man* (N.S.), *4,* 177–202.

Powell, Harry A. (1980). Review: Woman of value by A. Weiner. *American Anthropologist,* 700–702.

Schiefenhövel, S., & Schiefenhövel, W. (1996). Am evolutionaren Modell -Stillen und fruhe sexualization bei den Trobriandern (Erin Caddy, Trans.). In Christine E. Gottschaulk-Batschkus & Judith Sculer (Eds.), *Ethnomedizinische Perpectiven zur fruhen Kindheit Curare im Auftrag der Arbeitsgemeinschaft Ethnomedizin.*

Scoditti, Giancarlo. (1996). *Kitawa oral poetry: An example from Melanesia* (Pacific Linguistics Series D 87). Canberra: Australian National University Department of Linguistics.

Seligmann, C. G. (1910). *The Melanesians of British New Guinea.* Cambridge: Cambridge University Press.

Senft, Gunter. (1986). *Kilivila: The language of the Trobriand Islanders.* Berlin: Mouton de Gruyter.

Silas, Ellis. (1926). *A primitive Arcadia: Being the impressions of an artist in Papua.* London: T. Fisher Unwin.

Spiro, Melford. (1968). Virgin birth, parthenogenesis, and physiological paternity: An essay on cultural interpretation. *Man, 3,* 242–261.

Spiro, Melford. (1972). Correspondence: Reply to Montague. *Man, 7,* 315.

Spiro, Melford. (1973). Copulation in Kaduwaga. *Man, 8,* 631.

Stocking, George. (1977). Contradicting the doctor: Billy Hancock and the problem of the Baloma. *History of Anthropology Newsletter, 4*(1), 11–12.

Tambiah, S. J. (1968). The magical power of words. *Man, 3,* 175–208.

Tambiah, S. J. (1983). On flying witches and flying canoes: The coding of male and female values. In J. W. Leach & E. R. Leach (Eds.), *The Kula: New perspectives on Massim exchange* (pp. 249–276). Cambridge: Cambridge University Press.

Villeminot, Jacques et Paule. (1967). *Les Seigneurs des Mers du Sud. La vie ancestrale et paradisque de habitants de iles Trobriand.* Paris: P. Laffont.

Wayne, Helena. (Ed.). (1995). *The story of a marriage: Volume I. The letters of Bronislaw Malinowski and Elsie Masson.* London: Routledge.

Weiner, Annette B. (1976). *Women of value, men of renown: New perspective on Trobriand exchange.* Austin: University of Texas Press.

Weiner, Annette. (1987). *The Trobriand Islanders of Papua New Guinea.* New York: Holt, Rinehart and Winston.

Young, Michael. (1983). The Massim: An introduction. *The Journal of Pacific History, 18,* 4–10.

Young, Michael. (1986). The "Worst Disease": The cultural definition of hunger in Kalauna. In L. Manderson (Ed.), *Shared wealth and symbol: Food, culture and society in Oceania and Southeast Asia* (pp. 111–126). Cambridge: Cambridge University Press.

Tuareg

Susan J. Rasmussen

ALTERNATIVE NAMES

Kel Tamajaq ("People who speak Tamajaq"); Kel Tagelmust ("People of the Veil"); Targui; Touareg; and also names designating groups from the different regions and confederations, for example, Kel Air ("People of the Air region"); Kel Ewey ("People of the Bull", a political confederation), descent groups or clans ("Kel Nabarro"; "Kel Tafidet"), and the pre-colonial social strata, for example, Imajeghen (designating the aristocracy); Inaden (smith-artisans); and Imghad (tributaries).

LOCATION AND LINGUISTIC AFFILIATION

Most Tuareg today live in Saharan and Sahelian regions of West and North Africa: in the present-day countries of Mali, Niger, Burkina Faso, Algeria, and Libya. They are believed to have originated in the central Saharan Fezzan region of Libya. Some Tuareg from Mali and Niger have fled from recent drought and war, to refugee communities in Mauritania. In addition, many men travel extensively on caravan trade to Nigeria and labor migration to Nigeria and other countries, where they reside long term. Language spoken: a Berber language, Tamajaq (alternative spelling: Tamacheq), with several regional dialects: for example, Tamahac in the Ahaggar Mountain region of southern Algeria and Tayrt in the Air Mountain region of northern Niger.

OVERVIEW OF THE CULTURE

Estimated population: between approximately 750,000 and one million (Decalo, 1996; Rasmussen, 2001).

History

A number of sources—oral traditions recorded by Rodd (1926), Bernus (1981), and Nicolaisen & Nicolaisen (1997), Saharan rock art, and Arab chronicles—relate the early origins and migrations of various Tuareg confederations. As early as the 7th century AD, there were extensive migrations of pastoral Berber peoples, including two important groups related to many contemporary Tuareg: the Lemta and the Zarawa. Invasions of the Beni Hilal and Beni Sulaym Arabs into Tuareg regions of Tripolitania and Fezzan pushed the Tuareg more southward toward the Air Mountain region of present-day Niger. By the 14th century, the Tuareg had become prominent as stock-breeders and caravanners on Saharan and Sahelian trading routes that led to the salt, gold, ivory, and slave markets in North Africa, the Middle East, and Europe. Later, however, caravans declined when much trans-Saharan trade was diverted to the West African coast, and 19th-century European exploration and military expeditions in the Sahara and along the Niger River led to the incorporation of the region into French West Africa by the early 20th century (Claudot-Hawad, 1993; Decalo, 1996; Porch, 1984; Rasmussen, 2001).

Economy and Occupations

French colonial domination, as well as recurrent droughts and intermittent conflict with central state governments of the independent nations where Tuareg reside, have profoundly affected local subsistence. The French colonial administrators disrupted many local systems of adaptation and the natural ecological balance of the Sahara and Sahel. In some regions, they destroyed traditional irrigation and well systems, replacing them with artesian springs which attracted mosquitoes, and installed gasoline-powered pumps which altered the distances between herds, pasture, and water. Colonial policies imposed limitations on the trans-Saharan caravan trade, taxed subject populations, and encouraged sedentarization and the growing of cash crops which were hard on the soil and displaced many populations farther into the pastoral nomadic zone. Some Tuareg, in particular nobles, initially resisted secular schools as threats to local culture and magnets for census and taxation counts. The traditional aristocracy at first sent the children of their slaves to these schools. Consequently, until recently Tuareg of noble origins tended to be underrepresented in the new occupations of the governmental and urban infrastructures. Recently, however, attitudes toward education have been more favorable.

Most Tuareg today practice a mixed economy of livestock herding, oasis gardening, caravan and other itinerant trading expeditions, and labor migration. A series of recurrent droughts, however, has diminished the livestock of many herders, pressuring them toward greater sedentarization and more intensive gardening. In addition, there are more specialized artisan activities traditionally performed by smiths, an endogamous, hereditary occupational social stratum. Recently, some Tuareg have also become active in tourism.

Social and Political Conditions

Pre-colonial Tuareg society was organized into hereditary, hierarchical, specialized occupational groups or social strata who were also, in principle, endogamous: at the top of the social pyramid were nobles (called *imajeghen*), who controlled large livestock, managed the caravan trade, and collected tribute from peoples of varying degrees of tributary (*imghad*) and servile (*ighawalen* and *iklan*) status, and practiced mutual client-patron rights and obligations with their attached smith/artisan (*inaden*) families. These relationships featured some pollution beliefs, for example, belief in the activation of destructive powers by smiths upon nobles' neglect of obligations toward them. In addition, Islamic scholars (*ineslemen*, also popularly called "marabouts") served as scribes, ritual specialists, legal councils, and Quranic healers. Despite its marked stratification, this social organization has always been characterized by considerable flexibility and negotiability (Keenan, 1976; Nicolaisen, 1997; Rasmussen, 1999). Leaders in the recent nationalist/separatist Tuareg rebellion have called for wider identification beyond kinship and social stratum, on the basis of the Tamajaq language, its Tifinagh script, and Tuareg cultural identity. In many rural areas today, the most intact relationship is between persons of noble origins and smith/artisans: the latter continue to act as political go-betweens, assist at arranging noble marriages, perform praise-songs, serve food and tea at noble weddings, manufacture jewelry and household tools for noble patron families, and dress nobles' hair, in exchange for remuneration. Nowadays, however, many nobles are impoverished and experience difficulty fulfilling their obligations toward smiths. Tuareg political structure included local drum chiefs who headed noble clans or lineages, who elected the sultan (*amenukal*) or their larger regional confederation. In many groups, the drum chief office was inherited through matrilineal descent, although personal qualities of the chief were also important. Under the domination of first, the French colonial administration and later, the central governments of independent nation-states, Tuareg political leaders experienced modifications in their powers. Some chiefs' powers were diminished, others' were increased (Claudot-Hawad, 1993). In rural communities, elders and Islamic scholars adjudicate local-level dispute cases.

Family and Kinship

Most communities are predominantly semi-nomadic nowadays, and many household compounds are enclosed by either a fence or adobe wall, and contain adobe mud houses, usually owned by men, and the more traditional nomadic tent, owned by married women. In most Tuareg groups, descent is now bilateral: vestiges of ancient matrilineal institutions have become submerged within patrilineal institutions introduced upon conversion to Islam (Claudot-Hawad, 1993; Murphy, 1967; Nicoliasen, 1997). Many clans trace their descent to founding female ancestresses/culture heroines. In some groups, succession to the chiefship passes from maternal uncle to sister's son. In many groups, there are prominent symbols of matriliny and pre-Islamic cosmology alongside those of patriliny and Islam in rites of passage and healing rituals. There are alternative forms of property transmission, in pre-inheritance gifts called "living milk" (*akh ihuderan*), in which some property (herds, date palms) is reserved for sisters, daughters, and nieces. This is intended to compensate women for Quranic inheritance (*takachit*), in which brothers receive twice the amount sisters receive. While living milk property once constituted women's primary source of economic independence, its future is now uncertain since the advent of nation-state laws and Quranic rulings which tend to favor patrilineal inheritance and droughts which have diminished many women's livestock herds.

Religion

The local "pre-Islamic" or "popular Islamic" belief system interweaves with more "official" Islamic beliefs and practices. Tuareg converted to Islam under the influence of Sufism and Almoravid marabouts between the 7th and

11th centuries; (Norris, 1975, 1990). Some Tuareg initially resisted Islam, and many Arab explorers disapproved of them for some "laxness" in Islamic observances, in particular for not secluding women, who among Tuareg may travel, receive visitors, and interact freely with unrelated men. In many groups, Islamic scholars are very respected and play important roles in ritual, healing, and politics. Offerings of dates and stones are made to tombs of marabouts and ancient ruins of the People of the Night or the People of the Past (*Kel Nad* or *Kel Arou*). During daytime many people pray at these tombs and occasionally consult them in divination. At night they are believed to be haunted by evil spirits. There are spirit pantheons integrating Islamic and local cultural cosmologies: for example, spirits called *djinn* (mentioned in the Quran) and spirits called the People of Solitude or the Wild (*Kel Essuf*) figure prominently in local folktales, rituals, and healing. Persons may become possessed by these spirits, and undergo special rituals to cure them. There are frequent practices to ward off evil spirits, malevolent humans such as thieves, and other misfortunes such as birth defects: for example, use of amulets made by both smith/artisans and Islamic scholars/marabouts, and observance of many ritual restrictions or "taboos." There are diverse "witchcraft"-like powers or forces, each designated by a distinct Tamajaq term, believed to operate upon violation of various "taboos," for example, those of jealousy or covetousness upon greedy consumption of (often scarce) food, or upon too ostentatious display of possessions (Rasmussen, 1998a, 2001). Additional cultural values in moral conduct, such as *takarakit* or shame/reserve, and *imojagh* or dignity, are also significant as constraints in interpersonal relations and limitations on conduct: too-openly boasting, for example, invites catastrophe. Many taboos, therefore, serve as "leveling mechanisms," limiting accumulation of wealth or power and moderating consumption by any single individual or social stratum. Ideally, they serve to restrain undisciplined or selfish conduct in an environment that requires, ideally at least, balanced reciprocity, sharing of resources, and mutual aid.

THE CONTEXT OF HEALTH: ENVIRONMENTAL, ECONOMIC, SOCIAL, AND POLITICAL FACTORS

Most Tuareg still live in rural communities, in the Sahara desert and in the Sahel savanna areas along its fringe. Except in the mountain massifs (Air in Niger, Ahaggar in Algeria, and Adragh n Ifoghas in Mali), which offer somewhat milder conditions, the climate is among the harshest in the world. It is subject to a very short and unpredictable rainy season (approximately July–August), recurrent droughts, and temperature extremes: for example, temperatures in the Sahara may reach 130°F in the hot season (April–July), and may plunge to freezing (32°F) at night during the cold season (December–March), when there are also high winds sometimes reaching 80 miles an hour, and sandstorms, locally called the Harmattan. In the northern regions of Mali and Niger above the agricultural line, the soils require daily irrigation.

Many diseases and environmental dangers are regional and season-specific. Malaria mosquitoes, while less prevalent in the desert regions than further South, pose a threat in the desert near oases gardens and standing pools of water. Local pests such as scorpions are ubiquitous in the rainy season. During the cold dry Harmattan season, conjunctivitis, streptococci, and meningitis are common. Intestinal parasites and wound infections tend to occur more rarely than they do in the forest zones.

Tuareg diets vary according to degree of sedentarization and urbanization, and have changed over the past half-century. In more nomadic camps, the principal foods consumed are millet and dates from caravan trading, and, when rains and pasture are sufficient, dairy products such as milk and cheese from goats, sheep, and occasionally camels. Meat tends to be eaten only at festivals, religious holidays, and rites of passage. In semi-sedentarized villages and on oases, in addition to these more traditional items, grains (maize, wheat, barley) and vegetables (potatoes, onions, tomatoes) and a few fruits (usually citrus) are consumed. Additional products obtained less regularly from trade and, more recently, food relief aid agencies and some small shops, include macaroni, manioc flour, rice, peanuts, and beans. Since the early 1990s, millet has become more expensive because of droughts in regions of Niger and Mali where it is grown; thus an important cereal source of high-protein has been somewhat more difficult to obtain, in some cases now supplemented or replaced by store-bought manioc or refined flour. If this trend continues, it has ominous implications for local nutrition. In urban areas, the diet reflects more multi-ethnic influences: for example, rice is a status food, and if funds permit, it is eaten not solely at festivals (as in the countryside), but more often, nearly on a daily basis. In some towns with large livestock markets, meat is consumed much more frequently than in the countryside, in sauces over rice, millet, or maize. In the towns, many dishes from the countryside—for example, the nutritious beverage made of millet, dates, and goat-cheese and several wheat dishes—while valued by urban Tuareg, are more seldom prepared since some ingredients are more difficult to obtain.

According to many elderly persons, food was scarcer and less varied in the past, yet healthier. It was necessary to move about more often in order to find food (in nomadism and some hunting and gathering). They indicated that, although foods today are more abundant since one can find many sold in shops, such processed foods are more expensive and less nutritious. In the past, many people ate wild plants and grasses nowadays consumed only as herbal medicines. Periodic droughts threaten many herbal medicinal plants and trees, which traditionally constitute a rich local pharmacopoeia, particularly in more mountainous regions. Some foods are famine foods: for example, the core of the doum palm and thorns of various species of trees, pounded and grilled.

Most countries where Tuareg reside today are poor. For example, Niger has one of the lowest per capita incomes in the world, estimated at US$260 in 1987 (World Bank, *World development report*, 1989, p. 14). This income has declined in the past decade from World Bank-imposed economic austerity measures. Life expectancy in many countries where Tuareg reside is low; in Niger, for example, this was estimated at 45 years in 1987, and the infant mortality rate was estimated to be 135 deaths per 1,000 live births (World Bank, *World development report*, 1989, p. 226). More recently, infant mortality was estimated at 123 per 1,000 births (U.S. Department of State, *Background notes, Niger*, July 1994). The central government budget in 1994 was estimated at US$291.4 million (adjusted for devaluation in 1993 of the French West Africa C.F.A.). The 1994 investment budget (capital and development expenditures) was at $190 million (U.S. Department of State, *Background notes, Niger*, July 1994).

Private medical insurance and other benefits are not available to the vast majority of patients in many countries where Tuareg reside. Governments do not reimburse hospitals directly for the care of their employees, so the programs for government employees are in effect exemptions from payment. Throughout Niger, for example, public facilities administered by the Ministry of Public Health provide most of the biomedical health-care. There are two tiers of prices: private sector patients pay higher fees than public sector patients for private hospital rooms, diagnostic exams, and surgical procedures (Weaver, Wong, Sako, Simon, & Lee, 1994, p. 566). Only a small percentage of the population is employed by large companies and receive insurance benefits from them (Rasmussen, 2001, p. 13).

MEDICAL PRACTITIONERS

Types of traditional medical practitioners in Tuareg society include herbalists; bone-setters; Islamic scholars or marabouts; diviners; and spirit possession ritual exorcism specialists. Some of these specializations overlap, many practitioners refer patients to each other, and many patients consult more than one practitioner for a given illness. Herbalists are predominantly though not exclusively women; many are called "medicine women" (*tinesmegelen*). Many inherit their profession in clans and apprentice with an older female relative. These practitioners do diagnostic, healing, and referral work and cure mostly stomach afflictions with leaves, barks, and roots and also, sometimes, with medicines purchased in markets or brought into the region by trade. They diagnose through massage with special focus upon the stomach. Herbalists also practice some psycho-social, particularly marital, counseling and, along with Islamic scholars/marabouts, are often consulted to treat women's fertility problems. Many herbal medicine women emphasize their complementary, rather than competitive relationship to Islamic scholars/marabouts in healing, describing these healers as being "like husband and wife," and referring some patients to marabouts (Rasmussen, 1998b). Some herbal medicine women perform non-Quranic divination, through dreaming; this specialty is called *asawad*, denoting "to look or see," and its practitioners called *imaswaden* (sing. *amaswad* or *amanai*, fem. *tamaswad* or *tamanai*) (Rasmussen, 2001). Some herbalists also know bone-setting, while other practitioners set bones, but do not practice herbal medicine: they are called *imadasen* (sing. *amadas*, fem. *tamadas*). Other non-Quranic diviners are called by the Hausa-derived term *bokaye* (sing. *boka*); these diviners tend to be male and practice more often in the towns, whereas the herbalist/diviners tend to be female and practice more often in the countryside. *Bokaye* diviners work with plants as well, but supplement them with scents (usually perfumes, but also some herbal medicines inhaled through the nose) and other ritual paraphernalia such as cowrie shells. They are believed to work with a tutelary spirit, in a pact with the *Kel Essuf* spirits of the Wild, whom they must propitiate at intervals with sacrifices (Nicolaisen, 1961, 1997). Their position among the Tuareg is ambiguous, and some residents express ambivalence toward them, suggesting that this specialty originated from outside Tuareg society, perhaps from the neighboring Hausa people (Rasmussen, 2001). Islamic scholars or marabouts (*ineslemen*) are active in healing organic and non-organic illnesses, many of which are defined as caused by spirits. They also do psychosocial counseling. Marabouts cure with the Quran, which has special verses that cure diverse illnesses. Marabouts make amulets from these verses, to be worn around the neck or against the skin. Many men and some women see marabouts, although the latter sometimes feel intimidated by them, or find them unsympathetic and undergo alternate cures, seeing additional healers when their illnesses do not respond to marabouts' healing, explained locally as caused by spirits who do not respond to Quranic verses. In such cases, patients are referred to other healers, such as exorcism specialists who preside over a musical spirit possession ceremony called *tende n goumaten*, featuring drumming and singing believed to "please" and placate certain non-Quranic spirits alternately called *goumaten* or *Kel Essuf*, which predominantly afflict women (Rasmussen, 1995). These practitioners consist of musicians, who are often relatives and close friends of the possessed: namely, a female chorus who perform songs addressing the spirits and also containing critical social commentary; a woman who strikes the *asakalabo*, a calabash floating in water; and a drummer, usually a smith/artisan man or a woman of any social origin, who strikes the *tende* drum, constructed from a mortar with goat-hide stretched across its top.

CLASSIFICATION OF ILLNESS, THEORIES OF ILLNESS, AND TREATMENT OF ILLNESS

Despite some influence of hospitals and clinics, particularly in the towns, Tuareg cultural understandings concerning illness, particularly in the countryside, include many alternatives to the biomedical paradigm. Fundamental to the local paradigm is a continuum, rather than rigid opposition, between body and mind and between organic and non-organic illnesses. Much Tuareg medicine features counteractive theories of balance and harmony, for example, "hot" versus "cold" states of the body and diseases caused by imbalance of these forces. These states are gender-linked; for example, women should ideally be cool, and men should ideally be warm, but these states should not become too pronounced or intensified, for example, a man can become too hot and ill (Figueiredo in Claudot-Hawad, 1996, pp. 113–137). Too intense cold or heat, or conversely, accumulation of the opposite of the ideally-dominant quality brings illness, and requires a cure. These conditions of "hot/cold" and associated afflictions are sometimes literal, sometimes non-literal or metaphorical in connotations, for example, "hot" illnesses (*tuksi*) are believed to be caused by too much heat, from "hot" foods (dates, tomatoes), sunlight, or moon-beams (these latter may cause illness from direct contact, as in sun-stroke, or, alternately, by sitting on warm mats or from their reflections inside doorways). *Tuksi* may also result from anger and other strong sentiments. These require counteractive treatments with herbal medicines, ritual precautions, or dietary remedies (such as "cold" foods, e.g., millet). Many of these illnesses include stomach ailments. "Cold" illnesses (*tessmat*) are the counterpart of "hot" illnesses. These include urinary tract problems and STDs. Treatment is sought from herbalists, who often prescribe plant remedies and ritual bathing (Rasmussen, 1998b). Other diseases may be caused by the wind and aromas; covering the bodily orifices is important to prevent them.

There are additional conditions that defy neat classification into organic and non-organic, and have only approximate, rather than exact, translations into the English language and Euro-American established biomedical paradigms: for example, *anoughou* refers to a condition caused by a sudden change in routine with subsequent deprivation of the usual nourishment or habit; *tamazai* refers to a condition approximating depression, in which one suffers from a long-term hidden wish or resentment that cannot be directly expressed. This latter relates to important Tuareg cultural values that discourage direct or explicit speech and encourage indirect expression by allusion. Sometimes, many residents believe, this condition results from unrequited love or other love problems (*tarama*). It often provokes *goumaten* spirit possession. A stomach ailment called *karambaza*, usually diarrhea, is believed to be caused by the mystic ritual powers of smith/artisans (*tezma* or *ettama*), activated automatically when they are refused a present or denied a request. *Karambaza* attacks children or livestock of the offending party, often a noble patron. Its remedies involve seeing an herbalist, and also gathering up the sand in the smith/ artisans' footprints and throwing it into the fire.

Other illnesses may result from negative gossip (*togerchet*) by anyone, behind one's back (Casajus, 2000; Rasmussen, 2002). More lethal is an affliction believed to be caused by sorcery, called *ark echaghel* (literally denoting "bad work"). This is practiced surreptitiously and considered very dangerous and anti-social, for it requires the assistance of a marabout who is willing to misuse his powers destructively (Nicolaisen, 1961; Rasmussen, 2002). Sorcery almost always causes the death of the targeted victim. It is often transmitted through contamination of food, burying of harmful amulets beneath the ground where one walks, or through contact with the victims' clothing or other possessions or bodily fluids. Sometimes, an animal such as a dog may be sent to harm the intended victim.

Prominent in theories of health and illness among the Tuareg are concepts of the body and soul. The head (*eghef*) is considered the place where spirits reside, once they have entered the liver (*tessa*) and stomach (*tedis*). The stomach is also the symbol of the matriline, whereas the back (*aghuri*) is the symbol of the patriline. If herbal medicine women diagnose spirits "dancing" in a patient's stomach, they refer the patient to a marabout, who treats the patient in seclusion with counseling and the Quran, often also divining with the Quran and various cabalistic formulas. The liver is considered the seat of sentiments such as anger and love, and is the place where non-Quranic spirits often enter; these require the *tende n goumaten* spirit possession ritual, whose public curing with music and a large, diverse audience who encourage the possessed person in trance in her dancing, constitute an effective form of group therapy. Sometimes, patients see diverse healers in succession for the same ailment, for example, *bokaye* non-Quranic diviners after seeing herbalists and marabouts. *Bokaye* treat non-organic illness through non-Quranic divination methods; they diagnose with scents and oils. The soul (*iman*, also denoting life and breath), is believed to leave the body during sleep and walk about. Upon death, the soul may be contacted through dreams or through offerings to graves. The heart (*ewel*) is connected to sentiments such as generosity and compassion, and is closely associated with the soul and the liver in psycho-social and non-organic connotations; many non-organic illnesses such as spirit possession are referred to as "illnesses of the heart and soul."

SEXUALITY AND REPRODUCTION

Tuareg differ from some of their neighbors in the marked degree of free social interaction between the sexes, and in the generally high prestige and economic independence of Tuareg women. Unrelated men and women may visit each other, flirt, and conduct courtship. Conversation between the sexes is considered extremely important. Women may travel, visit, and receive male visitors before and following marriage (Murphy, 1964, 1967). Women may initiate divorce, and own the tent and inherit and manage livestock.

There is some variation among the different Tuareg confederations concerning pre- and extramarital sexual mores. In some groups, women and men may conduct pre- and extramarital affairs freely; in others, particularly among the more devoutly Muslim clans of Islamic scholars/ marabouts and in some chiefly families, there is a tendency to frown on this. Illegitimate children bring shame to the mother, and are often hidden and raised in distant regions.

Most women in semi-nomadic rural communities bear approximately six to eight children. There are indications that with increased sedentarization, many families prefer more children to assist with oasis gardening. Precise statistics from large samples are not available, but many local residents indicate that in the past, nomadic lifestyles discouraged having many children and there were efforts to space children. Although Tuareg women's status is not solely dependent upon childbearing, and Tuareg recognize that men as well as women may be responsible for childlessness, nowadays wives feel greater pressure than in the past to bear children, and fear that husbands may contract polygynous marriages if they are infertile.

HEALTH THROUGH THE LIFE CYCLE

Pregnancy and Birth

Local beliefs surrounding reproduction include the idea, expressed in local slang, that during menstruation, a woman's eggs (pl. *chikikaten*) are "broken"; at other times, they are whole. Some beliefs appear gender-based; for example, men describe women as being like "containers" or "leather sacks" during pregnancy; whereas women describe female reproductive physiology as centered on the stomach (*tedis*) and womb, called *ehan n barar* ("the child's tent"), and tend to place greater emphasis upon love as important in conception. The male is believed to transmit "heat" (*tarraf*) during conception. Iblis, the Devil, is considered to be the ultimate source of reproductive force.

Women must protect themselves from malevolent spirits believed to cause birth defects and infant mortality. During menstruation, there are ritual restrictions against praying, touching Islamic amulets, handling animal hides, and dressing hair. These measures are believed to protect *al baraka* in living things, and by extension, the yet-to-be born child (Rasmussen, 1991). Conception should not occur outside the tent under the moonlight, nor before the end of the week-long wedding. At childbirth, the earth opens up and threatens the woman in labor (Worley, 1992, pp. 54–64). In rural communities, babies are born within the mother's tent, with elderly female relatives attending. A baby should not be left in a doorway (at a crossroads); this action is believed to cause mental disabilities. New mothers are secluded for 40 days following birth, to protect them from jealous spirits threatening mother and child. A knife (metal is believed to ward off spirits) is stuck in the sand near the mother's bed, and Quranic amulets are prominently displayed. There are also ritual practices to promote fertility, for example, animal sacrifice followed by the married couple's consumption of specified foods: the animal's lungs, fresh rather than dried (symbol of life and breath), and eggs with Quranic verses written in vegetal ink.

Infancy

In the countryside, infant mortality is approximately 60%. Spirits are believed to sometimes mistake babies for goat-hide water-bags, and pull them back into the spirit world. Efforts are made to keep newborns in the world of humans. The firstborn male wears a tuft of hair on top of the head "in order to be pulled by the Prophet up into paradise" in the event of death. If a baby is stillborn, no condolence or nameday rituals are held; if a baby dies after crying, a condolence ritual is held as for deceased adults. Babies are often called "stranger" until they are 1 week old, then are given several names, Quranic by the father and marabout, and non-Quranic by the maternal grandmother, at the nameday ritual. The marabout writes Quranic verses in vegetal ink for the baby's mother to drink, to protect from malevolent spirits and humans. Special names are given to babies when a previously-born sibling died, in order to distract the spirits, for example, Tekle (denoting slave). Twins are regarded as somewhat demanding of scarce resources, but efforts are made not to make them angry. Mothers are supposed to treat each twin in identical ways: for example, they are given alliterative names (e.g., Alhassane and Alhouseini), are breast-fed at the same time, and arranged in the same sleeping positions. Babies are breast-fed for approximately two years. They sleep with their mother, and parents practice a postpartum sexual taboo for approximately two years; a child born before this time is considered shameful, and some birth defects are attributed to a lapse in practicing this taboo. There is no transitional baby food; small children are encouraged to eat adult food. Small children are toilet-trained gradually and casually: mothers encourage them to go far from the household to relieve themselves, with other children.

Childhood

Children are encouraged to assist with adult chores. In some regions, toddlers are given chickens to raise as pets to practice for future herding of their parents' livestock. Small children are encouraged to explore their physical environment freely, and corporal punishment is rare, although adults in rural communities keep a watchful eye on children's activities. Toddlers begin to wear protective religious amulets around weaning age. At around eight years, young girls begin to assist their mothers in cooking and fetching water. At this age, boys traditionally start to accompany their fathers on caravan trading expeditions, help in the oasis gardens, and smith/artisan boys begin apprenticeships at the forge with an older male relative. Children of both sexes herd livestock. Nowadays, more girls and boys of diverse social origins go to both Quranic and secular schools.

Adolescence

Despite children's full participation in adult tasks, Tuareg recognize adolescence as a distinct phase, and a Tamajaq term designates an adolescent: *ekabkab* (fem. *tekabkab*). Local concepts of the life course are, nonetheless, not strictly linear, chronological, or biological, but rather these phases are socially and ritually defined (Rasmussen, 1997). For example, females are considered to "become women" at marriage, rather than upon menstruation; males are considered marriageable upon taking up the men's face-veil, at approximately 18–20 years of age. Adolescents of either sex, but particularly women, are believed to be vulnerable to jealousy of spirits and humans at life transitions.

Adulthood and Middle-Age

Women and men among Tuareg are considered full social adults upon becoming parents; children are extremely important to mature, adult status. Informal adoption is often practiced in cases of childlessness, but rituals promoting fertility convey the importance of having children, and eventually, children-in-law (Rasmussen, 1997). Sons-in-law contribute important economic resources to their affinal household through bridewealth and groom-service. There is an extreme respect/reserve relationship between a man and his parents-in-law, particularly the mother-in-law. Thus, while Tuareg women generally enjoy high social prestige and some economic independence throughout life, these ideally increase upon their children's, particularly daughters' marriages.

There is equal access to medical care, both traditional and established bio-medical, for Tuareg men and women throughout life; the problem of unequal care principally affects Tuareg generally, regardless of gender, and arises from the predicament of most Tuareg as a marginalized group within the nation-state, from the geographic inaccessibility of rural communities, and general poverty of those nation-states where most Tuareg reside.

The Aged

Aged persons are ideally respected. Children are supposed to care for ailing parents. Elderly persons of either sex reside next door to children, and participate prominently in Islamic and pre-Islamic rituals. Many elderly persons continue to work (herding, caravanning, and gardening) as long as their health and energy permit, although more arduous labor (fetching well water and firewood) is usually performed by younger relatives.

Physical problems of older women are attributed to aging processes in general, rather than specifically from the cessation of menstruation (Rasmussen, 2000, pp. 91–116). Herbal medical specialists and other women report few symptoms associated with the "menopause" model in western established bio-medicine. Post-childbearing women's ritual and social roles apparently compensate for any perceived physical problems, and many such problems merge with other age-related problems in local medicine.

Dying and Death

Mortuary beliefs and practices interweave local Tuareg and Islamic cosmoligies and rituals. Dead souls are believed to wander in the vicinity of graves, and some communication with them is possible through dreams and divination. Funerals (called *iwichken* or *iban*) are held within 3 days of death, and consist principally of condolences at the home of relatives, during which special foods are served and the marabout comforts the bereaved, ritually spitting in a goblet in order to "calm" them. The body is prepared for burial by elderly same-sex living persons: it is washed, wrapped in a white shroud, and carried to the cemetery, accompanied by chanting from the Quran. Men and marabouts officiate at the burial. Two lines are formed in order to allow the angel of death to pass through. Thereafter, the name of the deceased is not mentioned. Graves are not individually marked, although stones are piled higher on tombs of prominent marabouts and chiefs, and offerings made to them. In keeping with Islamic beliefs, marabouts emphasize judgment day following death, when angels measure the relative weights of bad and good deeds of the deceased in life. While ancestor cults among Tuareg appear less elaborated than in some other African societies, there are commemorative rituals (*takote*) for deceased at intervals following death, which feature animal sacrifice, alms-giving, feasts, and reading from the Quran.

CHANGING HEALTH PATTERNS

Many rural Tuareg initially feared hospitals because they appeared implicated in colonial and post-colonial schemes to dominate and control: for example, many medicine distributions were accompanied by census counts, taxation records, and political speeches. Some viewed hospital and clinic staffs as unsympathetic and hostile to local culture (Rasmussen, 1994, 2001). More recently, many Tuareg are less fearful of hospitals and clinics due to their staffing, since the Peace Pact of 1995, with more local and Tamajaq-speaking personnel. Nevertheless, most rural people still go to traditional local medical practitioners before clinics and hospitals because of difficult access to hospitals, located only in the major towns—and travel there is often difficult. Prescription medicines are irregularly stocked and expensive. Some pills are now sold at cheaper prices at vending tables on the street; these are, however, often dangerous, uncontrolled, and unlabeled. Recently, in the wake of economic austerity and privatization policies initiated by the International Monetary Fund and the World Bank, private practitioners have sprung up in capital cities, but most rural residents cannot afford their fees.

References

Bernus, E. (1981). *Touaregs Nigeriens: Unite d'un peuple pasteur*. Paris: Editions de l'Office de la Recherche Scientifique et technique de l'Outre-Mer.

Casajus, D. (2000). *Gens de Parole: Langage, poesie et politique en pays touareg*. Paris: Editions la decouverte textes a l'appui/anthropologie.

Claudot-Hawad, H. (1993). *Touareg: Portrait en fragments*. Aix-en-Provence: Edisud.

Claudot-Hawad, H. (1996). *Touareg et Autres Sahariens entre Plusieurs Mondes*. Aix-en-Provence: Edisud.

Decalo, S. (1996). *Historical dictionary of Niger*. Lanham, MD, London: The Scarecrow Press.

Figueiredo, C. (1996). Identite et concitoyennete. La reelaboration des relations entre hommes et femmes aux marges de la societe Kel Adagh (Mali). In H. Claudot-Hawad (Ed.), *Touareg et Autres Sahariens entre Plusieurs Mondes* (pp. 113–137). Aix-en-Provence: Edisud.

Keenan, J. (1976). *Tuareg: People of Ahaggar*. New York: St. Martins Press.

Murphy, R. (1964). Social distance and the veil. *American Anthropologist, 66*, 1257–1274.

Murphy, R. (1967). Tuareg kinship. *American Anthropologist, 69*, 163–170.

Nicolaisen, J. (1961). Essaie sur la religion et la magie touaregues. *Folk, 3*, 113–160.

Nicolaisen, I. and Nicolaisen, J. (1997). *The pastoral Tuareg*. Rhodos, Copenhagen: The Carlsberg Foundation.

Norris, H. T. (1975). *The Tuareg: Their Islamic legacy and its Diffusion in the Sahel*. Wilts, England: Aris and Phillips.

Norris, H. T. (1990). *Sufi mystics of the Niger desert*. Oxford: Clarendon Press.

Porch, D. (1984). *The Conquest of the Sahara*. New York: Knopf (Random House).

Rasmussen, S. (1991). Lack of prayer: Ritual restrictions, social experience, and the anthropology of menstruation among the Tuareg. *American Ethnologist, 18*, 751–769.

Rasmussen, S. (1994). Female Sexuality, Social Reproduction, and the Politics of Medical Interventions in Niger: Kel Ewey Tuareg Perspectives. *Culture, Medicine, and Pshchiatry, 18*, 433–462.

Rasmussen, S. (1995). *Spirit possession and personhood among the Kel Ewey Tuareg*. Cambridge: Cambridge University Press.

Rasmussen, S. (1997). *The poetics and politics of Tuareg aging: Life course and personal destiny in Niger*. DeKalb: Northern Illinois University Press.

Rasmussen, S. (1998a). Ritual powers and social tensions as moral discourse among the Tuareg. *American Anthropologist, 100*, 458–468.

Rasmussen, S. (1998b). Only women know trees: Medicine women and the role of herbal healing in Tuareg culture. *Journal of Anthropological Research, 54*, 147–171.

Rasmussen, S. (1999). The slave narrative in life history and myth. *Ethnohistory, 46*, 67–108.

Rasmussen, S. (2000). From childbearers to culture-bearers: Transition to postchildbearing among Tuareg women. *Medical Anthropological Quarterly, 14*, 242–270.

Rasmussen, S. (2001). *Healing in community: Medicine, contested terrains, and cultural encounters among the Tuareg*. Westport, CT: Bergin & Garvey.

Rasmussen, S. (2002). Betrayal or affirmation? Transformation in witchcraft technologies of power, danger, and agency among the Tuareg. In H. Moore & T. Sanders (Eds.), *Magical interpretations, material realities* (pp. 136–160). London: Routledge.

Rodd, F., Lord of Rennell. (1926). *The people of the veil*. London: Anthropological Publications.

U.S. Department of State. (1994, July). *Background notes. Niger*. Washington, DC: U.S. Government Printing Office.

Weaver, M., Wong, H., Sako, A. S., Simon, R., & Lee, F. (1994). Prospects for reform of hospital fees in sub-Saharan Africa: A case-study of Niamey National Hospital in Niger. *Social Science and Medicine, 38*, 565–574.

World Bank. (1989). *World development report*. New York: Oxford University Press.

Worley, B. (1992). Where all the women are strong. *Natural History, 10*, 54–64.

Wape

William E. Mitchell

Alternative Names

Wapei, Wapi.

Location and Linguistic Affiliation

The Wape live on the large island of New Guinea located just north of Australia. Their country, Papua New Guinea, occupies the eastern half of the island; Sandaun Province, where the Wape reside, is situated in its northwest corner bordering Irian Jaya and is one of the country's poorest and the least developed. "Wape" is a term given by Westerners to the culturally similar Olo speakers. It is derived from the Olo word *metane wape*, which denotes a human in contrast to a spirit. Olo is one of the 47 languages of the Torricelli Phylum that is provisionally divided into 7 stocks and 13 families. Olo is classified within the Wapei Family, 23,378 speakers, and the Wapei-Palei Stock, 31,770 speakers (Foley, 1986; Laycock, 1975; McGregor & McGregor, 1982).

Most of the men, many children, and an increasing number of the women speak Tok Pisin (Melanesian Pidgin), the *lingua franca* (Mihalic, 1971). Tok Pisin is a post-contact language, the vocabulary of which is 80% English origin (Foley, 1986). Tok Pisin and English are the country's two official languages. English is the language of instruction of the village primary schools, but as attendance is not compulsory, only a minority of the Wape are fluent.

OVERVIEW OF THE CULTURE

The approximately 10,000 Wape live in the rain forests of the Torricelli Mountains located between New Guinea's north coast and the sprawling swamps of the Sepik River to the south. The average yearly rainfall is 264.16 cm. They occupy 55 villages between 396 and 853 m above sea level with a population density of approximately 50 people per square mile. Germany claimed the Wape area in 1885 but there is no evidence that they had any direct contact with the people. After World War I, the area became part of a mandated territory under the aegis of the League of Nations and administered by Australia. The first government patrols probably entered the Wape area from the coast in the early 1920s, but no stations were established. In 1926, when the first scientist, zoologist E. A. Briggs (1928, 1929) from the University of Sydney explored the area, it had received only a few visits from labor recruiters and prospectors.

Foreigners, including Brigg's student and explorer–writer A. J. Marshall as well as gold and oil prospectors, continued sporadic incursions into the area during the 1930s and with few exceptions were received peacefully. In 1935, a disastrous earthquake ripped through the Wape area taking an unknown number of lives. It destroyed houses and gardens, and created landslides that dammed rivers and streams (Marshall, 1938). During World War II, both the Allies and Japanese sent patrols into the area impressing local men as carriers. The Allies also used local labor to build a small military airstrip near Lumi village in the heart of the Wape area. Some Wape men went to work as laborers for the Allies at their base on the coast near Aitape. When dysentery broke out on the coast in 1944, many men fled back to their villages carrying the disease with them and the resulting epidemic took many lives. After World War II, in 1946, the eastern half of New Guinea became a trust territory of the United Nations under Australian administration. In 1947, two Franciscan priests from the coastal Aitape mission opened a mission station by the abandoned airstrip in Lumi and the mission remains active today. In 1948, the government established a patrol post nearby and a small hospital was built soon afterwards. In 1973, the territory was given internal self-government status and in 1975, gained full independence as Papua New Guinea.

Christian Brethren missionaries have been active in the area since 1951 and in the 1980s an indigenous evangelical church began winning some adherents. Although many Wape are nominal Christians, most continue to follow the rituals of their ancestors. Throughout the years of foreign contact, the government and missions have introduced numerous economic schemes for the Wape (Taru, 1977). None have enjoyed much success and the Wape remain subsistence farmers practicing slash and burn agriculture with limited access to money. To obtain money to buy coveted imported commodities, Wape men formerly worked as indentured laborers in other parts of the country. With this work no longer available and no town in the immediate area, there is an increase in the number of families moving to the coastal towns to find work. Unlike large parts of the highlands of Papua New Guinea where tribal warfare is still practiced, the Wape live at peace with each other and neighboring groups. In the past as at present, villages have no internal government except that of kinship although they participate in regional council, provincial, and parliamentary elections.

Villages traditionally were situated on ridges for easier protection and, as today, population is usually several hundred people. In the village center is a dirt plaza where children play and community rituals are performed. Wape houses are made of forest materials but where they formerly had a single earthen floor room, today they are more often built on stilts with several rooms serving different domestic functions in the Western fashion. Parents, today as traditionally, live together with their children in their own home. At puberty, sons move into a separate dwelling with other youths and unmarried men. Some villages still have a separate men's ceremonial house where ritual objects are stored and rituals are enacted. Wape do not enter each other's homes unless they are close kin, but most dwellings have a front verandah or an area beneath a raised house where neighbors and relatives gather to visit.

Dwellings have no electricity, telephone, or plumbing, and access to most villages is by narrow footpaths. In the 1980s, the government built a rough dirt road connecting the Wape to the coastal town of Wewak, but the heavy rains, scarcity of transport, and occasional blockades by landowners seeking toll along the route makes its use problematical. Regarding work, women cook, fetch water, and firewood, make string bags, process sago, and forage. Men hunt, prepare gardens, cut down the sago palms, and make tools and ritual objects. Men and women both participate in childcare, weeding, and harvesting.

The most important social and economic unit in Wape society is the patrilineage whose members usually live contiguously in a single village along with several other patrilineages. At birth a child becomes a member of her or his father's patrilineage and remains so until death. Patriclans are composed of several patrilineages from a number of villages. Clan ties provide access to others in times of hardship but the proviso to assist is not as stringent as among patrilineage mates. Inheritance of land and food trees is patrilineal. Marriage with a member of one's patrilineage and patriclan is forbidden, but it is sometimes violated among patriclan members. Today, women usually marry by choice and the groom's family pays bridewealth in the form of money to her patrilineage. A woman goes to live with her husband at marriage and, in the unlikely event of divorce, she returns to her village while the children should, but do not always, stay in their father's village. Plural marriage is permitted but uncommon.

One of the most striking aspects of village life is its placidness, a feature also observed by early visitors to the area (Marshall, 1938). Physical aggression is disapproved and seldom occurs and while verbal aggression is allowed, it too occurs infrequently. The customary response to an aggressive act is simply to turn away and ignore it. Men and woman live relatively harmoniously together and spousal abuse is unusual (Mitchell, 1999). Male ethos is egalitarian and the elaborate and pervasive reciprocal exchange system of pig meat, sago, and valuables among relatives in different villages mitigates against the accumulation of individual wealth (Mitchell, 1978). The gambling dice game *satu* is played by many men and is another important way of leveling wealth (Mitchell, 1988).

Wape traditional religion is animistic, with beliefs in ghosts, demons, witches, and sorcery that are intimately involved with ideas about health, sickness, and healing.

Unlike many New Guinea societies that center their ritual and ceremonial life around harvests or male initiation, the Wape center theirs on healing.

THE CONTEXT OF HEALTH: ENVIRONMENTAL, ECONOMIC, SOCIAL AND POLITICAL FACTORS

The Wape live in very broken forested country with generally thin and poor soils subject to landslides by the heavy rains, and the many steeply banked streams and rivers, which bisect the forest, contain few fish. The men are avid hunters but population density and the introduction of the shotgun (Mitchell, 1973) has depleted the wild game, for example, pigs and cassowaries. Because of the crime problem in some parts of Papua New Guinea since Independence, shot guns, always limited to one or two in a village, are now difficult to obtain and there are reports that wild game is increasing in number. As the Wape area is economically undeveloped, the people have restricted access to money for the purchase of food commodities such as tinned fish or rice from a local trade store. Some villagers keep a pig or two who, foraging for themselves, are fed just enough to prevent them from becoming feral. As pigs are raised primarily for ceremonial exchanges, the meat is distributed only periodically and is not an enduring source of protein. Men to the south occasionally bring smoked fish and pig into the area, but its availability is also random and undependable.

Most of what villagers eat is raised in their unfenced slash and burn gardens, for example, bananas, coconuts, sweet potatoes, taro, breadfruit, and *Gnetum gnemon* tree leaves, or processed from sago palms planted in swampy areas. As a result their diet consists primarily of sago starch, a notoriously poor nutrient, augmented with garden produce. With the absence of important amounts of fish and game in the diet, vegetables are their main source of protein. Consequently, the Wape diet has a negative effect on their health and maturation resulting, in part, in low birth weights, delayed growth, and malnutrition.

The introduction of health services by the missions and government in the 1940s was an important accomplishment of the colonial era. However, after Independence many of the expatriate health workers returned home severely shrinking the country's health resources. Unfortunately, the local training of replacement workers continues to lag. As a Third World country with limited economic resources, Papua New Guinea currently concentrates its health expenditures in the towns where the new indigenous elite live. The comparatively isolated Wape suffer accordingly with limited access to health as well as other services. Although a service may exist on governmental paper, staff and supplies are frequently inadequate to meet urgent rural needs.

Medical Practitioners

Indigenous

The Wape recognize three types of part-time indigenous practitioners who diagnose and/or treat sicknesses, none of which use trance or an altered state of consciousness (Mitchell, 1990). A *numoin* (Olo), the most feared and powerful, is a male shaman-witch who receives his healing prowess by magically killing persons whose ghosts give him his healing powers. In Tok Pisin he is called a *sangumaman* and is found in other parts of New Guinea as well as among some of the Australian aborigines. A *sangumaman* supposedly can fly and make himself invisible and his services are used only when a person is seriously ill. If the patient recovers, the *sangumaman* receives the credit. If not, further treatment is sought. Today there are few, if any, *sangumaman* practitioners among the Wape, but the less contacted groups to the south still have them and they are sometimes resorted to.

A *wobif* (Olo) does not have the supernatural powers of a *sangumaman* and is not feared by villagers. He learns his healing skills from an another *wobif* and is given a small fee or gift for his services. The third type of practitioner is a *glasman* (Tok Pisin) whose skills were brought into the area by men returning from contract labor work on the coastal plantations. He is solely a diagnostician and does not offer treatment. As a clairvoyant he determines the cause of the illness by looking "into" his patient and asking questions. Having specified the cause on the basis of an indigenous differential diagnosis he then prescribes an appropriate treatment that may include consulting a *wobif* or *sangumaman*. He too receives a small fee or gift for his services.

Introduced

The grassroots practitioner of modern medicine is the aide post orderly or *doktaboi* (Tok Pisin). These men have training in first aid procedures, hygiene, and the treatment of common illnesses and injuries including acute respiratory and alimentary tract infections, malaria, common skin diseases, burns, and wounds. Aid posts are found in strategically located villages and are the most important medical service in terms of the number of patients served. If a health problem is beyond the aid post orderly's competence, he refers the patient to the small regional hospital situated in the midst of Wape territory. The hospital wards are coed except for one reserved for mothers and babies. As patients must provide their own food, they are usually accompanied by relatives who cook for them. Hospital orderlies provide basic patient care and are supervised by several nurses and the medical officer. Some of the most frequent presenting cases include acute upper respiratory tract infections, malnutrition, skin infections, birth confinements, malaria, anemia, gastroenteritis, tropical ulcers, and wounds. Only minor surgery is performed. If transport can be arranged, major surgery cases and other cases demanding sophisticated medical skills and technology are taken out to the coastal hospital in Wewak.

At one time nurses regularly held maternal and child health clinics in each village but staffing shortages have made these more episodic. The work of these clinics is to examine pregnant and nursing mothers as well as infants and small children, give immunization shots, and provide instruction on diet and hygiene. When the clinics are active, the hospital census of childhood malnutrition cases increases.

Classification of Illness, Theories of Illness, and Treatment of Illness

Although many Wape have a rudimentary understanding of the biomedical theory of sickness in terms of, for example, a penicillin injection's attack on invading bacteria, it is seldom persuasive. It is the theories of their own culture intimately conveyed in myth, ceremonial, and everyday events that offer a more satisfying understanding of sickness and dying. The Wape believe in a complex set of unseen spirit forces that affect them in both positive and negative ways (Mitchell, 1987; Waisi, 1982). While these can, for example, cause an earthquake or inflict illness and death, with the proper rituals they can enhance life as well.

The spirit forces most frequently causing illness are ghosts and demons. Each patrilineage owns tracts of land wherein reside the dreaded ghosts of recently dead relatives who will attack their closest kin as well as others. Also in residence are ancestral ghosts who can bring sickness to trespassing individuals or to those in conflict with their descendants. Finally, there are the vengeful and unpredictable demons residing in places of unusual or strange appearance like a waterfall, landslide, or still pond. Any of these forces may cause sickness or injury by entering into an individual and seizing a vital organ, shooting foreign objects into the body or, for example, may cause an injurious fall when climbing a coconut tree. Such attacks may be of the malevolent force's own volition or at the direction of a member of the related patrilineage. Some of these forces also may be appealed to for help, for example, the *mani* demon when hunting or a long dead father's ghost is asked to protect the home while the family is visiting kin in another village.

An attack by a *sangumaman* is the most feared cause of illness; once a person is assaulted and made ill, rectification is considered rare. A *sangumaman* may attack on his own volition or is hired to attack one's personal enemy. Although the details of the form of the assault vary with the informant, the procedure is roughly as follows. The victim is attacked and rendered unconscious by choking after which small incisions are made in the skin and bits of flesh removed. These are later put in little packets for sale to facilitate hunting. The wounds are magically closed without scaring and the victim is brought back to consciousness. Confused and dazed, the victim is told to return to her village, that she will have no memory of the attack and will die after a specified number of days. Consequently, any person returning to the village in a dazed or feverish state is considered a possible victim of *sangumaman*.

Sorcery is another cause of illness. Many villagers know spells for affecting the physical and emotional life of others. To bring sickness or death, a spell is performed with a bit of the victim's exuviae. For this reason most individuals are careful about when and where they defecate and take care to destroy their hair and fingernail cuttings.

In the treatment of illness, the Wape have recourse to a variety of therapists and therapies including their own pharmacopoeia consisting mostly of plants. For a persisting malady they are likely to try a number of therapies before they return to health or die. Self-treatment is usually the first form of treatment intervention. For a headache, a tight band may be tied around the forehead; an aching leg may be superficially cut to let out the hot blood; a boil may receive a leaf poultice or, if ripe, lanced. Judged by the premises of Wape culture these actions, like those by biomedical practitioners, are palliative, not curative, in nature. They treat the noxious symptoms but in etiological terms do not treat the *cause* of the symptoms. If self-treatment does not help and the suffering increases, the afflicted person will probably seek the diagnostic cultural skills of a *sangumaman, wobif,* or *glasman.*

A *sangumaman* is adept at the magical removal of small objects, for example, stones, thorns, and slivers of bamboo, that a forest demon or another *sangumaman* allegedly shoots into their victim causing sickness. After probing his patient's body to locate the offending object, he magically removes it, sometimes shows it to his patient, and is paid a small fee. The treatment specialty of the *wobif* is the use of massage to mend skeletal breaks caused by malevolent spirits and magically sucking from the body bad blood and bits of taboo food eaten by the victim that caused the illness.

The cause of an illness is frequently attributed to one of the taboo demons living in the forest, rivers, and streams. As a particular demon is related to a patrilineage's land, the afflicted person must go to one of the lineage mates who know the demon's secret name and ask to be exorcised. Such requests are never refused. These exorcisms are a common occurrence in the village and often one of the first therapies attempted. The exorcism is informal and simple. Lightly brushing the afflicted area with some ginger leaves, the exorcist inaudibly appeals to the demon by his secret name to leave his victim in peace. Embarrassed by the chastisement, the spirit readily departs. Of course if the illness worsens, then the diagnosis was wrong and a new diagnosis must be made and another treatment tried.

Some of these demons, for example, *niyl, poril,* and *mani,* are the center of curing societies whose exclusively male membership cuts across lineage lines. Men initiated into a society learn the requisite secret name and are responsible for performing exorcisms and carrying the demon's mask in a curing festival. The *niyl* curing festival is the largest and is months in preparation. Its climax features a parade of the sacred masks, all-night traditional dancing, and a dawn exorcism rite. All of these activities are intimately articulated with the Wape's elaborate exchange system. Although a curing festival is held to exorcise specific victims, others previously afflicted by the demon also may participate. Rubbing their skin against the sweaty body of a mask carrier, they receive a kind of booster immunization shot as an inoculation against future attacks.

The Wape tolerate erratic and strange behavior in fellow villagers as long as it is not threatening or, if aggressive, episodic in nature. Some men, but not women, are especially prone to episodes of threatening manic behavior. If the behavior becomes chronic, a man will be forced to leave the village and made to live in the forest where his actions are not a constant threat to village tranquility. Whether an individual is benignly mad or berserk, the cause is the same; possession by a demon or ghost, an event sometimes precipitated by the chewing of betel nut. A person exhibiting crazy behavior is not blamed for her or his crazy actions but, being possessed, is perceived as similar to a horse controlling its rider.

There are a number of ways to cure a person from her or his madness. The most common is to exorcise the demon or ghost by appealing for it to leave its victim as previously described. If this is unsuccessful, two other treatments are intended to drive the spirit out. A possessed man acting aggressively is grabbed from behind and, as he is held, other men rub and beat his body with a special type of ginger plant to force the invasive spirit out of him. The other treatment is performed when a victim is not agitated and can cooperate with the treatment. The victim goes to a bamboo grove with family members and, lying down, is completely wrapped in several dried palm flower sheaths. The bamboo is lighted and explodes in loud reports whereby the victim breaks loose from his confinement and, running away, the offending spirit runs away too.

Villagers view the two therapeutic systems, one indigenous and the other introduced, as complementary, not competing and may utilize both when ill or injured. The indigenous one offers culturally compelling explanations of illness and related treatments. While the introduced biomedical system's explanation of illness is exotic and conceptually irrelevant, nevertheless it provides some powerful forms of therapy they gladly take advantage of.

SEXUALITY AND REPRODUCTION

Sexual intercourse between a man and a woman is not a carefree activity. Like childbirth and menstruation, it is associated with female sexuality and is ritually polluting. A woman's vaginal area, especially when she is menstruating, is contaminating to men. During her period she eats and sleeps by a separate fire in the back of the house and, if her husband eats her food, his hunting will be luckless. A husband who wishes to continue to hunt will prepare the family's food. Sexual intercourse may occur in the house or in the forest but not in the gardens; a demon angered with smelling the sexual secretions could ruin the plants. For the same reason a menstruating woman cannot go into the forest or a demon would make her sick. Her period over, she washes her vaginal area and resumes her regular activities.

If a man intends to maintain his strength and health, sexual intercourse must be carefully regulated. If he has intercourse too frequently—about every four days is appropriate—he will lose weight, his skin will droop in folds like an old man's, and he will become soft like an overripe banana. Semen is conceived as energizing; by its discharge into a woman, a man becomes weaker with its loss as she gains strength. Although today some Wape are aware of the scientific explanation of conception, the traditional belief is that it takes many acts of intercourse to create a baby. Regarding family size, there is no special desire to have a large family. Many couples are content with two or three children because finding sufficient food for a large family, including the meat children cry for, is a losing battle.

HEALTH THROUGH THE LIFE CYCLE

Pregnancy and Birth

According to one study, Wape women have an average of 4.1 live births with 2.7 remaining alive (Wark & Malcolm, 1969). A newly pregnant woman wishing to abort the fetus can perform magical spells to expel it. A pregnant woman continues her regular tasks best she can until parturition. Birth usually occurs in her house while attended by one or two female kin. Delivery is in a squatting position and after the baby is cleaned, the umbilical cord and placenta are buried in or near the house. If there are complications during the birth, the husband, kin, and friends gather and various exorcisms are performed and indigenous practitioners consulted. In an intractable case, she might be transported or carried to the area's hospital (J. Mitchell, 1973). Within a day or two after birth, she appears publicly with the baby carried in a sling.

A wife and husband should refrain from intercourse from the time she is pregnant until their new baby can walk, usually about 18 months. An admittedly difficult ideal to achieve, if they violate the taboo, the child may be too weak to stand up.

Infancy

The mean weight of the Wape baby at birth, 2.40 kg, is one of the lowest on record (Malcolm, 1973). Malaria is holoendemic and one of the main causes of high infant mortality. From birth until about a year old, an infant subsists on its mother's milk and premasticated sago paste. Nursing babies usually double their birth weight within 3 months and treble it just over a year. From between about 8 months and 1 year, infants may be offered premasticated taro and a green shoot when seasonally available. However green vegetables in the form of boiled *Gnetum gnemon* leaves are not offered until a child is walking and completely toilet trained, about 2 years old. As a result, by the age of 6 months, a baby is most likely deficient in both caloric and protein intake. After a year, more solids are occasionally added to the diet including sago grubs and breadfruit providing additional protein. As the Wape also have little access to meat, and fresh fish is taboo to babies, malnutrition is not an uncommon problem among babies and toddlers.

Although a mother has the main responsibility for her infant, the father will remain in the village with the baby in a sling on his body when his wife is in the forest making sago flour. If a baby cries to nurse, it is not unusual for a father to offer his own breast as a pacifier. Other children and relatives also carry the baby from time to time; until it can crawl it is in almost constant physical contact with one attentive caregiver or another. Weaning is a gradual process with the toddler returning occasionally to the breast until it is dry.

Childhood

Unless there is a school in the village or close by, children are left much on their own to play with their age mates in the central plaza or shadow the activities of their parents. Boys, more than girls, go into the forest and forage for small game and other food. Children sleep in the same house as their parents and take their meals there too. Dining is a private family affair. Overt signs of malnutrition are not as common among children as they are among toddlers and infants. Parents are affectionate to both their male and female children and to viciously strike or whip a child is almost unheard of.

When a toddler does not get its way, he may lie on the ground and kick and scream but the response is to ignore him. A sibling may be posted nearby to keep the child from hurting himself but he is left to cry it out, learning at an early age that tempestuous behavior has no rewards. Children are susceptible to the usual childhood diseases, especially upper respiratory tract infections; some younger children appear to have a constant cold, and mucous smeared noses are common. Just as infants are unusually small, children in comparison to their European cohorts are too. At some point while still a child, every Wape boy and girl is taken to a river for a ritual washing by the mother's brother to assure its growth. The mother's brothers, by cultural definition a succoring person, performs the ritual during a *niyl* curing festival.

Adolescence

Chronic undernutrition appears to be a factor in the comparatively late onset of secondary sexual characteristics. One study (Wark & Malcolm, 1969) notes the onset of menarche in girls at 18.4 years. The onset of puberty in boys including the lowering of the voice is correspondingly delayed in comparison to European populations. Unlike some New Guinea societies, the Wape have no special ceremonies marking puberty. While toddlers and younger children play together, adolescent boys and girls tend to separate into gender-determined groups. Girls take an important role in helping their mothers in the family's food quest, but the small birds and animals obtained by boys are usually reserved for themselves.

Adulthood

Premature aging has been observed in many New Guinea societies and the Wape are no exception. Measurements of adults show a marked decline in height with age from 2 to 3 cm and there is a concomitant loss of weight. The average age of death for both men and women is the mid-forties. With physical maturity coming late and death coming early, Wape adulthood is of limited duration for many.

The Aged

There are few aged in Wape society. It is an exception for an adolescent to have grandparents. An old person usually lives with her or his son's family. The elderly help with looking after babies in ways that their health and strength permit. Unlike others who move actively in and out of the village, the aged, thin and fragile, are more likely to spend most of their time in the village. Their knowledge about the past is valued by adults who consult them, for example, regarding past exchanges, kinship relations, and ownership of land and food trees.

Dying and Death

Most individuals die in the village under the care of their immediate family members. As death is not considered natural but due to a malevolent cause, various indigenous diagnoses and treatments are attempted until a person is near death. Recourse to the aide post's therapies are utilized but unless a patient is strong enough to walk to the hospital, it is usually considered too costly to have him or her transported by a group of carriers. The almost obsessive last minute attempts at treatments in some cases may be attributed to the family's affection for the patient as well as an attempt to insure against the predations of a vengeful ghost. At death, a person's spirit departs the body via the anus and becomes a dangerous homicidal ghost. Only in time is it becalmed and joins the patrilineage's ancestral ghosts and demons deep in the forest.

CHANGING HEALTH PATTERNS

A general point about Wape culture is that it is defensively oriented. Ever on the defense against a world of hostile spirits, the Wape have centered their ceremonial life around pacifying malevolent ghosts and demons who strike them down with sickness and death. A deeply religious and conservative people concerned with culture stability and continuity, they have dedicated much of their intellectual, affective, and behavioral energy to the social elaboration of a therapeutic system that features curing practitioners, male curing societies, and curing rites and festivals with extensive social and economic intervillage ties. This passionate concern with health and curing is probably one of the main reasons for the Wape's easy adoption of the biomedical therapies introduced by the government and missions in the 1940s. These new secular therapies do not challenge their own ritual ones in terms of explaining sickness and death but, perceived as effective palliatives, significantly augment their therapeutic armamentarium.

Whereas sickness may be alarmingly disruptive in some societies, the Wape, by placing it at the core of their culture have, as it were, turned sickness on its head and organized a way of life around it. This may, however, be changing. Recent reports allege that some villages are abandoning their curing festivals. What this purports for the future of Wape health patterns, only further research can reveal.

REFERENCES

Briggs, E. A. (1928). New Guinea—land of the devil devil. *Australian Museum Magazine, 3,* 265–273.
Briggs, E. A. (1929). The black heart of New Guinea. *Australian Geographer, 1,* 38–40.
Foley, W. A. (1986). *The Papuan languages of New Guinea.* Cambridge: Cambridge University Press.
Laycock, D. C. (1975). The Torricelli phylum. In S. A. Wurm (Ed.), *Papuan languages and the New Guinea linguistic scene, 1,* 767–780.
Malcolm, L. A. (1973). *Growth and development patterns and human differentiation in Papua New Guinean communities.* Prepared for the IXth International Congress of Anthropological and Ethnological Sciences, August–September 1973, at Chicago, IL.
Marshall, A. J. (1938). *The men and birds of paradise: Journeys through equatorial New Guinea.* London: William Heinemann.
McGregor, D. E. (1982). *The fish and the cross* (2nd ed.). Goroko: Melanesian Institute.
McGregor, D. E., & McGregor, A. R. F. (1982). *Olo language materials.* Canberra: Australian National University (Pacific Linguistics, Series D, No. 42).
Mihalic, F. (1971). *The Jacaranda dictionary and grammar of Melanesian Pidgin.* Brisbane: Jacaranda Press.
Mitchell, J. (1973). Life and birth in New Guinea. *Ms. Magazine, 1*(11), 21–23.
Mitchell, W. E. (1973). A new weapon stirs up old ghosts. *Natural History, 82,* 74–84.
Mitchell, W. E. (1978). On keeping equal: Polity and reciprocity among the New Guinea Wape. *Anthropological Quarterly, 51,* 5–15.
Mitchell, W. E. (1987). *The bamboo fire: Fieldwork with the New Guinea Wape* (2nd ed.). Prospect Heights, IL: Waveland Press.
Mitchell, W. E. (1988). The defeat of hierarchy: Gambling as exchange in a Sepik society. *American Ethnologist, 15,* 638–657.
Mitchell, W. E. (1990). Therapeutic systems of the Taute Wape. In N. Lutkehaus, C. Kaufmann, W. E. Mitchell, D. Newton, L. Osmundsen, & M. Schuster (Eds.), *Sepik heritage: Tradition and change in Papua New Guinea* (pp. 428–438). Durham: Carolina Academic Press.
Mitchell, W. E. (1999). Why Wape men don't beat their wives: Constraints toward domestic tranquility in a New Guinea society. In D. A. Counts, J. K. Brown, & J. C. Campbell (Eds.), *To have and to hit: Cultural perspectives on wife beating* (pp. 100–109). Urbana, IL: University of Illinois Press.
Taru, L. (1977). Causes of lack of development in the Lumi Sub-Province of the West Sepik Province between 1945 and 1955. *Yagl-Ambu, 4,* 314–328.

Waisi, P. (1982). *The Laufis world-view: An attempt to locate its metaphysical base*. A sub-thesis submitted to the University of Papua New Guinea as a partial requirement for the Bachelor of Arts Honours Degree in Philosophy. Papua New Guinea: Waigani.

Wark, L., & Malcolm, L. A. (1969). Growth and development of the Lumi child in the Sepik District of New Guinea. *Medical Journal of Australia, 2*, 129–136.

Yanomamö

Jennifer Kuzara and Raymond Hames

ALTERNATIVE NAMES

Yanomami, Waika, Waica, Guaica, Shori, Yanoama, Yanomama, Shiriana, Xiriana, Shidishana, and Guajaribo. The names Sanema, Sanumá, and Sanima are auto denominations of Yanomamö people to the north and east of the main tribal distribution. They are culturally and genetically very closely related to the larger Yanomamö groups and their dialect is partially intelligible.

The name Yanomamö means person, individual, or human. Alternative names such as Shamatari or Waica (Waika) are relative terms used by some Yanomamö to refer to other Yanomamö living to the south or north, respectively.

LOCATION AND LINGUISTIC AFFILIATION

The Yanomamö are located in the extreme southeastern corner of the Venezuelan state of Amazonas and in the northern portions of the Brazilian states of Roraima and Amazonas. In Venezuela, the northern extension of the Yanomamö is delimited to the north by headwaters of Erebato and Caura rivers, east along the Parima Mountains, and west along the Padamo and Mavaca in a direct line to the Brazilian border. In Brazil, they concentrate themselves in the headwaters of the Demini, Catrimani, Araca, Padauari, Uraricoera, Parima, and Mucajai rivers. In Brazil and Venezuela there are about 20,000 Yanomamö in 200–250 villages, covering an area of approximately 192,000 sq km. Dense tropical forest covers most of the land but sparse savannas may be found in higher elevations. The topography is flat to gently rolling with elevations ranging from 250 to 1,200 m.

The Yanomamö language has not been confidently associated with any other South American language group. Linguists have attempted to provisionally place the language in Macro-Chibchan (Greenberg, 1960), Carib (Migliazza, 1972), or Panoan (Speilman, 1979). Migliazza divides Yanomamö into four major dialectical groups known as Sanema (3,262 speakers), Yanam (856 speakers), Yanomam (5,331 speakers), and Yanomamö (11,752 speakers). The last two dialects, accounting for 81% of the total, are mutually intelligible while the others may not be.

OVERVIEW OF THE CULTURE

The Yanomamö are horticultural people with a strong dependence on foraging (Hames, 1989; Lizot, 1977). Hunting is still accomplished with bow and arrow and swidden cultivation of plantains and bananas (manioc in some places) is central to the subsistence economy. Except for villages associated with missions, no locally produced goods are traded with outsiders. Although steel goods such as machetes and axes are common, they have had little impact on subsistence other than to make it more efficient (Hames, 1979).

Village size ranges from about 40 to more than 250 with a mode of between 60 and 100 (Chagnon, 1997; Early & Peters, 2000). Where raiding is intense, villages tend to be large, and where raid is relatively uncommon, they are smaller (Chagnon, 1997). Villages tend to be located about a day's walk apart and in areas between major rivers. Over the last 30 years some villages have begun to concentrate near missions on large rivers. All villagers inhabit a circular communal lean-to (*shabono*), and in higher elevations fully enclosed communal houses are built. These houses begin to deteriorate within 2–3 years. As the thatch and structural members deteriorate, the house becomes infested with cockroaches, crickets, sand fleas (*Tungans penetrans*) (Hagen, Hames, Craig, Lauer, & Price, 2001), and fecal and soil-born parasites. Sand fleas can be a serious health problem for children, especially orphans. This unsanitary and vexatious situation is resolved by building a new communal structure nearby.

Yanomamö social organization could be characterized as tribal and egalitarian (*sensu*; Service, 1975). The headman and the shaman are the only two socially differentiated roles and, in some villages, there may be more than one headman. Descent is patrilineal and residence is patrilocal. Lineages are local, of shallow genealogical depth, and lineage members do not own communal property. The major role of lineages is in dispute settlement, vengeance and the regulation of marriage. Each village is an independent political entity. Weak and ephemeral alliances between several villages are frequently established for mutual defense and aggression. In some areas, warfare is chronic while in others it is episodic and rare.

Most Yanomamö live in simple nuclear families. Anywhere from 20% to 25% of all families are polygynous (Hames, 1996) Polyandry is not uncommon in some areas comprising 2% of all marriages (Early & Peters, 2000). Each household owns one or more garden and is responsible for realizing its subsistence requirements. Sharing of food resources is common but families tend to restrict their reciprocity to only a few other families (Hames, 1999).

THE CONTEXT OF HEALTH: ENVIRONMENTAL, ECONOMIC, SOCIAL, AND POLITICAL FACTORS

Disease patterns among the Yanomamö are affected by a variety of physical and social environmental factors that are only now beginning to be understood. Malaria appears to be endemic throughout much of their territory but there appears to be considerable variation in the species of *Plasmodium* involved and in the intensity and prevalence of infection. Other diseases may be dependent on geography; for example, higher elevations appear to have greater incidences of river blindness (*Onchocerciasis*). Contact by non-Yanomamö is by far the most serious health problem faced by the Yanomamö. Influenza, measles, and hepatitis have been introduced by nonnative peoples who visit the area, or, in the case of gold miners in Brazil, have illegally invaded Yanomamö land.

Patterns of Disease

There has been a considerable amount of biomedical and epidemiological research done on the Yanomamö since 1966 that we cannot hope to effectively detail here. Elsewhere (Hames & Kuzara, in press) we present a much more complete review of this research, especially as it relates to the complexities of disease processes, origins, interactions, epidemiology, genetic susceptibilities, and the role of contact and acculturation. The pattern of diseases that affects the Yanomamö has been changing rapidly since first contact with Europeans and others.

It is difficult to know with certainty what the disease pressures acting on the Yanomamö prior to contact would have been; it is likely that many of the infectious diseases affecting the Yanomamö today were absent before the last century. Presently, in addition to introduced infectious diseases, the Yanomamö suffer widely from macroparasite infections. The humoral immune defenses that are most effective against macroparasites preclude the T-cell driven immune defenses that would be of most aid against many introduced diseases, such as tuberculosis. Thus, the matrix of both diseases to which the Yanomamö have been exposed since before their first contact, exacerbated by the conditions of sedentism in which many Yanomamö now find themselves, and those to which they are newly exposed, is complex. Because first contact occurred so recently, many diseases which have been introduced can be traced to their original source or point of introduction. Others, however, seem to have preceded actual contact with Europeans, either through contacts between the Yanomamö and other Amerindians of the area, or through infected animal vectors.

One of the most serious parasitic diseases now afflicting the Yanomamö is malaria, a serious infection that can result in high fevers, hemolytic anemia, and potentially dangerous enlargement of the spleen and liver. Of the two species of *Plasmodium*, *P. vivax* is the most common species, while *P. falciparum* results in the most severe infection. In Yanomamö territories, *P. falciparum* accounts for most infections (68.6%, Torres, Magris, Villegas, Torres, & Dominguez, 2000; 57.1%, Mato, 1998).

The patterns of the distribution and prevalence of *P. falciparum* among the Yanomamö seem to be much different than in other regions, even in Venezuela itself: the average Venezuelan distribution of infections in 1992 showed that 76% were *P. vivax* and only 24% *P. falciparum*. Additionally, a study by Laserson et al. (1999a) showed that two entire Yanomamö villages were infected with a single genotype of *P. falciparum*, suggesting that the epidemic in these villages was introduced by a single carrier.

Malaria is related to several long-term health problems: hepatomegaly (enlargement of the liver) and splenomegaly (enlargement of the spleen). Both conditions can be severe and sometimes chronic, and may result in portal hypertension and eventual cirrhosis. Malaria can also result in acute hemolytic anemia, as infected blood cells are killed by the exiting parasites or by the immune system. Anemia, including hemolytic anemia, has been shown to affect nearly the entire population of Yanomamö, and may have severe health consequences for these people (Torres et al., 1988).

In one study of the Yanomamö villages of Ocamo and Mavaca (Torres, Noya, Mondolfi, Peceno, & Botto, 1988), however, 44% of individuals were subject to some degree of splenomegaly. In some cases a syndrome known as hyperreactive malarial splenomegaly (HMS) results in those who have suffered repeated infection. In the same study, 23% of malaria cases resulted in the syndrome. In some of these, subjects' spleens occupied up to two thirds of their abdominal cavities.

One explanation for this may be related to the previously mentioned characteristic of Yanomamö immune reactions that seems to favor humoral rather than T-cell driven immune responses. Research has shown a link between a decreased number of T-lymphocytes and the development of HMS in other study populations. Related overproduction of IgM seems to be implicated in the development of HMS.

Torres et al. (2000) established a link between HMS and hemolysis in the Ocamo and Mavaca villages in Venezuela. In the year prior to the study, 38 of the 550 inhabitants of Ocamo and Mavaca had to be evacuated for emergency transfusions because of severe hemolytic anemia. In the study, 26 patients exhibiting severe hemolysis were studied, all of whom fulfilled the diagnostic criteria for HMS. The condition seems to be related to a cold agglutinin-mediated autoimmune response, one of the B-cell mediated immune responses. Whatever the cause of the syndrome, it seems to have a higher prevalence among the Yanomamö than among many other populations living with endemic and hyperendemic malaria, and the health consequences, in the form of hemolysis and anemia, can be severe.

Onchocerciasis, also known as river blindness, is caused by a parasitic infection with filarial worms of the genus *Onchocerca*, often *O. volvulus*. The parasite was introduced to the Americas via European and African immigrants, and may not have affected the Yanomamö prior to contact. Studies have shown the disease to be endemic to varying degrees among the Yanomamö of Brazil (Grillet et al., 2001; Rassi, Laurda, & Giuaimaraes, 1976), from a prevalence of 240 cases per 1,000 in one village to 606 cases per 1,000 in another.

The prevalence of onchocerciasis seems to be greater at greater altitudes, thus highland Yanomamö may be more at risk than lowland. In a study of biting rates of three species of black fly (the vectors), the biting rates of two species increased with altitude (Grillet et al., 2001). This may be related to the fact that the larval stage of the fly requires fast-flowing, highly oxygenated water, which is more likely to be found at higher altitudes.

While onchocerciasis is not fatal, it can be severely disfiguring and debilitating. Blindness and disorders of lymphoid tissue are potential results. The most adverse consequences of this infection are caused by the inflammatory immune response to the microfilariae by the host, rather than the infection itself. This can sometimes involve intense itching, cracking and thickening of the skin, related bacterial infection, and eventual fibrosis. Inflammation of lymph nodes can also ensue, resulting in a condition known as hanging groin. Of 75 Yanomamö subjects who had tested positive in skin biopsies, only 17 had palpable nodules (Rassi et al., 1976). Lymphatic involvement was not found in any of the Yanomamö subjects of this study. In the study conducted by Rassi et al. (1976), vision problems were only found in two Yanomamö subjects.

Yanomamö are widely infected with various macroparasites, with varying degrees of pathogenicity. Lawrence et al. (1980) conducted a study surveying several types of parasite in two newly contacted Yanomamö villages, as well as in three native, though non-Yanomamö, villages of the same region that are in the process of acculturation.

One of the main routes through which parasites such as nematode worms are spread is oral-fecal contact. Among swidden agriculturalists such as the Yanomamö, periodic relocation is common, which helps to mitigate the effects of a lack of sanitary means of defecation, which generally takes place outside, but near, the village or *shabono*. When populations become more sedentary, such as in areas near missions, parasites can build up in the soil and infection can thus become more prevalent, more severe, or both, if the settlement does not also include some form of sanitation, such as designated pit latrines.

Ascaris lumbricoides, among the more common intestinal helminthes infecting the Yanomamö, is passed through ingestion of feces. The primary complication, particularly in children, is intestinal obstruction. The two Yanomamö villages had infection rates of 90% and 100%, higher than in the acculturating villages.

The Yanomamö villages had hookworm infection rates of 76% and 80%. Hookworm enters a new host through the skin, particularly that of bare feet. The primary complication of infection with hookworm is iron-deficiency anemia, through the loss of blood; no information was available on the prevalence specifically of this type of anemia among the Yanomamö; considering the prevalence of other types of anemia, as well as the prevalence of hookworm infection, this should be the subject of future investigations into Yanomamö health.

Infection rates of *Trichuris trichiura,* also a nemathelminth, in the Yanomamö villages were 66% and 92%. *T. trichiura*, like *A. lumbricoides*, requires a period of incubation in the soil before eggs are infectious. Once ingested, they reside in the intestine and can cause diarrhea and abdominal pain, and potentially rectal prolapse.

Strongyloides stercoralis, a roundworm, involves infection through the skin. In some individuals, particularly those who are immunosuppressed, fever, abdominal pain and even shock may occur. The prevalence of this infection was 4–12% in the two Yanomamö villages, similar to acculturating villages.

Several species of amebic parasites were surveyed in the same study. Two parasites had higher infection rates in Yanomamö villages than in acculturating villages. *Entamoeba histolytica*, the parasite that is responsible for amebic dysentery and tropical liver abscess, had a prevalence of 28% in one village and 78% in another. *Chilomastix mesnili*, which can cause diarrhea, was more prevalent among the Yanomamö, almost half of whom in one village were infected.

Escherichia coli was nearly ubiquitous in the Yanomamö villages, although it is most often nonpathogenic. *E. hartmanni* was far less common than other microparasites and did not differ from the acculturating villages. Other generally nonpathogenic parasites were studied. *Endolimax nana* and *Iodamoeba buschlii* were both more than twice as prevalent among Yanomamö as among non-Yanomamö.

Giardia lamblia, infection with which can cause severe symptoms, such as diarrhea, vomiting, and weight loss that may last for several months, was not present in one Yanomamö village, and present among 5% of the population of the other, but infected about a quarter of the residents of acculturating villages. The low rates among the Yanomamö are surprising, considering that much of the fresh water of the Americas is infected with this parasite. *G. lamblia* infected about a quarter of the population of the acculturating villages, meaning that this may be a health concern for the Yanomamö in the future. While not generally fatal, the severity of the symptoms and especially the duration of the infection make it an important concern.

One must be careful in interpreting these results; in acculturating villages, 14 different species of intestinal parasites were found, while only 11 species were found in Yanomamö villages. However, in the Yanomamö villages, a higher average number of species were found per person, ranging from 4.2 to 6.8. In acculturating villages, the average number of species per person was much lower, from 1.3 to 6.0. Whether this fact points to the health of the Yanomamö as being more precarious than that of Amerindian groups who are farther along on the road to acculturation is dependent on two things: the severity of each individual infection and the pathogenicity of each species.

It is possible that Yanomamö are better able to deal with certain parasitic infections than some other groups. Many of these infections are dangerous, not because of the parasite itself, but because of the body's reaction to the parasite during some stage of its development in the body. Allergic reactions, for example, are the primary cause for the most severe symptoms of onchocerciasis. These reactions, ranging from mild itch to shock, are caused by an overabundance of immunoglobulin E (IgE). Some studies have suggested, however, that South American Indians can have levels of IgE higher than many Europeans would be able to cope with. Indians living with such high levels of IgE show no signs of ill-health as a result, however, which may make for a reduced negative reaction to the presence of certain parasites in the body.

In addition to parasites, viruses present a great health risk to the Yanomamö. Particularly threatening are the hepatitis viruses, as well as the polyoma viruses and the measles virus. Hepatitis B (HBV) and hepatitis Delta (HDV) may both be found among the Yanomamö. The severity of the health risks of these viruses varies, but the primary threat is the damage done to the liver, which can be very serious, and often fatal, in chronic cases.

HBV is the most common form of the virus among the Yanomamö. In general, this virus can be asymptomatic or associated with mild liver disease and jaundice, and the body's defenses can overcome it in a few months. After this, the infected person becomes immune, and can no longer transmit the infection. In about 10% of adult cases, and 30% of cases in children, however, the disease is not overcome. Chronic HBV lasts through the person's lifetime, potentially causing jaundice, cirrhosis, ascites (abdominal distention and discomfort), and eventual hepatic failure. The chronically infected person may transmit the infection, primarily through blood contact, as well as through contact with saliva and sexual secretions. HDV, first described in 1977, is more acute and often fatal, particularly in individuals who already carry HBV, even those who were previously asymptomatic.

A 1986 study (Torres & Mandolfi, 1991) of Ocamo and Mavaca villages showed that of samples assayed, only 16.2% of the population had not been infected and 53.7% had developed immunity to HBV. A full 30.3% showed current infection with HBV, meaning that they were either chronic or newly infected. As the virus was introduced 18 years prior, and the prevalence of infection was so high, it is likely that many of these were chronic cases. These rates seem abnormally high, since generally only 10% of adult infections are chronic. However, 39.7% of those infected with HBV also tested positive for HDV, which can greatly exacerbate the HBV infection. This may explain why the rates of chronic HBV were so much higher than average.

In the same study, an examination of serum samples that had been collected in 1975 and preserved, 52.7% showed infection with HBV. This rate is likely so high because this was near the time of the first introduction of the disease, by a missionary in 1968 who had been infected and had been know to reuse needles in the administration of vitamin complexes to himself and to the Yanomamö of the area. In this case, a full 96.5% of the samples showed that the individual had been exposed to HBV at some point. Although only six samples were tested for HDV, all six of them were positive for the presence of anti-delta antigens, showing that this particularly severe form of hepatitis was present among the Yanomamö prior even to its description in the medical literature. However, these results may not be representative of the entire population, as the samples from 1975 were collected from individuals who showed some sign of liver disease.

Hepatitis poses a severe health risk for the Yanomamö who engage in many practices that make the wide and rapid spread of this blood- and saliva-borne disease likely. Breast-feeding, which can last 2–3 years, the premastication of children's food, the sharing of instruments used to pierce the body, the sharing of chewed tobacco, and the early onset of sexual activity are only some of the examples of this.

Venereal disease has been present among the Yanomamö since the 1960s, and gained greater prevalence since then (Peters, 1998, p. 247). HIV is also a concern, as Boa Vista, one of the larger settlements in close proximity to the Yanomamö, has one of the highest rates of HIV in Brazil.

Another kind of virus with long-term effects that is now a concern for the Yanomamö is the polyoma virus. Working in the field in 1969, James Neel and Arthur Bloom showed high rates among the Yanomamö of cytogenetic damage similar to that caused by SV40, the simian polyoma virus. The polyoma virus is a type of papova virus, or DNA tumor virus, a class of viruses which have been implicated in carcinogenesis. These viruses can also produce kidney, neurological, and lymphoid disease. A study by Major and Neel (1998) showed that among individuals from 33 villages, 43.06% had significant titers of antibody to JCV polyoma virus, and 40.65% had significant titers of antibody to BKV polyoma virus. The distribution and heterogeneity suggested that the viruses were of recent introduction to the Yanomamö, and that there were probably two points of entry of the virus into the group, with introduction from the eastern (Brazilian) portion of their territory more significant than that from the western (Venezuelan) side. Other studies have shown unexpected chromosomal damage among the Yanomamö (Neel, 1971). The Yanomamö were chosen for a study of chromosomal damage as an example of a population who had been exposed to minimum levels of toxic agents in the environment. The results were surprising, showing that compared to the controls (members of the expedition) and a group of Japanese examined in a similar study, the Yanomamö had very high rates of chromosomal damage. Particularly surprising was the number of dicentric chromosomes, in some cases several in a single cell, a condition which is exceedingly rare. The reasons for this have not been discovered, but may be related to congenital defect, the presence of a virus or other pathogen, or possibly an environmental agent such as one of the hallucinogenic plant preparations used by the adult males of the group (although this latter possibility is less likely, as similar damage was shown among female Yanomamö, who are prohibited from using these preparations).

Measles, although generally not fatal among urban peoples, is a severe health risk to the Yanomamö. A serological study conducted in 1966 and 1967 (Neel, Centralwall, Chagnon, & Casey, 1970) showed that all but a handful of Yanomamö had not before been exposed to measles; in only two villages of 18 did a significant number of individuals test positive, one of which was located near a mission and had been known to have experienced an epidemic. In the other village, none of the positive responders was younger than 28 years, suggesting that this village had sustained an epidemic of measles prior to contacts with people of European descent.

As a consequence of this discovery, Neel secured measles vaccine and was about to begin a vaccine campaign when he learned of the outbreak of measles in Brazil and Venezuela. He quickly entered the field, treated the infected and vaccinated a cordon around those who had not contracted measles.

Again, although measles is generally not fatal in western populations, fatality rates in non-western populations can be quite high The primary complications are bronchopneumonia and a high febrile response, as well as encephalitis in severe cases. In this epidemic, at least 36% of those suffering from measles were estimated to have developed pneumonia as well, and it is thought to be the primary cause of death associated with the disease.

The other complicating factor in the spread of a disease such as measles among naïve populations is the incapacitation of the population. Measles incurs an acute febrile response, as well as prostration and dehydration. Adequate water, nutrition, and care are essential to recovery. When an entire village is stricken, however, there is often no caretaker available; hunting and garden work are suspended, breastfeeding is suspended by ill mothers. It is important to note that Neel and others believe that the complete breakdown of the economic and care system among New World peoples is largely responsible for the devastating effects of Old World diseases on these populations (Neel et al., 1970). In this epidemic, most villages were visited by a government team or missionaries, and antibiotics were supplied to help prevent secondary infections. For some, especially distant, villages, treatment was provided late in the epidemic, however. Neel et al. (1970) estimate that of about 170 cases, 29, or 17.7%, proved fatal. This number, while disheartening, would certainly have been staggering had the team not been able to vaccinate so many. This would likely have been the case had Neel not predicted the fact that the Yanomamö were likely naïve to measles and tested them for resistance, and then worked to procure vaccine. It also shows what a devastating effect such an outbreak could potentially have had if health care workers had not been vigilant.

Finally, tuberculosis must be considered a significant threat to the health of the Yanomamö. The disease may be treated with some success; however, drug resistant strains have developed with an associated 80% mortality. More effective than treatment is vaccination with bacille Calmette-Guerin (BCG), a live attenuated *Mycobacterium bovis* vaccine. Sousa et al. (1997) examined individuals from five villages. Serum samples were taken and tuberculin skin tests were administered. Subjects were also examined for BCG scars and they were found in 76% of the sample. The results were somewhat surprising. Prevalence of active tuberculosis was high, at 6.5%. Reactivity to the skin test was found among only 42% of Yanomamö who had received the BCG vaccine just 3 years earlier. The control group, European Brazilian army recruits, had a reactivity of 72% after having received the vaccine a full 18 years previously.

While analysis of T-cell response in this study was not possible, this study suggested that the Yanomamö had a reduced T-cell responsiveness to the disease, with an elevated humoral response. This is based in part on the low reactivity to the tuberculin skin test. Also noted was the fact that among the Yanomamö, production of antibodies, particularly IgM, but also IgG, was elevated in comparison with the control group.

Most disturbing in this study was the fact that of 28 new cases of tuberculosis, 82% occurred in individuals who had received BCG vaccine 3 years earlier. This incidence is higher than that seen in many other populations, and should be of concern, considering that the prevalence of tuberculosis is 100 times higher among Brazilian

Yanomamö than the average for Amazonas State, Brazil. If the vaccine is less effective for the Yanomamö, other types of treatments will need to be readily available, and health care workers will need to be especially vigilant in detecting tuberculosis early, if the 50% mortality associated with the untreated disease is to be avoided.

Diet and Nutrition

Aside from the effects of the infections described above, the general health of the Yanomamö is good. In terms of diet and by many standard measures of cardiac and circulatory health, they fit an ideal profile. The traditional diet of the Yanomamö consists primarily of manioc and plantains, supplemented with wild fruits, fish, game, and insects. Except near missions, domestic animals and their products are not consumed by the Yanomamö, and there is very limited access to processed or refined sugars and grains (Lizot, 1977). The resulting diet, similar to that of some other isolated groups, is low in fat and sugar, and high in protein, fiber, and complex carbohydrates.

The effects on their health of this diet, as well as a lifestyle necessarily including a great deal of physical activity, are evidenced in a variety of measures. Mancilha-Carvalho and Douglas Crews (1990) measured anthropometric traits and serum lipids among a group of Yanomamö of the Surucucu area in Brazil. The study showed that total serum cholesterol levels among Yanomamö adult men were half that of American men, and among Yanomamö adult women, were two thirds that of American women. Mean levels among the Yanomamö for all age groups were 121.9 mg/dl for men and 142.5 mg/dl for women.

They also showed that both height and weight were lower among Yanomamö of both sexes than among Americans. Mean Body Mass Index (BMI) was 20.4 kg/m^2 for Yanomamö men, compared to 25.3 kg/m^2 for comparable samples of American men, and 21.3 kg/m^2 for Yanomamö women, compared to 25.0 kg/m^2 for comparable samples of American women. BMI, as well as lipid level, was higher for Yanomamö women than for Yanomamö men. This is likely due to the fact that about half of the women in the sample were either breastfeeding or pregnant at the time of the examination. Additionally, total serum cholesterol increased with age among the Yanomamö in this study. This must be interpreted with the consideration that the sample size for each age group was small, ranging from 1 to 17 individuals. The increase was, however, statistically significant, suggesting that age-related increase in cholesterol is not solely a product of a high-cholesterol, Western-type diet.

A study of sodium metabolism (Oliver, Cohen, & Neel, 1975), close to the time of first contact for those studied, showed that the mean blood pressure for Yanomamö was low, from 93.2 to 108.4 systolic and 58.6 to 69.4 diastolic for men, and 95.7 to 105.7 systolic and 61.6 to 64.5 diastolic for women. In contrast to most Western and acculturating groups, the Yanomamö did not seem to experience any age-related increase in blood pressure. These results certainly implicate diet as a factor; likely the fact that the Yanomamö diet includes no significant source of sodium chloride. Urine excretions of sodium were also measured, and were quite low.

In this study, excretions of potassium, found in high quantities in plantains and bananas, were much higher for the Yanomamö than for the control group. Aldosterone excretions, for which potassium is an important stimulus, were also high. Yet plasma renin suppression was not affected. The authors suggest that given these results, sodium deficiency is a more important stimulus upon renin activity than potassium suppression.

Mortality among the Yanomamö remains high, in spite of this picture of overall physical well-being. This is largely a result of infectious and parasitic diseases, which accounts for about 70% of Yanomamö mortality. The second leading cause of death is physical violence, accounting for about 13% of Yanomamö mortality (a figure that varies greatly depending on area). Accidental trauma and degenerative disease account for 7% and 6%, respectively (Melancon, 1982). Thus it is obvious that while the resources available to the Yanomamö are adequate, infectious disease poses by far the greatest threat to Yanomamö health and life than any other factor, even warfare.

MEDICAL PRACTITIONERS

Shamans (*shabori*) are the major medical practitioners among the Yanomamö. Young men who wish to become shamans are apprenticed by an experienced shaman who is usually a senior agnate (Taylor, 1976). They are taught appropriate chants to call spirits (or *hekura*, who are causes and cures for illnesses and exist in the form of tiny humanoids) and supernatural lore. The training is demanding often lasting more than a year and during this time, trainees are required to obey a series of taboos such as dietary restrictions and refraining from sex. In some villages, up to a quarter of all men may have done through shaman training but only a few distinguish themselves as consistent and effective healers.

A shaman may have dozens of *hekura* living in his body. All have animal counterparts, different personalities, and habits and may be male or female, with male *hekura* more powerful than the females. *Hekura* dwell inside the shaman's chest going about their daily activities until called upon.

Recent research among Brazilian Yanomamö suggests that contrary to previous observations on Yanomamö health practices, the Yanomamö use a wide variety of medicinal plants and that women were the main practitioners (Miliken & Albert, 1996). This research shows that male shamanic cures are typically followed by the use of medicinal plants administered by women. This same research reveals that the Yanomamö employ 113 medicinal plants, 11 of which are cultivated. Many of the plants used are also used by other Amazonian groups to treat many of the same illnesses. Knowledge of the uses of these plants is being rapidly lost through acculturation. It is possible that one of these plants (*Aspidosperma nitidum*) may have anti-malaria properties. Pharmacologists in Brazil are currently investigating this possibility.

Classification of Illness, Theories of Illness, and Treatment of Illness

The Yanomamö believe that nearly all illness (and most misfortunes) are caused by either spirits (*hekura*) or the ghosts (*bore*) of deceased Yanomamö. The Yanomamö have a complex conception of the soul. One of these, the *möamo*, can be attacked or lured out of the body by *hekura* sent by a malevolent shaman (Chagnon, 1997, p. 113). If counter magic is not preformed by a curing shaman, death will ensue.

There are two kinds of spirits that can cause illness: *bore* or sprits of dead Yanomamö and *hekura*. A *bore* is that part of the soul that travels to a distant area of forest when a person dies. If a kinsman offended the deceased during his lifetime, the ghost may avenge himself by causing sickness. An important offense is an improper mortuary ceremony which includes, among other things, destruction of the deceased's property and cremation. Alternatively, a warrior who kills an enemy is in danger from the ghost of the slain if he does not carefully follow the *unokaimou* ceremony to cleanse himself (Chagnon, 1997).

A much more common cause of illness is an attack by a *hekura* sent by an enemy shaman. Shamans have a variety of *hekura* at their disposal, which they send to destroy the souls of Yanomamö. Some are especially powerful and can only be counteracted by equally powerful shamans.

If someone falls ill, a shaman is called for a diagnosis. The shaman begins by taking hallucinogenic snuff so he may easily communicate with his *hekura*.

Sexuality and Reproduction

The Yanomamö are relatively permissive of premarital sex. Men aggressively seek sex with women whenever they have the opportunity and women acquiesce, sometimes, out of fear. Extramarital sex leads to fights among men over sexual rights to particular women. The most significant reproductive outcome of an illicit affair is abortion or infanticide (Peters, 1980). A strongly enforced postpartum sex taboo lasts for the 2-year duration of breast-feeding. The effectiveness of this taboo is reflected in the inter-birth intervals.

Ideas about Conception

If a woman fails to become pregnant, she is deemed to be at fault and is stigmatized (Peters, 2000). A shaman will be called to help a woman overcome this problem. In some cases, a shaman discovers that a man other than the woman's husband is using magic to cause her to be sterile or abort her fetuses. Women are expected to produce numerous children and to assiduously care for them. If twins are born, the weaker or the female is killed. The Yanomamö believe that a woman cannot adequately care for twins.

Health through the Life Cycle

Pregnancy and Birth

The Yanomamö believe that each act of sexual intercourse contributes to the formation of the fetus: a man makes the child while a woman delivers it. In effect, the uterus is a passive receptacle for the formation of the fetus which grows with each contribution of semen. Consequently, as many men who contributed to the formation of the fetus are fractional fathers. This belief in "partible paternity" is not uncommon in Amazonia and elsewhere (Beckerman et al., 1998).

Abortion is common among the Yanomamö: one researcher estimates that by age 30 most women are likely to have had at least one abortion (Early & Peters, 2000). There are a number of reasons for abortion which include: the woman is too young to bear a child; a current child is nursing; or the husband believes that the child his wife is bearing is not his own. Abortion is both crude and probably dangerous for the mother. It is accomplished by direct pressure or blows to the uterus to damage the fetus or placenta.

Birth takes place immediately outside of the village where a woman is assisted by several other women, usually her mother or sisters. No men are allowed to be present. The expectant mother elevates her self by sitting on a log or other object while other women hold her from behind. The infant is allowed to fall onto a bed of fresh plantain leaves. After delivery the umbilicus is severed with a bamboo knife, the infant is inspected and washed by the assistants, and returned to the mother.

Infancy

Infanticide occurs for all the reasons listed above for abortion and, in addition, if the child has a congenital defect. Preferential female infanticide is widely reported for the Yanomamö. The Yanomamö show a decided preference for males (as in the case of twin births described above) and in many villages there is an infantile and juvenile sex ratio strongly skewed towards boys. To some extent, this skewing is a consequence of preferential female infanticide. Early and Peters (2000, pp. 210–212) in a careful analysis of infanticide for Brazilian Yanomamö show that 8% of all births results in infanticide and of these 32% are preferential female infanticides. However, in nearly all cases the decision to kill a child is made prior to birth.

During an infant's first 6 months of life, it is in the exclusive care of his mother or a close female relative when the mother is indisposed. During this time fathers are believed to lack the competence to care for infants. Mothers are expected to provide attentive care for the child and nurse on demand. Throughout most of the day the infant is attached to the mother via a sling that runs diagonally from the mother's shoulder to her hip where the child is suspended. From this position the child has easy access to the breast. A Yanomamö infant travels with its mother as she journeys to forest or garden to collect food. If the infant is out of the sling, it is never more than a few feet from the mother. One of the most traumatic events a child faces is when it is weaned from the sling. Even though a child is able to walk, it expects to be carried when the mother takes it to the forest or garden. On the trail the mother sets the child down and encourages it to follow her. The child will follow for a short distance but ultimately refuses to walk and cries or throws a tantrum until the mother picks it up. Over a period of several days, the mother repeats the process until she ultimately refuses to carry the child and it now walks on its own.

Children nurse until they are about 3 years old or until a new child is born. Supplemental feeding begins at about 6 months with soft food such as plantains. During the period of breast-feeding, a mother should not have intercourse with her husband to prevent the birth of a new child who would deprive the infant of milk.

Childhood

During the first 3 years of life children are seldom disciplined. However, if discipline is meted out, it can be very severe. If the child cries or becomes upset, it is given the breast or its genitals are gently rubbed, or it is pacified in other ways. Between the ages of 4 and 8 children roam immediately about the village in mixed age and sex groups. However, during this time boys and girls begin to segregate in same-sex groups with boys having more freedom to range further from the village and the girls are encouraged to help their mothers or care for younger siblings.

Adolescence

During this period boys and girls begin to take on labor activities that ultimately mirror adults patterns. Girls begin this process much earlier. By the time they are 9 years of age, they help carry firewood, fetch water, and prepare food. By the time they are 12–13, they have mastered the rudiments of nearly all of a woman's tasks. Throughout this period they are expected to avoid boys and young men when they are bathing or otherwise out of sight of their parents. At the sign of their first menses, they are confined to a small enclosure near their parents' hearth where they enter their puberty confinement. While in confinement they are subject to a variety of taboos and rituals that mark their transition to adulthood. When a girl emerges from confinement, she is recognized as a mature woman eligible for marriage. Frequently, she takes up residence with her betrothed.

There is no rite of passage ceremony that marks the transition from boyhood to manhood. During adolescence boys are expected to hone their hunting skills and assist their fathers in heavy work such as house construction and garden clearance. They tend to associate with other males of their age: they hunt and fish, seek female companionship, and visit other villages.

Adulthood

There are few Yanomamö women who do not bear the scars of their husband's violence. Women may be jabbed in the buttocks or thigh by arrow points, burned with firewood, or cracked across the head with firewood splitting the skin on the skull. Sometimes these injuries lead to significant disability or death. There are probably no Yanomamö men who do not bear injuries from dueling or warfare. Most injuries are on the head in the form of split skin. In fact, the men will shave their tonsure to show these dueling injuries. Death through violence accounts for 5–30% of all adult male mortality (Hames, 2001).

The Aged

The aged are treated well especially if they are surrounded by numerous kin and can still make economic contributions to families with whom they are associated. Common ailments include arthritis and old injuries that limit mobility.

Dying and Death

As mentioned above, the Yanomamö believe that nearly all deaths are supernaturally caused. They are products of *hekura* sent by long distance by enemy shamans to weaken and destroy the soul. Alternatively, death may be the result of the blowing of charms (*oka*) by enemy Yanomamö from the edges of the settlement. An inquest is usually made by a local shaman in an attempt to identify the likely source of the death magic.

Immediately after death, the body is placed on a scaffold in the forest where the bones are picked clean by scavengers. After the flesh is removed, the bones are brought back to the village where they are ceremonially incinerated. Alternatively, the whole body may be cremated in the village. In either case, the bones are removed from the fire and crushed to powder, and stored in a gourd. Later, amid much ceremony, the bone ash is mixed into a puree of plantains and/or bananas and is consumed by close kin as a final mortuary rite. In many cases, the dead are mourned on a regular basis just after sunset. The mourners cry and sob while they speak of their sadness and bitterness at the loss of a loved one. Women often lead in these evening rituals and will smear their cheeks black with ashes to mark their mourning status. Mourning may last for years. Its duration and intensity is dependent on a complex of factors such as the deceased' social status (e.g., infants are mourned the least while prominent men are mourned the most), how deceased, and the cause of death.

CHANGING HEALTH PATTERNS

Centers for contact such as mission stations, government installations (frequently they are in the same place), and mining installations influence Yanomamö morbidity and mortality patterns. Chagnon (1997, pp. 241–254) claims a higher rate of mortality for Yanomamö villages at an intermediate distance from missions. Early and Peters (2000, pp. 187–188) cast reasonable doubt on this assertion. With regard to mission contact in Brazil, Early and Peters (2000, pp. 188–190) find no consistent variation in mortality patterns. Nevertheless, it is abundantly clear that HIV, measles, influenza, tuberculosis, and various forms of hepatitis have been introduced through contact with devastating consequences for the Yanomamö.

While the missions might be implicated in the introduction of potentially life-threatening diseases to the Yanomamö, as was the case with the introduction of hepatitis and measles, the missions do not constitute the only contact the Yanomamö have, especially in Brazil, where gold miners and even farmers living in relatively close proximity can present an even greater hazard to the Yanomamö. A case is cited by Peters (1998, pp. 245–246) of what is believed to be the first case of tuberculosis among the Yanomamö, which the individual likely contracted while in the employ of Brazilian farmers, in 1966. Beginning in 1973, the Perimetral Norte, a spur of the Transamazon Highway, was constructed through the southern edge of the Brazilian Yanomamö region in the Catrimani area. Contact with road construction crews and travelers led to outbreaks of influenza and measles that killed between 30% and 50% of the people in three different villages in 1977 (Saffirio & Hames, 1983, p. 11).

There is little doubt that the invasion of gold miners beginning in 1980 has had and continues to have a devastating impact on the Yanomamö. Ramos (1995, pp. 278–279) documents huge increases in malaria, anemia, splenomegaly, respiratory infections, and tuberculosis in the Surucucu and Paapiú regions inhabited by the Yanomamö and invaded by gold miners.

John Peters (1998), a long-time missionary and ethnographer among the Yanomamö, describes the health care and facilities available to the Brazilian Yanomamö with whom he has worked. Missions and government health centers, such as *Fundacão Nacionál do Índio* (FUNAI) and *Fundacão Nacionál de Saude* (FNS) provide medical care for the Yanomamö. For the Brazilian Yanomamö at least, the *Casa do Índio* and *Casa da Hekula* in Boa Vista are available to meet many health needs, and hospital care is available in Boa Vista as well, in cases of dire emergency. The FNS makes transport available via airplane. Finally, the Commisão Pró Yanomami (CCPY) is an umbrella organization devoted to enhancing the quality of medical care and health monitoring (as well as assisting the Yanomamö politically) for the Yanomamö. Their web site at *http://www.proyanomami.org.br/index_fl.htm* is an important source of current information on the variety of health crises facing the Yanomamö.

Obviously, the availability of proper medications, such as tuberculosis treatments and antibiotics is crucial to maintain the health of the Yanomamö. Until 1995 these were available from FUNAI and FNS personnel, by missionaries, and by field nurses (Yanomamö trained to provide basic care). Currently, medications are available from FUNAI and FNS at specific centers.

Thus, some care and treatments are available to the Yanomamö; however, these remain problematic. Transportation to Boa Vista for emergency care may not be fast enough in many cases. Also, many Yanomamö continue to move often, or to spend a great deal of time away from their villages: ensuring that all receive treatment or vaccination is difficult. Better sanitation, facilities (e.g., latrines), and practices especially with sedentism becoming more common near mission posts, must be implemented if conditions are to improve.

Nonetheless, disease being the greatest threat to Yanomamö life and health, these conditions must be augmented by access to medical resources, such as medicines and supplies, and vigilant health care, whether from missionaries, government posts or others, if these conditions are to be ameliorated for the Yanomamö.

REFERENCES

Basanez, M. G., Yarzabal, L., Takaoka, H., Suzuki, H., Noda, S., & Tada, I. (1988). The vectoral role of several blackfly species Diptera simuliidae in relation to human onchocerciasis in the Sierra Parima and Upper Orinoco regions of Venezuela. *Annals of Tropical Medicine and Parasitology, 82*, 597–612.

Beckerman, S., Lizzaralde, R., Ballew, C., Sissel, S., Fingelto, C., Garrison, A., & Smith, H. (1998). The Bari partible paternity project: Preliminary results. *Current Anthropology, 39*, 164–167.

Chagnon, N. (1997). *Yanomamö* (5th ed.). New York: Harcourt Brace Jovanovich.

Confalonieri, U. E., Araujo, A. J., & Ferreira, L. F. (1989). Intestinal parasites among Yanomami indians. *Memorias do Instituto Oswaldo Cruz, 84*, 111–114.

Crews, D. E., & Mancilha, C. J. J. (1991). Correlates of blood pressure in Yanomami indians of northwestern Brazil. *Ethnicity and Disease, 3*, 362–371.

Donnelly, C., Thomson, L., Stiles, H., Brewer, C., Neel, J. V., & Brunelle, J. (1977). Plaque, caries, periodontal diseases, and acculturation among Yanomamö Indians, Venezuela. *Community Denistry and Oral Epidemiology, 5*, 30–39.

Early, J., & Peters, J. (2000). *The Xiliana Yanomami of the Amazon.* Gainesville, FL: University Press of Florida.

Eveland, W., Oliver, C. W. J., & Neel, J. V. (1971). Characteristics of *Escherichia coli* serotypes in the Yanomamma, a primitive Indian tribe of South America. *Infection and Immunity, 4*, 753–756.

Franco, M. H. L. P., Brennan, S. O., Chua, E. K. M., Kragh, H. U., Callegari, J. S. M., Bezerra, M. Z. P. J., & Salzano, F. M. (1999). Albumin genetic variability in South America: Population distribution and molecular studies. *American Journal of Human Biology, 11*, 359–366.

Franco, R. F., Araujo, A. G., Zago, M. A., Guerreiro, J. F., & Figueiredo, M. S. (1997). Factor IX gene haplotypes in Amerindians. *Human Biology, 69*, 1–9.

Freeman, J., Laserson, K. F., Petralanda, I., & Spielman, A. (1999, May). Effect of chemotherapy on malaria transmission among Yanomami Amerindians: Simulated consequences of placebo treatment. *American Journal of Tropical Medicine and Hygiene, 60*, 774–780.

Gershowitz, H., & Neel, J. V. (1978) The immunoglobulin allotypes (Gm and Km) of twelve Indian tribes of Central and South America. *American Journal of Physical Anthropology, 49*, 289–301, maps, table.

Godoy, G. A., Volcan, G. S., Medrano, C., & Guevara, R. (1989). Onchocerciasis endemic in the state of Bolivar Venezuela. *Annals of Tropical Medicine and Parasitology, 83*, 405–410.

Greenberg, J. (1960). The general classification of Central and South American languages. In *Men and cultures. Selected papers of the Fifth Internation Congress of Anthropological and Ethnological Sciences, Philadelphia, 1956* (pp 791–794). Philadelphia.

Grillet, M. E., Basanez, M. G., Vivas, M. S., Villamizar, N., Frontado, H., Cortez, J., Coronel, P., & Botto, C. (2001, July). Human onchocerciasis in the Amazonian area of Southern Venezuela: Spatial and temporal variations in biting and parity rates of black fly (Diptera: Simuliidae) vectors. *Journal of Medical Entomology, [print] 38*, 520–530.

Hagen, E., Hames, R. , Craig, N., Lauer, M., & Price, M. (2001). Parental investment and child health in a Yanomamö village suffering short-term food stress. *Journal of Biosocial Science, 99*, 1–33.

Hames, R. (1979). A comparison of the efficiencies of the shotgun and bow in neotropical forest hunting. *Human Ecology, 7*, 219–252.

Hames, R. (1989). Time, efficiency, and fitness in the Amazonian protein quest. *Research in Economic Anthropology, 11*, 43–85.

Hames, R. (1996). Costs and benefits of monogamy and polygyny for Yanomamö women. *Ethology and Sociobiology, 17*, 181–199.

Hames, R. (1999). Reciprocal altruism in Yanomamö food exchange. In L. Cronk, N. Chagnon, & W. Irons (Eds.), *Human behavior and adaptation: An anthropological Perspective* (pp. 226–252). New York: Aldine de Gruyter.

Hames, R., & Kuzara, J. (in press). The nexus of Yanomamö growth, health, and demography. In Francisco Salzano & Magdalena Hurtado (Eds.), *Lost Paradises and the Ethics of Research and Publication*. Oxford University Press.

Hurtado, M., Hill, K., Kaplan, H., & Lancaster, J. (2001, June). The epidemiology of infectious diseases among South American Indians: A call for guidelines for ethical research. *Current Anthropology, [print] 42*, 425–432.

Intersalt, C. R. G. (1988). Intersalt: An international study of electrolyte excretion and blood pressure results for 24 hour urinary sodium and potassium excretion. *British Medical Journal, 297*, 319–328.

Jansson, B. (1990). Dietary total body and intracellular potassium to sodium ratios and their influence on cancer. *Cancer Detection and Prevention, 14*, 563–566.

Laserson, K. F., Petralanda, I., Almera, R., Barker, Jr., R. H., Spielman, A., Maguire, J. H., & Wirth, D. F. (1999a, December). Genetic characterization of an epidemic of Plasmodium falciparum malaria among Yanomami Amerindians. *Journal of Infectious Diseases, 180*, 2081–2085.

Laserson, K. F., Wypij, D., Petralanda, I., Spielman, A., & Maguire, J. H. (1999b, May). Differential perpetuation of malaria species among Amazonian Yanomami Amerindians. *American Journal of Tropical Medicine and Hygiene, 60*, 767–773.

Lawrence, D. N., Neel, J. V., Abadi, S. H., Moore, L., Adams, G., Healy, G., & Kagan, I. (1980). Epidemiologic studies among Amerindian populations of Amazonia. III intestinal parasitoses in newly contacted and acculturating villages. *American Journal of Tropical Medicine and Hygiene, 29*, 530–537.

Lizot, J. (1977). Population, resources, et Guerre chez les Yanomami. *Libre, 2*, 111–145.

Major, E. O., & Neel, J. V. (1998). The JC and BK human polyoma viruses appear to be recent introductions to some South American Indian tribes: There is no serological evidence of cross-reactivity with the simian polyoma virus SV40. *Proceedings of the National Academy of Sciences of the United States of America, 95*, 15525–15530.

Mancilha-Carvalho,. J. J., & Crews, D. E. 1990. Lipid profiles of Yanomamö Indians of Brazil. *Preventive Medicine, 19*, 66–75.

Melancon, T. (1982). *Marriage and reproduction among the Yanomamö of Venezuela*. Thesis, Pennsylvania State University.

Migliazza, E. (1972). *Yanomamö grammar and intelligibility*. Doctoral, Indiana.

Milliken, W., & Albert, B. (1996). The use of medicinal plants by the Yanomami Indians of Brazil. *Economic Botany, 50*, 10–25.

Neel, J. V. (1971). Genetic aspects of the ecology of disease in the American Indian. In F. M. Salzano (Ed.), *The ongoing evolution of Latin American populations* (pp. 561–590). Springfield, IL: Thomas.

Neel, J. V., Centerwall, W., Chagnon, N., & Casey, H. (1970). Notes on the effect of measles and measles vaccine in a virgin soil population of South American Indians. *American Journal of Epidemiology, 91*, 418–429.

Oliver, W. J., Cohen, E. L., & Neel, J. V. (1975). Blood pressure, sodium intake, and sodium related hormones in the Yanomamö Indians, a "no-salt" culture. *Circulation, 52*, 146–151.

Perez, M. S. (1998). Anemia and malaria in a Yanomami Amerindian population from the southern Venezuelan Amazon. *American Journal of Tropical Medicine and Hygiene, 59*, 998–1001.

Peters, J. F. (1980). The Shirishana of the Yanomami Brazil a demographic study. *Social Biology, 27*, 272–285.

Peters, J. F. (2000). *Life among the Yanomami*. Peterborough, Ontario, Canada: Broadview Press.

Ramos, A. R. (1995a). Papel político das epidemias o caso Yanomami. *Ya No Hay Lugar para Cazadores*, 55–89.

Ramos, A. R. (1995b). *Sanuma memories*. Madison, WI: University of Wisconsin Press.

Rassi, E., Laurda, B., & Giuaimaraes, J. (1976). Study of the area affected by onchocerciasis in Brazil: Survey of local residents. *Bulletin of the Pan American Health Organization, 10*, 33–45.

Saffirio, G., & Hames, R. (1983). In K. Kensinger & J. Clay (Eds.), *The Forest and the Highway*. Working Papers on South American Indians #6 and Cultural Survival Occasional Paper #11 (joint publication) (pp. 1–52). Cambridge, MA: Cultural Survival.

Service, E. (1975). *Origins of the state and civilization*. New York: W. W. Norton.

Sousa, A. O., Salem, J. I., Lee, F. K., Vercosa, M. C., Cruaud, P., Bloom, B. R., Lagrange, P. H., & David, H. L. (1997). An epidemic of tuberculosis with a high rate of tuberculin anergy among a population previously unexposed to tuberculosis, the Yanomami Indians of the Brazilian Amazon. *Proceedings of the National Academy of Sciences of the United States of America, 94*, 13227–13232.

Spielman, R. S. (1979). The evolutionary relationships of two populations: A study of the Guaymí and the Yanomama. *Current Anthropology, 20*, 377–388.

Taylor, K. (1976). Body and spirit among the Sanuma (Yanoama) of North Brazil. In F. Grollig & H. Haley (Eds.), *Medical Anthropology* (pp. 27–48). The Hague: Mouton.

Torres, J. R., & Mondolfi, A. (1991). Protracted outbreak of severe delta hepatitis experience in an isolated Amerindian population of the Upper Orinoco Basin Venezuela. *Reviews of Infectious Diseases, 13*, 52–55.

Torres, R. J., Noya, G. O., Mondolfi, G. A., Peceno, C., & Botto, A. C. (1988). Hyperreactive malarial splenomegaly in Venezuela. *American Journal of Tropical Medicine and Hygiene, 39*, 11–14.

Torres, R. J. R., Magris, M., Villegas, L., Torres, V. M. A., & Dominguez, G. (2000). Spur cell anaemia and acute haemolysis in patients with hyperreactive malarious splenomegaly. Experience in an isolated Yanomamö population of Venezuela. *Acta Tropica, [print] 77*, 257–262.

Vivas, M. S., Basanez, M. G., Botto, C., Rojas, S., Garcia, M., Pacheco, M., & Curtis, C. F. (2000, November). Amazonian onchocerciasis: Parasitological profiles by host-age, sex, and endemicity in southern Venezuela. *Parasitology, [print] 121*, 513–525.

Yoruba

Norma H. Wolff

Alternative Names

In the past the Yoruba have been identified by outsiders as the Anago, Olukumi, and Aku (Adediran, 1998) Major subgroups include the Egba, Egbado, Ekiti, Igbomina, Ijebu, Ijesa, Ife, Kabba, Ondo, Owo, and Oyo.

LOCATION AND LINGUISTIC AFFILIATION

The Yoruba are located in the tropical rain forest and guinea savanna zones of coastal West Africa, concentrated in southwestern Nigeria (Yorubaland) with smaller groupings in southeastern Benin, Togo, and Ghana. As a result of the diaspora, enclaves of Yoruba culture are evident in Cuba and Brazil. The Yoruba language, marked by dialect diversity, and in some cases unintelligibility between subgroups, is a tone language that is a member of the Kwa group of Niger–Congo languages.

OVERVIEW OF THE CULTURE

Twenty to twenty-five million people speaking dialects of the Yoruba language form one of the largest ethnic groups in Nigeria, making up about 20% of the population (Zeitlin & Babatunde, 1995). Yoruba culture emerged in a series of kingdoms and chiefdoms in the tropical forests of southwestern Nigeria between the 9th and 12th centuries AD. The forest environment, with wide biodiversity, an abundance of mineral resources, numerous north-south flowing rivers, and a climate conducive to year-round agriculture, encouraged surplus agriculture, leading to population growth, centralized political systems, large-scale craft production, and trade. The early concentration of populations into large indigenous towns which were political, economic, and residential centers prior to European contact set the Yoruba apart in sub-Saharan Africa.

The Yoruba are horticulturalists. Root crops, including a variety of yams, cocoyam, and cassava, are the most important food crops. Maize, rice, millet, beans, plantains, bananas, and a wide variety of vegetables including tomatoes, onions, okra, peppers, and greens are also grown. Kola nuts are harvested for trade, and palm oil is processed for sale. Cocoa, introduced during the colonial period, is a major cash crop. Agricultural surpluses are the rule and have long been a staple of trade between rural and urban areas.

Despite the economic importance of farming, Yoruba take great pride in being town dwellers. Long before European contact, Yoruba towns were residential centers and the foci for economic redistribution, political administration, ritual activities, and the production of crafts. With the rise of long-distance trade, they became important links in the major trans-Saharan and Atlantic coastal trade routes of the region. A typical walled Yoruba city (*ilu*) is made up of multi-familied lineage compounds with a centrally placed palace of the king to whom descent groups acknowledge allegiance. The town is the core of a kingdom, and the palace provides a central focus for administrative, ritual, and economic activities with a major market in front. Beyond the walls of a city, farmlands, corporately owned by descent groups, extend 2–30 miles outward. Farmers live in town compounds and commute to their farms on a daily or seasonal basis. Temporary small hamlets (*aba*) spring up at farm outposts and sometimes develop into larger more permanent villages (*abule*). Villagers remain loyal to their city-bound king and retain links to the city compounds of their lineages.

Men were full-time farmers in the past. Other indigenous male occupations include long-distance trade and part-time crafts such as iron-smithing, weaving, woodcarving, and beadworking, combined with farming. Men's crafts are hereditary occupations associated with specific lineages. While women farm and raise vegetables, the ultimate female occupation is trader. Virtually all women take part in local trade and the controlled distribution of foodstuffs and trade goods at periodic markets. Women's craft skills include pottery manufacture, weaving, dyeing, and decorating textiles with resist designs.

Political organization of a Yoruba subgroup revolves around an *oba*, a king with divine attributes. The oba embodies the kingdom; his health is an indicator of the strength and stability of the polity. Decisions concerning the maintenance of peace and order in the town, such as quarrels between lineages, land disputes and other local altercations, are made by ward or town chiefs. Palace chiefs, backed by the oba, are ultimate decision-makers concerning public affairs and outside threats that endanger the kingdom. Public morality and the maintenance of social order is supported by members of ancestral cults such as *Egungun* and *Oro* who take active roles in applying supernatural sanctions to encourage proper social behaviors and punish wrongdoing. Today, national and state sanctions overlay indigenous practices.

The concept of family begins with membership in a lineage that traces ancestry back to a known male ancestor five or six generations in the past. While individual social mobility is possible, lineage membership determines boundaries of possible achievement. Religious choice, rights to land and property, occupation and access to political office, in large part, are a prerogative of lineage membership. Living male descendants of a named progenitor, their wives and children and unmarried female descendants live together in a walled compound named for the ancestor. Today, with a changed social and economic climate, individuals feel more free to reside outside the family compound, but lineage loyalties remain strong, and visits and extended stays at family compounds are typical. Within the lineage, the polygynous family is the ideal. Beyond family, the Yoruba are group-oriented, and non-kin associations based on economic cooperation, occupation, age, and religion are common.

Religious belief reflects an ongoing relationship with the rain forest environment. Olorun (Olodumare), the supreme deity, created a multitude of lesser *orisha* (deities) who control the forces of nature and culture, including Osanyin, the deity of medicine. For each deity, there are ritual paraphernalia, taboos, praise names, symbols, sacrificial foods, and texts for worship and supplication. In contemporary "Yorubaland," indigenous belief is retained in varying degrees. Most Yoruba are Muslims or Christians. Islam is seen as more compatible by many, because polygyny is allowable and Islamic medicine shares characteristics with the Yoruba religio-medical system. Christianity's promotion of Western education has brought vast changes in Yoruba life and culture. However, Christian syncretic churches combine indigenous and Western beliefs and practices.

Social interactions are governed by two important Yoruba precepts: seniority and good character. Respect for age and seniority plays a part in every social interaction. From the palace to family compounds people are ranked by age and birth order with elders given positions of responsibility. Beyond the family, status accrues with lineage affiliation, individual industriousness, wealth, and achievement of honored positions such as chieftaincies. Personal character (*iwa*) is of extreme importance. To have good character one must honor and respect the elderly, show generosity and kindness, engage in selfless service to others, be truthful and moral, and show hospitality to all, especially strangers.

Contemporary Yoruba life and belief are shaped by multiple forces. Islam and Christianity influence belief, and Western-style education, industrialization, and modernization affect every aspect of life. With the introduction of universal primary education in the mid-1950s, awareness grew of opportunities to make one's own life decisions, enter new professions, and practice new health strategies. Today the Yoruba are identified as the most urbanized and industrialized ethnic group of sub-Saharan Africa. Nine Yoruba cities are estimated to have populations of over 100,000. With more than 50% of Nigeria's light manufacturing located in Yorubaland, the Yoruba have one of the highest population densities of sub-Saharan Africa and face some of the highest environmental pollution threats in the world (Adediran, 1998; Zeitlin & Babatunde, 1995).

THE CONTEXT OF HEALTH: ENVIRONMENTAL, ECONOMIC, SOCIAL, AND POLITICAL FACTORS

Environmental Factors

The Yoruba live in a tropical ecosystem where diseases, illnesses, and parasites are endemic. These include malaria, cholera, skin diseases, unspecific fevers, and guinea worm. Historically, global diseases such as smallpox and influenza took their toll. In a 1980 United Nations Report for Nigeria, infectious diseases, combined with malnutrition, accounted for about 70% of illnesses, with the leading causes of death being malaria, dysentery, pneumonia, measles, tuberculosis, gastroenteritis, and gonorrhea. Water-borne diseases such as cholera, diarrhoea, and other enteric infections are common due to the lack of reliable water supplies. Piped water is available only in urban areas. Villages depend on streams, wells, and rainwater stored in pots and drums for their water. Such water is often polluted. Another major factor in determining health status is diet. Ecological conditions are such that the major crops raised are high carbohydrate foods. They include cassava, yam, and maize. Protein sources are scarce and expensive to acquire. Children are fed a high carbohydrate diet, particularly *eko* or *ogi* (cornmeal pap). Malnutrition based on inadequate consumption of proteins and enteric infections caused by bad water and contaminated food, stored and cooked under unsanitary conditions, are common (Iyun, 1994). Aspects of the built environment also contribute to health problems. Particularly in the urban areas, overcrowding and inadequate public services contribute to deplorable sanitation and outbreaks of diseases in low income neighborhoods.

It is estimated that half of Nigeria's children die before the age of 6 from such causes as measles, gastroenteritis, malaria, tetanus, meningitis, pneumonia, diarrhoea, whooping cough, polio and tuberculosis (Osunwole, 1997; Oyeneye, 1991). In parts of Yorubaland, infant mortality rates remain high, 78 out of 1,000 children dying before the age of 3 (Folasade, 2000). The most common nonfatal chronic diseases are malaria and guinea worm infestation. In 1987, the World Health Organization estimated that Nigeria had the highest number of guinea worm cases (140 million) in the world. Complications of guinea worm infestation are the major cause of work and school absenteeism. At the end of the 20th century, epidemics of infectious diseases included cholera, yellow fever, cerebrospinal meningitis, and AIDS, as well as continuing chronic infections of malaria, guinea worm, and high rates of malnutrition (Metz, 1992). A newer trend that is affecting higher income groups is the prevalence of noninfectious social-stress diseases. Hypertension is a leading cause of illness of Yoruba adults over the age of 30 treated in hospitals in Ibadan, Yorubaland's largest city, with males over the age of 45 bearing the greatest risks of developing cardiac disease (Iyun, 1994, 1995). Upper income males are most likely to suffer from hypertension and myocarditis, while women run the risk of rheumatic and pulmonary heart disease (Iyun, 1995). The increase in heart disease in urban Westernized Africans has been attributed to changes in lifestyle associated with pursuit of the modern lifestyle. Risk factors leading to hypertension include dietary changes such as increases in alcohol and salt intake, smoking, physical inactivity, obesity, and continual psychological stress (Akinkugbe, 1995; Iyun, 1994, 1995). A four decade study revealed that 1 in every 10–15 Nigerian adults is hypertensive, leading to strokes, heart and kidney failure and visual impairment (Akinkugbe, 1995).

The Introduction and Impact of Biomedicine

Dual medical systems, indigenous and Western, have competed in Nigeria since European colonization. Yorubaland was a primary target for the establishment of Christianity in West Africa. European and American missionaries introduced biomedical care in the 19th century, and the British colonial government continued to develop modern healthcare facilities in the 20th century. In 1948, the British established Nigeria's first medical school as part of the University College of Ibadan, with a modern teaching hospital opening in 1957. By the time the British left Nigeria in 1960 they had laid the foundations for modern medical services with over 3,000 hospitals, all concentrated in urban areas (Falola, 1999). Following independence in 1960, a first priority of the new government was to control endemic diseases and improve medical care services, particularly building hospitals. However, in the early 1970s, it was felt that most of the Nigerian population still relied on traditional medicine; those who accepted modern medicine were the urban educated (Foreign Area Studies [FAS], 1972). In 1975, the Nigeria Primary Health Care program was established to control communicable diseases, improve environmental hygiene, deliver preventive and curative care to at-risk populations such as the elderly and handicapped, and provide family planning and mother–child health services. Problems persisted. In 1980, over 85% of the doctors in Nigeria trained in Western medicine were concentrated in urban areas, and only about 35% of Nigeria's population had access to modern health-care services (Oyeneye, 1991; Warren, Egunjobi, & Wahab, 1996). In the 1980s, Nigeria's first population policy was implemented (Riedmann, 1993), and a Primary Health Care program was launched to serve all urban and rural Nigerians. The goal was to increase health-care personnel, collect reliable health data, make essential drugs available to all, improve nutrition, promote health education, develop a family health program, promote family planning, introduce improved therapy for childhood diarrheal diseases, and implement vaccination campaigns against major childhood diseases (Metz, 1992). In the early 1990s, challenges remained, including continuing disparities between rural–urban availability of healthcare facilities and personnel. These problems persist in the 21st century due to continuing economic tensions and lack of foreign exchange limiting the availability of medical supplies, drugs, and equipment.

Medical Practitioners

Yorubas have a wide variety of options in seeking health care. Choices are determined by the nature and history of the illness and the availability of services. Treatment often starts in the family with herbal remedies made with recipes passed through generations or after consulting locally printed indigenous medical formula books (Maclean, 1971). Indigenous practitioners who tap nature's unseen powers to treat a full range of human misfortunes and illnesses are often the next recourse. *Onisegun* are herbalist-healers who heal illness and combat misfortune. *Oloogun* are more likely to both cure and cause illness and misfortune (Wolff, 1979, 2000). When supernatural causation of illness is suspected, the close ties between the medical and religious systems draw the afflicted to *babalawo*, divination priest-healers of Ifa, or *olorisa*, priests of deities. Additional specialists include midwives, bone setters, circumcisers who cut tribal marks and incisions for medicines placed under the skins, and specialists in the treatment of mental illnesses.

Gender is not a problem to practicing indigenous medicine. Women practitioners (*iya isegun*) specialize in women's and children's health, as well as treatment of conditions such as asthma, diabetes, jaundice, hypertension, and fractures (Osunwole, 1997). As midwives, female practitioners manage childbirth and gynecological problems that affect fertility. Women favor female healers because of the bonds of similar experience and their expertise on children's diseases, including whooping cough, measles, neo-natal tetanus, tuberculosis, and polio. Women do not treat mental illnesses and other conditions that necessitate using powerful supernaturally charged medicines for fear that exposure to such forces causes infertility or birthing of deformed children by the practitioner. Because of the importance placed on women treating women in indigenous context, there is a modern focus on training Traditional Birth Attendants (TBAs) who combine indigenous and biomedical practices to oversee pregnancies and act as educators on women's and children's health. TBAs provide both pre- and postnatal care and take an active role in Primary Health Care programs including family planning and safe mother programs. They give advice on curing infertility, preventing unwanted pregnancies, and consult with pregnant women on diet, activity levels, and sexual relations during pregnancy. As midwives, they provide indigenous herbal therapies to speed birth, stop postpartum bleeding and stimulate breast milk, as well as perform childbirth rituals (Osunwole, 1997; Oyeneye, 1991).

In the urban areas, health-care options are evidence of the contemporary indigenous and biomedical mix. Clinics and hospitals provide access to biomedical specialists, including doctors with a wide range of specialties, nurses, interns, midwives, and other medical workers. On the streets and in the markets, every sort of pharmaceutical drug is for sale, specialists offer indigenous medicinal ingredients including plants, animal parts, and minerals; peddlers hawk patent medicines and home remedies, and injectionists provide shots for every ailment. Contemporary religious beliefs also play a part in urban healing practices. Christian faith healers of the syncretic churches, such as Aladura, cure and drive out harmful spirits with prayers and holy water. Muslim scholar-healers cure with Islamic medicines and charms and the power of the holy word.

Biomedical doctors in Yorubaland have found that cooperation with indigenous healers is feasible, cost-effective, and welcomed by patients, particularly in the rural areas. Blending indigenous with scientific medical practices makes them more acceptable and sustainable in the Yoruba worldview. Adeoye Lambo, a Western-trained Yoruba psychiatrist who served as the head of the World Health Organization, was a pioneer in this effort. As director of Aro Psychiatric Hospital in southwestern Nigeria, Lambo found that Yoruba mental patients recovered more quickly when indigenous practitioners were involved in their treatment. He developed the therapeutic village approach where the mentally ill and their relatives live together in a nearby settlement where the patient receives treatment from both hospital and indigenous practitioners.

CLASSIFICATION OF ILLNESS, THEORIES OF ILLNESS, AND TREATMENT OF ILLNESS

The Yoruba take a holistic approach to health or *alafia* (lit. "peace"), a philosophy of life, health, and death tied to the indigenous religion. *Alafia* encompasses physical, social, emotional, psychological, and spiritual well-being in the total environmental setting (Ademuwagun, 1978). Physical health (*ilera*), a prerequisite to achieving life goals, is negatively impacted by *aisan*, minor illnesses, such as fevers, headaches, diarrhoea, and vomiting that interfere with daily activities, and *arun*, which includes serious pathological conditions, communicable diseases such as smallpox and venereal diseases, infirmities, chronic tiredness, debilitating mental illnesses, and unexplained misfortunes (Ademuwagun, 1978; Warren, Egunjobi, & Wahab, 1997a).

The indigenous medical system takes into account both natural and supernatural causation of illness and misfortune (Awolalu, 1979; Odebiyi, 1989). It is a "personalistic medical system" where causation includes super-sensory forces directed toward afflicted individuals in acts of "active purposeful intervention" by human or nonhuman beings (Foster, 1976). Giving offense to potentially dangerous human beings such as witches (*aje*) or sorcerers (*osho*) and to entities such as evil spirits (*anjonu*), deities (*orisha*), and ancestors (*egungun*) is an ever-present danger. The belief that misfortunes of all kinds result from power attacks directed by malevolent beings is firmly entrenched in Yoruba thought. Every stage of life is fraught with possibilities of supernatural intervention that negatively affect advancement. Medicines or *oogun* are designed to combat these forces.

Oogun takes tangible or intangible form to either cure, cause, or counter illness and misfortune. In medicines, *ashe*, a potent universal power that the creator god put into all nature, is focused and energized for good or evil purposes through the combination of ingredients, preparation procedures, and incantations or words (Abiodun, 1994). To make medicine is to transform the raw materials of nature such as leaves, roots, fruits, woods, bird and animal parts, minerals and soils, and sometimes artifacts, into power objects according to specific formulas. The potential danger in the powers incorporated into medicines necessitates taboos and precautions in the making. *Osanyin*, the deity of medicine, is often appealed to for success. Sexual abstinence and the banning of women from the area during preparation may be necessary because of the perceived polluting effect of females (Buckley, 1985; Wolff, 2000).

While there are dialectal differences, a complex lexicon identifies medical conditions, symptoms, and treatments. Distinctions are made between diseases by sex, age, location in or on the body, natural or spiritual causation, and harshness of effect (Warren, Osunwole, & Wolff, 1997b). There is also a sense of hot and cold illnesses and the need to balance the body with foods and medicines of the opposite nature. The hot diseases (*arun gbona*), usually characterized by high fevers, include epidemic diseases such as cholera, yellow fever, and smallpox, as well as severe mental illnesses; they may be caused by deities and are considered serious and difficult to cure. Cold diseases (*arun tutu*) are less serious, associated with natural causation such as exposure to cold, and may manifest themselves as pains and aches in the body and mucous discharges (Odebiyi, 1989).

Abnormal behavior and states of personality disintegration that are identified as psychoses or neuroses in the West are recognized. *Were* labels chronic psychoses and persons manifesting symptoms of schizophrenia, where the afflicted poses no threat to the social order. It is the most common condition treated by indigenous and biomedical practitioners. *Were alaso* ("*were* who wear clothes") refers to less severe cases where the personality is intact. *Were agba* ("elderly *were*") labels conditions such as Alzheimer's disease and senile dementia. *Were d'ile* ("*were* of the lineage") refers to hereditary psychosis. Withdrawal into paranoid uncommunicative psychosis is known as *dindinrin*. Yorubas are generally tolerant of the mentally ill who are allowed to roam if they are not violent. Severe mental illness, characterized by violent acute psychotic episodes involving mania, catatonic excitement, or agitated delirium, is called *isinwin*. Neuroses are also recognized and labeled; they are thought of as physical diseases. Persons diagnosed with *ori ode* ("hunter's head") have sensations of burning or insects crawling in their brains, coupled with vision problems, dizziness, trembling, and insomnia. Organic and hysterical paralysis (*aluro, egba, ategun*), posturing and tics (*aiyiperi*), and epilepsy (*warapa*) are also distinguished (Prince, 1961).

Medicines are categorized by function, techniques used in preparation, and the physical form of the resultant products (e.g., Buckley, 1985; Verger, 1995; Warren et al.,

1997b). Types commonly identified include *agunmu* (pounded medicines), where ingredients are pounded to a pulp, dried, and ground to powder to eat or mix in a drink; *oogun jijo or etu* (burnt medicines) made by charring ingredients in a dry pot and grinding them to powder to ingest dry or add to liquids to drink or rub into skin incisions (*gbere*); *agbo* (infusions) prepared by steeping fresh ingredients to ingest or rub on body; and *oshe* (black soap) made by pounding burnt or fresh ingredients into palm kernel oil soap used to wash. Charms (*onde*) are a popular form of medicine. The most requested charms are *awure* that bring power, wealth, and good luck in life's ventures (Verger, 1995). *Afose* are charms that cause misfortune to enemies if placed in their proximity. Medicines can also be nonmaterial. The importance placed on the efficacy of words is exhibited in incantations (*ofo*) and curses (*epe*) that call directly upon the powers of nature, spirits, deities, witches, and wizards for a desired effect. A particularly malevolent spirit (*shigidi*) is summoned by sorcerers to attack people as they sleep (Wolff, 2000). The prescription and proscription of medicinal foods (*aseje*) is another medicine category (Buckley, 1985; Odebiyi, 1989). Some food proscriptions are widely held. Sweets, in particular, are considered to cause health problems and discomfort caused by worms. In a commonsense cure, healers forbid the eating of vegetables when patients have dysentery or gastroenteritis, because eating vegetables as roughage is affective in the treatment of constipation. The magical principle of similarity is applied when okra and other slimy foods are banned during treatment of infected wounds or skin lesions of measles and smallpox. Thermally hot food and alcoholic beverages are prescribed to alter body balances to combat cold diseases. Food and behavior proscriptions are also associated with avoiding offence to deities. For example, *Soponna*, the deity of smallpox, may punish people for roasting and eating melon seeds in the dry season or whistling in the hot part of the day.

In treatment, practitioners take natural and supernatural causation into consideration. They observe patients by examining their skin, eyes, and afflicted body parts and may look at blood, urine, and stool samples. Before arriving at a diagnosis, there are questions about social relationships between patient, relatives, and neighbors to determine the likelihood of supernatural attack. Once causation is ascertained, the healer prepares an appropriate medicine. Healers have many remedies for any one condition that are tried serially until a cure is affected.

When stubborn conditions do not respond to treatment, divination may be performed to determine ultimate causes and treatment is adjusted. A common finding is that the ill person or someone close to the patient has broken a taboo associated with a deity. Dealing with such transgressions can involve sacrifices and joining cults dedicated to the deity.

SEXUALITY AND REPRODUCTION

Traditional Yoruba attitudes favor "excessive fertility" (Olusanya, 1969) in a society that stresses ancestry and descent. The economic importance of children's labor in horticultural farming practices, a desire for lineage continuity, and economic and emotional security for parents in old age, coupled with high child mortality rates, have favored a high birth rate. A Yoruba proverb asserts that a man's wealth is counted in terms of money, wives, and children. This strong desire for children shapes Yoruba sexuality and promotes men's desire to control women's fertility. While Yoruba women enjoy a surprising amount of economic independence, their lives still center upon childbearing. In a typical village surveyed in the mid-1990s, Total Fertility Rates averaged over 8 births per woman and the overall fertility rate of 6.5 is among the world's highest (Lawoyin & Onadeko, 1997; Riedmann, 1993). Education has modified but not eliminated the desire for large families. In a study of fertility attitudes of Nigerian university students in the United States, ideal family size was five to six children (Adebayo & Adamchak, 1991). Today, among university-trained Yoruba, four children per monogamous family is common. It is generally held that men are the decision-makers about how many children a wife should bear. However, when women wield significant economic power, men may have to negotiate their own preferences regarding family size. With literacy and the increased sense of self worth associated with education, coupled with lower infant mortality rates, women are more outspoken about reproductive goals.

By the late 1980s Nigeria's official policy was to encourage four children per woman, and biomedical contraceptives became increasingly available. Pills, condoms, spermicides, injections, diaphragms, and IUDs are widely available in urban and rural areas. However, attempts to introduce birth control technology such as IUDS have met with resistance from men (Renne, 1993).

Both illiterate and educated men consider that women using contraceptives without the consent of their husbands contradicts male authority and is morally wrong; their wives might be unfaithful if given access to such technology. In 1992, 43% of married Yoruba women were using some form of modern birth control (1993). Because women who use contraceptives risk detection from a disapproving husband, they often depend on patent medicine abortifacients or abortion at clinics or alternatively leave their husbands if they do not want more children. A man who wants more children may take another wife or threaten to do so; if he wants fewer children, he can abstain from sexual relations, withhold financial support, or refuse paternity for an unwanted child (Olusanya, 1969; Renne, 1993).

Abstinence is the major form of birth control, and men practice coitus interruptus or withdrawal (*adaye*). Postpartum abstinence for 2–4 years until a child is weaned is an indigenous form of birth control. During this period, the prohibition against sexual intercourse is reinforced by the prevailing belief that such activity spoils the breast milk and causes life-threatening infant diarrhea (Olusanya, 1969). The practice regulates the number of children a woman bears in a lifetime, but does not limit a man's desire for a large family of children in a polygynous society. The period of abstinence has declined as educated working women seek to complete their families by an earlier age (Osunwole, 1997). Additional indigenous birth control methods used by women include wearing magic iron rings (*oruka irin*) and amulets (*onde, igbada*) or drinking extremely salty water immediately after intercourse. Women employ abortifacients. For example, in the first month of pregnancy, women may drink a mixture of potash and lime juice to terminate a pregnancy or insert herbal suppositories into the vagina (Olusanya,1969).

Yoruba attitudes toward infertile individuals are harsh, with blame usually placed on women. The social disapproval directed toward barren women is indicated in the term of reference *agon*, from the verb *gon* meaning "to despise" or "hold in contempt" (Abraham, 1958). Initially, a woman who does not conceive after marriage is treated with indigenous medicines and given advice on sexual relations by midwives (Oyeneye, 1991). If barrenness persists, it is thought to be either preordained, a condition of destiny (*iponri*), or due to supernatural intervention. Deities and ancestors may have been neglected or offended so that sacrifices must be made. If a woman remains barren despite treatment, she may be pointed out as a witch so imbued with evil that her own body is corrupted (Wolff, 1979).

HEALTH THROUGH THE LIFE CYCLE

Pregnancy and Birth

The Yoruba believe that pregnancy and birth are natural phenomena that create no health problems unless taboos are broken or witchcraft, sorcery, or other forms of spiritual attack occur. To avoid such incidents pregnant women observe behavioral and dietary prescriptions and prohibitions to ensure the safety of mother and foetus (Maclean, 1971; Odebiyi & Togonu-Bickersteth, 1987). The cooperation of deities, particularly when they have been appealed to for children, is sought through appeasement rituals, sacrifices, and the consumption or avoidance of special foods (Odebiyi, 1989; Osunwole, 1997). Beliefs about the properties of foods also determine diet. Expectant mothers consume starchy foods while avoiding nutritional foods such as milk which is thought to create large babies that are difficult to birth (Odebiyi, 1989). Postpartum depression is recognized, and if severe symptoms emerge it is identified as *abisinwin*, a supernaturally caused mental illness to be treated by indigenous healers (Prince, 1961).

When a child is born, precautions are taken to ensure an easy delivery and the future welfare of the child. Men are not normally present. The husband's presence is thought to delay the delivery. Both male and female practitioners can act as indigenous birth attendants, but male specialists are called only when complications develop, particularly those requiring spiritual interventions. Traditionally umbilical cords are ritually buried in the family compound to link the child to the father's lineage and to protect the mother from future barrenness. To prevent infertility in the child's adulthood, no word is spoken until it cries. The newborn is held by the feet and shaken three times to ensure that it will be brave and not have spasms; its head is touched to the ground so that future accidental falls will not result in injuries (Bascom, 1969; Osunwole, 1997).

Children born with striking physical traits or birth circumstances, such as twins and triplets, those born with the umbilical cord wrapped round the neck, an unruptured caul, extra fingers or toes, or abundant curly hair, are given special names and treatment. Multiple births are common. The Yoruba have the highest rate of fraternal twinning (4.4% of all births) in the world (Leroy, Olaleye-Oruene, Koeppen-Schomerus, & Bryan, 2002). Twins are both desired and feared as they have special powers. In the past twin infanticide was practiced, but it was replaced by elaborate ritual activities that harness the powers, particularly when a twin dies (Renne & Bastian, 2001).

It is still common for children to be born in lineage compounds. In urban areas only about 40% of women use modern health institutions at delivery (Oyeneye, 1991). A reason many women prefer home delivery is that modern health workers discard umbilical cords. Harmful effects of home birthing are unsanitary conditions and remedies administered during the birthing. Maternal mortality is primarily due to malnutrition, contemporary shortness of pregnancy intervals, and absence of adequate obstetric services.

Infancy

Male and female circumcision is performed a few days after birth, although the Egba and Ijebu sub-groups do not practice female excision (Caldwell, Orubuloye, & Caldwell, 1997). Men report that circumcision enhances sexual performance and reproductive potential, as well as reduces the female libido and promiscuity. Women support the practice of female circumcision by claiming that it makes child delivery easier (Osunwole, 1997). There is also a belief that the clitoris tip is charged with dangerous power that can kill the birthing child if it touches its head (Caldwell et al., 1997). It is only recently with global campaigns condemning female circumcision that Yoruba women identified the practice as an act of male domination to reduce female sexuality. In a 1997 survey, male circumcision remained universal while female circumcision, still prevalent in the rural areas (98%), had slightly declined in the urban areas (94%). Education, religion, and socioeconomic status were significant factors in the decline. Only 87% of girls born to mothers with secondary school education were circumcised. Christians were less likely than Muslims to circumcise females, and there was a decrease in families where fathers were in professional, managerial, or clerical occupations. The most significant change has been the medicalization of circumcision beginning in the 1970s. With involvement by biomedical professionals, circumcision-related deaths are rarely reported, and there has been a major shift from excision to clitoridectomy in female circumcisions (Caldwell et al., 1997).

Infants are carried on the mother's back and breast-fed on demand. From birth breastmilk is supplemented with herbal infusions (*agbo*) thought to guard against a variety of ailments and ward off evil influences. Solid food in the form of cornstarch pap is introduced as early as 2 months. While seen as health-enhancing, these practices increase the likelihood of death-threatening diarrhea and other illnesses. Malnutrition is also a strong possibility if breast-feeding is not continued, because pap is a poor source of protein and kwashiorkor may develop (Davies-Adetugbo, 1997). In addition, the tonic purgatives (*agbo*) given to infant early each morning in the first 6 months and to older children whenever they fall sick may be detrimental to health as ingredients, including cow urine, are variable and unregulated (Maclean, 1971).

Some infants are singled out for special treatment. Tied to high infant mortality rates is the belief in *abiku* ("born-to-die") children, troublesome spirits that die before adulthood and return again and again to the same mother. Such children are treated with special medicines and given names such as Malomo ("Do not go again") (Maclean, 1971).

Childhood

There are many safeguards to ensure the health of children and protect them from evil influences and accidents. Home remedies are used to treat minor illnesses and indigenous and biomedical healers are consulted for more serious ailments. From birth, children wear charms and have imposed food proscriptions associated with attracting the favor of deities. However, children born with physical and mental defects are stigmatized. Congenital malformations in children are perceived as supernatural punishment for the parents and can be grounds for a husband to desert his wife (Odebiyi & Togonu-Bickersteth, 1987). There is little tolerance for individuals with physical abnormalities, despite the indigenous belief that they are favored children of Obatala (Orisha-nla), the "sculptor-divinity" who shapes children in the womb. Albinos (*afin*), dwarfs (*irara*), hunchbacks (*abuke*), cripples (*aro*), and the dumb (*odi*), for example, are sacred to Obatala (Awolalu, 1979). In the real world, handicapped children are ridiculed and avoided and often exhibit withdrawal behavior. For example, in studies of attitudes toward deaf children, no efforts were made at home to communicate with them, they were not taken out socially, and they were disciplined more harshly than normal siblings. Attempts at curing deafness and other handicaps focuses on placating witches or offended deities (Odebiyi & Togonu-Bickersteth, 1987).

Adolescence

While there are no elaborate prolonged puberty rituals, the importance of women's childbearing role is marked by special ceremonies with sacrifices to the deities to thank them when a girl reaches menarche (Togonu-Bichersteth, 1988).

Adulthood

The Yoruba have no ritual acknowledgment of adulthood. Adulthood is recognized by taking on the expected male and female roles defined by the division of labor. Men take up economic activities, such as farming and craft production, and women become childbearers and rearers. It is marriage and parenthood that fully identify individuals as adults. Men marry between ages 25 and 30 and women between 17 and 25. An integral part of marriage negotiations between families is investigations to reveal any family history of undesirable characteristics such as the presence of leprosy, insanity, epilepsy, barrenness, birth anomalies, or social digressions. It is expected that the bride be in good health and be able to bear strong healthy children. While expectations of virginity have altered in contemporary context, the unquestioned health of wife and future children is imperative (Eades, 1980; Olusanya, 1969).

The Aged

Yoruba old age occurs when the person can no longer effectively carry out primary gender roles associated with the division of labor. For men, the critical sign is decreased stamina, so that active farming and labor become difficult and economic productivity dwindles. The critical indicator for women is menopause when childbearing ceases. Three major factors contribute to a sense of well-being and happiness in old age. These are having good health, responsible children nearby, and sufficient money. Security for the elderly comes from within the family where adult children, particularly sons, provide for aged parent's economic and emotional needs. Without such support, old age is a period of sadness and dependancy upon distant relatives or strangers (Togunu-Bichersteth, 1988). Urbanization, industrialization, migration, and rural change have altered the context of aging. With national and local inflation and recurring economic crises, adult children have more difficulty in meeting their obligations, and the number of elderly destitutes has grown, specifically in urban areas (Togunu-Bichersteth, 1988). In a study of 706 urban women aged 55–102, it was found that 70% continued to work to meet their sustenance needs and suffered a higher rate of depression than rural elderly women (Udegbe, 1995).

Dying and Death

At death, funeral rites and burial are guided by the circumstances of death. Age at death, reputation, status, and the predominant religion of the family all play a part (Adelowo, 1988). In the case of elders, the funeral and burial is a time for descendants and relatives to come together and celebrate the good life, with the funeral used by the family as an opportunity to display affluence and filial status. It is common for burial to be postponed until funds can be collected for the feasting.

Rites are carried out to move the spirit away from the living world but enable it to return as ancestor, or through reincarnation, in the future. The first child of the right gender born after a death is named either Babatunde ("father returns") or Iyabo ("mother returns"). The corpse is buried under the floor of the house or outside the door in the lineage compound. The often unmarked grave site becomes a focal point where the ancestral spirit can be summoned by pouring libations and calling out praise names (*oriki*) of the deceased to encourage the spirit to return to the household of the living when help is needed. It was only with the adoption of Christianity and Islam that designated burial grounds were established.

The bodies of those whose deaths are considered unnatural are treated differently (Adelowo, 1988). There is no mourning allowable at the death of abiku ("born-to-die") children. The corpses of infants and abiku children who die young or in an untimely fashion, lepers, hunchbacks, albinos, pregnant women, persons killed by lightning, those who hang themselves and others who die unnatural deaths are buried in the bush outside town or in special sacred groves. Suspected witches and wizards may be denied burial, and their bodies are thrown into bush areas inhabited by wicked spirits.

Changing Health Patterns

The Yoruba are one of the most studied ethnic groups of sub-Saharan Africa. International and national research projects have provided continuing information on health status, the prevalence of diseases, fertility, and the interface of biomedicine with indigenous cultural belief. A current focus of research is AIDS. The presence of AIDS in Nigeria was officially announced in 1987. In 1990, it was estimated to affect less than 1% of the population (Metz, 1992). In early 2002, it was estimated by the Federal Ministry of Health of Nigeria that the HIV population in Nigeria was over three million (National HIV/Syphilis Seroprevalence Survey).

References

Abiodun, R. (1994). Understanding Yoruba art and aesthetics: The concept of *ase*. *African Arts, XXVII*(3), 68–78.

Abraham, R. C. (1958). *Dictionary of Modern Yoruba*. London: University of London Press.

Adebayo, A., & Adamchak, D. J. (1991). Ethnic affiliation and fertility attitudes of Nigerian university students. *College Student Journal, 1*, 470–477.

Adediran, 'B. (1998). Yorubaland up to the emergence of the states. In A. D. Ogunremi & 'B. Adediran (Eds.), *Culture and society in Yorubaland* (pp. 1–14). Ibadan: Rex Charles Publication in association with Connel Publications.

Adelowo, E. D. (1988). A comparative look at the phenomena of death (*iku*), burial and funerary rites (*isinku*) in Yorubaland. *Odu, 33*, 163–188.

Ademuwagun, Z. A. (1978). Alafia—the Yoruba concept of health: Implications for health education. *International Journal of Health Education, XXI*(2), 89–97.

Akinkugbe, O. O. (1995). Lifestyle and cardiovascular diseases: Emerging problems of health development in the Third World. In B. F. Iyun, Y. Verhasselt, & J. A. Hellen (Eds.), *The health of nations: Medicine, disease and development in the Third World* (pp. 177–183). Aldershot: Avebury.

Awolalu, J. O. (1979). *Yoruba beliefs and sacrificial rites*. London: Longman.

Bascom, W. (1969). *The Yoruba of southwestern Nigeria*. New York: Holt, Rinehart & Winston.

Buckley, A. D. (1985). *Yoruba medicines*. Oxford: Clarendon Press.

Caldwell, J. C., Orubuloye, I. O., & Caldwell, P. (1997). Male and female circumcision in Africa from a regional to a specific Nigerian examination. *Social Science and Medicine, 44*, 1181–1193.

Davies-Adetugbo, A. A. (1997). Sociocultural factors and the promotion of exclusive breast-feeding in rural Yoruba communities of Osun, State, Nigeria. *Social Science and Medicine, 45*, 113–125.

Eades, J. S. (1980). *The Yoruba today*. Cambridge: Cambridge University Press.

Falola, T. (1999). *The history of Nigeria*. Westport, CN, London: Greenwood Press.

Folasade, I. B. (2000). Environmental factors, situation of women and child mortality in southwestern Nigeria. *Social Science and Medicine, 51*, 1473–1189.

Foreign Area Studies (FAS) of The American University. (1972). *Area handbook for Nigeria*. Washington, DC: U.S. Government Printing Office.

Foster, G. M. (1976). Disease etiologies in non-western medical systems. *American Anthropologist, 78*, 773–782.

Iyun, B. F. (1994). Health problems in Ibadan region. In M. O. Filani, F. O. Akintola, & C. O. Ikporukpo (Eds.), *Ibadan region* (pp. 256–268). Ibadan: Rex Charles.

Iyun, B. F. (1995). Cardiac morbidity in Nigerian society: Trends and inter-relationships. In B. F. Iyun, Y. Verhasselt, & J. A. Hellen (Eds.), *The health of nations: Medicine, disease and development in the Third World* (pp. 139–151). Aldershot: Avebury.

Lawoyin, T. O., & Onadeko, M. O. (1997). Fertility and childbearing practices in a rural African community. *West African Journal of Medicine, 16*, 204–207.

Leroy, F., Olaleye-Oruene, T., Koeppen-Schomerus, G., & Bryan, E. (2002). Yoruba customs and beliefs pertaining to twins. *Twin Research: The Official Journal of the International Society for Twin Studies, 5*, 132–136.

Maclean, U. (1971). *Magical medicine: A Nigerian case-study*. London: Allan Lane, The Penguin Press.

Metz, H. C. (Ed.). (1992). *Nigeria: A country study*. (5th ed.). Washington, DC: Federal Research Division, Library of Congress.

Odebiyi, A. I. (1989). Food taboos in maternal and child health: The views of traditional healers in Ile-Ife, Nigeria. *Social Science and Medicine, 28*, 985–996.

Odebiyi, A. I., & Togonu-Bickersteth, F. (1987). Concepts and management of deafness in the Yoruba medical system: A case study of traditional healers in Ile-Ife, Nigeria. *Social Science and Medicine, 24*, 645–649.

Olusanya, P. O. (1969). Nigeria: Cultural barriers to family planning among the Yorubas. *Studies in Family Planning, 37*, 3–17.

Osunwole, S. A. (1997). The role of women in traditional Yoruba medicine. In D. M. Warren, L. Egunjobi, & B. Wahab (Eds.), *Studies of the Yoruba therapeutic system in Nigeria* (pp. 118–128). Ames: Iowa State University, Center for Indigenous Knowledge for Agriculture and Rural Development. Studies in Technology and Social Change, No. 28.

Oyeneye, O. Y. (1991). Family health in Nigeria. *International Journal of Sociology of the Family, 21*, 189–99.

Prince, R. (1961). Indigenous Yoruba psychiatry. In A. Kiev (Ed.), *Magic, faith, and healing: Studies in primitive psychiatry today* (pp. 84–120). New York: Free Press of Glencoe.

Renne, E. P. (1993). Gender ideology and fertility strategies in an Ekiti Yoruba village. *Studies in Family Planning, 24*, 343–353.

Renne, E. P. (1997). Local and institutional interpretations of IUDs in southwestern Nigeria. *Social Science and Medicine, 44*, 1141–1148.

Renne, E., & Bastian, M. L. (2001). Reviewing twinship in Africa. *Ethnology, 40*, 1–11.

Riedmann, A. (1993). *Science that colonizes: A critique of fertility studies in Africa*. Philadelphia: Temple University Press.

References

Togonu-Bichersteth, F. (1988). Perception of old age among Yoruba aged. *Journal of Comparative Family Studies, 19*, 113–122.

Udegbe, I. B. (1995). Current adjustment patterns of elderly Yoruba women and their health implications. In B. Iyun, B. Folasade, U. Verhasselt, & J. A. Hellen (Eds.), *The health of nations: Medicine, disease and development in the Third World* (pp. 217–223). Aldershot, England: Avebury.

Verger, P. F. (1995). *Ewe: The use of plants in Yoruba society.* Sao Paulo: Copmpanhia das Letras.

Warren, D. M., Egunjobi, L., & Wahab, B. (1996). The Yoruba concept of health and well-being: Implications for Nigerian national health policy. In F. Fairfax III, B. W. Wahab, L. Egunjobi, & D. M. Warren (Eds.), *Alaafia: Studies of Yoruba concepts of health and well-being in Nigeria* (pp. 4–9). Ames: Iowa State University Center for Indigenous Knowledge for Agriculture and Rural Development. Studies in Technology and Social Change, No. 25.

Warren, D. M., Egunjobi, L., & Wahab, B. (Eds.). (1997a). *Studies of the Yoruba therapeutic system in Nigeria.* Ames: Iowa State University Center for Indigenous Knowledge for Agriculture and Rural Development. Studies in Technology and Social Change, No. 28.

Warren, D. M., Osunwole, S. A., & Wolff, N. H. (1997b). Yoruba lexicon for therapeutics. In D. M. Warren, L. Egunjobi, & B. Wahab (Eds.), *Studies of the Yoruba therapeutic system in Nigeria* (pp. 44–72). Ames: Iowa State University Center for Indigenous Knowledge for Agriculture and Rural Development. Studies in Technology and Social Change, No. 28.

Wolff, N. H. (1979). Concepts of causation and treatment in the Yoruba medical system: The special case of barrenness. In Z. A. Ademuwagun, J. A. A. Ayoade, I. E. Harrison, & D. M. Warren (Eds.), *African therapeutic systems* (pp. 125–131). Waltham: Crossroads Press.

Wolff, N. H. (2000). The use of human images in Yoruba medicines. *Ethnology, 39*, 205–224.

Zeitlin, M. F., & Babatunde, E. D. (1995). The Yoruba family: Kinship, socialization, and child development. In M. F. Zeitlin, R. Megawangi, E. M. Kramer, N. D. Colletta, E. D. Babatunde, & D. Garman (Eds.), *Strengthening the family: Implications for international development* (pp. 142–181). Tokyo, NY, Paris: United Nations University Press.

Subject Index

A-TI *see* ataxia telangiectasia
AA *see* Alcoholics anonymous
AAMR *see* American Association of Mental Retardation
Abortion
    as birth control, 278
    China, 278
    and contraceptives, 88–89
    Greeks, 685
    Jamaican Maroons, 761–762
    selective, 278
    United States, 278, 284
Abuse, child *see* Child abuse/neglect
Abuse, domestic *see* Domestic abuse
Accumulating medical systems, 7
Achondroplasia (ACH), genetic disease, 411
Achromatopsia, genetic disease, 408, 411
Acromegaly, genetic disease, 411–412
Acupuncture, Han, 714
AD *see* Alzheimer's disease
ADA *see* Adenosine deaminase deficiency
Adaptation, disasters, 158
Adenomatous polyposis, familial (FAP), genetic disease, 412
Adenosine deaminase deficiency (ADA), genetic disease, 412
Adolescence
    African-Americans, 553–554
    Amish, 562
    Argentine Toba, 569–570
    Badaga, 577
    Baliem Valley Dani, 596–597
    Bangladeshis, 586
    British, 604
    Burmese, 612
    Cree, 620
    Datoga, 636
    Fore, 644–645
    French, 654
    Fulani, 662
    Garhwali, 671
    Garifuna, 678
    Greeks, 687
    growth events/duration, 237–239
    Hadza, 694
    Han, 716
    Hausa, 726
    Hmong, 739–740
    Iroquois, 749
    Jamaican Maroons, 762–763
    Japanese, 774–775
    Jat, 781–782
    Malagasy, 801, 801–802
    Malays, 847–848
    Maori, 809
    Matsigenka, 824
    Maya of Highland Mexico, 835
    Mongolia, 858
    Nahua, 869–870
    Navajo, 879–880
    Nepal, 889
    Northwest Coast, 899–900
    Ojibwa, 912
    Oklahoma Choctaw, 921
    Roma of the United States and Europe, 927
    Samoa, 935
    Saraguros, 944
    Shipibo, 955
    Sotho, 961–962
    Sudanese, 970
    Thai, 979
    Tongans, 987
    Trobriand, 996–997
    Tuareg, 1007
    Wape, 1015
    Yanomamö, 1025–1026
    Yoruba, 1037
Adrenoleukodystrophy (ALD), genetic disease, 412–413
Adulthood
    African-Americans, 554
    Amish, 562–563
    Argentine Toba, 570
    Badaga, 577
    Baliem Valley Dani, 597–598
    Bangladeshis, 586–587
    British, 604–605
    Burmese, 612–613
    Cree, 620–621
    Datoga, 636–637
    Fore, 645
    Fulani, 662

Garhwali, 671
Garifuna, 678
Greeks, 687
Hadza, 694
Han, 716
Hausa, 726–727
Hmong, 740
Iroquois, 749–750
Jamaican Maroons, 763–764
Jat, 782
Malays, 848
Maori, 809–810
Matsigenka, 824–825
Maya of Highland Mexico, 835
Mongolia, 858
Nahua, 869–870
Navajo, 880
Nepal, 889
Ojibwa, 912–913
Oklahoma Choctaw, 921
Roma of the United States and Europe, 927
Samoa, 935–936
Saraguros, 944
Shipibo, 955
Sotho, 962
Sudanese, 970
Thai, 979
Tuareg, 1008
Wape, 1015
Yanomamö, 1026
Yoruba, 1037
Agammaglobulinemia, genetic disease, 413
Age studies, phenomenology, 131–132
Aged
  African-Americans, 554–555
  Amish, 563
  Argentine Toba, 570
  Badaga, 578
  Bangladeshis, 587
  British, 605
  Burmese, 613
  Cree, 621
  Fore, 645
  French, 654–655
  Fulani, 662–663
  Garhwali, 671–672
  Garifuna, 678
  Greeks, 687
  Hadza, 694–695
  Han, 716–717
  Hausa, 727
  Hmong, 740
  Iroquois, 750
  Jamaican Maroons, 764
  Japanese, 775
  Jat, 782–783
  Malagasy, 802
  Malays, 848
  Maori, 810
  Maya of Highland Mexico, 835
  Mongolia, 858
  Nahua, 870
  Navajo, 880–881
  Nepal, 889
  Northwest Coast, 900
  Ojibwa, 913
  Oklahoma Choctaw, 921–922
  Roma of the United States and Europe, 927
  Samoa, 936
  Saraguros, 944–945
  Shipibo, 955–956
  Sotho, 962
  Sudanese, 970
  Thai, 979
  Tuareg, 1008
  Wape, 1015–1016
  Yanomamö, 1026
  Yoruba, 1038
Agency for International Development (USAID), 166
Aging, 217–223
  AD, 314–315
  anthropological perspectives, 312
  anthropology of, 217
  chronic diseases of, 311–318
  cultural construction, 218
  and disease, 311
  elderhood, 217–218
  gender and, 219
  global context, 220–221
  identifying, 218
  impact, population health, 311–312
  longevity, 435
  Matsigenka, 824–825
  menopause, 219
  old age, 217–218
  societal transformation, 220
  status, 219–220
  witchcraft, 220
  see also Population aging
AIDS *see* HIV/AIDS
AIS *see* Androgen insensitivity syndrome
Akaptonuria, genetic disease, 414
Albinism, genetic disease, 409, 413–414
Alcohol use, 293–301
  AA, 20, 298
  alcohol abuse, 296
  'alcoholic personality', 295
  alternatives, 298–299
  anthropology and alcohol studies, 294–296
  anthropology's contribution to understanding, 300
  culture 'window', 295–296
  Datoga, 636
  discourse, universalizing, 299–300
  education as prevention, 298
  epidemiology, 297–298
  Han, 706

Subject Index

milestones, 297
Mongolia, 853–854
practical implications, 296–300
problems, related, 296
public policies, 297–298
ritual setting hypothesis, 378
treatment, 298–299
trends, 296–300
WHO, 297, 299–300
see also Drug use
Alcoholics anonymous (AA), 298
illness narratives, 20
Alcoholism, Native Americans, 202
ALD see Adrenoleukodystrophy
ALS see Amyotrophic lateral sclerosis, familial
Altered states of consciousness (ASC), shamanism, 146–148
Alzheimer's disease (AD), 314–315
    African-Americans, 554–555
    cross-cultural distribution, 314–315
    genetic disease, 414–415
    prevention, 315
    risk factors, 314–315
    sociocultural studies, 315
    treatment, 315
Ambiguous genitalia, genetic disease, 415
American Academy of Pediatrics, breast milk substitutes, 234
American Association of Mental Retardation (AAMR), 494–495
Amok, Malaysia, 324
Amuk, Malays, 844–845
Amyotrophic lateral sclerosis, familial (FALS), genetic disease, 415
Androgen insensitivity syndrome (AIS), genetic disease, 415
Angelman syndrome (AS), genetic disease, 416
Anishinaabe sickness, Ojibwa, 910–911
Anosmia, genetic disease, 416
Anthropological health research, history, 4–5
Anthropological interviewing, psychoanalysis, 63–64
Anthropological research, comparative nature, 158–159
Anticipation, genetic disease, 397–398
Applied issues, cross-cultural health research, 3–11
AS see Angelman syndrome
ASC see Altered states of consciousness
Asthma, genetic disease, 416
Ataxia telangiectasia (A-TI), genetic disease, 416
Australia, refugee health, 194–195
Autism, genetic disease, 408, 417
Autosomal inheritance, genetic disease, 395–397
Ayurvedic system, Jat, 779

Balneology, Czechs, 626–627
Bariba, birth, 227
Beginning-of-life decision-making, bioethics, 77–78
Behavior disability/difference, 360–363
Behavioral intervention, HIV/AIDS, 465–469
Benefits, breast-feeding, 231
Benin, birth, 227
Beta thalassemia, Greeks, 683
Biafra, post-colonial development and health, 186
Bilateral donors, economic development, 166

Biocultural models, child growth, 236–237
Bioethics, 73–86
    anthropology importance, 73
    applied anthropological work, 79–80
    beginning-of-life decision-making, 77–78
    cultural diversity, 79–80
    cultural domain of inquiry, 75–78
    cultural process, 75–78
    death, 75–76
    defining, 74
    EBM, 81
    end-of-life care, 77
    ethics consultation, 79–80
    ethno-ethnic traditions, 78–79
    future research areas, 80–81
    genetic disease, 404–405
    genetics, 78
    historical development, 73–74
    identity, 78
    identity, bioethicists, 79
    informed consent, 80
    justice, 80–81
    cf. medical ethics, 74
    organ transplantation/donation, 76
    prognosis disclosure, 76–77
    public health, 80–81
    'race', 78
    reproductive ethics, 78
    research ethics, 80
    terminal diagnosis disclosure, 76–77
    theoretical issues, 75
    topical issues, 75–78
    truth-telling, 76–77
Biological warfare, population control, 277
Biomedical practitioners, Tongans, 983–984
Biomedical technologies, 86–95
    contraceptives, 87–89
    organ transplantation/donation, 91–92
    pharmaceuticals, 87–88
    reproductive technologies, 89–91, 102
Biomedicine, 95–109
    anthropology and, 96–97
    childbirth, 104
    disease construction, 101
    early studies, 95–96
    and humanism, 99
    knowledge, 97–99
    Local Biology, 99–100
    practice, 97–99
    principle of separation, 98
    realities, 99–101
    social stratification, 200–201
    stance of practice, 103
    translating, 104–105
    trends, 101–103
    worldview, 97–99
    Yoruba, 1032
Biomedicine cultures, emerging infectious diseases, 388–389

Biopolitics, death/dying, 248–249
Biosocial traits, childhood, 239–240
Biotechnology, genetic disease, 399–405
Bipolar affective disorder, genetic disease, 417
Birth, 224–230
   African-Americans, 552
   Amish, 561
   anthropology of, 225–226
   Argentine Toba, 567–568
   authoritative knowledge, 228–229
   Badaga, 575–576
   Baliem Valley Dani, 595–596
   Bangladeshis, 585
   Bariba, 227
   Benin, 227
   biomedicine, 104
   British, 603
   Burmese, 611
   contextualising birthing systems, 226–228
   Cree, 619
   cross-cultural analysis, 225–226
   cultural patterning, 225–226
   Czechs, 628
   Datoga, 634–635
   early ethnographies, 224–225
   early surveys, 224–225
   Fore, 644
   French, 653
   Fulani, 660
   Garhwali, 669–670
   Garifuna, 677–678
   global perspectives, 226–228, 229
   Greeks, 685–686
   Hadza, 693
   Han, 715–716
   Hausa, 725
   Hmong, 737–738
   Iroquois, 748
   Jamaican Maroons, 762
   Japanese, 773
   Jat, 780
   local perspectives, 226–228, 229
   Malagasy, 800
   Malays, 846–847
   Malaysia, 227
   Maori, 808
   Matsigenka, 823
   Maya of Highland Mexico, 834
   midwifery, 225–226
   Mongolia, 857
   multiple births, 438–439
   Nahua, 867–868
   Navajo, 878
   Nepal, 888
   Northwest Coast, 899
   Ojibwa, 911–912
   Oklahoma Choctaw, 920
   placenta disposal, 224–225
   Roma of the United States and Europe, 926
   Samoa, 934–935
   Saraguros, 942–943
   Shipibo, 954
   Sierra Leone, 227–228
   Sotho, 960–961
   Sudanese, 969
   Thai, 978
   Tongans, 986
   Trobriand, 995
   Tuareg, 1006–1007
   Wape, 1014–1015
   western, anthropology of, 228
   witchcraft, 227
   Yanomamö, 1024–1025
   Yoruba, 1036
   see also Midwifery
Birth control
   abortion as, 278
   population control, 273–274, 278
   see also Contraceptives; Family planning
BL *see* Burkitt lymphoma
Black Death *see* Bubonic plague
Blindness, disability/difference, 363
Blood pressure
   cultural consonance, 333–334
   differences, societies', 328–330
   stress, 333–334
Bloom syndrome, genetic disease, 417
BMD *see* Bone mineral density
Body disability/difference, 360–363
Bolivian *altiplano*, medical pluralism, 112–113
Bombay phenotype, genetic disease, 417
Bone diseases
   OA, 56
   osteoporosis, 312–314
Bone infections, paleopathology, 53–54
Bone mineral density (BMD), 312–313
Bone tumours, paleopathology, 56–57
'Bones', illness of, Datoga, 632
Bonesetters
   Jamaican Maroons, 756
   Nahua, 865
Bovine Spongiform Encephalitis (BSE), 390
Brain-death, 248–249
   organ transplantation/donation, 92–93
Brazil
   cholera, 308
   HIV/AIDS, 469
   possession and trance, 140
   psychoanalytic movements, 65–66
   sufferer experience, 28
Breast cancer
   African-Americans, 551
   demographic transition, 399–400
   genetic disease, 408, 417–418
   illness narratives, 20–21
Breast-feeding, 230–235

Subject Index

benefits, 231
costs, 231–232
cross species perspective, 230–232
cultural practices, 232–233
disease protection, 231
evolutionary perspective, 230–232
health benefits, 231–232
immunologic protection, 231
initiation, 232–233
La Leche League, 234–235
maintenance, 232–233
oxytocin, 231
physiological perspective, 230–232
prolactin, 231
substitutes, 234–235
supplementation, 233–234
weaning, 233–234
western practices, 234–235
BSE *see* Bovine Spongiform Encephalitis
Bubonic plague
    paleopathology, 54
    population control, 270
Buddhist medicine, Mongolia, 854
pre-Buddhist Traditions, Mongolia, 854, 855
Burial, Jamaican Maroons, 764
Burkitt lymphoma (BL), genetic disease, 418

CAH *see* Congenital adrenal hyperplasia
California, holistic/new age healing center, 114
CAM *see* Complementary and alternative medicine
Cancer, paleopathology, 56–57
Cannabis, drug use, 378–379
Cannibalism, Fore, 640
Cardiovascular disease, 328–335
    African-Americans, 553
    blood pressure, 328–330
Caries, paleopathology, 57
Case studies
    disasters, 160–162
    medical pluralism, 111–114
Cattle, Fulani, 656
CCR5 gene mutation *see* Chemokine (C-C) receptor 5
CDC *see* Centers for Disease Control and Prevention
CDD *see* Chronic degenerative disease
Centers for Disease Control and Prevention (CDC), 27
CGH *see* Hypertrichosis, congenital generalized
Chagas' disease, epidemiology, 479–481
Charcot-Marie-Tooth disease 1A (CMT1), genetic disease, 418
Chemokine (C-C) receptor 5 (CCR5), genetic disease, 408, 418–419
Chiapas area, Maya of Highland Mexico, 828–829
Child abuse/neglect, 242, 301–305
    breaking the cycle, 303–304
    categories, 301–302
    consequences, 304
    cross-cultural perspective, 303
    definitions, 301–302
    demographics, 302–303
    etiology, 303
    incidence, 302–303
    intervention, 304
    prevention, 304
Child care practices, SIDS, 510–511
Child growth, 235–244
    biocultural models, 236–237
    continuity, 239
    cross-cultural examples, 240–241
    DNA regulation, 236–237
    events/duration, 237–239
    life history theory, 237–239
    medical anthropology, 236
    plasticity, 239
    pygmies, 241
    reasons, 239
    risks, 241–242
    study reasons, 235–236
Childbirth *see* Birth
Childhood
    African-Americans, 552–553
    Amish, 562
    Argentine Toba, 569
    Badaga, 577
    Baliem Valley Dani, 596
    Bangladeshis, 586
    biosocial traits, 239–240
    British, 604
    Burmese, 612
    Cree, 620
    Czechs, 628–629
    Datoga, 636
    Fore, 644
    French, 654
    Fulani, 661–662
    Garhwali, 671
    Garifuna, 678
    Greeks, 687
    growth events/duration, 237–239
    Hadza, 693–694
    Han, 716
    Hausa, 725–726
    Hmong, 739
    Iroquois, 749
    Jamaican Maroons, 762–763
    Japanese, 774
    Jat, 781
    Malagasy, 800–801
    Malays, 847
    Maori, 808–809
    Matsigenka, 824
    Maya of Highland Mexico, 835
    Mongolia, 857–858
    Nahua, 869
    Navajo, 879
    Nepal, 888
    Ojibwa, 912
    Oklahoma Choctaw, 921
    Roma of the United States and Europe, 926–927

Samoa, 935
Saraguros, 944
Shipibo, 955
Sotho, 961
Sudanese, 970
Thai, 978–979
themes, 239–240
Tongans, 987
Trobriand, 996
Tuareg, 1007
Wape, 1015
Yanomamö, 1025
Yoruba, 1037
Children's Vaccine Initiative (CVI), 263
China
    abortion, 278
    diabetes mellitus, 343–344
    family planning, 89
    psychoanalytic movements, 65–66
Chinese Medicine, Traditional (TCM), Lijiang Naxi, 786–787
Cholera, 305–311
    Brazil, 308
    history, 305–306
    prevalence, 306
Chromosomal mutations, genetic disease, 394–395
Chronic degenerative disease (CDD), 34
Chronic diseases of aging, 311–318
    AD, 314–315
    osteoporosis, 312–314
CHRs *see* Community Health Representatives
Circulatory abnormalities, paleopathology, 56
Circumcision, female *see* Female genital cutting
CJD *see* Creutzfeldt-Jakob disease
CKN1 *see* Cockayne syndrome type 1
Class, social stratification, 203
Clinical medicine, illness narratives in, 45–46
Cloning, stem cell, 403
CMA *see* Critical medical anthropology
CMT1 *see* Charcot-Marie-Tooth disease 1A
Cockayne syndrome type 1 (CKN1), genetic disease, 419
Cognitive-ethnographic studies, illness treatment decisions, 16–18
Cognitive medical anthropology, 12–23
Colon cancer, genetic disease, 419
Colonial era, Haitians, 699
Colonialism *see* Post-colonial development and health
Color blindness, genetic disease, 419–420
Community-based interventions, diabetes mellitus, 347
Community context, Ojibwa, 907–908
Community Health Representatives (CHRs), Oklahoma Choctaw, 918
Complementary and alternative medicine (CAM)
    Amish, 560
    Czechs, 626
Conditions, medicalized, 120–121
Congenital adrenal hyperplasia (CAH), genetic disease, 408, 420
Congenital defects, paleopathology, 51–52
Congo, post-colonial development and health, 186
Consent, informed *see* Informed consent
Contact, Northwest Coast, 893–894

Contagions, Northwest Coast, 893–894
Contemporary paradigm, Samoa, 933–934
Contraception
    and abortion, 88–89
    biographies, 87–88
    biomedical technologies, 87–89
    Hmong, 737
    Jamaican Maroons, 761–762
    population control, 274
    reproductive health, 283–284
    social aspects, 283–284
Contragestin, population control, 274
Contranatals, population control, 274–275
Controversies, cross-cultural health research, 3–11
Conventional medicine, cf. vernacular medicine, 8
Costs, breast-feeding, 231–232
Cot death *see* Sudden infant death syndrome
Counseling, genetic disease, 403–404
Creutzfeldt-Jakob disease (CJD), 390, 399, 401–402, 409–410
    genetic disease, 420
    cf. *kuru*, 401
Cri-du-chat syndrome, genetic disease, 420
Critical medical anthropology (CMA), 23–30, 102
    critical perspective, 24–25
    disease, defining, 26
    health, defining, 26
    impact, 29–30
    key concepts, 26–29
    medical hegemony, 28–29
    medical pluralism, 29, 111
    medicalization, 28
    origin, 25–26
    sufferer experience, 27–28
    tobacco, 525–527
Critical perspective, health social sciences, 24–25
Crohn disease, genetic disease, 421
Cross-cultural distribution
    AD, 314–315
    osteoporosis, 313
Cross-cultural health research
    controversies, 3–11
    key concepts, 3–11
    vs. mono-cultural health research, 8–9
    theoretical issues, 3–11
Cross-cultural medical systems, 6–8
Cross-cultural perspectives
    child abuse/neglect, 303
    shamanism, 146
Cross-cultural research, schizophrenia, 489–491
Cross-cultural review, disability/difference, 360–363, 366–367
Cross species perspective, breast-feeding, 230–232
Cultural consensus theory, 14–15
Cultural consonance, blood pressure, 333–334
Cultural diversity, bioethics, 79–80
Cultural expression
    neuroses, 62–63
    psychoses, 62–63
Cultural factors, Malays, 840–842

Subject Index

Cultural history, African-Americans, 547–548
Cultural knowledge, comparing, 15–16
Cultural models, illness narratives, 18–21
Cultural settings, variability, 14–15
Cultural variation, evolutionary theory, 35
Culture and environment, Northwest Coast, 891–892
Culture-bound syndromes, 319–327
   amok, 324
   defining, 319–321
   Japan, 320–321
   koro, 320, 323–324
   kuru, 324–325
   latah, 325
   Malaysia, 320, 323–324
   'nerves', 320, 322
   social change, 325–326
   susto, 322–323
   types, 321–322
Culture, defining, 12
Culture of poverty, urban poor, 210
Culture overview
   African-Americans, 545–546
   Amish, 557–558
   Argentine Toba, 564–565
   Badaga, 572–573
   Baliem Valley Dani, 591–592
   Bangladeshis, 579–580
   British, 599–600
   Burmese, 607–609
   Cree, 615–616
   Czechs, 622–623
   Datoga, 629–630
   Fore, 638–641
   French, 646–647
   Fulani, 656–657
   Garhwali, 665
   Garifuna, 673–674
   Greeks, 681–682
   Hadza, 689–690
   Haitians, 696–699
   Han, 703–704
   Hausa, 718–719
   Hmong, 729–731
   Iroquois, 744–745
   Jamaican Maroons, 754–755
   Japanese, 766–768
   Jat, 777–778
   Lijiang Naxi, 784–785
   Malagasy, 795–796
   Malays, 840
   Maori, 804–805
   Matsigenka, 812
   Maya of Highland Mexico, 828
   Mongolia, 850–851
   Nahua, 863–864
   Navajo, 873
   Nepal, 883–884
   Northwest Coast, 891–894
   Ojibwa, 903–905
   Oklahoma Choctaw, 916
   Roma of the United States and Europe, 923–924
   Samoa, 929–931
   Saraguros, 937–938
   Shipibo, 947–948
   Sotho, 957–958
   Sudanese, 965
   Thai, 971–972
   Tongans, 981–982
   Trobriand, 990–991
   Tuareg, 1001–1003
   Wape, 1010–1011
   Yanomamö, 1017–1018
   Yoruba, 1030–1031
Culture shock, psychoanalysis, 59, 64–65
Culture 'window', alcohol use, 295–296
Current era, Haitians, 699–700
Curses, Datoga, 632
CVI *see* Children's Vaccine Initiative
Cystic fibrosis, genetic disease, 421
Cytogenic conditions, genetic disease, 394–395

DALY *see* Disability Life Years
Dancers, Jamaican Maroons, 756
Darwin, Charles
   genetic disease, 393
   population control, 269
Deafness, genetic disease, 421
Death control, population control, 272–273
Death/dying, 244–252
   bioethics, 75–76
   biopolitics problem, 248–249
   brain-death, 92–93, 248–249
   critique and innovation, 246
   defining, 75–76, 92–93, 248–249
   experience, 246–248
   hospice movement, 247–248
   organ transplantation/donation, 92–93
   as problem of life, 248–249
   process, 246–248
   religious studies, 245–246
   ritual studies, 245–246
   social inequality, disease distribution, 249
   social studies, 245–246
   studies, 246–248
   Uniform Determination of Death Act (1981), 248
   violent death, 249–251
   see also Sudden infant death syndrome
Death/dying, Jamaican Maroons, 764
Death/rebirth, shamanism, 147–148
Decision-making, cognitive-ethnographic studies, 16–18
Degenerative disease, paleopathology, 56
Demographic data, Czechs, 623–624
Demographic transition, genetic disease, 399–400
Demographics
   Japanese, 766
   Saraguros, 937–938

Demographics, child abuse/neglect, 302–303
Demography
　　Haitians, 696–697, 699
　　population control, 270–271
　　transition theory, 271
Demonic forces, Greeks, 683
Dengue fever, 306
　　epidemiology, 479–481
Dental paleopathology, 57
Depression
　　genetic disease, 422
　　mental disorder, 488–489
DGS see DiGeorge syndrome
DI see Diego blood group
Diabetes insipidus, genetic disease, 422
Diabetes mellitus, 335–353
　　African-Americans, 344, 346
　　China, 343–344
　　community-based interventions, 347
　　complications, 336–337
　　definitions, 335–336
　　diagnostic criteria, 336
　　dietary transitions, 341–343
　　educational interventions, 347
　　epidemiology, 337–338
　　evolution, 340–341
　　genetic disease, 422–423
　　genetic thriftiness, 340–341
　　Hispanic traditional medicine, 345
　　IDF, 337
　　India, 343–344
　　Iroquois, 751
　　Korea, 343–344
　　Latino traditional medicine, 345
　　lifestyle factors, 341–343
　　Maya of Highland Mexico, 829–830
　　medical anthropology, 339
　　Native Americans, 342, 345, 346
　　obesity, 335, 336, 338, 341
　　Oklahoma Choctaw, 919–920
　　patient-provider interactions, 346
　　Pima Indians, 342
　　populations studied, 338–339
　　prevalence, 337–338
　　risk factors, 335, 336, 338, 341
　　traditional healthcare beliefs, 343–345
　　type 1;, 335, 422
　　type 2;, 335, 341, 343–345, 422–423
Diachronicity, disasters, 158
Diarrhea, 353–360
　　biomedical diagnosis, 354–355
　　biomedical perspectives, 353
　　cause, 355
　　classification, 353–355
　　diagnostic categories, 353–354
　　environmental relationships, 357
　　epidemiologic cycles, 355–357
　　ethnomedical care/treatment, 357

　　ethnomedical classes, 353–354
　　ethnomedical perspectives, 353
　　ethnomedical transmission, 355
　　implications, theoretical/applied, 358–359
　　Mexico, 359
　　pathogens, 356–357
　　transmission routes, 355–357
　　treatment, 357–358
　　treatment, home, 357–358
　　treatment, plants, 359
　　urban poor, 209
　　web sites, 359–360
Diego blood group (DI), genetic disease, 423
Diet
　　Czechs, 624–625
　　Hadza, 690–691
　　Han, 704–705, 713
　　Malagasy, 797
　　Maya of Highland Mexico, 829–830, 836
　　Mongolia, 852–853
　　Trobriand, 992
　　Yanomamö, 1023
Dietary transitions, diabetes mellitus, 341–343
Diffusing medical systems, 7
Diffusion of Innovations Theory (DIT), HIV/AIDS, 467–468
DiGeorge syndrome (DGS), genetic disease, 423
Disability, 103
Disability/difference, 360–373
　　anthropology of, 366–369
　　behavior, 360–363
　　blindness, 363
　　body, 360–363
　　cross-cultural review, 360–363, 366–367
　　defining, 368
　　illness and, 368–369
　　neurofibromatosis, 363
　　research review/critique, 363–366
　　responses, 360–363
　　theory review/critique, 363–366
　　WHO, 368
Disability Life Years (DALY), depression, 488
Disaccharide intolerance, genetic disease, 423–424
Disasters, 157–164
　　adaptation, 158
　　anthropology perspectives, 158–159
　　case study, 160–162
　　classifying, 157
　　defining, 158
　　diachronicity, 158
　　earthquake case study, 160–162
　　ethnography, 160
　　history of research, 157–158
　　research methodologies, 159–160
　　resilience, 159
　　social aspects, 157
　　stress, 157
　　vulnerability, 159

Subject Index

Disease
    and aging process, 311
    anthropological perspectives, 307–308
    classification, paleopathologic, 51
    CMA, 26
    construction, 101
    defining, 3–4, 26
    ecological/evolutionary perspective, 307
    ecology, 31–33, 36
    emerging infectious, 383–391
    equilibrium and change, 32
    evolution, 31–32, 33–36
    genetic *see* Genetic disease
    insect vector *see* Vector-borne diseases
    Interpretive perspective, 308
    natural selection, 307
    political/economic perspective, 307
    vector-borne *see* Vector-borne diseases
    water-borne, 305–311
Disease distribution, social inequality, 249
Disease patterns, Yanomamö, 1018–1023
Disease protection, breast-feeding, 231
DIT *see* Diffusion of Innovations Theory
Divorce
    Hadza, 694
    Hmong, 736–737
DM *see* Myotonic dystrophy
DMD *see* Muscular dystrophy Duchenne type
DNA
    child growth regulation, 236–237
    mutations, 395
Domestic abuse
    Czechs, 625–626
    Datoga, 637
    Malagasy, 801–802
Double-Y syndrome, 395
    genetic disease, 424
Down syndrome, genetic disease, 394, 409, 424–425
Drug use, 374–382
    after 1960, 378–379
    anthropology of, 375
    background, 374–377
    cannabis, 378–379
    consequences, 378
    contributions, 380–381
    early accounts, 375
    ethnography, street drug use, 379
    hazards, 376
    history, 374–377
    HIV/AIDS, 379–380
    impact, 380–381
    methods, 377–380
    models, 377–380
    prohibition, 376–377
    western moralism, 376
    see also Alcohol use
Dwarfism, genetic disease, 408, 425, 431–432
Dying/death
    African-Americans, 555
    Amish, 563
    Argentine Toba, 570–571
    Badaga, 578
    Baliem Valley Dani, 598
    Bangladeshis, 587–588
    British, 605–606
    Burmese, 613–614
    Cree, 621
    Czechs, 629
    Datoga, 637
    Fore, 645
    French, 655
    Fulani, 663
    Garhwali, 672
    Garifuna, 679
    Greeks, 687–688
    Hadza, 694
    Han, 717
    Hausa, 727–728
    Hmong, 740
    Iroquois, 750
    Japanese, 775
    Jat, 783
    Malagasy, 802–803
    Malays, 848
    Maori, 810–811
    Matsigenka, 825
    Maya of Highland Mexico, 835–836
    Mongolia, 858
    Nahua, 870
    Navajo, 881
    Nepal, 889
    Ojibwa, 913
    Oklahoma Choctaw, 922
    Roma of the United States and Europe, 927–928
    Samoa, 936
    Saraguros, 945
    Sotho, 962–963
    Sudanese, 970
    Thai, 979–980
    Tongans, 987–988
    Trobriand, 998–999
    Tuareg, 1008
    Wape, 1016
    Yanomamö, 1026
    Yoruba, 1038
Dyslexia, genetic disease, 425

Earthquake case study, disasters, 160–162
EBM *see* Evidence-based medicine
Ebola, 389
Ecological/evolutionary perspective, disease, 307
Ecological perspectives, 31–37
    controversies, 33
    reproductive ecology, 33
Economic development, 164–170
    bilateral donors, 166

defining, 164–165
international agencies, 165–166
medical anthropologists in, 167–168
private foundations, 166–167
United Nations, 165
USAID, 166
WHO, 165–166
World Bank, 165–166
Economic factors
African-Americans, 546–548
Amish, 558–560
Argentine Toba, 565–566
Badaga, 573
Baliem Valley Dani, 592–593
Bangladeshis, 580–581
British, 600–601
Burmese, 609–610
Cree, 616–617
Czechs, 623–626
Datoga, 630–631
Fore, 641–643
French, 647
Fulani, 657–658
Garhwali, 665–666
Garifuna, 675
Hadza, 690–691
Haitians, 699–700
Han, 704–707
Hausa, 719–720
Iroquois, 745–746
Japanese, 768–771
Jat, 778
Lijiang Naxi, 785–787
Malagasy, 796–797
Malays, 840–842
Maori, 805–807
Matsigenka, 812–814
Maya of Highland Mexico, 828–831
Mongolia, 851–854
Nahua, 864
Navajo, 874
Nepal, 885
Northwest Coast, 894–897
Ojibwa, 905–908
Oklahoma Choctaw, 916–917
Roma of the United States and Europe, 924
Saraguros, 938–939
Shipibo, 948–951
Sotho, 958–959
Sudanese, 965–966
Thai, 972–975
Tongans, 982–983
Trobriand, 991–992
Tuareg, 1003–1004
Wape, 1011–1012
Yanomamö, 1018–1023
Yoruba, 1031–1032
Economy

Czechs, 623
Datoga, 630
Haitians, 697–698
Hmong, 730
Japanese, 767
Samoa, 930
Tuareg, 1001–1002
Ectodermal dysplasia, anhidrotic (EDA), genetic disease, 426
EDS *see* Energy dispersive spectrometer
Educational interventions, diabetes mellitus, 347
Edward syndrome, 395
  genetic disease, 426
Efficacy theories, Navajo, 877
Elderhood, aging, 217–218
Emerging infectious diseases, 383–391
  biomedicine cultures, 388–389
  culture, 386–387
  epidemiology, 386–387
  examples, 383–384
  future directions, 390
  molecular biology, 387–388
  social change, 387–388
  study methods, 386–388
  theoretical heritage, 384–386
Emphysema, congenital, genetic disease, 426
End-of-life care, bioethics, 77
Energy dispersive spectrometer (EDS), forensic anthropology, 39
Enteritis necroticans, Fore, 641–642
Environmental factors
  African-Americans, 546–548
  Amish, 558–560
  Argentine Toba, 565–566
  Badaga, 573
  Baliem Valley Dani, 592–593
  Bangladeshis, 580–581
  British, 600–601
  Burmese, 609–610
  Cree, 616–617
  Czechs, 623–626
  Datoga, 630–631
  Fore, 641–643
  French, 647
  Fulani, 657–658
  Garhwali, 665–666
  Garifuna, 674–675
  Greeks, 682–683
  Hadza, 690–691
  Haitians, 699–700
  Han, 704–707
  Hausa, 719–720
  Iroquois, 745–746
  Japanese, 768–771
  Jat, 778
  Lijiang Naxi, 785–787
  Malagasy, 796–797
  Malays, 840–842
  Maori, 805–807
  Matsigenka, 812–814

Subject Index

Maya of Highland Mexico, 828–831
Mongolia, 851–854
Nahua, 864
Navajo, 874
Nepal, 885
Northwest Coast, 894–897
Ojibwa, 905–908
Oklahoma Choctaw, 916–917
Roma of the United States and Europe, 924
Saraguros, 938–939
Shipibo, 948–951
Sotho, 958–959
Sudanese, 965–966
Thai, 972–975
Tongans, 982–983
Trobriand, 991–992
Tuareg, 1003–1004
Wape, 1011–1012
Yanomamö, 1018–1023
Yoruba, 1031–1032
Environmental factors, mental retardation, 498–500
Environmental relationships, diarrhea, 357
EPI *see* Expanded Program on Immunization
Epidemics *see* Emerging infectious diseases
Epidemiology
    alcohol use, 297–298
    emerging infectious diseases, 386–387
    genetic, 407–409
    mental retardation, 496–498
    SIDS, 507–509
    trends, Maori, 805
    vector-borne diseases, 479–481
Epilepsy, progressive myoclonic 2 (EPM2A), genetic disease, 426
EPM2A *see* Epilepsy, progressive myoclonic 2
Erythroblastosis fetalis, genetic disease, 426
Essential tremor (ETM1), genetic disease, 426
Ethics *see* Bioethics; Medical ethics
Ethnic identity, Garifuna, 673–674
Ethnicity, social stratification, 201–202
Ethnographic research vs. ethnological research, 8–9
Ethnography, disasters, 160
Ethnological research vs. ethnographic research, 8–9
Ethnomedical studies, vector-borne diseases, 481–482
Etiologies
    Hmong, 733–735
    Jamaican Maroons, 758–759
ETM1 *see* Essential tremor
Eugenics
    genetic disease, 393–394
    population control, 277–278
Evidence-based medicine (EBM), bioethics, 81
Evil eye
    Datoga, 632
    Garhwali, 667–668
    Greeks, 684
Evolutionary perspectives, 31–37
    breast-feeding, 230–232
    controversies, 35

    cultural variation, 35
    disease, 31–32, 33–36
    see also Paleopathology
Exorcism, possession and trance, 142
Expanded Program on Immunization (EPI), 263
Externalizing illness, 6–7
Extrusion illness, 7

Factor V Leiden thrombophilia, genetic disease, 426
FALS *see* Amyotrophic lateral sclerosis, familial
Familial fatal insomnia (FFI), genetic disease, 426–427
Family and kinship
    Czechs, 623
    Datoga, 630
    Haitians, 698
    Japanese, 767
    Tuareg, 1002
Family planning, 88–89
    population control, 279
Family, Saraguros, 938
FAP *see* Adenomatous polyposis, familial
Fatalism, Nepal, 886–887
Female genital cutting (FGC), 252–262
    elimination strategies, 257–260
    Islam, 256
    medicalization, 259
    Prevention of Discrimination and the Protection of Minorities, 254, 258
    Sierra Leone, 255
    Special Report, 254, 258
    Sudan, 255
    trivialization of culture, 253–257
Female pollution, Fore, 639–640
Fertility
    Han, 714–715
    Mongolia, 856
    Saraguros, 942
    Sudanese, 968–969
Fertility enhancement, reproductive health, 285
Fertility regulation, reproductive health, 283–284
Fetal hemoglobin, hereditary persistence, heterocellular (HPFH), genetic disease, 427
Fever, Datoga, 632
FFI *see* Familial fatal insomnia
FGC *see* Female genital cutting
FHC *see* Hypercholesterolemia, familial
Fires of life, Trobriand, 994
Five Elements, Han, 708–709
FMF *see* Mediterranean fever, familial
Food
    homelessness, 172–173
    see also Nutritional anthropology
Food therapy, Han, 713
Forensic anthropology, 37–42
    applications, 40
    definitions, 37
    EDS, 39
    goals, 38–39

history, 37–38
methodology, 39
related disciplines, interface with, 40–41
SEM, 39
techniques, new, 39–40
training, 38
Fractures, paleopathology, 52–53
Fragile X syndrome (FRAX), genetic disease, 427
Fructose intolerance, genetic disease, 427
Funerals, psychoanalysis, 60–61
Future directions
 emerging infectious diseases, 390
 HIV/AIDS, 473–474
 mental disorders, 491
 tuberculosis, 536–537
Future research areas, bioethics, 80–81

G6PD deficiency, Greeks, 683
Galactosemia, genetic disease, 428
Garrod, Archibald Edward, 393
Gastronomy, French, 649–650
Gaucher disease, genetic disease, 428
Gender
 aging and, 219
 social stratification, 203–204
Gender relations, Fore, 639
Gender research, vector-borne diseases, 483
Gendered body, phenomenology, 130
Genetic disease, 391–462
 achromatopsia, 408
 albinism, 409
 anticipation, 397–398
 antiquity, 392–393
 autism, 408
 autosomal inheritance, 395–397
 bioethics, 404–405
 biotechnology, 399–405
 breast cancer, 408
 CCR5 gene mutation, 408
 CJD, 390, 399, 401–402, 409–410
 congenital adrenal hyperplasia, 408
 cytogenic conditions, 394–395
 Darwin, Charles, 393
 demographic transition, 399–400
 distributions, 409–453
 Down syndrome, 394, 409
 dwarfism, 408
 etiology facets, 392
 eugenics, 393–394
 foundation, genetics, 393
 gene therapy, 403–404
 genetic counseling, 403–404
 genetic epidemiology, 407–409
 genomic imprinting, 397–398
 Herodotus, 392
 history, 391–394
 Human Genome Project, 400–401
 'inborn errors of metabolism', 393
 kuru, 409
 lactose persistence, 408
 MCAs, 394–395
 maple syrup urine disease, 408
 Maupertuis, 392
 mechanisms, 394–395
 Mendel, Gregor, 393
 mitochondrial inheritance, 397
 modes of inheritance, Mendelian, 395–397
 modes of inheritance, non-Mendelian, 397–399
 molecular genetic conditions, 395
 molecular genetics, 394
 monogenic factors, 397
 multifactorial disorders, 398–399
 mutations, chromosomal, 394–395
 mutations, DNA, 395
 New World syndrome, 409
 phenocopies, 399
 polydactyly, 408
 polygenic factors, 398–399
 porphyria variegata, 408
 SCD, 392, 408
 sex-linked inheritance, 397
 sources, current information, 399
 stem cell research/cloning, 403
 synopsis, syndromes, 411–453
 Tay-Sachs disease, 408, 409
 uniparental disomy, 397–398
 web sites, 399
 Weissmann, August, 393
 X-linked inheritance, 392, 397
 Y-linked inheritance, 397
 see also Named diseases
Genetic–environmental interactions, mental retardation, 501–502
Genetic factors, mental retardation, 498
Genetic thriftiness, diabetes mellitus, 340–341
Genetics, bioethics, 78
Genitalia, ambiguous, genetic disease, 415
Genomic imprinting, genetic disease, 397–398
Germs, Jamaican Maroons, 757–758
Gerstmann-Straussler-Sheinker disease (GSSD), genetic disease, 428
GGM *see* Glucose–galactose malabsorption
Ghettos *see* Urban poor
Gilles de la Tourette syndrome (GTS), genetic disease, 429
Glaucoma, genetic disease, 429
Global context, aging, 220–221
Global immunization efforts, 263
Global perspectives
 birth, 226–228, 229
 urban poor, 211
Glucose-6-phosphate dehydrogenase (G6PD) deficiency, genetic disease, 429
Glucose–galactose malabsorption (GGM), genetic disease, 429
Glycogen storage disease, genetic disease, 429
Government, Czechs, 623
Growth *see* Child growth
GSSD *see* Gerstmann-Straussler-Sheinker disease
GTS *see* Gilles de la Tourette syndrome

Subject Index

Guatemala earthquake, case study, 160–162
Guevedoces, genetic disease, 430
Gusii practices, weaning, 233–234

Haiti, medical pluralism, 113
HBM *see* Health Belief Model
HCV *see* Hepatitis C virus
HD *see* Huntington disease
HDN *see* Hemolytic disease of the newborn
Healers
    Hausa, 720–722
    Hmong, 732–733
    Tongans, 983
Healers, shamanism, 148–149
Healing
    and possession/trance, 141–142
    shamanism, 152–153
Healing behaviors, evolution, 34–35
Health
    African-Americans, 550
    CMA, 26
    defining, 3–4, 26
    homelessness and, 174–175
Health Belief Model (HBM), HIV/AIDS, 466–467
Health care system, Japanese, 768–769
Health conditions
    Hmong, 731–732
    Maya of Highland Mexico, 836–838
Health context, Jamaican Maroons, 755–756
Health development, 164–170
    medical anthropologists in, 167–168
    see also Economic development
Health, historical context, Ojibwa, 906–907
Health indicators, Samoa, 931–932
Health infrastructure
    Czechs, 625
    Haitians, 700
    Samoa, 932
    Thai, 974–975
Health insurance, Czechs, 625
Health patterns
    Argentine Toba, 571
    Badaga, 578
    Bangladeshis, 588
    Fulani, 663
    Garhwali, 672
    Garifuna, 679–680
    Greeks, 687–688
    Hadza, 694
    Iroquois, 750–752
    Maya of Highland Mexico, 836
    Mongolia, 858–861
    Nahua, 870–871
    Nepal, 889–890
    Northwest Coast, 900–901
    Oklahoma Choctaw, 922
    Roma of the United States and Europe, 928
    Saraguros, 945–946

    Sudanese, 970
    Tongans, 988
    Tuareg, 1008
    Wape, 1016
    Yanomamö, 1026–1027
    Yoruba, 1038
Health profile, Saraguros, 938–939
Health research, history, 4–5
Health-seeking process, 5–6
Health situation, Argentine Toba, 566
Health social sciences, critical perspective, 24–25
Hegemony, medical *see* Medical hegemony
Hemochromatosis, hereditary (HH), genetic disease, 430
Hemoglobin (Hb) alleles, genetic disease, 430
Hemolytic disease of the newborn (HDN), genetic disease, 431
Hemophilia, genetic disease, 431
Hepatitis C virus (HCV), 383, 389
Hepatitis, Yanomamö, 1021
Herbal practitioners, Han, 707–708
Herbal treatments, Han, 711–714
Herbalists
    Hmong, 732
    Jamaican Maroons, 756
    Navajo, 874
    Northwest Coast, 897
Hermansky-Pudlak syndrome (HPS), genetic disease, 431
Hermaphroditism, genetic disease, 431
Herodotus, genetic disease, 392
HGP *see* Human Genome Project
HH *see* Hemochromatosis, hereditary
Hispanic traditional medicine, diabetes mellitus, 345
History
    Czechs, 622–623
    Haitians, 697
    Hmong, 729–730
    Japanese, 766–767
    Samoa, 930
    Saraguros, 937–938
    Tuareg, 1001
History, anthropological health research, 4–5
HIV/AIDS, 383, 386–387, 389
    Baliem Valley Dani, 593
    Bangladeshis, 588
    behavioral intervention, 465–469
    behavioral research, anthropology in, 469–473
    Brazil, 469
    condom use, 465
    DIT, 467–468
    drug use, 379–380
    epidemiological features, 464–465
    ethnography and, 470–472
    future directions, 473–474
    Han, 705–706
    HBM, 466–467
    illness narratives, 20
    immune system effects, 464–465
    Lijiang Naxi, 791
    Malagasy, 799

medical features, 464–465
morbidity, 463
mortality, 463
Nepal, 889–890
post-colonial development and health, 189
prevention, 462–479
Puerto Rico, 469
reproductive health, 286
research, 462–479
research methods, 472–473
sexuality and, 286
social stratification, 202–203
Sotho, 958–959
syndemics, 27
Thai, 977–978, 979
TRA, 467
transmission, 464
treatment, 465
urban poor, 208, 209
Holistic/new age healing center, California, 114
Homelessness, 170–177
daytime respites, 176
defining, 170–171
doubling-up, 174
ending, 175–176
food, 172–173
health and, 174–175
leadership, 176
mental illness, 174–175
methodologies, study, 171
panhandling, 172
physical health, 174
skid row studies, 171–172
street life, 172
substance abuse, 175
supportive housing, 176
survival strategies, 171–174
theories, 170–171
transitional housing, 176
tuberculosis, 174
urban poor, 208
Homeopathic practitioners, Bangladeshis, 582
Homeopathy, French, 651–652
Hormone replacement therapy (HRT), 605
British, 605
Hospice movement, 247–248
Hospitals, Hausa, 722–723
HPFH *see* Fetal hemoglobin, hereditary persistence, heterocellular
HPS *see* Hermansky-Pudlak syndrome
HRAFs *see* Human Relations Area Files
HRT *see* Hormone replacement therapy
HTC2 *see* Hypertrichosis, congenital generalized
Human Genome Project (HGP), 98, 399
genetic disease, 400–401
Human Relations Area Files (HRAFs), 9
Humanism, and biomedicine, 99
Humoral pathology, Greeks, 683–684
Hunter syndrome, genetic disease, 432

Huntington disease (HD), genetic disease, 432
Hurler syndrome, 432
Hydrops fetalis, genetic disease, 432
Hypercholesterolemia, familial (FHC), genetic disease, 432
Hyperlipoproteinemia, genetic disease, 432–433
Hypertrichosis, congenital generalized (CGH) (HTC2), genetic disease, 433
Hypophosphatasia, familial, genetic disease, 433

IBD *see* Inflammatory bowel disease
Ideal household, Mongolia, 856–857
Identities, medicalized, 120–121
Identity, bioethics, 78
Identity, Fulani, 657
IDF *see* International Diabetes Federation
IHS *see* Indian Health Service
Illness
defining, 3–4
disability/difference and, 368–369
externalizing, 6–7
extrusion, 7
internalizing, 6–7
intrusion, 7
Illness classification
African-Americans, 550–551
Amish, 560
Argentine Toba, 566–567
Badaga, 574–575
Baliem Valley Dani, 594
Bangladeshis, 582–583
British, 602
Burmese, 610–611
Cree, 618–619
Czechs, 627–628
Datoga, 631–632
Fore, 643
French, 648–652
Fulani, 658–659
Garhwali, 666–669
Garifuna, 676–677
Greeks, 683–684
Hadza, 692
Haitians, 700–701
Han, 709–710
Hausa, 723–724
Hmong, 733–736
Iroquois, 747–748
Jamaican Maroons, 757–760
Japanese, 772–773
Jat, 779
Lijiang Naxi, 788–791
Malagasy, 798–799
Malays, 843–846
Maori, 807
Matsigenka, 815–822
Maya of Highland Mexico, 832–833
Mongolia, 855–856
Nahua, 865–867

Subject Index

Navajo, 875
Nepal, 886–887
Northwest Coast, 898–899
Ojibwa, 909–911
Oklahoma Choctaw, 919–920
Roma of the United States and Europe, 925
Samoa, 932–934
Saraguros, 940–941
Shipibo, 952–953
Sotho, 959–960
Sudanese, 966–967
Thai, 976–977
Tongans, 984–985
Trobriand, 992–993
Tuareg, 1005–1006
Wape, 1012–1014
Yanomamö, 1024
Yoruba, 1033–1035
Illness concept, Han, 709
Illness defined, African-Americans, 550
Illness diagnosis, Han, 710
Illness narratives, 42–49
    in anthropology, 44–45
    in clinical medicine, 45–46
    cultural models, 18–21
    illness autobiography, 42–43
    narratologists, 46–47
Illness theories
    African-Americans, 550–551
    Amish, 560
    Argentine Toba, 566–567
    Badaga, 574–575
    Baliem Valley Dani, 594
    Bangladeshis, 582–583
    British, 602
    Burmese, 610–611
    Cree, 610–611
    Czechs, 627–628
    Datoga, 631–632
    Fore, 643
    French, 648–652
    Fulani, 658–659
    Garhwali, 666–669
    Garifuna, 676–677
    Greeks, 683–684
    Hadza, 692
    Haitians, 700–701
    Han, 708–709
    Hausa, 723–724
    Hmong, 733–736
    Iroquois, 747–748
    Jamaican Maroons, 757–760
    Japanese, 772–773
    Jat, 779
    Lijiang Naxi, 788–791
    Malagasy, 798–799
    Malays, 843–846
    Maori, 807
    Matsigenka, 815–822
    Maya of Highland Mexico, 832–833
    Mongolia, 855–856
    Nahua, 865–867
    Navajo, 875
    Nepal, 886–887
    Northwest Coast, 898–899
    Ojibwa, 909–911
    Oklahoma Choctaw, 919–920
    Roma of the United States and Europe, 925
    Samoa, 932–934
    Saraguros, 940–941
    Shipibo, 952–953
    Sotho, 959–960
    Sudanese, 966–967
    Thai, 976–977
    Tongans, 984–985
    Trobriand, 993–994
    Tuareg, 1005–1006
    Wape, 1012–1014
    Yanomamö, 1024
    Yoruba, 1033–1035
Illness treatment
    African-Americans, 550–551
    Amish, 560
    Argentine Toba, 566–567
    Badaga, 574–575
    Baliem Valley Dani, 594
    Bangladeshis, 582–583
    British, 602
    Burmese, 610–611
    Cree, 610–611
    Czechs, 627–628
    Datoga, 631–632
    Fore, 643
    French, 648–652
    Fulani, 658–659
    Garhwali, 666–669
    Garifuna, 676–677
    Greeks, 683–684
    Hadza, 692
    Haitians, 700–701
    Han, 710–711
    Hausa, 723–724
    Hmong, 733–736
    Iroquois, 747–748
    Jamaican Maroons, 757–760
    Japanese, 772–773
    Jat, 779
    Lijiang Naxi, 788–791
    Malagasy, 798–799
    Malays, 843–846
    Maori, 807
    Matsigenka, 815–822
    Maya of Highland Mexico, 832–833
    Mongolia, 855–856
    Nahua, 865–867
    Navajo, 875–877

Nepal, 886–887
Northwest Coast, 898–899
Ojibwa, 909–911
Oklahoma Choctaw, 919–920
Roma of the United States and Europe, 925
Samoa, 932–934
Saraguros, 940–941
Shipibo, 952–953
Sotho, 959–960
Sudanese, 966–967
Thai, 976–977
Tongans, 984–985
Trobriand, 994
Tuareg, 1005–1006
Wape, 1012–1014
Yanomamö, 1024
Yoruba, 1033–1035
Illness treatment decisions, cognitive–ethnographic studies, 16–18
Immigration, population control, 276
Immunization, 262–268
　acceptance, 265–266
　barriers, 265
　British, 603–604
　CVI, 263
　demand, 265–266
　EPI, 263
　global efforts, 263
　history, 262–263
　Jamaican Maroons, 763
　media reports, 267
　Netherlands, 266
　Philippines, 266–267
　resistance, 266–267
　vaccination programs, 264–265
Immunologic diseases, paleopathology, 56
Immunologic protection, breast-feeding, 231
In vitro fertilization (IVF), 91
'Inborn errors of metabolism', genetic disease, 393
India
　diabetes mellitus, 343–344
　medical pluralism, 111
　psychoanalytic movements, 65–66
Indian Act (1876), Ojibwa, 905
Indian Health Service (IHS), Oklahoma Choctaw, 917–919
Indonesia, post-colonial development and health, 185–187, 189
Industrial development, Han, 704
Infancy
　African-Americans, 552
　Amish, 561–562
　Argentine Toba, 568
　Badaga, 576–577
　Baliem Valley Dani, 596
　Bangladeshis, 585–586
　British, 603–604
　Burmese, 612
　Cree, 619–620
　Czechs, 628
　Datoga, 635–636

Fore, 644
French, 653–654
Fulani, 660–661
Garhwali, 670–671
Garifuna, 678
Greeks, 686–687
growth events/duration, 237–239
Hadza, 693
Han, 716
Hausa, 725
Hmong, 738–739
Iroquois, 748–749
Jamaican Maroons, 762
Japanese, 773–774
Jat, 781
Malagasy, 800
Malays, 847
Maori, 808
Matsigenka, 823–824
Maya of Highland Mexico, 834–835
Mongolia, 857
Nahua, 868–869
Navajo, 878–879
Nepal, 888
Ojibwa, 912
Oklahoma Choctaw, 921
Roma of the United States and Europe, 926
Samoa, 935
Saraguros, 943–944
Shipibo, 954–955
Sotho, 961
Sudanese, 969–970
Thai, 978
Tongans, 986–987
Trobriand, 996
Tuareg, 1007
Wape, 1015
Yanomamö, 1025
Yoruba, 1036–1037
Infectious disease
　paleopathology, 53–54
　urban poor, 209
Infectious diseases, emerging *see* Emerging infectious diseases
Infertility
　reproductive health, 285
　reproductive technologies, 89–91, 102
Inflammatory bowel disease (IBD), genetic disease, 433–434
Informed consent, bioethics, 80
Infrastructure
　Han, 706–707
　Mongolia, 853
Inherited disease *see* Genetic disease
Inner-cities *see* Urban poor
Insect vector diseases *see* Vector-borne diseases
Internalizing illness, 6–7
International Diabetes Federation (IDF), 337
International relations, Haitians, 698
Internet sites *see* Web sites

Subject Index

Interpretive perspective, disease, 308
Interviewing, anthropological *see* Anthropological interviewing
Intrusion illness, 7
Iran
    family planning, 89
    menstruation, 282
Islam, FGC, 256
IVF *see* In vitro fertilization

Japan
    culture-bound syndromes, 320–321
    family planning, 89
    medical pluralism, 113
    psychoanalytic movements, 65–66
Justice, bioethics, 80–81
Juvenile stage, growth events/duration, 237–239

Key concepts, cross-cultural health research, 3–11
Kinship, Saraguros, 938
Klinefelter syndrome, genetic disease, 434
Korea, diabetes mellitus, 343–344
Koro, culture-bound syndrome, 320, 323–324
KRAS *see* Ras oncogene
Kuru
    cf. CJD, 401
    culture-bound syndrome, 324–325, 386, 387
    genetic disease, 399, 409, 434
Kuru, Fore, 642–643

La Leche League, breast-feeding, 234–235
Lactose persistence, genetic disease, 408
Language of Wider Communication (LWC), African-Americans, 545
Laos, Hmong in, 731–732
Latah, culture-bound syndrome, 325
Latah, Malays, 844–845
Latino traditional medicine, diabetes mellitus, 345
Leber's hereditary optic neuropathy (LHON), genetic disease, 434–435
Legislation
    Uniform Determination of Death Act (1981), 248
    violent death, 249–251
Leprosy, paleopathology, 54
Lesch-Nyhan syndrome, genetic disease, 435
LHON *see* Leber's hereditary optic neuropathy
Life cycle, growth events/duration, 237–239
Life cycle, health through, Lijiang Naxi, 792–793
Life history theory
    child growth, 237–239
    principles, 237–239
    trade-offs, 237–239
Lifestyle factors, diabetes mellitus, 341–343
Linguistic affiliation
    African-Americans, 545
    Amish, 557
    Argentine Toba, 564
    Badaga, 572
    Baliem Valley Dani, 591
    Bangladeshis, 579

British, 599
Burmese, 607
Cree, 614–615
Czechs, 622
Datoga, 629
Fore, 638
French, 646
Fulani, 656
Garhwali, 665
Garifuna, 673
Greeks, 681
Hadza, 689
Haitians, 696
Han, 703
Hausa, 718
Hmong, 729
Iroquois, 743–744
Jamaican Maroons, 754
Japanese, 765–766
Jat, 777
Lijiang Naxi, 783–784
Malagasy, 794–795
Malays, 839–840
Maori, 804
Maya of Highland Mexico, 827–828
Mongolia, 850
Nahua, 863
Navajo, 873
Nepal, 883
Northwest Coast, 890–891
Ojibwa, 903
Oklahoma Choctaw, 915
Roma of the United States and Europe, 923
Samoa, 929
Saraguros, 937
Shipibo, 947
Sotho, 957
Sudanese, 964–965
Thai, 971
Tongans, 980–981
Trobriand, 990
Tuareg, 1001
Wape, 1009–1010
Yanomamö, 1017
Yoruba, 1029–1030
Local Biology, biomedicine, 99–100
Local perspectives
    birth, 226–228, 229
    urban poor, 211
Location
    African-Americans, 545
    Amish, 557
    Argentine Toba, 564
    Badaga, 572
    Baliem Valley Dani, 591
    Bangladeshis, 579
    British, 599
    Burmese, 607

Cree, 614–615
Czechs, 622
Datoga, 629
Fore, 638
French, 646
Fulani, 656
Garhwali, 665
Garifuna, 673
Greeks, 681
Hadza, 689
Haitians, 696
Han, 703
Hausa, 718
Hmong, 729
Iroquois, 743–744
Jamaican Maroons, 754
Japanese, 765–766
Jat, 777
Lijiang Naxi, 783–784
Malagasy, 794–795
Malays, 839–840
Maori, 804
Matsigenka, 812
Maya of Highland Mexico, 827–828
Mongolia, 850
Nahua, 863
Navajo, 873
Nepal, 883
Northwest Coast, 890–891
Ojibwa, 903
Oklahoma Choctaw, 915
Roma of the United States and Europe, 923
Samoa, 929
Saraguros, 937
Shipibo, 947
Sotho, 957
Sudanese, 964–965
Thai, 971
Tongans, 980–981
Trobriand, 990
Tuareg, 1001
Wape, 1009–1010
Yanomamö, 1017
Yoruba, 1029–1030
Long-QT syndrome (LQT), genetic disease, 435
Longevity, 435
  see also Aging
Low-income countries, nutritional anthropology, 181–182
LQT see Long-QT syndrome
LWC see Language of Wider Communication
Lymphatic filariasis, epidemiology, 479–481

MCAs see Major chromosome anomalies
MAD1 see Major affective disorder 1
'Magical' healers, Hmong, 732–733
Major affective disorder 1 (MAD1), genetic disease, 436
Major chromosome anomalies (MCAs), 394–395
  genetic disease, 419

Malaria, 479–485
  Baliem Valley Dani, 593
  Bangladeshis, 588
  epidemiology, 479–481
  Greeks, 683
  Yanomamö, 1018–1019
  see also Vector-borne diseases
Malaysia
  amok, 324
  birth, 227
  culture-bound syndromes, 320, 323–324, 325
  family planning, 89
  koro, 320, 323–324
  latah, 325
Male initiation, Fore, 640
Malnutrition, 241–242
  Baliem Valley Dani, 592–593
  low-income countries, 181–182
  Malagasy, 797
  urban poor, 208–209
Malthus, Thomas Robert, population control, 269, 272
MAOA see Monoamine oxidase A deficiency
Maple syrup urine disease, genetic disease, 408, 436
Marfan syndrome, genetic disease, 436
Marriage, Saraguros, 938
Maupertuis, genetic disease, 392
MDRTB see Multi-drug-resistant tuberculosis
MDS see Multidimensional scaling
Measles, Yanomamö, 1022
Media reports, immunization, 267
Medical anthropologists, in economic development, 167–168
Medical anthropology
  child growth, 236
  diabetes mellitus, 339
Medical decision making, Hmong, 735–736
Medical ethics
  cf. bioethics, 74
  defining, 74
Medical expenses, Czechs, 625
Medical hegemony, CMA, 28–29
Medical pluralism, 109–116
  Bolivian *altiplano*, 112–113
  case studies, 111–114
  CMA, 29, 111
  conceptions, 110–111
  Haiti, 113
  India, 111
  Japan, 113
  power relations, 114–115
  United States, 113–114
  Zaire, 111–112
Medical pluralism, post-socialist, Mongolia, 860–861
Medical practitioners
  African-Americans, 548–550
  Amish, 560
  Argentine Toba, 566
  Badaga, 573–577
  Baliem Valley Dani, 593–594

Subject Index

Bangladeshis, 581–582
British, 601–602
Czechs, 626–627
Datoga, 631
Fore, 643
French, 647–648
Fulani, 658
Garhwali, 666
Garifuna, 676
Greeks, 683
Hadza, 692
Haitians, 700
Han, 707–708
Hausa, 720–723
Hmong, 732–733
Iroquois, 746–747
Jamaican Maroons, 756–757
Japanese, 771–772
Jat, 778–779
Lijiang Naxi, 787–788
Malagasy, 797–798
Malays, 842–843
Maori, 807
Matsigenka, 814–815
Maya of Highland Mexico, 832
Mongolia, 854–855
Nahua, 864–865
Navajo, 874
Nepal, 885–886
Northwest Coast, 897
Ojibwa, 908–909
Oklahoma Choctaw, 917–919
Roma of the United States and Europe, 925
Samoa, 932
Saraguros, 940
Shipibo, 951–952
Sotho, 959
Sudanese, 966
Thai, 975–976
Tongans, 983–984
Trobriand, 992
Tuareg, 1004–1005
Wape, 1012
Yanomamö, 1023–1024
Yoruba, 1032–1033
Medicalization
    CMA, 28
    conditions, 120–121
    critique, 118–120
    FGC, 259
    identities, 120–121
    menopause, 122
    modernization, 116–118
    naturalization of social control, 116–125
    reproduction, 119
    risk as self-governance, 122–123
    social stratification, 200–201
    wellness, 121–122

    women's responses, 118–119
Medicinal drugs, Czechs, 627
Medicinal plants, Maya of Highland Mexico, 836–838
Medicine classification, Han, 711–714
Mediterranean fever, familial (FMF) (MEF), genetic disease, 436–437
Melanoma, malignant, genetic disease, 437
MELAS see Mitochondrial encephalopathy lactic acidosis syndrome
Memory, embodied, phenomenology, 131–132
Mendel, Gregor, genetic disease, 393
Menkes syndrome, genetic disease, 437
Menopause, 282
    aging and, 219
    Greeks, 687
    medicalization, 122
Menstruation
    Iran, 282
    reproductive health, 281–282
Mental disorders, 486–493
    background, 486–487
    culture-specifics, 487–488
    defining, 486
    depression, 488–489
    future directions, 491
    kinds of disorders, 488–491
    schizophrenia, 489–491
    universals, 487–488
Mental health
    Greeks, 684
    Samoa, 934
Mental illness
    Czechs, 627–628
    homelessness, 174–175
    Japanese, 771
Mental retardation, 493–505
    AAMR, 494–495
    background, 493–494
    cultural factors, 499–500
    defining, 494–496
    diagnosis, 494–496
    environmental factors, 498–500
    epidemiological data, 498–499
    epidemiology, 496–498
    etiology, 497
    family supports, 500–501
    future perspectives, 502–503
    genetic-environmental interactions, 501–502
    genetic factors, 498
    international data, 501
    intervention issues, 500–502
    population differences, 497
    services issues, 500–502
    X-linked non-specific, 437
Metabolic disorders, paleopathology, 55
Methicillin-resistant staphylococcia (MRSA), 383, 389
Mexico, diarrhea treatment, 359
Middle-age, Tuareg, 1008
Midwifery, 225–226
    Jamaican Maroons, 756

Nahua, 865, 867–868
Netherlands, 228–229
see also Birth
Migration, population control, 276
Mineral water, French, 650
Miscarriage, Datoga, 634
Mitochondrial encephalopathy lactic acidosis syndrome (MELAS), genetic disease, 437
Mitochondrial inheritance, genetic disease, 397
Mitochondrial myopathies, genetic disease, 437
Mobile pastoralism, Mongolia, 852
Modernization, medicalization, 116–118
Molecular biology, emerging infectious diseases, 387–388
Mono-cultural health research, vs. cross-cultural health research, 8–9
Monoamine oxidase A deficiency (MAOA), genetic disease, 438
Morality, possession and trance, 140
Morbidity
   Japanese, 769–771
   Maya of Highland Mexico, 830–831
Mortality
   Japanese, 769–771
   Matsigenka, 813
   Maya of Highland Mexico, 830–831
Mosaicism, chromosomal, genetic disease, 438
Mourning, psychoanalysis, 60–61
Moxibustion, Han, 714
MPS1/2 *see* Mucopolysaccharidosis type 1/2
MRSA *see* Methicillin-resistant staphylococciae
MTb *see* Mycobacterium tuberculosis
Mucopolysaccharidosis type 1/2 (MPS1/2), 438
Multi-drug-resistant tuberculosis (MDRTB), 387–388, 389
Multidimensional scaling (MDS), 13–14
Multifactorial disorders, genetic disease, 398–399
Multiple births, 438–439
Mummification, paleopathology, 49–50
Muscular dystrophy Duchenne type (DMD), genetic disease, 439
Mutations, chromosomal, 394–395
Mutations, DNA, 395
Mycobacterium tuberculosis (MTb), syndemics, 27
Myopia 2 (MYP2), genetic disease, 439
Myotonic dystrophy (DM), genetic disease, 439

Narratives, illness *see* Illness narratives
National Health Service (NHS), British, 601, 605
National Strategies for Sustainable Development (NSSD), population control, 272
Native Americans
   alcoholism, 202
   diabetes mellitus, 342, 345, 346
Natural causes, illness, Garhwali, 667
Natural selection, disease, 307
Naturalistic health problems, Maya of Highland Mexico, 832–833
Naturalistic medicine, 7
Naturalization of social control, medicalization, 116–125
'Nerves', culture-bound syndrome, 320, 322
Netherlands
   immunization, 266
   midwifery, 228–229
Neurofibromatosis
   disability/difference, 363
   genetic disease, 440
Neuroses, cultural expression, 62–63
New Guinea, *kuru*, 324–325, 386, 387
New reproductive technologies (NRTs), 89–91, 102
New World syndrome, genetic disease, 409, 440
NHS *see* National Health Service
NRTs *see* New reproductive technologies
NSSD *see* National Strategies for Sustainable Development
Nutrition
   Han, 704–705, 713
   Yanomamö, 1023
Nutritional anthropology, 178–184
   history, 179
   low-income countries, 181–182
   methods, 179–180
   themes, 181–183
   theoretical orientations, 179–180
   today's health, 182–183
Nutritional disorders
   overnutrition, 242
   paleopathology, 55
   undernutrition, 241–242
   see also Malnutrition

Obesity
   African-Americans, 553
   diabetes mellitus, 335, 336, 338, 341
   genetic disease, 440
   Navajo, 880
   Tongans, 985
Occupations
   Czechs, 623
   Datoga, 630
   Haitians, 697–698
   Japanese, 767
   Tuareg, 1001–1002
Ombiasa, shamanism, 797–798
Onchocerciasis, epidemiology, 479–481
Onchocerciasis, Yanomamö, 1019
Oral rehydration salts (ORS), 88
Organ transplantation/donation
   bioethics, 76
   biomedical technologies, 91–92
   brain-death, 92–93
   exploitation, 92–93
ORS *see* Oral rehydration salts
Osteoarthritis (OA), paleopathology, 56
Osteogenesis imperfecta, genetic disease, 440–441
Osteoporosis, 312–314
   cross-cultural distribution, 313
   diagnosis, 312–313
   prevention, 313–314
   risk factors, 313

Subject Index

treatment, 313–314
WHO, 313
Osteoporosis, French, 648
Oxytocin, breast-feeding, 231

P53 tumor suppressor protein (TP53), genetic disease, 441
Paleopathology, 49–58
   bone tumours, 56–57
   cancer, 56–57
   circulatory abnormalities, 56
   congenital defects, 51–52
   degenerative disease, 56
   dental, 57
   disease classification, 51
   findings, 51–57
   history, 50
   immunologic diseases, 56
   infectious disease, 53–54
   metabolic disorders, 55
   mummification, 49–50
   nutritional disorders, 55
   osteoarthritis (OA), 56
   pneumoconiosis, 55
   rheumatoid arthritis, 56
   technical considerations, 50–51
   traumatic injury, 52–53
   vitamin abnormalities, 55
   see also Evolutionary perspectives
Parasitic infections, Yanomamö, 1019–1020
Parkinson disease, genetic disease, 441
Pastoralism, Fulani, 656
Patau syndrome, 395
   genetic disease, 441
Pendred syndrome, genetic disease, 441
Penicillin-resistant neisseriae gonorrhoea (PRNG), 383
Periodontal disease, paleopathology, 57
Personal symbols, psychoanalysis, 61
Personalistic health problems, Maya of Highland Mexico, 833
Personalistic medicine, 7
Pharmaceuticals
   biographies, 87–88
   biomedical technologies, 87–88
Phenocopies, genetic disease, 399
Phenomenology, 125–136
   age studies, 131–132
   of the body, 127–128
   body, ethnographic uses, 128–129
   body in distress, 129–130
   body of difference, 130–131
   gendered body, 130
   memory, embodied, 131–132
   methodological perspectives, 126–127
   suffering, politics of, 132
   theoretical perspectives, 126–127
Phenylketonuria, genetic disease, 441–442
Philadelphia chromosome, 395
   genetic disease, 442–443

Philippines, immunization, 266–267
Philosophical underpinnings, Japanese, 767
Phocomelia, genetic disease, 443
Physical handicaps, Datoga, 632
Pigbel, Fore, 641–642
Pill, 88–89
   social aspects, 283–284
Pima Indians, diabetes mellitus, 342
PKD1 *see* Polycystic kidney disease, adult
Placenta, disposal, 224–225
Pneumoconiosis, paleopathology, 55
Polio, paleopathology, 54
Political conditions
   Datoga, 630
   Haitians, 698
   Tuareg, 1002
Political/economic perspective, disease, 307
Political factors
   African-Americans, 546–548
   Amish, 558–560
   Argentine Toba, 565–566
   Badaga, 573
   Baliem Valley Dani, 592–593
   Bangladeshis, 580–581
   British, 600–601
   Burmese, 609–610
   Cree, 616–617
   Czechs, 623–626
   Datoga, 630–631
   Fore, 641–643
   French, 647
   Fulani, 657–658
   Garhwali, 665–666
   Garifuna, 676
   Greeks, 682–683
   Hadza, 690–691
   Haitians, 699–700
   Han, 704–707
   Hausa, 719–720
   Iroquois, 745–746
   Japanese, 768–771
   Jat, 778
   Lijiang Naxi, 785–787
   Malagasy, 796–797
   Malays, 840–842
   Maori, 805–807
   Matsigenka, 812–814
   Maya of Highland Mexico, 828–831
   Mongolia, 851–854
   Nahua, 864
   Navajo, 874
   Nepal, 885
   Northwest Coast, 894–897
   Ojibwa, 905–908
   Oklahoma Choctaw, 916–917
   Roma of the United States and Europe, 924
   Saraguros, 938–939

Shipibo, 948–951
Sotho, 958–959
Sudanese, 965–966
Thai, 972–975
Tongans, 982–983
Trobriand, 991–992
Tuareg, 1003–1004
Wape, 1011–1012
Yanomamö, 1018–1023
Yoruba, 1031–1032
Political identity, Garifuna, 673–674
Political organization
    Fore, 639
    Hmong, 730
Political position, Maori, 806
Pollution, environmental, Czechs, 624
Polycystic kidney disease, adult (PKD1), genetic disease, 443
Polydactyly, genetic disease, 408, 443
Polygenic factors, genetic disease, 398–399
Polyoma virus, Yanomamö, 1021–1022
Polyposis of colon, familial, genetic disease, 443
Poor, urban *see* Urban poor
Population
    Czechs, 622
    Datoga, 629–630
    Samoa, 929–930
Population aging
    health impact, 311–312
    United Nations, 221
    see also Aging
Population control, 269–280
    biological warfare, 277
    birth control, 273–274
    bubonic plague, 270
    contraceptives, 274
    contragestin, 274
    contranatals, 274–275
    controversies, 275–279
    cultural revolution, 269–270
    Darwin, 269
    death control, 272–273
    demography, 270–271
    ecological models, 271–272
    eugenics, 277–278
    family planning, 279
    historical background, 269–270
    immigration, 276
    issues, 275–279
    Malthus, 269, 272
    medicine developments, 272–275
    methods, 270–275
    migration, 276
    NSSD, 272
    organizations, governmental/private, 275
    other species, 279
    overconsumption, 275–276
    overpopulation, 275
    responses, 269–270
    responsibilities, procreative/reproductive, 279
    rights, procreative/reproductive, 279
    sterilization programs, 277–278
    theories, 270–275
    warfare, 276–277
    ZPG, 275
Population differences, mental retardation, 497
Population distribution, Han, 704
Population history, Northwest Coast, 894–897
Population statistics, African-Americans, 546
Populations studied, diabetes mellitus, 338–339
Porphyria variegata, genetic disease, 408, 443–444
Possession and trance, 137–145
    behavioral manifestations, 137–138
    belief sources, 137
    belief types, 137
    Brazil, 140
    exorcism, 142
    geographic distributions, 138–139
    and healing, 141–142
    morality, 140
    religions, 140, 142–143
    sociocultural correlates, 138–139
    western thought, 139–140
    women, 140
Post-colonial development and health, 184–191
Postcolonial era, Haitians, 699
Prader-Willi syndrome (PWS), genetic disease, 444
Pre-contact paradigm, Samoa, 932–933
Pregnancy
    African-Americans, 552
    Amish, 561
    Argentine Toba, 567–568
    Badaga, 575–576
    Baliem Valley Dani, 595–596
    Bangladeshis, 585
    British, 603
    Burmese, 611
    Cree, 619
    Czechs, 628
    Datoga, 633–635
    Fore, 644
    French, 653
    Fulani, 660
    Garhwali, 669–670
    Garifuna, 677–678
    Greeks, 685–686
    growth events/duration, 237–239
    Hadza, 693
    Han, 715–716
    Hausa, 725
    Hmong, 737–738
    Iroquois, 748
    Jamaican Maroons, 762
    Japanese, 773
    Jat, 780
    Malagasy, 800
    Malays, 846–847

Subject Index

Maori, 808
Matsigenka, 823
Maya of Highland Mexico, 834
Mongolia, 857
Nahua, 867–868
Navajo, 878
Nepal, 888
Northwest Coast, 899
Ojibwa, 911–912
Oklahoma Choctaw, 920
outcomes, 284–285
Roma of the United States and Europe, 926
Samoa, 934–935
Saraguros, 942–943
Shipibo, 954
Sotho, 960–961
Sudanese, 969
Thai, 978
Tongans, 986
Trobriand, 995
Tuareg, 1006–1007
Wape, 1014–1015
Yanomamö, 1024–1025
Yoruba, 1036
see also Reproductive health
Preventive measures, Czechs, 628
Principle of separation, biomedicine, 98
Prion protein (PRNP)
  genetic disease, 444
  see also Creutzfeldt-Jakob disease; Familial fatal insomnia; Gerstmann-Straussler-Sheinker disease; *Kuru*
Private foundations, economic development, 166–167
PRNG *see* Penicillin-resistant neisseriae gonorrhoea
PRNP *see* Prion protein
Production, Saraguros, 938
Progerias, genetic disease, 444
Prognosis disclosure, bioethics, 76–77
Prolactin, breast-feeding, 231
Prophylaxis, psychoanalysis, 60–61
Prostate cancer, British, 605
Proteomics, 98
Proteus syndrome, genetic disease, 444–445
Pseudohermaphroditism, genetic disease, 445
Psychoanalysis, 58–69
  anthropological interviewing, 63–64
  culture shock, 59, 64–65
  funerals, 60–61
  mourning, 60–61
  personal symbols, 61
  prophylaxis, 60–61
  psychoanalytic movements, 65–66
  psychoses, 62–63
  ritual treatments, 59–61
  shamanism, 59–61
Psychoanalytic movements, Brazil, 65–66
Psychoses
  cultural expression, 62–63
  treatment, 62

PTC tasting, genetic disease, 445
Public health, bioethics, 80–81
Public health infrastructure, Lijiang Naxi, 786
Public policies, alcohol use, 297–298
Puerto Rico
  family planning, 89
  HIV/AIDS, 469
Pulaaku, Fulani, 657
PWS *see* Prader-Willi syndrome
Pygmies, 241
  see also Dwarfism

Qigong, Han, 713
Quadratic Assignment Program, 15, 16

'Race', bioethics, 78
Ras oncogene (KRAS), genetic disease, 445–446
RB1 *see* Retinoblastoma
Refugee health, 191–198
  healthcare provision, 195–196
  implications, conflict/displacement, 193–194
  third country resettlement, 194–195
Religions
  African-Americans, 548, 549–550
  Burmese, 608–609
  Czechs, 623
  Datoga, 630
  FGC, 256
  Fulani, 657
  Greeks, 682
  Haitians, 698–699, 701
  Han, 707
  Hmong, 730–731
  Islam, 256
  possession and trance, 140, 142–143
  Samoa, 930–931
  Saraguros, 938
  Tongans, 981
  Tuareg, 1002–1003
Religious beliefs, Yoruba, 1031
Religious healing, Greeks, 684
Religious ideology, Jamaican Maroons, 759–760
Religious studies, death/dying, 245–246
Religious traditions, Japanese, 767
Reproduction
  Amish, 560–561
  Argentine Toba, 567
  Badaga, 575
  Baliem Valley Dani, 595
  Bangladeshis, 583–584
  British, 602–603
  Burmese, 611
  Czechs, 628
  Datoga, 632–633
  Fore, 643–644
  French, 652–653
  Fulani, 659–660
  Garhwali, 669

Garifuna, 677
Greeks, 684–685
Hadza, 692–693
Haitians, 701–702
Han, 714–715
Hausa, 724
Hmong, 736–737
Iroquois, 748
Jamaican Maroons, 760–762
Japanese, 772–773
Jat, 779–780
Lijiang Naxi, 791–792
Malagasy, 799
Malays, 846
Maori, 808
Matsigenka, 822–823
Maya of Highland Mexico, 834
medicalization, 119
Mongolia, 856, 857
Nahua, 867
Navajo, 878
Nepal, 887
Oklahoma Choctaw, 920
Roma of the United States and Europe, 926
Samoa, 934
Saraguros, 942
Shipibo, 953–954
Sotho, 960
Sudanese, 967–969
Thai, 977–978
Tongans, 986
Trobriand, 994–995
Tuareg, 1006
Wape, 1014
Yanomamö, 1024
Yoruba, 1035–1036
Reproductive ecology, 33
Reproductive ethics, bioethics, 78
Reproductive health, 280–290
    fertility enhancement, 285
    fertility regulation, 282–284
    HIV/AIDS, 286
    infections, 285–286
    infertility, 285
    menstruation, 281–282
    pregnancy outcomes, 284–285
    reproductive life cycle, 281–282
    STDs, 285–286
    see also Pregnancy
Reproductive technologies, 89–91, 102
Research ethics, bioethics, 80
Research interests, 9–10
Reservations, Iroquois, 743–744
Resilience, disasters, 159
Resort patterns, health-seeking process, 5–6
Retinoblastoma (RB1), genetic disease, 446
Rett syndrome (RTT), genetic disease, 446
Rhesus isoimmunization, genetic disease, 446–447

Rheumatoid arthritis, paleopathology, 56
Risk as self-governance, medicalization, 122–123
Risk factors
    AD, 314–315
    diabetes mellitus, 335, 336, 338, 341
    osteoporosis, 313
    SIDS, 507–509
    stress, 331–333
Risks, child growth, 241–242
Ritual studies, death/dying, 245–246
Ritual treatments, psychoanalysis, 59–61
RNA viruses, 385
RTT see Rett syndrome
Russia
    MDRTB, 387–388, 389
    psychoanalytic movements, 65–66
Russian modernist traditions, Mongolia, 854–855, 856

St. Lucia, stress, 332
Samoan community, stress, 332–333
Scanning electron microscope (SEM), forensic anthropology, 39
SCD see Sickle cell disease
Schizophrenia (SCZD)
    cross-cultural research, 489–491
    genetic disease, 447
    mental disorder, 489–491
'Science men', Jamaican Maroons, 756–757
Self-care, African-Americans, 551
Self-governance, risk as, 122–123
SEM see Scanning electron microscope
Seripigari, Matsigenka, 814–815
Severe combined immunodeficiency disease, X-linked (XSCID), genetic disease, 447
Sex-determining region of the Y chromosome (SRY), genetic disease, 448
Sex-linked inheritance, genetic disease, 397
Sexual orientation, social stratification, 202–203
Sexual pleasure, Hmong, 737
Sexuality
    Amish, 560–561
    Argentine Toba, 567
    Badaga, 575
    Baliem Valley Dani, 595
    Bangladeshis, 583–584
    British, 602–603
    Burmese, 611
    Czechs, 628
    Datoga, 632–633
    Fore, 643–644
    French, 652–653
    Fulani, 659–660
    Garhwali, 669
    Garifuna, 677
    Greeks, 684–685
    Hadza, 692–693
    Haitians, 701–702
    Han, 714–715
    Hausa, 724

Subject Index

Hmong, 736–737
Iroquois, 748
Jamaican Maroons, 760–762
Japanese, 772–773
Jat, 779–780
Lijiang Naxi, 791–792
Malagasy, 799
Malays, 846
Maori, 808
Matsigenka, 822–823
Maya of Highland Mexico, 834
Mongolia, 856
Nahua, 857
Navajo, 878
Nepal, 887
Oklahoma Choctaw, 920
Roma of the United States and Europe, 926
Samoa, 934
Saraguros, 942
Shipibo, 953–954
Sotho, 960
Sudanese, 967–969
Thai, 977–978
Tongans 986
Trobriand, 994–995
Tuareg, 1006
Wape, 1014
Yanomamö, 1024
Yoruba, 1035–1036
Sexuality, HIV/AIDS and, 286
Sexually transmitted diseases (STDs), 285–286
Shamanism, 145–154
   animism, 151–152
   ASC, 146–148
   ASC bases, therapeutic processes, 149–150
   ASC, psychobiological structures, 148–149
   community relations, 150
   contemporary, 152–153
   cross-cultural perspective, 146
   death and rebirth, 147–148
   Han, 707
   healers, 148–149
   healing, contemporary, 152–153
   Hmong, 732, 734
   illness, contemporary, 152–153
   Malagasy, 797–798
   Matsigenka, 814–815
   meaning and emotions, 150–151
   Nahua, 865, 866
   Nepal, 885–886, 887
   Northwest Coast, 897
   ombiasa, 797–798
   psychoanalysis, 59–61
   role-taking, 152
   soul journey, 147
   soul loss, 148
   spirit, 151–152
   therapeutic processes, 148, 149–152

   universals, 146–148
   Yanomamö, 1023–1024
Sickle cell disease (SCD), 392, 408
   genetic disease, 448–449
Sickness, defining, 3–4
SIDS *see* Sudden infant death syndrome
Sierra Leone
   birth, 227–228
   FGC, 255
Situs invertus viscerum, genetic disease, 449
Sleep physiology research, SIDS, 509–510
Slums *see* Urban poor
SMA *see* Society for Medical Anthropology
Smallpox, Northwest Coast, 895, 897
Social change
   culture-bound syndromes, 325–326
   emerging infectious diseases, 387–388
Social conditions
   Datoga, 630
   Haitians, 698
   Tuareg, 1002
Social factors
   African-Americans, 546–548
   Amish, 558–560
   Argentine Toba, 565–566
   Badaga, 573
   Baliem Valley Dani, 592–593
   Bangladeshis, 580–581
   British, 600–601
   Burmese, 609–610
   Cree, 616–617
   Czechs, 623–626
   Datoga, 630–631
   Fore, 641–643
   French, 647
   Fulani, 657–658
   Garhwali, 665–666
   Garifuna, 675
   Greeks, 682–683
   Hadza, 690–691
   Haitians, 699–700
   Han, 704–707
   Hausa, 719–720
   Iroquois, 745–746
   Japanese, 768–771
   Jat, 778
   Lijiang Naxi, 785–787
   Malagasy, 796–797
   Malays, 840–842
   Maori, 805–807
   Matsigenka, 812–814
   Maya of Highland Mexico, 828–831
   Mongolia, 851–854
   Nahua, 864
   Navajo, 874
   Nepal, 885
   Northwest Coast, 894–897
   Ojibwa, 905–908

Oklahoma Choctaw, 916–917
Roma of the United States and Europe, 924
Saraguros, 938–939
Shipibo, 948–951
Sotho, 958–959
Sudanese, 965–966
Thai, 972–975
Tongans, 982–983
Trobriand, 991–992
Tuareg, 1003–1004
Wape, 1011–1012
Yanomamö, 1018–1023
Yoruba, 1031–1032
Social inequality, disease distribution, 249
Social organization
    Fore, 639
    Fulani, 656–657
    Hmong, 730
    Northwest Coast, 892–893
    Samoa, 930
    Saraguros, 938
Social sciences, health *see* Health social sciences
Social stratification, 198–206
    biomedicine, 200–201
    class, 203
    constructing hierarchy, 201–204
    defining, 198–200
    ethnicity, 201–202
    gender, 203–204
    health and, 200
    HIV/AIDS, 202–203
    medicalization, 200–201
    negotiating health, 201
    sexual orientation, 202–203
    western context, 198–206
Social studies, death/dying, 245–246
Socialism, Mongolia, 859
post-socialism, Mongolia, 859–861
Society for Medical Anthropology (SMA), 4–5
Socio-economic circumstances, Maori, 805–806
Sociocultural norms, Japanese, 767–768
Sorcery
    Fore, 640–641
    Garhwali, 667–668
    Hmong, 734
    Matsigenka, 814–815
    Trobriand, 993
    Wape, 1013–1014
Soul loss, shamanism, 148
Souls, Hmong, 734
Spasmophilia, French, 648
Spirits
    Hmong, 734
    Trobriand, 993
Spiritual intervention, Jamaican Maroons, 759–760
Spiritual life, Garifuna, 674
Spirituality, African-Americans, 548, 549–550
Spoken Soul, African-Americans, 545

Squatter settlements *see* Urban poor
SRY *see* Sex-determining region of the Y chromosome
STDs *see* Sexually transmitted diseases
Stem cell research/cloning, genetic disease, 403
Sterilization programs, population control, 277–278
Stigma
    tuberculosis, 533–534
    vector-borne diseases, 483
Stress, 328–335
    blood pressure, 328–330, 333–334
    disasters, 157
    models, 331–333
    risk factors, 331–333
    St. Lucia, 332
    Samoan community, 332–333
Substance abuse, homelessness, 175
Sudan, FGC, 255
Sudden infant death syndrome (SIDS), 506–518
    child care practices, 510–511
    epidemiology, 507–509
    history, 507–509
    prevention, 513–515
    rates across cultures, 511–513
    risk factors, 507–509
    sleep physiology research, 509–510
Sufferer experience, Brazil, 28
Sufferer experience, CMA, 27–28
Suicide, Japanese, 771
Supernatural causes, illness, Garhwali, 667–668
Supplementation, breast-feeding, 233–234
Susto, culture-bound syndrome, 322–323
Symbols, personal, psychoanalysis, 61
Syndemics
    CMA, 26–27
    urban poor, 211
Syphilis, paleopathology, 54

Taboos, Garhwali, 668
Tauva'u, Trobriand, 993
Tay-Sachs disease (TSD), genetic disease, 408, 409, 449
TCM *see* Traditional Chinese Medicine
Temporomandibular joint (TMJ), illness narratives, 20
Terminal diagnosis disclosure, bioethics, 76–77
Testicular feminization (TF), genetic disease, 449
Thalassemia, alpha, genetic disease, 449–450
Thalassemia, beta, genetic disease, 450
Thalassemia, beta, Greeks, 683
Theoretical issues, cross-cultural health research, 3–11
Theory of Reasoned Action (TRA), HIV/AIDS, 467
Therapeutic practices, Lijiang Naxi, 786–787
Tibetan Buddhist medicine, Mongolia, 858–859
Tibetan medicine, Mongolia, 855–856
TMJ *see* Temporomandibular joint
Tobacco, 518–528
    anthropological studies, 522–523
    CMA, 525–527
    and the Colonies, 521
    as a critical commodity, 521–522

Subject Index

cultural studies, 523–525
Han, 706
and health, 518–519
impact on Europe, 520–521
Mongolia, 853–854
New World origins, 519–520
social history, 519–521
Tourette syndrome *see* Gilles de la Tourette syndrome
TP53 *see* P53 tumor suppressor protein
TRA *see* Theory of Reasoned Action
Traditional Chinese Medicine (TCM), Lijiang Naxi, 786–787
Trance *see* Possession and trance
Transforming liquids, Trobriand, 997–998
Traumatic injury
　illness narratives, 43
　paleopathology, 52–53
Treatment decisions, cognitive-ethnographic studies, 16–18
Trends
　alcohol use, 296–300
　biomedicine, 101–103
Triplo-X syndrome, 395
　genetic disease, 450
TSD *see* Tay-Sachs disease
Tuberculosis, 383
　adherence issues, 532–533
　care-seeking behaviours, 531–532
　control, 528–542
　cultural research, 531–535, 536–537
　current approaches, 535–536
　current tools, 535–536
　diagnosis, 529–530
　epidemiological features, 529
　future directions, 536–537
　history, 530–531
　homelessness, 174
　MDRTE, 387–388, 389
　medical features, 529
　MTb, 27
　program structure, 535
　provider behavior, 534–535
　research, 528–542
　service delivery, 534–535
　social research, 531–535, 536–537
　stigma, 533–534
　treatment, 529–530
　Yanomamö, 1022–1023
Tuina, Han, 713
Tuina, Han, 713–714
Turner syndrome, genetic disease, 450–451
Tuskegee experiment, African-Americans, 547–548
Twins/twinning, 451
　see also Multiple births
Tyrosinemia, genetic disease, 451

UNHCR *see* United Nations High Commissioner for Refugees
UNICEF, breast milk substitutes, 234–235
Uniform Determination of Death Act (1981), 248
Uniparental disomy, genetic disease, 397–398

United Nations
　economic development, 165
　population aging, 221
　refugee health, 191–193
　UNHCR, 191–193, 195
United Nations High Commissioner for Refugees (UNHCR), 191–193, 195
United States
　abortion, 278, 284
　diabetes mellitus, 344–346
　Hmong in, 731–732, 733
　medical pluralism, 113–114
Urban poor, 207–213
　culture of poverty, 210
　diarrhea, 209
　global perspectives, 211
　HIV/AIDS, 209
　homelessness, 208
　infectious disease, 209
　local perspectives, 211
　malnutrition, 208–209
　representing, 209–210
　structural violence, 211
　syndemics, 211
　threats to health, 207–209
　violence, 208
USAID *see* Agency for International Development

Vaccination *see* Immunization
Vancomycin-resistant enterococciae (VRE), 383, 389
Variability, measuring, 14–15
Vector-borne diseases, 479–485
　community participation, health programs, 482–483
　ecological studies, 480–481
　epidemiology, 479–481
　ethnomedical studies, 481–482
　gender research, 483
　health programs, 482–483
　stigma, 483
Vernacular medicine, cf. conventional medicine, 8
Violence, urban poor, 208
Violent death, 249–251
Vitamin abnormalities, paleopathology, 55
Voodoo
　African-Americans, 550
　Haitians, 700
VRE *see* Vancomycin-resistant enterococciae
Vulnerability, disasters, 159

Waardenburg syndrome (WS1), genetic disease, 451
Warfare
　biological, 277
　population control, 276–277
Water-borne diseases, 305–311
　anthropological contributions to study of, 306–309
　anthropological perspectives, 307–308
　cholera, 305–311
　dengue fever, 306

determinants, 309
typology, 306
water insecurity, 308–309
web sites, 309–310
Water therapies, French, 650
Weaning, 233–234
  Gusii practices, 233–234
  see also Breast-feeding
Web sites
  diarrhea, 359–360
  genetic disease, 399
  water-borne diseases, 309–310
Weissmann, August, genetic disease, 393
Werner syndrome (WRN), genetic disease, 451
West Nile Virus, 385, 389
Western medicine, Han, 708
White man's sickness, Ojibwa, 910
WHO *see* World Health Organization
Wilson disease (WND), genetic disease, 452
Witchcraft
  aging and, 220
  birth, 227
  Datoga, 632
  Garhwali, 667–668
  Hausa, 721
WND *see* Wilson disease
Women, possession and trance, 140
Women's responses, medicalization, 118–119
World Bank, economic development, 165–166

World Health Organization (WHO)
  alcohol use, 297, 299–300
  breast milk substitutes, 234–235
  cholera, 305
  disability/difference, 368
  economic development, 165–166
  health, defining, 26
  osteoporosis, 313
Worm infections, paleopathology, 54–55
Worm infections, Yanomamö, 1019–1020
WRN *see* Werner syndrome
WS1 *see* Waardenburg syndrome

X-linked inheritance, genetic disease, 392, 397
Xeroderma pigmentosum (XP), genetic disease, 452
XSCID *see* Severe combined immunodeficiency disease, X-linked
XX male syndrome, genetic disease, 452
XY female syndrome, genetic disease, 452

Y-linked inheritance, genetic disease, 397
Yangsheng, Han, 713
Yunani system, Bangladeshis, 582

Zaire
  medical pluralism, 111–112
  post-colonial development and health, 185, 188
Zellweger syndrome, genetic disease, 452
Zero Population Growth (ZPG), 275

Cultural and Alternative Names Index

Afnu *see* Hausa
African-Americans, 545–557
Afro-Americans *see* African-Americans
Afro-Caribbean *see* African-Americans
Aku *see* Yoruba
Algonquin *see* Ojibwa
Alsean *see* Northwest Coast
Amish, 557–564
Anago *see* Yoruba
Anishinaabe *see* Ojibwa
Argentine Toba, 564–572
Athapaskans *see* Northwest Coast
Ayisyen see Haitians
Ayiti see Haitians
Aztec *see* Nahua

Badaga, 572–578
Baliem Valley Dani, 591–599
Bangladeshis, 579–590
Basotho *see* Sotho
Basuto *see* Sotho
Bella Bella *see* Northwest Coast
Bella Coola *see* Northwest Coast
Bengalis *see* Bangladeshis
Betsileo *see* Malagasy
Black Americans *see* African-Americans
Black Caribs *see* Garifuna
Bohemians *see* Czechs
Bowoyan *see* Trobriand
British, 599–606
Bungi *see* Ojibwa
Burgher *see* Badaga
Burmese, 607–614
Bweyowa *see* Trobriand

Cayuga *see* Iroquois
Chama *see* Shipibo
Chattah *see* Oklahoma Choctaw
Chinese *see* Han
Chinookans *see* Northwest Coast
Chippewa *see* Ojibwa
Chocs *see* Oklahoma Choctaw
Choctaw Nation of Oklahoma *see* Oklahoma Choctaw
Choctaws *see* Oklahoma Choctaw

Coast Salish *see* Northwest Coast
Conibo *see* Shipibo
Coos *see* Northwest Coast
Cree, 614–622
Czechs, 622–629

Dani *see* Baliem Valley Dani
Datoga, 629–638
Datoog *see* Datoga
Diné *see* Navajo

Eeyou see Cree
Eeyouch see Cree
Egba *see* Yoruba
Egbado *see* Yoruba
Ekiti *see* Yoruba
England *see* British
Eyak *see* Northwest Coast

Fore, 638–645
France *see* French
French, 646–655
French Republic *see* French
The Friendly Islands *see* Tongans
Fula see Fulani
Fulani, 656–664
   see also Hausa
FulBe see Fulani

Garhwali, 664–672
Garifuna, 672–681
Grand Valley Dani *see* Baliem Valley Dani
Great Britain *see* British
Greeks, 681–688
Guaica *see* Yanomamö
Guajaribo *see* Yanomamö
Gypsy *see* Roma of the United States and Europe

Habasha *see* Hausa
Hadza, 689–696
Hadzabe *see* Hadza
Hadzapi *see* Hadza
Haida *see* Northwest Coast
Haihais *see* Northwest Coast

Haisla *see* Northwest Coast
Haïti see Haitians
Haitians, 696–702
Haïtien see Haitians
Han, 703–717
Han-ren see Han
Hatsa *see* Hadza
Hausa, 718–729
Hausa-Fulani *see* Hausa
Haytian *see* Haitians
Heiltsuk *see* Northwest Coast
Hli-khin *see* Lijiang Naxi
Hmong, 729–743
Hodénosaunee *see* Iroquois
Hua-ren see Han

Ife *see* Yoruba
Igbomina *see* Yoruba
Ijebu *see* Yoruba
Ijesa *see* Yoruba
Imajeghen *see* Tuareg
Imghad *see* Tuareg
Inaden *see* Tuareg
Iroquois, 743–754

Jamaican Maroons, 754–765
Japanese, 765–776
Jat, 777–783

Kabba *see* Yoruba
Kangeju *see* Hadza
Kel Air *see* Tuareg
Kel Ewey *see* Tuareg
Kel Nabarro *see* Tuareg
Kel Tafidet *see* Tuareg
Kel Tagelmust *see* Tuareg
Kel Tamajaq *see* Tuareg
Kilivilan *see* Trobriand
Kiriwinian *see* Trobriand
Kogapakori *see* Matsigenka
Kugapakori *see* Matsigenka
Kwakiutl *see* Northwest Coast
Kwakwaka'wakw *see* Northwest Coast

Leeward Maroons *see* Jamaican Maroons
Lesotho *see* Sotho
Lijiang Naxi, 783–794

Machiguenga *see* Matsigenka
Madagascar *see* Malagasy
Makah *see* Northwest Coast
Malagasy, 794–804
Malays, 839–849
Mangati *see* Datoga
Maori, 804–811
Maroons *see* Jamaican Maroons
Matsigenka, 812–827
Matsiguenka *see* Matsigenka

Maya of Highland Mexico, 827–839
Mei-ji Hua-ren see Han
Meo *see* Hmong
Mexicano *see* Nahua
Mexijcatl *see* Nahua
Miao *see* Hmong
Mississauga Ojibwa *see* Ojibwa
Mississippi Choctaw *see* Oklahoma Choctaw
Mohawk *see* Iroquois
Mongol *see* Mongolia
Mongolia, 850–863
Mongolian *see* Mongolia
Moravians *see* Czechs
Morenos *see* Garifuna
Mosuo *see* Lijiang Naxi
Myanmarese *see* Burmese
Myanmars *see* Burmese

Na *see* Lijiang Naxi
Nahua, 863–872
Nahuat *see* Nahua
Nahuatl *see* Nahua
Naru *see* Lijiang Naxi
Navajo, 873–883
Naxi *see* Lijiang Naxi
Naze *see* Lijiang Naxi
Nepal, 883–890
New Zealand Maori *see* Maori
Nihon see Japanese
Nipissing *see* Ojibwa
Nippon see Japanese
Nootka *see* Northwest Coast
Northern Ireland *see* British
Northern Ojibwa *see* Ojibwa
Northwest Coast, 890–902
Nuu-chah-nulth *see* Northwest Coast
Nuxalk *see* Northwest Coast
Nyankimpong Pickibo *see* Jamaican Maroons

Oji-Cree *see* Ojibwa
Ojibwa, 903–915
Ojibway *see* Ojibwa
Ojibwe *see* Ojibwa
Ojicree *see* Ojibwa
Oklahoma Choctaw, 915–923
Old Order Amish *see* Amish
Olukumi *see* Yoruba
Ondo *see* Yoruba
Oneida *see* Iroquois
Onondaga *see* Iroquois
Oowekeeno *see* Northwest Coast
Owo *see* Yoruba
Oyo *see* Yoruba

Pacific Islanders *see* Tongans
Pahari see Garhwali
Peul see Fulani
Peulh see Fulani

Cultural and Alternative Names Index

Plain People *see* Amish
Plains Ojibwa *see* Ojibwa
Polynesians *see* Tongans
Porekina see Fore
Pullo see Fulani

Quichua language *see* Saraguros
Quileute/Chemakum *see* Northwest Coast

République Française *see* French
Roma of the United States and Europe, 923–929
Romanies *see* Roma of the United States and Europe

Samoa, 929–937
Sanema *see* Yanomamö
Sanima *see* Yanomamö
Sanumá *see* Yanomamö
Saraguros, 937–947
Saulteaux *see* Ojibwa
Saulteurs *see* Ojibwa
Scotland *see* British
Seneca *see* Iroquois
Shamatari *see* Yanomamö
Shidishana *see* Yanomamö
Shipibo, 947–956
Shiriana *see* Yanomamö
Shori *see* Yanomamö
Siamese *see* Thai
Siuslaw *see* Northwest Coast
Sotho, 957–964
Southeastern Ojibwa *see* Ojibwa
Southern-Sotho *see* Sotho
Southwestern Chippewa *see* Ojibwa
Sudanese, 964–971

Takelma *see* Northwest Coast
Tanala *see* Malagasy
Tang-ren see Han
Tangata Whenua o Aotearoa *see* Maori
Targui *see* Tuareg
Tatog *see* Datoga
Tatoga *see* Datoga

Thai, 971–980
Tindiga *see* Hadza
Tinkers *see* Roma of the United States and Europe
Tlingit *see* Northwest Coast
Toba *see* Argentine Toba
Tongans, 980–990
Touareg *see* Tuareg
Travelers *see* Roma of the United States and Europe
Trobriand, 990–1000
Trobriand Islanders *see* Trobriand
Trobrianders *see* Trobriand
Tsimshian *see* Northwest Coast
Tuareg, 1001–1009
Tuscarora *see* Iroquois

Union of Myanmar *see* Burmese
United Kingdom *see* British

Waica *see* Yanomamö
Waika *see* Yanomamö
Wakindiga *see* Hadza
Wales *see* British
Wape, 1009–1017
Wapei *see* Wape
Wapi *see* Wape
Watindiga *see* Hadza
Windward Maroons *see* Jamaican Maroons

Xiriana *see* Yanomamö

Yamato see Japanese
Yanoama *see* Yanomamö
Yanomama *see* Yanomamö
Yanomami *see* Yanomamö
Yanomamö, 1017–1029
Yindunixiya Huaren see Han
Yongning Naxi see Lijiang Naxi
Yoruba, 1029–1039

Zhong Huar Re Min Gong He Guo see Han
Zhongguo-ren see Han
Zutt see Jat